Semiconductor Physics and Applications

M. BALKANSKI

Université Pierre et Maire Curie
Paris, France

and

R. F. WALLIS

University of California
Irvine, California, USA

OXFORD
UNIVERSITY PRESS

OXFORD
UNIVERSITY PRESS

Great Clarendon Street, Oxford OX2 6DP

Oxford University Press is a department of the University of Oxford.
It furthers the University's objective of excellence in research, scholarship,
and education by publishing worldwide in

Oxford New York

Athens Auckland Bangkok Bogotá Buenos Aires Calcutta
Cape Town Chennai Dar es Salaam Delhi Florence Hong Kong Istanbul
Karachi Kuala Lumpur Madrid Melbourne Mexico City Mumbai
Nairobi Paris São Paulo Singapore Taipei Tokyo Toronto Warsaw

with associated companies in Berlin Ibadan

Oxford is a registered trade mark of Oxford University Press
in the UK and in certain other countries

Published in the United States
by Oxford University Press Inc., New York

First published 2000

A catalogue record for this book is available from the British Library

Library of Congress Cataloging in Publication Data

Balkanski, Minko, 1927-
Semiconductor physics and applications/M. Balkanski and R.F. Wallis.
p. cm.
Includes bibliographical references and index.
1. Semiconductors. I. Wallis, R. F. (Richard Fisher), 1924-II. Title. III. Series.

QC611 .B185 2000 537.6′22—dc21 00-038489

ISBN 0 19 851741 6 Hardback
ISBN 0 19 851740 8 Paperback

Typeset by Newgen Imaging Systems (P) Ltd, Chennai, India
Printed in Great Britain
on acid-free paper by The Bath Press, Avon

SEMICONDUCTOR PHYSICS AND APPLICATIONS

To Nelly and Camilla

Preface

In recent years there have been remarkable advances in high technology that have revolutionized telecommunications, computing, and information storage and retrieval. Devices such as field effect transistors, integrated circuits, microprocessors, and electromagnetic radiation detectors and emitters have played key roles in this revolution. In large measure these devices are based upon semiconductors as the active materials. It is therefore essential for the further development of high technology that young workers entering the field be well trained in the physics and applications of semiconductors and that an up-to-date, well-written textbook covering these topics be available. It is our hope and intention that the present volume satisfies this need.

The material included in this book falls into two categories:

1. the basic physics of semiconductors;
2. applications of semiconductors to practical devices with emphasis on the basic physical principles upon which the devices operate.

The part on the basic physics starts with a discussion of the composition and structure of semiconductors. Next come chapters on the basic theory of electronic energy bands and on the detailed characteristics of energy bands in pure, bulk semiconductors. The effects of impurities on electronic states are then discussed. Chapters on semiconductor statistics and lattice vibrations in semiconductors precede a chapter on charge-carrier scattering and transport. In the next chapter the effects of surfaces on semiconductor properties are treated. Optical properties form the subject matter of Chapter 10. The part on basic physics concludes with a chapter on magneto-optical phenomena and nonlinear electro-optical effects.

The part on applications is not an entity separate from the part on basic physics. The various applied topics are accompanied by references to the basic topics to which they are most closely related. The first applied chapter covers p–n junctions and their use as rectifiers and capacitors. Then follow chapters on bipolar junction transistors and on semiconductor lasers and other photodevices. The importance of semiconductor interfaces is expanded upon in chapters on the electronic, lattice dynamical, optical, and transport properties of heterostructures. The book concludes with a chapter on metal-oxide-semiconductor field effect transistors and one on device applications of heterostructures.

This book is intended to serve as a textbook for a course on semiconductor physics and applications at the advanced undergraduate and beginning graduate levels. It is assumed that the reader has command of the basic material in quantum mechanics, statistical mechanics, and electromagnetic theory. It is desirable, but not essential, that the reader be acquainted with elementary solid state physics. Examples and exercises are included to emphasize particular points and provide an appreciation of the magnitudes of the physical quantities encountered. Extensive use is made of figures to enhance the clarity of the presentation and to establish contact with experimental results. At the beginning of each chapter there is a summary of key ideas developed in that chapter. This summary enables the reader to get a quick impression of the material covered in the chapter and the important qualitative results. Problem sets appear at the ends of the chapters.

In writing this book we have benefited from a number of prior works. They include *Introduction to Solid State Physics* by C. Kittel, *Solid State Physics* by H. Ibach and H. Lüth, *Electronic Structure and the Properties of Solids* by W. A. Harrison, *Survey of Semiconductor Physics* by K. W. Böer, *Introduction to Applied Solid State Physics* by R. Dalven, *Semiconductor Physics* by K. Seeger, *Semiconductors* by R. A. Smith, *Physique des Semiconducteurs et des Composants Electroniques* by H. Mathieu, *Physics and Technology of Semiconductor Devices* by A. S. Grove, *Physics of Semiconductor Devices* by S. M. Sze, *Fundamentals of Solid-State Electronics* by C.-T. Sah, *Solid State Electronic Devices* by B. G. Streetman, *Quantum Processes in Semiconductors* by B. K. Ridley, *Wave Mechanics Applied to Semiconductor Heterostructures* by G. Bastard, *Quantum Semiconductor Structures* by C. Weisbuch and B. Vinter, and *Principles of Optics* by M. Born and E. Wolf.

Finally, the authors wish to express their appreciation to Jeannie M. Brown whose diligence and patience in typing the manuscript made possible the production of this volume.

Paris M. B.
Irvine R. F. W.
December 1999

Contents

Basic characteristics of semiconductors

1

Key ideas

A *semiconductor* is neither a good insulator nor a very good conductor of electric current.

The *electrical resistivity* of semiconductor generally decreases strongly with increasing temperature, whereas the resistivity of a metal generally increases weakly with increasing temperature.

A semiconductor can be an *element*, such as Si or Ge, or a *compound*, such as GaAs or InSb.

Solid semiconductors can exist in the *crystalline* or *amorphous* form.

A *lattice* is a periodic array of points in space.

A *basis* is a set of atoms associated with each *lattice point*.

Semiconductors can crystallize in the *diamond*, *zincblende*, *wurtzite*, or other *structures*.

The *bonding* between atoms in a semiconductor can be *covalent*, *ionic*, or *mixed*.

Layered semiconductors can *intercalate* foreign atoms or molecules.

Very pure semiconductors can be produced by the *zone refining method*.

General Properties

The history of semiconductors dates back to the nineteenth century. The decrease in resistance of silver sulfide with increasing temperature was noticed by Michael Faraday (1833). Selenium was found to be photoconducting by W. Smith (1873), and the rectifying properties of lead sulfide were observed by F. Braun (1874). During World War II silicon found application as a rectifier in radar. By far the most important event, however, was the discovery of transistor action in germanium by J. Bardeen and W. H. Brattain in 1948, which ushered in the explosive development of high technology based on semiconductors that continues to the present day.

1.1 Qualitative properties

Semiconductors can be distinguished from other materials by a number of physical properties, one of the most important of which is the electric resistivity ρ or the difficulty with which an electric current can pass through

Table 1.1 Room temperature resistivities for the categories of materials

Category	ρ (Ω cm)
Insulators	10^{12}
Semiconductors	10^{6}–10^{-3}
Metals	10^{-6}

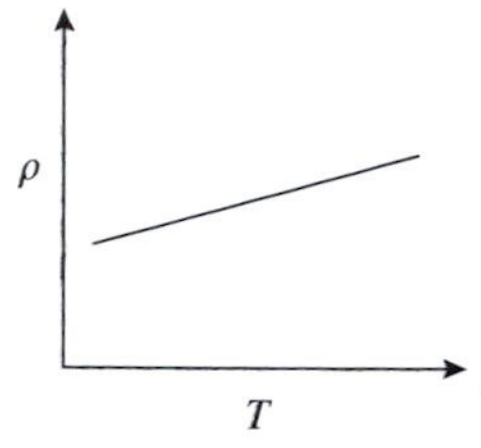

Fig. 1.1
Resistivity versus temperature for a metal.

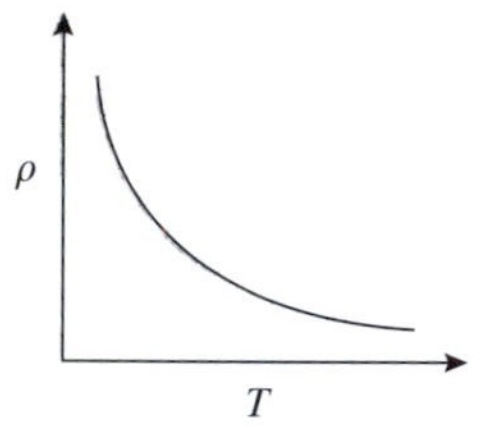

Fig. 1.2
Resistivity versus temperature for a semiconductor.

the material under the influence of an electric field. Materials can be classified into three categories: insulators, for which the resistivity is very high; metals, for which the resistivity is very low; and semiconductors, for which the resistivity is intermediate in value and is highly temperature dependent. Typical values for the room temperature resistivities of these three types of materials are given in Table 1.1. The temperature dependences of the resistivities of metals and semiconductors are quite different. For metals the temperature dependence is typically weak and the resistivity increases with increasing temperature (Fig. 1.1). For semiconductors the opposite is typically the case. The temperature dependence is strong and the resistivity for the most part decreases with increasing temperature (Fig. 1.2). A number of other aspects of the electrical properties of semiconductors should be mentioned:

1. non-ohmic behavior and rectifying effects;
2. large impurity effects;
3. both positive and negative charge carriers;
4. high thermoelectric power;
5. sensitivity to light—production of photovoltage and change in resistance.

Details concerning these effects will be found in later chapters.

1.2 Composition of semiconductors

Which solids are semiconductors? Among the most important are the elements silicon (Si) and germanium (Ge), which appear in group IV of the periodic table of the elements. Carbon in various forms can be a semiconductor. Other elemental semiconductors are selenium (Se) and tellurium (Te), which appear in group VI.

The list of compounds which are semiconductors is very large. It includes the III–V compounds, formed from group III and group V elements, such as GaAs and InSb; the II–VI compounds such as CdS, CdSe, and ZnSe; the III–VI compounds such as GaS and InSe; and the IV–VI compounds such as PbS and PbTe. Many ternary compounds such as $CuFeS_2$ (chalcopyrite) are also semiconductors. Organic compounds which are semiconductors include anthracene, $C_{14}H_{10}$, and polyacetylene, $(CH)_x$.

Many properties of semiconductors are drastically modified by the presence of impurities. In silicon, for example, the electrical and optical properties can be significantly changed by the addition of impurities such as boron or arsenic. The process of deliberately adding known impurities in a controlled manner is known as **doping**. In compound semiconductors, deviations from stoichiometry can affect their properties.

1.3 Structure of solids

1.3.1 Crystalline and amorphous forms

Solids can exist in the crystalline or the amorphous form. In the **crystalline form**, the atoms or ions are arranged in a periodic array, i.e., there is long range order in the system. In the **amorphous form** there is no long range order. A **glass** is an amorphous solid formed by the rapid cooling of a viscous liquid and has a viscosity higher than 10^{13} poise. Whether a semiconductor sample is crystalline or amorphous depends on how it was prepared. In the following discussion we shall focus on crystalline semiconductors; amorphous semiconductors will be discussed in a later chapter.

1.3.2 Lattice and basis

The arrangement of the atoms in an ideal crystal is specified by the **crystal structure**. The structure is characterized by two elements: a **lattice** and a **basis**. A lattice is a periodic array of points in space. A basis is a set of points attached identically to each **lattice point**. Choosing one point of the lattice as the origin, the position vector $\boldsymbol{R}(\ell)$ of an arbitrary lattice point can be expressed as

$$\boldsymbol{R}(\ell) \equiv \boldsymbol{R}(\ell_1\ell_2\ell_3) = \ell_1\boldsymbol{a}_1 + \ell_2\boldsymbol{a}_2 + \ell_3\boldsymbol{a}_3 \tag{1.1}$$

where $\boldsymbol{a}_1, \boldsymbol{a}_2, \boldsymbol{a}_3$ are noncoplanar vectors called the **primitive translation vectors** and ℓ_1, ℓ_2, ℓ_3 are integers that take on all integer values and are referred to collectively as ℓ. The primitive translation vectors are the shortest translation vectors in their respective directions that carry one point of the lattice into another point of the lattice. They form the edges of a parallelepiped known as a **primitive unit cell** whose periodic repetition generates the entire lattice. If $\boldsymbol{a}_1, \boldsymbol{a}_2, \boldsymbol{a}_3$ are chosen to form a right handed triple, i.e., the scalar triple product $\boldsymbol{a}_1 \cdot (\boldsymbol{a}_2 \times \boldsymbol{a}_3)$ is positive, then this triple product is the volume Ω_0 of the primitive unit cell:

$$\Omega_0 = \boldsymbol{a}_1 \cdot (\boldsymbol{a}_2 \times \boldsymbol{a}_3). \tag{1.2}$$

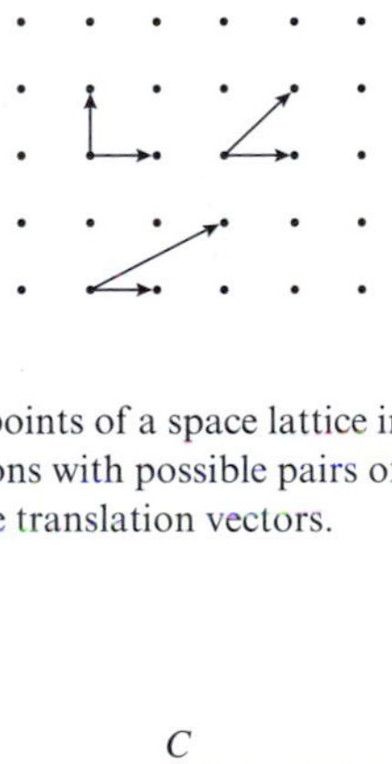

Fig. 1.3
Lattice points of a space lattice in two dimensions with possible pairs of primitive translation vectors.

It should be emphasized that neither the primitive translation vectors nor the primitive unit cell are unique for a given lattice. Three possible pairs of primitive translation vectors for a two-dimensional Bravais lattice are shown in Fig. 1.3. Furthermore, for cubic lattices, the primitive translation vectors may not be orthogonal, and the primitive unit cell may not have the shape of a cube. This is the case for both the fcc and bcc lattices. In general, there exist unit cells which are not primitive and whose volume is larger than that of a primitive unit cell. Of particular interest is the **conventional unit cell** which has the symmetry of the **crystal system** to which the lattice belongs. Thus, for fcc and bcc lattices the conventional unit cell is a cube, frequently referred to as the **elemental cube**. The conventional unit cell is characterized by three axes $\boldsymbol{a}, \boldsymbol{b}, \boldsymbol{c}$ that coincide with three noncoplanar edges of the cell that meet at a corner and three angles α, β, γ between pairs of axes as shown in Fig. 1.4. The magnitudes a, b, c of the axes are the **lattice constants**.

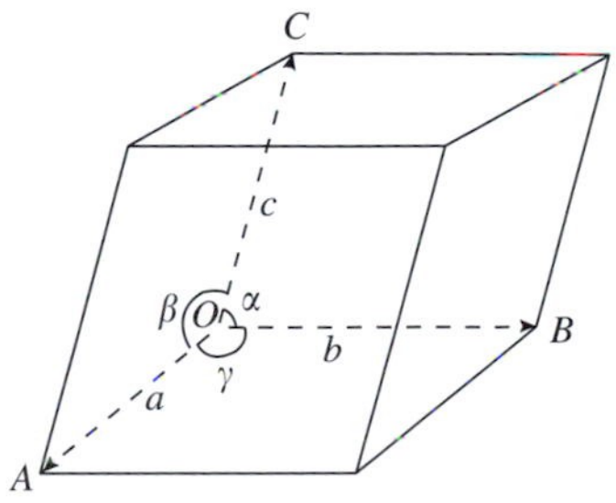

Fig. 1.4
Crystal axes $\boldsymbol{a}, \boldsymbol{b}, \boldsymbol{c}$ and angles α, β, γ.

1.3.3 Bravais lattices

It was shown by Bravais that there are 14 possible lattices in three-dimensional space. A tabulation of the Bravais lattices and crystal systems

Table 1.2 The 14 Bravais lattices for the 7 crystal systems

System	Bravais lattice	Restrictions on conventional cell
cubic	simple or primitive	$a = b = c$ $\alpha = \beta = \gamma = 90°$
	face-centered	
	body-centered	
tetragonal	primitive	$a = b \neq c$
	body-centered	$\alpha = \beta = \gamma = 90°$
orthorhombic	primitive	$a \neq b \neq c$
	base-centered	$\alpha = \beta = \gamma = 90°$
	face-centered	
	body-centered	
hexagonal	primitive	$a = b \neq c$ $\alpha = \beta = 90°,\ \gamma = 120°$
trigonal	primitive	$a = b = c$ $\alpha = \beta = \gamma < 120° \neq 90°$
monoclinic	primitive	$a \neq b \neq c$
	base-centered	$\alpha = \gamma = 90° \neq \beta$
triclinic	primitive	$a \neq b \neq c$ $\alpha \neq \beta \neq \gamma$

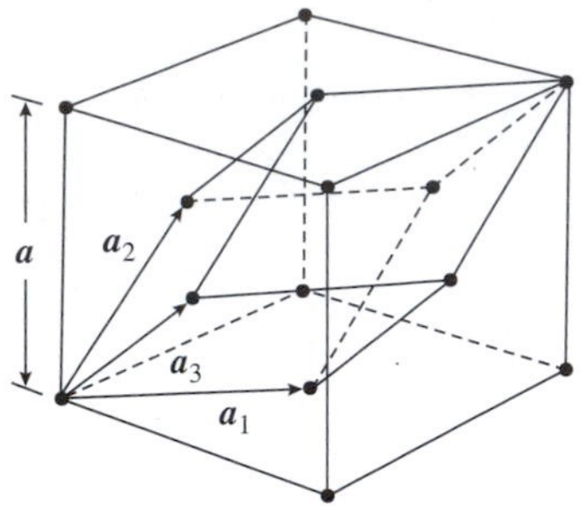

Fig. 1.5
Conventional and primitive unit cells of the face-centered cubic lattice.

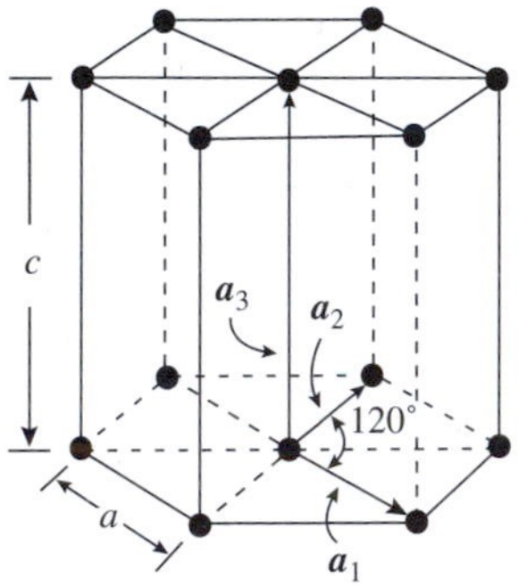

Fig. 1.6
Conventional and primitive unit cells of the hexagonal lattice.

is given in Table 1.2. Three of the Bravais lattices are cubic: simple cubic (sc), face-centered cubic (fcc), and body-centered cubic (bcc). Another one is hexagonal. The face-centered cubic lattice is shown in Fig. 1.5 and the hexagonal lattice in Fig. 1.6.

Example 1.1: Body-centered cubic lattice
Describe the positions of the atoms in a body-centered cubic lattice and specify one or more sets of primitive translation vectors.
Solution. The **body-centered cubic lattice** (bcc) consists of a simple cubic lattice with additional points, one at the center of each lattice cube. The additional points constitute a second simple cubic lattice (Fig. 1.7).

The corner points of the original simple cubic lattice are center points of the second simple cubic lattice. Hence, each point of a body-centered cubic lattice has surroundings that are identical to those of every other point, which is a property characteristic of a Bravais lattice. A possible set of primitive translation vectors is given by: $\boldsymbol{a}_1 = a(1, 0, 0)$, $\boldsymbol{a}_2 = a(\frac{1}{2}, \frac{1}{2}, \frac{1}{2})$, and $\boldsymbol{a}_3 = a(0, 0, 1)$ as shown in Fig. 1.8. A more symmetric set is given by $\boldsymbol{a}_1 = a(\frac{1}{2}, -\frac{1}{2}, -\frac{1}{2})$, $\boldsymbol{a}_2 = a(\frac{1}{2}, \frac{1}{2}, \frac{1}{2})$, and $\boldsymbol{a}_3 = a(-\frac{1}{2}, -\frac{1}{2}, \frac{1}{2})$ as shown in Fig. 1.9. The angle between a pair of the latter primitive translation vectors is $109°28'$.

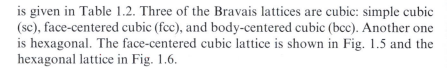

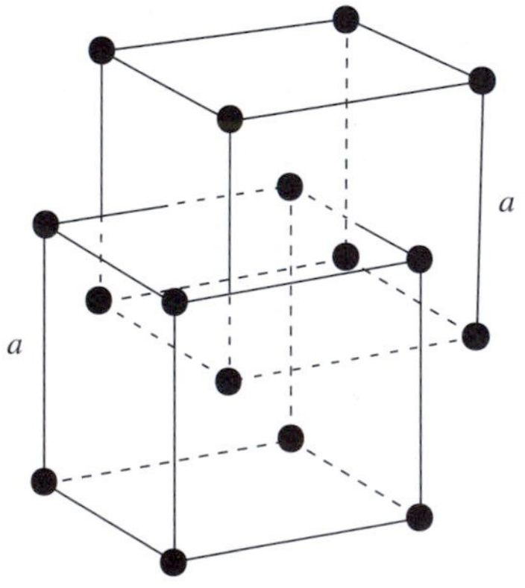

Fig. 1.7
Two interpenetrating cubes forming part of a body-centered cubic Bravais lattice.

In simple crystals such as those of the metallic elements copper and iron, there is only one atom associated with a given lattice point. In more complicated crystals, including those of all semiconductors, there is more than one atom associated with a given lattice point, i.e., the crystal structure has a basis. A primitive unit cell contains the same number of atoms s as does the basis. The atoms of the basis are identified by an index κ which takes the values $1, 2, \ldots, s$. The position vector of the κth atom in the basis can be written as

$$\boldsymbol{R}(\kappa) = d_1^\kappa \boldsymbol{a}_1 + d_2^\kappa \boldsymbol{a}_2 + d_3^\kappa \boldsymbol{a}_3 \tag{1.3}$$

where $d_1^\kappa, d_2^\kappa, d_3^\kappa$ can be chosen such that $0 \leq d_1^\kappa, d_2^\kappa, d_3^\kappa \leq 1$. The complete position vector for the κth atom in unit cell ℓ is specified by

$$\boldsymbol{R}(\ell\kappa) = \boldsymbol{R}(\ell) + \boldsymbol{R}(\kappa). \tag{1.4}$$

crystal structure = lattice + basis

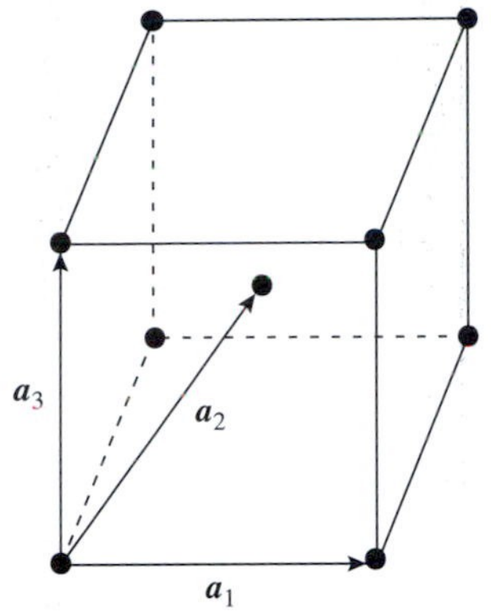

Fig. 1.8
Primitive translation vectors for the body-centered cubic Bravais lattice which connect the lattice point at the origin to lattice points at two cube corners and one cube center.

1.3.4 Crystallographic terminology

The vector $\boldsymbol{R}(\ell)$ is a **translation vector** of the lattice. The crystal as viewed from any point $\boldsymbol{r}$ appears exactly the same as viewed from another point $\boldsymbol{r}'$ given by

$$\boldsymbol{r}' = \boldsymbol{r} + \boldsymbol{R}(\ell). \tag{1.5}$$

In other words, translating the crystal by $\boldsymbol{R}(\ell)$ leaves the crystal invariant.

A **lattice direction** can be defined in terms of a translation vector

$$\boldsymbol{T}(n) = n_1\boldsymbol{a} + n_2\boldsymbol{b} + n_3\boldsymbol{c} \tag{1.6}$$

where $\boldsymbol{a}, \boldsymbol{b}, \boldsymbol{c}$ were introduced at the end of Section 1.3.2 and n_1, n_2, n_3 are integers replacing ℓ_1, ℓ_2, ℓ_3. The lattice direction from the origin to the endpoint of $\boldsymbol{T}(n)$ is designated by enclosing in square brackets the values of n_1, n_2, n_3 for the shortest translation vector parallel to $\boldsymbol{T}(n)$: $[n_1 n_2 n_3]$. If a coefficient n_i is negative, it is designated by placing a bar over it: $\bar{n}_i$.

Certain lattice directions can be transformed into one another by symmetry operations of the crystal. Such directions are called **equivalent**. In a cubic crystal, the directions [100], [010], [001], $[\bar{1}00]$, $[0\bar{1}0]$, $[00\bar{1}]$ are equivalent and are designated collectively by $\langle 100 \rangle$. For a general set of equivalent directions, one chooses the n_1, n_2, n_3 values for one of the directions and encloses them in angular brackets.

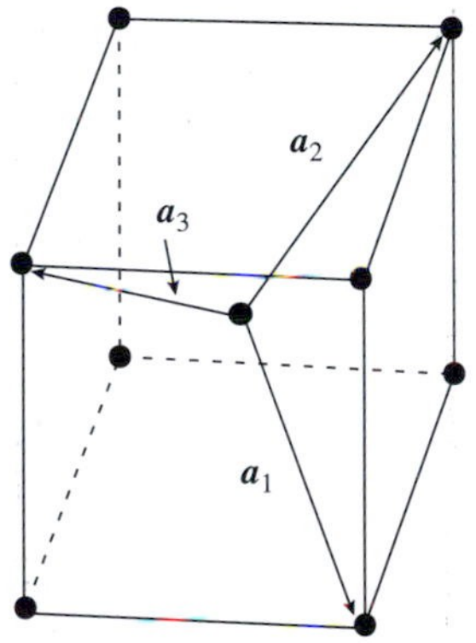

Fig. 1.9
Primitive translation vectors of the body-centered cubic lattice connecting the lattice point at the origin to lattice points at three cube centers.

A **lattice plane** is determined by the intercepts of the plane on the crystallographic axes and is specified by the **Miller indices**. Ordinarily, one uses the axes that coincide with the edges of the conventional unit cell and expresses the intercepts in units n_1, n_2, n_3 of the lattice parameters a, b, c. The Miller indices h, k, l are obtained by taking the reciprocals of n_1, n_2, n_3 and multiplying them by the smallest number that clears the fractions. The lattice plane is designated by the Miller indices enclosed in parentheses: $(hk\ell)$. If an index is negative, a bar is placed over it. It should be noted that here ℓ is not the same as in Eq. (1.1).

A set of crystallographically equivalent planes is identified by the Miller indices for one of the planes enclosed in curly brackets $\{hk\ell\}$. Thus, the set of equivalent planes (100), (010), (001), $(\bar{1}00)$, $(0\bar{1}0)$, $(00\bar{1})$ of a cubic lattice

Table 1.3 Characteristic parameters of face-centered cubic and hexagonal lattices

	Face-centered	Hexagonal
Conventional cell volume	a^3	$(3\sqrt{3}/2)a^2c$
Lattice points per cell	4	3
Primitive cell volume	$\frac{1}{4}a^3$	$(\sqrt{3}/2)a^2c$
Lattice points per unit volume	$4/a^3$	$2/\sqrt{3}a^2c$
Numbers of nearest neighbors	12	$6\ (c > a)$
Nearest neighbor distance	$a/\sqrt{2}$	$a(c > a)$
Number of second neighbors	6	$2(c < \sqrt{3}a)$
Second neighbor distance	a	$c(c < \sqrt{3}a)$

is represented collectively by $\{100\}$. The smallest distance between a pair of parallel $(hk\ell)$ planes in a cubic crystal of lattice constant a is given by

$$d_{hk\ell} = \frac{a}{\sqrt{h^2 + k^2 + \ell^2}}. \tag{1.7}$$

The Bravais lattices that occur most frequently among semiconductors are the face-centered cubic lattice and the hexagonal lattice. The primitive and conventional unit cells of these two lattices are shown in Figs. 1.5 and 1.6, respectively. Their characteristic parameters are given in Table 1.3.

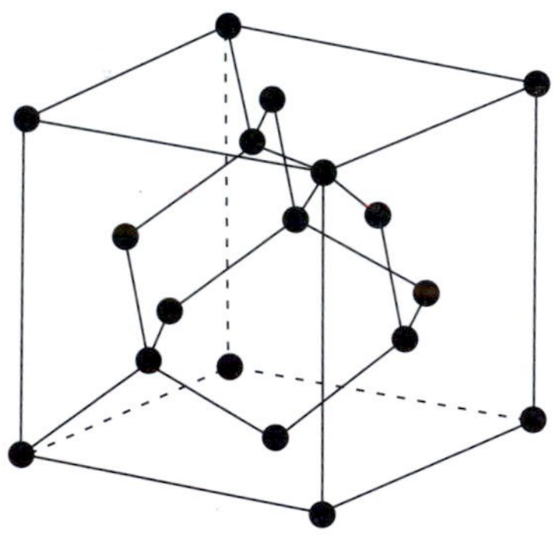

Fig. 1.10
Conventional unit cell of the diamond structure.

1.3.5 Structures of semiconductors

With these preliminaries out of the way, we can now discuss the structures of the important types of semiconductors.

1.3.5.1 Group IV semiconductors

Group IV semiconductors such as Si and Ge crystallize in the diamond structure which consists of two interpenetrating fcc lattices displaced with respect to each other along the (111) direction by one quarter the body diagonal of the elemental cube. There are two atoms per primitive unit cell and there is a center of inversion midway between two nearest neighbor atoms. The diamond structure is illustrated in Fig. 1.10.

Example 1.2: Crystal structure of diamond
Describe the diamond structure and its Bravais lattice.
Solution. The Bravais lattice of the diamond structure is face-centered cubic. A symmetric set of primitive translation vectors can be taken to be $\boldsymbol{a}_1 = a\left(0, \frac{1}{2}, \frac{1}{2}\right)$, $\boldsymbol{a}_2 = a\left(\frac{1}{2}, 0, \frac{1}{2}\right)$, $\boldsymbol{a}_3 = a\left(\frac{1}{2}, \frac{1}{2}, 0\right)$. The angle between pairs of vectors is 60°. The volume of the primitive unit cell is $a^3/4$. The diamond structure consists of two interpenetrating face-centered cubic lattices, displaced with respect to one another along the body diagonal of the cubic cell by one quarter of the length of the diagonal. The basis consists of two atoms located at $a(0, 0, 0)$ and $a\left(\frac{1}{4}, \frac{1}{4}, \frac{1}{4}\right)$. The coordination number, which is the number of the nearest neighbors of an atom in the structure, is four. The number of next-nearest neighbors is 12. The maximum proportion of available volume which can be filled by hard spheres is 0.34, which is significantly smaller than the value of 0.74 for a single fcc lattice. The nearest neighbors of the first and second

basis atoms are at $a(\frac{1}{4}, \frac{1}{4}, \frac{1}{4})$, $a(-\frac{1}{4}, -\frac{1}{4}, \frac{1}{4})$, $a(-\frac{1}{4}, \frac{1}{4}, -\frac{1}{4})$, $a(\frac{1}{4}, -\frac{1}{4}, -\frac{1}{4})$ and at $a(0, 0, 0)$, $a(\frac{1}{2}, \frac{1}{2}, 0)$, $a(\frac{1}{2}, 0, \frac{1}{2})$, $a(0, \frac{1}{2}, \frac{1}{2})$, respectively. The angle between nearest neighbor bonds is $109°28'$.

The diamond structure is not a Bravais lattice because the environments of the two atoms in the basis differ in orientation. This structure is a natural consequence of covalent bonding.

Exercise. Silicon crystallizes in the diamond structure. The silicon lattice constant is $a = 5.43$ Å. Calculate the distance between nearest neighbors r_1 and between next nearest neighbors r_2.
Answer. $r_1 = \sqrt{3}a/4 = 2.35$ Å; $r_2 = a/\sqrt{2} = 3.84$ Å

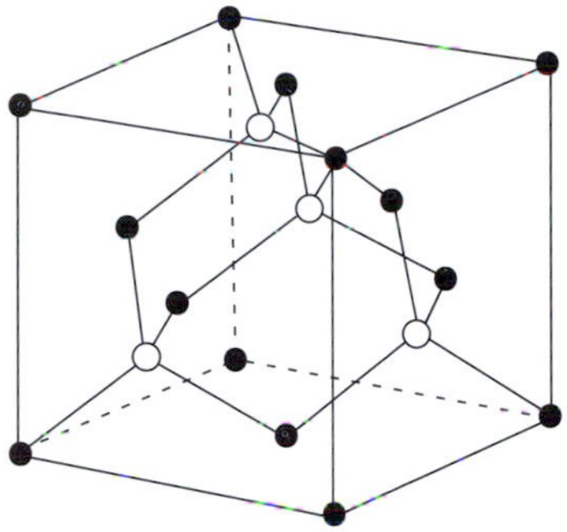

Fig. 1.11
Conventional unit cell of the zincblende structure.

1.3.5.2 III–V semiconductors

III–V semiconductors such as GaAs and InSb crystallize in the zincblende structure which is derived from the diamond structure by making the two interpenetrating fcc lattices different as shown in Fig. 1.11. For example, in GaAs one fcc lattice contains the Ga atoms and the other fcc lattice contains the As atoms. There is no center of inversion between nearest neighbor atoms because these two atoms are different. About each atom of a given species, say Ga, there are four equidistant atoms of the other species, As, arranged at the corners of a tetrahedron. In each elemental cube there are four molecular units of GaAs.

Example 1.3: Crystal structure of zincblende
Describe the crystal structure of zincblende (cubic zinc sulfide) and its Bravais lattice.
Solution. The Bravais lattice of the zincblende structure is face-centered cubic. The face-centered cubic lattice is formed by adding to the simple cubic lattice one point in the center of each square face. Any point can be considered to be either a corner point or a face-centered point of any of the three kinds of faces (bottom, top and side) of a simple cube. Hence, the face-centered cubic lattice is a Bravais lattice.

The primitive unit cell of the face-centered cubic crystal is rhombohedral as seen in Fig. 1.5 and has three of its edges coinciding with the primitive translation vectors. The conventional unit cell is a cube. The primitive cell has one quarter of the volume of the conventional cell and has less symmetry.

The zincblende structure consists of two interpenetrating fcc lattices displaced from each other by one-quarter of a body diagonal. The S atoms are placed on one fcc lattice and the Zn atoms on the other fcc lattice, as in Fig. 1.7. There are four molecules of ZnS per conventional cell. The basis attached to each point of the lattice is ZnS, with respective coordinates of the atoms: $S = a(0, 0, 0)$ and $Zn = a(\frac{1}{4}, \frac{1}{4}, \frac{1}{4})$. The ZnS structure does not have inversion symmetry.

The diamond structure is obtained from the zincblende structure by making all atoms the same species.

Exercise. Describe the crystal structure of InAs, which has a lattice constant $a = 6.048$ Å (see Fig. 1.11).

(a) What are the volume Ω_0 of the primitive unit cell and the number of atoms in it?

Answer. $\Omega_0 = |\boldsymbol{a}_1 \cdot \boldsymbol{a}_2 \times \boldsymbol{a}_3| = a^3/4 = 55.3\,\text{Å}^3$, where $\boldsymbol{a}_1 = a(0, \frac{1}{2}, \frac{1}{2})$, $\boldsymbol{a}_2 = a(\frac{1}{2}, 0, \frac{1}{2})$, $\boldsymbol{a}_3 = a(\frac{1}{2}, \frac{1}{2}, 0)$; two atoms

(b) Calculate the angle θ between two tetrahedral bonds knowing that the angle is that between two body diagonals of a cube.

Answer. Two body-diagonal vectors, taking the cube center as origin, are $(1, 1, 1)$ and $(-1, -1, 1)$, each with length $\sqrt{3}$ in units of $\frac{a}{2}$. The distance between their endpoints is $2\sqrt{2}$. From trigonometry, $c^2 = a^2 + b^2 - 2ab\cos\theta$, $c = 2\sqrt{2}$, $a = b = \sqrt{3}$.
$8 = 3 + 3 - 2 \cdot 3\cos\theta$, $\cos\theta = -\frac{1}{3}$, $\theta = 109°28'$.

(c) How many atoms are in the conventional unit cell?

Answer. From Fig. 1.11, there are:

4 complete	∘ atoms
$6 \times \frac{1}{2}$	• atoms in faces
$8 \times \frac{1}{8}$	• atoms at corners
8 atoms	

(d) What are the number and separation r_1 of nearest neighbors?

Answer. 4 nearest neighbors; $r_1 = \sqrt{3}a/4 = 2.62\,\text{Å}$

(e) What are the number and separation r_2 of second neighbors?

Answer. 12 second neighbors; $r_2 = a/\sqrt{2} = 4.28\,\text{Å}$ (same as for nearest neighbors of the fcc lattice)

1.3.5.3 II–VI semiconductors

CdS, a II–VI compound, typically occurs in the wurtzite structure which consists of four interpenetrating hexagonal lattices, two occupied by Cd atoms and the other two by S atoms. There are four atoms per primitive unit cell and no center of inversion. Each atom is surrounded by four nearest neighbor atoms of the opposite species located at the corners of a tetrahedron as shown in Fig. 1.12.

As we have seen previously, the zincblende structure consists of two interpenetrating face-centered cubic sublattices, one sublattice being displaced with respect to the other by $\frac{1}{4}a, \frac{1}{4}a, \frac{1}{4}a$, where a is the lattice constant. The wurtzite structure (Fig. 1.12) can be viewed as consisting of two interpenetrating hexagonal close-packed (hcp) structures, one for each of the atomic components. The hcp structures are displaced with respect to one another along the a_3-axis by an amount uc where u is a parameter on the order of 3/8. It should be noted that the hcp structure is **not** a Bravais lattice, but consists of two interpenetrating hexagonal Bravais lattices displaced relative to one another by a vector $\boldsymbol{d}$ given by

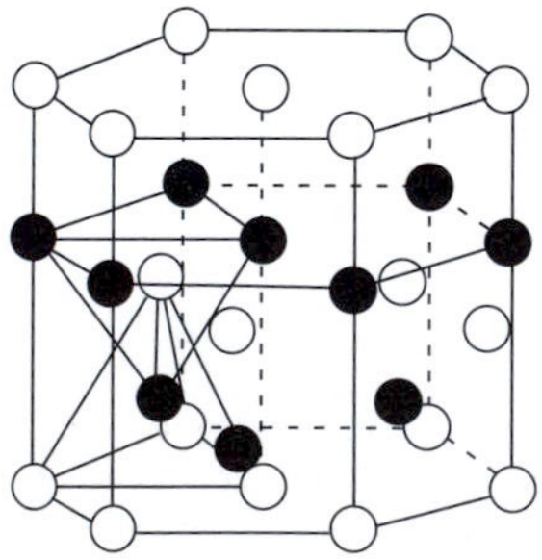

Fig. 1.12
Wurtzite structure showing the tetrahedral environment of each type of atom.

$$\boldsymbol{d} = \tfrac{1}{3}\boldsymbol{a}_1 + \tfrac{2}{3}\boldsymbol{a}_2 + \tfrac{1}{2}\boldsymbol{a}_3, \tag{1.8}$$

where $\boldsymbol{a}_1$, $\boldsymbol{a}_2$, and $\boldsymbol{a}_3$ are shown in Fig. 1.6. The wurtzite structure therefore contains four interpenetrating hexagonal Bravais lattices and four atoms per primitive unit cell.

1.3.5.4 III–VI semiconductors

III–VI semiconductors such as GaSe and InSe crystallize in layered structures, Fig. 1.13, with various **polytypes** possible depending on the stacking of the layers. The number of atoms per primitive unit cell can be four or eight according to the polytype.

In a layered semiconductor consisting of metal atoms and chalcogen atoms such as InSe, the metal atoms lie in two-dimensional sheets and are bound together by chemical bonds. Pairs of metal sheets are connected by chemical bonds between pairs of metal atoms, one atom of a pair being in one sheet and the other atom being in the other sheet. The chalcogen atoms are also arranged in two-dimensional sheets and are bound by chemical bonds to metal atoms in adjacent metal sheets. Each metal sheet is chemically bound to another metal sheet on one side and to a chalcogen sheet on the other. Each chalcogen sheet, on the other hand, is chemically bound only to the adjacent metal sheet. The result is a sandwich of four sheets, a pair of metal sheets on the inside and a pair of chalcogen sheets on the outside, as shown in Fig. 1.14a. Also shown (Figs. 1.14b and 1.14c) are views from above and from the side of the hexagonal array of chalcogen and metal atoms in adjacent sheets. Chemical bonds do not exist between sandwiches. Sandwiches are bound together by weak **van der Waals forces** to form the crystal. The spacing between successive sandwiches is called a **van der Waals gap.**

The physical properties of layered compounds are strongly anisotropic. These materials are not mechanically strong. If a shear stress is applied in a direction parallel to the sheets, the sandwiches will slide with respect to each other and, in fact, can be pulled off like onion peels.

The electrical conductivity is also anisotropic. The conductivity is typically high within the plane of the sheets and very low perpendicular to it.

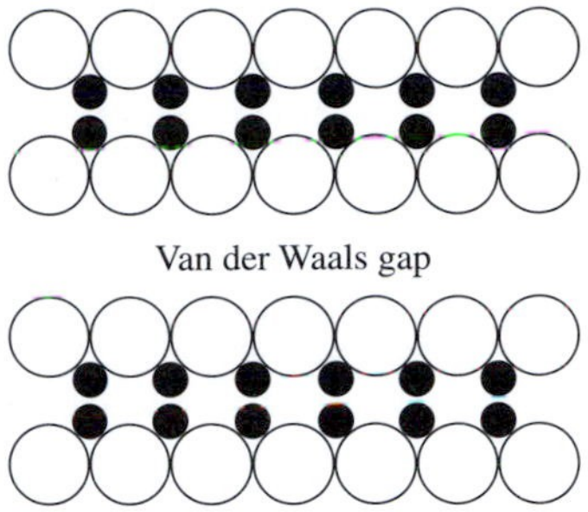

Fig. 1.13
Layered structure.

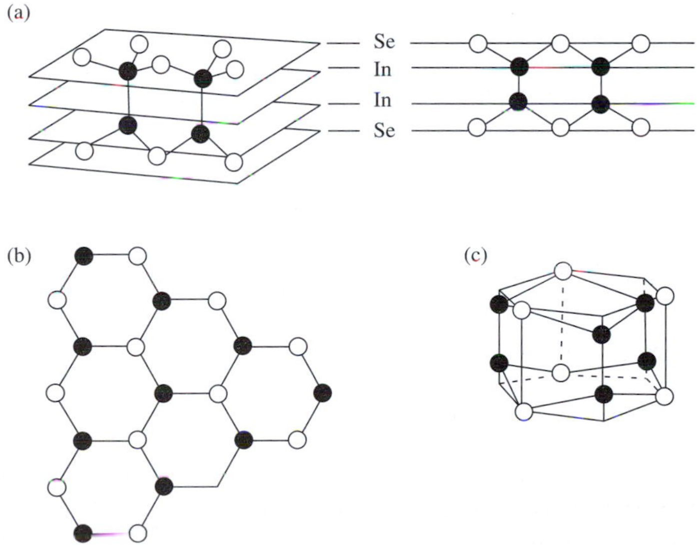

Fig. 1.14
Schematic representation of the InSe structure: (a) Structure of a sandwich composed of two In sheets between two Se sheets. (b) Hexagonal array of a subsandwich consisting of adjacent In and Se sheets viewed from above. (c) Chair-like deformation of the hexagonal array.

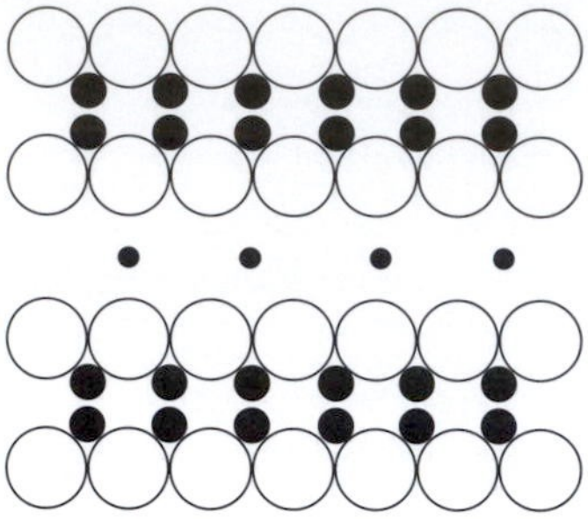

Fig. 1.15
Intercalated Li atoms (small full circles) in the van der Waals gap between sandwiches of InSe.

Table 1.4 Lattice constants a in Å for representative cubic semiconductors at room temperature

Material	Structure	a	Ref.	Material	Structure	a	Ref
Si	diamond	5.43	a				
Ge	diamond	5.65	a	ZnS	zincblende	5.423	b
AlP	zincblende	5.431	b	ZnSe	zincblende	5.661	b
AlAs	zincblende	5.631	b	ZnTe	zincblende	6.082	b
AlSb	zincblende	6.142	b	CdS	zincblende	5.832	b
GaP	zincblende	5.447	b	CdSe	zincblende	6.052	b
GaAs	zincblende	5.646	b	CdTe	zincblende	6.423	b
GaSb	zincblende	6.130	b	PbS	rocksalt	5.935	b
InAs	zincblende	6.048	b	PbSe	rocksalt	6.152	b
InSb	zincblende	6.474	b	PbTe	rocksalt	6.353	b

a Kittel (1986)
b Weißmantel and Hamann (1979)

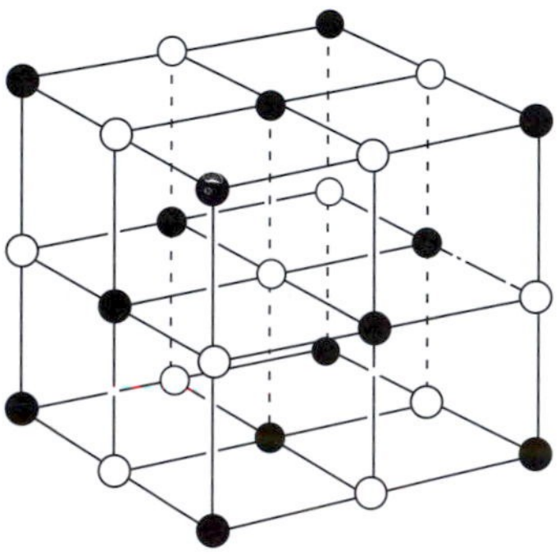

Fig. 1.16
Rock salt structure.

Comment. In the process of **intercalation**, foreign atoms and even molecules can easily be introduced into the van der Waals gap. The physical properties of the material can be drastically modified by intercalation. When InSe is intercalated with Li, as shown in Fig. 1.15, for example, its conductivity increases markedly as a result of free electrons donated by the Li atoms. The Li^+ ions which arise can diffuse easily within the van der Waals gap.

1.3.5.5 IV–VI semiconductors

IV–VI semiconductors such as PbS and PbTe crystallize in the rock salt structure, Fig. 1.16, which consists of two interpenetrating fcc lattices displaced along the (100) direction by one-half the edge of the elemental cube. There are two atoms per primitive unit cell and each atomic site is a center of inversion.

Lattice constants for representative semiconductors are given in Tables 1.4 and 1.5.

Table 1.5 Lattice constant a and the c/a ratio for representative wurtzite structure semiconductors at room temperature (after Weißmantel and Hamann 1979)

Material	a	c/a
ZnS	3.819	1.64
CdS	4.139	1.62
CdSe	4.309	1.63

1.4 Chemical bonding in semiconductors

1.4.1 Diamond structure semiconductors

The bonds between nearest neighbor atoms in Si and Ge are covalent, electron pair bonds as described by Pauling (1960). Each atom resides at the center of a tetrahedron and is bound to four nearest neighbors located at the corners of the tetrahedron. This tetrahedral bonding is the same type as that found in many carbon compounds and in the diamond form of carbon. It may be understood as follows. The electronic configurations of C, Si and Ge are

$$\mathrm{C}:\ (1s)^2\ (2s)^2\ (2p)^2$$

$$\mathrm{Si}:\ (1s)^2\ (2s)^2\ (2p)^6\ (3s)^2\ (3p)^2$$

$$\mathrm{Ge}:\ (1s)^2\ (2s)^2\ (2p)^6\ (3s)^2\ (3p)^6\ (4s)^2 (4p)^2.$$

In each case there are four electrons, two s-electrons and two p-electrons beyond closed shells. Each of these four electrons can be paired with a similar electron of opposite spin on a neighboring atom to form an electron pair bond. If we represent the electrons by dots, then the electron pair bond formation in silicon, for example, can be represented symbolically by

$$\cdot \dot{\underset{\cdot}{\mathrm{Si}}} \cdot + \cdot \dot{\underset{\cdot}{\mathrm{Si}}} \cdot \longrightarrow \cdot \dot{\underset{\cdot}{\mathrm{Si}}} : \dot{\underset{\cdot}{\mathrm{Si}}} \cdot$$

Quantum mechanically, the bond strength is determined by the overlap of the atomic orbitals of the two electrons on neighboring atoms. This overlap can be enhanced by "promoting" an s-electron to a p-state. For Si, the resulting electron configuration is $(1s)^2\,(2s)^2\,(2p)^6\,(3s)\,(3p)^3$. Let the $3s$ and $3p$ orbitals be denoted by s, p_x, p_y, p_z. One can "hybridize" these orbitals to form sp^3 **hybrid orbitals** or **tetrahedral orbitals** by constructing the following linear combinations (Pauling 1960):

$$\begin{aligned}
\varphi_{t1} &= \tfrac{1}{2}(s + p_x + p_y + p_z) \\
\varphi_{t2} &= \tfrac{1}{2}(s + p_x - p_y - p_z) \\
\varphi_{t3} &= \tfrac{1}{2}(s - p_x + p_y - p_z) \\
\varphi_{t4} &= \tfrac{1}{2}(s - p_x - p_y + p_z).
\end{aligned}$$

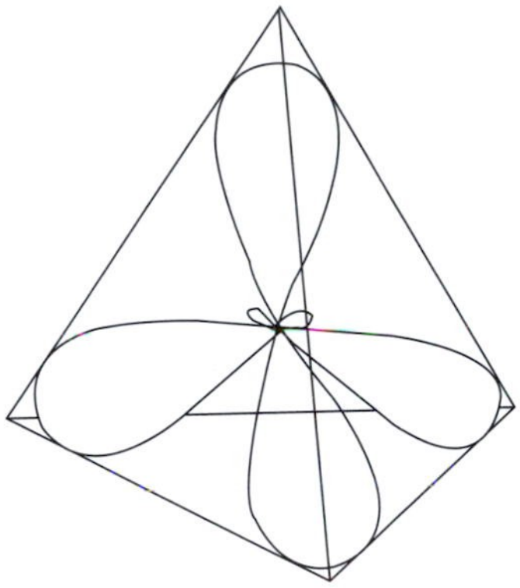

Fig. 1.17
sp^3 tetrahedral orbitals.

The tetrahedral orbitals have their maximum amplitudes along the directions to the corners of a tetrahedron whose center is at the origin of coordinates of the orbitals. If two such orbitals are centered on neighboring atoms and oriented along the line connecting the two atoms, a very large overlap of the orbitals occurs, which leads to a correspondingly large bond strength for the electron pair bond between the atoms. The angle between any pair of bonds of a given atom is $109°28'$. Such bonds are extremely strong, as indicated by the great hardness and resistance to shear of diamond.

> **Comment.** Starting with an isolated silicon atom in the $3s^2 3p^2$ ground state configuration, an energy of 3.5 eV is required to promote an s electron to a p state yielding the $3s3p^3$ configuration. The directions of maximum amplitude of the sp^3 orbitals along the four tetrahedral axes are shown in Fig. 1.17. When electron pair bonds are formed between neighboring atoms with sp^3 orbitals, a substantial amount of energy is gained that offsets the promotional energy and yields a net binding energy. The bonds between a given atom and its four nearest neighbors are shown in Fig. 1.18. The angle θ between bonds of $109°28'$ is extremely resistant to deformation.

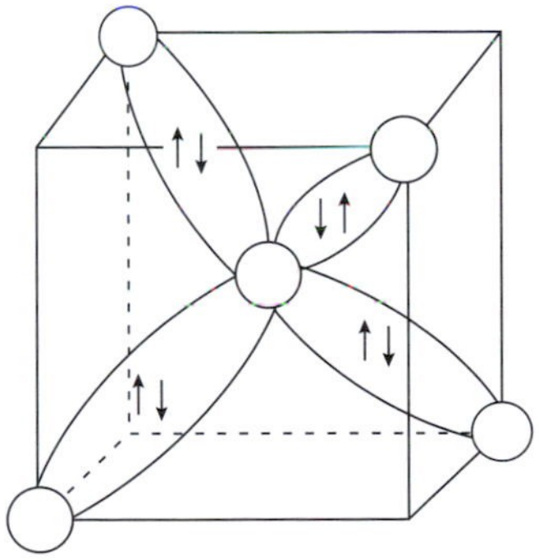

Fig. 1.18
Electron pair bonds between a Si atom and its four nearest neighbors.
The arrows indicate electron spin orientation.

1.4.2 Zincblende structure semiconductors

We already noted that the zincblende structure, like the diamond structure, contains two interpenetrating fcc lattices. This suggests that the bonding is primarily covalent with sp^3 orbitals participating in the electron pair bonds. Taking GaAs as an example, the Ga and As atoms have the electron configurations

$$\text{Ga}: (1s)^2\ (2s)^2\ (2p)^6\ (3s)^2\ (3p)^6\ (4s)^2\ (4p)$$
$$\text{As}: (1s)^2\ (2s)^2\ (2p)^6\ (3s)^2\ (3p)^6\ (4s)^2\ (4p)^3$$

which do not correspond to four electrons beyond closed shells. However, we can achieve the latter situation by transferring one electron from an As atom to a nearest neighbor Ga atom yielding an As^+ ion and a Ga^- ion. When this is done for all Ga–As pairs, each atom in the crystal has four electrons beyond closed shells and tetrahedral electron pair bonds can be constructed using sp^3 hybrid orbitals just as in the group IV semiconductors. We can therefore refer to the bond in Ga^-As^+ as a **covalent bond**. On the other hand, three electrons can be transferred from a Ga atom to an As atom to yield $Ga^{+++}As^{---}$. Each atom is an ion with only closed shells of electrons. Hence, the bond in $Ga^{+++}As^{---}$ can be regarded as an **ionic bond**. The actual bond is a **mixed bond** containing both covalent and ionic character.

The chemical bond between two atoms A and B can be described by a **molecular orbital** ψ given by (Coulson *et al.* 1962).

$$\psi = \varphi_A + \lambda\varphi_B, \tag{1.9}$$

where φ_A and φ_B are atomic orbitals centered on atoms A and B, respectively, and λ is a variational parameter. For the case that A and B refer to As and Ga atoms, respectively, φ_A and φ_B are sp^3 hybrid orbitals oriented along the line between nearest neighbor As and Ga atoms. The orbital part of the wave function Ψ for the two electrons of opposite spin forming the electron pair bond is

$$\begin{aligned}\Psi &= \psi(1)\psi(2)\\ &= [\varphi_A(1) + \lambda\varphi_B(1)][\varphi_A(2) + \lambda\varphi_B(2)].\end{aligned} \tag{1.10}$$

Multiplying out the right hand side of Eq. (1.9) gives

$$\begin{aligned}\Psi = \varphi_A(1)\varphi_A(2) + \lambda[\varphi_A(1)\varphi_B(2) + \varphi_A(2)(\varphi_B(1)]\\ + \lambda^2\varphi_B(1)\varphi_B(2).\end{aligned} \tag{1.11}$$

In the first term on the right hand side of Eq. (1.10) both electrons are localized on atom A, in the second term they are evenly distributed between atoms A and B, and in the last term they are both localized on atom B.

The value of λ lies in the range of $0 \leq \lambda \leq \infty$. For $\lambda = 0$, both electrons are localized on atom A, while for $\lambda = \infty$, they are localized on atom B. These cases correspond to purely ionic bonds. For $\lambda = 1$, each electron is equally shared between the A and B atoms, and we have a purely covalent bond.

If a semiconductor has the composition $A^N B^{8-N}$, where A is the anion (group IV, V or VI element) and B is the cation (group IV, III, or II element), the static effective charges Q_A and Q_B on atoms A and B are given

Table 1.6 Magnitude of effective charge Q and ionicity parameter λ for compound semiconductors (after Coulson *et al.* 1962)

	Q/e	λ		Q/e	λ
ZnS	0.47	0.49	AlAs	0.47	0.68
CdS	0.49	0.48	GaAs	0.46	0.68
BP	0.32	0.71	InAs	0.49	0.68
AlP	0.46	0.68	ZnTe	0.45	0.49
GaP	0.45	0.68	CdTe	0.47	0.49
InP	0.49	0.68	AlSb	0.44	0.69
ZnSe	0.47	0.49	GaSb	0.43	0.69
CdSe	0.49	0.48	InSb	0.46	0.68

by (Coulson *et al.* 1962)

$$Q_A = \frac{N\lambda^2 - (8 - N)}{1 + \lambda^2} e = -Q_B, \tag{1.12}$$

where e is the magnitude of the electron charge. The quantity N takes on the values 4 for group IV semiconductors, 5 for III–V semiconductors, and 6 for II–VI semiconductors. Since group IV semiconductors have pure covalent bonds, $\lambda = 1$ and $Q_A = Q_B = 0$. The III–V and II–VI semiconductors have $\lambda \neq 1$ and $Q_A = -Q_B \neq 0$. The bonds have a partial ionic character and are therefore mixed bonds. The magnitudes of the static effective charges for a number of compound semiconductors are given in Table 1.6. An alternative treatment of ionicity has been given by Phillips (1973).

1.4.3 III–VI layered semiconductors

In a layered compound such as InSe, each In atom is bonded to three nearest neighbor Se atoms which lie in a plane and one nearest neighbor In atom which lies on a line perpendicular to this plane and connecting the two In atoms. These four atoms are located at the corners of a distorted tetrahedron. Each Se atom, on the other hand, is bonded directly only to three nearest neighbor In atoms. This bonding can be understood by noting that In and Se atoms have three and six electrons, respectively, beyond closed shells:

$$\mathrm{In}: (1s)^2\,(2s)^2\,(2p)^6\,(3s)^2\,(3p)^6\,(4s)^2\,(4p)^6\,(5s)^2(5p)$$
$$\mathrm{Se}: (1s)^2\,(2s)^2\,(2p)^6\,(3s)^2\,(3p)^6(4s)^2\,(4p)^4.$$

By transferring one electron from an Se atom to a neighboring In atom according to the equation

$$\dot{\underset{\cdot}{\mathrm{In}}}\cdot + \cdot\dot{\underset{\cdot\cdot}{\mathrm{Se}}}: \longrightarrow \cdot\dot{\underset{\cdot}{\mathrm{In}}}\cdot^{-} + \cdot\dot{\underset{\cdot}{\mathrm{Se}}}:^{+}$$

we obtain four electrons on the In^+ ion and five electrons on the Se^+ ion. Four sp^3 hybrid orbitals can be constructed on each ion. In the case of In^-, four unpaired electrons can occupy the four orbitals and be used to form four electron-pair bonds with four neighboring atoms. In the case of Se^+, two electrons must pair up in a single sp^3 orbital and the other three electrons can occupy the remaining three sp^3 orbitals and be used to form three electron-pair bonds with three neighboring In atoms. It is the fact that one of the sp^3 orbitals of Se^+ cannot be used to form an electron-pair bond that provides a simple explanation for the layered structure of InSe.

1.4.4 Wurtzite structure semiconductors

The II–VI material ZnS occurs in both the zincblende and wurtzite structures. CdS is a II–VI semiconductor which occurs primarily in the wurtzite structure. The electronic configurations of the atoms are

$$\begin{aligned} \mathrm{Cd}&: (1s)^2\,(2s)^2\,(2p)^6\,(3s)^2\,(3p)^6\,(4s)^2\,(4p)^6(5s)^2 \\ \mathrm{S}&: (1s)^2\,(2s)^2\,(2p)^6\,(3s)^2\,(3p)^4 \end{aligned}$$

corresponding to two and six electrons beyond closed shells for Cd and S, respectively. By transferring two electrons from an S atom to a neighboring Cd atom according to

$$\cdot\mathrm{Cd}\cdot + \cdot\dot{\underset{\cdot\cdot}{\mathrm{S}}}: \longrightarrow \cdot\dot{\underset{\cdot}{\mathrm{Cd}}}\,\cdot^{2-} + \cdot\dot{\underset{\cdot}{\mathrm{S}}}\,\cdot^{2+}$$

we obtain four electrons on each ion and the possibility of four tetrahedral bonds per atom using sp^3 hybrid orbitals. The orientation of the tetrahedra in successive layers of the wurtzite structure differs from that of the zincblende structure.

1.4.5 IV–VI semiconductors

The rock salt structure in which the lead chalcogenides PbS, PbSe, and PbTe crystallize is typical of strongly ionic materials such as NaCl and KBr. This suggests that the chalcogen atoms acquire two electrons from neighboring Pb atoms and become doubly negatively charged with closed shell electronic configurations similar to those of the halide ions in the alkali halides. Since Pb has four electrons beyond closed shells, we have, for example,

$$\cdot\dot{\underset{\cdot}{\mathrm{Pb}}}\cdot + \cdot\dot{\underset{\cdot\cdot}{\mathrm{S}}}: \longrightarrow \cdot\mathrm{Pb}\,\cdot^{2+} + :\ddot{\underset{\cdot\cdot}{\mathrm{S}}}:^{2-}. \tag{1.13}$$

Although the bonding in the lead salts seems to be predominantly ionic, there is evidence (Smith 1968) that the bonding also has some covalent character. Indeed, in none of the compound semiconductors is the bonding purely covalent or purely ionic.

1.5 Growth of pure semiconductor crystals

The concentration of impurities is an essential factor concerning the quality and use of semiconductors. It is therefore important to have good control of the impurity content. Furthermore, the material should be available in single-crystal form with very few crystalline defects such as dislocations and stacking faults. At the present time silicon is the basis for the production of the vast majority of semiconductor devices. No other material can match it. A content of certain impurities of 10^{-9} or less is often required.

The starting point is pure polycrystalline silicon obtained by hydrogen reduction of $SiCl_4$, which is a liquid that can be distilled many times in order to purify it. The silicon thus produced is melted and a single crystal is grown from the melt. One of the main difficulties in achieving single crystals of large dimensions is controlling the temperature gradient necessary to obtain solidification. A widely used method is the **Czochralski method** in which a seed crystal is touched to the melt and the crystal is "pulled" from it by slowly withdrawing the seed as shown in Fig. 1.19. The small single crystal seed is mounted on a rotating axis and is put in contact with the surface of the liquid. The temperature gradients are adjusted in such a way that crystal growth occurs at the surface of the seed. Crystals of diameters up to 50 cm are obtainable by this method. At present the standard diameters used for microelectronic applications range from 10 to 15 cm.

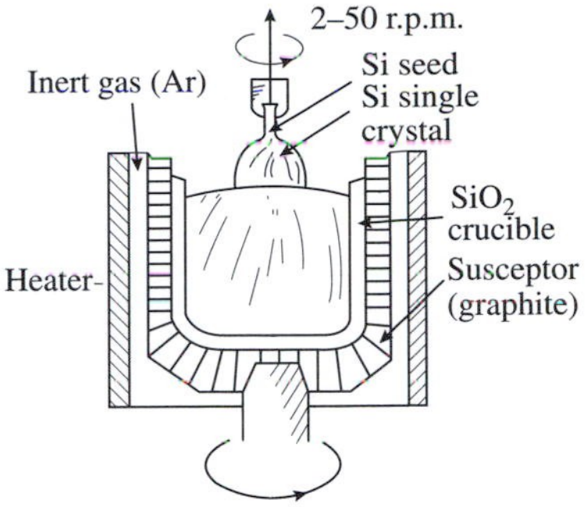

Fig. 1.19
Diagram of the Czochralski method of crystal growth.

In the Czochralski method a possible source of pollution is the crucible. At the melting point of silicon (1418 C), the crucible is always chemically attacked and pollutes the melt. An effective way of removing the impurities is to use the **zone refining method** or **floating zone method** developed by Pfann and shown in Fig. 1.20. One starts with a single crystal or polycrystalline silicon ingot. A narrow cylinder on the bottom of the ingot is heated by a coil attached to a high-frequency generator. The local induction heating forms a narrow molten zone. By slowly moving the heating coil upward, a slow displacement of the molten zone takes place followed by solidification. Once the whole polycrystalline ingot has been passed through by the molten zone, one obtains a single-crystal ingot. By repeating this process many times one can obtain extremely pure material. Silicon with impurity levels as low as 10^{-10} can be produced by zone refining. Such extraordinary purity has made possible the development of modern electronic and computer technology to the extent we know it today.

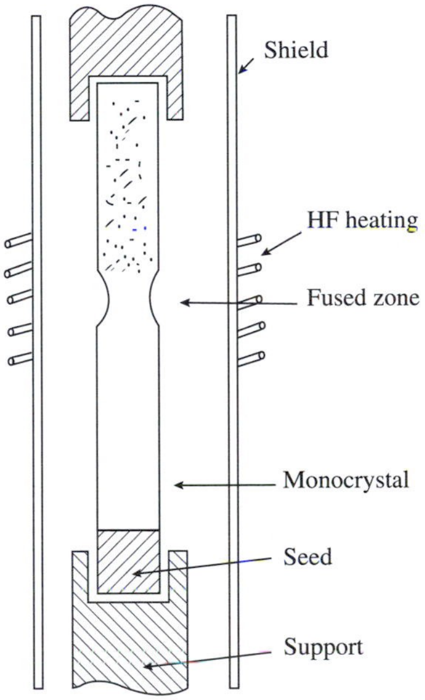

Fig. 1.20
Diagram of the zone refining method of crystal purification.

To understand the basis of zone refining consider a solid in equilibrium with its liquid phase. The impurity concentration is c_L in the liquid phase and c_S in the solid phase. The segregation coefficient K of the impurity is defined to be the ratio of c_L to c_S:

$$K = \frac{c_L}{c_S}. \tag{1.14}$$

In general K is much larger than unity, impurities being more soluble in the liquid than in the solid. Let the impurity concentration in the starting ingot be c_0. The initial molten zone also has impurity concentration c_0, but the first solid segment obtained after the passage of the molten zone has the concentration $c' = c_0/K$. If $K > 1$, the solid segment has a lower impurity concentration than initially. As the molten zone progresses, it becomes enriched with impurities until its impurity concentration reaches the value Kc_0. Further motion of the molten zone through the crystal gives no additional purification.

After the first passage of the molten zone is complete, the ingot is allowed to cool down. A second passage is then started from the bottom with the new impurity concentration of the ingot being c'. The first segment which recrystallizes after the second passage of the molten zone has the impurity concentration $c'' = c_0/K^2$. Many passages can be made starting always at the bottom of the ingot. Eventually the impurity concentration profile reaches a steady state which cannot be improved upon by further passages. If the ingot length is L and the zone length is ℓ, the final impurity concentration c_f at the bottom of the ingot is given by (Pfann 1957)

$$c_f = c_0/K^S,$$

where $S \simeq 1.4\,L/\ell$. Clearly, if both K and S are large, very significant purification can be achieved.

Problems

1. Prove that in a cubic crystal the direction [hkℓ] is perpendicular to the plane ($hk\ell$). Can this demonstration be generalized to all crystal systems?
2. Determine the direction perpendicular to the (111) plane of the tetragonal lattice.
3. Identify the plane with maximum density of atoms in a face-centered cubic lattice. Show that the maximum proportion of the available volume which can be filled by hard spheres in a face-centered cubic lattice is 0.74. What is this proportion for a body-centered cubic lattice?
4. The lattice constant of the diamond form of carbon is 3.567 Å. Given that the density is 3.516 g/cm^3, calculate the number of atoms in the elemental cube. Is your result consistent with the diamond structure?
5. Carbon also crystallizes as graphite which belongs to the hexagonal system. If the lattice constants are a = 2.41 Å, and c = 6.70 Å, and the density is 2.25 g/cm^3, how many atoms does the primitive unit cell contain? How many interpenetrating hexagonal Bravais lattices does the graphite structure have?
6. For CdS in the wurtzite structure the lattice constants are specified by a = 4.14 Å and $c = 1.62\,a$. Calculate the density of the crystal and the volume of the primitive unit cell. What is the number of nearest neighbors and the distance between them? How many atoms does the primitive unit cell contain?

References

C. A. Coulson, L. B. Redei, and D. Stocker, *Proc. Roy. Soc.* (London) A**270**, 357 (1962).

C. Kittel, *Introduction to Solid State Physics*, Sixth edition (John Wiley, New York, 1986).

L. Pauling, *The Nature of the Chemical Bond*, Third edition (Cornell University Press, Ithaca, 1960).

W. G. Pfann, in *Solid State Physics*, Vol. 4, eds. F. Seitz and D. Turnbull (Academic Press, New York, 1957).

J. C. Phillips, *Bonds and Bands in Semiconductors* (Academic Press, New York, 1973).

R. A. Smith, *Semiconductors* (Cambridge University Press, Cambridge, 1968).

Ch. Weißmantel and C. Hamann, *Grundlagen der Festkörperphysik* (Springer-Verlag, Berlin, 1979).

Electronic energy bands: basic theory

2

Key ideas

Bands and gaps

The eigenstates of an electron moving in the periodic potential of a crystal are *Bloch states* whose energy eigenvalues can only take certain allowed values lying in *energy bands*.

Electrons move as if the ions are instantaneously at rest.

2.1 Schrödinger equation

The Fourier expansion of a *periodic potential* involves sums over wave vectors called *reciprocal lattice vectors*.

2.2 Electrons in a periodic potential

The electronic eigenstates are determined by solving the *Schrödinger equation* for a periodic potential.

2.3 Schrödinger equation for a periodic potential

The *density of electronic states* in $\boldsymbol{k}$-space for crystal volume Ω is

$$g(\boldsymbol{k}) = \frac{\Omega}{(2\pi)^3}.$$

2.4 Expansion of the eigenfunction in plane waves

Each eigenfunction of an electron moving in a periodic potential consists of a *plane wave factor* and a *periodic function factor*.

2.5 Bloch's theorem

A weak periodic potential perturbation leads to strong scattering of electrons with wave vectors $\boldsymbol{k}$ satisfying the *Bragg condition* $\boldsymbol{k} \cdot \left|\frac{\boldsymbol{G}}{2}\right| = \left|\frac{\boldsymbol{G}}{2}\right|^2$ where $\boldsymbol{G}$ is a reciprocal lattice vector.

2.6 Electrons in a weak periodic potential

The planes which bisect the reciprocal lattice vectors $\boldsymbol{G}$, the Bragg planes, enclose spaces called *Brillouin zones*.

2.7 Brillouin zones

The degeneracy of the zero order states $|\boldsymbol{k}\rangle$ and $|\boldsymbol{k}-\boldsymbol{G}\rangle$ at the Brillouin zone boundary is lifted in degenerate perturbation theory. The perturbed energies E_k are split at the Brillouin zone boundary by an amount $2|V_G|$, leading to a discontinuity or *gap* in E_k which separates two *energy bands*. In the *reduced zone scheme*, all of the energy bands can be represented within the range of wave vectors in the first Brillouin zone.

2.8 Energy bands and energy band gaps

When the electron–ion interaction is strong, the *tight binding* or *LCAO method* provides the electron wave functions as series of functions localized about the nuclei. The localized functions can be represented by *Wannier functions* or by *atomic orbitals*.

2.9 Tight binding method

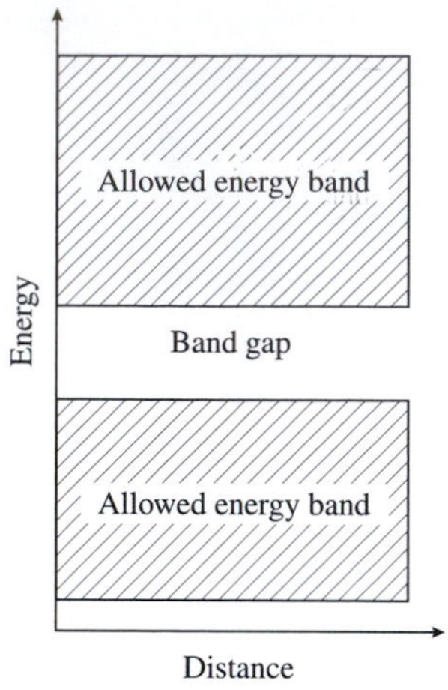

Fig. 2.1
Allowed energy bands and forbidden energy gaps of an electron in a periodic potential.

Although applications such as rectifiers that we now associate with semiconductors had been known as far back as the nineteenth century, it was not until after the development of quantum mechanics in the late 1920s that the physical basis for these properties became understood. A. H. Wilson showed in 1931 how the electronic band theory of crystals developed by F. Bloch can be applied to the understanding of semiconductors. Properties such as the negative temperature coefficient of electrical resistivity follow naturally from Wilson's theory. The key ingredients are the **Bloch states** which are the eigenstates of an electron moving in the periodic potential of a crystal. The energies of the Bloch states cannot take on all possible values, but are restricted to certain allowed regions or **bands** separated from one another by forbidden regions or **band gaps** as shown in Fig. 2.1. The allowed energy bands are analogous to the pass bands of an electric transmission line and are characteristic of any wave propagating through a periodic structure (Brillouin 1953).

In a pure semiconductor without defects and at 0 K, certain bands are completely full of electrons and the remaining bands are completely empty. As we shall see later, this situation corresponds to an insulator. If the temperature of the crystal is raised, electrons will be excited from filled bands to empty bands leading to partial occupancy of these bands. Under these circumstances, the crystal becomes a semiconductor with a conductivity that increases rapidly with increasing temperature as more and more electrons are excited into originally empty bands. The underlying basis for these statements will become apparent as we proceed.

2.1 Schrödinger equation

An ideal crystal consists of an infinite array of ions located at or near lattice sites and an assembly of electrons moving in the field of the ions. The basic properties of crystals follow from the dynamics of these interacting systems. In order to solve this dynamical problem, we first write down the complete Schrödinger equation that includes electron–electron, electron–ion, and ion–ion interactions. We then introduce the approximation that enables us to find solutions corresponding to the different aspects of the behavior of the crystal: electronic motion, ionic motion, and their interactions with various perturbations including external fields.

The complete Schrödinger equation for n electrons and N ions can be written in terms of the electronic coordinates $\boldsymbol{r}_1, \ldots, \boldsymbol{r}_n$ and the ionic coordinates $\boldsymbol{R}_1, \ldots, \boldsymbol{R}_N$ as

$$\Bigg[-\sum_{i=1}^{n} \frac{\hbar^2}{2m} \nabla_i^2 - \sum_{k=1}^{N} \frac{\hbar^2}{2M_k} \nabla_k^2 + \frac{1}{2} \sum_{i,j=1}^{n}{}' \frac{e^2}{r_{ij}} + V_{ii}(\boldsymbol{R}_1, \ldots, \boldsymbol{R}_N) + V_{ei}(\boldsymbol{r}_1, \ldots, \boldsymbol{r}_n; \boldsymbol{R}_1, \ldots, \boldsymbol{R}_N) \Bigg] \Psi = E\Psi, \tag{2.1}$$

where the first term on the left hand side is the kinetic energy of the electrons, the second is the kinetic energy of the ions, the third is the electron–electron interaction potential, the fourth is the ion–ion interaction potential, and the last is the electron–ion interaction potential. In the third term, r_{ij} is the distance between electrons i and j, and the prime on the sum means that the contribution for $i = j$ is excluded. The spin–orbit interaction and other relativistic effects are omitted for the present.

Since the electron mass m is smaller by a factor of at least 1/1800 than any ion mass M_k, the electron motion is in general much faster than that of the ions. The electron distribution adjusts continuously to the positions of the ions and at any instant of time can be assumed to be the same as though the ions were at rest. This assumption is called the **adiabatic approximation** (Born and Oppenheimer 1927). It can be expressed by writing the complete eigenfunction Ψ as a product of functions of the electronic coordinates $\boldsymbol{r}_1, \ldots, \boldsymbol{r}_n$, represented collectively by $\boldsymbol{r}$, and the ionic coordinates $\boldsymbol{R}_1, \ldots, \boldsymbol{R}_N$, represented collectively by $\boldsymbol{R}$:

$$\Psi(\boldsymbol{r}, \boldsymbol{R}) = \psi(\boldsymbol{r}, \boldsymbol{R})\varphi(\boldsymbol{R}). \tag{2.2}$$

We see that a partial separation of the electronic and ionic coordinates has been achieved by the adiabatic approximation.

The function $\psi(\boldsymbol{r}, \boldsymbol{R})$ is the **electronic eigenfunction** and is a function of the electronic coordinates $\boldsymbol{r}$ as variables with the ionic coordinates $\boldsymbol{R}$ appearing as fixed parameters. It satisfies the **electronic Schrödinger equation**

$$\left[-\sum_{i=1}^{n}\frac{\hbar^2}{2m}\nabla_i^2 + \frac{1}{2}\sum_{i,j=1}^{n}{}' \frac{e^2}{r_{ij}} + V_{ei}(\boldsymbol{r}, \boldsymbol{R})\right]\psi(\boldsymbol{r}, \boldsymbol{R}) = E_e(\boldsymbol{R})\psi(\boldsymbol{r}, \boldsymbol{R}), \tag{2.3}$$

where $E_e(\boldsymbol{R})$ is the electronic energy eigenvalue and is a function of the ionic coordinates $\boldsymbol{R}$. The function $\varphi(\boldsymbol{R})$ is the **ionic eigenfunction** and is taken to satisfy the approximate equation

$$\left[-\sum_{k=1}^{N}\frac{\hbar^2}{2M_k}\nabla_k^2 + \Phi(\boldsymbol{R})\right]\varphi(\boldsymbol{R}) = E\varphi(\boldsymbol{R}), \tag{2.4}$$

where $\Phi(\boldsymbol{R}) = E_e(\boldsymbol{R}) + V_{ii}(\boldsymbol{R})$ is the effective potential energy of the ions and E is the total energy eigenvalue. Equation (2.4) is the **ionic Schrödinger equation** and arises when Eq. (2.2) is substituted into Eq. (2.1), Eq. (2.3) is used, and the terms

$$-\sum_{k=1}^{N}\frac{\hbar^2}{2M_k}\left[\varphi(\boldsymbol{R})\nabla_k^2\psi(\boldsymbol{r}, \boldsymbol{R}) - \boldsymbol{\nabla}_k\varphi(\boldsymbol{R}) \cdot \boldsymbol{\nabla}_k\psi(\boldsymbol{r}, \boldsymbol{R})\right]$$

are neglected. The latter terms represent the **nonadiabatic coupling** of the electronic and ionic motions.

Up to this point the ionic coordinates $\boldsymbol{R}$ in Eq. (2.3) are arbitrary. If the ions occupy the sites $\boldsymbol{R}_0$ of the equilibrium crystal structure, the potential $V_{ei}(\boldsymbol{r}, \boldsymbol{R}_0)$ becomes a **periodic potential** with the periodicity of the crystal lattice.

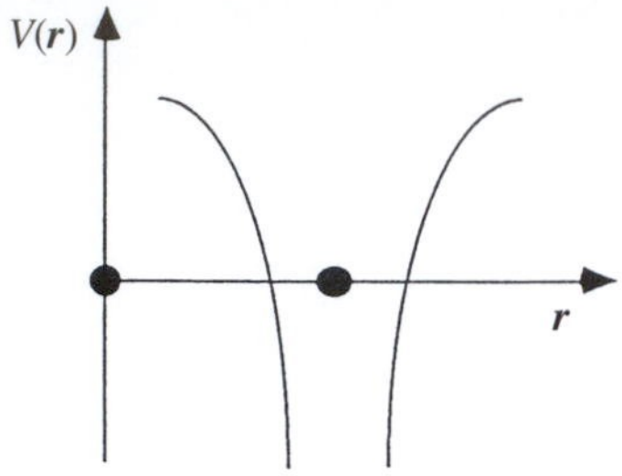

Fig. 2.2
Potential of a single isolated ion.

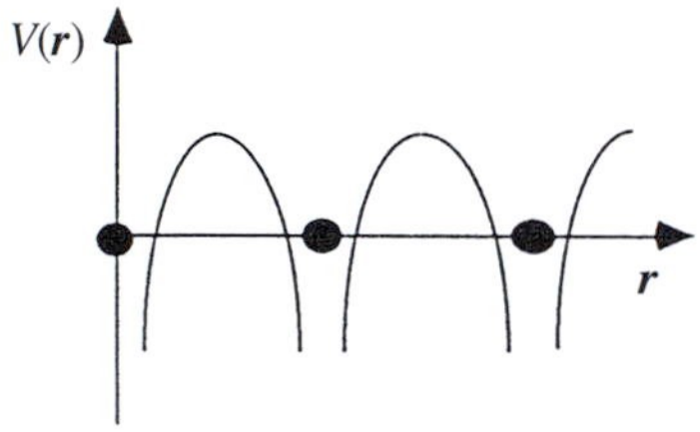

Fig. 2.3
Potential along a line of ions.

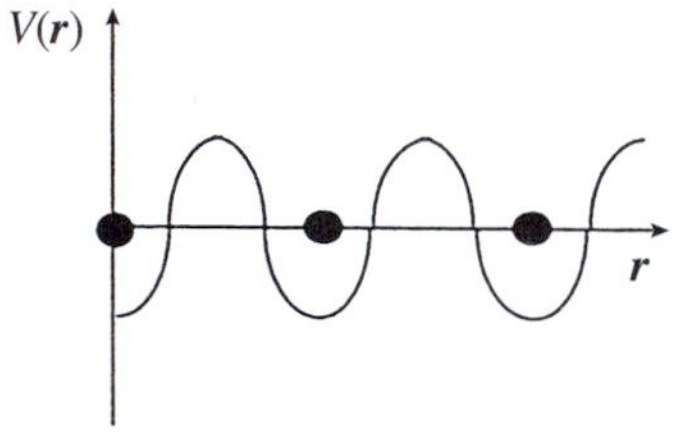

Fig. 2.4
Potential along a line between planes of ions.

2.2 Electrons in a periodic potential

In a real semiconductor crystal the periodicity of the perfect crystal is disrupted by impurities, vacancies, interstitials and other defects and by the thermal vibrations of the nuclei about their equilibrium positions. In well-prepared crystals at moderate temperatures these deviations from periodicity lead to changes in the electronic energies that are relatively small and can be neglected in first approximation. The primary problem is then the motion of the electrons in the field of the **periodic potential** of the positive ions, each of which provides an attractive potential as shown in Fig. 2.2. To simplify the problem even further, we neglect the Coulomb interactions between the electrons and focus on a single electron moving in a periodic potential $V(\boldsymbol{r})$ where $\boldsymbol{r}$ is the position vector of the electron. Schematic diagrams of periodic potentials along lines containing ions and between planes of ions are shown in Figs. 2.3 and 2.4, respectively.

The periodicity condition satisfied by $V(\boldsymbol{r})$ is

$$V(\vec{\boldsymbol{r}}+\vec{\boldsymbol{R}}(\ell)) = V(\boldsymbol{r}), \tag{2.5}$$

where $\boldsymbol{R}(\ell)$ is a lattice vector given by Eq. (1.1). Any periodic function can be expanded in a Fourier series. For $V(\boldsymbol{r})$ we can write

$$V(\boldsymbol{r}) = \sum_{\boldsymbol{G}} V_{\boldsymbol{G}} e^{i\boldsymbol{G}\cdot\boldsymbol{r}}, \tag{2.6}$$

where $\boldsymbol{G}$ is a **reciprocal lattice vector** which can be determined in the following way. Replace $\boldsymbol{r}$ by $\boldsymbol{r}+\boldsymbol{R}(\ell)$ in Eq. (2.6):

$$V(\boldsymbol{r}+\boldsymbol{R}(\ell)) = \sum_{\boldsymbol{G}} V_{\boldsymbol{G}} e^{i\boldsymbol{G}\cdot(\boldsymbol{r}+\boldsymbol{R}(\ell))} = \sum_{\boldsymbol{G}} V_{\boldsymbol{G}} e^{i\boldsymbol{G}\cdot\boldsymbol{r}} e^{i\boldsymbol{G}\cdot\boldsymbol{R}(\ell)}. \tag{2.7}$$

Now substitute Eqs. (2.6) and (2.7) into the periodicity condition, Eq. (2.5):

$$\sum_{\boldsymbol{G}} V_{\boldsymbol{G}} e^{i\boldsymbol{G}\cdot\boldsymbol{r}} e^{i\boldsymbol{G}\cdot\boldsymbol{R}(\ell)} = \sum_{\boldsymbol{G}} V_{\boldsymbol{G}} e^{i\boldsymbol{G}\cdot\boldsymbol{r}}. \tag{2.8}$$

In order for Eq. (2.8) to be valid for all values of $\boldsymbol{r}$, we must have

$$e^{i\boldsymbol{G}\cdot\boldsymbol{R}(\ell)} = 1. \tag{2.9}$$

Making use of Eq. (1.1), we obtain

$$e^{i\boldsymbol{G}\cdot\boldsymbol{a}_i} = 1, \quad i = 1, 2, 3 \tag{2.10}$$

from which follows

$$\boldsymbol{G}\cdot\boldsymbol{a}_i = 2\pi n_i, \quad i = 1, 2, 3 \tag{2.11}$$

where the quantities n_i are integers. It can be easily verified that the solution of Eqs. (2.11) is

$$\boldsymbol{G} = n_1\boldsymbol{b}_1 + n_2\boldsymbol{b}_2 + n_3\boldsymbol{b}_3, \tag{2.12}$$

where

$$b_1 = \frac{2\pi}{\Omega_0} a_2 \times a_3, \quad b_2 = \frac{2\pi}{\Omega_0} a_3 \times a_1, \quad b_3 = \frac{2\pi}{\Omega_0} a_1 \times a_2, \tag{2.13}$$

and Ω_0 is the volume of the primitive unit cell given by Eq. (1.2). The vectors $\boldsymbol{b}_1, \boldsymbol{b}_2, \boldsymbol{b}_3$ are the primitive translation vectors of the reciprocal lattice. They are related to the primitive translation vectors of the direct lattice by the equations

$$\boldsymbol{a}_i \cdot \boldsymbol{b}_j = 2\pi\delta_{ij}, \quad i, j = 1, 2, 3 \tag{2.14}$$

where δ_{ij} is the Kronecker delta ($\delta_{ij} = 1$ if $i = j$, 0 if $i \neq j$).

The Fourier coefficients of the periodic potential, $V_{\boldsymbol{G}}$, can be obtained by inverting the Fourier series in Eq. (2.6). The result is

$$V_{\boldsymbol{G}} = \frac{1}{\Omega_0} \int_{\text{cell}} d^3 r V(\boldsymbol{r}) e^{-i\boldsymbol{G}\cdot\boldsymbol{r}}. \tag{2.15}$$

We note that V_0 is simply the average potential over a primitive unit cell.

Example 2.1: Primitive unit cell of the reciprocal lattice
Derive the relation between the volumes of the primitive unit cells of the reciprocal and direct lattices.
Solution. The volume Ω_r of the primitive unit cell of the reciprocal lattice is given by

$$\Omega_r = |\boldsymbol{b}_1 \cdot (\boldsymbol{b}_2 \times \boldsymbol{b}_3)|.$$

Using the expressions

$$\boldsymbol{b}_1 = \frac{2\pi}{\Omega_0} \boldsymbol{a}_2 \times \boldsymbol{a}_3, \quad \boldsymbol{b}_2 = \frac{2\pi}{\Omega_0} \boldsymbol{a}_3 \times \boldsymbol{a}_1, \quad \boldsymbol{b}_3 = \frac{2\pi}{\Omega_0} \boldsymbol{a}_1 \times \boldsymbol{a}_2,$$

we obtain

$$\Omega_r = \frac{(2\pi)^2}{\Omega_0^2} |\boldsymbol{b}_1 \cdot [(\boldsymbol{a}_3 \times \boldsymbol{a}_1) \times (\boldsymbol{a}_1 \times \boldsymbol{a}_2)]|.$$

The theorem of vector analysis concerning the vector triple product gives

$$\begin{aligned}(\boldsymbol{a}_3 \times \boldsymbol{a}_1) \times (\boldsymbol{a}_1 \times \boldsymbol{a}_2) &= [(\boldsymbol{a}_3 \times \boldsymbol{a}_1) \cdot \boldsymbol{a}_2]\boldsymbol{a}_1 - [(\boldsymbol{a}_3 \times \boldsymbol{a}_1) \cdot \boldsymbol{a}_1]\boldsymbol{a}_2 \\ &= [(\boldsymbol{a}_3 \times \boldsymbol{a}_1) \cdot \boldsymbol{a}_2]\boldsymbol{a}_1.\end{aligned}$$

Combining this result with the condition

$$\boldsymbol{b}_1 \cdot \boldsymbol{a}_1 = 2\pi,$$

we obtain

$$\begin{aligned}\Omega_r &= \frac{(2\pi)^2}{\Omega_0^2} \cdot 2\pi |(\boldsymbol{a}_3 \times \boldsymbol{a}_1) \cdot \boldsymbol{a}_2| \\ &= \frac{(2\pi)^3}{\Omega_0^2} |\boldsymbol{a}_1 \cdot (\boldsymbol{a}_2 \times \boldsymbol{a}_3)|.\end{aligned}$$

Substituting

$$|\boldsymbol{a}_1 \cdot (\boldsymbol{a}_2 \times \boldsymbol{a}_3)| = \Omega_0$$

from Eq. (1.2) yields the desired result

$$\Omega_r = \frac{(2\pi)^3}{\Omega_0}$$

which states that the volume of the primitive unit cell of the reciprocal lattice is inversely proportional to the volume of the primitive unit cell of the direct lattice.

2.3 Schrödinger equation for a periodic potential

In dealing with the dynamics of electrons in crystals, we shall be primarily concerned with electron energies on the order of a few eV. The **de Broglie wavelength** λ is related to the energy E by

$$E = \frac{h^2}{2m\lambda^2}, \tag{2.16}$$

where h is Planck's constant and m is the electron mass. For the energies of interest, λ is on the order of a lattice spacing, so diffraction effects will be important and the wave picture of the electron must be employed through the **Schrödinger equation.**

In the coordinate representation, the momentum operator $\boldsymbol{p}$ is related to the gradient operator $\boldsymbol{\nabla}$ by $\boldsymbol{p} = -i\hbar\boldsymbol{\nabla}$, where $\hbar = h/2\pi$. The Schrödinger equation for an electron moving in a periodic potential $V(\boldsymbol{r})$ is

$$H\psi(\boldsymbol{r}) = E\psi(\boldsymbol{r}), \tag{2.17}$$

where

$$H = H_0 + V(\boldsymbol{r}), \tag{2.18}$$

H_0 is the kinetic energy operator given by

$$H_0 = \frac{p^2}{2m} = -\frac{\hbar^2\nabla^2}{2m}, \tag{2.19}$$

∇^2 is the Laplacian operator, and $\psi(\boldsymbol{r})$ is the eigenfunction. Only for a few special cases of the periodic potential can the Schrödinger equation be solved analytically. One can, however, obtain a formal solution by expanding $\psi(\boldsymbol{r})$ in a series of functions forming a complete set. A particularly convenient set for this purpose consists of the eigenfunctions $\chi(\boldsymbol{r})$ of the free electron Hamiltonian H_0,

$$H_0\chi(\boldsymbol{r}) = E_0\chi(\boldsymbol{r}), \tag{2.20}$$

where E_0 is the free electron energy eigenvalue. With H_0 given by Eq. (2.19), one sees that plane waves constitute solutions to Eq. (2.20),

$$\chi(\boldsymbol{r}) = e^{i\boldsymbol{k}\cdot\boldsymbol{r}}, \tag{2.21}$$

with

$$E_0 = \frac{\hbar^2 k^2}{2m}. \tag{2.22}$$

The wave vector $\boldsymbol{k}$ has constant components that are conveniently specified by **periodic boundary conditions**. One starts with the crystal in the form of a parallelepiped with edge lengths L_1, L_2, L_3 and periodically reproduces the crystal until all of the space is filled. The original crystal serves as a unit cell for a supercrystal, and one requires that the eigenfunction $\chi(\boldsymbol{r})$ satisfy the periodicity conditions

$$\chi(\boldsymbol{r} + N_i \boldsymbol{a}_i) = \chi(\boldsymbol{r}), \quad i = 1, 2, 3 \tag{2.23}$$

where $N_i|\boldsymbol{a}_i| = L_i$, the N_i are integers, and the $\boldsymbol{a}_i$ are the primitive translation vectors.

Equation (2.23) is the periodic boundary condition of Born and von Kármán. Substituting Eq. (2.21) into this equation yields

$$e^{iN_i \boldsymbol{k}\cdot\boldsymbol{a}_i} = 1. \tag{2.24}$$

If we write $\boldsymbol{k}$ in the form

$$\boldsymbol{k} = p_1 \boldsymbol{b}_1 + p_2 \boldsymbol{b}_2 + p_3 \boldsymbol{b}_3, \tag{2.25}$$

where the $\boldsymbol{b}_i$ are the primitive translation vectors of the reciprocal lattice, and use Eq. (2.14), we obtain

$$e^{2\pi i N_i p_i} = 1 \tag{2.26}$$

or

$$p_i = \frac{m_i}{N_i}, \tag{2.27}$$

where the m_i are integers. Equation (2.25) can now be written as

$$\boldsymbol{k} = \frac{m_1}{N_1}\boldsymbol{b}_1 + \frac{m_2}{N_2}\boldsymbol{b}_2 + \frac{m_3}{N_3}\boldsymbol{b}_3 \tag{2.28}$$

which is the specification of $\boldsymbol{k}$ according to periodic boundary conditions.

It should be noted that the volume associated with one $\boldsymbol{k}$ value in reciprocal space is

$$\Delta\Omega_{\boldsymbol{k}} = \frac{1}{N_1 N_2 N_3}|[\boldsymbol{b}_1 \cdot (\boldsymbol{b}_2 \times \boldsymbol{b}_3)]| \tag{2.29}$$

and that $N = N_1 N_2 N_3$ is the number of primitive unit cells in the crystal. Since $|\boldsymbol{b}_1 \cdot (\boldsymbol{b}_2 \times \boldsymbol{b}_3)|$ is the volume of the primitive unit cell of the reciprocal lattice, the number of $\boldsymbol{k}$ values in the latter volume is given by

$$\frac{|\boldsymbol{b}_1 \cdot (\boldsymbol{b}_2 \times \boldsymbol{b}_3)|}{\Delta\Omega_{\boldsymbol{k}}} = N. \tag{2.30}$$

In other words, the number of allowed wave vectors $\boldsymbol{k}$ in a primitive unit cell of the reciprocal lattice is equal to the number of lattice sites in the crystal. For a crystal of macroscopic size, this is a very large number, on the order of Avogadro's number.

It is clear from the way periodic boundary conditions are introduced that surface effects are neglected. If one is only interested in the bulk properties of macroscopic crystals, the error associated with the use of periodic boundary conditions is extremely small and is more than compensated by the enormous simplification of the mathematics relative to that required when surface effects are treated explicitly.

Example 2.2: Density-of-states in reciprocal space
Derive an expression for the density-of-states in reciprocal space.
Solution. The quantity $\Delta\Omega_k$ is the volume in reciprocal space associated with one $\boldsymbol{k}$ value as specified by periodic boundary conditions. The inverse of $\Delta\Omega_k$ is the number of $\boldsymbol{k}$ values associated with unit volume in reciprocal space, or in other words, the density-of-states in reciprocal space $g(\boldsymbol{k})$. From Eq. (2.30) we have

$$g(\boldsymbol{k}) = \frac{1}{\Delta\Omega_k} = \frac{N}{|\boldsymbol{b}_1 \cdot \boldsymbol{b}_2 \times \boldsymbol{b}_3|}.$$

Since the volume of the primitive unit cell of the reciprocal lattice Ω_r is $|\boldsymbol{b}_1 \cdot \boldsymbol{b}_2 \times \boldsymbol{b}_3|$,

$$g(\boldsymbol{k}) = \frac{N}{\Omega_r}.$$

But we saw in Example 2.1 that

$$\Omega_r = \frac{(2\pi)^3}{\Omega_0},$$

so

$$g(\boldsymbol{k}) = \frac{N\Omega_0}{(2\pi)^3} = \frac{\Omega}{(2\pi)^3},$$

where Ω is the volume of the crystal. If we introduce a factor of 2 to account for the two possible orientations of electron spin, we obtain the density-of-states including spin

$$g_s(\boldsymbol{k}) = \frac{\Omega}{4\pi^3}.$$

2.4 Expansion of the eigenfunction in plane waves

The plane wave states that constitute the eigenfunctions of H_0 form a complete set in terms of which we can expand an eigenfunction of the full Hamiltonian H. Such an expansion is equivalent to expressing the eigenfunction in terms of a Fourier series appropriate to periodic boundary conditions and takes the form

$$\psi(\boldsymbol{r}) = \sum_{\boldsymbol{k}} C(\boldsymbol{k}) e^{i\boldsymbol{k}\cdot\boldsymbol{r}}, \tag{2.31}$$

where the quantities $C(\boldsymbol{k})$ are the expansion coefficients and $\boldsymbol{k}$ is a wave vector that is specified by periodic boundary conditions, $\boldsymbol{k} = (m_1/N_1)\boldsymbol{b}_1 + (m_2/N_2)\boldsymbol{b}_2 + (m_3/N_3)\boldsymbol{b}_3$, in Eq. (2.28). Substituting Eq. (2.31) into the Schrödinger equation, Eq. (2.17), and using the expansion of the periodic potential in Fourier series, Eq. (2.6), yield

$$\sum_{\boldsymbol{k}} (\hbar^2 k^2/2m) C(\boldsymbol{k}) e^{i\boldsymbol{k}\cdot\boldsymbol{r}} + \sum_{\boldsymbol{G}} \sum_{\boldsymbol{k}} V_{\boldsymbol{G}} C(\boldsymbol{k}) e^{i(\boldsymbol{G}+\boldsymbol{k})\cdot\boldsymbol{r}} = E \sum_{\boldsymbol{k}} C(\boldsymbol{k}) e^{i\boldsymbol{k}\cdot\boldsymbol{r}}. \tag{2.32}$$

In the second term on the left hand side of this equation, we make the substitution $\boldsymbol{k}' = \boldsymbol{G} + \boldsymbol{k}$ and obtain

$$\begin{aligned} \sum_{\boldsymbol{G}} \sum_{\boldsymbol{k}} V_{\boldsymbol{G}} C(\boldsymbol{k}) e^{i(\boldsymbol{G}+\boldsymbol{k})\cdot\boldsymbol{r}} &= \sum_{\boldsymbol{G}} \sum_{\boldsymbol{k}'} V_{\boldsymbol{G}} C(\boldsymbol{k}' - \boldsymbol{G}) e^{i\boldsymbol{k}'\cdot\boldsymbol{r}} \\ &= \sum_{\boldsymbol{G}} \sum_{\boldsymbol{k}} V_{\boldsymbol{G}} C(\boldsymbol{k} - \boldsymbol{G}) e^{i\boldsymbol{k}\cdot\boldsymbol{r}}, \end{aligned} \tag{2.33}$$

where the last form results from the replacement of the dummy summation variable $\boldsymbol{k}'$ by the dummy summation variable $\boldsymbol{k}$.

The Schrödinger equation can now be rewritten in the form

$$\sum_{\boldsymbol{k}} \left\{ (\hbar^2 k^2/2m) C(\boldsymbol{k}) + \sum_{\boldsymbol{G}} V_{\boldsymbol{G}} C(\boldsymbol{k} - \boldsymbol{G}) - E C(\boldsymbol{k}) \right\} e^{i\boldsymbol{k}\cdot\boldsymbol{r}} = 0. \tag{2.34}$$

For this equation to be valid for an arbitrary value of $\boldsymbol{r}$, we must set the coefficient of each factor $e^{i\boldsymbol{k}\cdot\boldsymbol{r}}$ equal to zero:

$$(\hbar^2 k^2/2m) C(\boldsymbol{k}) + \sum_{\boldsymbol{G}} V_{\boldsymbol{G}} C(\boldsymbol{k} - \boldsymbol{G}) - E C(\boldsymbol{k}) = 0. \tag{2.35}$$

Additional equations involving the coefficients $C(\boldsymbol{k} - \boldsymbol{G})$ can be obtained by replacing $\boldsymbol{k}$ by $\boldsymbol{k} - \boldsymbol{G}$ and $\boldsymbol{G}$ by $\boldsymbol{G}'$ in Eq. (2.35) and considering all values of $\boldsymbol{G}$. Since we have infinitely many reciprocal lattice vectors $\boldsymbol{G}$, we have an infinite set of linear homogeneous algebraic equations in the coefficients $C(\boldsymbol{k} - \boldsymbol{G})$. For a nontrivial solution to exist, the determinant of the coefficients $C(\boldsymbol{k} - \boldsymbol{G})$ must be zero. This **secular equation** specifies the allowed eigenvalues of the energy E for a given value of $\boldsymbol{k}$. In practice, the determinant is made finite by restricting the values of $\boldsymbol{G}$ employed to those with magnitude less than some specified value. The nature of the solutions to the secular equation is discussed in Section 2.9.

2.5 Bloch's theorem

We note from Eq. (2.35) that a given $\boldsymbol{k}$ is coupled to only those $\boldsymbol{k}$'s that differ from it by a reciprocal lattice vector $\boldsymbol{G}$. We can therefore pick a particular $\boldsymbol{k}$

as a starting point and rewrite Eq. (2.31) in the form

$$\begin{aligned}\psi_k(r) &= \sum_G C(k-G)e^{i(k-G)\cdot r} \\ &= e^{ik\cdot r}\left[\sum_G C(k-G)e^{-iG\cdot r}\right]. \end{aligned} \tag{2.36}$$

The quantity in square brackets on the right hand side of Eq. (2.36) is a periodic function of r with the periodicity of the crystal lattice, as may be verified by replacing r by $r + R(\ell)$ and using Eq. (2.9). We call this periodic function $u_k(r)$,

$$u_k(r) = \sum_G C(k-G)e^{-iG\cdot r}, \tag{2.37}$$

and write

$$\psi_k(r) = e^{ik\cdot r}u_k(r). \tag{2.38}$$

Equation (2.38) constitutes **Bloch's theorem**, which states that the eigenfunction of an electron moving in a periodic potential consists of a **plane wave factor** and a **periodic function factor**. An eigenfunction of the form (2.38) is called a **Bloch function** (Bloch 1928).

Comment. If the position vector r is augmented by a lattice translation vector $R(\ell)$, the Bloch function $\psi_k(r)$ is changed only by a phase factor. To demonstrate this, let us replace r by $r + R(\ell)$ on both sides of Eq. (2.38) which states Bloch's theorem:

$$\psi_k(r + R(\ell)) = e^{ik\cdot(r+R(\ell))}u_k(r + R(\ell)).$$

Since $u_k(r)$ is periodic with the periodicity of the lattice,

$$u_k(r + R(\ell)) = u_k(r),$$

and we can write

$$\begin{aligned}\psi_k(r + R(\ell)) &= e^{ik\cdot(r+R(\ell))}u_k(r) \\ &= e^{ik\cdot R(\ell)}e^{ik\cdot r}u_k(r) \\ &= e^{ik\cdot R(\ell)}\psi_k(r).\end{aligned}$$

Thus, $\psi_k(r + R(\ell))$ is simply the original Bloch function $\psi_k(r)$ multiplied by the phase factor $e^{ik\cdot R(\ell)}$ which is independent of r.

2.6 Electrons in a weak periodic potential

For semiconductors and many other solids, the proper representation of the one-electron eigenfunctions by means of the plane wave expansion as expressed in Eq. (2.36) requires hundreds of plane waves. This means that the corresponding Schrödinger equation, Eqs. (2.35), must be solved on a computer, and it becomes difficult to extract the essential physics in a simple way. To avoid this complication we consider a weak potential that we treat by perturbation theory. We thereby obtain analytic results from which some important physical principles can be deduced.

Treating the potential $V(\boldsymbol{r})$ as a perturbation, the zero-order eigenfunctions of the Schrödinger equation, Eq. (2.17), are plane waves

$$\psi_{\boldsymbol{k}}^{(0)}(\boldsymbol{r}) = \frac{1}{\sqrt{\Omega}} e^{i\boldsymbol{k}\cdot\boldsymbol{r}} \equiv |\boldsymbol{k}\rangle, \tag{2.39}$$

where Ω is the volume of the crystal, and the zero-order energy eigenvalues are given by

$$E_{\boldsymbol{k}}^{(0)} = \frac{\hbar^2 k^2}{2m}. \tag{2.40}$$

Assuming for the moment that the state of the wave vector $\boldsymbol{k}$ is not coupled by the potential to another state of equal energy, we can write the perturbed energy eigenvalue to second order as

$$E_{\boldsymbol{k}} = E_{\boldsymbol{k}}^{(0)} + \langle \boldsymbol{k}|V(\boldsymbol{r})|\boldsymbol{k}\rangle + \sum_{\boldsymbol{k}'} \frac{\langle \boldsymbol{k}|V(\boldsymbol{r})|\boldsymbol{k}'\rangle\langle \boldsymbol{k}'|V(\boldsymbol{r})|\boldsymbol{k}\rangle}{E_{\boldsymbol{k}}^{(0)} - E_{\boldsymbol{k}'}^{(0)}}, \tag{2.41}$$

where

$$\langle \boldsymbol{k}|V(\boldsymbol{r})|\boldsymbol{k}'\rangle = \frac{1}{\Omega}\int d^3r e^{-i\boldsymbol{k}\cdot\boldsymbol{r}} V(\boldsymbol{r}) e^{i\boldsymbol{k}'\cdot\boldsymbol{r}}. \tag{2.42}$$

Substituting Eq. (2.6) into Eq. (2.42) yields the relation

$$\langle \boldsymbol{k}|V(\boldsymbol{r})|\boldsymbol{k}'\rangle = \frac{1}{\Omega}\int d^3r\, e^{i(\boldsymbol{k}'-\boldsymbol{k})\cdot\boldsymbol{r}} \sum_{\boldsymbol{G}} V_{\boldsymbol{G}} e^{i\boldsymbol{G}\cdot\boldsymbol{r}} = \sum_{\boldsymbol{G}} V_{\boldsymbol{G}} \delta_{\boldsymbol{k}',\boldsymbol{k}-\boldsymbol{G}}. \tag{2.43}$$

Using this result, Eq. (2.41) becomes

$$E_{\boldsymbol{k}} = E_{\boldsymbol{k}}^{(0)} + V_0 + \sum_{\boldsymbol{G}} \frac{V_{\boldsymbol{G}} V_{-\boldsymbol{G}}}{E_{\boldsymbol{k}}^{(0)} - E_{\boldsymbol{k}-\boldsymbol{G}}^{(0)}}. \tag{2.44}$$

Taking the complex conjugate of Eq. (2.6) and noting that $V(\boldsymbol{r})$ is real, we have

$$V(\boldsymbol{r}) = V^*(\boldsymbol{r}) = \sum_{\boldsymbol{G}} V_{\boldsymbol{G}}^* e^{-i\boldsymbol{G}\cdot\boldsymbol{r}}$$

which, upon replacing the dummy summation variable $\boldsymbol{G}$ by $-\boldsymbol{G}$, becomes

$$V(\boldsymbol{r}) = \sum_{\boldsymbol{G}} V_{-\boldsymbol{G}}^* e^{i\boldsymbol{G}\cdot\boldsymbol{r}}. \tag{2.45}$$

Comparing Eqs. (2.6) and (2.45) gives the result

$$V_G = V^*_{-G}. \tag{2.46}$$

Equation (2.44) can then be rewritten as

$$E_k = E_k^{(0)} + V_0 + \sum_G \frac{|V_G|^2}{E_k^{(0)} - E_{k-G}^{(0)}}, \tag{2.47}$$

which is correct to $O(V_G^2)$ as long as $E_k^{(0)} \neq E_{k-G}^{(0)}$.

The first order corrected eigenfunction has the form

$$\begin{aligned}\psi_k(r) &= \psi_k^{(0)}(r) + \sum_{k'} \frac{\langle k'|V(r)|k\rangle}{E_k^{(0)} - E_{k'}^{(0)}} \psi_{k'}^{(0)}(r) \\ &= \psi_k^{(0)}(r) + \sum_G \frac{V_G^*}{E_k^{(0)} - E_{k-G}^{(0)}} \psi_{k-G}^{(0)}(r).\end{aligned} \tag{2.48}$$

Making use of the fact that $\psi_k^{(0)}(r)$ is a plane wave as given by Eq. (2.39), we can rewrite Eq. (2.48) as

$$\psi_k(r) = \frac{1}{\sqrt{\Omega}} e^{ik\cdot r} \left\{ 1 + \sum_G \frac{V_G^*}{E_k^{(0)} - E_{k-G}^{(0)}} e^{-iG\cdot r} \right\} \tag{2.49}$$

which has the Bloch form of a plane wave modulated by a periodic function $u_k(r)$ given by

$$u_k(r) = \frac{1}{\sqrt{\Omega}} \left\{ 1 + \sum_G \frac{V_G^*}{E_k^{(0)} - E_{k-G}^{(0)}} e^{-iG\cdot r} \right\}. \tag{2.50}$$

2.7 Brillouin zones

The periodicity of the system in real space leads to a periodic division of wave vector space into Brillouin zones. Within a given zone the energy of an electron is a continuous function of energy, but undergoes a discontinuity in crossing a zone boundary. The Brillouin zone is a most useful concept in the representation of electronic states in crystals.

We now discuss the above results in detail. The perturbed energy eigenvalue given by Eq. (2.47) contains a constant shift due to the first order term V_0 and a second order term that is wave vector dependent. The latter term is negative at $k = 0$ and becomes large in magnitude as $E_k^{(0)}$ approaches $E_{k-G}^{(0)}$. Since there is a sum over G in Eq. (2.47), there is an infinite number of singularities in the expression for the second order perturbed energy. A similar situation applies to the perturbed eigenfunction given by Eq. (2.49). If a singularity occurs for $G = G_0$, the amplitude of the plane wave state $k - G_0$ in the perturbed eigenfunction diverges corresponding to strong scattering of the electron from the state k to the state $k - G_0$.

The condition for the singularity,

$$E_{\boldsymbol{k}}^{(0)} = E_{\boldsymbol{k}-\boldsymbol{G}_0}^{(0)}, \tag{2.51}$$

can be evaluated by using the expression for the unperturbed energy given by Eq. (2.40). The result is

$$\boldsymbol{k} \cdot \left(\frac{\boldsymbol{G}_0}{2}\right) = \left|\frac{\boldsymbol{G}_0}{2}\right|^2. \tag{2.52}$$

Equation (2.52) is the equation for the plane in wave vector space that bisects the reciprocal lattice vector $\boldsymbol{G}_0$. It is the famous **Bragg condition** for the coherent scattering of plane waves by a periodic structure. The perpendicular bisector plane is known as a **Bragg plane**. We note that $\boldsymbol{k} = \boldsymbol{G}_0/2$ satisfies Eq. (2.52) and that $E_{\boldsymbol{G}_0/2}^{(0)} = E_{-\boldsymbol{G}_0/2}^{(0)}$ from Eq. (2.51).

Since there is a Bragg condition for each $\boldsymbol{G}$, the Bragg planes divide wave vector space into zones which are called **Brillouin zones.** The **first Brillouin zone** encloses the origin and is the smallest volume entirely enclosed by the Bragg planes of reciprocal lattice vectors drawn from the origin. The second Brillouin zone consists of those portions of volume just outside and contiguous with the first zone which, when translated by appropriate reciprocal lattice vectors, will exactly fill the first zone. Third and higher zones can be described in a similar manner. It can be shown (Ziman 1972) that the volume of each Brillouin zone is the same as that of the primitive unit cell of the reciprocal lattice. Consequently, we deduce from Eq. (2.30) that each zone contains N values of the wave vector $\boldsymbol{k}$ allowed by periodic boundary conditions, where N is the number of primitive unit cells in the crystal. The first three Brillouin zones for a two-dimensional square lattice are shown in Fig. 2.5.

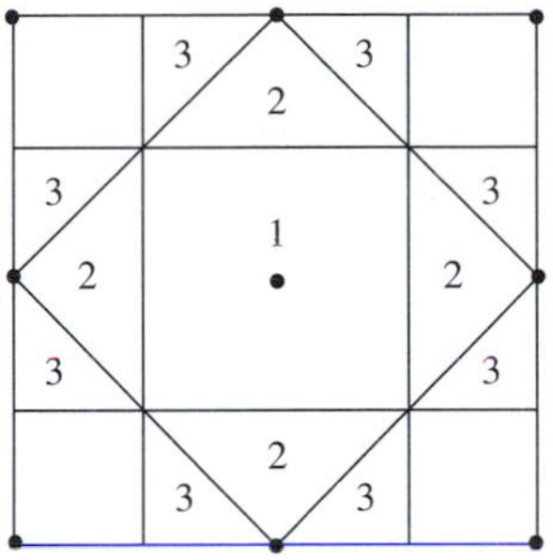

Fig. 2.5
The first three Brillouin zones of a two-dimensional square lattice.

Example 2.3: Brillouin zone
Describe the construction of the first Brillouin zone for a face-centered cubic lattice.

Solution. The primitive translation vectors of the face-centered cubic lattice with lattice constant a are given by

$$\boldsymbol{a}_1 = a\left(\tfrac{1}{2}, \tfrac{1}{2}, 0\right), \quad \boldsymbol{a}_2 = a\left(\tfrac{1}{2}, 0, \tfrac{1}{2}\right), \quad \boldsymbol{a}_3 = a\left(0, \tfrac{1}{2}, \tfrac{1}{2}\right).$$

They can also be written in terms of orthogonal unit vectors $\hat{x}, \hat{y}, \hat{z}$ as

$$\boldsymbol{a}_1 = \tfrac{1}{2}a(\hat{x} + \hat{y}), \quad \boldsymbol{a}_2 = \tfrac{1}{2}a(\hat{x} + \hat{z}), \quad \boldsymbol{a}_3 = \tfrac{1}{2}a(\hat{y} + \hat{z}).$$

The volume of the primitive unit cell is

$$\Omega_0 = |\boldsymbol{a}_1 \cdot (\boldsymbol{a}_2 \times \boldsymbol{a}_3)| = \tfrac{1}{4}a^3.$$

These primitive translation vectors can be combined with Eqs. (2.12) and (2.13) to give the reciprocal lattice vectors $\boldsymbol{G}$,

$$\boldsymbol{G} = \frac{2\pi}{a}[\ell_1(-\hat{x} - \hat{y} + \hat{z}) + \ell_2(-\hat{x} + \hat{y} - \hat{z}) + \ell_3(\hat{x} - \hat{y} - \hat{z})],$$

where each of ℓ_1, ℓ_2, ℓ_3 spans all integer values. From the form of these vectors we see that the reciprocal lattice of a face-centered cubic lattice is a body-centered cubic lattice. The two sets of $\boldsymbol{G}$'s with smallest nonzero magnitude are:

$$\left.\begin{array}{ll} \pm\frac{2\pi}{a}(-\hat{x}-\hat{y}+\hat{z}), & \pm\frac{2\pi}{a}(-\hat{x}+\hat{y}-\hat{z}), \\ \pm\frac{2\pi}{a}(\hat{x}-\hat{y}-\hat{z}), & \pm\frac{2\pi}{a}(\hat{x}+\hat{y}+\hat{z}) \end{array}\right\} |\boldsymbol{G}| = \sqrt{3}\left(\frac{2\pi}{a}\right)$$

$$\pm\frac{2\pi}{a}2\hat{x},\ \pm\frac{2\pi}{a}2\hat{y},\ \pm\frac{2\pi}{a}2\hat{z}, |\boldsymbol{G}| = 2\left(\frac{2\pi}{a}\right).$$

The perpendicular bisector planes of these two sets of $\boldsymbol{G}$'s are specified by the equations

$$\pm(-k_x-k_y+k_z) = \frac{3\pi}{a}, \quad \pm(-k_x+k_y-k_z) = \frac{3\pi}{a},$$

$$\pm(k_x-k_y-k_z) = \frac{3\pi}{a}, \quad \pm(k_x+k_y+k_z) = \frac{3\pi}{a},$$

and

$$\pm k_x = \frac{2\pi}{a}, \quad \pm k_y = \frac{2\pi}{a}, \quad \pm k_z = \frac{2\pi}{a}.$$

The boundary of the first Brillouin zone of the face-centered cubic lattice has contributions from all 14 of these perpendicular bisector planes as can be seen in Fig. 2.6.

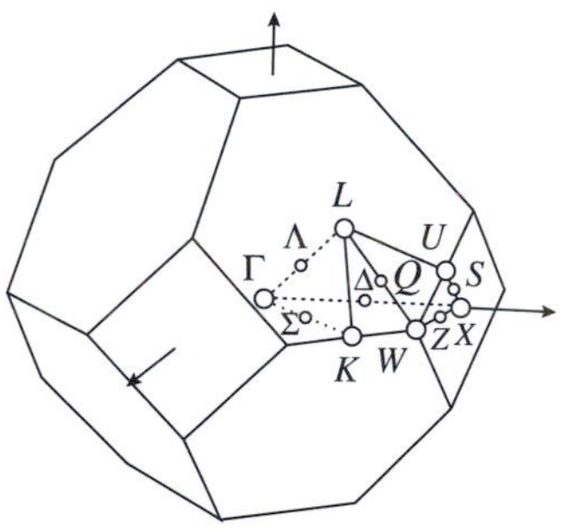

Fig. 2.6
The first Brillouin zone of the face-centered cubic lattice with some points of high symmetry indicated by letters.

The first Brillouin zone is the smallest volume centered at the origin that is bounded by perpendicular bisector planes. It has the form of the **Wigner–Seitz primitive unit cell** for a body-centered cubic lattice and is uniquely specified by the symmetry operations of the crystal lattice. Points of high symmetry denoted by Γ, L, X, etc., are shown in Fig. 2.6.

2.8 Energy bands and energy band gaps

In order to get further insight into the effect of the periodic potential on the energy eigenvalues of an electron, we need to modify our procedure in order to handle the degeneracy that occurs in zero order at the Brillouin zone boundary where Eq. (2.51) is satisfied. We use degenerate perturbation theory and focus on the two degenerate states $\boldsymbol{k}$ and $\boldsymbol{k}-\boldsymbol{G}_0$, while neglecting all others. The two equations of the set of Eqs. (2.35) that involve these two states are

$$(E_{\boldsymbol{k}}^{(0)} + V_0 - E_{\boldsymbol{k}})C(\boldsymbol{k}) + V_{\boldsymbol{G}_0}C(\boldsymbol{k}-\boldsymbol{G}_0) = 0 \tag{2.53a}$$

$$V_{\boldsymbol{G}_0}^{*}C(\boldsymbol{k}) + (E_{\boldsymbol{k}-\boldsymbol{G}_0}^{(0)} + V_0 - E_{\boldsymbol{k}})C(\boldsymbol{k}-\boldsymbol{G}_0) = 0. \tag{2.53b}$$

The secular equation for a nontrivial solution is

$$\begin{vmatrix} E_{\boldsymbol{k}}^{(0)} + V_0 - E_{\boldsymbol{k}} & V_{\boldsymbol{G}_0} \\ V_{\boldsymbol{G}_0}^* & E_{\boldsymbol{k}-\boldsymbol{G}_0}^{(0)} + V_0 - E_{\boldsymbol{k}} \end{vmatrix} = 0. \tag{2.54}$$

Solving this equation gives

$$E_{\boldsymbol{k}} = V_0 + \frac{1}{2}\left\{ E_{\boldsymbol{k}}^{(0)} + E_{\boldsymbol{k}-\boldsymbol{G}_0}^{(0)} \pm \left[(E_{\boldsymbol{k}}^{(0)} - E_{\boldsymbol{k}-\boldsymbol{G}_0}^{(0)})^2 + 4|V_{\boldsymbol{G}_0}|^2 \right]^{\frac{1}{2}} \right\}. \tag{2.55}$$

A very important consequence of this result is that, as $E_{\boldsymbol{k}}^{(0)}$ and $E_{\boldsymbol{k}-\boldsymbol{G}_0}^{(0)}$ become equal, the two solutions for $E_{\boldsymbol{k}}$ come no closer together than $2|V_{\boldsymbol{G}_0}|$. In other words there is a discontinuity or **forbidden energy gap** in the energy versus $\boldsymbol{k}$ curve that separates two allowed **energy bands**. States of electrons propagating through the crystal do not have energies in the forbidden energy gap.

To describe the behavior of the energy eigenvalue more precisely near the degeneracy point $\boldsymbol{k} = \frac{1}{2}\boldsymbol{G}_0$, we let $\boldsymbol{q} = \boldsymbol{k} - \frac{1}{2}\boldsymbol{G}_0$ and use Eq. (2.40) to eliminate $\boldsymbol{k}$ from Eq. (2.55). The result is

$$E_{\boldsymbol{k}} = V_0 + \frac{\hbar^2}{2m}\left(\tfrac{1}{4}G_0^2 + q^2\right) \pm \left[\frac{\hbar^4}{4m^2}(\boldsymbol{G}_0 \cdot \boldsymbol{q})^2 + |V_{\boldsymbol{G}_0}|^2 \right]^{\frac{1}{2}}. \tag{2.56}$$

As $\boldsymbol{q} \to 0$, $E_{\boldsymbol{k}} \to V_0 + \frac{\hbar^2}{8m}G_0^2 \pm |V_{\boldsymbol{G}_0}|$. For small q, we expand the square root in Eq. (2.56) in Taylor series and obtain

$$E_{\boldsymbol{k}} \simeq V_0 + \frac{\hbar}{8m}G_0^2 + \frac{\hbar^2}{2m}\left(1 \pm \frac{\hbar^2 G_0^2 \cos^2\theta}{4m|V_{\boldsymbol{G}_0}|}\right)q^2 \pm |V_{\boldsymbol{G}_0}|, \tag{2.57}$$

where θ is the angle between $\boldsymbol{G}_0$ and $\boldsymbol{q}$.

The behavior of $E_{\boldsymbol{k}}$ as a function of $\boldsymbol{k}$ is illustrated schematically in Fig. 2.7 for a particular value of θ. Two bands of allowed energies are separated by an energy gap E_g whose magnitude is $2|V_{\boldsymbol{G}_0}|$. The wave vector at which the gap occurs is specified by the Bragg condition contained in Eq. (2.52), and it therefore lies on a Brillouin zone boundary. Note that the q^2 dependence forces $E_{\boldsymbol{k}}$ to be normal to the zone boundary itself.

Each nonzero reciprocal lattice vector can give rise to an energy gap. The first three allowed energy bands and the gaps between them are shown in Fig. 2.8 for a one-dimensional system. Except for the lowest energy band, the energy bands are split into two parts, one lying in the domain of negative wave vectors and the other in the domain of positive wave vectors. This behavior is a feature of the **extended zone scheme** depicted in Fig. 2.8.

An alternative procedure, which in some ways is more attractive, is the **reduced zone scheme**. It is obtained by using appropriate reciprocal lattice vectors to translate the parts of an energy band that lie in an outer Brillouin zone into the first Brillouin zone as shown in Fig. 2.9. In this way, all of the

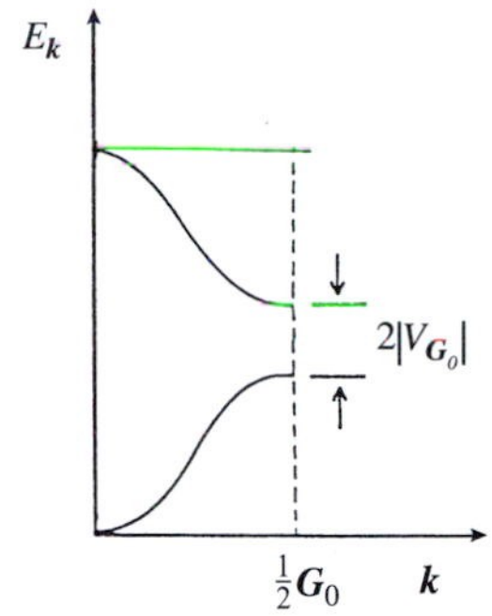

Fig. 2.7
Energy $E_{\boldsymbol{k}}$ as a function of $\boldsymbol{k}$ near the degeneracy point $\frac{1}{2}\boldsymbol{G}_0$.

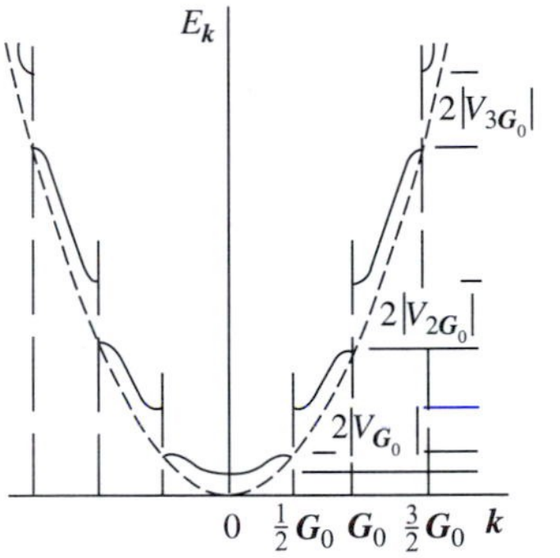

Fig. 2.8
The first three allowed energy bands in the extended zone scheme.

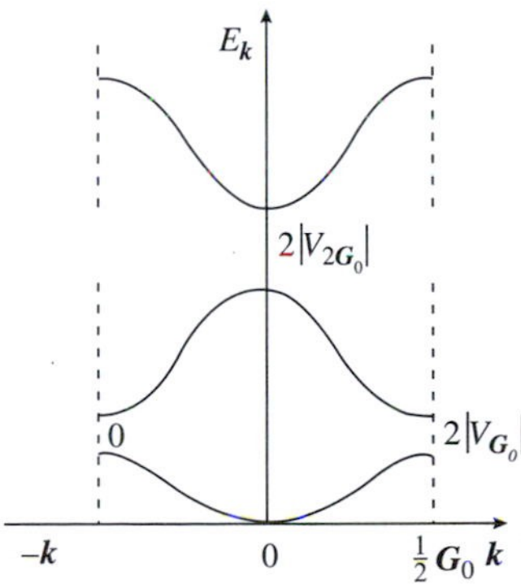

Fig. 2.9
The first three allowed energy bands in the reduced zone scheme.

energy bands can be represented within the same range of wave vectors. The reduced zone scheme can be justified by noting that if the wave vector $\boldsymbol{k}'$ of the Bloch function $\psi_{\boldsymbol{k}'}(\boldsymbol{r})$ differs from another wave vector $\boldsymbol{k}$ by a reciprocal lattice vector $\boldsymbol{G}$,

$$\boldsymbol{k}' = \boldsymbol{k} + \boldsymbol{G}, \tag{2.58}$$

then by using Eq. (2.38), we see that

$$\psi_{\boldsymbol{k}'}(\boldsymbol{r}) = e^{i\boldsymbol{k}'\cdot\boldsymbol{r}}u_{\boldsymbol{k}'}(\boldsymbol{r}) = e^{i\boldsymbol{k}\cdot\boldsymbol{r}}e^{i\boldsymbol{G}\cdot\boldsymbol{r}}u_{\boldsymbol{k}'}(\boldsymbol{r}). \tag{2.59}$$

The factor $e^{i\boldsymbol{G}\cdot\boldsymbol{r}}$ has the periodicity of the crystal and so does the function $u_{\boldsymbol{k}}(\boldsymbol{r})$ defined by

$$u_{\boldsymbol{k}}(\boldsymbol{r}) \equiv e^{i\boldsymbol{G}\cdot\boldsymbol{r}}u_{\boldsymbol{k}'}(\boldsymbol{r}) = e^{i\boldsymbol{G}\cdot\boldsymbol{r}}u_{\boldsymbol{k}+\boldsymbol{G}}(\boldsymbol{r}). \tag{2.60}$$

Consequently,

$$\psi_{\boldsymbol{k}}(\boldsymbol{r}) \equiv e^{i\boldsymbol{k}\cdot\boldsymbol{r}}u_{\boldsymbol{k}}(\boldsymbol{r}) = e^{i(\boldsymbol{k}+\boldsymbol{G})\cdot\boldsymbol{r}}u_{\boldsymbol{k}+\boldsymbol{G}}(\boldsymbol{r}) = \psi_{\boldsymbol{k}+\boldsymbol{G}}(\boldsymbol{r}). \tag{2.61}$$

The Bloch functions $\psi_{\boldsymbol{k}}(\boldsymbol{r})$ and $\psi_{\boldsymbol{k}+\boldsymbol{G}}(\boldsymbol{r})$ are therefore equivalent.

To complete the justification of the reduced scheme, we show that the energy eigenvalue is a periodic function of $\boldsymbol{k}$ with the periodicity of the reciprocal lattice. If the form of the Schrödinger equation in Eq. (2.35) is modified by replacing $\boldsymbol{G}$ by $\boldsymbol{G}'$ and $\boldsymbol{k}$ by $\boldsymbol{k}-\boldsymbol{G}$, the result is

$$\frac{\hbar^2}{2m}(\boldsymbol{k}-\boldsymbol{G})^2C(\boldsymbol{k}-\boldsymbol{G}) + \sum_{\boldsymbol{G}'} V_{\boldsymbol{G}'}C(\boldsymbol{k}-\boldsymbol{G}-\boldsymbol{G}') = E_{\boldsymbol{k}}C(\boldsymbol{k}-\boldsymbol{G}), \tag{2.62}$$

where the dependence of the energy eigenvalue on $\boldsymbol{k}$ is explicitly indicated. Since Eq. (2.62) stands for an infinite number of equations corresponding to the infinite number of values of $\boldsymbol{G}$, there is an infinite number of solutions for $E_{\boldsymbol{k}}$ and the associated coefficients $C(\boldsymbol{k}-\boldsymbol{G})$ which we label by an index n having the values $1, 2, 3, \ldots$:

$$\frac{\hbar^2}{2m}(\boldsymbol{k}-\boldsymbol{G})^2C_n(\boldsymbol{k}-\boldsymbol{G}) + \sum_{\boldsymbol{G}'} V_{\boldsymbol{G}'}C_n(\boldsymbol{k}-\boldsymbol{G}-\boldsymbol{G}') = E_{n\boldsymbol{k}}C_n(\boldsymbol{k}-\boldsymbol{G}). \tag{2.63}$$

The index n is the **band index**, and the values of $E_{n\boldsymbol{k}}$ for given n and various $\boldsymbol{k}$ form the nth energy band. The corresponding Bloch functions specified by Eq. (2.36) are designated $\psi_{n\boldsymbol{k}}(\boldsymbol{r})$.

To establish the periodicity of $E_{n\boldsymbol{k}}$, we modify the first term on the left hand side of Eq. (2.63) by introducing $\delta_{\boldsymbol{G},\boldsymbol{G}'}$ and summing over $\boldsymbol{G}'$ and the second term by letting $\boldsymbol{G}'' = \boldsymbol{G}+\boldsymbol{G}'$. The result is

$$\sum_{\boldsymbol{G}'}\frac{\hbar^2}{2m}(\boldsymbol{k}-\boldsymbol{G}')^2C_n(\boldsymbol{k}-\boldsymbol{G}')\delta_{\boldsymbol{G},\boldsymbol{G}'} + \sum_{\boldsymbol{G}''} V_{\boldsymbol{G}''-\boldsymbol{G}}C_n(\boldsymbol{k}-\boldsymbol{G}'') = E_{n\boldsymbol{k}}C_n(\boldsymbol{k}-\boldsymbol{G}). \tag{2.64}$$

Now replace G'' by G' to give

$$\sum_{G'}\left[\frac{\hbar^2}{2m}(k-G')^2\delta_{G,G'}+V_{G'-G}\right]C_n(k-G')$$
$$=E_{nk}C_n(k-G) \quad (2.65)$$

which can be rewritten in the equivalent form

$$\sum_{G'}\left[\frac{\hbar^2}{2m}(k-G')^2\delta_{k-G,\,k-G'}+V_{k-G-(k-G')}\right]C_n(k-G')$$
$$=E_{nk}C_n(k-G). \quad (2.66)$$

If we decrease k in Eq. (2.66) by the reciprocal lattice vector G'' and set $K=G+G''$, $K'=G'+G''$, the result can be expressed as

$$\sum_{K'}\left[\frac{\hbar^2}{2m}(k-K')^2\delta_{k-K,\,k-K'}+V_{k-K-(k-K')}\right]C_n(k-K')$$
$$=E_{nk-G''}C_n(k-K). \quad (2.67)$$

Since both G and K in Eqs. (2.66) and (2.67) span the space of reciprocal lattice vectors and since G' and K' are dummy summation variables spanning the same space, we see that $E_{nk-G''}$ and E_{nk} are eigenvalues of the same matrix. If there is no degeneracy of bands at k, then

$$E_{nk-G''}=E_{nk} \quad (2.68)$$

for any value of G'' and therefore E_{nk} is periodic in reciprocal space. This result can be extended without difficulty to the case where k is a point of degeneracy.

We have thus established the justification for the reduced zone scheme, which is the scheme most commonly used in the discussion of electronic energy bands. In Section 2.5 we noted that the number of values of the wave vector k allowed by periodic boundary conditions is equal to N, the number of primitive unit cells in the crystal. Since the Bloch functions in a given energy band are distinguished by their wave vectors, it follows that there are N Bloch functions in an energy band. Each Bloch state can be occupied by two electrons of opposite spin in accordance with the Pauli principle, so $2N$ electrons can occupy a given energy band.

Especially important for semiconductor physics are the face-centered cubic and hexagonal lattices. The first Brillouin zone of the former is given in Fig. 2.6 and that of the latter in Fig. 2.10. The zone-center point ($k=0$) is denoted by Γ. Particular points of high symmetry on the zone boundary and in certain directions within the zone are labeled by letters as shown.

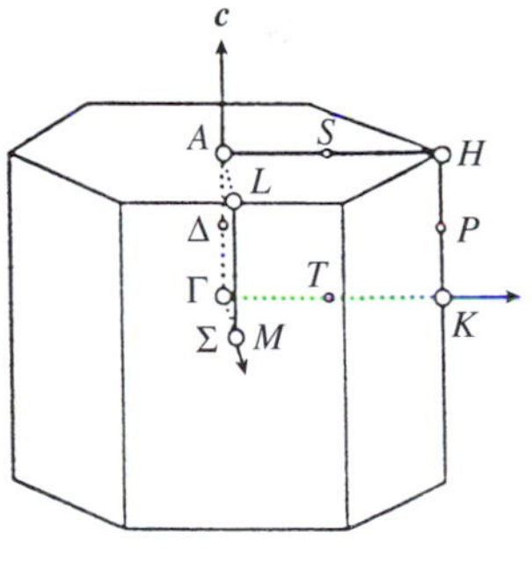

Fig. 2.10
Brillouin zone for the hexagonal lattice.

Example 2.4: The Kronig–Penney Model
Develop analytic solutions to the Schrödinger equation for a periodic potential consisting of an array of square-well potentials.
Solution. Consider a one-dimensional periodic potential, $V(x)=V(x+a)$, having lattice constant a and represented by the

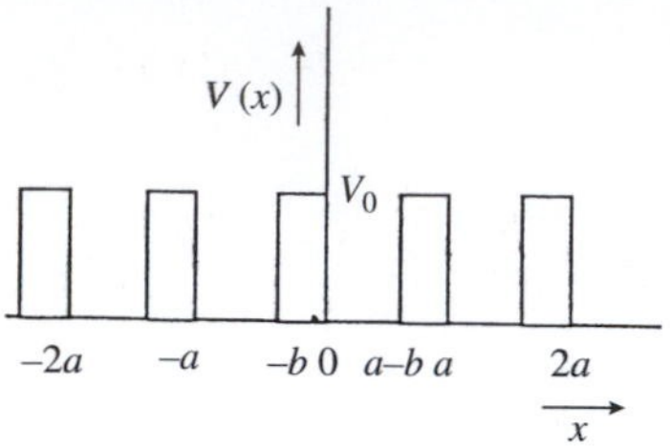

Fig. 2.11
Potential energy of the Kronig–Penney model.

square-well array shown in Fig. 2.11. The one-dimensional Schrödinger equation for this case is

$$-\frac{\hbar^2}{2m}\frac{d^2\psi(x)}{dx^2} + V(x)\psi(x) = E\psi(x),$$

where E is the energy eigenvalue and $\psi(x)$ is the eigenfunction.

In the region $0 < x < a - b$ where $V(x) = 0$, the eigenfunction is a linear combination of plane waves travelling to the right and to the left:

$$\psi_1(x) = Ae^{i\beta x} + Be^{-i\beta x}.$$

The energy eigenvalue is related to β by $E = \hbar^2\beta^2/2m$ or $\beta = \sqrt{2mE}/\hbar$.

In the other region $-b < x < 0$, within the barrier, the solution is of the form

$$\psi_2(x) = Ce^{\alpha x} + De^{-\alpha x},$$

where $V_0 - E = \hbar^2\alpha^2/2m$ or $\alpha = \sqrt{2m(V_0 - E)}/\hbar$. At each discontinuity of potential, the eigenfunction must satisfy the boundary conditions that it is continuous and has a continuous first derivative with respect to the coordinate. For the discontinuity at $x = 0$, we have

$$\psi_1(0) = \psi_2(0)$$
$$\psi_1'(0) = \psi_2'(0),$$

where the primes indicate first derivatives with respect to argument. For the discontinuity at $x = -b$, the boundary conditions are

$$\psi_1(-b) = \psi_2(-b)$$
$$\psi_1'(-b) = \psi_2'(-b).$$

The eigenfunction must also satisfy Bloch's theorem which for present purposes can be written as

$$\psi_1(a - b) = e^{ika}\psi_1(-b)$$
$$\psi_1'(a - b) = e^{ika}\psi_1'(-b).$$

The boundary conditions at $x = -b$ then become

$$e^{-ika}\psi_1(a - b) = \psi_2(-b)$$
$$e^{-ika}\psi_1'(a - b) = \psi_2'(-b).$$

If one now utilizes the expansions of $\psi_1(x)$ and $\psi_2(x)$ in plane waves, one obtains

$$A + B = C + D$$
$$i\beta(A - B) = \alpha(C - D)$$
$$e^{-ika}(Ae^{i\beta(a-b)} + Be^{-i\beta(a-b)}) = Ce^{-\alpha b} + De^{\alpha b}$$
$$i\beta e^{-ika}(Ae^{i\beta(a-b)}) - Be^{-i\beta(a-b)}) = \alpha(Ce^{-\alpha b} - De^{\alpha b}).$$

For a non-trivial solution to these equations, we set the determinant of the coefficients of A, B, C, D to zero:

$$\begin{vmatrix} 1 & 1 & -1 & -1 \\ i\beta & -i\beta & -\alpha & \alpha \\ e^{-ika+i\beta(a-b)} & e^{-ika-i\beta(a-b)} & -e^{-\alpha b} & -e^{\alpha b} \\ i\beta e^{-ika+i\beta(a-b)} & -i\beta e^{-ika-i\beta(a-b)} & -\alpha e^{-\alpha b} & \alpha e^{\alpha b} \end{vmatrix} = 0.$$

Evaluating the determinant gives the equation

$$\cos(ka) = \frac{\alpha^2 - \beta^2}{2\alpha\beta} \sinh(\alpha b) \sin[\beta(a-b)] + \cosh(\alpha b) \cos[\beta(a-b)].$$

Since α and β are functions of the energy eigenvalue E, this equation specifies the dependence of E on the wave vector k.

A simple result arises if one passes to the limit $b \to 0$, $V_0 \to \infty$ in such a way that the quantity $mV_0ba/\hbar^2$ approaches the finite value p:

$$\cos(ka) = \frac{p}{\beta a} \sin(\beta a) + \cos(\beta a).$$

This result corresponds to a potential $V(x)$ consisting of a periodic array of delta functions. Since the left hand side lies in the interval -1 to $+1$ for real k, the allowed values of E are those that cause the right hand side to lie in that same interval. They correspond to the heavily drawn ranges of βa in Fig. 2.12. Note that the forbidden ranges become smaller as E increases. Figure 2.13 shows the allowed bands of energy $E(k)$ separated by gaps at the zone boundaries.

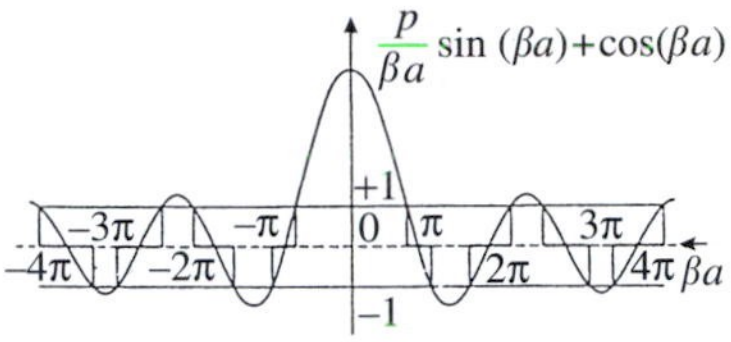

Fig. 2.12
The dependence of the function $(p/\beta a)\sin(\beta a) + \cos(\beta a)$ on βa with $p = 3\pi/2$ (after Kronig and Penney 1931).

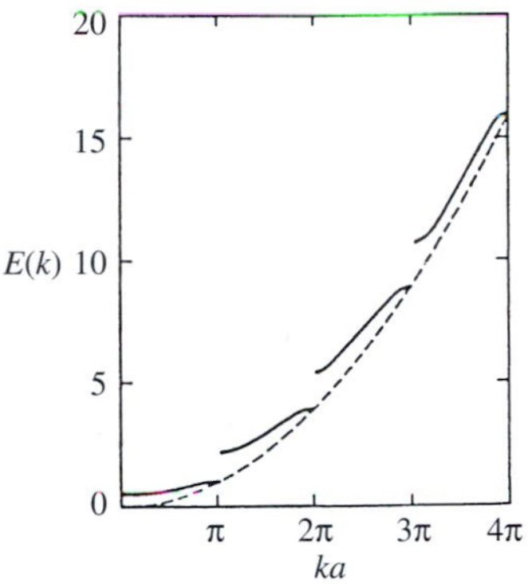

Fig. 2.13
Plot of energy in units of $\pi^2\hbar^2/2ma^2$ versus reduced wave vector for the Kronig–Penney model with $p = 3\pi/2$ (after Kittel 1986).

2.9 Tight binding method

The expansion of Bloch wave functions in a series of plane waves converges well if the electron–ion interaction is weak, i.e., the electron energy undergoes only a small change as a result of the interaction. It is of interest to consider an alternative method, the **tight binding method**, which is particularly appropriate when the electron–ion interaction is strong and the electrons are tightly bound to the positive ions. In this method the electron wave function is expanded in a series of functions that are localized about the various nuclei of the crystal.

2.9.1 Wannier functions

One can construct a set of localized functions by taking suitable linear combinations of Bloch functions $\psi_{n\boldsymbol{k}}(\boldsymbol{r})$. The resulting functions, called **Wannier functions**, are defined by

$$w_n(\boldsymbol{r} - \boldsymbol{R}(\ell)) = N^{-\frac{1}{2}} \sum_{\boldsymbol{k}} e^{-i\boldsymbol{k}\cdot\boldsymbol{R}(\ell)} \psi_{n\boldsymbol{k}}(\boldsymbol{r}) \tag{2.69}$$

for the case of one atom per unit cell. The fact that w_n is a function of $\boldsymbol{r} - \boldsymbol{R}(\ell)$ only can be seen by using Bloch's theorem and writing

$$w_n(\boldsymbol{r} - \boldsymbol{R}(\ell)) = N^{-\frac{1}{2}} \sum_{\boldsymbol{k}} e^{-i\boldsymbol{k}\cdot\boldsymbol{R}(\ell)} e^{i\boldsymbol{k}\cdot\boldsymbol{r}} u_{n\boldsymbol{k}}(\boldsymbol{r})$$
$$= N^{-\frac{1}{2}} \sum_{\boldsymbol{k}} e^{i\boldsymbol{k}\cdot(\boldsymbol{r}-\boldsymbol{R}(\ell))} u_{n\boldsymbol{k}}(\boldsymbol{r} - \boldsymbol{R}(\ell)), \tag{2.70}$$

where we have used the periodicity of $u_{n\boldsymbol{k}}(\boldsymbol{r})$. One can use the orthonormality of the Bloch functions $\psi_{n\boldsymbol{k}}(\boldsymbol{r})$ to show that Wannier functions centered on different lattice sites are orthogonal:

$$\int d^3 r w_n^*(\boldsymbol{r} - \boldsymbol{R}(\ell)) w_n(\boldsymbol{r} - \boldsymbol{R}(\ell')) = \delta_{\ell,\ell'}. \tag{2.71}$$

Each Wannier function is localized about the lattice site entering into its definition. This can be easily demonstrated for the special case of a simple cubic lattice with $u_{n\boldsymbol{k}}(\boldsymbol{r})$ independent of $\boldsymbol{k}$: $u_{n\boldsymbol{k}}(\boldsymbol{r}) = u_{n0}(\boldsymbol{r})$. Then

$$w_n(\boldsymbol{r} - \boldsymbol{R}(\ell)) = N^{-\frac{1}{2}} u_{n0}(\boldsymbol{r}) \sum_{\boldsymbol{k}} e^{i\boldsymbol{k}\cdot(\boldsymbol{r}-\boldsymbol{R}(\ell))}$$
$$= N^{-\frac{1}{2}} u_{n0}(\boldsymbol{r}) \prod_{\alpha} \frac{\sin[\pi(r_\alpha - R_\alpha(\ell))/a]}{\pi(r_\alpha - R_\alpha(\ell))/a}. \tag{2.72}$$

The function

$$f(x) = \frac{\sin x}{x} \tag{2.73}$$

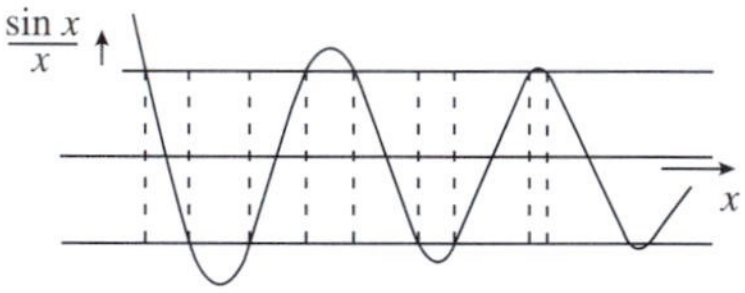

Fig. 2.14
The function $(\sin x)/x$ plotted versus x.

is a localized function that has its maximum at $x = 0$ and decays in an oscillatory fashion as $x \to \pm\infty$. This behavior is shown in Fig. 2.14.

Although the orthogonality of Wannier functions localized on different lattice sites is a desirable property, they are of limited use in energy band calculations, because the desired quantities, the Bloch states, must be known in order to calculate the Wannier functions. Recourse is therefore made to alternative forms of localized functions.

2.9.2 LCAO method

A convenient set of **localized functions** is composed of **atomic orbitals** $\varphi_{\kappa i}\boldsymbol{r}$, where κ designates the type of atom and i the particular atomic state of that atom ($1s$, $2s$, $2p_x$, etc.). A tight binding Bloch function $\psi_{\boldsymbol{k},\kappa i}(\boldsymbol{r})$ is constructed having the form

$$\psi_{\boldsymbol{k},\kappa i}(\boldsymbol{r}) = \frac{1}{\sqrt{N}} \sum_{\ell} e^{i\boldsymbol{k}\cdot\boldsymbol{R}(\ell\kappa)} \varphi_{\kappa i}(\boldsymbol{r} - \boldsymbol{R}(\ell\kappa)), \tag{2.74}$$

where $\boldsymbol{R}(\ell\kappa)$ is the position vector of the κth nucleus in the ℓth primitive unit cell. In the terminology of theoretical chemistry, the expansion contained in Eq. (2.74) is known as a **linear combination of atomic orbitals** or **LCAO**.

The contributions with $\boldsymbol{k} = 0$ to several Bloch functions of orbitals centered on adjacent atomic sites are shown in Fig. 2.15.

The eigenfunctions and energy eigenvalues of the Hamiltonian are determined by taking a linear combination of the Bloch functions $\psi_{\boldsymbol{k},\kappa i}\boldsymbol{r}$ for various values of κ and i and choosing the coefficients to minimize the expectation value of the Hamiltonian subject to the normalization condition on the eigenfunction. The condition for achieving this result is that the following determinantal equation be satisfied:

$$|\underline{\mathsf{H}} - E_{\boldsymbol{k}}\underline{\mathsf{S}}| = 0. \tag{2.75}$$

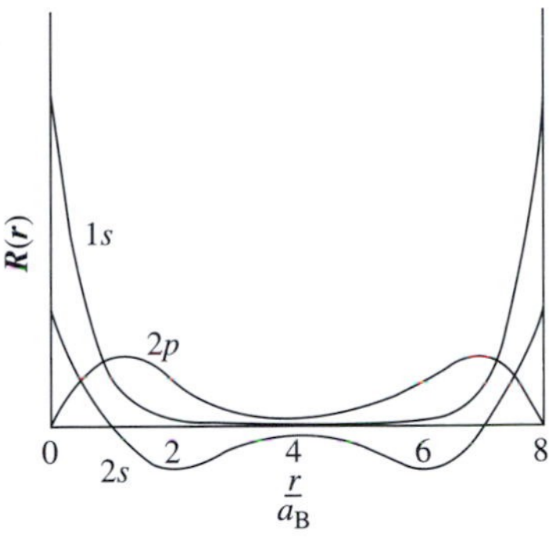

Fig 2.15
Contributions to several $\boldsymbol{k} = 0$ Bloch functions of $1s$, $2s$, and $2p_z$ orbitals on adjacent atomic sites.

The elements of the **Hamiltonian matrix** $\underline{\mathsf{H}}$ in Eq. (2.75) are designated by $H_{\kappa i,\kappa' i'}$ and are given by

$$H_{\kappa i,\kappa' i'} = \int \psi^*_{\boldsymbol{k},\kappa i}(\boldsymbol{r}) H \psi_{\boldsymbol{k},\kappa' i'}(\boldsymbol{r}) d^3 r, \tag{2.76}$$

where the Hamiltonian for H is given by Eq. (2.18). The elements of the **overlap matrix** $\underline{\mathsf{S}}$ are designated by $S_{\kappa i,\kappa' i'}$ and are given by

$$S_{\kappa i,\kappa' i'} = \int \psi^*_{\boldsymbol{k},\,\kappa i}(\boldsymbol{r}) \psi_{\boldsymbol{k},\,\kappa' i'}(\boldsymbol{r}) d^3 r. \tag{2.77}$$

It should be emphasized that in general $S_{\kappa i,\kappa' i'}$ is not the Kronecker $\delta_{\kappa i,\kappa' i'}$ because atomic orbitals $\varphi_{\kappa i}(\boldsymbol{r} - \boldsymbol{R}(\ell\kappa))$ and $\varphi_{\kappa' i'}(\boldsymbol{r} - \boldsymbol{R}(\ell'\kappa'))$ centered on different nuclei are not orthonormal. In principle, the size of the matrices $\underline{\mathsf{H}}$ and $\underline{\mathsf{S}}$ is infinite, but in practice, one truncates the matrices by including only a finite number of occupied and nearby unoccupied atomic states.

The expression for $H_{\kappa i,\kappa' i'}$ in Eq. (2.76) can be rewritten using Eq. (2.74):

$$H_{\kappa i,\kappa' i'} = \frac{1}{N} \int \left[\sum_{\ell} e^{-i\boldsymbol{k}\cdot\boldsymbol{R}(\ell\kappa)} \varphi^*_{\kappa i}(\boldsymbol{r} - \boldsymbol{R}(\ell\kappa)) \right] H \times \left[\sum_{\ell'} e^{i\boldsymbol{k}\cdot\boldsymbol{R}(\ell'\kappa')} \varphi_{\kappa' i'}(\boldsymbol{r} - \boldsymbol{R}(\ell'\kappa')) \right] d^3 r. \tag{2.78}$$

Assuming periodic boundary conditions, we change the variable of integration from $\boldsymbol{r}$ to $\boldsymbol{r}' = \boldsymbol{r} - \boldsymbol{R}(\ell\kappa)$:

$$\begin{aligned} H_{\kappa i,\kappa' i'} &= \frac{1}{N} \int \left[\sum_{\ell} e^{-i\boldsymbol{k}\cdot\boldsymbol{R}(\ell\kappa)} \varphi^*_{\kappa i}(\boldsymbol{r}') \right] H \\ &\quad \times \left[\sum_{\ell'} e^{i\boldsymbol{k}\cdot\boldsymbol{R}(\ell'\kappa')} \varphi_{\kappa' i'}(\boldsymbol{r}' - \boldsymbol{R}(\ell'\kappa') + \boldsymbol{R}(\ell\kappa)) \right] d^3 r' \\ &= \frac{1}{N} \sum_{\ell} \sum_{\bar{\ell}} e^{i\boldsymbol{k}\cdot[\boldsymbol{R}(\bar{\ell}) + \boldsymbol{R}(\kappa') - \boldsymbol{R}(\kappa)]} \\ &\quad \times \int \varphi^*_{\kappa i}(\boldsymbol{r}') H \varphi_{\kappa' i'}(\boldsymbol{r}' - \boldsymbol{R}(\bar{\ell}\kappa') + \boldsymbol{R}(0\kappa)) d^3 r', \end{aligned}$$

where we have set $\bar{\ell} = \ell' - \ell$ and have made use of Eq. (1.4). The sum over ℓ can be done immediately to yield a factor of N which cancels the factor $1/N$. Dropping the prime on the dummy integration variable $\boldsymbol{r}'$, we obtain

$$H_{\kappa i,\kappa' i'} = \sum_{\bar{\ell}} e^{i\boldsymbol{k}\cdot[\boldsymbol{R}(\bar{\ell}\kappa')-\boldsymbol{R}(0\kappa)]} \times \int \varphi^*_{\kappa i}(\boldsymbol{r}) H \varphi_{\kappa' i'}(\boldsymbol{r} - \boldsymbol{R}(\bar{\ell}\kappa') + \boldsymbol{R}(0\kappa)) d^3r. \tag{2.80}$$

An analysis similar to that just given can be applied to the overlap integrals $S_{\kappa i,\kappa' i'}$ defined by Eq. (2.77). The result is

$$S_{\kappa i,\kappa' i'} = \sum_{\bar{\ell}} e^{i\boldsymbol{k}\cdot[\boldsymbol{R}(\bar{\ell}\kappa')-\boldsymbol{R}(0\kappa)]} \times \int \varphi^*_{\kappa i}(\boldsymbol{r}) \varphi_{\kappa' i'}(\boldsymbol{r} - \boldsymbol{R}(\bar{\ell}\kappa') + \boldsymbol{R}(0\kappa)) d^3r. \tag{2.81}$$

We see that $S_{\kappa i,\kappa' i'}$ consists of a sum of terms containing either one-center or two-center integrals depending on whether $\boldsymbol{R}(\bar{\ell}\kappa') = \boldsymbol{R}(0\kappa)$ or $\boldsymbol{R}(\ell\kappa') \neq \boldsymbol{R}(0\kappa)$, respectively. Their evaluation can be conveniently carried out using spherical coordinates for the one-center integrals and elliptic coordinates for the two-center integrals. The latter can be reduced to certain basic two-center integrals using the relations developed by Slater and Koster (1954). Since the two-center integrals decrease exponentially with increasing separation of the centers, it is customary to neglect these integrals for centers with separation greater than some chosen value.

Consistent with the spirit of the tight binding procedure, one can express the potential energy $V(\boldsymbol{r})$ in the Hamiltonian H as a sum of atomic potentials:

$$V(\boldsymbol{r}) = \sum_{\ell\kappa} v_\kappa(\boldsymbol{r} - \boldsymbol{R}(\ell\kappa)), \tag{2.82}$$

where $v_\kappa(\boldsymbol{r})$ is the potential due to nucleus κ. We can then group the terms of H in the following way after making the replacement $\boldsymbol{r} - \boldsymbol{R}(\ell\kappa) \to \boldsymbol{r}$:

$$H = H_0 + v_\kappa(\boldsymbol{r}) + \sum_{\bar{\ell}''\kappa''\neq 0\kappa} v_{\kappa''}(\boldsymbol{r} - \boldsymbol{R}(\bar{\ell}''\kappa'') + \boldsymbol{R}(0\kappa)). \tag{2.83}$$

If we now assume that the atomic orbital $\varphi_{\kappa i}(\boldsymbol{r})$ is an eigenfunction of the atomic Hamiltonian $H_0 + v_\kappa(\boldsymbol{r})$ with energy eigenvalue ϵ_i^κ, then Eq. (2.79) for $H_{\kappa i,\kappa' i'}$ can be rewritten as

$$H_{\kappa i,\kappa' i'} = \epsilon_i^\kappa S_{\kappa i,\kappa' i'} + V_{\kappa i,\kappa' i'}, \tag{2.84}$$

where

$$V_{\kappa i,\kappa' i'} = \sum_{\bar{\ell}} e^{i\boldsymbol{k}\cdot[\boldsymbol{R}(\bar{\ell}\kappa')-\boldsymbol{R}(0\kappa)]}$$
$$\times \sum_{\bar{\ell}''\kappa''\neq 0\kappa} \int \varphi_{\kappa i}(\boldsymbol{r}) v_{\kappa''}(\boldsymbol{r} - \boldsymbol{R}(\bar{\ell}''\kappa'') + \boldsymbol{R}(0\kappa))$$
$$\times \varphi_{\kappa' i'}(\boldsymbol{r} - \boldsymbol{R}(\bar{\ell}\kappa') + \boldsymbol{R}(0\kappa)) d^3 r. \tag{2.85}$$

The sum over $\bar{\ell}''\kappa''$ contains three-center integrals, in which $\bar{\ell}''\kappa''$, $\bar{\ell}\kappa'$ and 0κ are all different, and two-center integrals in which either $\bar{\ell}''\kappa'' = \bar{\ell}\kappa' \neq 0\kappa$ or $\bar{\ell}''\kappa'' \neq \bar{\ell}\kappa' = 0\kappa$. Since the three functions $\varphi_{\kappa i}(\boldsymbol{r})$, $\varphi_{\kappa' i'}(\boldsymbol{r})$, and $v_\kappa(\boldsymbol{r})$ are all localized functions, the three-center integrals are typically small compared to the two-center integrals, where two of the three functions strongly overlap on the same center. Consequently the three-center integrals are frequently neglected. The results of Slater and Koster can be used to reduce the two-center integrals appearing in $V_{\kappa i,\kappa' i'}$ to certain basic integrals.

In order to calculate the energy band structure for a given material, it is necessary to have values for the basic integrals entering into $V_{\kappa i,\kappa' i'}$ and $S_{\kappa i,\kappa' i'}$. These values are determined in the **empirical** method by choosing them so that experimental data reflecting the band structure are reproduced. In the **ab initio** method, the basic integrals are calculated from tabulated atomic orbitals and atomic potentials such as those of Herman and Skillman (1963) and Clementi and Roetti (1974). Typically, the results of the *ab initio* method do not agree well with experimental data for semiconductors, so recourse is then made to a semi-empirical method in which the basic integrals are multiplied by scale factors to produce agreement with experiment. In many cases, the introduction of just a few scale factors leads to good results.

Example 2.5: Calculation of a simple energy band
Consider a one-dimensional, periodic array of hydrogen atoms, all in the $1s$ state, with lattice constant a. Calculate the energy as a function of the wave vector $\boldsymbol{k}$.
Solution. Since there is only one atom per primitive unit cell, the basis index κ and atomic state index i each take on only one value. We can drop κ and take $i = 1s$. The determinantal equation given by Eq. (2.75) reduces to the simple equation

$$\sum_{\ell} e^{ik\ell a} \int \varphi_{1s}^*(x,y,z)[H_0 + \sum_{\ell'} v(x - \ell' a, y, z)]\varphi_{1s}(x - \ell a, y, z) d^3 r$$
$$= E_k \sum_{\ell} e^{ik\ell a} \int \varphi_{1s}^*(x,y,z)\varphi_{1s}(x - \ell a, y, z) d^3 r.$$

Utilizing the result

$$[H_0 + v(x,y,z)]\varphi_{1s}(x,y,z) = \epsilon_{1s}\varphi_{1s}(x,y,z)$$

and separating out the terms with $\ell = 0$, we obtain

$$\epsilon_{1s} + \int \varphi^*_{1s}(x,y,z) \sum_{\ell' \neq 0} v(x - \ell' a, y, z)\varphi_{1s}(x,y,z)d^3r$$
$$+ \sum_{\ell \neq 0} e^{ik\ell a} \int \varphi^*_{1s}(x,y,z) \left[\epsilon_{1s} + \sum_{\ell' \neq 0} v(x - \ell' a, y, z)\right]$$
$$\times \varphi_{1s}(x - \ell a, y, z)d^3r$$
$$= E_k \left[1 + \sum_{\ell \neq 0} e^{ik\ell a} \int \varphi^*_{1s}(x,y,z)\varphi_{1s}(x - \ell a, y, z)d^3r\right],$$

where we have assumed that the atomic orbital $\varphi_{1s}(x,y,z)$ is normalized to unity. We now restrict ourselves to nearest neighbor overlaps. Letting

$$J_{1s} = \int \varphi^*_{1s}(x,y,z)v(x \pm a, y, z)\varphi_{1s}(x,y,z)d^3r$$
$$K_{1s} = \int \varphi^*_{1s}(x,y,z)v(x + a, y, z)\varphi_{1s}(x + a, y, z)d^3r$$
$$= \int \varphi^*_{1s}(x,y,z)v(x - a, y, z)\varphi_{1s}(x - a, y, z)d^3r$$
$$S_{1s} = \int \varphi^*_{1s}(x,y,z)\varphi_{1s}(x \pm a, y, z)d^3r,$$

and solving for E_k, we obtain

$$E_k = \epsilon_{1s} + \frac{2}{1 + 2S_{1s}\cos ka}[J_{1s} + K_{1s}\cos ka].$$

Taking $v(x,y,z)$ to be the Coulomb potential $-e^2/r$, we can evaluate the integrals to give (Pauling and Wilson 1935)

$$J_{1s} = \frac{e^2}{a_B}\left\{-\frac{1}{D} + e^{-2D}\left(1 + \frac{1}{D}\right)\right\},$$
$$K_{1s} = -\frac{e^2}{a_B}e^{-D}(a + D),$$
$$S_{1s} = e^{-D}\left(1 + D + \frac{1}{3}D^2\right),$$

where $D = a/a_B$ and a_B is the Bohr radius.

Let us consider the particular case $D = 5$. The integrals then have the values $J_{1s} = -0.3999$ Ry, $K_{1s} = -0.0809$ Ry, and $S_{1s} = 0.0966$. With $\epsilon_{1s} = -1.0$ Ry, the behavior of E_k as a function of ka is shown in Fig. 2.16. Note that the band width is $|4(K_{1s} - 2J_{1s}S_{1s}/(1 - 4S^2_{1s})| = 0.015$ Ry. It is interesting that the energy of a free electron is parabolic in wave vector for all wave vectors, whereas in

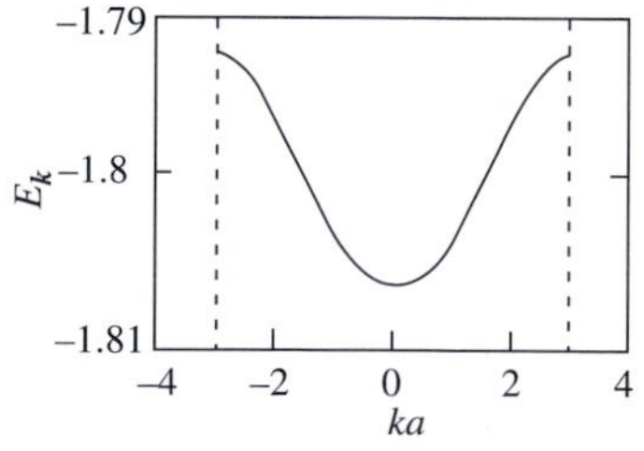

Fig. 2.16
Energy band for a linear chain of 1s hydrogen atoms.

the tight binding model parabolic behavior occurs only near the zone center and zone boundaries.

The tight binding procedure provides a valuable picture of how energy bands form as the isolated atoms are brought together to form the crystal. When the atoms are infinitely far apart, the S_{1s}, J_{1s} and K_{1s} integrals are zero and $E_k = \epsilon_i^{\kappa}$. The degeneracy of the eigenvalue ϵ_i^{κ} is equal to the number of atoms of type κ in the crystal, i.e., the number of primitive unit cells N. As the atoms approach each other, the overlap of the atomic orbitals commences and the integrals S_{1s}, J_{1s}, and K_{1s} increase in magnitude as the lattice constant a decreases. The degeneracies are then removed, and each atomic level broadens out into a band. The widths of the bands increase as the lattice constant decreases. Since there is one state for each atom of type κ, the total number of orbital states in a band is N. Therefore, taking into account spin, $2N$ electrons can be accommodated in a band, just as in the nearly free electron picture.

A schematic representation of the energy bands as a function of atomic separation for tetrahedrally bound semiconductors such as Si and Ge is shown in Fig. 2.17. At the equilibrium separation r_0 there is a forbidden energy gap E_g between the occupied and the unoccupied states that results from the sp^3 hybridization.

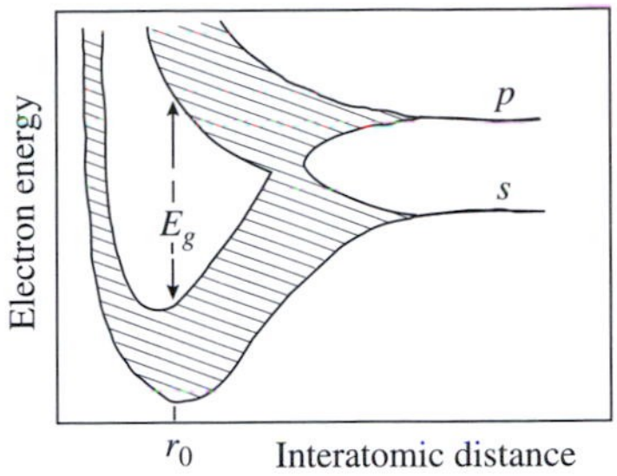

Fig. 2.17
Energy bands as a function of atomic separation for tetrahedrally bound semiconductors like Si and Ge.

Problems

1. Show that $\exp(i\boldsymbol{k}\cdot\boldsymbol{R}(\ell))$ is an eigenvalue and the Bloch function $\psi_{\boldsymbol{k}}(\boldsymbol{r})$ is an eigenfunction of the crystal translation operator T defined by $T\boldsymbol{r} = \boldsymbol{r} + \boldsymbol{R}(\ell)$.
2. Find the energy gap at the corner point $(\pi/a, \pi/a)$ of the Brillouin zone for a square lattice in two dimensions with the crystal potential

$$V(x, y) = -4V_0 \cos(2\pi x/a)\cos(2\pi y/a).$$

3. Consider the diamond structure and its reciprocal lattice associated with the conventional cubic cell. Let $\boldsymbol{b}_c$ be a primitive translation vector of this reciprocal lattice.
   (a) Show that the Fourier component V_G of the crystal potential seen by an electron is zero for $\boldsymbol{G} = 2\boldsymbol{b}_c$.
   (b) Demonstrate that the energy gap vanishes at the zone boundary plane normal to the end of the vector $\boldsymbol{b}_c$ in the first order approximation to the solutions of the wave equation in a periodic lattice.
   (c) Using Eqs. (2.35), where the only equations to be retained are those that contain both coefficients $C(\frac{1}{2}G)$ and $C(-\frac{1}{2}G)$, show that the potential energy $2V_0 \cos Gx$ creates an energy gap of exactly $2V_0$ at the zone boundary.
4. Find the energy of the lowest energy band at $k = 0$ for the Kronig–Penney model when the potential energy is a delta function and $p \ll 1$. For this case find the band gap at $k = \frac{\pi}{a}$.
5. By applying successive approximations, estimate the width of the first allowed band ΔE and the forbidden gap E_g for the Kronig–Penney model with $p = \frac{3\pi}{2}$ and $a - 3$ Å.
6. Prove that the Wannier functions centered at different lattice sites are orthogonal.

References

F. Bloch, *Z. Physik* **52**, 555 (1928).

M. Born and J. R. Oppenheimer, *Ann. Phys.* **84**, 457 (1927); M. Born and K. Huang, *Dynamical Theory of Crystal Lattices* (Oxford University Press, Oxford, 1954).

L. Brillouin, *Ann. Phys.* **17**, 88 (1922); *Wave Propagation in Periodic Structures*, Second edition (Dover Publications, New York, 1953).

E. Clementi and C. Roetti, *Atomic Data and Nuclear Data Tables* **14**, 177 (1974).

F. Herman and S. Skillman, *Atomic Structure Calculations* (Prentice-Hall, Englewood Cliffs, N. J., 1963).

C. Kittel, *Introduction to Solid State Physics*, Sixth edition (John Wiley, New York, 1986).

R. de L. Kronig and W. G. Penney, *Proc. Roy. Soc.* (London) **130**, 499 (1931).

L. Pauling and E. B. Wilson, Jr., *Introduction to Quantum Mechanics* (McGraw-Hill, New York, 1935).

J. C. Slater and G. F. Koster, *Phys. Rev.* **94**, 1478 (1954).

A. H. Wilson, *Proc. Roy. Soc.* (London) **133**, 458 (1931).

J. M. Ziman, *Principles of the Theory of Solids*, Second edition (Cambridge University Press, Cambridge, 1972).

Electronic energy bands: semiconductors

3

Key ideas

A perfect semiconductor at 0 K has its *valence bands* completely full of electrons and its *conduction bands* completely empty. Between these bands is the *fundamental band gap*.

Orbital and *spin angular momenta* of electrons interact to modify their energy levels.

The effective electron–ion potential can be represented by a *pseudopotential* consisting of attractive and repulsive parts.

Electrons interact with one another through the *Coulomb interaction*. In the *Hartree–Fock method* the Pauli exclusion principle leads to the *exchange interaction* between electrons. In the *density functional method* the energy of interacting electrons is a functional of the electron density $n(\mathbf{r})$.

Excited state energies can be calculated with the aid of the *electron self-energy operator* that combines the effects of exchange and correlation.

The $\mathbf{k} \cdot \mathbf{p}$ *method* provides the energy $E_{n\mathbf{k}}$ as a function of wave vector $\mathbf{k}$ near a *band extremum* without calculating the entire band structure.

A *nondegenerate band* with extremum at $\mathbf{k} = 0$ in a cubic crystal has an *isotropic effective mass*. A *degenerate band* with extremum at $\mathbf{k} = 0$ in a cubic crystal exhibits *warping* of the constant energy surfaces.

A band with extremum at $\mathbf{k} \neq 0$ in a cubic crystal has an *anisotropic effective mass*.

Silicon and germanium have *indirect gaps* with extrema of valence and conduction bands at different $\mathbf{k}$ values. The majority of III–V, II–VI, and IV–VI semiconductors have *direct gaps* with extrema of valence and conduction bands at $\mathbf{k} = 0$.

Energy band gaps can be modified by changing the *temperature*, by applying *pressure* or *stress*, or by forming an *alloy*.

Disorder in the atomic arrangement produces an *amorphous semiconductor*.

Energy bands in semiconductors

Pure semiconductors without deviations from periodicity have the important characteristic that at the absolute zero of temperature all energy bands up to and including a certain band are completely full of electrons and bands of higher energy are completely empty. There is a gap between the top edge of the highest filled band and the bottom edge of the lowest empty band. This gap is called the **fundamental gap**. The electrons occupying the lowest filled bands are strongly localized about individual nuclei and are termed **core electrons**. The electrons in the higher filled bands are less localized than core electrons and participate in covalent bonds between atoms. These electrons are **valence electrons**, and the bands they occupy are **valence bands**. For reasons that will become clear later, the bands that are empty at 0 K are **conduction bands**.

The general discussion of energy bands in Chapter 2 presents the basic concepts that underlie energy band theory, but omits a number of effects that have significant impact on the energy bands of real semiconductors. The effects that we discuss in this chapter are the spin–orbit interaction, the electron–ion interaction and pseudopotentials, and the electron–electron interaction. The characteristics of the energy bands of typical semiconductors are presented and analyzed with the aid of $\boldsymbol{k} \cdot \boldsymbol{p}$ perturbation theory.

3.1 Spin–orbit interaction

An electron possesses both orbital and spin angular momenta. These two forms of angular momentum interact through their associated magnetic moments to modify the energy of the electron. This interaction is the basis for the familiar LS coupling in atoms. For an electron of position vector $\boldsymbol{r}$ and momentum $\boldsymbol{p}$ moving in a central potential $V(\boldsymbol{r})$, the spin–orbit interaction energy is given by (Thomas 1926)

$$H_{so} = \xi(r)\boldsymbol{L} \cdot \boldsymbol{S}, \tag{3.1}$$

where $\boldsymbol{L}$ is the orbital angular momentum operator $\boldsymbol{r} \times \boldsymbol{p}$, $\boldsymbol{S}$ is the spin angular momentum operator $\frac{1}{2}\hbar\boldsymbol{\sigma}$, $\boldsymbol{\sigma}$ is the Pauli spin matrix vector, and

$$\xi(r) = \frac{1}{2m^2c^2}\frac{1}{r}\frac{dV}{dr}. \tag{3.2}$$

For a general potential the spin–orbit contribution to the Hamiltonian is given by

$$H_{so} = \frac{\hbar}{4m^2c^2}\boldsymbol{\sigma} \cdot \boldsymbol{\nabla} V \times \boldsymbol{p}, \tag{3.3}$$

where $\boldsymbol{\nabla}$ is the gradient operator. H_{so} should be added to the Hamiltonian given by Eq. (2.15). The presence of the speed of light c in H_{so} indicates that the spin–orbit interaction is a relativistic effect.

Comment. The spin–orbit interaction is one of several relativistic effects. When an electron moves in the strong local fields very near the nuclei, its speed can approach that of light, and one must take into account relativistic terms in the wave equation. The Schrödinger equation is replaced by the Dirac relativistic equation which, after eliminating the minor components, takes the form

$$\left[\left\{-\frac{\hbar^2}{2m}\nabla^2 + V(\mathbf{r}) - \frac{\hbar^4}{8m^3c^2}\boldsymbol{\nabla}^4 - \frac{\hbar^2}{4m^2c^2}\boldsymbol{\nabla}V(\mathbf{r})\cdot\boldsymbol{\nabla}\right\}\underline{1} - \frac{i\hbar^2}{4m^2c^2}\boldsymbol{\sigma}\cdot(\boldsymbol{\nabla}V(\mathbf{r})\times\boldsymbol{\nabla})\right]\underline{\psi} = E\underline{\psi},$$

where $\underline{\psi}$ is a 2-component spinor, $\underline{1}$ is the 2×2 unit matrix, and

$$\sigma_x = \begin{pmatrix} 0 & 1 \\ 1 & 0 \end{pmatrix}, \quad \sigma_y = \begin{pmatrix} 0 & -i \\ i & 0 \end{pmatrix}, \quad \sigma_z = \begin{pmatrix} 1 & 0 \\ 0 & -1 \end{pmatrix}.$$

The first two terms are the kinetic and potential energies, the third and fourth terms are the relativistic corrections to the kinetic energy and potential energy (the Darwin correction), and the fifth term represents the spin–orbit coupling. The relativistic corrections become important for heavy nuclei.

Since the magnitudes of both the potential and its gradient are very large near the nucleus, the spin–orbit interaction tends to reflect atomic characteristics. In particular, the energies of $j=\frac{3}{2}$ and $j=\frac{1}{2}$ atomic levels are split by the spin–orbit interaction. This situation carries over to energy bands in crystals, as may be readily seen in the tight-binding picture. For example, energy bands derived from $j=\frac{3}{2}$ and $j=\frac{1}{2}$ atomic levels are split.

Example 3.1: Spin–orbit splitting of valence bands
Discuss the spin–orbit splitting of the valence bands in Si, Ge, and InSb.

Solution. Semiconductors such as Si and Ge have the diamond structure. The structures of the valence bands are similar in diamond, Si, and Ge, with the point of maximum energy at $\boldsymbol{k}=0$. This point has energy E_V and is called the **band edge**. The valence band edge states are derived from p-like atomic states and would be threefold degenerate in the absence of spin. With spin the degeneracy is six-fold without spin–orbit interaction. Introducing spin–orbit interaction leads to a splitting of the valence band edge into two band edges: an upper band edge associated with fourfold degenerate $p_{\frac{3}{2}}$-like states and a lower band edge associated with two-fold degenerate $p_{\frac{1}{2}}$-like states.

When $\boldsymbol{k}\neq 0$, the $p_{\frac{3}{2}}$ bands split into two doubly degenerate bands called light and heavy hole bands. The energy difference between the $p_{\frac{3}{2}}$ and $p_{\frac{1}{2}}$ bands at $\boldsymbol{k}=0$ is denoted by Δ_{so}. A schematic representation of the valence band splittings is shown in Fig. 3.1. For diamond the value of Δ_{so} is ~ 0.006 eV, which is much less than the fundamental gap of 5.4 eV. As the atomic constituents become heavier, the spin–orbit splitting increases and the fundamental gap decreases. For Ge, $\Delta_{so}\sim 0.3$ eV and $E_g\simeq 0.74$ eV. InSb has a spin–orbit splitting of 0.82 eV which is significantly larger than its fundamental gap of

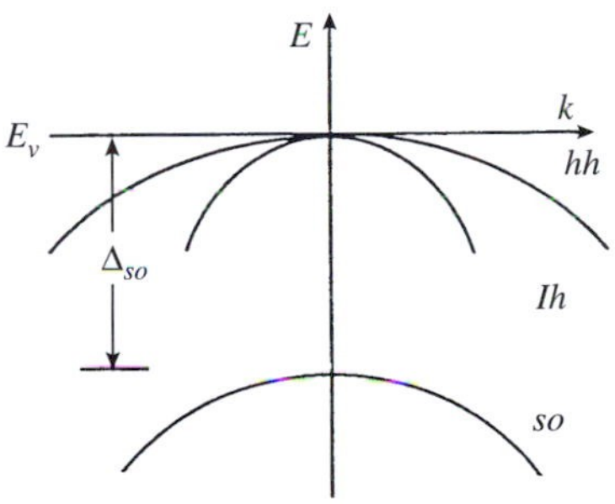

Fig. 3.1
Valence bands near the Brillouin zone center for diamond structure semiconductors with spin–orbit splitting Δ_{so}. The bands shown are the heavy hole (*hh*), light hole (*lh*), and split off (*so*) bands. Each band is doubly degenerate.

0.23 eV, in contrast to the situation in Si and Ge. In semiconductors composed of heavy elements, the spin–orbit splitting is one of the important factors determining the band structure.

3.2 Electron–ion interaction and pseudopotentials

If all the electrons of a crystal are dealt with explicitly in an energy band calculation, the electron–ion interaction is simply the Coulomb interaction due to the positive nuclei. Treating all the electrons, however, is very difficult analytically and very computer-intensive numerically. Part of the reason for this difficulty is that the wave functions of the loosely bound, more extended electronic states must be orthogonal to the wave functions of the tightly bound, more localized core states. Furthermore, the electron–ion potential is not weak in the core region, so plane waves are a poor approximation there. If the extended state wave functions are expanded in terms of plane waves, it may take an enormous number of plane waves to achieve core state orthogonalization and to properly account for the strong electron–ion potential in the core region.

3.2.1 Orthogonalized plane wave method

A way of handling the orthogonality problem was pointed out by Herring (1940) who proposed the **orthogonalized plane wave method** (OPW method). If $|\boldsymbol{k}\rangle$ denotes a plane wave of wave vector $\boldsymbol{k}$ and $|c\rangle$ a core state, the orthogonalized plane wave $|\boldsymbol{k}\rangle_{OPW}$ is given by

$$|\boldsymbol{k}\rangle_{OPW} = |\boldsymbol{k}\rangle - \sum_c \langle c|\boldsymbol{k}\rangle |c\rangle, \tag{3.4}$$

where the summation is over all core states. It is easily seen that $|\boldsymbol{k}\rangle_{OPW}$ is orthogonal to any core state $|c'\rangle$:

$$\begin{aligned}\langle c'|\boldsymbol{k}\rangle_{OPW} &= \langle c'|\boldsymbol{k}\rangle - \sum_c \langle c|\boldsymbol{k}\rangle\langle c'|c\rangle \\ &= \langle c'|\boldsymbol{k}\rangle - \sum_c \langle c|\boldsymbol{k}\rangle \delta_{cc'} \\ &= \langle c'|\boldsymbol{k}\rangle - \langle c'|\boldsymbol{k}\rangle = 0.\end{aligned} \tag{3.5}$$

The exact wave function for an extended electron state can be expanded in a series of OPWs of the form

$$|\phi_{\boldsymbol{k}}^{OPW}\rangle = \sum_{\boldsymbol{G}} C(\boldsymbol{k}-\boldsymbol{G})|\boldsymbol{k}-\boldsymbol{G}\rangle_{OPW}, \tag{3.6}$$

where $\boldsymbol{G}$ is a reciprocal lattice vector. Substitution of Eq. (3.6) into the Schrödinger equation, Eq. (2.17), and use of Eq. (3.4) gives

$$\begin{aligned}&\sum_{\boldsymbol{G}} C(\boldsymbol{k}-\boldsymbol{G})[(H_0+V)|\boldsymbol{k}-\boldsymbol{G}\rangle - \sum_c \langle c|\boldsymbol{k}-\boldsymbol{G}\rangle \epsilon_c |c\rangle] \\ &\quad = \sum_{\boldsymbol{G}} C(\boldsymbol{k}-\boldsymbol{G})[E\,|\boldsymbol{k}-\boldsymbol{G}\rangle - \sum_c \langle c|\boldsymbol{k}-\boldsymbol{G}\rangle E\,|c\rangle],\end{aligned} \tag{3.7}$$

where we assume the core states are sufficiently localized so that

$$(H_0 + V)|c\rangle = \epsilon_c|c\rangle \tag{3.8}$$

and ϵ_c is independent of $\boldsymbol{k}$. Noting that H_0 is the kinetic energy operator defined by Eq. (2.19) and taking the scalar product of Eq. (3.7) with the plane wave state $\langle \boldsymbol{k} - \boldsymbol{G}'|$ we obtain

$$\begin{aligned}\sum_{\boldsymbol{G}} C(\boldsymbol{k}-\boldsymbol{G})\Bigg[\frac{\hbar^2}{2m}(\boldsymbol{k}-\boldsymbol{G})^2\delta_{\boldsymbol{GG}'} + \langle \boldsymbol{k}-\boldsymbol{G}'|V|\boldsymbol{k}-\boldsymbol{G}\rangle \\ - \sum_c \epsilon_c\langle c|\boldsymbol{k}-\boldsymbol{G}\rangle\langle \boldsymbol{k}-\boldsymbol{G}'|c\rangle\Bigg] \\ = \sum_{\boldsymbol{G}} EC(\boldsymbol{k}-\boldsymbol{G})[\delta_{\boldsymbol{GG}'} - \sum_c\langle c|\boldsymbol{k}-\boldsymbol{G}\rangle\langle \boldsymbol{k}-\boldsymbol{G}'|c\rangle].\end{aligned} \tag{3.9}$$

Exploiting the Kronecker deltas and rearranging terms, Eq. (3.9) becomes

$$\begin{aligned}\frac{\hbar^2}{2m}(\boldsymbol{k}-\boldsymbol{G})^2C(\boldsymbol{k}-\boldsymbol{G}') + \sum_{\boldsymbol{G}}\langle \boldsymbol{k}-\boldsymbol{G}'|V+V_R||\boldsymbol{k}-\boldsymbol{G}\rangle C(\boldsymbol{k}-\boldsymbol{G}) \\ = EC(\boldsymbol{k}-\boldsymbol{G}'),\end{aligned} \tag{3.10}$$

where

$$V_R = \sum_c (E - \epsilon_c)|c\rangle\langle c|. \tag{3.11}$$

In coordinate space, we have

$$V_R(\boldsymbol{r}, \boldsymbol{r}') = \sum_c (E - \epsilon_c)\varphi_c^*(\boldsymbol{r})\varphi_c(\boldsymbol{r}'),$$

where $\varphi_c(\boldsymbol{r})$ is a core state orbital.

The quantity V_R is a potential operator that is nonlocal and energy-dependent. It is repulsive in character, because for extended electronic states, $E > \epsilon_c$. It therefore tends to cancel the electron–ion potential V which is attractive in character. Consequently, the effective potential $V + V_R$ appearing in Eq. (3.10) tends to be weak, a fact which was exploited by Phillips and Kleinman (1959) in their development of the **pseudopotential method**.

3.2.2 Pseudopotential method

The effective potential $V + V_R$ is known as the **pseudopotential** V_{ps}:

$$V_{ps} = V + V_R. \tag{3.12}$$

In terms of the pseudopotential, Eq. (3.10) can be rewritten as

$$\sum_{\boldsymbol{G}'} H_{\boldsymbol{G},\boldsymbol{G}'}(k)C(\boldsymbol{k}-\boldsymbol{G}') = EC(\boldsymbol{k}-\boldsymbol{G}), \tag{3.13}$$

where

$$H_{G,G'}(\mathbf{k}) = \frac{\hbar^2}{2m}(\mathbf{k}-\mathbf{G})^2\delta_{G,G'} + \langle \mathbf{k}-\mathbf{G}|V_{ps}|\mathbf{k}-\mathbf{G}'\rangle \tag{3.14}$$

and we have interchanged the roles of $\mathbf{G}$ and $\mathbf{G}'$. The fact that V_{ps} is weak compared to V means that the Fourier components of V_{ps} are small, except for the first few reciprocal lattice vectors, and the convergence of the series in Eq. (3.6) for $|\phi_{\mathbf{k}}^{OPW}\rangle$ should be rapid.

Following Phillips and Kleinman, we take the coefficients $C(\mathbf{k}-\mathbf{G})$ that satisfy Eq. (3.13) and define the function

$$\chi_{\mathbf{k}}(\mathbf{r}) = \sum_{\mathbf{G}} C(\mathbf{k}-\mathbf{G})|\mathbf{k}-\mathbf{G}\rangle. \tag{3.15}$$

We can then re-express $|\phi_{\mathbf{k}}^{OPW}\rangle$ as

$$|\phi_{\mathbf{k}}^{OPW}\rangle = |\chi_{\mathbf{k}}\rangle - \sum_{c}\langle c|\chi_{\mathbf{k}}\rangle|c\rangle. \tag{3.16}$$

Substituting this form for $|\phi_{\mathbf{k}}^{OPW}\rangle$ into the Schrödinger equation $H|\phi_{\mathbf{k}}^{OPW}\rangle = E\,|\phi_{\mathbf{k}}^{OPW}\rangle$ yields

$$H|\chi_{\mathbf{k}}\rangle - \sum_{c}\langle c|\chi_{\mathbf{k}}\rangle H|c\rangle = E\,|\chi_{\mathbf{k}}\rangle - \sum_{c}\langle c|\chi_{\mathbf{k}}\rangle E\,|c\rangle \tag{3.17}$$

Using Eq. (3.8) and rearranging, we get

$$\left[H + \sum_{c}(E-\epsilon_c)|c\rangle\langle c|\right]|\chi_{\mathbf{k}}\rangle = E\,|\chi_{\mathbf{k}}\rangle \tag{3.18}$$

or, using Eqs. (2.18), (3.11), and (3.12),

$$[H_0 + V_{ps}]|\chi_{\mathbf{k}}\rangle = E\,|\chi_{\mathbf{k}}\rangle. \tag{3.19}$$

We see that $|\chi_{\mathbf{k}}\rangle$ satisfies a Schrödinger-like equation with the pseudopotential as the effective potential and the eigenvalue given by the exact eigenvalue E. $|\chi_{\mathbf{k}}\rangle$ is known as the **pseudowavefunction**. Since V_{ps} is weak, $|\chi_{\mathbf{k}}\rangle$ can be expanded in a rapidly convergent series of plane waves. This is a computationally much less demanding task than the solution of Eq. (2.17) in which the full potential V appears and for which a plane wave expansion is much less rapidly convergent.

There still remains the nontrivial problem of actually obtaining the pseudopotential. Not only is the pseudopotential nonlocal and dependent on the energy eigenvalue E, it is also not unique (Ziman 1972). In practice much use has been made of local forms of the pseudopotential that contain parameters chosen to fit experimental data, but recently nonlocal forms have become common.

3.2.2.1 Empirical pseudopotential method

A particularly simple approach (Cohen and Bergstresser 1966) assumes that the pseudopotential can be expressed as a sum of atomic potentials $v_\kappa(\boldsymbol{r})$,

$$V_{ps}(\boldsymbol{r}) = \sum_{\ell\kappa} v_\kappa(\boldsymbol{r} - \boldsymbol{R}(\ell\kappa)), \tag{3.20}$$

just as in the tight binding method. The Fourier transform of $V_{ps}(\boldsymbol{r})$ then takes the form

$$V_{ps}(\boldsymbol{G}) = \frac{1}{\Omega}\int d^3r \sum_{\ell\kappa} v_\kappa(\boldsymbol{r} - \boldsymbol{R}(\ell\kappa))e^{-i\boldsymbol{G}\cdot\boldsymbol{r}}, \tag{3.21}$$

where Ω is the volume of the crystal. This expression can be recast as

$$\begin{aligned} V_{ps}(\boldsymbol{G}) &= \frac{1}{\Omega}\sum_{\ell\kappa} e^{-i\boldsymbol{G}\cdot\boldsymbol{R}(\ell\kappa)} \int d^3r v_\kappa(\boldsymbol{r} - \boldsymbol{R}(\ell\kappa))e^{-i\boldsymbol{G}\cdot(\boldsymbol{r}-\boldsymbol{R}(\ell\kappa))} \\ &= \frac{1}{\Omega}\sum_{\ell} e^{-i\boldsymbol{G}\cdot\boldsymbol{R}(\ell)} \sum_{\kappa} e^{-i\boldsymbol{G}\cdot\boldsymbol{R}(\kappa)} \int d^3r' v_\kappa(\boldsymbol{r}')e^{-i\boldsymbol{G}\cdot\boldsymbol{r}'}, \end{aligned} \tag{3.22}$$

where we have assumed periodic boundary conditions. The sum over ℓ gives the number of unit cells N by Eq. (2.9). With the introduction of

$$v_\kappa(\boldsymbol{G}) = \frac{n_c}{\Omega_0}\int d^3r v_\kappa(\boldsymbol{r})e^{-i\boldsymbol{G}\cdot\boldsymbol{r}}, \tag{3.23}$$

Eq. (3.22) becomes

$$V_{ps}(\boldsymbol{G}) = \frac{1}{n_c}\sum_{\kappa} e^{-i\boldsymbol{G}\cdot\boldsymbol{R}(\kappa)} v_\kappa(\boldsymbol{G}), \tag{3.24}$$

where n_c is the number of atoms in the primitive unit cell.

The Fourier transformed atomic potential $v_\kappa(\boldsymbol{G})$ is known as the **pseudopotential form factor** for atom κ. For semiconductors having the diamond or zincblende structure, it is convenient to choose the origin of coordinates midway between the two atoms in the primitive unit cell at the point $a(\frac{1}{8}, \frac{1}{8}, \frac{1}{8})$. Then $\boldsymbol{R}(1) = -a(\frac{1}{8}, \frac{1}{8}, \frac{1}{8})$ and $\boldsymbol{R}(2) = +a(\frac{1}{8}, \frac{1}{8}, \frac{1}{8})$. Substitution of these values into Eq. (3.24), setting $\boldsymbol{\tau} = a(\frac{1}{8}, \frac{1}{8}, \frac{1}{8})$, and taking $n_c = 2$ gives the result

$$V_{ps}(\boldsymbol{G}) = \{v_S(\boldsymbol{G})S^S(\boldsymbol{G}) + iv_A(\boldsymbol{G})S^A(\boldsymbol{G})\}, \tag{3.25}$$

where $v_S(\boldsymbol{G}) = \frac{1}{2}[v_1(\boldsymbol{G}) + v_2(\boldsymbol{G})]$ and $v_A(\boldsymbol{G}) = \frac{1}{2}[v_1(\boldsymbol{G}) - v_2(\boldsymbol{G})]$ are the symmetric and antisymmetric form factors, respectively. The quantities $S^S(\boldsymbol{G}) = \cos\boldsymbol{G}\cdot\boldsymbol{\tau}$ and $S^A(\boldsymbol{G}) = \sin\boldsymbol{G}\cdot\boldsymbol{\tau}$ are the symmetric and antisymmetric **structure factors**, respectively. Since the two atoms in the unit cell of the diamond structure are the same species, $v_1(\boldsymbol{G}) = v_2(\boldsymbol{G})$ and $v_A(\boldsymbol{G}) = 0$. Consequently,

$$V_{ps}(\boldsymbol{G}) = v_S(\boldsymbol{G})\cos\boldsymbol{G}\cdot\boldsymbol{\tau}. \tag{3.26}$$

Cohen and Bergstresser have calculated the band structures near the fundamental energy gap for a large number of semiconductors having the diamond or zincblende structure. They chose the form factors to fit optical data and found that the smallest three or four nonzero values of $|\boldsymbol{G}|$ were sufficient to give good convergence of the reciprocal lattice vector sums. Their results, however, do not include the effect of the spin–orbit interaction or nonlocal effects.

Example 3.2: Band structure of ZnSe
Discuss the determination of the band structure of ZnSe by the empirical pseudopotential method (EPM).
Solution. The pseudopotential Hamiltonian is

$$H_{ps} = -\frac{\hbar^2}{2m}\nabla^2 + V_{ps}(\boldsymbol{r}).$$

Making use of Eq. (3.25), the pseudopotential $V_{ps}(\boldsymbol{r})$ for a zincblende-structure semiconductor can be expressed as a Fourier series involving symmetric and antisymmetric pseudopotential form factors and structure factors,

$$V_{ps}(\boldsymbol{r}) = \sum_{|\boldsymbol{G}|\leqq \boldsymbol{G}_0} [S^S(\boldsymbol{G})v_S(\boldsymbol{G}) + iS^A(\boldsymbol{G})v_A(\boldsymbol{G})]e^{i\boldsymbol{G}\cdot\boldsymbol{r}},$$

where $\boldsymbol{G}_0$ is a cutoff reciprocal lattice vector. Both the symmetric and antisymmetric structure factors contribute for a zincblende structure material.

For the determination of the form factors, only the first five reciprocal lattice vectors which have squared amplitudes of 0, 3, 4, 8, and 11 in reduced units are considered (Chelikowsky and Cohen 1976). Larger reciprocal lattice vectors are assumed to give negligible contributions to the pseudopotential. The symmetric structure factor for $G^2 = 4$ is zero and the antisymmetric structure factor for $G^2 = 0$ and $G^2 = 8$ is zero. One can take $v_S(0) = 0$, since it is a constant contribution to all energy levels. Thus, there are three symmetric and three antisymmetric form factors to be specified. Approximate form factors are determined by comparison of experimental optical data with the calculated band structure.

The band structure is obtained by calculating the roots of the secular equation derived from the Hamiltonian matrix. The matrix for the zincblende structure is complex, since the matrix elements of the antisymmetric potential are pure imaginary. The procedure consists of adjusting the band structure in successive steps to make it consistent with the experimental results. One starts with a diamond-structure semiconductor for which the matrix involved is real and 20×20 in size, for example, Ge. Reflectivity and photo-emission measurements are used to estimate the energy difference between various electronic states in the Brillouin zone. A few levels near the band gap are fit by determining the form factors to thousandths of a rydberg. Having found the potential for Ge, one next considers GaAs by keeping the same symmetric potential and adding a small antisymmetric potential. To obtain the potential for ZnSe, one keeps the same symmetric potential and increases the antisymmetric form factor of GaAs by a factor of ~ 2.3 to within 0.01 Ry. The band structure of ZnSe calculated according to this procedure is shown in Fig. 3.2.

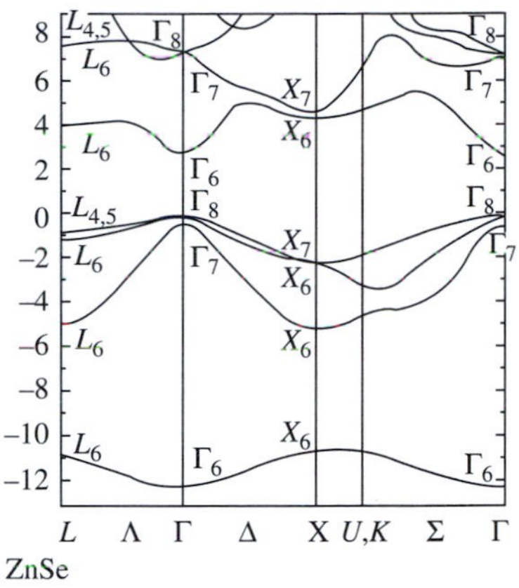

Fig. 3.2
Band structure of ZnSe by the empirical pseudopotential method (after Chelikowski and Cohen 1976).

3.2.2.2 Nonlocal pseudopotential method

A more sophisticated treatment of semiconductor energy bands has been given by Chelikowsky and Cohen (1976) who included both the spin–orbit interaction and nonlocal terms in the pseudopotential. The matrix elements of the spin–orbit Hamiltonian given by Eq. (3.3) were calculated with respect to OPWs. In order to include spin effects, the state indices that appear in Eq. (3.4) must be augmented by spin state indices s as follows: $|\boldsymbol{k}\rangle \rightarrow |\boldsymbol{k}s\rangle = |\boldsymbol{k}\rangle|\,s\rangle$, $|c\rangle \rightarrow |cs\rangle = |c\rangle|\,s\rangle$. The matrix elements of H_{so} then take the form

$$\begin{aligned}H^{so}_{\boldsymbol{G}s,\boldsymbol{G}'s'} &= \langle \boldsymbol{k}-\boldsymbol{G}s|H_{so}|\boldsymbol{k}-\boldsymbol{G}'s'\rangle \\ &\quad - \sum_{cs''}\langle \boldsymbol{k}-\boldsymbol{G}s|H_{so}|cs''\rangle\langle cs''|\boldsymbol{k}-\boldsymbol{G}'s'\rangle \\ &\quad - \sum_{cs''}\langle cs''|H_{so}|\boldsymbol{k}-\boldsymbol{G}'s'\rangle\langle \boldsymbol{k}-\boldsymbol{G}s|cs''\rangle \\ &\quad + \sum_{cs''}\sum_{c's'''}\langle \boldsymbol{k}-\boldsymbol{G}s|cs''\rangle\langle cs''|H_{so}|c's'''\rangle\langle c's'''|\boldsymbol{k}-\boldsymbol{G}'s'\rangle. \end{aligned} \tag{3.27}$$

Since the spin–orbit interaction is strong only near the nucleus of an atom, the dominant contribution to the matrix element comes from the last term involving the double sum. Writing

$$H_{so} = \boldsymbol{\Lambda}\cdot\boldsymbol{\sigma}, \tag{3.28}$$

where

$$\boldsymbol{\Lambda} = \frac{\hbar}{4m^2c^2}\boldsymbol{\nabla}V \times \boldsymbol{p}, \tag{3.29}$$

we have to a good approximation

$$H^{so}_{\boldsymbol{G}s,\boldsymbol{G}'s'} \simeq \langle s|\boldsymbol{\sigma}|s'\rangle \cdot \sum_{c}\sum_{c'}\langle \boldsymbol{k}-\boldsymbol{G}|c\rangle\langle c|\boldsymbol{\Lambda}|c'\rangle\langle c'|\boldsymbol{k}-\boldsymbol{G}'\rangle. \tag{3.30}$$

The extreme localization of the spin–orbit interaction has the additional consequence that we can write $\boldsymbol{\Lambda}$ as a sum of atomic contributions,

$$\boldsymbol{\Lambda} = \sum_{\ell\kappa}\frac{\hbar}{4m^2c^2}\boldsymbol{\nabla}v_{\kappa}(\boldsymbol{r}-\boldsymbol{R}(\ell\kappa)) \times \boldsymbol{p}. \tag{3.31}$$

Since the core index c denotes both the electronic state t and the lattice site $\ell\kappa$ of the core state atom, the only significant contributions to $\langle c|\boldsymbol{\Lambda}|c'\rangle$ are those for which c and c' have the same lattice site as one of the terms in Eq. (3.31) for $\boldsymbol{\Lambda}$. We can therefore rewrite Eq. (3.30) for $H^{so}_{\boldsymbol{G}s,\boldsymbol{G}'s'}(\boldsymbol{k})$ as

$$\begin{aligned}H^{so}_{\boldsymbol{G}s,\boldsymbol{G}'s'}(\boldsymbol{k}) &\simeq \langle s|\boldsymbol{\sigma}|s'\rangle \cdot \sum_{t}\sum_{t'}\sum_{\ell\kappa}\langle \boldsymbol{k}-\boldsymbol{G}|t\ell\kappa\rangle\langle t\kappa|\boldsymbol{\Lambda}|t'\kappa\rangle \\ &\quad \times \langle t'\ell\kappa|\boldsymbol{k}-\boldsymbol{G}'\rangle. \end{aligned} \tag{3.32}$$

Let us consider the sum S given by

$$S = \sum_{\ell} \langle \boldsymbol{k} - \boldsymbol{G} | t\ell\kappa \rangle \langle t'\ell\kappa | \boldsymbol{k} - \boldsymbol{G}' \rangle$$
$$= \sum_{\ell} \frac{1}{\Omega^2} \int e^{-i(\boldsymbol{k}-\boldsymbol{G})\cdot\boldsymbol{r}} \varphi_{t\kappa}(\boldsymbol{r} - \boldsymbol{R}(\ell\kappa)) d^3r$$
$$\times \int e^{i(\boldsymbol{k}-\boldsymbol{G}')\cdot\boldsymbol{r}'} \varphi^*_{t'\kappa}(\boldsymbol{r}' - \boldsymbol{R}(\ell\kappa)) d^3r'$$
$$= \sum_{\ell} \frac{1}{\Omega^2} e^{i(\boldsymbol{G}-\boldsymbol{G}')\cdot\boldsymbol{R}(\ell\kappa)} I^*_{t\kappa}(\boldsymbol{k} - \boldsymbol{G}) I_{t'\kappa}(\boldsymbol{k} - \boldsymbol{G}'), \tag{3.33}$$

where

$$I_{i\kappa}(\boldsymbol{k}) = \int e^{i\boldsymbol{k}\cdot\boldsymbol{r}} \varphi^*_{i\kappa}(\boldsymbol{r}) d^3r, \quad i = t, t' \tag{3.34}$$

is the Fourier transform of the orbital $\varphi_{i\kappa}(\boldsymbol{r})$ for core state i of atom κ. The sum over ℓ can now be carried out with the aid of Eqs. (1.4) and (2.5) to yield

$$S = \frac{N}{\Omega^2} e^{i(\boldsymbol{G}-\boldsymbol{G}'\cdot\boldsymbol{R}(\kappa)} I^*_{t\kappa}(\boldsymbol{k} - \boldsymbol{G}) I_{t'\kappa}(\boldsymbol{k} - \boldsymbol{G}'). \tag{3.35}$$

At this point we restrict ourselves to core states that are p-states, which are the most important core states for semiconductor band structure calculations (Chelikowsky and Cohen 1976). It can then be shown (Weisz 1966, Bloom and Bergstresser 1968) that the integral $I_{i\kappa}(\boldsymbol{k})$ is proportional to $|\boldsymbol{k}|$. Combining the magnitudes of $\boldsymbol{k} - \boldsymbol{G}$ and $\boldsymbol{k} - \boldsymbol{G}'$ with angular contributions from $\langle \boldsymbol{k} - \boldsymbol{G} | c \rangle$, $\langle c | \boldsymbol{\Lambda} | c' \rangle$, and $\langle c' | \boldsymbol{k} - \boldsymbol{G}' \rangle$ gives a factor $(\boldsymbol{k} - \boldsymbol{G}) \times (\boldsymbol{k} - \boldsymbol{G}')$ in the expression for $H^{so}_{\boldsymbol{G}s,\boldsymbol{G}'s'}(\boldsymbol{k})$. Utilizing this result together with Eq. (3.35), we can rewrite Eq. (3.32) in the form (Chelikowsky and Cohen 1976)

$$H^{so}_{\boldsymbol{G}s,\boldsymbol{G}'s'}(\boldsymbol{k}) \simeq (\boldsymbol{k} - \boldsymbol{G}) \times (\boldsymbol{k} - \boldsymbol{G}') \langle s | \boldsymbol{\sigma} | s' \rangle$$
$$\{-i\lambda^S[\cos(\boldsymbol{G} - \boldsymbol{G}') \cdot \tau] + \lambda^A \sin[(\boldsymbol{G} - \boldsymbol{G}') \cdot \tau]\}, \tag{3.36}$$

where we have restricted ourselves to diamond and zincblende structures, $\boldsymbol{\tau}$ is defined just before Eq. (3.25), and

$$\lambda^S = \tfrac{1}{2}(\lambda_1 + \lambda_2), \quad \lambda^A = \tfrac{1}{2}(\lambda_1 - \lambda_2). \tag{3.37}$$

The quantities λ_1 and λ_2 contain the contributions from the radial integrals involving atoms 1 and 2 in the primitive unit cell. In principle they may be calculated from tabulated atomic orbitals, but in practice one usually needs to adjust them to yield good agreement with experiment.

Nonlocal effects can be included by adding to the local atomic pseudopotential a term of the form (Chelikowsky and Cohen 1976)

$$V_{NL}(\boldsymbol{r}, E) = \sum_{\ell=0} A_\ell(E) f_\ell(r) \mathcal{P}_\ell, \tag{3.38}$$

where $A_\ell(E)$ is an energy-dependent well depth, $f_\ell(r)$ is a function simulating the effect of core states, and $\mathcal{P}_\ell$ is a projection operator for the

ℓth angular momentum component. A convenient form for $f_\ell(r)$ is $f_\ell(r) = 1$ for $r < R_\ell$ and $f_\ell(r) = 0$ for $r \geq R_\ell$. For semiconductor applications, ℓ can usually be restricted to the values 0, 1, 2. The quantities $A_\ell(E)$ contain parameters that are chosen to fit experimental data. The matrix elements of $V_{NL}(\boldsymbol{r}, E)$ with respect to plane wave states have the form

$$V_{NL}(\boldsymbol{k}-\boldsymbol{G}, \boldsymbol{k}-\boldsymbol{G}') = \frac{4\pi}{\Omega_a} \sum_{\ell,i} A^i_\ell(E)(2l+1)P_\ell(\cos\theta_{\boldsymbol{k}-\boldsymbol{G},\boldsymbol{k}-\boldsymbol{G}'}) \times S^i(\boldsymbol{G}'-\boldsymbol{G})F^i_\ell(\boldsymbol{k}-\boldsymbol{G}, \boldsymbol{k}-\boldsymbol{G}'), \qquad (3.39)$$

where Ω_a is the volume per atom, $P_\ell(x)$ is a Legendre polynomial, $\theta_{\boldsymbol{k}-\boldsymbol{G},\boldsymbol{k}-\boldsymbol{G}'}$ is the angle between $\boldsymbol{k}-\boldsymbol{G}$ and $\boldsymbol{k}-\boldsymbol{G}'$, $S^i(\boldsymbol{G}'-\boldsymbol{G})$ is the structure factor for species i and $F^i_\ell(\boldsymbol{k}-\boldsymbol{G}, \boldsymbol{k}-\boldsymbol{G}')$ is a somewhat complicated function of spherical Bessel functions.

The nonlocal pseudopotential method has proved to be a very powerful procedure for calculating accurate energy band structures of semiconductors. Numerous specific examples are presented in Section 3.5.

3.2.2.3 *Ab initio* pseudopotentials

The pseudopotentials that have been discussed so far have involved parameters that are determined by fitting experimental data for the crystal being studied. A conceptually more satisfactory procedure is to avoid data for the crystal and use only data for the individual constituent atoms. Pseudopotentials generated in this way are known as **ab initio pseudopotentials**. They can be viewed as producing truly predictive calculations for crystals. A set of *ab initio* pseudopotentials (Bachelet *et al.* 1982) has been developed for every atom from H to Pu. They can be used advantageously in the nonlocal pseudopotential method and in procedures to be discussed in the next section which include the electron–electron interaction explicitly.

We now have in hand the ingredients for a one-electron theory of semiconductor band structure. By one-electron theory we mean a theory in which the eigenfunction for N electrons is approximated by an antisymmetrized product of one-electron eigenfunctions. However, the electron–electron interaction has not yet been taken explicitly into account.

3.3 Electron–electron interaction

Up to this point we have treated only a single electron and have obtained a Hamiltonian, $H_1(\boldsymbol{r})$, which involves the coordinates of only that one electron. We know, however, that electrons interact with one another through the Coulomb interaction. It is therefore necessary to deal explicitly with the coordinates of all N electrons in the system $\boldsymbol{r}_1, \boldsymbol{r}_2, \boldsymbol{r}_3, \ldots, \boldsymbol{r}_N$. The Hamiltonian then takes the form

$$H = \sum_{i=1}^{N} H_1(\boldsymbol{r}_i) + \frac{1}{2}\sum_{i,j=1}^{N}{}' \frac{e^2}{4\pi\epsilon_0 r_{ij}}, \qquad (3.40)$$

where the prime on the second summation means that the terms for $i = j$ are excluded. The presence of the Coulomb interaction terms enormously complicates the problem, because the Hamiltonian is no longer separable into a sum of terms, each term involving the coordinates of only one electron. Recourse must then be made to approximation methods, some of which we now describe.

3.3.1 Hartree method

In the Hartree method, the N-electron eigenfunction is expressed as a product of one-electron eigenfunctions $\varphi_n(\boldsymbol{r})$:

$$\Psi_N = \varphi_1(\boldsymbol{r}_1)\varphi_2(\boldsymbol{r}_2)\ldots\varphi_N(\boldsymbol{r}_N). \tag{3.41}$$

Each electron is assumed to move in the average field due to the other electrons. The eigenfunction $\varphi_i(\boldsymbol{r})$ then satisfies the **Hartree equation**

$$\left\{H_0 + V_{ps} + \frac{e^2}{4\pi\epsilon_0}\sum_j \int d^3r' |\varphi_j(\boldsymbol{r}')|^2 \frac{1}{|\boldsymbol{r}-\boldsymbol{r}'|}\right\}\varphi_i(\boldsymbol{r}) = E_i\varphi_i(\boldsymbol{r}), \tag{3.42}$$

where the sum over j need not be restricted to $j \neq i$, because the volume element of the integral cancels the singular term. There is an equation of the form (3.42) for each occupied state $\varphi_i(\boldsymbol{r})$, so we have a set of coupled nonlinear integro-differential equations. They can be solved by iteration on a computer by inserting an initial guess for the $\varphi_j(\boldsymbol{r})$ into the Coulomb term.

3.3.2 Hartree–Fock method

There is a fundamental difficulty with the N-electron function given by the simple product in Eq. (3.41). It does not satisfy the **Pauli exclusion principle** that a given electronic state can be occupied by at most one electron. In the **Hartree–Fock method**, this difficulty is remedied by writing the N-electron eigenfunction as a **Slater determinant** of one-electron functions

$$\Psi_N(\boldsymbol{r}_1 s_1 \ldots \boldsymbol{r}_N s_N) = \frac{1}{\sqrt{N}}\begin{vmatrix} \varphi_1(\boldsymbol{r}_1)\alpha_1(s_1) & \ldots & \varphi_N(\boldsymbol{r}_1)\alpha_N(s_1) \\ \varphi_1(\boldsymbol{r}_2)\alpha_1(s_2) & \ldots & \varphi_N(\boldsymbol{r}_2)\alpha_N(s_2) \\ \vdots & & \vdots \\ \varphi_1(\boldsymbol{r}_N)\alpha_1(s_N) & \ldots & \varphi_N(\boldsymbol{r}_N)\alpha_N(s_N) \end{vmatrix}, \tag{3.43}$$

where $\alpha_j(s_i)$ is a spin eigenfunction. If two electrons occupy the same state $\varphi_j(\boldsymbol{r})\alpha_j(s)$, then two columns become identical, and the determinant vanishes. The Pauli principle is therefore obeyed.

The one-electron functions $\varphi_i(\boldsymbol{r})$ can be determined by minimizing the expectation value of the Hamiltonian $\langle\Psi_N|H|\Psi_N\rangle$ with respect to variations in the $\varphi_i(\boldsymbol{r})$. The variational calculation leads to an Euler equation of the form

$$\left\{H_0 + V_{ps} + \frac{e^2}{4\pi\epsilon_0}\sum_j \int d^3r|\varphi_j(\boldsymbol{r}')|^2 \frac{1}{|\boldsymbol{r}-\boldsymbol{r}'|}\right\}\varphi_i(\boldsymbol{r})$$
$$-\frac{e^2}{4\pi\epsilon_0}\sum_j \int d^3r'\varphi_j^*(\boldsymbol{r}')\varphi_i(\boldsymbol{r}')\frac{1}{|\boldsymbol{r}-\boldsymbol{r}'|}\varphi_j(\boldsymbol{r}) = E_i\varphi_i(\boldsymbol{r}). \qquad (3.44)$$

This equation is one of the Hartree–Fock equations that determine the one-electron functions $\varphi_i(\boldsymbol{r})$. The last term on the left hand side is a new term called the **exchange term**. It arises from the direct Coulomb term preceding it by the rearrangement $\varphi_j(\boldsymbol{r}')\varphi_i(\boldsymbol{r}) \to -\varphi_i(\boldsymbol{r}')\varphi_j(\boldsymbol{r})$. In other words, the subscripts i and j are interchanged and a minus sign is added.

The exchange term can be rewritten in the form

$$\int d^3r' V_{ex}(\boldsymbol{r},\boldsymbol{r}')\varphi_i(\boldsymbol{r}'), \qquad (3.45)$$

where

$$V_{ex}(\boldsymbol{r},\boldsymbol{r}') = -\sum_j \varphi_j^*(\boldsymbol{r}')\frac{e^2}{4\pi\epsilon_0|\boldsymbol{r}-\boldsymbol{r}'|}\varphi_j(\boldsymbol{r}) \qquad (3.46)$$

is the **exchange operator**. It is both a nonlinear and **nonlocal operator**, a fact which complicates the solution of the Hartree–Fock equations. They can be solved exactly for a system of free electrons, the one-electron functions being plane waves. Exchange leads to an effective repulsion between electrons of parallel spin (Seitz 1940). Some of the physical consequences of the Hartree–Fock method, however, are not satisfactory (Kittel 1987, Ashcroft and Mermin 1976).

3.3.3 Density functional method

A significant advance in the treatment of the electron–electron interaction was made by Hohenberg and Kohn (1964) and by Kohn and Sham (1965). Hohenberg and Kohn showed that the energy of an interacting electron gas in an external potential is a functional of the electron density $n(\boldsymbol{r})$. Minimization of the energy functional with respect to variations in $n(\boldsymbol{r})$ leads to the correct ground state energy. Kohn and Sham showed that, if $n(\boldsymbol{r})$ for an N-electron system is expressed in terms of one-electron functions $\varphi_i(\boldsymbol{r})$ by

$$n(\boldsymbol{r}) = \sum_{i=1}^{N} |\varphi_i^*(\boldsymbol{r})|^2, \qquad (3.47)$$

the minimization of the energy functional leads to a set of equations, the **Kohn–Sham equations**, that determine the one-electron functions. By treating exchange in the **local density approximation** (LDA) and introducing the correlation energy ϵ_c, which accounts for the difference between the Hartree–Fock energy and the exact energy, Kohn and Sham obtained the equations

$$\left\{H_0 + V_{ei} + \frac{e^2}{4\pi\epsilon_0}\int d^3r' \frac{n(\mathbf{r}')}{|\mathbf{r}-\mathbf{r}'|} + \mu_{xc}(n(\mathbf{r}))\right\}\varphi_i(\mathbf{r}) = E_i\varphi_i(\mathbf{r}), \quad (3.48)$$

where V_{ei} is the electron–ion potential,

$$\mu_{xc}(n) = \frac{d(n\epsilon_{xc}(n))}{dn} \quad (3.49)$$

and $\epsilon_{xc}(n)$ is the one-electron exchange and correlation energy. The exchange part of $\mu_{xc}(n)$ as derived by Kohn and Sham has the form

$$\mu_x(n) = -\frac{1.22}{r_s(n)} \text{ rydbergs.} \quad (3.50)$$

The quantity $r_s(n)$ is the radius of a sphere containing one electron expressed in units of the Bohr radius a_B and given by

$$r_s(n) = \left[\frac{3}{4\pi n(\mathbf{r})}\right]^{1/3} \cdot \frac{1}{a_B}. \quad (3.51)$$

The correlation part has been estimated to be (Wigner 1934)

$$\mu_c = -\frac{0.88}{r_s(n) + 7.8}\left[1 + \frac{r_s(n)}{3(r_s(n) + 7.8)}\right] \text{ rydbergs.} \quad (3.52)$$

Other forms for both the exchange and correlation potentials have been suggested.

The density functional formalism of Hohenberg, Kohn, and Sham gives, in principle, exact results for the electronic ground state energy and for structural and lattice dynamical properties that are determined by that energy. Computational difficulties generally force the use of the LDA (Wendel and Martin 1979, Yin and Cohen 1982) or Quantum Monte Carlo methods (Fahy *et al.* 1990). Representing the electron–ion interaction by *ab initio* pseudopotentials and using the LDA, Yin and Cohen obtained values for the lattice constant a, bulk modulus B_0, and cohesive energy E_{coh} of silicon and germanium that are presented in Table 3.1 together with the

Table 3.1 Calculated and experimental lattice constant, bulk modulus and cohesive energy for silicon and germanium (after Yin and Cohen 1982)

	Lattice constant a (Å)		Bulk modulus B_0 (Mbar)		Cohesive energy E_{coh} (eV/atom)	
	Calc.	Exp.	Calc.	Exp.	Calc.	Exp.
Si	5.451	5.429	0.98	0.99	4.84	4.63
Ge	5.655	5.652	0.73	0.77	4.26	3.85

experimental values. The agreement between theory and experiment is very good for a and B_0, and reasonably good for E_{coh}.

3.3.4 Excited electronic states

Of great importance to the properties of semiconductors are **excited electronic states**. They play a crucial role in properties such as electrical conductivity and optical absorption. Although the density functional method in the LDA gives quite good results for ground state properties such as the lattice constant and bulk modulus, as we have seen in the previous section, it does not give reliable results for energy gaps and other properties involving excited states. For example, the LDA calculations of Yin and Cohen yield minimum energy gaps of 0.48 eV and 0.47 eV for Si and Ge, respectively, which are significantly less than the experimental values of 1.17 eV and 0.74 eV.

An alternative approach to the excited state problem is based on the determination of **quasiparticle** energies E_{nk} through the solution of the Schrödinger-like equation

$$
\begin{aligned}
&[H_0 + V_{ei} + V_H + \int d^3r' \sum(\mathbf{r}, \mathbf{r}'; E_{nk})]\psi_{nk}(\mathbf{r}') \\
&\quad = E_{nk}\psi_{nk}(\mathbf{r}), \qquad (3.53)
\end{aligned}
$$

where V_H is the average Coulomb (Hartree) potential due to the electrons, and $\sum(\mathbf{r}, \mathbf{r}'; E_{nk})$ is the electron self-energy operator. The latter operator contains the effects of exchange and correlation. It has been evaluated with sufficient accuracy to yield significant results for semiconductors by Hybertsen and Louie (1986). They obtained the values of 1.29 eV and 0.75 eV for the minimum energy gaps of Si and Ge, respectively, which are in much better agreement with the experimental values than the LDA results of Yin and Cohen. A number of interband transition energies were calculated that are in good overall agreement with experiment for both Si and Ge. Similar results have been obtained for the III–V compounds AlP, GaP, InP, AlAs, GaAs, InAs, AlSb, GaSb, and InSb by Zhu and Louie (1991). The energy eigenvalues as functions of wave vector were calculated by Hybertsen and Louie for the valence bands of Ge. The results are in good agreement with experiment.

3.4 The $k \cdot p$ method

It was stated in the introduction to this chapter that the top of the valence band and the bottom of the conduction band are separated by the fundamental gap. Electrons in the conduction band and holes in the valence band tend to accumulate in the vicinity of their respective band extrema in accordance with statistical mechanical distribution laws. It therefore proves useful to have a procedure for developing the energy versus wave vector relationship E_{nk} near a band extremum without calculating the entire band structure. The $\mathbf{k} \cdot \mathbf{p}$ method is such a procedure.

3.4.1 Nondegenerate bands

As we have seen in Chapter 2, an electron in free space has an energy-wave vector relationship that is isotropic,

$$E_0 = \frac{\hbar^2 k^2}{2m}, \tag{3.54}$$

where $k^2 = k_x^2 + k_y^2 + k_z^2$. By differentiating this expression twice with respect to k, one obtains

$$\frac{1}{m} = \frac{1}{\hbar^2}\frac{d^2 E_0}{dk^2}. \tag{3.55}$$

On the other hand, the results that will be presented for energy bands in semiconductors frequently exhibit an anisotropic relationship between energy and wave vector, so that the energy is not simply a function of the magnitude of $\boldsymbol{k}$, but depends separately on k_x, k_y, k_z. One must then define an **inverse effective mass tensor** whose elements for band n are given by

$$\left(\frac{1}{m_n^*}\right)_{\mu\nu} = \frac{1}{\hbar^2}\frac{\partial^2 E_{n\boldsymbol{k}}}{\partial k_\mu \partial k_\nu}. \tag{3.56}$$

In general, these elements are not constants, but are functions of k_x, k_y, k_z. Of particular interest are the values of the elements at band extrema, as discussed above. A derivation of Eq. (3.56) will be given in Chapter 4.

Let the energy band of interest have an extremum at wave vector $\boldsymbol{k}_0$. At a nearby wave vector $\boldsymbol{k}$, the Bloch function can be written as

$$\psi_{n\boldsymbol{k}}(\boldsymbol{r}) = e^{i\boldsymbol{k}_0\cdot\boldsymbol{r}} e^{i\Delta\boldsymbol{k}\cdot\boldsymbol{r}} u_{n\boldsymbol{k}_0+\Delta\boldsymbol{k}}(\boldsymbol{r}), \tag{3.57}$$

where $\Delta\boldsymbol{k} = \boldsymbol{k} - \boldsymbol{k}_0$. Substituting Eq. (3.54) into the Schrödinger equation,

$$\left[\frac{p^2}{2m} + V(\boldsymbol{r})\right]\psi_{n\boldsymbol{k}}(\boldsymbol{r}) = E_{n\boldsymbol{k}}\psi_{n\boldsymbol{k}}(\boldsymbol{r}), \tag{3.58}$$

where $\boldsymbol{p}$ is the momentum operator and $V(\boldsymbol{r})$ is the periodic potential, we obtain

$$\left[\frac{p^2}{2m} + \frac{\hbar\boldsymbol{k}_0\cdot\boldsymbol{p}}{m} + \frac{\hbar\Delta\boldsymbol{k}\cdot\boldsymbol{p}}{m} + \frac{\hbar^2(\boldsymbol{k}_0+\Delta\boldsymbol{k})^2}{2m} + V(\boldsymbol{r})\right] u_{n\boldsymbol{k}_0+\Delta\boldsymbol{k}}(\boldsymbol{r})$$
$$= E_{n\boldsymbol{k}_0+\Delta\boldsymbol{k}} u_{n\boldsymbol{k}_0+\Delta\boldsymbol{k}}(\boldsymbol{r}). \tag{3.59}$$

Let us assume that the problem for $\boldsymbol{k} = \boldsymbol{k}_0$ has been solved for all n, i.e., we know the functions $u_{n\boldsymbol{k}_0}(\boldsymbol{r})$ and the eigenvalues $E_{n\boldsymbol{k}_0}$. The terms involving $\Delta\boldsymbol{k}$ in Eq. (3.59) can then be treated by perturbation theory. Note that the terms $\hbar^2\boldsymbol{k}_0\cdot\Delta\boldsymbol{k}/m$ and $\hbar^2(\Delta\boldsymbol{k})^2/2m$ are constants and can be combined with $E_{n\boldsymbol{k}_0}$, so the only term that involves an operator and needs to be dealt with is $\hbar\Delta\boldsymbol{k}\cdot\boldsymbol{p}/m$.

For the case that the state $n\boldsymbol{k}_0$ is nondegenerate, the perturbed energy can be written to second order as

$$E_{n\boldsymbol{k}_0+\Delta\boldsymbol{k}} = E_{n\boldsymbol{k}_0} + \frac{\hbar}{m}\Delta\boldsymbol{k}\cdot\langle n\boldsymbol{k}_0|\boldsymbol{p}|n\boldsymbol{k}_0\rangle + \frac{\hbar^2}{m}\boldsymbol{k}_0\cdot\Delta\boldsymbol{k} + \frac{\hbar^2(\Delta\boldsymbol{k})^2}{2m} + \frac{\hbar^2}{m^2}\sum_{\alpha\beta}\Delta k_\alpha\Delta k_\beta\sum_{n'}{}' \frac{\langle n\boldsymbol{k}_0|p_\alpha|n'\boldsymbol{k}_0\rangle\langle n'\boldsymbol{k}_0|p_\beta|n\boldsymbol{k}_0\rangle}{E_{n\boldsymbol{k}_0}-E_{n'\boldsymbol{k}_0}}, \quad (3.60)$$

where the prime on the sum means the term $n' = n$ is excluded. If the energy $E_{n\boldsymbol{k}}$ has an extremum at $\boldsymbol{k} = \boldsymbol{k}_0$, the term linear in $\Delta\boldsymbol{k}$ must vanish, so that

$$E_{n\boldsymbol{k}_0+\Delta\boldsymbol{k}} = E_{n\boldsymbol{k}_0} + \frac{\hbar^2(\Delta\boldsymbol{k})^2}{2m} + \frac{\hbar^2}{m^2}\sum_{\alpha\beta}\Delta k_\alpha\Delta k_\beta\sum_{n'}{}' \frac{\langle n\boldsymbol{k}_0|p_\alpha|n'\boldsymbol{k}_0\rangle\langle n'\boldsymbol{k}_0|p_\beta|n\boldsymbol{k}_0\rangle}{E_{n\boldsymbol{k}_0}-E_{n'\boldsymbol{k}_0}}. \quad (3.61)$$

From the definition of the inverse effective mass tensor,

$$\left(\frac{1}{m_n^*}\right)_{\mu\nu} = \frac{1}{\hbar^2}\frac{\partial^2 E_{n\boldsymbol{k}}}{\partial k_\mu \partial k_\nu}\bigg|_{\boldsymbol{k}=\boldsymbol{k}_0}, \quad (3.62)$$

and using Eq. (3.61), we obtain

$$\left(\frac{m}{m_n^*}\right)_{\mu\nu} = \delta_{\mu\nu} + \frac{1}{m}\sum_{n'}{}'\left\{\frac{\langle n\boldsymbol{k}_0|p_\mu|n'\boldsymbol{k}_0\rangle\langle n'\boldsymbol{k}_0|p_\nu|n\boldsymbol{k}_0\rangle}{E_{n\boldsymbol{k}_0}-E_{n'\boldsymbol{k}_0}} + \frac{\langle n\boldsymbol{k}_0|p_\nu|n'\boldsymbol{k}_0\rangle\langle n'\boldsymbol{k}_0|p_\mu|n\boldsymbol{k}_0\rangle}{E_{n\boldsymbol{k}_0}-E_{n'\boldsymbol{k}_0}}\right\}. \quad (3.63)$$

The principal contributions to the sum over n' typically arise from those states with $E_{n'\boldsymbol{k}_0}$ closest to $E_{n\boldsymbol{k}_0}$.

Example 3.3: Effective mass for a spherical, parabolic band
Evaluate the inverse effective mass ratio for a spherical, parabolic band.
Solution. Take the semiconductor to be cubic with band n derived from an s-like state at $\boldsymbol{k}_0 = 0$ and bands n' from p-like states at $\boldsymbol{k}_0 = 0$. From symmetry, $\langle s0|p_x|x'0\rangle = \langle s0|p_y|y'0\rangle = \langle s0|p_z|z'0\rangle = P_{sp'}(0)$, and all other matrix elements of p_μ are zero. Then,

$$\left(\frac{m}{m_s^*}\right)_{\mu\nu} = \left[1 + \frac{2}{m}\sum_{p'}{}'\frac{|P_{sp'}(0)|^2}{E_{s0}-E_{p'0}}\right]\delta_{\mu\nu}. \quad (3.64)$$

If the states $s0$ and $p0$ correspond to the conduction band edge and the valence band edge, respectively, and we neglect all other p-like states at $\boldsymbol{k}_0 = 0$, we obtain the approximate result (Kane 1957)

$$\left(\frac{m}{m_s^*}\right)_{\mu\nu} = \left[1 + \frac{2}{m}\frac{|P_{sp}(0)|^2}{E_g}\right]\delta_{\mu\nu}, \quad (3.65)$$

where $E_g = E_{s0} - E_{p0}$. Equation (3.65) is particularly useful for narrow gap semiconductors such as InSb. The energy as a function of wave vector takes the form characteristic of a spherical, parabolic band:

$$E_{s\boldsymbol{k}} = E_{s0} + \frac{\hbar^2 k^2}{2m^*}, \tag{3.66}$$

where

$$\frac{1}{m^*} = \frac{1}{m^*_{xx}} = \frac{1}{m^*_{yy}} = \frac{1}{m^*_{zz}}$$

and $k^2 = k_x^2 + k_y^2 + k_z^2$. As the energy gap increases the effective mass also increases.

3.4.2 Valence bands of Si and Ge

Turning now to the case for which band n is degenerate at its extremum $\boldsymbol{k}_0$, we label these bands by $n = \delta_1, \delta_2, \ldots$ and the corresponding eigenvalue energy by $E_{\delta\boldsymbol{k}_0}$. We focus on the valence bands of Si and Ge whose maxima are at $\boldsymbol{k}_0 = 0$, so $\Delta\boldsymbol{k} = \boldsymbol{k}$. The bands are triply degenerate in the absence of spin considerations. Applying degenerate perturbation theory, we must deal with nondiagonal second order matrix elements having the form

$$H_{ij} = \frac{\hbar^2}{m^2} \sum_{\ell}{}' \frac{\langle \delta_i \boldsymbol{k}_0 | \boldsymbol{k} \cdot \boldsymbol{p} | \ell \boldsymbol{k}_0 \rangle \langle \ell \boldsymbol{k}_0 | \boldsymbol{k} \cdot \boldsymbol{p} | \delta_j \boldsymbol{k}_0 \rangle}{E_{\delta\boldsymbol{k}_0} - E_{\ell\boldsymbol{k}_0}}. \tag{3.67}$$

The perturbed energy eigenvalues $E_p(\boldsymbol{k})$ are then specified by the secular equation

$$\begin{vmatrix} H_{11} - \lambda & H_{12} & H_{13} \\ H_{21} & H_{22} - \lambda & H_{23} \\ H_{31} & H_{32} & H_{33} - \lambda \end{vmatrix} = 0, \tag{3.68}$$

where $\lambda = E_p(\boldsymbol{k}) - \hbar^2 k^2/2m$. Utilizing the symmetry properties of Si and Ge, one can reduce Eq. (3.68) to the form (Dresselhaus *et al.* 1955)

$$\begin{vmatrix} Lk_x^2 + M(k_y^2 + k_z^2) - \lambda & Nk_x k_y & Nk_x k_z \\ Nk_x k_y & Lk_y^2 + M(k_x^2 + k_z^2) - \lambda & Nk_y k_z \\ Nk_x k k_z & Nk_y k_z & Lk_z^2 + M(k_x^2 + k_y^2) - \lambda \end{vmatrix} = 0, \tag{3.69}$$

where the quantities L, M, N involve particular intermediate states in Eq. (3.67).

To properly treat the valence bands of Si and Ge as well as those of compound semiconductors, it is necessary to take into account the spin–orbit interaction as given by Eq. (3.3). With the inclusion of spin, but excluding the spin–orbit interaction, the 6 p-like valence bands at $\boldsymbol{k} = 0$ are degenerate. The spin–orbit interaction splits this degeneracy into a $j_{3/2}$ set of four-fold degenerate states and a $j_{1/2}$ set of two-fold degenerate states. The secular equation now takes the form (Dresselhaus *et al.* 1955)

$$\begin{vmatrix}
\frac{H_{11}+H_{22}}{2}-\lambda & -\frac{H_{13}-iH_{23}}{\sqrt{3}} & -\frac{H_{11}-H_{22}-2iH_{12}}{2\sqrt{3}} & 0 & -\frac{H_{13}-iH_{23}}{\sqrt{6}} & -\frac{H_{11}-H_{22}-2iH_{12}}{\sqrt{6}} \\
-\frac{H_{13}+iH_{23}}{\sqrt{3}} & -\frac{4H_{33}+H_{11}+H_{22}}{6}-\lambda & 0 & -\frac{H_{11}-H_{22}-2iH_{12}}{2\sqrt{3}} & -\frac{H_{11}+H_{22}-2H_{33}}{3\sqrt{2}} & \frac{H_{13}-iH_{23}}{\sqrt{2}} \\
-\frac{H_{11}-H_{22}+2iH_{12}}{2\sqrt{3}} & 0 & \frac{4H_{33}+H_{11}+H_{22}}{6}-\lambda & \frac{H_{13}-iH_{23}}{\sqrt{3}} & \frac{H_{13}+iH_{23}}{\sqrt{2}} & \frac{H_{11}+H_{22}-2H_{33}}{3\sqrt{2}} \\
0 & -\frac{H_{11}-H_{22}+2iH_{12}}{2\sqrt{3}} & \frac{H_{13}+iH_{23}}{\sqrt{3}} & \frac{H_{11}+H_{22}}{2}-\lambda & \frac{H_{11}-H_{22}+2iH_{12}}{\sqrt{6}} & -\frac{H_{13}+iH_{23}}{\sqrt{6}} \\
-\frac{H_{13}+iH_{23}}{\sqrt{6}} & -\frac{H_{11}+H_{22}-2H_{33}}{3\sqrt{2}} & \frac{H_{13}-iH_{23}}{\sqrt{3}} & \frac{H_{11}-H_{22}-2iH_{12}}{3} & \frac{H_{11}+H_{22}+H_{33}}{3}-\Delta_{so}-\lambda & 0 \\
-\frac{H_{11}-H_{22}+2iH_{12}}{\sqrt{6}} & \frac{H_{13}+iH_{23}}{\sqrt{2}} & \frac{H_{11}+H_{22}-2H_{33}}{3\sqrt{2}} & -\frac{H_{13}-iH_{23}}{\sqrt{6}} & 0 & \frac{H_{11}+H_{22}+H_{33}}{3}-\Delta_{so}-\lambda
\end{vmatrix} = 0, \tag{3.70}$$

where Δ_{so} is the spin–orbit splitting between the $j_{3/2}$ and $j_{1/2}$ states at $\boldsymbol{k} = 0$. In addition the momentum operator $\boldsymbol{p}$ must be replaced by the operator $\pi = \boldsymbol{p} + (\hbar/4mc^2)\boldsymbol{\sigma} \times \boldsymbol{\nabla} V$ in the matrix elements in Eq. (3.67) and the term $\hbar^2 k^2/2m$ must be added to λ to give the energy eigenvalue $E_p(\boldsymbol{k})$.

The solutions of the secular equation to order k^2 may be obtained by considering only the 4×4 block in the upper left corner and the 2×2 block in the lower right corner. The two 2×4 strips neglected in this approximation affect the solutions only to order k^4/Δ_{so}. Solving the 4×4 determinantal equation gives the result

$$E_{3/2}(\boldsymbol{k}) = Ak^2 \pm [B^2k^4 + C^2(k_x^2k_y^2 + k_y^2k_z^2 + k_z^2k_x^2)]^{\frac{1}{2}}, \tag{3.71}$$

where

$$A = \tfrac{1}{3}(L + 2M) + \hbar^2/2m \tag{3.72a}$$

$$B = \tfrac{1}{3}(L - M) \tag{3.72b}$$

$$C^2 = \tfrac{1}{3}[N^2 - (L - M)^2]. \tag{3.72c}$$

Each solution specified by Eq. (3.71) occurs twice for each value of $\boldsymbol{k}$. This double degeneracy (Kramers degeneracy) is a consequence of time-inversion symmetry and the fact that crystals of the diamond structure possess a center of inversion (Kittel 1987).

The constant energy surfaces described by Eq. (3.71) are nonspherical for $C \neq 0$ and are referred to as fluted or warped surfaces. The intersections of these surfaces with the (100) plane in $\boldsymbol{k}$-space are plotted in Fig. 3.3 for silicon and Fig. 3.4 for germanium.

Turning now to the 2×2 determinantal equation, its solution can be written as

$$E_{1/2}(\boldsymbol{k}) = -\Delta_{so} + Ak^2, \tag{3.73}$$

where A is given by Eq. (3.72a) provided Δ_{so} is small compared to E_g, and a two-fold degeneracy due to time-reversal symmetry is present. Since the band described by Eq. (3.73) is split off at $\boldsymbol{k} = 0$ from the bands described by Eq. (3.71) by the spin–orbit interaction, it is referred to as the **split-off band**.

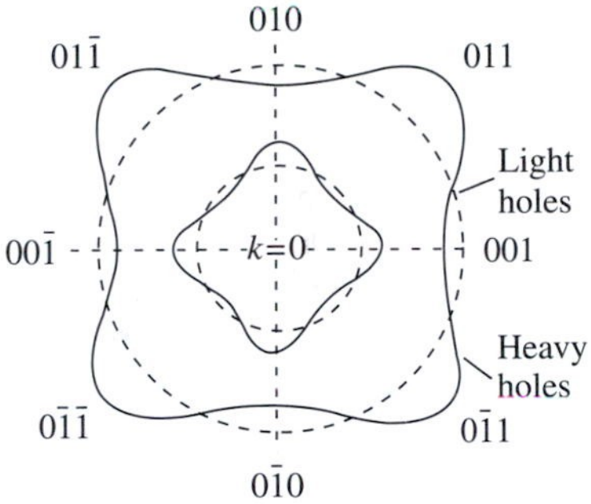

Fig. 3.3
Constant energy contours in the (100) plane in $\boldsymbol{k}$-space for the two fluted energy surfaces of the valence band of Si.

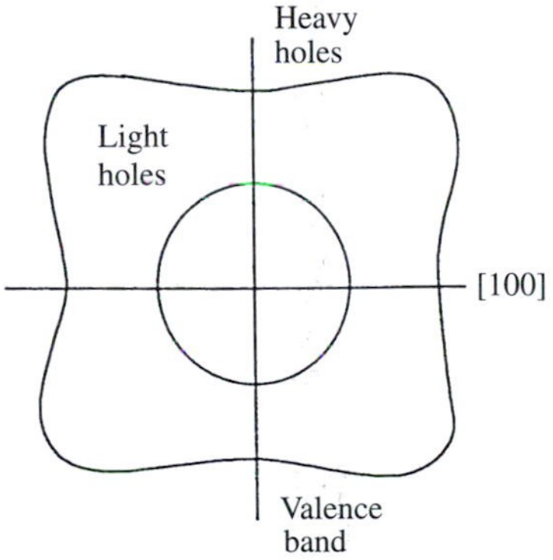

Fig. 3.4
Constant energy contours in the (100) plane in $\boldsymbol{k}$-space for the two fluted energy surfaces of the valence band of Ge (after Dresselhaus *et al.* 1955).

Table 3.2 Valence band parameters for silicon and germanium in units of $\hbar^2/2m$ (after Kittel 1986)

	A	$\|B\|$	$\|C\|$
Si	−4.29	0.68	4.87
Ge	−13.38	8.48	13.15

The values of the valence band constants A, B, C for Si and Ge have been determined from cyclotron resonance measurements (Dresselhaus *et al.* 1955; Dexter *et al.* 1956). Recent values are given in Table 3.2. We note that A is negative, consistent with the fact that the curvature of the valence band is negative at $\boldsymbol{k} = 0$. One can also show that $B < 0$. The plus sign in Eq. (3.71) corresponds to heavy holes, whereas the minus sign corresponds to light holes. The value of the spin–orbit split-off parameter Δ_{so} can be determined by spectroscopic measurements. It is found to be 0.045 eV for Si and 0.297 eV for Ge (Böer 1990). A schematic plot of the heavy hole, light hole, and split-off bands is given in Fig. 3.5.

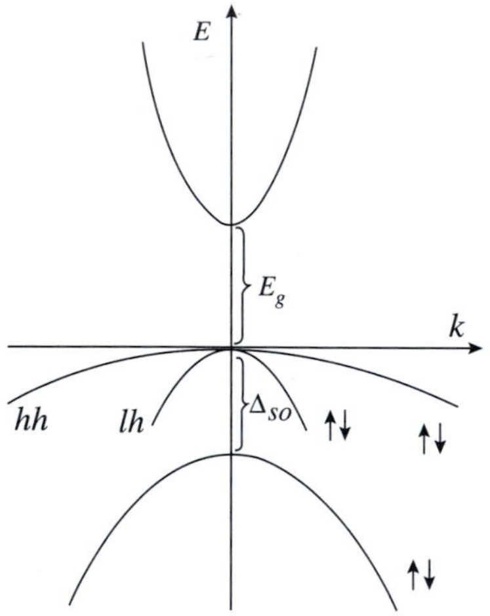

Fig. 3.5
Heavy hole (*hh*), light hole (*lh*), and split-off (*so*) energy bands for diamond structure semiconductors. The Kramers degeneracy is indicated by the up and down arrows.

3.4.3 Conduction bands of Si and Ge

Theoretical calculations (Herman 1954) show that the conduction band minima of Si and Ge do not occur at $\boldsymbol{k} = 0$. In the case of Si, there are six minima of the same energy at $\boldsymbol{k}_0 = (\pm k_0, 0, 0)$, $(0, \pm k_0, 0)$, and $(0, 0, \pm k_0)$, where k_0 is about 0.8 of the way to the Brillouin zone boundary in the ΓX direction. Such minima are often referred to as valleys. One can then say that the Si conduction band has six valleys. Even though Si is cubic, a change $\Delta\boldsymbol{k}$ parallel to $\boldsymbol{k}_0$ is not equivalent to a change perpendicular to $\boldsymbol{k}_0$, since, if $\boldsymbol{k}_0 = (\pm k_0, 0, 0)$, $\langle n\boldsymbol{k}_0|p_x|n'\boldsymbol{k}_0\rangle \neq \langle n\boldsymbol{k}_0|p_y|n'\boldsymbol{k}_0\rangle = \langle n\boldsymbol{k}_0|p_z|n'\boldsymbol{k}_0\rangle$. Consequently, the expressions for the inverse effective mass ratios become

$$\frac{m}{m_\ell^*} \equiv \frac{m}{m_{xx}^*} = 1 + \frac{2}{m}{\sum_{n'}}' \frac{|\langle n\boldsymbol{k}_0|p_x|n'\boldsymbol{k}_0\rangle|^2}{E_{n\boldsymbol{k}_0} - E_{n'\boldsymbol{k}_0}} \tag{3.74}$$

$$\frac{m}{m_t^*} \equiv \frac{m}{m_{yy}^*} = \frac{m}{m_{zz}^*} = 1 + \frac{2}{m}{\sum_{n'}}' \frac{|\langle n\boldsymbol{k}_0|p_y|n'\boldsymbol{k}_0\rangle|^2}{E_{n\boldsymbol{k}_0} - E_{n'\boldsymbol{k}_0}}. \tag{3.75}$$

The relation of energy to wave vector is

$$E_{n\boldsymbol{k}} = E_{n\boldsymbol{k}_0} + \frac{\hbar^2}{2}\left[\frac{(k_x - k_{0x})^2}{m_\ell^*} + \frac{(k_y - k_{0y})^2 + (k_z - k_{0z})^2}{m_t^*}\right]. \tag{3.76}$$

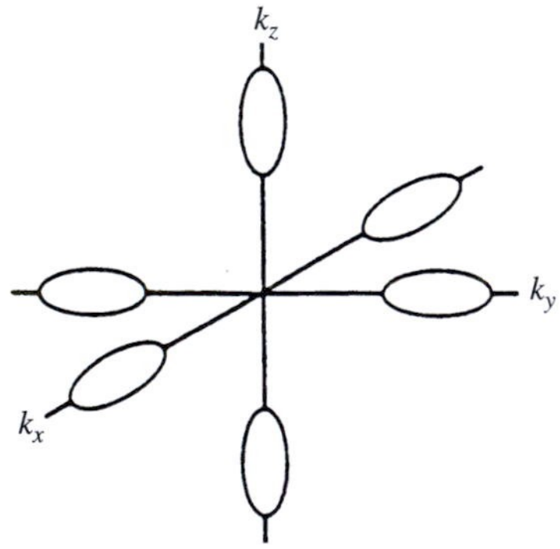

Fig. 3.6
Constant energy surfaces for the conduction band of Si.

The surfaces of constant energy are ellipsoids of revolution with major and minor axes proportional to $\sqrt{m_\ell^*}$ and $\sqrt{m_t^*}$, respectively. The set of constant energy surfaces is shown schematically in Fig. 3.6.

In the case of Ge, there are four conduction band ellipsoids that lie along $\langle 111\rangle$ directions as shown in Fig. 3.7. The band edges are at the Brillouin zone boundary (L point). By rotating the coordinate axes so that one axis lies along a [111] direction, one can show that the energy-wave vector relation of Eq. (3.76) is valid for the conduction band of Ge for appropriate values of m_ℓ^* and m_t^*.

For Si and Ge, $m_\ell^* > m_t^*$, so the ellipsoids are prolate and resemble a rugby ball. The major-to-minor axis ratio $\sqrt{m_\ell^*/m_t^*}$ is a measure of the anisotropy of the ellipsoids.

3.4.4 Zincblende structure semiconductors

Analysis of the band structures of zincblende structure semiconductors using the $\boldsymbol{k} \cdot \boldsymbol{p}$ method leads to results that are qualitatively similar to those just described for diamond structure materials. However the lack of a center of inversion in the zincblende structure leads to certain effects on the band structure that are not found in the diamond structure. For example, the nature of the Kramers degeneracy is different in the two cases. If we denote the two Kramers degenerate states by up and down arrows (loosely, but not precisely related to spin direction), the Kramers degeneracy for the diamond structure is specified by $E(\uparrow, \boldsymbol{k}) = E(\downarrow, \boldsymbol{k})$ and $E(\uparrow, -\boldsymbol{k}) = E(\downarrow, -\boldsymbol{k})$. These relations do not hold for the zincblende structure, for which we have $E(\uparrow, \boldsymbol{k}) = E(\downarrow, -\boldsymbol{k})$ and $E(\downarrow, \boldsymbol{k}) = E(\uparrow, -\boldsymbol{k})$. We therefore arrive at the qualitative picture shown in Fig. 3.8 for the valence bands. The behavior of the bands shown implies that there is a linear splitting near $\boldsymbol{k} = 0$. This is, in fact, the case. The splitting is specified by the contributions (Dresselhaus 1955)

$$\Delta E(\boldsymbol{k}) = \pm\bar{C}\{k^2 + [3(k_x^2k_y^2 + k_y^2k_z^2 + k_z^2k_x^2)]^{\frac{1}{2}}\}^{\frac{1}{2}} \tag{3.77a}$$

$$\Delta E(\boldsymbol{k}) = \pm\bar{C}\{k^2 - [3(k_x^2k_y^2 + k_y^2k_z^2 + k_z^2k_x^2)]^{\frac{1}{2}}\}^{\frac{1}{2}} \tag{3.77b}$$

for the light and heavy hole bands with $\bar{C}$ a constant. The effect of these contributions is to shift the maximum of the valence band slightly away from $\boldsymbol{k} = 0$. Interband magneto-optical experiments on InSb (Pidgeon and Groves 1969) have shown that the valence band edge is raised by $\sim 10^{-5}$eV corresponding to a very small value of $\bar{C}$. This shift is not evident in the energy band figures for zincblende structure semiconductors shown in subsequent figures. Henceforth, we shall neglect $\bar{C}$ in our analysis.

It has become customary in discussing the valence bands of zincblende structure semiconductors to use the dimensionless Luttinger band parameters $\gamma_1, \gamma_2, \gamma_3$ which are related to the Dresselhaus parameters A, B, C by

$$\gamma_1 = -\frac{2m}{\hbar^2}A, \quad \gamma_2 = -\frac{m}{\hbar^2}B, \quad \gamma_3 = \frac{m}{\hbar^2}\left[\tfrac{1}{3}C^2 + B^2\right]^{\frac{1}{2}}. \tag{3.78}$$

The energies of the light and heavy hole bands are specified by the expression

$$E_{\ell,h}(\boldsymbol{k}) = \frac{\hbar^2}{2m}\{\gamma_1 k^2 \pm [4\gamma_2^2k^4 + 12(\gamma_3^2 - \gamma_2^2) \times (k_x^2k_y^2 + k_y^2k_z^2 + k_z^2k_x^2)]^{\frac{1}{2}}\}, \tag{3.79}$$

where the + and − signs refer to light and heavy holes, respectively. We note that the condition for spherical bands is $\gamma_2 = \gamma_3$. Values of the Luttinger parameters are presented in Table 3.3.

In the case of III–V semiconductors such as GaAs and InSb, the minimum of the conduction band is at $\boldsymbol{k} = 0$, and there is no linear splitting such

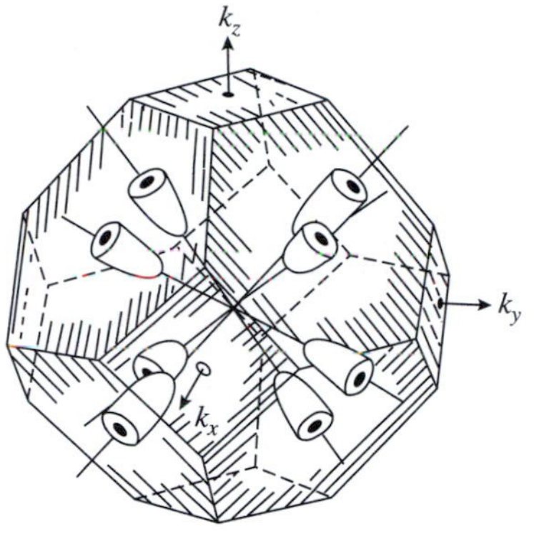

Fig. 3.7
Constant energy surfaces for the conduction band of Ge.

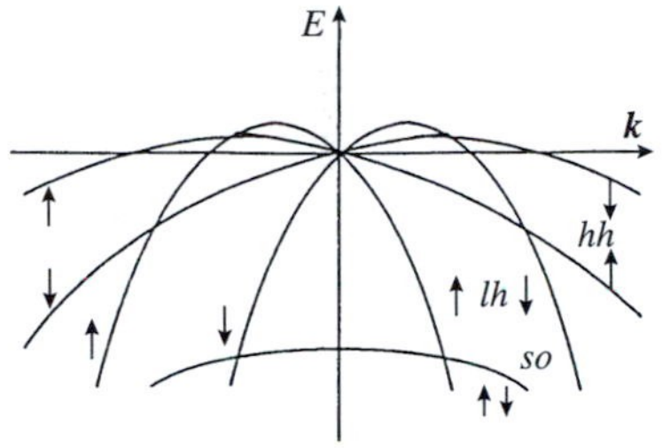

Fig. 3.8
Heavy hole (*hh*), light hole (*lh*), and split-off (*so*) energy bands in the [110] direction for zincblende structure semiconductors. The arrows distinguish different Kramers states.

Table 3.3 Luttinger parameters γ_1, γ_2, γ_3 for III–V and II–VI zincblende structure semiconductors (after Pidgeon and Groves 1969, Lawaetz 1971, Böer 1990)

III–V	γ_1	γ_2	γ_3	II–VI	γ_1	γ_2	γ_3
AlP	3.47	0.06	1.15	ZnS	2.54	0.75	1.09
AlAs	4.04	0.78	1.57	ZnSe	3.77	1.24	1.67
AlSb	4.15	1.01	1.75	ZnTe	3.74	1.07	1.64
GaP	4.05	0.49	1.25	CdTe	5.29	1.89	2.46
GaAs	6.95	2.25	2.86	HgS	−41.3	−21.0	−20.7
GaSb	13.3	4.4	5.7	HgSe	−26.0	−13.7	−13.2
InP	5.15	0.94	1.62	HgTe	−18.7	−10.2	−9.6
InAs	20.4	8.3	9.1				
InSb	33.5	14.5	15.6				

as in the valence band. A splitting does appear in third order in k and is given by (Dresselhaus 1955)

$$\Delta E(\boldsymbol{k}) = \pm C_1[k^2(k_x^2k_y^2 + k_y^2k_z^2 + k_z^2k_x^2) - 9k_x^2k_y^2k_z^2]^{\frac{1}{2}}, \tag{3.80}$$

where C_1 is a constant related to the spin–orbit interaction. This splitting is also too small to be observable in the energy band figures of this chapter. The conduction bands of GaAs and InSb are therefore very nearly spherical. GaP, on the other hand, has conduction band minima along the six $\langle 100 \rangle$ directions as in Si. The minima are so close to the zone boundary at X that a **camel's back structure** appears.

3.4.5 Extended $\boldsymbol{k} \cdot \boldsymbol{p}$ method

The $\boldsymbol{k} \cdot \boldsymbol{p}$ method outlined in the preceding sections is valid only in the vicinity of an energy band extremum. The method can be modified, however, so that it provides accurate results over the entire Brillouin zone. The starting point is the observation that the periodic functions $u_{n\boldsymbol{k}_0}(\boldsymbol{r})$ form a complete set in terms of which *any* periodic function with the periodicity of the crystal lattice can be expanded. In particular, one can expand $u_{n\boldsymbol{k}}(\boldsymbol{r})$ in a series of the $u_{n\boldsymbol{k}_0}(\boldsymbol{r})$,

$$u_{n\boldsymbol{k}}(\boldsymbol{r}) = \sum_{n'} c_{n'}(\boldsymbol{k}_0) u_{n'\boldsymbol{k}_0}(\boldsymbol{r}), \tag{3.81}$$

where the coefficients $c_{n'}(\boldsymbol{k}_0)$ are the elements of the eigenvectors of the Hamiltonian matrix

$$\begin{bmatrix} E_{n_0\boldsymbol{k}_0} & \langle n_0\boldsymbol{k}_0|\frac{\hbar}{m}\Delta\boldsymbol{k}\cdot\boldsymbol{p}|n_1\boldsymbol{k}_0\rangle & \langle n_0\boldsymbol{k}_0|\frac{\hbar}{m}\Delta\boldsymbol{k}\cdot\boldsymbol{p}|n_2\boldsymbol{k}_0\rangle & \cdots \\ \langle n_1\boldsymbol{k}_0|\frac{\hbar}{m}\Delta\boldsymbol{k}\cdot\boldsymbol{p}|n_0\boldsymbol{k}_0\rangle & E_{n_1\boldsymbol{k}_0} & \langle n_1\boldsymbol{k}_0|\frac{\hbar}{m}\Delta\boldsymbol{k}\cdot\boldsymbol{p}|n_2\boldsymbol{k}_0\rangle & \cdots \\ \langle n_2\boldsymbol{k}_0|\frac{\hbar}{m}\Delta\boldsymbol{k}\cdot\boldsymbol{p}|n_0\boldsymbol{k}_0\rangle & \langle n_2\boldsymbol{k}_0|\frac{\hbar}{m}\Delta\boldsymbol{k}\cdot\boldsymbol{p}|n_1\boldsymbol{k}_0\rangle & E_{n_2\boldsymbol{k}_0} & \cdots \\ \vdots & \vdots & \vdots & \end{bmatrix}. \tag{3.82}$$

If a sufficiently large number of bands is included in the Hamiltonian matrix, the diagonalization of the latter leads to accurate energy bands over the entire Brillouin zone.

This procedure has been exploited by Cardona and Pollak (1966) for Si. Using optical data, they determined interband matrix elements of $\boldsymbol{p}$ for up to 15 bands. They then constructed a 15×15 Hamiltonian matrix whose diagonalization gave quite good results for the valence and conduction bands over the entire zone.

3.4.6 Nonparabolic bands: the Kane model

A simple example of the exact $\boldsymbol{k}\cdot\boldsymbol{p}$ method is the **Kane model** (Kane 1957) which is based on the observation from perturbation theory that the bands having the greatest influence on a given band are those that are closest in energy to that band. In the Kane model one focuses on a particular band and includes in the Hamiltonian matrix only that band and those other bands closest to it in energy. Exact diagonalization of the resulting matrix gives results for the energy bands that include **nonparabolic effects**, i.e., corrections to the energy of higher order than k^2.

Example 3.4: Kane model for a two-band system
Evaluate $E_{n\boldsymbol{k}}$ for two bands $n = c$ and $n = v$, assuming that the band extrema are at $\boldsymbol{k} = 0$.
Solution. The secular equation for the determination of $E_{n\boldsymbol{k}}$ takes the form

$$\begin{vmatrix} E_{c0} - \lambda & \frac{\hbar}{m}\boldsymbol{k}\cdot\langle c0|\boldsymbol{p}|v0\rangle \\ \frac{\hbar}{m}\boldsymbol{k}\cdot\langle v0|\boldsymbol{p}|c0\rangle & E_{v0} - \lambda \end{vmatrix} = 0, \tag{3.83}$$

where $\lambda = E_{n\boldsymbol{k}} - \hbar^2 k^2/2m$. Expanding the determinant yields a quadratic equation in λ whose solution is

$$\lambda = \frac{1}{2}\left\{E_{c0} + E_{v0} \pm \left[(E_{c0} - E_{v0})^2 + 4\left(\frac{\hbar}{m}\boldsymbol{k}\cdot|\langle c0|\,\boldsymbol{p}|v0\rangle|\right)^2\right]^{\frac{1}{2}}\right\}. \tag{3.84}$$

Expansion of the square root in Eq. (3.84) leads to the series solutions for the conduction and valence band energies:

$$\lambda_c = E_{c0} + \frac{\hbar^2}{m^2}\frac{(\boldsymbol{k}\cdot|\langle c0|\boldsymbol{p}\,|v0\rangle|)^2}{E_{c0} - E_{v0}} - \frac{\hbar^4}{m^4}\frac{(\boldsymbol{k}\cdot|\langle c0|\boldsymbol{p}|v0\rangle|)^4}{(E_{c0} - E_{v0})^3} + \cdots \tag{3.85a}$$

$$\lambda_v = E_{v0} - \frac{\hbar^2}{m^2}\frac{(\boldsymbol{k}\cdot|\langle c0|\boldsymbol{p}|v0\rangle|)^2}{E_{c0} - E_{v0}} + \frac{\hbar^4}{m^4}\frac{(\boldsymbol{k}\cdot|\langle c0|\boldsymbol{p}|v0\rangle|)^4}{(E_{c0} - E_{v0})^3} - \cdots. \tag{3.85b}$$

The first two terms on the right hand sides of these expressions give the energy to order k^2 and constitute the **parabolic approximation.**

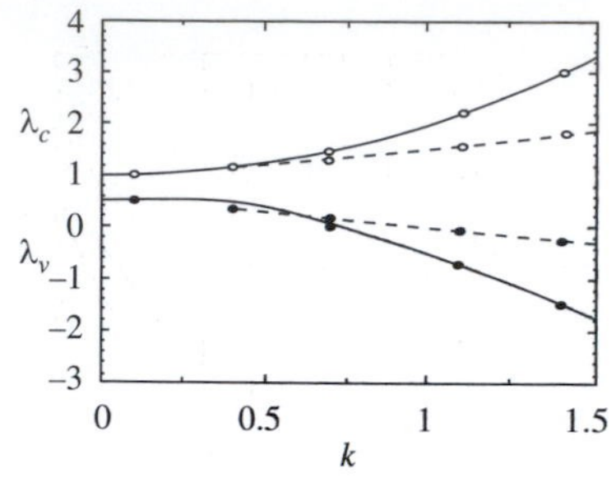

Fig. 3.9
Band energies λ_c and λ_v versus k. Solid curves: parabolic case; dashed curves: nonparabolic case.

The third terms on the right hand sides are of order k^4 and are **nonparabolic corrections**. Note from Eq. (3.84) that as $\boldsymbol{k} \to \infty$, $\lambda \sim |\boldsymbol{k}|$, a distinctly nonparabolic behavior. Plots of the band energies λ_c and λ_v for the parabolic and nonparabolic cases are shown in Fig. 3.9.

The Kane model can be expanded to include the conduction band, the light and heavy hole bands and the spin–orbit split-off band (Kane 1957). A 4×4 Hamiltonian matrix is diagonalized to yield the following band energies to order k^2:

$$E_c(\boldsymbol{k}) = E_g + \frac{\hbar^2 k^2}{2m}\left[1 + \frac{2P^2}{3m}\left(\frac{2}{E_g} + \frac{1}{E_g + \Delta_{so}}\right)\right] \tag{3.86a}$$

$$E_{vh} = \frac{\hbar^2 k^2}{2m} \tag{3.86b}$$

$$E_{vl} = \frac{\hbar^2 k^2}{2m}\left(1 - \frac{4P^2}{3mE_g}\right) \tag{3.86c}$$

$$E_{so} = -\Delta_{so} + \frac{\hbar^2 k^2}{2m}\left[1 - \frac{2P^2}{3m(E_g + \Delta_{so})}\right]. \tag{3.86d}$$

In Eqs. (3.86), P is the momentum matrix element between conduction and valence bands. The quantities P^2/mE_g and $P^2/m(E_g + \Delta_{so})$ are typically large compared to unity for materials such as InSb. The effective masses of the $v2$ and split-off (so) bands then have the proper negative signs. The neglect of higher band interactions leads to an incorrect sign for the $v1$ band effective mass.

3.5 Energy band structures for specific semiconductors

Detailed energy band structures have been calculated for a large number of semiconductors. In this section we present band structures for representative elemental and compound semiconductors and discuss the qualitative features of the individual structures.

3.5.1 Elemental semiconductors

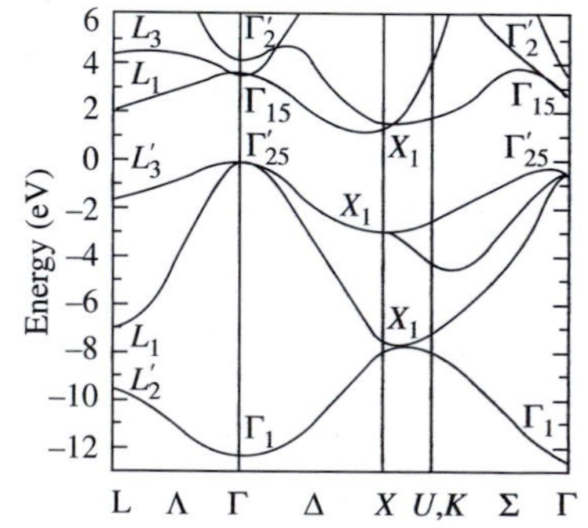

Fig. 3.10
Band structure of Si (after Chelikowsky and Cohen 1976).

The energy band structure of Si for the principal directions in the Brillouin zone is presented in Fig. 3.10. The maximum of the uppermost filled (valence) band occurs at the Γ point ($\boldsymbol{k} = 0$), and the minimum of the lowest empty (conduction) band occurs along the ΓX direction about 0.8 of the way to the zone boundary. This is an example of an **indirect gap** characterized by the extrema of the valence and conduction bands being at different points in $\boldsymbol{k}$-space. The magnitude of the energy gap between valence and conduction bands at room temperature is 1.14 eV. The **direct gap** at the Γ point is 3.45 eV at 5 K.

It is evident from Fig. 3.10 that the valence band at Γ'_{25} is degenerate. This band is derived from atomic $3p$ states which are not split by the cubic field at Γ. They are split by the spin–orbit interaction, but the latter is too small to give a splitting that is evident on the scale of the figure. As one moves away from the Γ point, the symmetry is reduced from cubic and the p-like bands split, as is evident for the ΓX and ΓL directions. Every band, however, is at least doubly degenerate because of the Kramers degeneracy due to time-reversal symmetry and the presence of a center of inversion (Kittel 1987).

The energy band structure of Ge is shown in Fig. 3.11. It is qualitatively similar to that of Si, but there are some significant differences. As in the case of Si, the valence band maximum occurs at the Γ point. However, the conduction band minimum is at the L-point, i.e., at the Brillouin zone boundary along the [111] direction. The order of the Γ_{15} and Γ'_2 bands is reversed compared to Si, but the fundamental energy gap is still indirect and is equal to 0.67 eV at room temperature. The spin–orbit splitting of the uppermost valence band is 0.3 eV and is clearly seen in Fig. 3.11. The minimum direct gap at Γ between valence and conduction bands is 0.90 eV, slightly larger than the indirect gap.

In the sequence of homopolar materials such as in column IV of the periodic table, the fundamental gap is largest at the top for diamond ($E_g \simeq 5.4$ eV) and decreases going down the column until α-Sn is reached, which has zero fundamental gap and whose band structure is shown in Fig. 3.12.

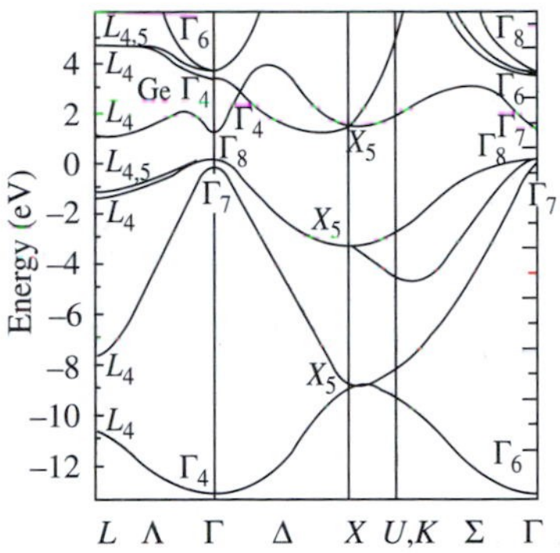

Fig. 3.11
Band structure of Ge (after Chelikowsky and Cohen 1976).

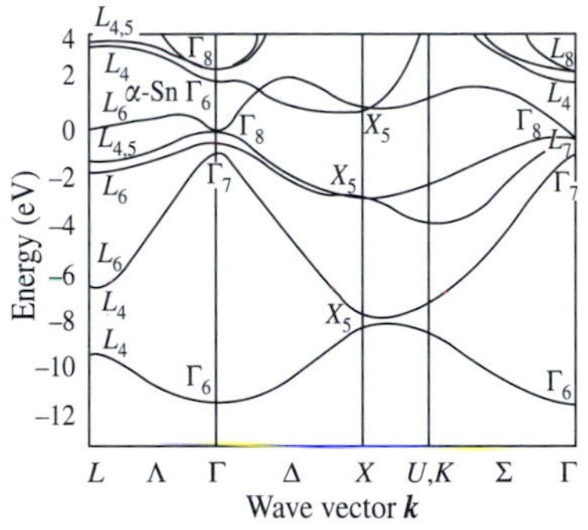

Fig. 3.12
Band structure of α-Sn (after Chelikowsky and Cohen 1976).

3.5.2 III–V semiconductors

Chelikowsky and Cohen have calculated the energy band structure of the III–V compounds corresponding to all possible pairings of Ga and In with P, As, and Sb. The band structure of GaP is shown in Fig. 3.13. It is similar to that of Si in that the minimum energy gap is indirect and occurs between the Γ point of the valence band and a point near the Brillouin zone boundary in the ΓX direction of the conduction band. The band structures of the other five compounds are very similar to one another. Specifically, the minimum energy gap is direct and occurs at the Γ point for each of them. As an example, the band structure of GaAs is shown in Fig. 3.14. In addition to the principal conduction band minimum at Γ, there is a secondary minimum at the L-point. The order of these two minima is the reverse of that found in Ge (Fig. 3.11).

The spin–orbit splitting of the valence bands of GaAs at Γ is clearly evident in Fig. 3.14 as the separation of the Γ_7 and Γ_8 bands. The splitting of the upper valence band Γ_8 as $\boldsymbol{k}$ becomes different from zero can be seen in the ΓL and ΓX directions. This splitting is a consequence of the lowered symmetry when $\boldsymbol{k}$ is finite. The band structures of other III–V semiconductors such as GaSb, InP, InAs, and InSb are qualitatively similar to that of GaAs. Just as with the group IV materials, the fundamental gaps of III–V compounds become smaller and smaller as one goes down the periodic table.

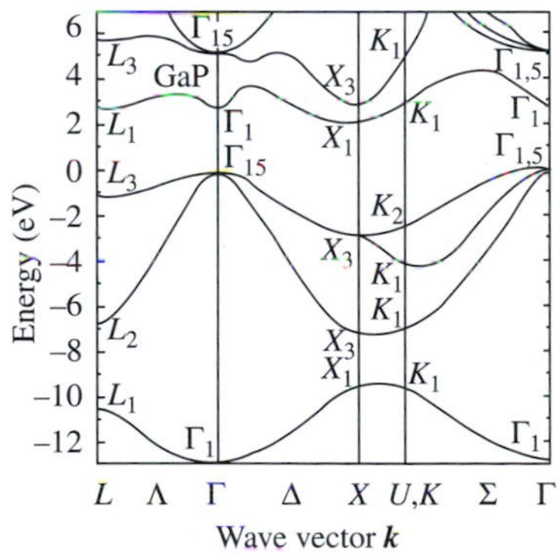

Fig. 3.13
Band structure of GaP (after Chelikowsky and Cohen 1976).

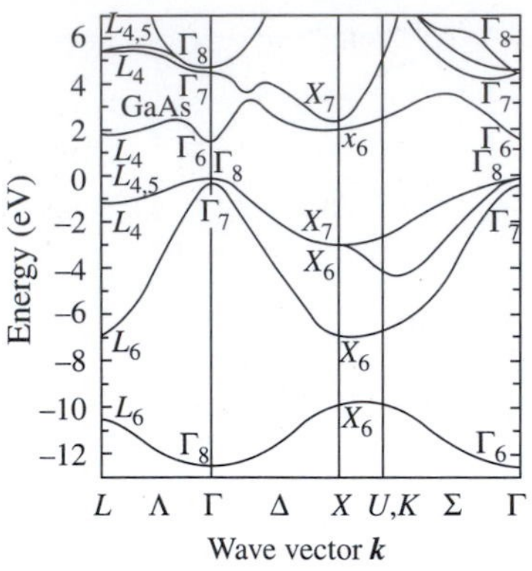

Fig. 3.14
Band structure of GaAs (after Chelikowsky and Cohen 1976).

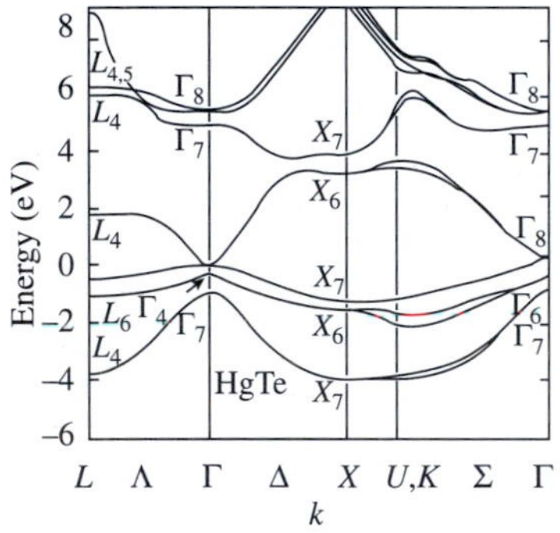

Fig. 3.15
Band structure of HgTe (after Cohen and Chelikowski 1988).

Table 3.4 Direct and indirect band gaps and spin–orbit splittings in eV of group IV, III–V, II–VI, and IV–VI semiconductors at low temperature (after Burstein and Egli 1955, Cardona 1969, Harrison 1980, Kittel 1986, Böer 1990)

IV and III–V	E_g^{ind}	E_g^{dir}	Δ_{so}	II–VI and IV–VI	E_g^{dir}	Δ_{so}
C (diamond)	5.4		0.006	ZnS (cubic)	3.78	0.067
Si	1.17	3.45	0.045	ZnS (hex)	3.91	0.086
Ge	0.74	0.90	0.297	ZnSe	2.82	0.403
α-Sn		0	0.73	ZnTe	2.39	0.97
AlP	~ 3.0			CdS	2.58	0.062
AlAs	~ 2.2	2.77		CdSe	1.84	0.416
AlSb	1.65		0.75	CdTe	1.61	0.811
GaP	2.35	2.90	0.08	HgS	≤ 0	0.32
GaAs	1.82	1.52	0.34	HgSe	≤ 0	0.58
GaSb	1.22	0.81	0.76	HgTe	≤ 0	1.05
InP	2.19	1.42	0.11	PbS	0.29	
InAs	1.53	0.42	0.43	PbSe	0.16	
InSb	1.03	0.23	0.82	PbTe	0.19	
				SnTe	0.3	

3.5.3 II–VI and IV–VI semiconductors

The II–VI and IV–VI compounds essentially comprise the Zn, Cd, Hg, Pb, and Sn chalcogenides and some oxides. The II–VI materials crystallize in the zincblende or wurtzite structure, whereas the IV–VI materials crystallize in the NaCl structure. An example of band structure is that of ZnSe shown in Fig. 3.2. It closely resembles that of GaAs. The band structure of HgTe is given in Fig. 3.15 and is seen to be like that of α-Sn. However, the mercury chalcogenides may have overlapping valence and conduction bands and be semimetals as discussed in Section 3.6.1. A tabulation of valence-conduction band gaps and spin–orbit splittings of the valence band is given in Table 3.4 for a variety of semiconductors.

3.6 Modification of energy band gaps

We have seen that under the same experimental conditions the energy band gap varies from one semiconductor to another. For a given material the band gap can be changed by heavy doping with impurities or by alloying with another semiconductor or by varying the temperature or applying stress. In this section we examine these methods of modifying the band gap.

3.6.1 Semiconductor alloys

Semiconductor alloys can be formed in various ways. Binary alloys are made by mixing similar elements such as Si and Ge, for example, to form Ge_xSi_{1-x}. Ternary alloys of compound semiconductors can be obtained by substituting an element such as Al for the homologous element Ga in GaAs to form $Ga_{1-x}Al_xAs$. Alternatively, one can mix the nonmetal constituents by substituting S for Se in ZnSe to give ZnS_xSe_{1-x}. Quaternary alloys result if both the metal and nonmetal constituents are mixed to give, for example,

$Ga_xAl_{1-x}P_yAs_{1-y}$. The band gap of an alloy is a continuous function of composition. In a ternary alloy, the variation with composition is linear if the two constituents that are varied have nearly the same atomic radii and the same bonding strength to the third constituent. For two constituents A and B,

$$\bar{E}_g(x) = E_g(A) + x[E_g(B) - E_g(A)], \tag{3.87}$$

where the bar over $E_g(x)$ designates an average over different microscopic configurations consistent with a given composition x. An example of linear variation of the band gap with composition is given in Fig. 3.16 for the system $ZnSe_xS_{1-x}$. When the characteristic parameters of the two constituents are significantly different, **bowing** of the bandgap occurs as shown in Fig. 3.17 for the system $ZnTe_xSe_{1-x}$. Bowing can be described by the empirical relation

$$\bar{E}_{gb}(x) = \bar{E}_g(x) - bx(1-x), \tag{3.88}$$

where $\bar{E}_{gb}(x)$ is the band gap with bowing and b is the bowing parameter. For many semiconductor alloys, b is positive (Böer 1990). The band gap of the alloy for a given composition is then smaller than the linear gap $\bar{E}_g(x)$ at that composition.

In some semiconductor alloys with narrow gaps, more profound changes occur in the band structure as a function of composition. Of particular interest is $Hg_{1-x}Cd_xTe$ whose band structure for three different compositions is shown in Fig. 3.18. If a band gap is defined by $\Delta E = E(\Gamma_6) - E(\Gamma_8)$ at $\boldsymbol{k} = 0$, then ΔE for pure CdTe is $+1.6\,\text{eV}$ corresponding to a semiconductor. At a certain intermediate composition ΔE is zero, corresponding to a gapless semiconductor, and for pure HgTe, ΔE is negative. ΔE corresponds to the fundamental gap defined as the difference in energy of the lowest empty state and the highest filled state only if $\Delta E \geq 0$. If $\Delta E < 0$, it is no longer the fundamental gap. The latter is zero for compositions with $\Delta E < 0$, because the lowest empty band and highest filled band are the Γ_8 bands that touch at $\boldsymbol{k} = 0$. Thus, HgTe appears to be a gapless semiconductor like α-Sn. However, there is a fine point that must be mentioned. Since HgTe lacks a center of inversion, there are terms linear in k that raise the valence band edge above its value at $k = 0$ and produce a small overlap of the valence and conduction bands. The result is a **semimetal** with both electrons in the conduction band and holes in the valence band present.

Another interesting system is the alloy $Pb_{1-x}Sn_xTe$. In PbTe the conduction and valence bands are L_6^- and L_6^+, respectively, whereas in SnTe they are reversed, L_6^+ and L_6^-. As x increases, the gap decreases until it reaches zero at $x = 0.62$. For $x > 0.62$ the system is again a semiconductor (Dimmock *et al.* 1966). This behavior is shown in Fig. 3.19. It provides a basis for designing materials with specific band gaps.

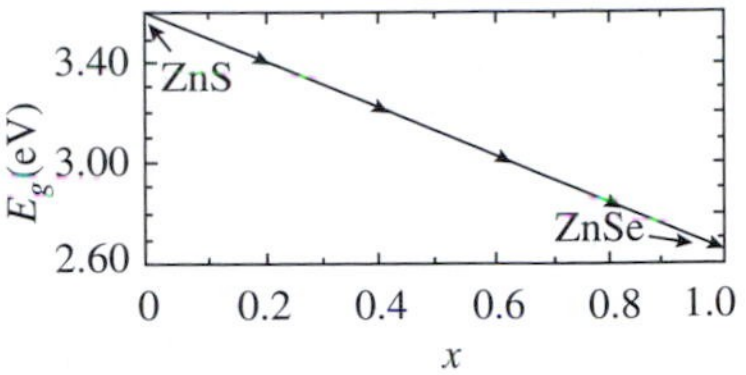

Fig. 3.16
Dependence of band gap on composition for $ZnSe_xS_{1-x}$ (after Larach *et al.* 1957).

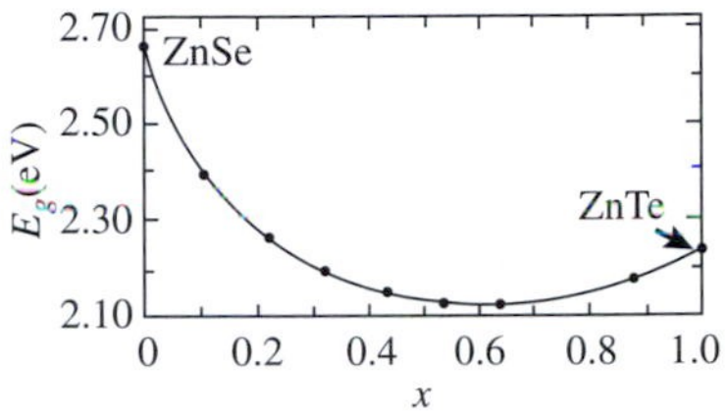

Fig. 3.17
Dependence of band gap on composition for $ZnTe_xSe_{1-x}$ (after Larach *et al.* 1957).

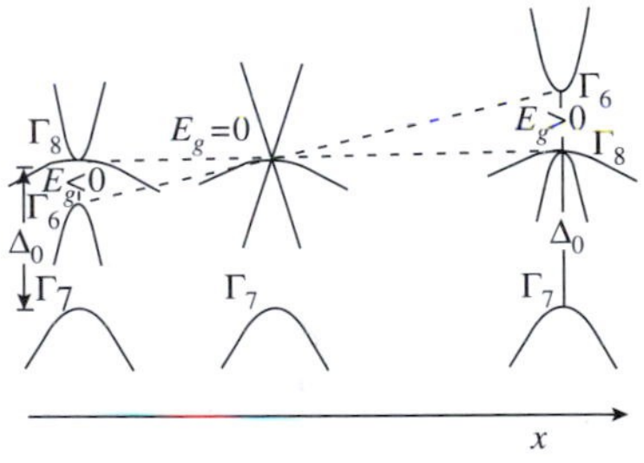

Fig. 3.18
Dependence of band structure on composition x or pressure p for $Hg_{1-x}Cd_xTe$ (after Böer 1990).

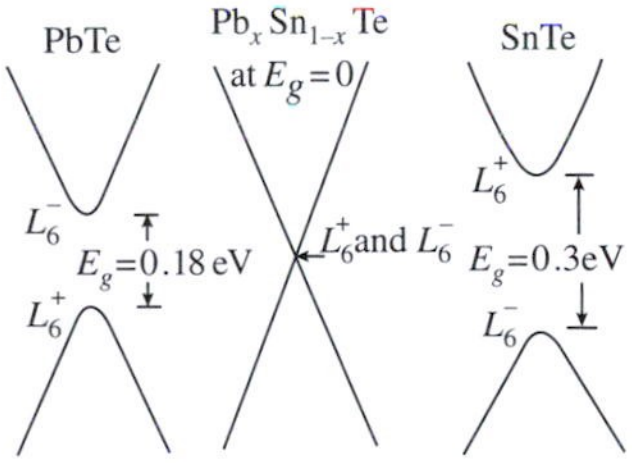

Fig. 3.19
Dependence of band gap on composition for $Pb_{1-x}Sn_xTe$ (after Dimmock *et al.* 1966).

3.6.2 Temperature and pressure dependence of band gaps

The band gap in semiconductors is dependent on crystal bonding and lattice constant. Both of these parameters change when the temperature T

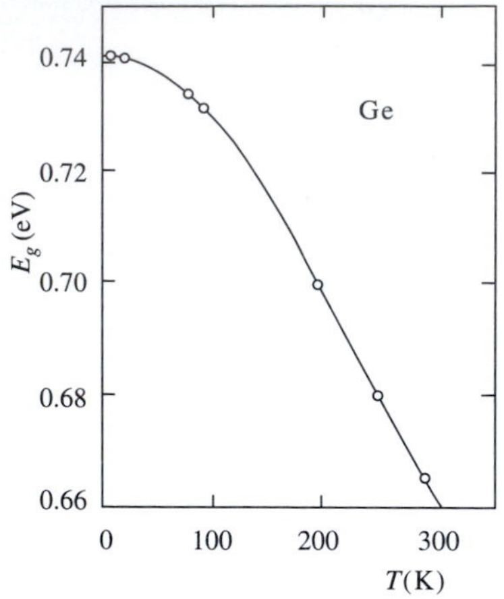

Fig. 3.20
Dependence of band gap on temperature for Ge (after MacFarlane *et al.* 1957).

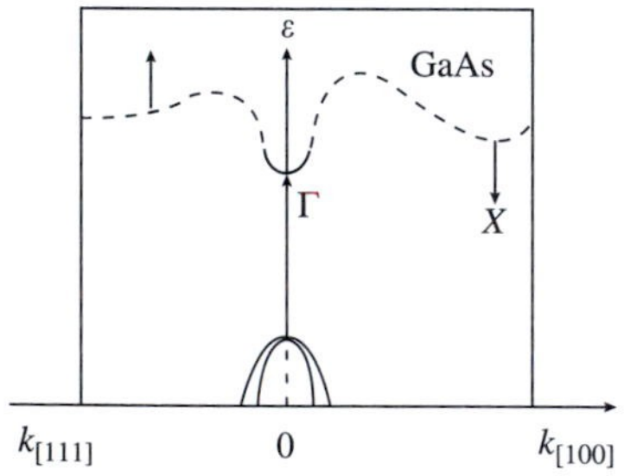

Fig. 3.21
Dependence of band edges on pressure for GaAs (after Böer 1990).

or pressure p changes. The change in band gap is described by

$$\Delta E_g = \left(\frac{\partial E_g}{\partial T}\right)_p \Delta T + \left(\frac{\partial E_g}{\partial p}\right)_T \Delta p. \tag{3.89}$$

The values of $(\partial E_g/\partial T)_p$ are typically $\sim -40\,\text{meV/K}$ and those of $(\partial E_g/\partial p)_T \sim 10\,\text{meV/kbar}$.

Considering first the effect of temperature, one finds that the band gap decreases quadratically with T at very low T and linearly with T at room temperature. This behavior is illustrated for Ge in Fig. 3.20 and is related to effects on the band structure due to thermal expansion and lattice vibrations. Empirically, the band gap as a function of temperature can be represented by (Varshni 1967)

$$E_g(T) = E_g(0) - \frac{AT^2}{B+T}, \tag{3.90}$$

where A and B are constants.

The dependence of the band gap on pressure can be complicated if the band edges at different symmetry points move in opposite directions as the pressure increases. An example is GaAs, which is shown in Fig. 3.21. The band edge at Γ moves upward relative to the valence band edge, whereas the edge at X moves downward. As a result, the minimum band gap is direct at pressures below 8×10^4 atm and indirect at pressures above 8×10^4 atm. The resulting dependence of the minimum band gap on pressure for GaAs is given in Fig. 3.22.

Application of uniaxial stress to a cubic crystal reduces the symmetry and splits the light and heavy hole bands at $\boldsymbol{k} = 0$. The number of equivalent conduction band minima is reduced if the stress is in a suitable crystallographic direction.

3.7 Amorphous semiconductors

In addition to the crystalline state a semiconductor such as Si can exist in the **amorphous state** in which the atoms do not form a periodic array, but are disordered. The absence of periodicity means that Bloch's theorem no

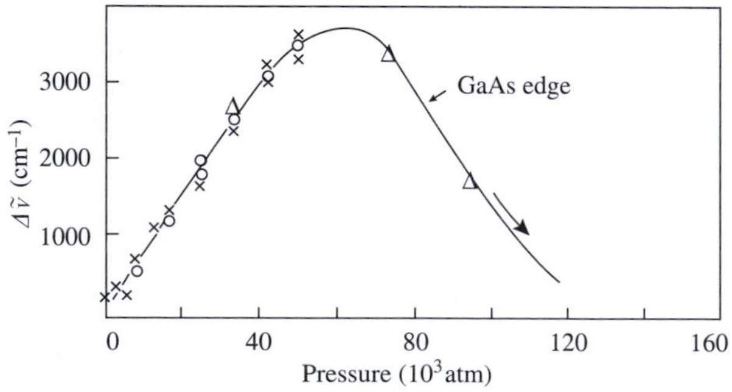

Fig. 3.22
Dependence of band gap on pressure for GaAs (after Edwards *et al.* 1959).

longer holds, and that the wave vector $\boldsymbol{k}$ is no longer a good quantum number. Nevertheless, sufficient short range order remains that the electronic density-of-states retains its gross features. Band edges, however, are smeared out, and band gaps are no longer clearly defined. States whose energies have moved into what had been the forbidden gap may have a localized character rather than the extended character of Bloch states.

α-Si:H is one of the most common amorphous semiconductors. Creating disorder in an Si lattice, which is tetrahedrally bonded, causes relaxation of he requirement for a fourfold coordination of nearest neighbors and produces dangling bonds. These bonds need to be saturated and combine easily with monovalent atoms such as H to yield α-Si:H.

Amorphous semiconductor alloys are frequently used in applications. In particular, amorphous Si–Ge alloys are useful in solar cells in view of the possibility to maximize performance by varying the composition. The dependence of the band gap of hydrogenated Si–Ge alloys (α-$Si_{1-x}Ge_x$:H) on composition is shown in Fig. 3.23.

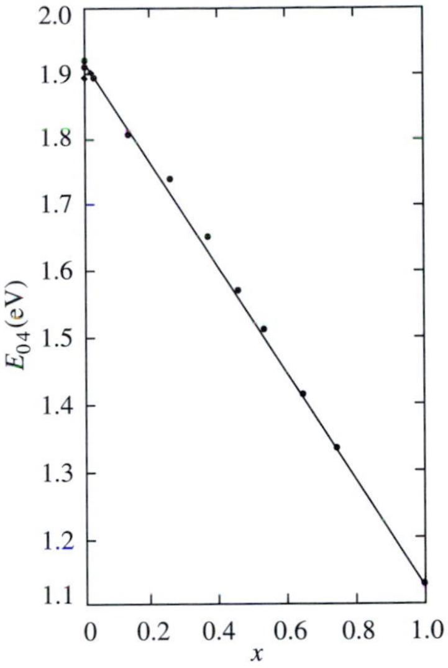

Fig. 3.23
Dependence of band gap on composition for α-$Si_{1-x}Ge_x$:H (after Mackenzie *et al.* 1985).

Problems

1. Determine the expectation value of the Hamiltonian, Eq. (3.40), for a state with wave function given by Eq. (3.41) under the assumption that the $\varphi_i(r)$ satisfy the condition of orthonormality:

$$\int \varphi_i^*(\mathbf{r})\varphi_j(\mathbf{r})d^3r = \delta_{ij}.$$

2. Show that the stationary condition

$$\delta \int \Psi_N^* H \Psi_N d^3r_1 \cdots d^3r_N = 0$$

together with the orthonormality condition handled by Lagrange multipliers leads to the Hartree equation, Eq. (3.42).

3. Repeat the calculations of problems 1 and 2 using the wave function of Eq. (3.43).
4. Prove that the only nonzero contributions to the exchange term in the Hartree–Fock equations come from electrons of parallel spin.
5. Calculate the energies of the conduction, valence, and split-off bands of GaAs as functions of k using Eq. (3.86). Take $P = 8\hbar/a$, where a is the lattice constant, and take $E_g(E_g^{dir})$ and Δ_{so} from Table 3.4. Plot the energies versus k and compare the results qualitatively to those for GaAs in Fig. 3.14. Explain any significant differences.
6. Extend the tight binding treatment of Chapter 2 to the case of Ge with two atoms per unit cell and taking into account the 4s and 4p atomic states. The result specifies the valence band energies (Reference: Chadi and Cohen 1975).

References

N. W. Ashcroft and N. D. Mermin, *Solid State Physics* (Holt, Rinehart, and Winston, New York, 1976).
G. B. Bachelet, D. R. Hamann, and M. Schlüter, *Phys. Rev.* **B26**, 4199 (1982).
S. Bloom and T. K. Bergstresser, *Solid State Commun.* **6**, 465 (1968).
K. W. Böer, *Survey of Semiconductor Physics* (Van Nostrand Reinhold, New York, 1990).

E. Burstein and P. Egli, in *Advances in Electronics and Electron Physics*, Vol. VII, ed. H. Brooks (Academic Press, New York, 1955).
M. Cardona, *Modulation Spectroscopy* (Academic Press, New York, 1969).
M. Cardona and F. H. Pollak, *Phys. Rev.* **142**, 530 (1966).
D. J. Chadi and M. L. Cohen, *Phys. Stat. Sol.* B**68**, 405 (1975).
J. R. Chelikowsky and M. L. Cohen, *Phys. Rev.* B**14**, 556 (1976).
M. L. Cohen and T. K. Bergstresser, *Phys. Rev.* **141**, 789 (1966).M. L. Cohen and J. R. Chelikowsky, *Electronic Structure and Optical Properties of Semiconductors* (Springer-Verlag, Berlin, 1988).
R. N. Dexter, H. J. Zeiger, and B. Lax, *Phys. Rev.* **104**, 637 (1956).
J. O. Dimmock, I. Melngailis, and A. J. Strauss, *Phys. Rev. Lett.* **16**, 1193 (1966).
G. Dresselhaus, *Phys. Rev.* **100**, 580 (1955).
G. Dresselhaus, A. F. Kip, and C. Kittel, *Phys. Rev.* **98**, 368 (1955).
A. L. Edwards, T. E. Slykhouse, and H. G. Drickamer, *J. Phys. Chem. Solids* **11**, 140 (1959).
S. Fahy, X. W. Wang, and S. G. Louie, *Phys. Rev.* B**42**, 3503 (1990).
W. A. Harrison, *Electron Structure and the Properties of Solids* (W. H. Freeman, San Francisco, 1980).
F. Herman, *Phys. Rev.* **93**, 1214 (1954).
C. Herring, *Phys. Rev.* **57**, 1169 (1940).
P. Hohenberg and W. Kohn, *Phys. Rev.* **136**, B**864** (1964).
M. S. Hybertsen and S. G. Louie, *Phys. Rev.* B**34**, 5390 (1986).
E. O. Kane, *J. Phys. Chem. Solids* **1**, 249 (1957).
C. Kittel, *Introduction to Solid State Physics*, Sixth edition (John Wiley, New York, 1986).
C. Kittel, *Quantum Theory of Solids*, Second revised printing (John Wiley, New York, 1987), p. 92.
W. Kohn and L. J. Sham, *Phys. Rev.* **140**, A1133 (1965).
S. Larach, R. E. Schrader, and C. F. Stocker, *Phys. Rev.* **108**, 587 (1957).
P. Lawaetz, *Phys. Rev.* B**4**, 3460 (1971).
G. G. MacFarlane, T. P. McLean, J. E. Quarrington, and V. Roberts, *Phys. Rev.* **108**, 1377 (1957).
K. D. Mackenzie, J. R. Eggert, D. J. Leopold, Y. M. Li, S. Lin, and W. Paul, *Phys. Rev.* B**31**, 2198 (1985).
J. C. Phillips and L. Kleinman, *Phys. Rev.* **116**, 287 (1959).
C. R. Pidgeon and S. H. Groves, *Phys. Rev.* **186**, 824 (1969).
F. Seitz, *Modern Theory of Solids* (McGraw-Hill, New York, 1940), p. 242.
L. H. Thomas, *Nature* **117**, 514 (1926).
Y. P. Varshni, *Phys. Stat. Sol.* **19**, 459 and **20**, 9 (1967).
G. Weisz, *Phys. Rev.* **149**, 504 (1966).
H. Wendel and R. M. Martin, *Phys. Rev.* B**19**, 5251 (1979).
E. Wigner, *Phys. Rev.* **46**, 1002 (1934).
M. T. Yin and M. L. Cohen, *Phys. Rev.* B**26**, 5668 (1982).
X. Zhu and S. G. Louie, *Phys. Rev.* B**43**, 14142 (1991).
J. M. Ziman, *Principles of the Theory of Solids*, Second edition (Cambridge University Press, Cambridge, 1972), p. 102.

Kinematics and dynamics of electrons and holes in energy bands

4

Key ideas

A free electron in an energy band can be represented semiclassically by a *wave packet* of Bloch states that is *spatially localized*. The velocity of the center of the wave packet is the *group velocity* $\boldsymbol{v}_g$.

The curvature of an energy band is proportional to the *inverse effective mass*. The *inverse effective mass tensor* is defined by

$$\left(\frac{1}{m^*}\right)_{\alpha\beta} = \frac{1}{\hbar^2}\frac{\partial^2 E_{nk}}{\partial k_\alpha \partial k_\beta}.$$

An *external force* $\boldsymbol{F}$ acting on a band electron produces a change of its wave vector $\boldsymbol{k}$ with time,

$$\boldsymbol{F} = \hbar\frac{d\boldsymbol{k}}{dt},$$

which is analogous to the classical relation of force to time rate of change of momentum $\boldsymbol{p}$:

$$\boldsymbol{F} = \frac{d\boldsymbol{p}}{dt}.$$

One therefore defines the *crystal momentum* to be $\hbar\boldsymbol{k}$.

An electric field $\boldsymbol{\mathcal{E}}$ produces an accelerated electron wave packet:

$$\frac{d\boldsymbol{v}_g}{dt} = -e\left(\frac{1}{m^*}\right)\cdot\boldsymbol{\mathcal{E}}.$$

A *hole* is an *empty state* in an otherwise filled band. Both the *electric charge* of a hole and its *effective mass* are *positive*.

The effective mass of a charge carrier can be measured by *cyclotron resonance*.

The concentration and charge sign of a charge carrier can be measured by the *Hall effect*.

Electrons and holes in energy bands

One of the features that distinguishes semiconductors from metals is that the current carriers can include not only electrons in the conduction band, but also empty electron states or holes in the valence band. The dynamical response of these current carriers to external electric and magnetic fields is strongly affected by the energy band structure of the semiconductor and is characterized by their effective mass and charge. These parameters can be determined by means of cyclotron resonance and Hall effect, respectively.

4.1 Group velocity

The wave functions corresponding to Bloch states are wave-like in character and are non-localized. In order to discuss the motion of an electron from one point to another, it is necessary to form a localized wave packet, so that the electron can be assigned a particular coordinate at a particular time.

A wave packet can be created by setting up a linear superposition of time-dependent Bloch functions of various wave vectors $\boldsymbol{k}$ with coefficients $a_{n\boldsymbol{k}}$ that peak at a particular wave vector $\boldsymbol{k}_0$:

$$f_{n\boldsymbol{k}_0}(\boldsymbol{r}, t) = \int d^3k a_{n\boldsymbol{k}} \psi_{n\boldsymbol{k}}(\boldsymbol{r}, t). \tag{4.1}$$

The time-dependent Bloch functions $\psi_{n\boldsymbol{k}}(\boldsymbol{r}, t)$ can be written as

$$\psi_{n\boldsymbol{k}}(\boldsymbol{r}, t) = e^{i\boldsymbol{k}\cdot\boldsymbol{r}} u_{n\boldsymbol{k}}(\boldsymbol{r}) e^{-i(E_{n\boldsymbol{k}}/\hbar)t}, \tag{4.2}$$

where $E_{n\boldsymbol{k}}$ is the energy eigenvalue of the Bloch state. We now exploit the sharply peaked nature of the quantities $a_{n\boldsymbol{k}}$ by setting $\boldsymbol{k} = \boldsymbol{k}_0 + \Delta\boldsymbol{k}$ and carrying out expansions in powers of $\Delta\boldsymbol{k}$:

$$E_{n\boldsymbol{k}} = E_{n\boldsymbol{k}_0} + \Delta\boldsymbol{k} \cdot \boldsymbol{\nabla}_{\boldsymbol{k}_0} E_{n\boldsymbol{k}_0} + \cdots \tag{4.3}$$

$$u_{n\boldsymbol{k}}(\boldsymbol{r}) = u_{n\boldsymbol{k}_0}(\boldsymbol{r}) + \Delta\boldsymbol{k} \cdot \boldsymbol{\nabla}_{\boldsymbol{k}_0} u_{n\boldsymbol{k}_0}(\boldsymbol{r}) + \cdots. \tag{4.4}$$

In the case of $a_{n\boldsymbol{k}}$, its rapidly varying character requires us to retain its full dependence on $\Delta\boldsymbol{k}$:

$$a_{n\boldsymbol{k}} = a_{n\Delta\boldsymbol{k}}. \tag{4.5}$$

Substituting Eqs. (4.2–4.5) into Eq. (4.1) gives

$$f_{n\boldsymbol{k}_0}(\boldsymbol{r}, t) = e^{i\boldsymbol{k}_0\cdot\boldsymbol{r}} u_{n\boldsymbol{k}_0}(\boldsymbol{r}) e^{-i(E_{n\boldsymbol{k}_0}/\hbar)t} \times \int d^3\Delta k a_{n\Delta\boldsymbol{k}} e^{i\Delta\boldsymbol{k}\cdot[\boldsymbol{r}-(\nabla_{\boldsymbol{k}_0} E_{n\boldsymbol{k}_0}/\hbar)t]}, \tag{4.6}$$

where, for our purposes, it is sufficiently accurate to retain only the leading term in the expansion of $u_{n\boldsymbol{k}}(\boldsymbol{r})$.

The expression we have obtained for $f_{n\boldsymbol{k}_0}(\boldsymbol{r}, t)$ represents a Bloch function of wave vector $\boldsymbol{k}_0$ modulated by an envelope function given by the integral over $\Delta\boldsymbol{k}$. If we consider the value of the envelope function for given values of $\boldsymbol{r}$ and t, we note that the envelope function has the same value for all $\boldsymbol{r}$ and t satisfying

$$\boldsymbol{r} - (\boldsymbol{\nabla}_{\boldsymbol{k}_0} E_{n\boldsymbol{k}_0}/\hbar)t = \text{constant}. \tag{4.7}$$

Equation (4.6) can therefore be regarded as specifying simple rectilinear motion of the wave packet for the electron that is characterized by a velocity $\boldsymbol{v}_g$ given by

$$\boldsymbol{v}_g = \frac{1}{\hbar}\boldsymbol{\nabla}_{\boldsymbol{k}_0} E_{n\boldsymbol{k}_0}. \tag{4.8}$$

The velocity $\boldsymbol{v}_g$ is the **group velocity** of the electron wave packet. It is a measure of the rate of transport of mass, charge, and energy of the electron. Since $\boldsymbol{v}_g$ is proportional to the $\boldsymbol{k}$-gradient of the Bloch state energy, it is determined by the slope of the $E_{n\boldsymbol{k}}$ versus $\boldsymbol{k}$ curve. Figure 4.1 gives a graphical representation of $\boldsymbol{v}_g$ as a function of $\boldsymbol{k}$.

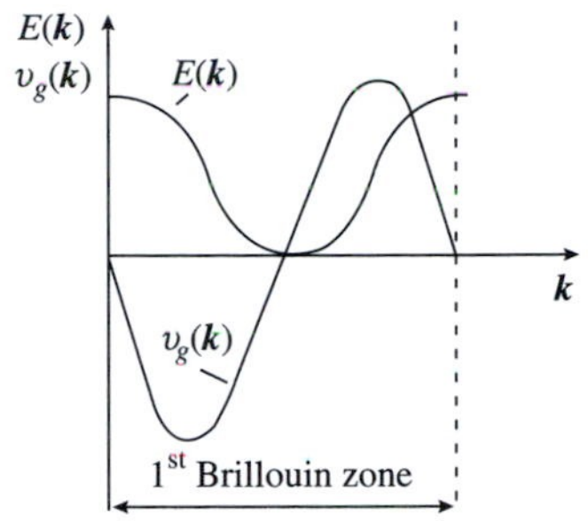

Fig. 4.1
Graphical representation of $\boldsymbol{v}_g$ and $E_{n\boldsymbol{k}}$ as functions of wave vector $\boldsymbol{k}$.

Comment. Equations (4.6) and (4.8) specify the time evolution of a wave packet for an electron characterized by a group velocity $\boldsymbol{v}_g$. Figure 4.2 gives a schematic representation of the wave packet in real space. At a particular time t, the wave packet has its maximum amplitude at some value of $\boldsymbol{r}$, say $\boldsymbol{r}_0$. The spread Δr of the wave packet and the width Δk of the region over which $a_{n\boldsymbol{k}}$ is appreciable are subject to the limitation $\Delta r \Delta k \simeq 1$. The wave vector interval Δk should be small compared to the dimensions of the Brillouin zone, so that $E_{n\boldsymbol{k}}$ varies little over the principal levels appearing in the wave packet, i.e., $\Delta k \ll 1/a$ where a is the lattice constant. The above mentioned limitation then gives

$$\Delta r \simeq \frac{1}{\Delta k} \gg a. \tag{4.9}$$

Irrespective of the value of $\boldsymbol{r}_0$, the wave packet of Bloch states characterized by $\Delta k \ll 1/a$ is spread out in real space over many primitive unit cells. This behavior is shown in Fig. 4.2.

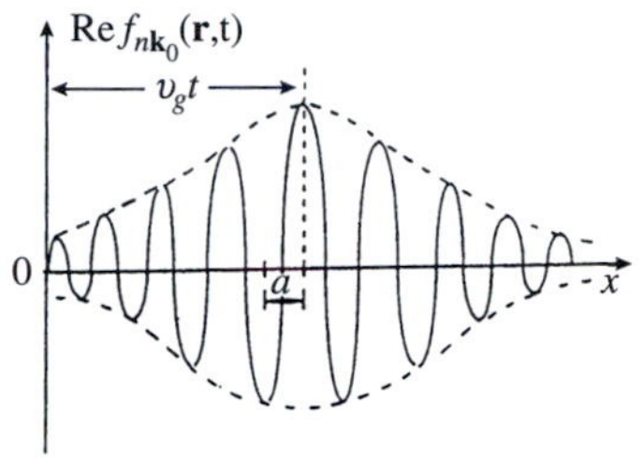

Fig. 4.2
Representation in real space of a wave packet of an electron in the conduction band.

4.2 Inverse effective mass tensor

Since a wave packet can be regarded as a semiclassical description of electron motion, we can make use of various classical relations between velocity, force, and energy. Consider the time derivative of $\boldsymbol{v}_g$:

$$\begin{aligned}\frac{d\boldsymbol{v}_g}{dt} &= \frac{d}{dt}\left(\frac{1}{\hbar}\boldsymbol{\nabla}_{\boldsymbol{k}} E_{n\boldsymbol{k}}\right) \\ &= \frac{1}{\hbar}\boldsymbol{\nabla}_{\boldsymbol{k}}\frac{dE_{n\boldsymbol{k}}}{dt}.\end{aligned} \tag{4.10}$$

Using the classical relation

$$\frac{dE}{dt} = \boldsymbol{F}\cdot\boldsymbol{v}_g \tag{4.11}$$

and introducing the force $\boldsymbol{F}$ acting on the particle, we can rewrite Eq. (4.10) as

$$\frac{d\boldsymbol{v}_g}{dt} = \frac{1}{\hbar}(\boldsymbol{\nabla}_k \boldsymbol{v}_g) \cdot \boldsymbol{F}, \tag{4.12}$$

where we have assumed that $\boldsymbol{F}$ is independent of $\boldsymbol{k}$. Replacing $\boldsymbol{v}_g$ on the right hand side of Eq. (4.12) by its definition in Eq. (4.8), we obtain

$$\frac{d\boldsymbol{v}_g}{dt} = \frac{1}{\hbar^2}(\boldsymbol{\nabla}_k \boldsymbol{\nabla}_k E_{nk}) \cdot \boldsymbol{F}. \tag{4.13}$$

Comparison of Eq. (4.13) to Newton's second law of motion shows that the quantity $\hbar^{-2}\nabla_k\nabla_k E_{nk}$, which is a dyadic, has the dimensions of inverse mass. We defined the elements of the **inverse effective mass tensor** for band n by Eq. (3.62). For a simple parabolic band,

$$\frac{1}{m_n^*} = \frac{1}{\hbar^2}\frac{\partial^2 E_{nk}}{\partial k^2}. \tag{4.14}$$

We see that the **curvature** of the energy band is proportional to the inverse effective mass. The larger the curvature, the smaller the effective mass.

4.3 Force equation

Let us re-examine the classical relation given by Eq. (4.11). Applying the chain rule for differentiation to the left hand side yields the result

$$\frac{d\boldsymbol{k}}{dt} \cdot \boldsymbol{\nabla}_k E_{nk} = \boldsymbol{F} \cdot \boldsymbol{v}_g.$$

Eliminating the gradient of the band energy with the aid of Eq. (4.8) and equating the coefficients of $\boldsymbol{v}_g$ on each side of the resulting equation, we obtain

$$\boldsymbol{F} = \hbar\frac{d\boldsymbol{k}}{dt}. \tag{4.15}$$

The quantity $\hbar\boldsymbol{k}$ is known as the **crystal momentum**. Equation (4.15) is therefore the analogue of the classical relation

$$\boldsymbol{F} = \frac{d\boldsymbol{p}}{dt}, \tag{4.16}$$

where $\boldsymbol{p}$ is the classical momentum.

4.4 Dynamics of electrons

Consider the special case in which the force arises from an applied electric field $\boldsymbol{\mathcal{E}}$. Then,

$$\boldsymbol{F} = -e\boldsymbol{\mathcal{E}} \tag{4.17}$$

and

$$\hbar \frac{d\boldsymbol{k}}{dt} = -e\boldsymbol{\mathcal{E}}. \tag{4.18}$$

An electric field therefore causes the wave vector $\boldsymbol{k}$ to change with time. Alternatively, we can combine Eqs. (4.13), (4.14), and (4.17) to yield

$$\frac{d\boldsymbol{v}_g}{dt} = -e\left(\frac{1}{m^*}\right) \cdot \boldsymbol{\mathcal{E}} \tag{4.19}$$

which shows that the electron wave packet is accelerated by the electric field and can give rise to an electric current.

The qualitative picture of electrical conduction just described is valid for a partially filled energy band. An electron of wave vector $\boldsymbol{k}$ can make a transition to a nearby empty state of different wave vector under the influence of a weak electric field. However, if the band is completely filled with electrons, there are no empty states into which an electron can make a transition, and consequently, the conductivity of a filled band is zero.

We thus arrive at a simple distinction between **insulators** and **conductors**. An insulator has all bands up to and including a certain band completely filled with electrons at the absolute zero of temperature. All bands above the uppermost filled band are completely empty, and the lowest empty band is separated from the uppermost filled band by an energy gap $E_g \gg k_B T_r$, where T_r is room temperature. In practice, E_g must be several electron volts as, for example, in the alkali halides.

A conductor contains at least one band that is partially filled with electrons. If these electrons arise by thermal excitation across a forbidden energy gap from an otherwise filled band, the material is a **semiconductor**. Note that under this definition, a semiconductor is simply an insulator with a relatively small energy gap. If one or more bands remain partially filled down to the absolute zero of temperature, the material is a **metal** or a **semimetal**.

4.5 Dynamics of holes

If an electron is excited from a filled energy band to an unfilled energy band, the empty state left in the otherwise filled band is called a **hole**. Of great importance are holes lying near the top of the uppermost filled band (valence band) of a semiconductor. In assessing the properties of such holes we examine first their wave vector. Since the reciprocal lattice has a center of inversion, the full valence band has a total wave vector of zero. The wave vector $\boldsymbol{k}_h$ to be associated with a hole is the difference between the total wave vector of the filled band and that of a band with an electron of wave vector $\boldsymbol{k}_e$ missing:

$$\begin{aligned} \boldsymbol{k}_h &= \boldsymbol{0} - \boldsymbol{k}_e \\ &= -\boldsymbol{k}_e. \end{aligned} \tag{4.20}$$

Note that the hole wave vector is not that of the missing electron, but is the negative of it.

Turning now to the energy of a hole, we observe that as $|\boldsymbol{k}_e|$ increases, the vacant state moves lower in the valence band, and the energy $E_e(\boldsymbol{k}_e)$ of that state decreases. However, the total energy of the system of electrons occupying the valence band increases by the same amount, because as the vacant state drops from its initial higher energy state to its final lower energy state, an occupied state makes the reverse transition. This scenario leads us to define the energy of the hole, $E_h(\boldsymbol{k}_h)$, to be the negative of $E_e(\boldsymbol{k}_e)$:

$$E_h(\boldsymbol{k}_h) = -E_e(\boldsymbol{k}_e). \tag{4.21}$$

Since $\boldsymbol{k}_h = -\boldsymbol{k}_e$, we have

$$E_h(\boldsymbol{k}_h) = -E_e(-\boldsymbol{k}_h). \tag{4.22}$$

For every state with wave vector $\boldsymbol{k}$, there is another state of equal energy with wave vector $-\boldsymbol{k}$ (see Figs. 3.5 and 3.8). Consequently,

$$E_h(\boldsymbol{k}_h) = -E_e(\boldsymbol{k}_h). \tag{4.23}$$

As stated above, E_e is a decreasing function of its argument; hence, E_h is an increasing function of its argument. It characterizes a normal particle whose energy increases as its momentum increases.

In the case of a spherical parabolic valence band, we have

$$E_e(\boldsymbol{k}_e) = E_V + \frac{\hbar^2 k_e^2}{2m_e^*}, \tag{4.24}$$

where E_V is the energy of the valence band edge and the effective mass m_e^* is negative. The hole energy then becomes

$$E_h(\boldsymbol{k}_h) = -E_V - \frac{\hbar^2 k_h^2}{2m_e^*}. \tag{4.25}$$

It is convenient to rewrite this result as

$$E_h(\boldsymbol{k}_h) = -E_V + \frac{\hbar^2 k_h^2}{2m_h^*}, \tag{4.26}$$

where m_h^* is the effective mass of a hole. Clearly, one must have $m_h^* = -m_e^*$. Since m_e^* is negative, m_h^* is positive. Plots of $E_e(\boldsymbol{k}_e)$ and $E_h(\boldsymbol{k}_h)$ versus wave vector are presented in Fig. 4.3.

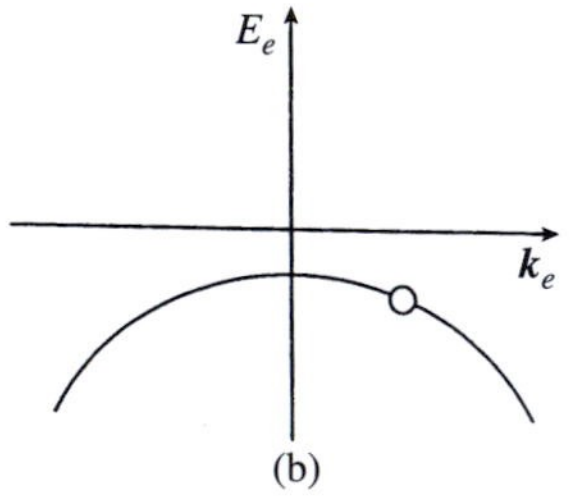

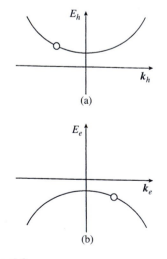

Fig. 4.3
(a) Hole energy E_h and (b) missing electron energy E_e as functions of wave vector. A pair of hole and missing electron is indicated by the circles.

The next question concerns the group velocity, v_{gh}, of a hole. It is given by

$$\boldsymbol{v}_{gh} = \frac{1}{\hbar}\nabla_{\boldsymbol{k}_h} E_h(\boldsymbol{k}_h), \tag{4.27}$$

while that of the missing electron is

$$\boldsymbol{v}_{ge} = \frac{1}{\hbar}\nabla_{\boldsymbol{k}_e} E_e(\boldsymbol{k}_e). \tag{4.28}$$

Replacing $\boldsymbol{k}_e$ by $-\boldsymbol{k}_h$,

$$\begin{aligned}\boldsymbol{v}_{ge} &= -\frac{1}{\hbar}\nabla_{\boldsymbol{k}_h} E_e(-\boldsymbol{k}_h) \\ &= -\frac{1}{\hbar}\nabla_{\boldsymbol{k}_h} E_e(\boldsymbol{k}_h). \end{aligned} \tag{4.29}$$

Making use of Eq. (4.23) leads to

$$\begin{aligned}\boldsymbol{v}_{ge} &= \frac{1}{\hbar}\nabla_{\boldsymbol{k}_h} E_h(\boldsymbol{k}_h) \\ &= \boldsymbol{v}_{gh}. \end{aligned} \tag{4.30}$$

Thus, the group velocities of the hole and missing electron are equal.

The last property of holes to be discussed is their charge e_h. The equation of motion, Eq. (4.18), applied to the missing electron is

$$\hbar\frac{d\boldsymbol{k}_e}{dt} = -e\boldsymbol{\mathcal{E}}. \tag{4.31}$$

Replacing $\boldsymbol{k}_e$ by $-\boldsymbol{k}_h$, one obtains the equation of motion of the hole

$$\hbar\frac{d\boldsymbol{k}_h}{dt} = e_h\boldsymbol{\mathcal{E}}, \tag{4.32}$$

where $e_h = +e$. Thus, the charge of a hole is positive. A hole near the top of an otherwise filled band can therefore be regarded as having both positive charge and positive effective mass. An electron near the bottom of a conduction band also has positive effective mass, but negative charge.

4.6 Experimental determination of effective masses: cyclotron resonance in semiconductors

Cyclotron resonance experiments in semiconductors exploit the classical motion of charged particles in a constant magnetic field. A particle with charge e_c and velocity $\boldsymbol{v}$ will experience a force under the influence of a magnetic field $\boldsymbol{\mathcal{B}}$ given by the Lorentz expression

$$\boldsymbol{F}_m = e_c\boldsymbol{v} \times \boldsymbol{\mathcal{B}}. \tag{4.33}$$

The magnitude of the force for $\boldsymbol{v}$ perpendicular to $\boldsymbol{\mathcal{B}}$ is

$$F_m = |e_c|v\mathcal{B}. \tag{4.34}$$

If the particle has mass m, it executes a uniform circular motion about the field with radius r, an acceleration v^2/r, and a centripetal force

$$F_c = m\frac{v^2}{r}. \tag{4.35}$$

Equating the two forces, we get for the radius of the circle $r = mv/|e_c|\mathcal{B}$. The period of the circular motion is $T = 2\pi r/v$. The angular frequency of the particle is called the **cyclotron frequency** and is given by

$$\omega_c = \frac{2\pi}{T} = \frac{v}{r} = \frac{|e_c|\mathcal{B}}{m}. \tag{4.36}$$

By analogy with a classical particle in free space, a charge carrier in a semiconductor with an effective mass m^* has a cyclotron frequency

$$\omega_c = \frac{e\mathcal{B}}{m^*}. \tag{4.37}$$

Subjected to an external radiation field, a carrier in a cyclotron orbit absorbs energy if the frequency of the radiation field approaches the cyclotron frequency. The electric field $\boldsymbol{\mathcal{E}}$ of the radiation should have a nonzero component in the plane of the cyclotron motion. The resonant absorption of energy at the cyclotron frequency is called cyclotron resonance. To determine the cyclotron resonance conditions let us consider the equation of motion for a free charge carrier subjected to an external electromagnetic field

$$m^*\left(\frac{d\boldsymbol{v}}{dt} + \frac{1}{\tau}\boldsymbol{v}\right) = e_c(\boldsymbol{\mathcal{E}} + \boldsymbol{v} \times \boldsymbol{\mathcal{B}}). \tag{4.38}$$

Here τ is the relaxation time of the carriers, $\boldsymbol{v}$ is the drift velocity under the influence of the electric field $\boldsymbol{\mathcal{E}}$, and the magnetic field of the electromagnetic radiation is neglected in comparison with the static magnetic field $\boldsymbol{\mathcal{B}}$.

For plane-polarized radiation with $\boldsymbol{\mathcal{E}} = (\mathcal{E}_x^{(0)}, 0, 0)e^{i\omega t}$, $\boldsymbol{v} = (v_x^{(0)}, v_y^{(0)}, 0)\times e^{-i\omega t}$ and $\boldsymbol{\mathcal{B}}$ taken in the z-direction, the equations of motion become

$$m^*\left(-i\omega + \frac{1}{\tau}\right)v_x = e_c\mathcal{E}_x + e_c v_y \mathcal{B} \tag{4.39a}$$

$$m^*\left(-i\omega + \frac{1}{\tau}\right)v_y = -e_c v_x \mathcal{B}. \tag{4.39b}$$

Solving for v_x and the current density component $j_x = n_c e_c v_x$, one calculates the complex conductivity according to

$$\sigma(\omega) = \frac{j_x}{\mathcal{E}_x} = \frac{n_c e_c v_x}{\mathcal{E}_x} = \sigma_0\left[\frac{1 + i\omega\tau}{1 + (\omega_c^2 - \omega^2)\tau^2 + 2i\omega\tau}\right], \tag{4.40}$$

where $\sigma_0 = n_c e_c^2 \tau / m^*$ is the static conductivity and n_c is the carrier concentration. The power absorbed is proportional to the real part of the conductivity given by

$$\mathrm{Re}\sigma(\omega) = \sigma_0\left[\frac{1 + (\omega_c^2 + \omega^2)\tau^2}{[1 + (\omega_c^2 - \omega^2)\tau^2]^2 + 4\omega^2\tau^2}\right]. \tag{4.41}$$

In Fig. 4.4 is shown the relative power absorption at constant frequency as a function of the static magnetic field intensity in units of ω_c/ω for two

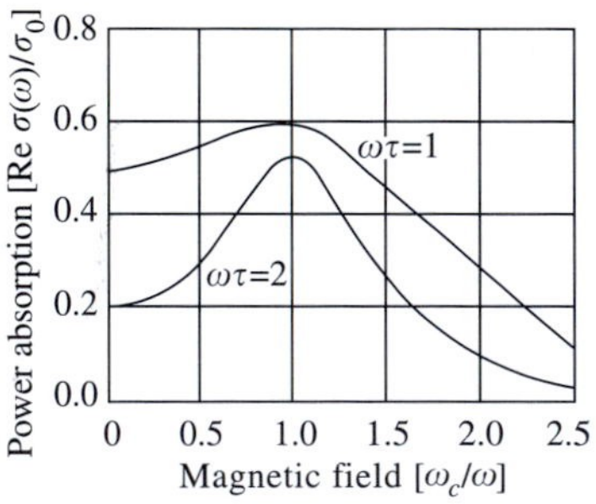

Fig. 4.4
Relative power absorption versus magnetic field in units of ω_c/ω for two different relaxation times in units of $\omega\tau$ (after Dresselhaus *et al.* 1955).

different relaxation times. This sets the conditions for observing sharp resonances.

The first experiments on cyclotron resonance of electrons and holes in Si and Ge were reported by G. Dresselhaus, A. F. Kip, and C. Kittel (1953, 1955), and by R. N. Dexter, H. J. Zeiger, and B. Lax (1954, 1956). The experimental conditions are determined by the effective mass of the carriers under investigation. For $m^*/m \simeq 0.1$, and $\omega = 1.5 \times 10^{11}$ rad s^{-1}, resonance occurs at $\mathcal{B} = 860\,G$ corresponding to $\omega = \omega_c$. The line width is determined by the collision relaxation time τ, which describes the effect of collisions of the carriers with imperfections. For cyclotron resonance to be observable it is necessary to have the mean free path of the charge carrier long enough to permit the carrier to sweep out one radian of arc around the cyclotron orbit between collisions. This condition is satisfied if $\omega_c\tau > 1$. To obtain long relaxation times it is advantageous to work with high-purity crystals at liquid helium temperatures. These requirements can be relaxed with the use of higher-frequency radiation and higher magnetic fields. The apparatus used in cyclotron resonance experiments is essentially the same as that for conventional paramagnetic resonance except that the geometry is modified so that the microwave electric field is perpendicular to the external applied magnetic field. Absorption of energy under cyclotron resonance conditions is determined by measuring the changes in the Q (quality factor) of a microwave cavity in which the sample is placed. The external magnetic field may be varied in order to obtain the dependence of the power absorption in the sample on the magnetic field. From the absorption spectrum this effective mass may be obtained. Typical cyclotron resonance spectra for Ge and Si are shown in Figs. 4.5 and 4.6, respectively.

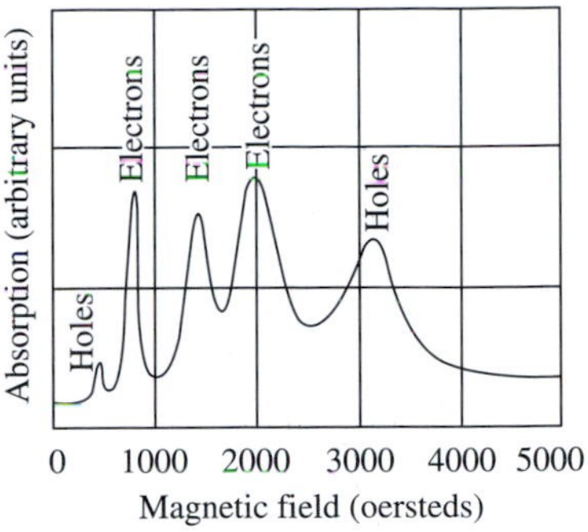

Fig. 4.5
Cyclotron resonance absorption versus magnetic field at 24000 Mc/s for Ge at 4 K (after Dresselhaus *et al.* 1955).

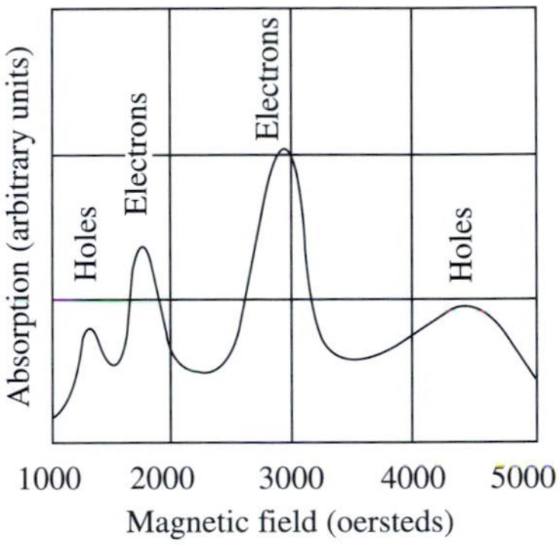

Fig. 4.6
Cyclotron resonance absorption versus magnetic field at 24000 Mc/s for Si at 4 K (after Dresselhaus *et al.* 1955).

4.6.1 Cyclotron resonance of conduction electrons in Ge and Si

The conduction bands near their extrema in Ge and Si consist of sets of ellipsoidal constant energy surfaces located at equivalent positions in $\boldsymbol{k}$-space at or near the zone boundary and are described by the equation

$$E(\boldsymbol{k}) = \hbar^2 \left(\frac{k_x^2 + k_y^2}{2m_t^*} + \frac{k_z^2}{2m_\ell^*} \right), \tag{4.42}$$

where m_t^* is the transverse effective mass and m_ℓ^* is the longitudinal effective mass.

To discuss the motion of a conduction electron in the presence of a uniform static magnetic field $\boldsymbol{\mathcal{B}}$, we consider the group velocity $\boldsymbol{v}_g = (1/\hbar)\nabla_{\boldsymbol{k}} E_{\boldsymbol{k}}$. Making use of Eq. (4.42), we find that

$$v_{gx} = \frac{\hbar k_x}{m_t^*},\ v_{gy} = \frac{\hbar k_y}{m_t^*},\ v_{gz} = \frac{\hbar k_z}{m_\ell^*}. \tag{4.43}$$

Combining the equation of motion given by Eq. (4.15) with the Lorentz force given by Eq. (4.33) and setting $\boldsymbol{v} = \boldsymbol{v}_g$ yields

$$\hbar \frac{d\boldsymbol{k}}{dt} = -e\boldsymbol{v}_g \times \boldsymbol{\mathcal{B}}. \tag{4.44}$$

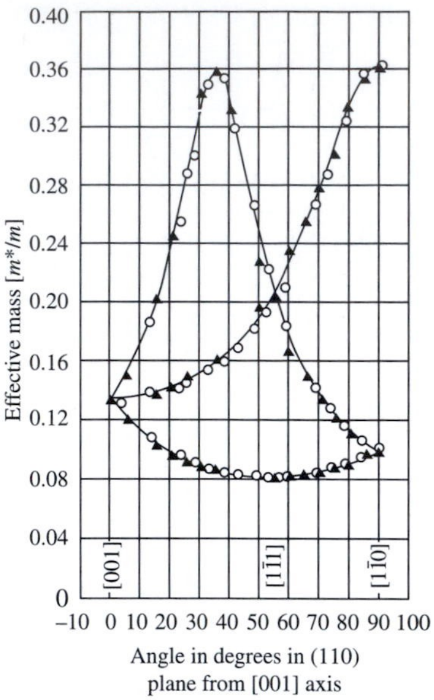

Fig. 4.7
Effective mass of conduction electrons in Ge at 4 K versus the angle between a magnetic field in the (110) plane and the [001] axis (after Dresselhaus *et al.* 1955).

Table 4.1 Electron and hole effective masses for silicon and germanium in units of free electron mass (after Kittel 1986)

	m_ℓ^*	m_t^*	$\bar{m}_{\ell h}^*$	$\bar{m}_{hh}^*$
Si	0.92	0.19	0.16	0.52
Ge	1.59	0.082	0.043	0.34

We take $\mathcal{B}$ to lie in the xz-plane, $\mathcal{B} = \mathcal{B}(\sin\theta, 0, \cos\theta)$. Eliminating $\boldsymbol{k}$ from Eq. (4.44) with the aid of Eq. (4.43), and assuming that $\boldsymbol{v}_g$ varies as $\exp(-i\omega t)$ result in the set of equations

$$i\omega v_{gx} - \omega_t \cos\theta v_{gy} = 0 \tag{4.45a}$$

$$\omega_t \cos\theta v_{gx} + i\omega v_{gy} - \omega_t \sin\theta v_{gz} = 0 \tag{4.45b}$$

$$\omega_\ell \sin\theta v_{gy} + i\omega v_{gz} = 0, \tag{4.45c}$$

where $\omega_t = e\mathcal{B}/m_t^*$ and $\omega_\ell = e\mathcal{B}/m_\ell^*$. Setting the determinant of the coefficients of v_{gx}, v_{gy}, v_{gz} equal to zero gives

$$\omega^2 = \omega_c^2 = \omega_t^2 \cos^2\theta + \omega_t\omega_\ell \sin^2\theta, \tag{4.46}$$

where ω_c is the cyclotron frequency. If we define the **cyclotron effective mass** m_c^* by

$$\omega_c = \frac{e\mathcal{B}}{m_c^*}, \tag{4.47}$$

we find from Eq. (4.46) that

$$\left(\frac{1}{m_c^*}\right)^2 = \frac{\cos^2\theta}{m_t^{*2}} + \frac{\sin^2\theta}{m_t^* m_\ell^*}. \tag{4.48}$$

The value of the cyclotron effective mass for conduction electrons in Ge at 4 K is plotted in Fig. 4.7 for various directions of the magnetic field in a (110) plane measured from the [001] direction. If one assumes that there is a set of equivalent constant energy surfaces oriented along the ⟨111⟩ directions in the Brillouin zone, the effective mass parameters can be derived from a fit of the theoretical expression of Eq. (4.48) to the experimental points. The results are presented in Table 4.1. In Fig. 4.8 are shown the results for the angular dependence of the effective mass of conduction electrons in Si. From these results one may conclude that there is a set of equivalent constant energy ellipsoids oriented along the ⟨100⟩ directions in the Brillouin zone. The values deduced for the mass parameters are listed in Table 4.1.

Exercise. Cyclotron resonance of electrons in silicon
The constant energy surfaces for the conduction band of Si are ellipsoids of revolution whose major axes are parallel to the six equivalent ⟨100⟩ directions in $\boldsymbol{k}$-space. Find the frequencies of all the cyclotron resonances with static magnetic field along the [100], [110], and [111] directions.

Answer. [100]: ω_t, $\sqrt{\omega_t\omega_\ell}$; [110]: $\sqrt{\omega_t\omega_\ell}$, $\sqrt{(\omega_t^2 + \omega_t\omega_\ell)/2}$; [111]: $\sqrt{(\omega_t^2 + 2\omega_t\omega_\ell)/3}$

4.6.2 Cyclotron resonance of holes in Ge and Si

The uppermost valence bands of Ge and Si are rather complicated as a consequence of their two-fold orbital degeneracy at $\boldsymbol{k} = 0$. They originate from $p_{3/2}$ atomic levels and are characterized by the energy-wave vector relationship given in Eq. (3.71):

$$E_{3/2}(\boldsymbol{k}) = Ak^2 \pm [B^2k^4 + C^2(k_x^2k_y^2 + k_y^2k_z^2 + k_z^2k_x^2)]^{\frac{1}{2}}. \qquad (3.71)$$

The minus sign in Eq. (3.71) gives the light hole band and the plus sign gives the heavy hole band. The warping terms with coefficient C^2 make the hole effective masses anisotropic. Warping effects can be included in an approximate way by means of the expression (Dresselhaus *et al.* 1955)

$$m_h^* = \frac{\hbar^2}{2}\frac{1}{|A \pm [B^2 + (C/2)^2]^{\frac{1}{2}}|} \times \left|1 \pm \frac{C^2(1 - 3\cos^2\theta)^2}{64[B^2 + (C/2)^2]^{\frac{1}{2}}\left\{A \pm [B^2 + (C/2)^2]^{\frac{1}{2}}\right\}}\right| \qquad (4.49)$$

which applies when the magnetic field lies in the (110) plane and makes an angle θ with the [100] direction. In Fig. 4.9 is shown a plot of experimental cyclotron resonance data for holes in Ge as a function of the angle θ. The constants A, B, C were evaluated by fitting Eq. (4.49) to the data. The values thus determined are given in Table 3.2. A similar procedure was applied to holes in Si with qualitatively similar results (Dresselhaus *et al.* 1955).

It is convenient in calculating various properties to have an isotropic approximation for the hole masses. Such an approximation is provided by the following expression for the valence band energies (Smith 1968):

$$E_v(\boldsymbol{k}) = \left\{A \pm [B^2 + (C^2/6)]^{\frac{1}{2}}\right\}k^2. \qquad (4.50)$$

The average light and heavy hole effective masses are specified by

$$\frac{m}{\bar{m}_{\ell h}^*} = \left\{-A + [B^2 + (C^2/6)]^{\frac{1}{2}}\right\}\left(\frac{2m}{\hbar^2}\right) \qquad (4.51a)$$

$$\frac{m}{\bar{m}_{hh}^*} = \left\{-A - [B^2 + (C^2/6)]^{\frac{1}{2}}\right\}\left(\frac{2m}{\hbar^2}\right). \qquad (4.51b)$$

The expression for the light hole effective mass is a rather good approximation, since the light hole band is nearly spherical, but that for the heavy hole is less satisfactory due to the significant warping of the heavy hole band. Using the values of A, B, and C from Table 3.2, one finds the values of $\bar{m}_{\ell h}^*$ and $\bar{m}_{hh}^*$ that are listed in Table 4.1.

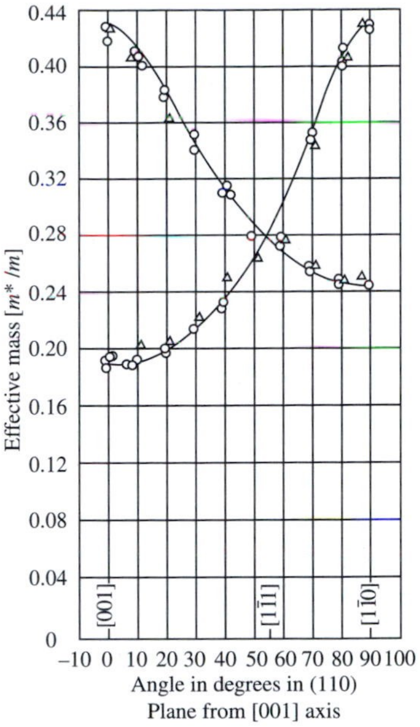

Fig. 4.8
Effective mass of conduction electrons in Si at 4 K versus the angle between a magnetic field in the (110) plane and the [001] axis (after Dresselhaus *et al.* 1955).

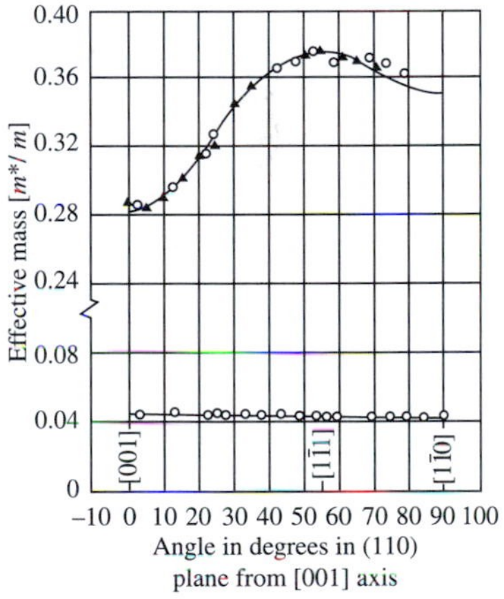

Fig. 4.9
Effective mass of holes in Ge at 4 K versus the angle between a magnetic field in the (110) plane and the [001] axis (after Dresselhaus *et al.* 1955).

Table 4.2 Conduction electron effective masses in units of the free electron mass for direct gap III–V and II–VI compounds (after Harrison 1980, Böer 1990)

III–V	m_e^*	II–VI	m_e^*
GaAs	0.066	ZnS_{cub}	0.34
GaSb	0.045	ZnS_{hex}	0.28
InP	0.077	ZnSe	0.16
InAs	0.024	ZnTe	0.12
InSb	0.014	CdS	0.21
		CdSe	0.11
		CdTe	0.096

4.6.3 Effective masses of carriers in compound semiconductors

Information concerning the effective masses of carriers in compound semiconductors has been obtained not only from cyclotron resonance, but also from other magneto-optical phenomena such as free carrier Faraday effect and interband magneto-absorption that are described in Chapter 11.

The conduction electron effective masses of III–V and II–VI compounds with direct gaps at $\boldsymbol{k} = 0$ are listed in Table 4.2. Comparing the effective masses with the corresponding band gaps in Table 3.2, we see that the effective mass decreases as the gap decreases. This behavior is consistent with the Kane model as expressed by Eq. (3.65). As pointed out in Chapter 3, GaP has an indirect gap with conduction band minima very close to the X-point on the Brillouin zone boundary in the $\langle 100 \rangle$ direction. The effective mass of conduction electrons is anisotropic with $m_\ell^* = 0.91\,m$ and $m_t^* = 0.25\,m$.

The effective masses of holes are determined by the Dresselhaus parameters A, B, C or the Luttinger parameters $\gamma_1, \gamma_2, \gamma_3$. As noted in Chapter 3 the effective mass depends on the direction of propagation of the Bloch wave due to warping of the constant energy surfaces. Average or sphericalized effective masses for light and heavy holes can be calculated using, for example, Eqs. (4.51). An alternative method (Lax and Mavroides 1955) uses the quantities

$$\bar{\gamma} = (2\gamma_2^2 + 2\gamma_3^2)^{\frac{1}{2}} \tag{4.52}$$

and

$$\gamma_a = \frac{6(\gamma_3^2 - \gamma_2^2)}{\bar{\gamma}(\gamma_1 - \bar{\gamma})} \tag{4.53}$$

to express the average light and heavy hole masses as

$$\begin{aligned} \bar{m}_{\ell h}^* &= (\gamma_1 + \bar{\gamma})^{-1} m \\ \bar{m}_{hh}^* &= (\gamma_1 - \bar{\gamma})^{-1} (1 + 0.05\gamma_a + 0.0164\gamma_a^2)^{2/3} m. \end{aligned} \tag{4.54}$$

Values of average effective hole masses are given in Table 4.3. Also shown are the spin–orbit split-off effective masses m_{so}^*.

Table 4.3 Average effective hole masses in units of the free electron mass for the valence bands in III–V and II–VI semiconductors (after Lawaetz 1971, Kittel 1986, Böer 1990)

III–V	$\bar{m}_{\ell h}^*$	$\bar{m}_{hh}^*$	m_{so}^*	II–VI	$\bar{m}_{\ell h}^*$	$\bar{m}_{hh}^*$	m_{so}^*
GaP	0.16	0.5	0.46	ZnS_{cub}	0.23	1.8	0.40
GaAs	0.08	0.5	0.15	ZnSe	0.15	1.4	0.30
GaSb	0.06	0.3	0.14	ZnTe	0.15	1.3	0.33
InP	0.08	0.4	0.15	CdTe	0.10	1.4	0.28
InAs	0.025	0.4	0.08				
InSb	0.021	0.4	0.11				

4.7 Experimental determination of carrier charge and concentration: Hall effect

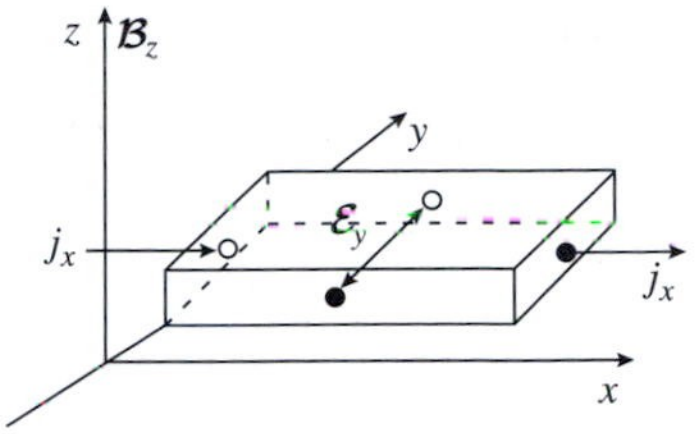

Fig. 4.10
Experimental configuration for the Hall effect.

Another phenomenon that provides useful information about current carriers is the **Hall effect**. Consider a bar of rectangular cross section in which a DC current flows in the presence of a uniform external magnetic field $\boldsymbol{\mathcal{B}}$ as shown in Fig. 4.10. For the steady state situation the equation of motion of a current carrier given by Eq. (4.38) reduces to

$$\boldsymbol{F} = \frac{m^* \boldsymbol{v}}{\tau} = e_c \boldsymbol{\mathcal{E}} + \boldsymbol{j}_1 \times \boldsymbol{\mathcal{B}}, \tag{4.55}$$

where $\boldsymbol{j}_1$ is the current density per unit carrier concentration, $e_c\boldsymbol{v}$, and $\boldsymbol{F}$ is the force on a carrier.

In the configuration shown in Fig. 4.10, let us first apply an electric field $\mathcal{E}_x$ in the x-direction parallel to the axis of the bar. A current will arise in the direction of the field with

$$j_{1x} = \left(\frac{\tau}{m^*}\right) e_c^2 \mathcal{E}_x. \tag{4.56}$$

Note that j_{1x} is in the same direction for both electrons and holes. Now let us turn on a magnetic field in the z-direction, $\mathcal{B}_z$. Carriers are deflected in the negative y-direction, whether electrons or holes, and accumulate on the front xz-face of the bar. At the same time an excess of carriers of opposite sign appears on the opposite wall creating an electric field component $\mathcal{E}_y$. Note that $\mathcal{E}_y$ has opposite signs for electrons and holes. In the steady state this transverse electric field, called the **Hall field**, gives rise to a force that just cancels the Lorentz force due to the magnetic field:

$$e_c \mathcal{E}_y = j_{1x} \mathcal{B}_z. \tag{4.57}$$

The total current density j_x is given by nj_{1x} for electrons and pj_{1x} for holes, where n and p are the concentrations of electrons and holes, respectively. The Hall field can now be expressed as

$$\mathcal{E}_y = R_H j_x \mathcal{B}_z, \tag{4.58}$$

where the **Hall coefficient** R_H is given by

$$R_H = -\frac{1}{ne} \tag{4.59a}$$

for electrons and by

$$R_H = \frac{1}{pe} \tag{4.59b}$$

for holes. Since $\mathcal{E}_y$, j_x, and $\mathcal{B}_z$ are experimentally measurable, an experimental value of R_H can be deduced and from it the concentration of free carriers. Furthermore, the sign of the Hall coefficient gives the sign of the

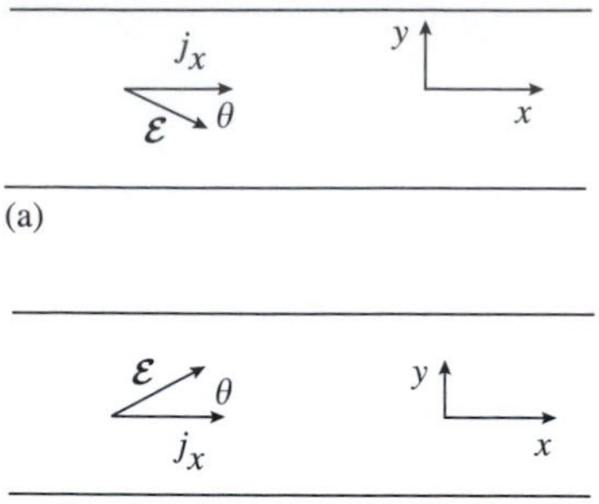

(a)

(b)

Fig. 4.11
Geometry of the Hall angle: (a) electrons, (b) holes.

carrier charge. The value of j_{1x} can now be obtained, and from a value of m^* determined by cyclotron resonance, one can calculate the scattering time τ with the aid of Eq. (4.53). It should be emphasized that Eqs. (4.56) are valid only for an energy-independent scattering time. The case of an energy-dependent τ is treated in Section 8.6.2.

The **Hall angle** θ is defined as the angle between the electric field $\boldsymbol{\mathcal{E}}$ and the current density $\boldsymbol{j}$ in the steady state. Since $\boldsymbol{j}$ is in the x-direction, θ is the angle between $\boldsymbol{\mathcal{E}}$ and the x-direction and is therefore specified by

$$\tan\theta = \frac{\mathcal{E}_y}{\mathcal{E}_x}. \tag{4.60}$$

The geometry is shown in Fig. 4.11 for electrons and holes. If we substitute Eq. (4.53) into Eq. (4.54), we obtain

$$\mathcal{E}_y = \frac{e_c\tau}{m^*}\mathcal{E}_x\mathcal{B}_y. \tag{4.61}$$

The quantity $e_c\tau/m^*$ is the proportionality constant between the velocity component v_x and the electric field component $\mathcal{E}_x$. We can write this relation as

$$v_x = \frac{e_c}{e}\mu\mathcal{E}_x, \tag{4.62}$$

where the **conductivity mobility** μ is $e\tau/m^*$ and is always positive. The equation specifying the Hall angle now takes the form

$$\tan\theta = \frac{e_c}{e}\mu\mathcal{B}_y. \tag{4.63}$$

Coupling an experimentally measured value of $\mathcal{E}_y$ with those of $\mathcal{E}_x$ and $\mathcal{B}_y$ enables one to determine both θ and μ.

> **Comment**. Hall effect measurements are the most frequently used procedure to obtain two of the basic characteristics of a semiconductor: carrier concentration and mobility. In order to simplify the analysis it is general practice to use a sample in the shape of a long parallelepiped. Consider as an example a bar of silicon whose dimensions are: length, 0.01 m; height, 0.001 m and width 0.001 m. Between the contacts at the ends of the bar a current $I_x = 10\,\text{mA}$ flows. A magnetic field $\mathcal{B}$ of 0.1 T is applied. The voltage applied between the ends of the bar is 5 V, and the potential difference measured in the transverse direction is 0.001 V. The electric field $\mathcal{E}_x$ is then 500 V/m, $\mathcal{E}_y$ is 1 V/m, and the current density j_x is $10^4\,\text{A/m}^2$. The Hall coefficient R_H is $10^{-3}\,\text{m}^3/\text{C}$ (these dimensions for R_H follow from Eqs. (4.59). The Hall angle is given by $\tan\theta = 1/500 = 0.002$. The mobility $\mu = 0.02\,\text{m}^2/(\text{V/s}) = 200\,\text{cm}^2/(\text{V/s})$. Since the Hall coefficient is positive, the carriers must be holes. The hole concentration p is $6.2\times10^{21}\,\text{m}^{-3} = 6.2\times10^{15}\,\text{cm}^{-3}$.

Problems

1. Consider a spherical nonparabolic energy band with the energy as a function of wave vector given by

$$E_k = \frac{\hbar^2 k^2}{2m_0^*} - \frac{\hbar^2 a^2 k^4}{4\pi^2 m_0^*}, \tag{4.64}$$

where m_0^* is the effective mass at $k = 0$ and a is the lattice constant. Calculate the group velocity v_g as a function of k and plot v_g versus k over the range $0 \leq k \leq \pi/a$ for $m_0^* = 0.1\,m$ and $a = 4\,\text{Å}$.

2. Using the same expression for E_k as in Problem 1, calculate the effective mass m^* as a function of k and plot m^* versus k over the range $0 \leq k \leq \pi/a$.
3. Consider an electron moving in an energy band under the influence of an applied electric field. Assuming no scattering of the electron, calculate the time required for the electron to cross the first Brillouin zone if $\mathcal{E} = 10\,\text{V/m}$ and $a = 4\,\text{Å}$. The inverse of this time is the frequency of a **Bloch oscillation**.
4. Compare the effective masses for light and heavy holes in Si and in Ge that result from Eqs. (4.49) and (4.51). Take the magnetic field to be in the [110] direction ($\theta = 90°$). Also compare with the experimental values of $m_{lh}^* = 0.16\,m$, $m_{hh}^* = 0.53\,m$ for Si and $m_{lh}^* = 0.044\,m$, $m_{hh}^* = 0.35\,m$ for Ge.

References

K. W. Böer, *Survey of Semiconductor Physics* (Van Nostrand Reinhold, New York, 1990).

R. N. Dexter, H. J. Zeiger, and B. Lax, *Phys. Rev.* **95**, 557 (1954); **104**, 637 (1956).

G. Dresselhaus, A. F. Kip, and C. Kittel, *Phys. Rev.* **92**, 827 (1953); **98**, 368 (1955).

W. A. Harrison, *Electronic Structure and the Properties of Solids* (W. H. Freeman, San Francisco, 1980).

C. Kittel, *Introduction to Solid State Physics*, Sixth edition (John Wiley, New York, 1986).

P. Lawaetz, *Phys. Rev.* B4, 3460 (1971).

B. Lax and J. G. Mavroides, *Phys. Rev.* **100**, 1650 (1955).

R. A. Smith, *Semiconductors* (Cambridge University Press, Cambridge, 1968).

Electronic effects of impurities

5

Key ideas

Donor impurities in *n-type semiconductors* provide free electrons to the conduction band and positively charged *donor ions*. *Acceptor impurities* in *p-type semiconductors* provide free holes to the valence band and negatively charged *acceptor ions*.

Shallow impurities have ionization energies that are small compared to the fundamental gap. Their energy levels and eigenfunctions are well described by *effective mass theory*.

The *anisotropic effective mass* of the conduction band in Si and Ge causes a splitting of energy levels associated with *p*-like hydrogenic states.

A set of donor levels is associated with each conduction band minimum: six minima for Si and four for Ge.

The wave vector dependence of the dielectric constant leads to coupling between impurity states associated with different extrema of an energy band and a splitting of degeneracies.

In materials such as InSb, the small effective mass and large dielectric constant lead to a large *effective Bohr radius* of *donor levels*.

Degeneracy and *warping* of the valence bands lead to complicated structure of *acceptor levels*.

Deep level centers have energy levels near the midpoint of the fundamental gap. The ground state ionization energy depends significantly on the nature of the impurity or defect. *Central cell corrections* are required in the impurity potential.

At sufficiently high impurity concentrations, the wave functions of neighboring impurities overlap sufficiently to produce an *impurity band*.

Impurities in semiconductors

Impurities can have a drastic effect on the properties of semiconductors, particularly the electrical properties. They can provide additional current carriers, such as conduction electrons or valence holes, and scattering centers which perturb the motion of the carriers. Impurities are conveniently classified according to the column of the periodic table in which they occur.

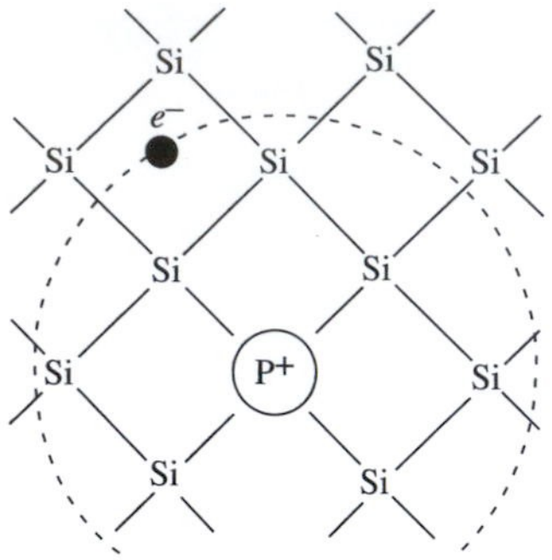

Fig. 5.1
Phosphorus atom introduced substitutionally into an Si lattice to form a donor center P^+ ion to which an extra electron is weakly bound.

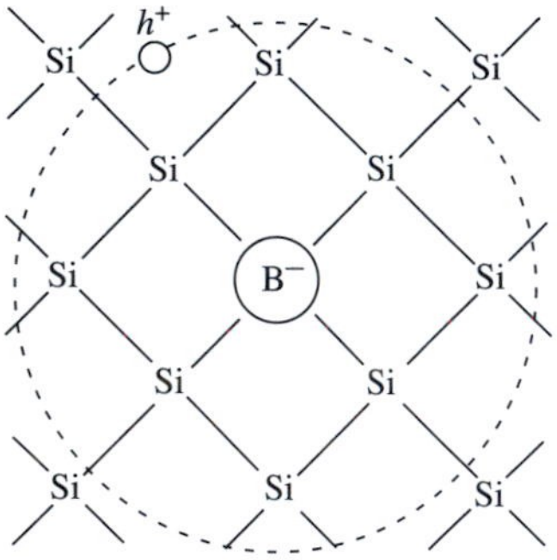

Fig. 5.2
Boron atom introduced substitutionally into a Si lattice to form an acceptor center B^- ion to which a hole is weakly bound.

5.1 Qualitative aspects of impurities

Let us consider first a group IV semiconductor such as Si. If a group V element such as P is added to Si, the P atoms can enter substitutionally into the crystal by replacing Si atoms in the diamond structure. Phosphorus contains five electrons beyond closed shells, of which four can be paired up with electrons on four neighboring Si atoms to form four electron-pair bonds. The fifth electron is bound to the P^+ ion at sufficiently low temperatures, but is thermally excited to the conduction band at higher temperatures and is therefore unbound to the P^+ ion. The situation is illustrated schematically in Fig. 5.1.

Phosphorus is called a **donor impurity**, because it "donates" electrons to the conduction band. Silicon doped with P is known as **n-type** Si, because the current carriers provided by P are negatively charged conduction electrons. Other group V elements such as As and Sb behave very much like P.

Let us now consider an impurity from group III of the periodic table such as B, which can also enter Si substitutionally. Boron contains three electrons beyond closed shells and therefore lacks one electron needed to form electron pair bonds with four nearest neighbor Si atoms. The missing electron on the B atom can be regarded as a hole. At sufficiently low temperatures, the hole is bound to the B atom. At high temperatures an electron from a far away electron pair bond can transfer to the B atom and, together with the other three electrons, form four electron-pair bonds between the B atom and its four nearest neighbor Si atoms. The hole is thereby transferred from the B atom to the far away electron pair bond and is no longer bound to the B atom. The B atom in this process has become negatively charged. The situation is illustrated schematically in Fig. 5.2.

Boron is referred to as an **acceptor** impurity, because it "accepts" electrons from elsewhere in the crystal. Silicon doped with B is known as **p-type** Si, because the current carriers provided by B are positively charged holes. Other group III elements such as Al and Ga behave much the same way as B.

Other impurities of interest are group VI elements such as S and Se, which behave as double donors and group II elements such as Zn and Cd, which behave as double acceptors. If we consider III–V semiconductors, replacing a group V atom by a group VI atom leads to a donor impurity, whereas replacing a group III atom by a group II atom yields an acceptor impurity.

5.2 Effective mass theory

A common feature of the impurity systems described in the preceding section is that one deals with a current carrier of a certain charge moving in the field of an impurity ion of the opposite charge. The impurity system thus corresponds to a hydrogen-like atom.

A problem of great importance is to determine the wave functions and energy eigenvalues of the impurity atom, properly taking into account the energy band structure and dielectric screening properties of the host crystal. Certain aspects simplify the problem:

1. The impurity potential is weak, because the impurity ion is strongly screened by the high dielectric constant of a typical semiconductor crystal.

2. The weak impurity potential is slowly varying over a lattice constant.
3. The impurity state is very spread out, and only wave vectors $\boldsymbol{k}$ near the band extremum are important in the Fourier decomposition of the impurity wave function.

A systematic development of the theory that incorporates these simplifications in a consistent way has been carried out by Luttinger and Kohn. The result (Luttinger and Kohn 1955) is the **effective mass theory** of impurity states. The starting point of the theory is the Schrödinger equation satisfied by the impurity wave function $\psi_i(\boldsymbol{r})$,

$$\left[-\frac{\hbar^2}{2m}\nabla^2 + V_i(\boldsymbol{r}) + V_p(\boldsymbol{r})\right]\psi_i(\boldsymbol{r}) = E_i\psi_i(\boldsymbol{r}), \tag{5.1}$$

where $V_i(\boldsymbol{r})$ is the impurity potential, $V_p(\boldsymbol{r})$ is the periodic potential of the host crystal, and E_i is the energy eigenvalue. Let us first consider donor impurities and a conduction band with minimum at $\boldsymbol{k} = 0$. Luttinger and Kohn expanded $\psi_i(\boldsymbol{r})$, not in terms of Bloch functions $\psi_{n\boldsymbol{k}}(\boldsymbol{r})$, but in terms of the functions

$$\chi_{n\boldsymbol{k}}(\boldsymbol{r}) = e^{i\boldsymbol{k}\cdot\boldsymbol{r}}u_{n0}(\boldsymbol{r}) \tag{5.2}$$

which are known as **Luttinger–Kohn functions**. These functions form a complete set. They enable one to use Fourier transform techniques unencumbered by the $\boldsymbol{k}$-dependence of $u_{n\boldsymbol{k}}(\boldsymbol{r})$.

By substituting the expansion

$$\psi_i(\boldsymbol{r}) = \sum_n \sum_{\boldsymbol{k}} A_{n\boldsymbol{k}}\chi_{n\boldsymbol{k}}(\boldsymbol{r}) \tag{5.3}$$

into the Schrödinger equation, Eq. (5.1), and making use of the simplifications (1)–(3) listed above, Luttinger and Kohn were able to show with the aid of canonical transformations that the impurity wave function can be written to a good approximation as

$$\psi_i(\boldsymbol{r}) \simeq F_c(\boldsymbol{r})u_{c0}(\boldsymbol{r}). \tag{5.4}$$

The function $F_c(\boldsymbol{r})$ is known as the **effective mass wave function** or **envelope function**. It satisfies the **effective mass equation**,

$$\{E_c(-i\boldsymbol{\nabla}) + V_i(\boldsymbol{r})\}F_c(\boldsymbol{r}) = E_{ci}F_c(\boldsymbol{r}), \tag{5.5}$$

where $E_c(-i\boldsymbol{\nabla})$ is obtained by replacing $\boldsymbol{k}$ in the band energy $E_{c\boldsymbol{k}}$ by the operator $-i\boldsymbol{\nabla}$. The impurity potential can be well approximated by the Coulomb potential

$$V_i(\boldsymbol{r}) = -\frac{Ze^2}{4\pi\epsilon_0\epsilon r}, \tag{5.6}$$

where Ze is the charge of the impurity ion and ϵ is the static dielectric constant of the semiconductor. The functions $F_c(\boldsymbol{r})$ and $\psi_i(\boldsymbol{r})$ are shown schematically in Fig. 5.3.

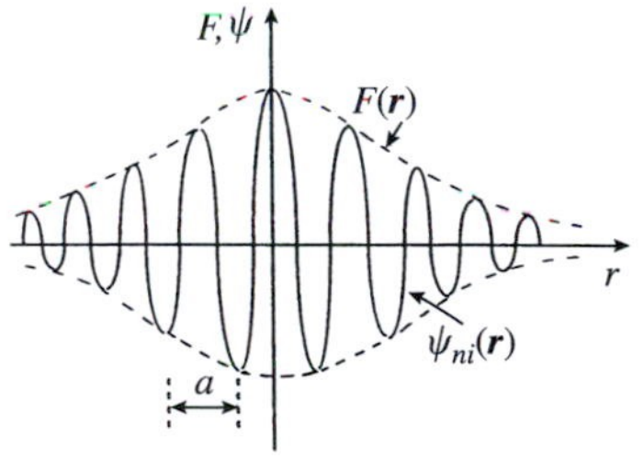

Fig. 5.3
Representation of the effective mass wave function or envelope function $F_c(\boldsymbol{r})$ and the impurity wave function $\psi_i(\boldsymbol{r})$. a is the lattice constant.

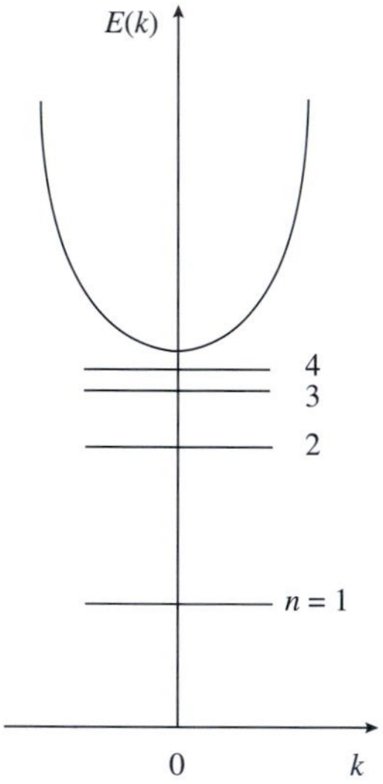

Fig. 5.4
Hydrogen-like energy spectrum of a shallow donor impurity.

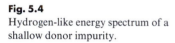

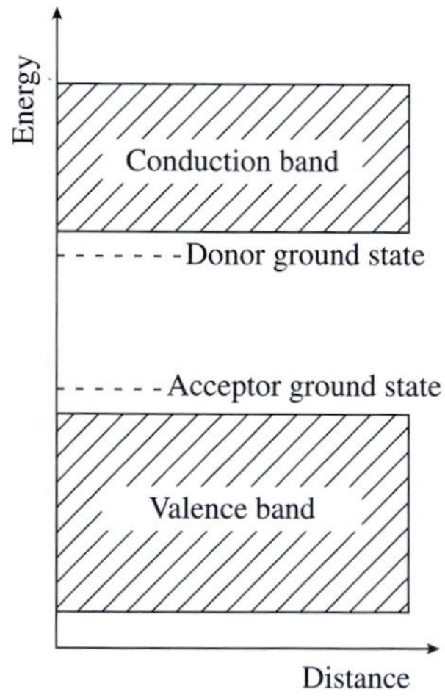

Fig. 5.5
Schematic representation of the position of the impurity ground states: Donors near the conduction band and acceptors near the valence band.

Example 5.1: Impurity states for a spherical parabolic band
Find the impurity eigenstates associated with a spherical parabolic band and a Coulomb potential.
Solution. The energy of a spherical parabolic band with effective mass m^* can be written as

$$E_{\mathbf{k}} = \frac{\hbar^2 k^2}{2m^*}, \tag{5.7}$$

if the band minimum is at $\mathbf{k} = 0$. The effective mass equation then takes the form

$$\left[-\frac{\hbar^2\nabla^2}{2m^*} - \frac{Ze^2}{4\pi\epsilon_0\epsilon r}\right] F_c(\mathbf{r}) = E_{ci} F_c(\mathbf{r}) \tag{5.8}$$

which is the Schrödinger equation for a hydrogen-like atom. The eigenfunctions are characterized by an effective Bohr radius a_B^*, which is a measure of the orbit size and is given by

$$a_B^* = \frac{4\pi\epsilon_0\hbar^2\epsilon}{m^* Ze^2}. \tag{5.9}$$

The binding energy of the impurity ground state is the effective Rydberg Ry* given by

$$\mathrm{Ry}^* = \frac{\hbar^2}{2m^* a_B^{*2}} = \frac{m^* Z^2 e^4}{2(4\pi\epsilon_0)^2\hbar^2\epsilon^2}. \tag{5.10}$$

The energy eigenvalues are those of a hydrogen-like atom given by

$$E(n_H) = -\frac{\mathrm{Ry}^*}{n_H^2}, \tag{5.11}$$

where n_H is the principal quantum number with the values 1, 2, 3, ... A schematic representation of the energy spectrum is given in Fig. 5.4. This example is applicable to donor impurities in GaAs.

A diagram showing the impurity ground state relative to the valence and conduction bands for a donor impurity is given in Fig. 5.5. We see that the donor level lies just below the conduction band edge. For an acceptor impurity the ground state lies just above the valence band edge. Impurity levels that lie close to the corresponding band edges are called **shallow levels**.

Exercise. Calculate the effective Bohr radius a_B^* and effective Rydberg Ry* for a donor impurity in GaAs with $m^* = 0.07m$, $Z = 1$, and $\epsilon = 12$.
Answer. 91 Å, 6.6 meV

The results of this exercise demonstrate that the effective Bohr radius is more than a factor of ten greater than the lattice constant (5.65 Å) of GaAs and that the ionization energy of the impurity electron is less than 1% of the energy gap (1.52 eV) of GaAs. The basic assumptions of the effective mass theory, that the impurity state is very spreadout and that the impurity potential is weak, are therefore confirmed.

5.3 Donor impurities in Si and Ge

5.3.1 Effects of ellipsoidal constant energy surfaces

A complication arises when one deals with donor impurities in Si and Ge, namely, the constant energy surfaces of the conduction band are not spheres, but ellipsoids of revolution as specified by Eq. (3.76). Consider a single ellipsoid centered at $\boldsymbol{k}_0$. The effective mass equation in the parabolic approximation then takes the form

$$\left[\sum_{\alpha=x,y,z} \frac{\hbar^2(k_{0\alpha} - i\nabla_\alpha)^2}{2m_\alpha^*} - \frac{Ze^2}{4\pi\epsilon_0\epsilon r}\right] F_c(\boldsymbol{r}) = E_{ci} F_c(\boldsymbol{r}). \quad (5.12)$$

The anisotropy of the kinetic energy operator renders it impossible to solve this equation exactly. Recourse must then be made to approximation methods such as the variational method (Faulkner 1969). Faulkner has carried out detailed variational calculations for donor impurities in Si and Ge. He determined the effective Rydberg to be 31.3 meV for Si and 9.8 meV for Ge.

The anisotropy of the kinetic energy operator in the effective mass equation given by Eq. (5.12) causes a splitting of excited states such as p-states. The $m = 0$ state is split off from the $m = \pm 1$ states. This splitting has been observed experimentally in Si (Picus *et al.* 1956, Aggarwal and Ramdas 1965) and in Ge (Reuszer and Fisher 1964) by means of infrared spectroscopy.

We have noted in Chapter 3 that the conduction band of Si has six equivalent minima or "valleys" along [100] directions in $\boldsymbol{k}$-space. Within the present theory, there is a set of donor levels associated with each valley. A given donor level therefore possesses a six-fold degeneracy due to the multiple valleys. In the case of Ge, there are four equivalent valleys located along [111] directions and centered at the L point on the Brillouin zone boundary. Each donor level has a four-fold degeneracy due to multiple valleys. A more sophisticated theory shows, however, that the multiple valley degeneracies are split by interactions between valleys, as will be discussed in the following section.

5.3.2 Valley–orbit interaction

The effective mass theory of impurity states presented thus far assumes that the dielectric constant is independent of wave vector. In fact, the dielectric constant is not a constant, but is a function of wave vector $\boldsymbol{k}$ and frequency ω. For present purposes the frequency dependence can be ignored, but the wave vector dependence cannot. The latter dependence leads to coupling between impurity states associated with different extrema of an energy band, as in the case of donor impurities in Si or Ge.

Let us consider a semiconductor with r equivalent minima in the conduction band located at wave vectors $\boldsymbol{k}_{0\mu}$, $\mu = 1, 2, 3, \ldots, r$. If $\epsilon(\boldsymbol{k})$ is the wave-vector-dependent dielectric function, the effective mass equation given by Eq. (5.12) must be generalized to a set of coupled differential equations

$$\sum_{\alpha=x,y,z} \frac{\hbar^2(k_{0\mu\alpha} - i\nabla_\alpha)^2}{2m^*_{\mu\alpha}} F_{c\mu}(\boldsymbol{r}) - \sum_{\nu=1}^{r} \frac{Ze^2}{4\pi\epsilon_0\epsilon(\boldsymbol{k}_{0\nu} - \boldsymbol{k}_{0\mu})r} F_{c\nu}(\boldsymbol{r}) = E_{ci}F_{c\mu}(\boldsymbol{r}), \tag{5.13}$$

where $m^*_{\mu\alpha}$ is the $\alpha\alpha$ component of the diagonal effective mass tensor of the μ-th minimum and $F_{c\mu}(\boldsymbol{r})$ is the effective mass function of the μ-th minimum. The terms in Eq. (5.13) with $\nu \neq \mu$ can be treated as perturbations with $\epsilon(\boldsymbol{k})$ taken from theoretical results for Si and Ge (Nara 1965). Using variational solutions to Eq. (5.13), one finds (Baldereschi 1970) that the six-fold degenerate ground state for donors in Si is split into a singlet (A_1) ground state, a doublet (E) excited state, and a triplet (T_2) excited state with relative energies given by

$$E(T_2) - E(A_1) = \Delta_1 = 10.6\,\text{meV}$$
$$E(E) - E(T_2) = \Delta_2 = 1.1\,\text{meV}.$$

The experimental values of the splitting parameters Δ_1 and Δ_2 derived from infrared spectroscopy (Aggarwal and Ramdas 1965) are $\Delta_1 = 11.85$, 21.15, and 9.94 meV and $\Delta_2 = 1.35$, 1.42, and 2.50 meV for P, As, and Sb impurities, respectively. The splittings of excited states derived from $2s$, $2p$, and higher hydrogenic states are very small, because the envelope functions are very spread out in real space and require only very small $\boldsymbol{k}$ in their Fourier decomposition.

For Ge, the four-fold degenerate ground state is split into a singlet (A_1) ground state and a triplet (T_2) excited state with calculated splitting

$$E(T_2) - E(A_1) = \Delta_1 = 0.6\,\text{meV}.$$

The experimental values of Δ_1 are 2.83, 4.23 and 0.32 meV for P, As, and Sb impurities, respectively.

The ionization energies for the lower lying states of donor impurities in Si and Ge are listed in Table 5.1. The agreement between theory and experiment for the excited states is very good, but is significantly less satisfactory for the $1s$ states. For the latter there is a clear dependence of the ionization energy on the nature of the impurity, in contradition with the theory so far developed.

5.4 Donor impurities in III–V semiconductors

III–V semiconductors such as GaAs and InSb have a single conduction band minimum at $\boldsymbol{k} = 0$. The conduction band to a good approximation is

Table 5.1 Ionization energies (meV) for donors in Si and Ge (after Reuszer and Fisher 1964, Aggarwal and Ramdas 1965, Faulkner 1969, Baldereschi 1970, Bassani *et al.* 1974)

System	State	Theory	P	As	Sb
Si	$1s(A_1)$	40.65	45.5	53.7	42.7
	$1s(T_2)$	30.05	33.9	32.6	32.9
	$1s(E)$	28.95	32.6	31.2	30.5
	$2p_0$	11.51	11.5	11.5	11.6
	$2p_\pm$	6.40	6.4	6.3	6.3
	$3p_0$	5.48	5.5	5.5	5.3
	$3p_\pm$	3.12	3.1	3.1	3.0
Ge	$1s(A_1)$	10.26	12.76	14.04	10.19
	$1s(T_2)$	9.66	9.93	9.81	9.87
	$2p_0$	4.74	4.74	4.73	4.74
	$2p_\pm$	1.73	1.73	1.73	1.73
	$3p_0$	2.56	2.56	2.56	2.57
	$3p_\pm$	1.03	1.05	1.02	1.03

spherical and parabolic. The donor states are therefore described by the isotropic effective mass equation, Eq. (5.8). Similar considerations apply to donor impurities in II–VI semiconductors. In the case of InSb the conduction band effective mass is so small ($m^* \simeq 0.014\,m$) and the dielectric constant is so large ($\epsilon \simeq 18$) that the effective Bohr radius is exceptionally large ($a_B^* \simeq 1000\,a_B$) and the ionization energy is exceptionally small ($E_I \simeq 0.59\,\text{meV}$).

Other III–V semiconductors such as GaP have multiple conduction band minima away from $\boldsymbol{k} = 0$. The donor impurity states are described by the effective mass equation given by Eq. (5.13) and are qualitatively similar to those of Si.

A tabulation of experimental and theoretical binding energies of shallow donors in III–V and II–VI semiconductors is given in Table 5.2. Effective mass theory is seen to give rather good results for shallow donors replacing either the cation or the anion of the host crystal.

Table 5.2 Binding energies (meV) for donors in III–V and II–VI semiconductors (after Watts 1977)

System	Donor	Theory	Experiment
GaAs	Si	5.72	5.84
	S	5.72	5.87
InSb	Te	0.59	0.6
CdTe	In	11.6	14
ZnSe	Al	25.7	26.3
	Cl	25.7	26.9

5.5 Acceptor impurities

The effective mass theory of acceptor impurity states in Si and Ge is complicated by the degeneracy of the valence band at $\boldsymbol{k} = 0$. The warping of the constant energy surfaces as described by Eq. (3.71) further complicates the problem. The effective mass formalism of Luttinger and Kohn, nevertheless, is able to handle these difficulties in a straightforward manner (Luttinger and Kohn 1955). The effective mass equation becomes a matrix equation obtained by first making the replacement $k_\alpha \to -i\nabla_\alpha$ in the matrix of Eq. (3.69) and adding the impurity potential to the diagonal elements. This matrix then operates on the column vector of components $F_\alpha(\boldsymbol{r})$ of the effective mass wave function, and the result is set equal to zero:

$$\begin{bmatrix} -L\frac{\partial^2}{\partial x^2} - M\left(\frac{\partial^2}{\partial y^2}+\frac{\partial^2}{\partial z^2}\right) + V_i - E_{vi} & -N\frac{\partial^2}{\partial x\partial y} & -N\frac{\partial^2}{\partial x\partial z} \\ -N\frac{\partial^2}{\partial x\partial y} & -L\frac{\partial^2}{\partial y^2} - M\left(\frac{\partial^2}{\partial x^2}+\frac{\partial^2}{\partial z^2}\right) + V_i - E_{vi} & -N\frac{\partial^2}{\partial y\partial z} \\ -N\frac{\partial^2}{\partial x\partial z} & -N\frac{\partial^2}{\partial y\partial z} & -L\frac{\partial^2}{\partial z^2} - M\left(\frac{\partial^2}{\partial x^2}+\frac{p^2}{\partial y^2}\right) + V_i - E_{vi} \end{bmatrix} \begin{bmatrix} F_1(\boldsymbol{r}) \\ F_2(\boldsymbol{r}) \\ F_3(\boldsymbol{r}) \end{bmatrix} = 0, \tag{5.14}$$

where $V_i = -Ze^2/4\pi\epsilon_0\epsilon r$.

Equation (5.14) represents a set of three coupled partial differential equations in the three functions $F_1(\boldsymbol{r})$, $F_2(\boldsymbol{r})$, and $F_3(\boldsymbol{r})$. The spin–orbit interaction at this point has not been included. To include it, one applies the procedure described above to the matrix of Eq. (3.70). The result is a set of six coupled partial differential equations. Neither set can be solved analytically. In those cases where the acceptor binding energy is small compared to the spin–orbit splitting of the valence band, the set of six coupled equations can be decomposed into a set of four coupled equations and a set of two coupled equations, which determine the acceptor states associated with the light and heavy hole bands and with the split-off band, respectively. The problem can be simplified by separating the Hamiltonian matrix into a spherically symmetric part and a cubic correction (Baldereschi and Lipari 1973).

Table 5.3 Experimental and theoretical ground state ionization energies (meV) for Group III acceptors in Ge (after Mendelson and James 1964)

Acceptor	Experimental	Theoretical
B	10.3	9.3
Ga	10.8	9.3
In	11.4	9.3
Tl	13.0	9.3

Variational methods have been applied to the solution of the acceptor impurity problem (Kohn and Schechter 1955, Mendelson and James 1964). The ground state is four-fold degenerate, has even parity, and belongs to the irreducible representation Γ_8, and has no radial nodes. It is conveniently represented by the symbol $(8+0)$. Excited states which are p-like belong to the Γ_6, Γ_7, or Γ_8 representations and have odd parity. The number of radial nodes varies. For a given symmetry type and a given number of nodes, there may be more than one eigenstate. These eigenstates are distinguished by adding another index, as in $(8-11)$ and $(8-12)$.

In Table 5.3 are given experimental and theoretical ionization energies for Group III acceptors in Ge. There are clearly significant differences in the ionization energies of the various acceptors, in contradiction with the effective mass result. The energies of the low-lying bound states of In in Ge are presented in Fig. 5.6 and identified with the corresponding hydrogenic levels. The parity of the levels according to effective mass theory should lead to selection rules for optical transitions. In fact, the tetrahedral symmetry of a substitutional site lacks a center of inversion, and transitions from the ground state to all the indicated excited states are allowed (Dickey and Dimmock 1967). Effective mass theory smooths out the host

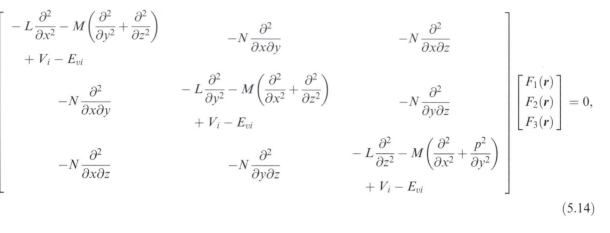

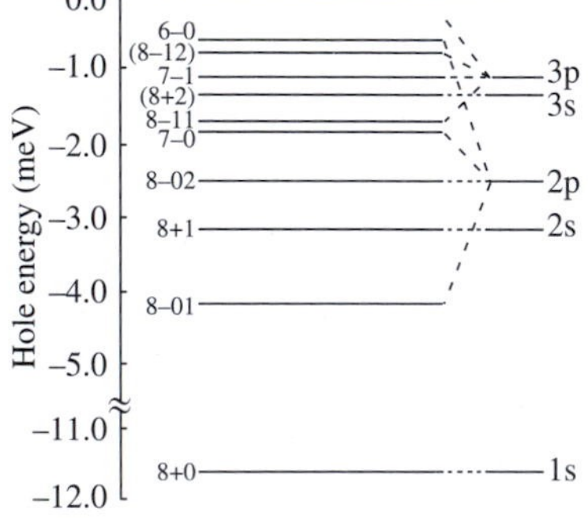

Fig. 5.6
Low-lying bound state energies of In in Ge (after Dickey and Dimmock 1967).

lattice into a dielectric continuum with the loss of the point group symmetry of individual lattice sites.

Acceptor impurities in III–V and II–VI semiconductors can be treated by the same theoretical approaches employed for Si and Ge. Effective mass theory gives a reasonable picture of the experimental results for the binding energies for a number of cases as shown in Table 5.4.

Table 5.4 Binding energies (meV) for acceptors in III–V and II–VI semiconductors (after Baldereschi and Lipari 1973)

System	Accepter	Theory	Experiment
GaP	Zn	47.5	64.0
GaAs	Zn	25.6	31.0
InSb	Cd	8.6	~10
ZnSe	Li	110.1	114

5.6 Central cell corrections and deep levels

We have seen in the preceding sections of this chapter that effective mass theory predicts the same ground state ionization energy for all impurities of a given type (donor or acceptor) in a given semiconductor, since the ionization energy depends only on the carrier effective mass and the dielectric constant of the host crystal. The latter properties are little affected by impurities at low concentration.

It is clear from the results presented in Tables 5.1 and 5.2 that the ground state ionization energy depends significantly on the nature of the impurity, in contradiction to the effective mass theory prediction. The reason for this discrepancy is that the impurity potential within the unit cell occupied by the impurity cannot be adequately described by a Coulomb potential screened by the bulk dielectric constant of the host crystal. The impurity potential tends to have a short range component superposed on the Coulomb potential. Since impurity ground states are typically derived from hydrogenic *s*-states and have wave functions that are nonzero at the impurity nucleus, they are strongly influenced by the short range part of the potential which varies significantly from one impurity to another. Excited states derived from hydrogenic *p*-states, on the other hand, have wave functions that are zero at the impurity nucleus and are much less affected by the short range potential, as is evident in the data of Table 5.1.

The short range contribution to the impurity potential is known as a **central cell correction**. The implementation of this correction may be achieved by means of more sophisticated theories based on pseudopotentials. A particularly simple case is that of **isocoric impurities** which have the same number of core electrons as the atom of the host crystal, e.g., P and S in Si (Pantelides and Sah 1974). For these impurities the perturbation potential is not responsible for binding additional core states. However, impurity and host atoms have different Coulomb and exchange contributions of core electrons to the potential. Taking this effect together with many-valley effects into account, a weak short range contribution to the effective impurity potential can be obtained and incorporated into the effective mass equations. The results for the ionization energies of P and S in Si are given in Table 5.5. The agreement between theory and experiment is quite good. Individual valley–orbit interaction splittings for P, As, Sb and other donor impurities have also been calculated.

Table 5.5 Ionization energies (meV) for P and S in Si. The S^+ and S^0 states refer to one and two electrons bound to the center, respectively (after Pantelides and Sah 1974)

Impurity	Theory	Experiment
P	42.4	45.5
$S(S^0)$	297.1	302.0
$S(S^+)$	659.3	613.6

For the case of nonisocoric impurities the perturbation potential binds additional core electrons in the impurity cell and cannot be regarded as weak. In other systems also, the perturbation potential is strong. These systems include vacancies and host atom interstitials, noble metal atom impurities, and transition metal atom impurities. Effective mass theory is no longer applicable, because the **defect levels** tend to lie toward the center of the fundamental gap and cannot be ascribed solely to either the valence or conduction band.

Impurity or defect levels lying toward the center of the gap are referred to as **deep levels**. They play an important role in catalyzing the recombination of electrons and holes, but contribute negligibly to the equilibrium concentrations of electrons and holes. Shallow donors and acceptors, on the other hand, control the concentration of current carriers and thereby the electrical conductivity.

As an example of deep levels we show in Table 5.6 the ionization energies of $3d$ transition elements in Si (Zunger 1986). The ionization energies are all considerably larger than the effective mass theory value of ~ 0.03 eV.

Table 5.6 Ionization energies (eV) of interstitial $3d$ transition metal dopants in Si (after Zunger 1986)

Ti	V	Cr	Mn	Fe
0.89	0.72	0.95	0.75	0.385
1.09	1.01	0.38	1.06	
0.25	0.30		0.25	

A number of approaches have been made to the theory of deep levels. They include tight binding methods (Hjalmarson *et al.* 1980), and Green's function methods (Baraff and Schlüter 1979, Zunger and Lindefelt 1983). Recent emphasis has been placed on the use of the density functional approach together with *ab initio* pseudopotentials (Zunger and Lindefelt 1983, Zunger 1986).

5.7 Impurity bands

Up to now our discussion of impurity states has treated the impurities as isolated entities. At a finite concentration n_I, however, the impurities have an average nearest neighbor separation of approximately $n_I^{-1/3}$. At sufficiently high n_I, $\sim 10^{18}\,\text{cm}^{-3}$ for Si, neighboring impurities are sufficiently close that their wave functions overlap enough to produce a significant perturbation of the energy levels. The energy levels broaden out into a band called an **impurity band**.

If the impurities were located on the sites of a lattice, the impurity band would be associated with reciprocal lattice vectors and an impurity Brillouin zone. For group V donor impurities in Si the impurity band would be half full if the two-fold spin degeneracy were taken into account. Such an impurity band therefore has metallic character and should exhibit finite conductivity even at very low temperatures. This behavior has in fact been observed in n-Ge (Fritzche 1955).

The assumption of periodicity of the impurity sites is, of course, not valid for real semiconductors. The impurities are **disordered**. The energies of impurity states are not confined by band edges, but are spread out over an extended range characterized by a density-of-states function. Furthermore, it has been shown (Anderson 1958) that disorder can lead to **localized states** and the appearance of insulating character. The localized and delocalized states are separated by the **mobility edge** (Mott 1967). If by changing the composition or by other means, the Fermi energy can be made to pass through the mobility edge, a **metal-insulator transition** occurs. The concentration at which the transition takes place is specified by the relation

$$n_I^{1/3} a_B^* = 0.27. \tag{5.15}$$

An approximate form of this relation can be derived by observing that free carriers screen the ion potential of an impurity and reduce the binding energy of a carrier. For the screened Coulomb potential

$$V(r) = -\frac{e^2}{4\pi\epsilon_0 r} e^{-q_s r}, \tag{5.16}$$

where q_s is the inverse screening length specified by $q_s^2 = 4(3n_I/\pi)^{1/3}/a_B^*$, an electron is bound only if $q_s < 1.19/a_B^*$ (Rogers *et al.* 1970) or

$$n_I^{1/3} a_B^* < 0.36. \tag{5.17}$$

Problems

1. Calculate the donor ionization energy, the radius of the ground state orbital, and the minimum concentration for which there will be sufficient orbital overlap to produce an impurity band in n-InSb. The dielectric constant ϵ is 18 and the electron effective mass m_e^* is $0.014\,m$.
2. Calculate the effect of nonparabolicity on the donor ionization energy in n-InSb using the effective mass equation

$$[p^2/2m^* + E_c^{(4)} p^4 + V_i(\mathbf{r})]F_c(\mathbf{r}) = E_{ci} F_c(\mathbf{r}),$$

where

$$E_c^{(4)} = -[(m/m^*) - 1]^2/4E_g m^2,$$

$\mathbf{p} = -i\hbar\nabla$, and $V_i(\mathbf{r})$ is the Coulomb potential. Take $F_c(\mathbf{r})$ to be a variational function of the form $A(\exp(-\alpha r)$, where α is the variational parameter. Is the effect of the nonparabolicity to increase or decrease the ionization energy?

3. Calculate the binding energy of a donor impurity in Si using the effective mass equation, Eq. (5.12), and a variational trial function of the form

$$F_c(\mathbf{r}) = (\alpha\beta^2/\pi a_B^{*3})^{\frac{1}{2}} \exp\{-[\alpha^2 x^2 + \beta^2(y^2 + z^2)]^{\frac{1}{2}}/a_B^*\}$$

with $a_B^* = 4\pi\epsilon_0\epsilon\hbar^2/me^2$, α and β variational parameters, and $\epsilon = 12$. Take the values of the m_α^* from Table 4.1, set $k_{0x} = k_{0y} = k_{0z} = 0$, and ignore the valley–orbit interaction. Compare your result with that for a spherical band with $m^* = (m_\ell^* m_t^{*2})^{1/3}$.

4. The electric field of an ionized impurity is screened by free carriers giving rise to the screened potential $V_i(\mathbf{r}) = -(e^2/4\pi\epsilon_0\epsilon r) \times \exp(-q_s r)$, where q_s is the inverse screening length. Using a trial function of the form $F_c(r) = A\exp(-\alpha r)$ in the variational method, calculate the binding energy as a function of q_s for a donor impurity in CdTe taking $\epsilon = 11$ and $m_e^* = 0.10\,m$. Is your result consistent with the remark just above Eq. (5.17)?

References

R. L. Aggarwal and A. K. Ramdas, *Phys. Rev.* **A137**, 602 (1965); **A140**, 1246 (1965).
P. W. Anderson, *Phys. Rev.* **109**, 1492 (1958).
A. Baldereschi, *Phys. Rev.* **B1**, 4673 (1970).

A. Baldereschi and N. O. Lipari, *Phys. Rev.* B**8**, 2697 (1973).
G. A. Baraff and M. Schlüter, *Phys. Rev.* B**19**, 4965 (1979).
F. Bassani, G. Iadonisi, and B. Pregiosi, *Rep. Prog. Phys.* **37**, 1099 (1974).
D. H. Dickey and J. O. Dimmock, *J. Phys. Chem. Solids* **28**, 529 (1967).
R. A. Faulkner, *Phys. Rev.* **184**, 713 (1969).
H. Fritzsche, *Phys. Rev.* **99**, 406 (1955).
H. P. Hjalmarson, P. Vogel, D. J. Wolford, and J. D. Dow, *Phys. Rev. Lett.* **44**, 810 (1980).
W. Kohn and D. Schechter, *Phys. Rev.* **99**, 1903 (1955).
J. M. Luttinger and W. Kohn, *Phys. Rev.* **97**, 869 (1955).
K. S. Mendelson and H. M. James, *J. Phys. Chem. Solids* **25**, 729 (1964).
N. F. Mott, *Adv. Phys.* **16**, 49 (1967).
H. Nara, *J. Phys. Soc. Jpn.* **20**, 778 (1965).
S. T. Pantelides and C. T. Sah, *Phys. Rev.* B**10**, 621, 638 (1974).
G. S. Picus, E. Burstein, and B. Henvis, *J. Phys. Chem.* Solids **1**, 75 (1956).
J. H. Reuszer and P. Fisher, *Phys. Rev.* A**135**, 1125 (1964).
F. J. Rogers, H. C. Graboske Jr., and D. J. Harwood, *Phys. Rev.* A**1**, 1577 (1970).
R. K. Watts, *Point Defects in Crystals* (John Wiley, New York, 1977).
A. Zunger, in *Solid State Physics*, Vol. 39, eds. H. Ehrenreich, F. Seitz, and D. Turnbull (Academic Press, New York, 1986), p. 275.
A. Zunger and U. Lindefelt, *Phys. Rev.* B**27**, 1191 (1983).

Semiconductor statistics

6

Key ideas

In an *intrinsic semiconductor* free charge carriers arise from the excitation of electrons from the valence band to the conduction band creating equal concentrations of free electrons in the conduction band and free holes in the valence band.

At thermal equilibrium the *Fermi–Dirac distribution function* specifies the *occupation number* of a state.

The *density-of-states* in an energy band is the number of states per unit volume per unit energy interval.

The *intrinsic carrier concentration* enters the *law of mass action* that relates the concentration of electrons and holes.

In *extrinsic semiconductors* the charge carriers arise primarily from impurities.

Donor impurities produce an *n-type semiconductor*. In the *freeze-out range* the free carrier concentration increases exponentially with temperature, but in the *saturation range* it is nearly constant.

Acceptor impurities produce a *p-type semiconductor*.

In a *compensated semiconductor* both donor and acceptor impurities are present.

In an n-type semiconductor electrons are the *majority carriers* and holes are the *minority carriers*. In a p-type semiconductor the roles of electrons and holes are reversed.

Statistics

In order to calculate properties such as electrical conductivity, it is necessary to know the concentration of electrons in the conduction band and holes in the valence band. These current carriers can arise by thermal or optical excitation of electrons from the valence band to the conduction band, by excitation of carriers from impurity states, or by injection from an outside source. In this chapter we focus on carrier concentrations in a semiconductor in thermal equilibrium.

Semiconductors can be classified according to the origin of their current carriers. If the carriers are due primarily to the excitation of electrons from the valence band to the conduction band, the semiconductor is **intrinsic**. This is the situation in an ideal, perfectly pure semiconductor, but it may also occur in an impure semiconductor if the temperature is sufficiently high. On the other hand, if the current carriers arise primarily from impurities, the semiconductor is **extrinsic**. Its electrical behavior depends on non-intrinsic properties of the material.

When both donor and acceptor impurities are present and have random distributions, the semiconductor is **compensated**. At equilibrium electrons are transferred from the donors to the acceptors, thus creating ionized donors and acceptors. If no neutral donors and acceptors remain, one has **ideal compensation**.

If a semiconductor contains impurities whose ground states lie far from the band edges, the impurities give rise to **deep traps**. Carriers originating from shallow impurities can be captured by deep traps. The result is a **semi-insulating semiconductor** in which the thermal excitation of carriers into the valence or conduction bands is difficult.

6.1 Intrinsic semiconductors

Let us consider an intrinsic semiconductor with an energy gap E_g between the valence and conduction band extrema. Since electrons are Fermi particles, the Fermi–Dirac distribution, $f_{FD}(\boldsymbol{k})$, specifies the occupation number of a state of energy $E_{\boldsymbol{k}}$ in thermal equilibrium at absolute temperature T,

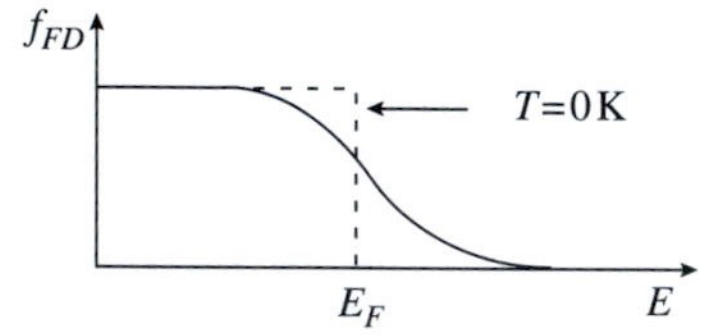

Fig. 6.1
Fermi–Dirac distribution function plotted against energy E for $T = 0$ K and $T \neq 0$ K.

$$f_{FD}(\boldsymbol{k}) = \frac{1}{e^{\beta[E_{\boldsymbol{k}}-E_F]}+1}, \tag{6.1}$$

where $\beta = 1/k_BT$, k_B is Boltzmann's constant and E_F is the Fermi energy or chemical potential. A plot of the Fermi–Dirac distribution function versus energy is given in Fig. 6.1.

When the temperature T is zero, $f_{FD}(\boldsymbol{k}) = 1$ for $E_{\boldsymbol{k}} < E_F^0$ and $f_{FD}(\boldsymbol{k}) = 0$ for $E_{\boldsymbol{k}} > E_F^0$. All of the states situated below the zero temperature Fermi energy E_F^0 are occupied and all of the states above E_F^0 are empty. The Fermi energy lies between the occupied and the unoccupied states of the system at $T = 0$.

When $E_{\boldsymbol{k}} - E_F$ becomes very large compared to k_BT, it is possible to neglect the quantity $+1$ after the exponential in the denominator of Eq. (6.1). The Fermi–Dirac distribution then reduces to the Boltzmann distribution

$$f_B(E) = e^{-(E_{\boldsymbol{k}}-E_F)/k_BT}.$$

The concentration of electrons in the conduction band n is given by

$$n = \frac{2}{\Omega}\sum_{\boldsymbol{k}} f_{FD}(\boldsymbol{k}) = \frac{2}{\Omega}\sum_{\boldsymbol{k}} \frac{1}{e^{\beta[E_{c\boldsymbol{k}}-E_F]}+1}, \tag{6.2}$$

where the sum is over all states in the conduction band with energies $E_{c\boldsymbol{k}}$, Ω is the volume of the system, and a factor of 2 has been included to account for the spin degeneracy. Using the prescription for converting the sum over

$\boldsymbol{k}$ to an integral,

$$\sum_{\boldsymbol{k}} \longrightarrow \frac{\Omega}{(2\pi)^3} \int d^3k,$$

we obtain

$$n = \frac{2}{(2\pi)^3} \int d^3k \frac{1}{e^{\beta[E_{c\boldsymbol{k}} - E_F]} + 1}. \tag{6.3}$$

The Fermi–Dirac integral in this expression cannot in general be evaluated in closed form. However, with certain approximations its analytic evaluation is possible.

6.1.1 Spherical parabolic energy bands

The first case we consider is a spherical parabolic conduction band with effective mass m_c^* and energy given by

$$E_{c\boldsymbol{k}} = E_C + \frac{\hbar^2 k^2}{2m_c^*}, \tag{6.4}$$

where E_C is the conduction band edge. We take the temperature sufficiently low that $k_B T \ll E_C - E_F$ and approximate the Fermi–Dirac distribution by

$$f_{FD}(\boldsymbol{k}) \simeq e^{\beta[E_F - E_{c\boldsymbol{k}}]}. \tag{6.5}$$

Equation (6.3) now becomes

$$n = \frac{2}{(2\pi)^3} \int d^3k e^{\beta(E_F - E_C - \hbar^2 k^2/2m_c^*)}. \tag{6.6}$$

Introducing spherical coordinates in $\boldsymbol{k}$-space and integrating over the angles gives

$$n = \frac{2}{2\pi^2} e^{\beta(E_F - E_C)} \int_0^\infty dk\, k^2 e^{-\beta\hbar^2 k^2/2m_c^*}. \tag{6.7}$$

Changing the variable of integration from k to $E = \hbar^2 k^2/2m_c^*$, we obtain

$$n = e^{\beta(E_F - E_C)} \int_0^\infty N_c(E) e^{-\beta E} dE, \tag{6.8}$$

where

$$N_c(E) = \frac{1}{2\pi^2} \left(\frac{2m_c^*}{\hbar^2}\right)^{3/2} E^{\frac{1}{2}} \tag{6.9}$$

is the **density-of-states in energy** for the conduction band, i.e., the number of states per unit volume per unit energy interval. This quantity is plotted in

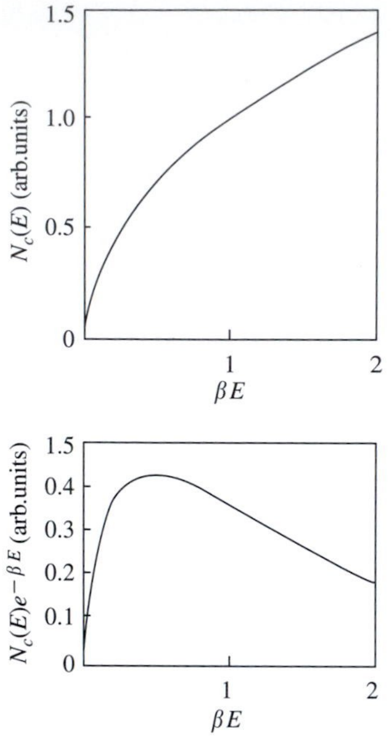

Fig. 6.2
Density-of-states $N_c(E)$ and the product $N_c(E)\exp(-\beta E)$ plotted against energy E.

Fig. 6.2 as a function of energy E. The free carrier concentration is also represented in this figure. It should be noted that Eq. (6.8) is valid for a more complicated dependence of $E_{ck} - E_C$ on k than that of Eq. (6.4) provided the proper expression for $N_c(E)$ is used.

The integral in Eq. (6.8) is proportional to $\Gamma\left(\frac{3}{2}\right) = \pi^{\frac{1}{2}}/2$. The result for n is

$$n = \bar{N}_c e^{\beta(E_F - E_C)}, \tag{6.10}$$

where

$$\bar{N}_c = 2(m_c^* k_B T/2\pi\hbar^2)^{3/2} \tag{6.11}$$

is the **effective density-of-states** for the conduction band.

The effective density-of-states $\bar{N}_c$ represents a weighted sum of all the states in the conduction band. It is, in fact, the partition function of the noninteracting electron gas. The weighting coefficient is the Boltzmann factor which expresses the fact that the higher the energy of the state in the conduction band the lower the probability that the state will be occupied by an electron. At a given temperature $\bar{N}_c$ represents the effective degeneracy of the conduction band regarded as a single level with energy E_C. Because the approximations employed lead to the use of the Boltzmann distribution, the carriers form a **nondegenerate system**. When the criterion $k_B T \ll E_C - E_F$ is not satisfied, the full Fermi–Dirac distribution must be used, and the carriers form a **degenerate system**. Nondegenerate semiconductors are characterized by a relatively low concentration of shallow impurities, whereas degenerate semiconductors have a relatively high concentration of shallow impurities. In the latter case the overlapping of atomic orbitals of neighboring impurities can lead to the formation of impurity bands as described in Chapter 5. The Fermi energy can then lie sufficiently close to E_C that degenerate behavior persists to low temperatures.

Example 6.1: Effective density-of-states: conduction band
Calculate the effective density-of-states $\bar{N}_c$ characterizing the conduction bands of semiconductors GaAs and InP.
Solution. Substituting the values $\hbar = 1.055 \times 10^{-27}$ erg s and $k_B = 1.38 \times 10^{-16}$ erg K^{-1} into the expression for the effective density-of-states in the conduction band, Eq. (6.11), one gets

$$\bar{N}_c = 2.5 \times 10^{19} \left(\frac{m_c^*}{m}\right)^{3/2} \left(\frac{T}{300}\right)^{3/2} \text{cm}^{-3}.$$

The conduction bands of GaAs and InP are to a good approximation spherical and parabolic with effective masses m_c^* equal to $0.066\,m$, and $0.077\,m$ respectively. The corresponding effective densities of states are 4.2×10^{17} cm^{-3} and 5.3×10^{17} cm^{-3}, respectively, at 300 K.

Turning now to holes in the valence band, we note that, since a hole is a state lacking an electron, the distribution function for holes is $1 - f_{FD}(\boldsymbol{k})$. The concentration of holes p is then given by

$$p = \frac{2}{\Omega}\sum_{\boldsymbol{k}}[1 - f_{FD}(\boldsymbol{k})] = \frac{2}{\Omega}\sum_{\boldsymbol{k}}\frac{1}{e^{\beta[E_F - E_v(\boldsymbol{k})]} + 1}, \tag{6.12}$$

where $E_v(\boldsymbol{k})$ is the energy of a valence band state. For the moment we consider a single valence band that is spherical and parabolic with band edge E_V and hole effective mass m_v^*. Then

$$E_v(\boldsymbol{k}) = E_V - \frac{\hbar^2 k^2}{2m_v^*}. \tag{6.13}$$

Making the low temperature approximation $k_B T \ll E_F - E_V$ and proceeding as in the case of conduction electrons, we find for the concentration of holes

$$p = \bar{N}_v e^{\beta(E_V - E_F)}, \tag{6.14}$$

where the effective density-of-states for holes is given by

$$\bar{N}_v = 2(m_v^* k_B T/2\pi\hbar^2)^{3/2}. \tag{6.15}$$

Example 6.2: Effective density-of-states: valence band
Calculate the effective density-of-states $\bar{N}_v$ for the valence bands of GaAs and InP.
Solution. Using the values of $\hbar$ and k_B given in Example 6.1, one gets

$$\bar{N}_v = 2.5 \times 10^{19} \left(\frac{m_v^*}{m}\right)^{3/2} \left(\frac{T}{300}\right)^{3/2} \text{cm}^{-3}.$$

The major part of the effective density-of-states comes from heavy holes whose masses are $0.5\,m$ and $0.4\,m$ for GaAs and InP, respectively. Neglecting the light hole contribution, one finds that $\bar{N}_v$ is $0.88 \times 10^{19}\,\text{cm}^{-3}$ and $0.63 \times 10^{19}\,\text{cm}^{-3}$ for GaAs and InP, respectively, at $T = 300\,K$. The contribution of light holes can be obtained by using the results of Section 6.1.3.

The expressions for the concentrations of electrons and holes given by Eqs. (6.10) and (6.14) involve the Fermi energy E_F which is as yet undetermined. We can eliminate it by multiplying the two expressions together to yield

$$np = \bar{N}_c \bar{N}_v e^{-\beta E_g} = n_i^2, \tag{6.16}$$

where $E_g = E_C - E_V$ and n_i is the **intrinsic carrier concentration**. Equation (6.16) is the **law of mass action** that relates the concentrations of conduction electrons and holes in thermal equilibrium, i.e., the product of the electron and hole concentrations in a nondegenerate semiconductor is constant at a given temperature. It must be emphasized that the law of mass action is very general. The basic approximation made is that the separation of the Fermi energy from the valence and conduction band edges is large compared to $k_B T$. It makes no assumption about the source or sources of electrons and holes, whether by excitation across the gap or by ionization of impurities. In chemistry, a similar law controls the concentrations of H^+ and OH^- ions in water.

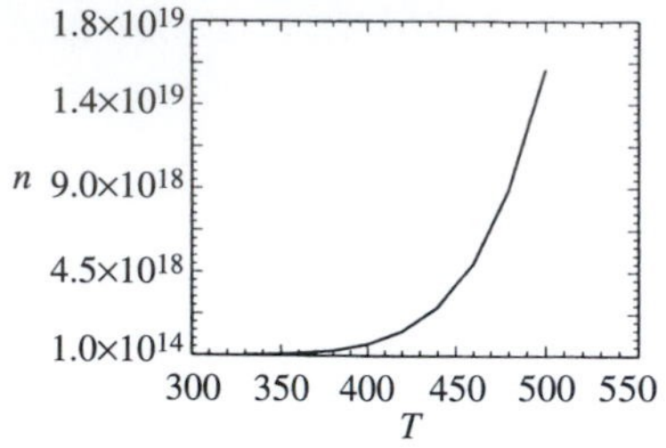

Fig. 6.3
Conduction electron concentration n versus temperature T for $E_g = 1.0\,\mathrm{eV}$ and $m_c^* = m_v^* = 0.1\,m$.

For the particular case of excitation across the gap in an intrinsic semiconductor, there are no impurities to be ionized, and charge conservation requires that $n = p$. From the law of mass action, we obtain

$$\begin{aligned} n = p = n_i &= (\bar{N}_c \bar{N}_v)^{\frac{1}{2}} e^{-\beta E_g/2} \\ &= 2(k_B T/2\pi\hbar^2)^{3/2} (m_c^* m_v^*)^{3/4} e^{-E_g/2k_B T}. \end{aligned} \tag{6.17}$$

The concentrations of conduction electrons and valence holes increase essentially exponentially with increasing temperature and with decreasing energy gap. This behavior is primarily responsible for the characteristic temperature dependence of the electrical conductivity of an intrinsic semiconductor. In Fig. 6.3 n is plotted as a function of temperature for $E_g = 1.0\,\mathrm{eV}$ and $m_c^* = m_v^* = 0.1\,m$.

Having obtained an expression that gives both n and p, we can now determine the Fermi energy E_{Fi} of an intrinsic semiconductor by equating the expressions for n in Eq. (6.10) and p in Eq. (6.14):

$$\bar{N}_v e^{\beta(E_V - E_{Fi})} = \bar{N}_c e^{\beta(E_{Fi} - E_C)}. \tag{6.18}$$

Solving for E_{Fi} yields the result

$$E_{Fi} = E_V + \tfrac{1}{2} E_g + \tfrac{3}{4} k_B T \log\left(\frac{m_v^*}{m_c^*}\right). \tag{6.19}$$

We see that at $T = 0\,\mathrm{K}$, the intrinsic Fermi energy lies at the center of the forbidden gap. As the temperature increases, E_{Fi} increases or decreases depending on whether $m_c^* < m_v^*$ or $m_c^* > m_v^*$, respectively. If $m_c^* = m_v^*$, E_{Fi} lies at the center of the gap for all temperatures.

6.1.2 Ellipsoidal energy bands

For the case of a semiconductor such as Si or Ge, we must take into account the fact that the constant energy surfaces of the conduction band are ellipsoids of revolution. The conduction band energy that appears in Eq. (6.6) must be replaced by the corresponding expression from Eq. (3.76):

$$n = \frac{2}{(2\pi)^3} \int d^3k e^{\beta\{E_F - E_C - (\hbar^2/2)[(k_x^2 + k_y^2)/m_t^* + (k_z - k_0)^2/m_\ell^*]\}}. \tag{6.20}$$

In order to evaluate the integral, we "sphericalize" the energy band by introducing the scale transformation

$$k_x = \left(\frac{m_t^*}{m}\right)^{\frac{1}{2}} k'_x, \quad k_y = \left(\frac{m_t^*}{m}\right)^{\frac{1}{2}} k'_y, \quad k_z - k_0 = \left(\frac{m_\ell^*}{m}\right)^{\frac{1}{2}} k'_z. \tag{6.21}$$

The conduction electron concentration then becomes

$$n = \frac{2}{(2\pi)^3} \left(\frac{m_\ell^* m_t^{*2}}{m^3}\right)^{\frac{1}{2}} \int d^3k' e^{\beta(E_F - E_C - \hbar^2 k'^2/2m)}. \tag{6.22}$$

Proceeding as in the case of a spherical parabolic band, we rewrite Eq. (6.22) in the form

$$n = e^{\beta(E_F - E_C)} \int_0^\infty N_c'(E) e^{-\beta E} dE, \tag{6.23}$$

where the density-of-states $N_c'(E)$ is given by

$$N_c'(E) = \frac{1}{2\pi^2} \left(\frac{2m_{DOS}^*}{\hbar^2} \right)^{3/2} E^{\frac{1}{2}}. \tag{6.24}$$

The density-of-states effective mass m_{DOS}^* is related to m_ℓ^* and m_t^* by

$$m_{DOS}^* = (m_\ell^* m_t^{*2})^{1/3}. \tag{6.25}$$

Carrying out the integration over E in Eq. (6.23) leads to the result

$$n = \bar{N}_c' e^{\beta(E_F - E_C)}, \tag{6.26}$$

where

$$\bar{N}_c' = 2(m_{DOS}^* k_B T / 2\pi\hbar^2)^{3/2}. \tag{6.27}$$

If there are r equivalent extrema in the conduction band, each extremum contributes the amount given by Eq. (6.26), so the total concentration of conduction electrons is then

$$n = r\bar{N}_c' e^{\beta(E_F - E_C)}. \tag{6.28}$$

It should be noted that Eqs. (6.14) and (6.28) are not restricted to pure semiconductors. They apply also to doped semiconductors, but E_F is no longer the intrinsic Fermi energy.

Comment. For Ge with density-of-states effective mass for the conduction band 0.22 m and heavy hole effective mass 0.34 m, one obtains for the effective density-of-states of the conduction band $\bar{N}_c = 0.26 \times 10^{19}\,\text{cm}^{-3}$ and of the valence band $\bar{N}_v = 0.50 \times 10^{19}\,\text{cm}^{-3}$ at $T = 300$ K. For Si with effective masses $m_{DOS}^* = 0.32\,m$ and $m_{hh}^* = 0.52\,m$, one has $\bar{N}_c = 0.45 \times 10^{19}\,\text{cm}^{-3}$ and $\bar{N}_v = 0.94 \times 10^{19}\,\text{cm}^{-3}$ at $T = 300$ K.

For the case of a spherical parabolic valence band, the Fermi energy can be eliminated from Eqs. (6.14) and (6.28). The concentrations of electrons and holes are then specified by Eq. (6.17) with m_c^* replaced by m_{DOS}^* and with a factor $r^{\frac{1}{2}}$ inserted:

$$n = p = 2r^{\frac{1}{2}} (k_B T / 2\pi\hbar^2)^{3/2} (m_{DOS}^* m_v^*)^{3/4} e^{-E_g/2k_B T}. \tag{6.29}$$

6.1.3 Multiple valence bands

In Si and Ge, the uppermost valence bands (light and heavy hole bands) have degenerate maxima at $\boldsymbol{k} = 0$. In the approximation that each of these bands is spherical and parabolic, their contributions to the concentrations of light and heavy holes are

$$p_\ell = \bar{N}_{v\ell} e^{\beta(E_V - E_F)}, \tag{6.30}$$

where

$$\bar{N}_{v\ell} = 2(m^*_{\ell h} k_B T/2\pi\hbar^2)^{3/2}, \tag{6.31}$$

and

$$p_h = \bar{N}_{vh} e^{\beta(E_V - E_F)}, \tag{6.32}$$

where

$$\bar{N}_{vh} = 2(m^*_{hh} k_B T/2\pi\hbar^2)^{3/2}. \tag{6.33}$$

The effective masses of light and heavy holes are $m^*_{\ell h}$ and m^*_{hh}, respectively. The total concentration of holes is given by

$$p = p_\ell + p_h = (\bar{N}_{v\ell} + \bar{N}_{vh}) e^{\beta(E_V - E_F)}. \tag{6.34}$$

Eliminating the Fermi energy from Eqs. (6.28) and (6.34) yields the law of mass action in the form

$$np = r\bar{N}_c'(\bar{N}_{v\ell} + \bar{N}_{vh}) e^{-\beta E_g}. \tag{6.35}$$

For intrinsic material, $n = p = n_i$. Then from Eq. (6.35)

$$n_i = n = p = [r\bar{N}_c'(\bar{N}_{v\ell} + \bar{N}_{vh})]^{\frac{1}{2}} e^{-\beta E_g/2}. \tag{6.36}$$

Using the expressions for $\bar{N}_c'$, $\bar{N}_{v\ell}$ and $\bar{N}_{vh}$ given in Eqs. (6.27), (6.31), and (6.33), respectively, we obtain

$$n_i = 2r^{\frac{1}{2}}(k_B T/2\pi\hbar^2)^{3/2}[m^{*3/2}_{DOS}(m^{*3/2}_{\ell h} + m^{*3/2}_{hh})]^{\frac{1}{2}} e^{-\beta E_g/2}. \tag{6.37}$$

As we have seen in Chapter 3, the light and heavy hole valence bands of Si and Ge are warped, not spherical and parabolic. Taking the low-temperature limit of the hole distribution function in Eq. (6.12) and converting the sum over $\boldsymbol{k}$ to an integral, we obtain

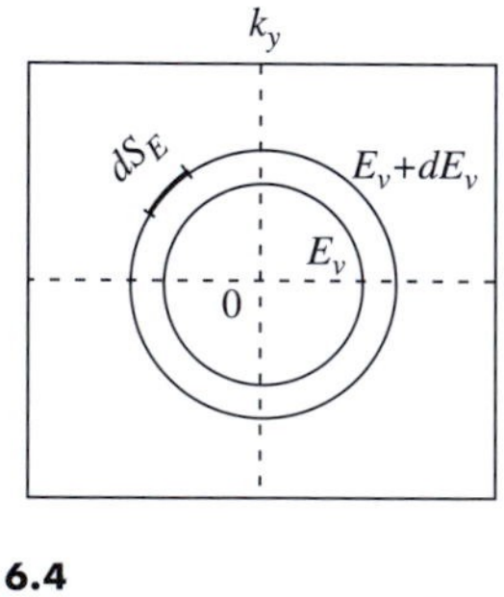

Fig. 6.4
Constant energy surfaces in $\boldsymbol{k}$-space.

$$p = \frac{2}{(2\pi)^3}\int d^3k e^{\beta[E_{v\boldsymbol{k}} - E_F]}, \tag{6.38}$$

where a factor of 2 has been included for spin and $E_{v\boldsymbol{k}}$ takes the forms appropriate for the light hole and heavy hole valence bands. The evaluation of the integral proceeds by considering two surfaces of constant energy in $\boldsymbol{k}$-space, one with energy E_v and the other with energy $E_v + dE_v$ as shown in Fig. 6.4. Let dS_E be the element of area at some point on the constant energy surface E_v. The element of volume between the constant energy surfaces E_v and $E_v + dE_v$ is a right cylinder of base dS_E and height $dk_\perp$: $d^3k = dS_E dk_\perp$.

The quantity $dk_\perp$ is the perpendicular distance between the constant energy surfaces. Also perpendicular to these surfaces is the gradient of the energy, $\nabla_k E_{vk}$. The energy difference dE_v is given by

$$dE_v = dk_\perp |\nabla_k E_{vk}|, \tag{6.39}$$

so the element of volume d^3k becomes

$$d^3k = dk_\perp dS_E = \frac{dE_v}{|\nabla_k E_{vk}|} dS_E. \tag{6.40}$$

Equation (6.38) then takes the form

$$p = \frac{2}{(2\pi)^3} \int dE_v \int_{\sum(E_v)} \frac{dS_{E_v}}{|\nabla_k E_{vk}|} e^{\beta[E_{vk} - E_F]}, \tag{6.41}$$

where $\sum(E_v)$ is the constant energy surface of energy E_v. In actual cases such as Si and Ge, the integrals in Eq. (6.41) must be done numerically.

6.2 Extrinsic semiconductors

As we have noted in Chapter 5, impurity atoms are an important source of charge carriers in a semiconductor. Since the electron or hole ionization energy may be small compared to the forbidden energy gap, there may be significant thermal production of free carriers at relatively low temperatures.

A very general expression for the ratio of electron and hole concentrations in doped semiconductors can be derived from Eqs. (6.10) and (6.14) by rewriting them as

$$n = \bar{N}_c e^{\beta(E_F - E_{Fi} + E_{Fi} - E_C)} = n_i e^{\beta(E_F - E_{Fi})} \tag{6.42a}$$

and

$$p = \bar{N}_v e^{\beta(E_V - E_{Fi} + E_{Fi} - E_F)} = n_i e^{\beta(E_{Fi} - E_F)}, \tag{6.42b}$$

where E_{Fi} is the intrinsic Fermi energy. Taking the ratio of Eqs. (6.42a) and (6.42b), we obtain

$$\frac{n}{p} = e^{2\beta(E_F - E_{Fi})}. \tag{6.43}$$

If the Fermi energy of the doped material is higher than that of the intrinsic material, the electron concentration is higher than the hole concentration, whereas if the Fermi energy is lower than that of the intrinsic material, the hole concentration is higher than the electron concentration.

The detailed statistical analysis of impurity levels is complicated by the fact that these states are localized. If there is already one carrier of a given spin in a localized impurity orbital, the addition of a second carrier of opposite spin to that orbital causes a large change in energy as a result of the

Coulomb repulsion of the two carriers. The binding energy of the second carrier is typically so weak that the bound state with two carriers can be ignored.

6.2.1 Donor impurities

It is convenient in the development of impurity statistics to use the **grand canonical ensemble** of statistical mechanics (Tolman 1938, Guggenheim 1953, Landsberg 1958, Teitler and Wallis 1960). The probability that a donor impurity state has $n\uparrow$ electrons with spin up and $n\downarrow$ electrons with spin down is given by

$$P(n\uparrow, n\downarrow) = Ce^{\beta[(n\uparrow+n\downarrow)E_F - E_I(n\uparrow,n\downarrow)]}, \tag{6.44}$$

where C is a normalization constant and $E_I(n\uparrow, n\downarrow)$ is the energy of the state. The occupation numbers $n\uparrow$ and $n\downarrow$ can take on the values 0, 1. In accordance with the discussion in the preceding paragraph, we shall exclude the possibility that both $n\uparrow$ and $n\downarrow$ take on the value 1 simultaneously.

As a specific example, we consider a donor impurity such as phosphorus in silicon or germanium. The treatment can be easily modified to apply to acceptor impurities such as boron. The phosphorus atom can have its extra electron, not needed for covalent band formation, in a localized state with its spin either up or down. As discussed in the preceding chapter, the ground state of phosphorus in either silicon or germanium is a singlet with the only degeneracy due to spin. In the present simplified discussion, we neglect the excited bound states and treat only the ground state and the continuum of unbound states. The probability that we have one electron localized on a phosphorus atom is therefore given by

$$P_1 = \frac{P(0,1) + P(1,0)}{P(0,0) + P(0,1) + P(1,0)}. \tag{6.45}$$

Using Eq. (6.44), we obtain

$$P_1 = \frac{2e^{\beta(E_F - E_I^d)}}{1 + 2e^{\beta(E_F - E_I^d)}} = \frac{1}{1 + \frac{1}{2}e^{\beta(E_I^d - E_F)}}, \tag{6.46}$$

where E_I^d is the energy of the donor impurity state given by $E_I^d = E_I(1,0) = E_I(0,1)$ and we have taken $E_I(0,0)$ (the energy of the crystal with the donor electron removed) to be zero. The quantity P_1 is the **occupation factor** of the impurity state. Note that it is not the same as the Fermi–Dirac distribution defined in Eq. (6.1).

The probability that the extra electron of the phosphorus atom is ionized into the conduction band is given by

$$1 - P_1 = \frac{\frac{1}{2}e^{\beta(E_I^d - E_F)}}{1 + \frac{1}{2}e^{\beta(E_I^d - E_F)}} = \frac{1}{1 + 2e^{\beta(E_F - E_I^d)}}. \tag{6.47}$$

If the concentration of donor impurities is n_d, the concentration of electrons in the conduction band due to the ionization of the donors is

$$n = n_d(1 - P_1) = \frac{n_d}{1 + 2e^{\beta(E_F - E_I^d)}}. \tag{6.48}$$

Before proceeding with the general analysis of Eq. (6.48), we examine the simple but important situation in which E_F is sufficiently below E_I^d that $E_I^d - E_F \gg k_B T$. Then the impurity level is essentially unoccupied, nearly all the donor impurities are ionized, and $n \simeq n_d$. A more precise appraisal of n is provided by taking into account the electrons excited from the valence band to the conduction band. The condition of charge neutrality is then

$$n = p + n_d^+, \tag{6.49}$$

where n_d^+ is the concentration of ionized donors. For the case under consideration, $n_d^+ \simeq n_d$. Eliminating the hole concentration using the law of mass action, we obtain

$$n = \frac{n_i^2}{n} + n_d. \tag{6.50}$$

The physically significant solution to this quadratic equation is

$$n = \frac{1}{2}\left[n_d + \left(n_d^2 + 4n_i^2\right)^{\frac{1}{2}}\right]. \tag{6.51}$$

Equation (6.51) is particularly useful when $n_d \simeq n_i$. If this is not so, we have the two limiting cases $n_d \gg n_i$, $n \simeq n_d$ and $n_d \ll n_i$, $n \simeq n_i$. The first case corresponds to an extrinsic semiconductor and the second to an intrinsic semiconductor.

Turning now to the general analysis of Eq. (6.48), we need a second relation involving the Fermi energy E_F. This relation is provided by Eq. (6.28) in which the approximation $E_C - E_F \gg k_B T$ has been made. However, we no longer require that $E_I^d - E_F \gg k_B T$. Hence, E_F can be close to E_I^d and can even be above it. Noting that the energy E_I^d of the impurity state is given by

$$E_I^d = E_C - E_d, \tag{6.52}$$

where E_d is the donor ionization energy, Eq. (6.48) becomes

$$n = \frac{n_d}{1 + 2e^{\beta(E_F - E_C + E_d)}}. \tag{6.53}$$

Eliminating $E_F - E_C$ from Eqs. (6.28) and (6.53) yields the relation

$$\frac{n^2}{n_d - n} = \tfrac{1}{2} r \bar{N}_c' e^{-\beta E_d}. \tag{6.54}$$

Two limiting cases of Eq. (6.54) are readily evaluated: $\beta E_d \gg 1 (k_B T \ll E_d)$ and $\beta E_d \ll 1 (k_B T \gg E_d)$. In the first case, $n \ll n_d$ and one has the approximate solution

$$n \simeq (r\bar{N}_c' n_d/2)^{\frac{1}{2}} e^{-\beta E_d/2} \tag{6.55}$$

according to which the conduction electron concentration increases essentially exponentially with increasing temperature. This is the **freeze-out range** of temperature. In the second case, $n \simeq n_d$, essentially all the donor impurities are ionized. This is the **saturation range** of temperature.

Example 6.3: Conduction electron concentration in n-Si
Calculate the conduction electron concentration as a function of temperature for group V donor impurities in Si taking $E_d = 0.030$ eV, $m_\ell^* = 0.92\,m$, $m_t^* = 0.19\,m$ and $r = 6$.
Solution. Equation (6.54) is a quadratic equation in n which can be solved to give the physically meaningful result

$$n = 2n_d \left[1 + \left(1 + 4\frac{n_d}{\bar{N}_c^{eff}} \right)^{\frac{1}{2}} \right]^{-1}, \tag{6.56}$$

where

$$\bar{N}_c^{eff} = \tfrac{1}{2} r\bar{N}_c' e^{-\beta E_d}. \tag{6.57}$$

The limiting cases discussed above follow from Eq. (6.56) for $n_d/\bar{N}_c^{eff} \gg 1$ and $n_d/\bar{N}_c^{eff} \ll 1$, respectively. Inserting the values of the material parameters and a range of values of the temperature into Eq. (6.56), one obtains the dependence of n on temperature shown in Fig. 6.5 for $n_d = 10^{16}\,\text{cm}^{-3}$.

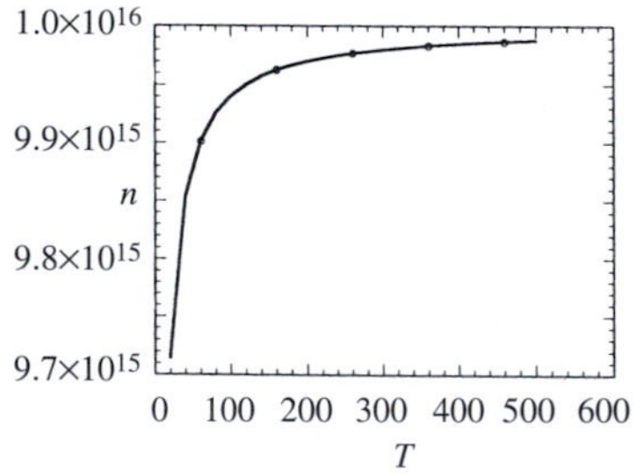

Fig. 6.5
Conduction electron concentration n versus temperature T for n-Si.

It is of interest to obtain an expression for the Fermi energy for the system discussed in Example 6.3. Substituting the result for n given by Eq. (6.56) into Eq. (6.28) and solving for E_F yields

$$E_F = E_C - k_B T \log\{(r\bar{N}_c'/2n_d)[1 + (1 + 4n_d/\bar{N}_c^{eff})^{\frac{1}{2}}]\}. \tag{6.58}$$

In the low-temperature limit, $n_d/\bar{N}_c^{eff} \gg 1$, and E_F is given by

$$E_F \simeq E_C - \tfrac{1}{2}E_d + \tfrac{1}{2}k_B T \log\left(\frac{n_d}{2r\bar{N}_c'}\right). \tag{6.59}$$

The Fermi energy starts midway between the impurity level and the conduction band edge at $T = 0$ and initially rises as T increases. As T continues to increase, it is necessary to take into account the holes generated thermally in the valence band in order to properly trace the further evolution of E_F as a function of T.

When holes are present, the **condition of electroneutrality** takes the form

$$n = n_d - n_d^o + p, \tag{6.60}$$

where n_d^0 is the concentration of neutral donors. The quantity $n_d - n_d^0$ is the electron concentration arising from donors alone and is given by Eq. (6.48).

Substituting the expressions for n and p given by Eqs. (6.28) and (6.34), respectively, we obtain

$$r\bar{N}'_c e^{\beta(E_F - E_C)} = \frac{n_d}{1 + 2e^{\beta(E_F - E_I^d)}} + (\bar{N}_{v\ell} + \bar{N}_{vh})e^{\beta(E_V - E_F)}. \quad (6.61)$$

This is a cubic equation in $\exp(\beta E_F)$ which can be solved numerically for E_F. The variation of E_F with temperature is displayed in Fig. 6.6 for n-Si. The Fermi energy lies about halfway between the donor level and the conduction band edge at very low temperatures and drops to about the midpoint of the energy gap at high temperatures for which $k_B T > E_g$.

The conduction electron concentration n is plotted versus inverse temperature in Fig. 6.7. Note the clearly defined intrinsic, saturation and freeze-out ranges of temperature. The slopes of the curves in the intrinsic and freeze-out ranges on the semilog plot provide values for E_g and E_d, respectively.

The foregoing treatment is applicable to donor impurities in semiconductors such as GaAs for which the conduction band minimum occurs at $\boldsymbol{k} = 0$, the constant energy surfaces are spherical, and $r = 1$. For those cases such as Si, Ge, and GaP in which the minimum occurs away from $\boldsymbol{k} = 0$, the constant energy surfaces are ellipsoids of revolution, $r > 1$, and a more general treatment is required that includes the contribution of excited bound states. This treatment is given in Section 6.2.5.

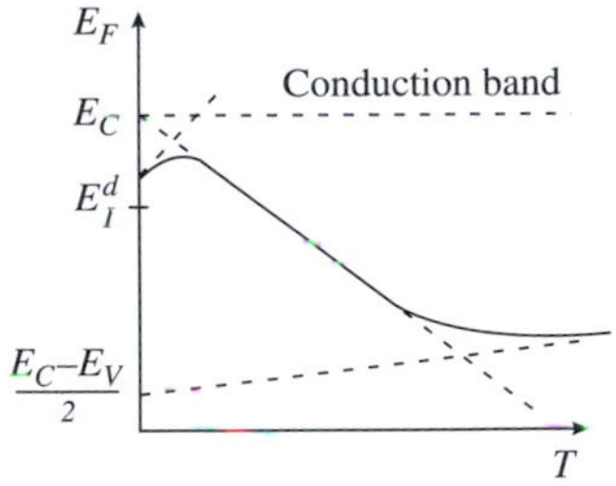

Fig. 6.6
Fermi energy E_F versus temperature T for n-Si.

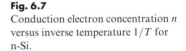

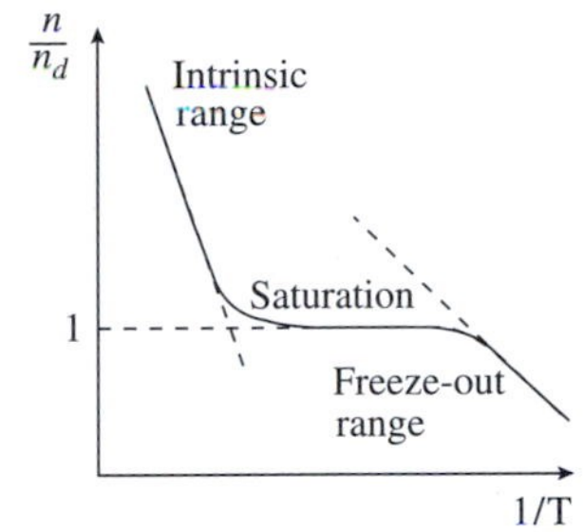

Fig. 6.7
Conduction electron concentration n versus inverse temperature $1/T$ for n-Si.

6.2.2 Acceptor impurities

The treatment of acceptor impurities is straightforward if the current carriers are regarded as holes. The major complication is the degeneracy of the valence band at $\boldsymbol{k} = 0$ associated with the light and heavy hole bands. We shall neglect warping effects and restrict our discussion to the case in which the light and heavy hole bands are spherical and parabolic.

We use the grand canonical ensemble and a procedure analogous to that employed for donors. Only those cases in which either no hole or one hole is trapped on an acceptor impurity are considered. For holes the energy scale must be inverted compared to electrons: $E_F \to -E_F$, $E_I \to -E_I$. The probability that an acceptor impurity state has $p_\ell \uparrow$ light holes with spin up, $p_\ell \downarrow$ light holes with spin down, $p_h \uparrow$ heavy holes with spin up, and $p_h \downarrow$ heavy holes with spin down can be written as

$$P(p_\ell \uparrow, p_\ell \downarrow, p_h \uparrow, p_h \downarrow) = Ce^{\beta[(p_\ell\uparrow + p_\ell\downarrow + p_h\uparrow + p_h\downarrow)(E_I^a - E_F)]}, \quad (6.62)$$

where E_I^a is the energy of the acceptor impurity state. The probability that a single hole is localized on an acceptor atom is

$$P_1 = \frac{P(0,0,0,1) + P(0,0,1,0) + P(0,1,0,0) + P(1,0,0,0)}{P(0,0,0,0) + P(0,0,0,1) + P(0,0,1,0) + P(0,1,0,0) + P(1,0,0,0)}. \quad (6.63)$$

Substituting Eq. (6.62) into Eq. (6.63) yields

$$P_1 = \frac{1}{1 + \frac{1}{4} e^{\beta(E_F - E_I^a)}}. \tag{6.64}$$

Following a procedure analogous to that used for donors, we find that the concentration of holes in the valence band is

$$p = \frac{n_a}{1 + 4e^{\beta(E_V + E_a - E_F)}}, \tag{6.65}$$

where n_a is the concentration of acceptor impurities, $E_I^a = E_V + E_a$, and E_a is the acceptor ionization energy. In Section 6.1.3 it was shown (Eq. (6.34)) that

$$p = (\bar{N}_{v\ell} + \bar{N}_{vh}) e^{\beta(E_V - E_F)}, \tag{6.66}$$

when the light and heavy hole bands are spherical and parabolic. Eliminating $E_V - E_F$ from Eqs. (6.65) and (6.66) gives

$$\frac{p^2}{n_a - p} = \frac{1}{4}(\bar{N}_{v\ell} + \bar{N}_{vh}) e^{-\beta E_a}, \tag{6.67}$$

which is analogous to Eq. (6.54) for donors. The same limiting cases can be analyzed for acceptors as were done for donors.

Comment. The case of an acceptor impurity such as boron can be analyzed in the following way (Guggenheim 1953). A neutral boron atom has an unpaired electron in the $2p$ shell. If a second electron is added to pair up with this $2p$ electron and one takes into account the pair of $2s$ electrons, the boron ion thus formed has the electron configuration necessary to form four tetrahedral bonds with four nearest neighbors. The procedure used for donors in Section 6.2.1 can be applied, but with emphasis on occupation numbers $n = 1$ and $n = 2$ rather than $n = 0$ and $n = 1$. For Group III acceptors, $n = 0$ corresponds to a state with the unpaired $2p$ electron removed. This state has an energy so high that it can be neglected.

Using the grand canonical ensemble one can express the probability of finding a pair of $2p$ electrons (or no hole) on a boron atom as

$$1 - P_1 \simeq \frac{e^{\beta(2E_F - E_2)}}{4e^{\beta(E_F - E_1)} + e^{\beta(2E_F - E_2)}} = \frac{1}{1 + 4e^{\beta(E_I^a - E_F)}}, \tag{6.68}$$

where E_1 and E_2 are the energies of the $n = 1$ and $n = 2$ states, respectively, $E_I^a = E_2 - E_1$, and the factor of 4 appears in the denominator because the acceptor ground state is four-fold degenerate. Since a bound pair of $2p$ electrons on a boron atom corresponds to a hole that is ionized into the valence band, the concentration of free holes is

$$p = (1 - P_1) n_a = \frac{n_a}{1 + 4e^{\beta(E_I^a - E_F)}}. \tag{6.69}$$

One should note the interchange of E_F and the impurity energy in Eq. (6.69) compared to their order in Eq. (6.48) for donor impurities. The present treatment thus justifies the inversion of signs of E_F and E_I that was introduced in the treatment of acceptors based on the hole picture.

The quantities P_1 defined by Eqs. (6.46) and (6.64) for donors and acceptors can be written in the general form

$$P_1 = \frac{1}{1 + \frac{1}{g} e^{\pm\beta(E_I - E_F)}}, \tag{6.70}$$

where the + and − signs refer to donors and acceptors, respectively, and g is the degeneracy of the impurity level. The corresponding expression for the free carrier concentration is

$$\left.\begin{matrix} n \\ p \end{matrix}\right\} = \frac{1}{1 + g e^{\pm\beta(E_F - E_I)}} \times \left\{\begin{matrix} n_d \\ n_a \end{matrix}\right. . \tag{6.71}$$

6.2.3 Compensated semiconductors

In a **compensated semiconductor** both donor and acceptor impurities are present and distributed randomly. Electrons from the donors will transfer to the acceptors thus creating ionized donors and acceptors, even at $T = 0$ K. Partial compensation means that not all donors and acceptors are ionized at $T = 0$ K. Ideal compensation means that all donors and acceptors are ionized at $T = 0$ K. In this section we treat an n-type semiconductor such as n-Si that is partially compensated by acceptor impurities.

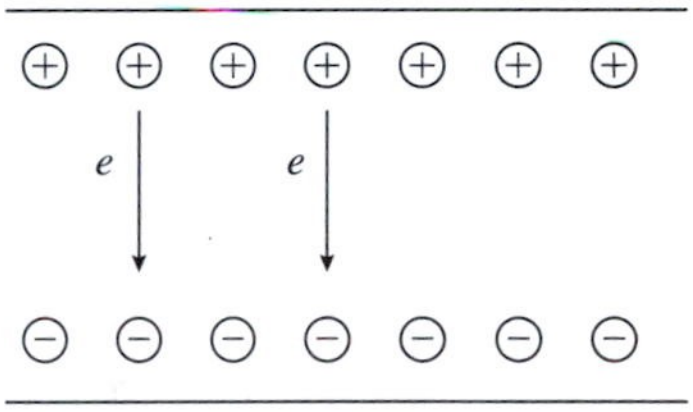

Fig. 6.8
Schematic representation of electron transfer from donor to acceptor impurities.

Let the concentrations of donor and acceptor impurities be n_d and n_a, respectively, with $n_d > n_a$. We assume, for simplicity, that the impurities are group V elements for the donors and group III elements for the acceptors. At the temperatures of interest, $k_B T \ll E_g$, all the acceptor levels are occupied by electrons transferred from donor impurities as indicated in the diagram of Fig. 6.8. The concentration of neutral donors n_d^0 is given by

$$n_d^0 = n_d - n_a - n, \tag{6.72}$$

where n is the concentration of electrons thermally excited to the conduction band.

Following the analysis based on the grand canonical ensemble given in Section 6.2.1, we can write the fraction of donor levels that are occupied by electrons as

$$\frac{n_d^0}{n_d} = \frac{1}{1 + \frac{1}{2} e^{\beta(E_I^d - E_F)}} = \frac{1}{1 + \frac{1}{2} e^{\beta}(E_C - E_d - E_F)}. \tag{6.73}$$

The energy difference $E_C - E_F$ can be eliminated using Eq. (6.28) and the result simplified with the aid of Eq. (6.57) to give

$$\frac{n_d}{n_d^0} = 1 + \frac{r\bar{N}_c'}{2n} e^{-\beta E_d} = 1 + \frac{\bar{N}_c^{eff}}{n}. \tag{6.74}$$

Using Eq. (6.65) and rearranging yields

$$\frac{n(n + n_a)}{n_d - n_a - n} = \bar{N}_c^{eff} \tag{6.75}$$

which is the generalization of Eq. (6.54) to include compensation by acceptor impurities. The solution to Eq. (6.75) is

$$n = \frac{2(n_d - n_a)}{1 + \frac{n_a}{\bar{N}_c^{eff}} + \left[\left(1 - \frac{n_a}{\bar{N}_c^{eff}}\right)^2 + 4\frac{n_d}{\bar{N}_c^{eff}}\right]^{\frac{1}{2}}}, \tag{6.76}$$

where n_c is given by Eq. (6.57).

Example 6.4: Conduction electron concentration with compensation present
Derive expressions for n with $n_d \gg n_a \neq 0$ and $k_B T \ll E_d$ or $k_B T \gg E_d$.
Solution. In the very low-temperature limit, n is much less than n_a and can be neglected compared to the latter. We then obtain directly from Eq. (6.75) the result

$$n \simeq \left(\frac{n_d - n_a}{n_a}\right)\bar{N}_c^{eff} = \frac{1}{2}\left(\frac{n_d - n_a}{n_a}\right) r\bar{N}_c' e^{-\beta E_d} \tag{6.77}$$

which shows clearly that n decreases as n_a increases. In the high-temperature limit with both n_a and n_d very small compared to n_c, Eq. (6.76) can be expanded in Taylor series to give

$$n \simeq n_d - n_a \tag{6.78}$$

corresponding to all noncompensated donors being ionized.

The Fermi energy can be calculated by eliminating n from Eqs. (6.28) and (6.76) and solving for E_F. The result is a straightforward generalization of Eq. (6.58).

6.2.4 Majority and minority carriers

In previous sections we have seen that electronic transitions from the valence band to the conduction band in intrinsic material lead to electrons in the conduction band and holes in the valence band. Furthermore, the excitation of donor impurities in n-type material and acceptor impurities in p-type material leads to conduction electrons and valence holes, respectively. Our discussion of the concentrations of free carriers in the latter two cases, however, did not develop explicit expressions for the concentrations of holes in n-type material and electrons in p-type material that arise from interband excitation. We now remedy this deficiency.

6.2.4.1 n-type material

Consider a semiconductor with donor concentration n_d, donor ionization energy E_d, and no compensating acceptor impurities: $n_a = 0$. At very low temperatures, $k_BT \ll E_d$, the vast majority of the impurity electrons are bound to the donor centers and the valence band is essentially completely full. The Fermi energy will then be situated between the donor energy level E_1^d and the bottom of the conduction band E_C as specified by Eq. (6.59).

With increasing temperature the donor impurities progressively ionize. The Fermi energy drops and eventually falls below the donor energy level. At this point essentially all donor impurities are ionized, and the concentration of conduction electrons is equal to the saturation range value:

$$n = n_d. \tag{6.79}$$

With further increase in the temperature, thermal excitation of carriers across the gap becomes significant and leads to the production of holes in the valence band.

The hole concentration p is given by the law of mass action $np = n_i^2$, where n_i is the intrinsic carrier concentration at the temperature of interest. Noting that $n_i \ll n, n_d$ we see from Eq. (6.50) that Eq. (6.79) remains a very good approximation. Then

$$p = \frac{n_i^2}{n_d}. \tag{6.80}$$

For Si at room temperature the intrinsic carrier concentration n_i is $1.5 \times 10^{10}\,\mathrm{cm}^{-3}$. If the concentration of donors n_d is $10^{16}\,\mathrm{cm}^{-3}$, the hole concentration p at room temperature is $2.2 \times 10^4\,\mathrm{cm}^{-3}$ which is very small compared to both n_d and n_i. For this reason, the conduction electrons in an n-type material are called the **majority carriers** and the holes are called the **minority carriers**. In the saturation range the conductivity of n-type material depends only on the donor impurity concentration; the minority holes give a negligible contribution to the conductivity.

6.2.4.2 p-type material

If the semiconductor is doped only with acceptor impurities, the conductivity is p-type, and there are no compensating donor impurities: $n_d = 0$. In the saturation range,

$$p = n_a. \tag{6.81}$$

As the temperature increases, interband excitation of conduction electrons and holes increases.

Proceeding as in the case of n-type material, we find under typical conditions that the conduction electron concentration is much less than that of the holes. Consequently, in a p-type material, the **majority carriers are holes** and the **minority carriers are conduction electrons**.

In Fig. 6.9 are presented schematic representations of the density-of-states, Fermi distribution, and free carrier concentrations as functions of energy for n- and p-type semiconductors.

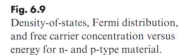

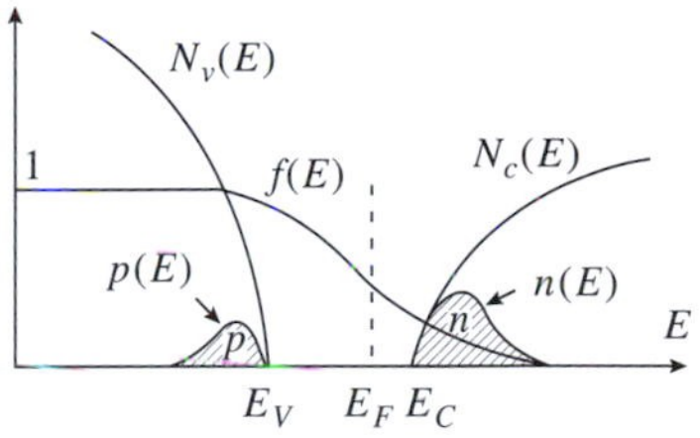

Fig. 6.9
Density-of-states, Fermi distribution, and free carrier concentration versus energy for n- and p-type material.

6.2.5 Contribution of excited impurity states

As discussed in Chapter 5, the donor ground state in Si is six-fold degenerate within effective mass theory if the degeneracy due to spin is neglected.

One state is associated with each of the six conduction band minima. Inclusion of the valley–orbit interaction splits the degeneracy to produce a singlet A_1 ground state, a triplet T_2 excited state, and a doublet E excited state. The A_1–T_2 splitting is ~ 10 meV and the T_2–E splitting is ~ 1 meV. If the temperature is such that the splittings are comparable to $k_B T$, Eqs. (6.48) and (6.53) must be modified to include the contributions of the excited states. Furthermore, there are additional excited bound states associated with $2s, 2p, 3s, \ldots$ levels. These levels are typically sufficiently close to the conduction band edge that they can be neglected if $n_d/\bar{N}_c \ll 1$.

In Ge the four-fold degeneracy of the donor ground state in effective mass theory is split by the valley–orbit interaction into a singlet ground state and a triplet excited state. Considerations similar to those for Si apply also to the contribution of excited donor states in Ge.

The grand canonical ensemble enables one to account for the occupancy of excited states very easily. If the system has impurity levels of energies $E_{Ik}, E_{I\ell}, \ldots$ and occupation numbers $n_k, n_\ell, \ldots$ the probability of the occurrence of this state in an ensemble of members containing all different possible occupation numbers may be written as (Tolman 1938)

$$P(n_k, n_\ell, \ldots) = Ce^{\beta[(n_k+n_\ell+\cdots)E_F-(n_k E_{Ik}+n_\ell E_{I\ell}+\cdots)]}, \tag{6.82}$$

where degenerate levels are not accounted for explicitly and we are focusing on donor impurities. Let us now assume that an impurity ion can trap only one carrier. Since the values of $n_k, n_\ell, \ldots$ are restricted to 0 or 1, the occupation numbers must satisfy one or the other of the constraints

$$n_k + n_\ell + \cdots = 0 \tag{6.83a}$$

$$n_k + n_\ell + \cdots = 1. \tag{6.83b}$$

In the latter case one of the occupation numbers is unity and all the others are zero. The occupation number that is unity can be any one of them, however. The normalized probability that one impurity level is occupied can be written as

$$P_1 = \frac{S_1}{1+S_1}, \tag{6.84}$$

where

$$S_1 = \sum_{n_k, n_\ell, \ldots} e^{\beta[E_F-(n_k E_{Ik}+n_\ell E_{I\ell}+\cdots)]} \tag{6.85}$$

and the constraint stated in Eq. (6.83b) applies.

If an impurity level is degenerate with energy E_{Is} and degeneracy factor g_s, each of these degenerate levels will have its occupation number take on the value unity in turn, thus giving rise to g_s identical terms in the sum S_1. Taking into account the degeneracies of all levels, we can express P_1 in the alternative form

$$P_1 = \frac{\sum_s g_s e^{\beta(E_F-E_{Is})}}{1+\sum_s g_s e^{\beta(E_F-E_{Is})}} = \frac{1}{1+[\sum_s g_s e^{\beta(E_F-E_{Is})}]^{-1}} \tag{6.86}$$

which is the generalization of Eq. (6.46) to an arbitrary number of degenerate states.

The conduction electron concentration is

$$n = n_d(1 - P_1) = \frac{n_d}{1 + \sum_s g_s e^{\beta(E_F - E_{Is})}}. \tag{6.87}$$

Combining this result with Eq. (6.28) gives a relation that determines the Fermi energy:

$$r\bar{N}_c' e^{\beta(E_F - E_C)} = \frac{n_d}{1 + \sum_s g_s e^{\beta(E_F - E_{Is})}}. \tag{6.88}$$

A numerical solution of Eq. (6.88) yields the Fermi energy as a function of temperature. The electron concentration can then be calculated from Eq. (6.28) for various temperatures. Electrons thermally excited from the valence band can be accounted for as in Eq. (6.61). The case of acceptor impurities can be handled in similar fashion.

Problems

1. Calculate the concentration of conduction electrons and holes in intrinsic Si at 300 K in the approximation of spherical, parabolic valence bands. The values of the required parameters are: $E_g = 1.14\,\text{eV}$, $r = 6$, $m_\ell^* = 0.92\,m$, $m_t^* = 0.19\,m$, $m_{\ell h}^* = 0.16\,m$, $m_{hh}^* = 0.52\,m$.
2. Consider a sample of Si doped with As to a concentration $n_d = 10^{17}\,\text{cm}^{-3}$.
   (a) Calculate the concentration of electrons occupying impurity levels at 300 K.
   (b) Deduce the concentrations of both electrons in the conduction band and holes in the valence band at that temperature.
   (c) Deduce the value of the Fermi energy.
   The value of the impurity ionization energy E_d is 0.03 eV.
3. For the case of donor impurities find explicit expressions for E_F in the two limiting cases
   (a) $\left(\frac{8n_d}{r\bar{N}_c'}\right) e^{E_d/k_B T} \ll 1$ (weak doping)
   (b) $\left(\frac{8n_d}{r\bar{N}_c'}\right) e^{E_d/k_B T} \gg 1$ (low temperature).
4. Consider n-Ge with As impurities and singlet A_1 ground state and triplet T_2 excited state ionization energies given in Table 5.1. For $n_d = 10^{16}\,\text{cm}^{-3}$ and neglecting higher bound states, calculate E_F as a function of temperature T over the range $0 \le T \le 150\,\text{K}$ and plot E_F versus T. Calculate n as a function of T and plot n versus T.

References

E. A. Guggenheim, *Proc. Phys. Soc. Lond.* A**66**, 121 (1953).
P. T. Landsberg, *Semiconductors and Phosphors* (Interscience, New York, 1958).
S. Teitler and R. F. Wallis, *J. Phys. Chem. Solids* **16**, 71 (1960).
R. C. Tolman, *Principles of Statistical Mechanics* (Oxford University Press, Oxford, 1938).

Lattice vibrations in semiconductors

7

Key ideas

In the *harmonic approximation* the equations of motion are linear in the displacement components of the atoms.

The *normal mode frequencies* of a monatomic linear chain are confined to a band between zero and a maximum frequency.

The normal mode frequencies of a diatomic linear chain lie in the *acoustic branch* or the *optical branch* with a *gap* between the branches.

Elastic continuum theory provides a simple treatment of long-wavelength modes of vibration.

Phonon dispersion curves are determined by *inelastic neutron scattering*. Short range interactions are insufficient to account for the experimental data. The deformability of the electron charge distribution is taken into account by the *shell model* and the *bond charge model*. The partial ionic character of the electron-pair bonds and the effective charge of the atom is taken into account by the *deformation dipole model*. The *linear response method* provides full phonon dispersion curves without fitting parameters to experimental curves.

In a *normal mode of vibration* all atoms vibrate with the same frequency.

The vibrational specific heat obeys the *Debye* T^3*-law* at low temperatures and the *Dulong–Petit law* at higher temperatures.

Anharmonic effects are responsible for *thermal expansion* and *diffusive thermal conductivity*.

Impurities and other defects can give rise to *localized modes*.

Piezoelectricity can increase the elastic moduli and the speed of sound.

Applied *stress* can cause shifts and splittings of electronic and vibrational energy levels.

Phonons

An important source of deviations from periodicity in a crystal is the displacement of an atom from its equilibrium position. These deviations arise naturally from the thermal energy of the atoms. If the crystal is in a stable configuration, the displacement of an atom leads to a force that tends to restore the atom to its equilibrium position. However, the kinetic energy of the atom causes it to overshoot the latter. As a result the atom vibrates about its equilibrium position. The forces of interaction between atoms couple the atomic vibrations together, giving rise to **lattice vibrations.**

7.1 Equations of motion

As developed in the Born–Oppenheimer approximation discussed in Chapter 2, the potential energy for the nuclear motion, $\Phi(\boldsymbol{R})$, is given by the sum of the electronic energy eigenvalue, $E_e(\boldsymbol{R})$, and the ion–ion interaction energy, $V_{ii}(\boldsymbol{R})$. At moderate temperatures for which the nuclear displacements are small, we can expand the nuclear potential energy in powers of their displacement components $u_\alpha(\ell\kappa)$,

$$\Phi(\boldsymbol{R}) = \Phi_0 + \sum_{\ell\kappa\alpha} \Phi_\alpha(\ell\kappa) u_\alpha(\ell\kappa) + \frac{1}{2}\sum_{\ell\kappa\alpha}\sum_{\ell'\kappa'\beta} \Phi_{\alpha\beta}(\ell\kappa;\ell'\kappa') \times u_\alpha(\ell\kappa)u_\beta(\ell'\kappa') + \sum_{\ell\kappa\alpha}\sum_{\ell'\kappa'\beta}\sum_{\ell''\kappa''\gamma} \Phi_{\alpha\beta\gamma}(\ell\kappa;\ell'\kappa';\ell''\kappa'') u_\alpha(\ell\kappa)u_\beta(\ell'\kappa')u_\gamma(\ell''\kappa'') + \cdots, \tag{7.1}$$

where

$$\boldsymbol{u} = \boldsymbol{R} - \boldsymbol{R}^{(0)} \tag{7.2}$$

and the superscript zero denotes the equilibrium value. The terms up to and including the quadratic terms are the **harmonic terms**, while those terms of higher order than quadratic are **anharmonic**. Since the right hand side of Eq. (7.1) is a Taylor series, the coefficients have the interpretation

$$\Phi_\alpha(\ell\kappa) = \left.\frac{\partial\Phi}{\partial u_\alpha(\ell\kappa)}\right|_{\boldsymbol{u}=0}, \quad \Phi_{\alpha\beta}(\ell\kappa;\ell'\kappa') = \left.\frac{\partial^2\Phi}{\partial u_\alpha(\ell\kappa)\partial u_\beta(\ell'\kappa')}\right|_{\boldsymbol{u}=0},$$
$$\Phi_{\alpha\beta\gamma}(\ell\kappa;\ell'\kappa';\ell''\kappa'') = \left.\frac{\partial^3\Phi}{\partial u_\alpha(\ell\kappa)\partial u_\beta(\ell'\kappa')\partial u_\gamma(\ell''\kappa'')}\right|_{\boldsymbol{u}=0}. \tag{7.3}$$

From the symmetry of partial derivatives, we have the relation

$$\Phi_{\alpha\beta}(\ell\kappa;\ell'\kappa') = \Phi_{\beta\alpha}(\ell'\kappa';\ell\kappa) \tag{7.4}$$

with similar relations for $\Phi_{\alpha\beta\gamma}(\ell\kappa;\ell'\kappa';\ell''\kappa'')$ obtained by permuting $\ell\kappa\alpha$, $\ell'\kappa'\beta$, and $\ell''\kappa''\gamma$.

The solution of the Schrödinger equation for the nuclei, Eq. (2.4), is facilitated by solving the classical problem of the motion of the nuclei. The classical equations of motion are given by a set of equations of which those for nucleus $\ell\kappa$ are

$$M_\kappa \frac{\partial^2 u_\alpha(\ell\kappa)}{\partial t^2} = -\frac{\partial\Phi(\boldsymbol{R})}{\partial u_\alpha(\ell\kappa)} \tag{7.5}$$

with $\alpha = x, y, z$. Restricting ourselves to the harmonic terms in Eq. (7.1), we evaluate the right hand side of Eq. (7.5) and obtain

$$M_\kappa \frac{\partial^2 u_\alpha(\ell\kappa)}{\partial t^2} = -\Phi_\alpha(\ell\kappa) - \sum_{\ell'\kappa'\beta} \Phi_{\alpha\beta}(\ell\kappa;\ell'\kappa')u_\beta(\ell'\kappa'). \tag{7.6}$$

If all displacement components $u_\beta(\ell'\kappa')$ are zero, so that all atoms are in their equilibrium positions, there can be no force acting on any atom; hence,

$$\Phi_\alpha(\ell\kappa) = 0 \tag{7.7}$$

for all $\ell\kappa\alpha$. Furthermore, if all atoms are displaced rigidly according to

$$u_\beta(\ell'\kappa') = \epsilon\delta_{\beta\gamma}, \tag{7.8}$$

where ϵ is a constant, no relative motion of the atoms occurs and no interatomic forces arise; hence,

$$\sum_{\ell'\kappa'\beta} \Phi_{\alpha\beta}(\ell\kappa;\ell'\kappa')\epsilon\delta_{\beta\gamma} = 0$$

or

$$\sum_{\ell'\kappa'\gamma} \Phi_{\alpha\gamma}(\ell\kappa;\ell'\kappa') = 0. \tag{7.9}$$

Equation (7.9) states the condition of **infinitesimal translational invariance** that must be satisfied by the force constants $\Phi_{\alpha\gamma}(\ell\kappa;\ell'\kappa')$. From this condition and replacing γ by β, we obtain the alternative form

$$\Phi_{\alpha\beta}(\ell\kappa;\ell\kappa) = -{\sum_{\ell'\kappa'}}' \Phi_{\alpha\beta}(\ell\kappa;\ell'\kappa'), \tag{7.10}$$

where the prime on the sum denotes that the term with $\ell'\kappa' = \ell\kappa$ is omitted.

We now rewrite the equations of motion in two forms that will prove to be useful. Using Eq. (7.7), Eq. (7.6) becomes

$$M_\alpha \frac{\partial^2 u_\alpha(\ell\kappa)}{\partial t^2} = -\sum_{\ell'\kappa'\beta} \Phi_{\alpha\beta}(\ell\kappa;\ell'\kappa')u_\beta(\ell'\kappa'). \tag{7.11}$$

Eliminating $\Phi_{\alpha\beta}(\ell\kappa;\ell\kappa)$ from the latter equation with the aid of Eq. (7.10) yields

$$M_\kappa \frac{\partial^2 u_\alpha(\ell\kappa)}{\partial t^2} = {\sum_{\ell'\kappa'\beta}}' \Phi_{\alpha\beta}(\ell\kappa;\ell'\kappa')[u_\beta(\ell\kappa) - u_\beta(\ell'\kappa')]. \tag{7.12}$$

Both forms of the equations of motion just obtained are linear homogeneous differential-difference equations. This suggests that we seek plane wave solutions of the form

$$u_\alpha(\ell\kappa;\boldsymbol{q},t) = M_\kappa^{-\frac{1}{2}} W_\alpha(\kappa)e^{i[\boldsymbol{q}\cdot\boldsymbol{R}(\ell)-\omega t]}, \tag{7.13}$$

in which $\boldsymbol{W}(\kappa)$ is a measure of the amplitude of the wave, $\boldsymbol{q}$ is the wave vector, and ω is the angular frequency. Substitution of this trial solution into Eq. (7.12) gives the linear, homogeneous algebraic equations

$$\sum_{\kappa'\beta}[D_{\alpha\beta}(\kappa\kappa';\boldsymbol{q}) - \omega^2\delta_{\alpha\beta}\delta_{\kappa\kappa'}]W_\beta(\kappa') = 0, \tag{7.14}$$

where

$$D_{\alpha\beta}(\kappa\kappa';\boldsymbol{q}) = (M_\kappa M_{\kappa'})^{-\frac{1}{2}} \sum_{\ell'} \Phi_{\alpha\beta}(\ell\kappa;\ell'\kappa') e^{i\boldsymbol{q}\cdot[\boldsymbol{R}(\ell')-\boldsymbol{R}(\ell)]} \tag{7.15}$$

is an element of the **dynamical matrix** for the crystal. If we impose periodic boundary conditions on the atomic displacements, the values of the wave vector $\boldsymbol{q}$ are specified by Eq. (2.28) with $\boldsymbol{k} \rightarrow \boldsymbol{q}$. Under periodicity, the force constants $\Phi_{\alpha\beta}(\ell\kappa;\ell'\kappa')$ depend only on the difference $\bar{\ell} = \ell' - \ell$ and not on ℓ and ℓ' separately. Consequently, we can rewrite Eq. (7.15) with the aid of Eq. (1.1) as

$$D_{\alpha\beta}(\kappa\kappa';\boldsymbol{q}) = (M_\kappa M_{\kappa'})^{-\frac{1}{2}} \sum_{\bar{\ell}} \Phi_{\alpha\beta}(0\kappa;\bar{\ell}\kappa') e^{i\boldsymbol{q}\cdot\boldsymbol{R}(\bar{\ell})}. \tag{7.16}$$

The nontrivial solution of Eqs. (7.14) requires that the determinant of the coefficients of the amplitudes $W_\beta(\kappa')$ be zero:

$$\left| D_{\alpha\beta}(\kappa\kappa';\boldsymbol{q}) - \omega^2 \delta_{\alpha\beta}\delta_{\kappa\kappa'} \right| = 0. \tag{7.17}$$

This is the **secular equation**. If the crystal is three-dimensional and has r atoms per primitive unit cell, the dynamical matrix has dimensions $3r \times 3r$, and the secular equation has $3r$ eigenvalues for ω^2 for each value of $\boldsymbol{q}$. These eigenvalues are distinguished by the **branch index** j. For a stable crystal the eigenvalues must be non-negative. The square roots of the eigenvalues are the frequencies of the **normal modes of vibration** of the crystal and are designated by $\omega_{\boldsymbol{q}j}$. They are conventionally taken to be non-negative. A plot of $\omega_{\boldsymbol{q}j}$ versus $\boldsymbol{q}$ constitutes a **phonon dispersion curve**. After the normal mode frequencies are calculated, they may be substituted into Eq. (7.14) and the corresponding amplitudes $W_\beta(\kappa';\boldsymbol{q}j)$ determined. The latter are the **eigenvectors** of the dynamical matrix.

The eigenvectors can be chosen to be orthonormal. A particularly convenient choice is designated by $e_{\alpha\kappa}(\boldsymbol{q}j)$, for which the orthonormality condition takes the form

$$\sum_{\kappa\alpha} e^*_{\alpha\kappa}(\boldsymbol{q}j) e_{\alpha\kappa}(\boldsymbol{q}j') = \delta_{jj'}. \tag{7.18}$$

The property of closure is also satisfied:

$$\sum_{j} e^*_{\alpha\kappa}(\boldsymbol{q}j) e_{\beta\kappa'}(\boldsymbol{q}j) = \delta_{\alpha\beta}\delta_{\kappa\kappa'}. \tag{7.19}$$

Of the $3r$ branches three of them have the property that their frequencies $\omega_{\boldsymbol{q}j}$ go to zero as the wave vector $\boldsymbol{q}$ goes to zero. These three branches are called **acoustic branches**, because in the small q limit (long wavelength limit) they correspond to sound waves. The remaining $3r - 3$ branches are called **optical branches**, because they strongly scatter or absorb electromagnetic radiation when q is very small. If we focus on the acoustic branches of cubic crystals and take $\boldsymbol{q}$ in a high-symmetry direction such as $\langle 100\rangle$, $\langle 110\rangle$, or $\langle 111\rangle$, one branch has all atomic displacements parallel to $\boldsymbol{q}$ and is called the

longitudinal acoustic branch. The other two branches have their atomic displacements perpendicular to $\boldsymbol{q}$ and are called **transverse acoustic branches**. The transverse branches can be chosen such that the displacements of the one are orthogonal to those of the other. In similar fashion the normal modes of the optical branches can be classified as longitudinal or transverse. If $\boldsymbol{q}$ is not in a high-symmetry direction, the branches have a mixed longitudinal-transverse character.

> **Comment.** A semiconductor of the diamond or zinc-blende structure has two atoms per unit cell. There are six phonon branches of which three are optical with two transverse and one longitudinal and three acoustic with two transverse and one longitudinal.

7.2 Monatomic linear chain

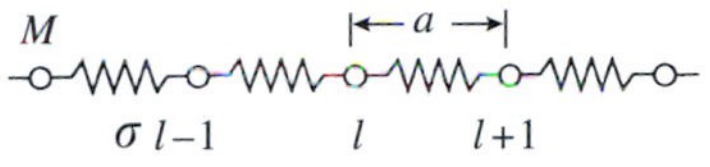

Fig. 7.1
Diagram of monatomic linear chain with nearest neighbor interactions.

We now illustrate the general procedures given in the preceding sections by some simple examples. The first case that we consider is the monatomic linear chain with nearest neighbor interactions. We assume only one atom per primitive unit cell. Both the κ index and the α index can be suppressed. A diagram of the system is shown in Fig. 7.1. Referring to the equations of motion given by Eq. (7.11), with nearest neighbor interactions only, the nonzero force constants are $\Phi(\ell, \ell+1)$, $\Phi(\ell, \ell)$, and $\Phi(\ell, \ell-1)$. From Eq. (7.4) and the periodicity of the chain, $\Phi(\ell, \ell+1)$ and $\Phi(\ell, \ell-1)$ must be equal and have the same value for all ℓ. If $u(\ell+1) > u(\ell)$, we see from Eq. (7.11) that the force constant $\Phi(\ell, \ell+1)$ must be negative in order to have the force on atom ℓ act in the direction of increasing ℓ, as it should from physical considerations. Setting

$$\Phi(\ell, \ell+1) = \Phi(\ell, \ell-1) = -\sigma, \tag{7.20}$$

we obtain from Eq. (7.10)

$$\Phi(\ell, \ell) = 2\sigma. \tag{7.21}$$

Using these results the equations of motion become

$$M\frac{\partial^2 u(\ell)}{\partial t^2} = \sigma[u(\ell+1) + u(\ell-1) - 2u(\ell)]. \tag{7.22}$$

The elements of the dynamical matrix specified by Eq. (7.16) are given by

$$\begin{aligned} D(\boldsymbol{q}) &= \frac{1}{M}[2\sigma - \sigma e^{iqa} - \sigma e^{-iqa}] \\ &= \frac{2\sigma}{M}(1 - \cos qa) \\ &= \frac{4\sigma}{M}\sin^2(qa/2), \end{aligned} \tag{7.23}$$

where a is the lattice constant. The secular equation, Eq. (7.17), yields the normal mode frequencies

$$\omega_{\hat{q}} = \left(\frac{4\sigma}{M}\right)^{\frac{1}{2}} \sin(qa/2). \tag{7.24}$$

The values of the wave vector q specified by the periodic boundary condition $u(N+\ell) = u(\ell)$ are $q = 2\pi n/Na$, where N is the number of atoms in the chain and n is an integer. The vectors of the reciprocal lattice G are given by $G = 2\pi n/a$. For the case under consideration, Eq. (7.13) for the displacements reduces to

$$u_q(\ell;t) = M^{-\frac{1}{2}} W e^{i(qa\ell - \omega t)}. \tag{7.25}$$

If we augment q by a reciprocal lattice vector G, we obtain

$$\begin{aligned} u_{q+G}(\ell;t) &= M^{-\frac{1}{2}} W e^{i\left[\left(q+\frac{2\pi n}{a}\right)a\ell - \omega t\right]} \\ &= u_{\hat{q}}(\ell;t). \end{aligned} \tag{7.26}$$

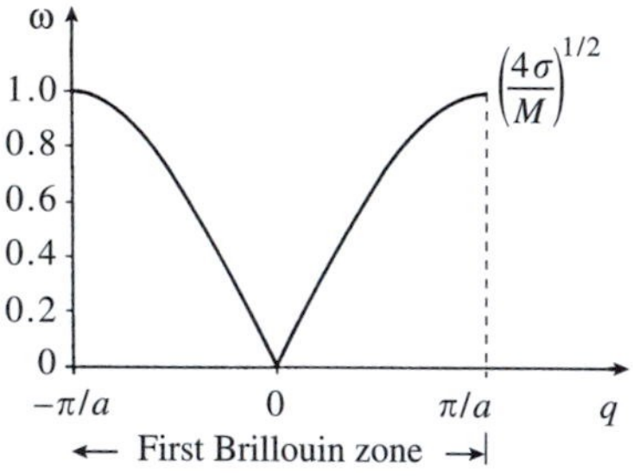

Fig. 7.2
Normal mode frequencies versus wave vector for a monatomic linear chain.

Thus, the physically distinct sets of displacements corresponding to the normal modes of vibration can be restricted to values of q in the first Brillouin zone. There are therefore N normal modes corresponding to the N values of q in the first Brillouin zone. One can have longitudinal modes with the $u(\ell)$ parallel to the chain axis or transverse modes with the $u(\ell)$ perpendicular to the axis. The force constants for these two cases would in general be different.

A plot of the normal mode frequencies versus q is shown in Fig. 7.2. For q small, ω_q increases linearly with q. At larger q, ω_q increases more slowly and reaches its maximum value at the Brillouin zone boundary. The normal mode at the zone boundary has atomic displacements that alternate in sign moving along the chain, but have the same magnitude.

7.3 Diatomic linear chain

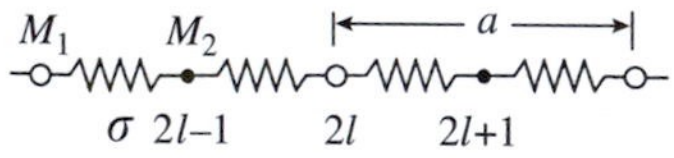

Fig. 7.3
Diagram of diatomic linear chain with alternating masses M_1 and M_2 and nearest neighbor interactions.

No real semiconductor has only one atom per primitive unit cell. Since group IV semiconductors as well as III–V semiconductors have two atoms per primitive cell, one can get a qualitative picture of their normal vibrational modes by considering linear chains with either alternating force constants or alternating atomic masses. We shall discuss the latter explicitly and refer the former to a problem.

Consider the diatomic linear chain with alternating masses M_1 and M_2 and nearest neighbor interactions as shown in Fig. 7.3. We assume periodic boundary conditions with N unit cells in the period. The nonzero elements of the force constant matrix are of the following form:

$$\Phi(\ell,1;\ell,2) = \Phi(\ell,1;\ell-1,2) = \Phi(\ell,2;\ell+1,1) = \Phi(\ell,2;\ell,1) = -\sigma \tag{7.27a}$$

$$\Phi(\ell,1;\ell,1) = \Phi(\ell,2;\ell,2) = 2\sigma. \tag{7.27b}$$

The equations of motion take the

$$M_1 \frac{\partial^2 u(\ell,1)}{\partial t^2} = \sigma[u(\ell,2) + u(\ell-1,2) - 2u(\ell,1)] \tag{7.28a}$$

$$M_2 \frac{\partial^2 u(\ell,2)}{\partial t^2} = \sigma[u(\ell+1,2) + u(\ell,1) - 2u(\ell,2)]. \tag{7.28b}$$

The elements of the dynamical matrix are

$$D(11;q) = M_1^{-1}(2\sigma), \quad D(22;q) = M_2^{-1}(2\sigma) \tag{7.29a}$$

$$D(12;q) = -(M_1 M_2)^{-\frac{1}{2}}\sigma(1 + e^{-iqa}) \tag{7.29b}$$

$$D(21;q) = -(M_1 M_2)^{-\frac{1}{2}}\sigma(1 + e^{iqa}) \tag{7.29c}$$

which, upon substitution into Eq. (7.17) give the secular equation

$$\begin{vmatrix} \dfrac{2\sigma}{M_1} - \omega^2 & -\dfrac{\sigma}{(M_1 M_2)^{\frac{1}{2}}}(1 + e^{-iqa}) \\ -\dfrac{\sigma}{(M_1 M_2)^{\frac{1}{2}}}(1 + e^{iqa}) & \dfrac{2\sigma}{M_2} - \omega^2 \end{vmatrix} = 0. \tag{7.30}$$

Evaluating the determinant and solving the resulting quadratic equation in ω^2 yield

$$\begin{aligned} \omega_q^2 &= \sigma\left\{\frac{1}{M_1} + \frac{1}{M_2} \pm \left[\left(\frac{1}{M_1} + \frac{1}{M_2}\right)^2 - \frac{4}{M_1 M_2}\sin^2\frac{qa}{2}\right]^{\frac{1}{2}}\right\} \\ &= \frac{\sigma}{\bar{M}}\left\{1 \pm \left[1 - \frac{4\bar{M}^2}{M_1 M_2}\sin^2\frac{qa}{2}\right]^{\frac{1}{2}}\right\}, \end{aligned} \tag{7.31}$$

where $\bar{M}$ is the reduced mass of the two types of atoms:

$$\frac{1}{\bar{M}} = \frac{1}{M_1} + \frac{1}{M_2}. \tag{7.32}$$

Consistent with the fact that there are two atoms in the primitive unit cell, there are two branches in the lattice vibration spectrum corresponding to the plus and minus signs in Eq. (7.31). The branch with the minus sign is the **acoustic branch** labeled by $j = 1$, while the branch with the plus sign is the **optical branch** labeled by $j = 2$. The normal mode frequencies ω_{q1} and ω_{q2} are the appropriate square roots of the right hand side of Eq. (7.31). They are plotted as functions of q in Fig. 7.4.

It is worth while to work out explicit results for certain limiting cases, namely, $q = 0$ and $q = \pi/a$ (Brillouin zone boundary). The normal mode frequencies are easily evaluated from Eq. (7.31) and are listed in Table 7.1 for the case $M_1 > M_2$.

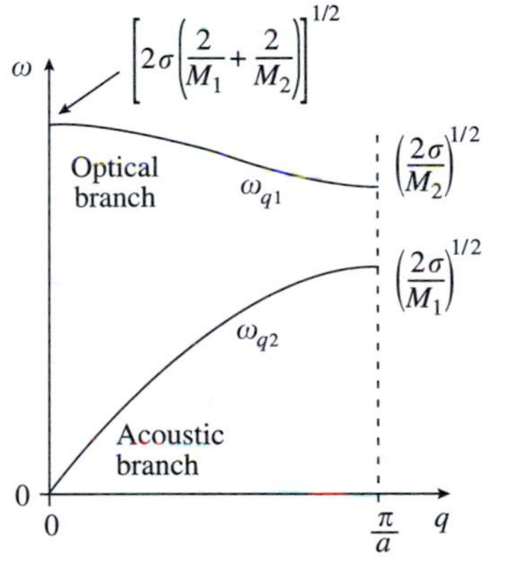

Fig. 7.4
Normal mode frequencies versus wave vector for a diatomic linear chain.

Table 7.1 Normal mode frequencies for the diatomic linear chain at two values of the wave vector q

q	ω_{q1}	ω_{q2}
0	0	$\left(\frac{2\sigma}{\bar{M}}\right)^{\frac{1}{2}}$
$\frac{\pi}{a}$	$\left(\frac{2\sigma}{M_1}\right)^{\frac{1}{2}}$	$\left(\frac{2\sigma}{M_2}\right)^{\frac{1}{2}}$

Also of interest are the amplitudes of the atomic displacements for these limiting cases. The amplitudes $W(1)$ and $W(2)$ may be obtained from Eq. (7.14) which take the form

$$\left(\frac{2\sigma}{M_1} - \omega^2\right) W(1) - \frac{\sigma}{(M_1 M_2)^{\frac{1}{2}}}(1 + e^{-iqa}) W(2) = 0 \tag{7.33a}$$

$$-\frac{\sigma}{(M_1 M_2)^{\frac{1}{2}}}(1 + e^{iqa}) W(1) + \left(\frac{2\sigma}{M_2} - \omega^2\right) W(2) = 0 \tag{7.33b}$$

for the case of interest. By substituting the appropriate values of the normal mode frequencies into either of these equations, one can determine the relationship between $W(1)$ and $W(2)$ for the wave vectors of interest. The amplitudes $U(1)$ and $U(2)$ of the physical displacements $u(\ell 1)$ and $u(\ell 2)$ are related to $W(1)$ and $W(2)$ by

$$U(1) = M_1^{-\frac{1}{2}} W(1), \quad U(2) = M_2^{-\frac{1}{2}} W(2), \tag{7.34}$$

as can be seen from Eq. (7.13).

To illustrate this procedure, let us consider the acoustic branch for $q = 0$. Then $\omega = 0$, and Eq. (7.33a) or Eq. (7.33b) yields

$$M_1^{-\frac{1}{2}} W(1) = M_2^{-\frac{1}{2}} W(2).$$

Eliminating $W(1)$ and $W(2)$ with the aid of Eq. (7.34), we obtain

$$U(1) = U(2) \tag{7.35}$$

which corresponds to a rigid translation of the crystal. The other limiting cases can be worked out in similar fashion. The results are presented in Table 7.2 for the case $M_1 > M_2$. We see that for the optical branch at $q = 0$, the two atoms in the unit cell move in opposite directions. For the acoustic branch at the Brillouin zone boundary, only the atoms of the heavier mass move, whereas for the optical branch, only the atoms of the lighter mass move.

Example 7.1: Vibrations of a square lattice
Derive an expression for the normal mode frequencies of a monatomic simple square lattice with nearest neighbor interactions.
Solution. The lattice is assumed to have lattice constant a with the atoms labeled by integers ℓ, m. The equations of motion have

Table 7.2 Relations between atomic displacement amplitudes for the diatomic linear chain at two values of the wave vector q for $M_1 > M_2$

q	Acoustic branch	Optical branch
0	$U(1) = U(2)$	$M_1 U(1) = -M_2 U(2)$
π/a	$U(1) \neq 0,\ U(2) = 0$	$U(1) = 0,\ U(2) \neq 0$

the form

$$M\ddot{u}_{\ell m} = \sigma\big(u_{\ell+1,m} + u_{\ell-1,m} - 2u_{\ell,m}\big) + \tau\big(u_{\ell,m+1} + u_{\ell,m-1} - 2u_{\ell m}\big)$$
$$M\ddot{v}_{\ell m} = \sigma\big(v_{\ell,m+1} + v_{\ell,m-1} - 2v_{\ell m}\big) + \tau\big(v_{\ell+1,m} + v_{\ell-1,m} - 2v_{\ell m}\big),$$

where u, v are the x, y components of displacement, M is the atomic mass, σ is the central force constant, and τ is the noncentral force constant. Since the equations of motion are uncoupled, we can take solutions of the form

$$u_{\ell m} = Ue^{i(q_x a\ell + q_y am - \omega t)}$$
$$v_{\ell m} = 0$$

and

$$u_{\ell m} = 0$$
$$v_{\ell m} = Ve^{i(q_x a\ell + q_y am - \omega t)}.$$

The equations of motion for the u-displacements yield normal mode frequencies specified by

$$\omega^2 = \frac{4\sigma}{M}\left[\sin^2\left(\frac{q_x a}{2}\right) + \frac{\tau}{\sigma}\sin^2\left(\frac{q_y a}{2}\right)\right],$$

while those for the v-displacements yield

$$\omega^2 = \frac{4\tau}{M}\left[\sin^2\left(\frac{q_x a}{2}\right) + \frac{\sigma}{\tau}\sin^2\left(\frac{q_y a}{2}\right)\right].$$

The wave vector components q_x, q_y lie in the range $-\pi/a$ to $+\pi/a$. For waves propagating in the x-direction, the u-displacement modes are longitudinal and the v-displacement modes are transverse. For propagation in the y-direction, the assignments are reversed. The model treated in this example is the **Rosenstock–Newell model** (Rosenstock and Newell 1953).

7.4 Three-dimensional crystals

7.4.1 Elastic continuum theory

Since real semiconductors are three-dimensional and have at least two atoms per primitive unit cell, the secular equation that must be solved to determine the normal mode frequencies, Eq. (7.17), involves a determinant that is at least 6×6 in size. Analytic solutions of this equation are feasible only in special cases, so one must use a computer for the general case. The situation is further complicated by the fact that the interatomic interactions are long range, even in homopolar semiconductors such as Si and Ge (Herman 1959). In heteropolar semiconductors such as GaAs, the atoms are electrically charged ions for which the long range Coulomb interaction must be taken into account.

There is one limiting case for which considerable progress can be made analytically, namely, the elastic continuum limit where the wavelength of the vibrational mode is much larger than the lattice constant. For a cubic crystal this condition takes the form $|\boldsymbol{q}|a \ll 1$. One can then expand $u_\alpha(\ell+\delta,\kappa)$ in powers of $R_\alpha(\delta)$:

$$u_\alpha(\ell+\delta,\kappa) \simeq u_\alpha(\ell\kappa) + \sum_\beta R_\beta(\delta)\frac{\partial u_\alpha(\ell\kappa)}{\partial x_\beta} + \frac{1}{2}\sum_{\beta\gamma} R_\beta(\delta)R_\gamma(\delta)\frac{\partial^2 u_\alpha(\ell\kappa)}{\partial x_\beta \partial x_\gamma} + \cdots. \tag{7.36}$$

Substituting this expansion into the equations of motion, Eq. (7.12), and retaining terms through second order, we obtain the equations of motion of elasticity theory

$$\rho\frac{\partial^2 u(\boldsymbol{r})}{\partial t^2} = C_{11}\frac{\partial^2 u(\boldsymbol{r})}{\partial x^2} + (C_{12}+C_{44})\left(\frac{\partial^2 v(\boldsymbol{r})}{\partial x \partial y} + \frac{\partial^2 w(\boldsymbol{r})}{\partial x \partial z}\right) + C_{44}\left(\frac{\partial^2 u(\boldsymbol{r})}{\partial y^2} + \frac{\partial^2 u(\boldsymbol{r})}{\partial z^2}\right), \tag{7.37}$$

plus two other equations obtained by cyclically permuting (u, v, w) and (x, y, z). We have expressed $\boldsymbol{u}$ as (u, v, w), ρ is the crystal density, and C_{11}, C_{12}, and C_{44} are the elastic moduli in the Voigt notation (Born and Huang 1954). The elastic moduli are linear combinations of the second order force constants $\Phi_{\alpha\beta}(\ell\kappa;\ell'\kappa')$. In Table 7.3 we present the elastic moduli for a number of cubic semiconductors.

The equations of motion of elasticity theory, as expressed by Eq. (7.37), are linear, homogeneous, partial differential equations which admit solutions of plane wave form

$$\boldsymbol{u}(\boldsymbol{r},t) = [u(\boldsymbol{r},t), v(\boldsymbol{r},t), w(\boldsymbol{r},t)] = (U, V, W)e^{i\boldsymbol{q}\cdot\boldsymbol{r}-i|\boldsymbol{q}|ct}, \tag{7.38}$$

where U, V, W are amplitudes and c is the speed of the wave. Substitution of this form into the equations of motion yields a set of three linear, homogeneous algebraic equations in U, V, W. The secular equation which must be solved to provide a nontrivial solution involves a 3×3 determinant and

Table 7.3 Elastic moduli (in 10^{11} dyne/cm^2 or 10^{10} N/m^2) for various cubic semiconductors (after Harrison 1980, Böer 1990)

Crystal	C_{11}	C_{12}	C_{44}	Crystal	C_{11}	C_{12}	C_{44}
C	107.6	12.50	57.7	InP	10.22	5.76	4.60
Si	16.57	6.39	7.96	InAs	8.33	4.53	3.96
Ge	12.89	4.83	6.71	InSb	6.67	3.65	3.02
AlAs	12.02	5.70	5.89	ZnS	10.40	6.50	4.62
GaP	14.12	6.25	7.05	ZnSe	8.10	4.88	4.41
GaAs	11.81	5.32	5.92	ZnTe	7.13	4.07	3.12
GaSb	8.84	4.03	4.32	CdTe	5.35	3.68	1.99

leads to a cubic equation in c^2. There are therefore three solutions for c^2 and three branches in the vibrational spectrum. All three are acoustic, since the frequency $\omega = |\boldsymbol{q}|c$ goes to zero as $|\boldsymbol{q}|$ goes to zero for each branch.

For certain high-symmetry directions of propagation, the secular equation can be factored and solutions of transverse or longitudinal polarization obtained. We list a few of these cases:

1. transverse wave, [100] propagation direction

$$\boldsymbol{u} = (0, v, 0) = (0, V, 0)e^{iq(x-c_t t)}$$
$$c_t = (C_{44}/\rho)^{\frac{1}{2}}$$

2. longitudinal wave, [100] propagation direction

$$\boldsymbol{u} = (u, 0, 0) = (U, 0, 0)e^{iq(x-c_\ell t)}$$
$$c_\ell = (C_{11}/\rho)^{\frac{1}{2}}$$

3. longitudinal wave, [110] propagation direction

$$\boldsymbol{u} = (u, v, 0) = (U, U, 0)e^{iq(x+y)/\sqrt{2}}e^{-iqc_\ell t}$$
$$c_\ell = [(C_{11} + C_{12} + 2C_{44})/2\rho]^{\frac{1}{2}}.$$

The speeds of acoustic waves such as these three can be measured experimentally using ultrasonic techniques (Truell *et al.* 1969). From the speeds and the density, the elastic moduli can be determined.

7.4.2 Three-dimensional lattices

The equations of motion for a three-dimensional lattice have the form specified by Eq. (7.12). The force constant matrices $\Phi_{\alpha\beta}(\ell\kappa, \ell'\kappa')$ are 3×3 matrices. For a crystal with periodic boundary conditions, the solutions to the equations of motion can be expressed as plane waves given by Eq. (7.13). If there are N unit cells in the crystal and r atoms per unit cell, the number of independent solutions is $3Nr$, i.e., the number of degrees of freedom.

An important quantity that enters into the calculation of many lattice-dynamical properties is the **frequency distribution function** $F(\omega)$. It is defined as the number of normal modes per unit frequency range and can be expressed in terms of the density-of-states in reciprocal space derived in Example 2.2,

$$g(\boldsymbol{q}) = \frac{\Omega}{(2\pi)^3}, \tag{7.39}$$

where the electron wave vector $\boldsymbol{k}$ has been replaced by the phonon wave vector $\boldsymbol{q}$. In order to obtain $F(\omega)$, we need to sum up the number of normal

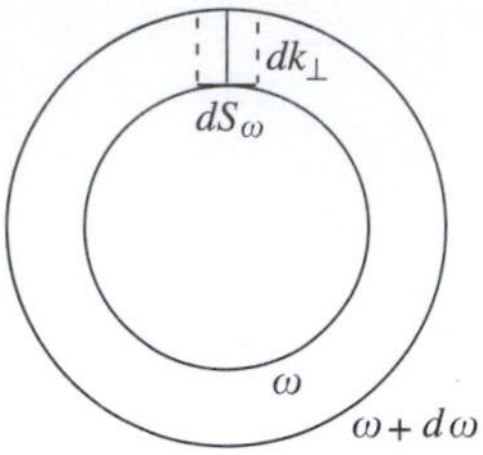

Fig. 7.5
Constant frequency surfaces in wave vector space.

modes with frequencies lying in a shell between two constant frequency surfaces at ω and $\omega + d\omega$ in $\boldsymbol{q}$-space, as shown in Fig. 7.5. This sum is given by

$$F(\omega)d\omega = \int_{\text{shell}} g(\boldsymbol{q})d^3q = \frac{\Omega}{(2\pi)^3}\int_{\text{shell}} d^3q.$$

The element of volume between the constant frequency surfaces has a base dS_ω and an altitude $dq_\perp$, so that

$$\int_{\text{shell}} d^3q = \int dS_\omega dq_\perp.$$

The element of frequency $d\omega$ is related to $dq_\perp$ by

$$d\omega = |\boldsymbol{\nabla}_{\boldsymbol{q}}\omega| dq_\perp,$$

where $\boldsymbol{\nabla}_{\boldsymbol{q}}\omega$ is the gradient of ω and is the group velocity $\boldsymbol{v}_g$ of the normal mode. One can then write for the element of volume

$$dS_\omega dq_\perp = dS_\omega \frac{d\omega}{v_g}.$$

A simple case is provided by the isotropic elastic continuum with constant speeds of transverse and longitudinal waves c_t and c_ℓ. The dispersion relations are $\omega = c_t q$ and $\omega = c_\ell q$, and the surfaces of constant frequency are spheres. Then for branch j,

$$dS_\omega = q^2 \sin\theta\, d\theta\, d\phi = \frac{\omega^2}{c_j^2}\sin\theta\, d\theta\, d\phi,$$

where θ, ϕ are polar angles. The frequency distribution function is given by

$$F(\omega) = \frac{\Omega}{(2\pi)^3}\int_0^\pi \sin\theta\, d\theta \int_0^{2\pi} d\phi \left(\frac{2}{c_t^3} + \frac{1}{c_\ell^3}\right)\omega^2 = \frac{\Omega\omega^2}{2\pi^2}\left(\frac{2}{c_t^3} + \frac{1}{c_\ell^3}\right) \quad (7.40)$$

and is proportional to ω^2.

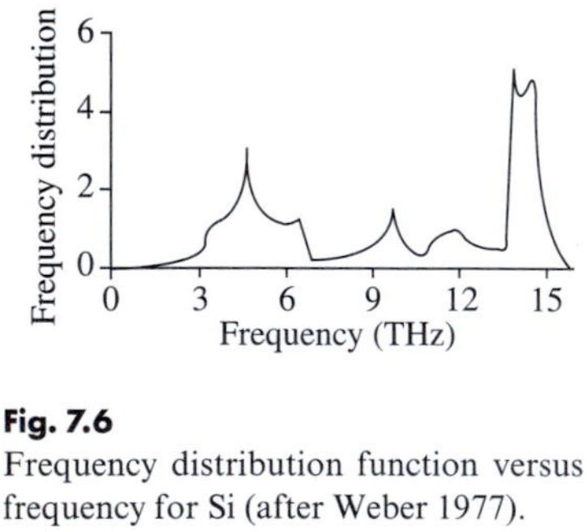

Fig. 7.6
Frequency distribution function versus frequency for Si (after Weber 1977).

Real semiconductors are elastically anisotropic with nonconstant phonon speeds. The evaluation of $F(\omega)$ requires accurate phonon frequencies over the entire Brillouin zone that are determined either experimentally or theoretically. An example is shown in Fig. 7.6 for Si. An interesting feature is the presence of **Van Hove singularities** in $F(\omega)$ at **critical points** characterized by $\partial\omega/\partial q_\alpha = 0$ (Van Hove 1953). There are various types of Van Hove singularities. In one-dimensional systems $F(\omega)$ typically varies as $|\omega_s - \omega|^{-\frac{1}{2}}$ near a singularity at ω_s and in two-dimensional systems as $\log|\omega_s - \omega|$. In three-dimensional systems $dF/d\omega$ has singularities behaving as $|\omega_s - \omega|^{-\frac{1}{2}}$. Critical points are isolated and finite in number. In accordance with Morse's theorem the periodicity of a crystal requires the existence of critical points in the surfaces of constant frequency in $\boldsymbol{q}$-space, and hence of singularities in the frequency distribution (Maradudin *et al.* 1971).

7.5 Lattice dynamical models for semiconductors

A **phonon** is the quantum of energy of a normal mode of given frequency. One of the principal tasks of lattice dynamical theory is to develop an understanding of experimental phonon dispersion curves, i.e., $\omega_{\boldsymbol{q}j}$ versus $\boldsymbol{q}$ curves. These curves can be determined by the technique of inelastic neutron scattering. A monoenergetic beam of neutrons of energy E_i and wave vector $\boldsymbol{k}_i$ is incident on the crystal of interest. The neutrons interact with the atoms of the crystal and are scattered into a state with energy E_s and wave vector $\boldsymbol{k}_s$. At the same time a phonon of energy $\hbar\omega_{\boldsymbol{q}j}$ is created (Stokes process) or destroyed (anti-Stokes process). Conservation of energy and crystal momentum require that

$$\hbar\omega_{\boldsymbol{q}j} = E_i - E_s \quad \text{Stokes} \tag{7.41a}$$

$$= E_s - E_i \quad \text{anti Stokes} \tag{7.41b}$$

$$\boldsymbol{q} = \boldsymbol{k}_i - \boldsymbol{k}_s \quad \text{Stokes} \tag{7.41c}$$

$$= \boldsymbol{k}_s - \boldsymbol{k}_i \quad \text{anti Stokes.} \tag{7.41d}$$

By measuring the energies and momenta of the incident and scattered neutrons, one can determine both $\boldsymbol{q}$ and $\omega_{\boldsymbol{q}j}$ for the phonon that is created or destroyed.

An alternative technique for obtaining phonon frequencies is based on light scattering. Photons replace neutrons as the scattered particles, but the same conservation laws apply. Highly monochromatic visible light is typically used, e.g., from an argon-ion laser. However, the wave vectors of such photons are so small compared to the Brillouin zone boundary that only phonons with wave vectors near the zone center can be studied in one-phonon processes. When the frequency of the scattered light differs from that of the incident light by the frequency of an optical phonon, the scattering process is called **Raman scattering**. An optical phonon frequency is thus determined. If, on the other hand, the difference in frequency of the scattered and incident light corresponds to the frequency of an acoustic phonon, the process is called **Brillouin scattering**, and an acoustic phonon frequency is determined.

Additional optical techniques for the measurement of phonon frequencies include the absorption, emission, and reflection of infrared radiation. A detailed treatment of optical techniques is presented in Chapter 10.

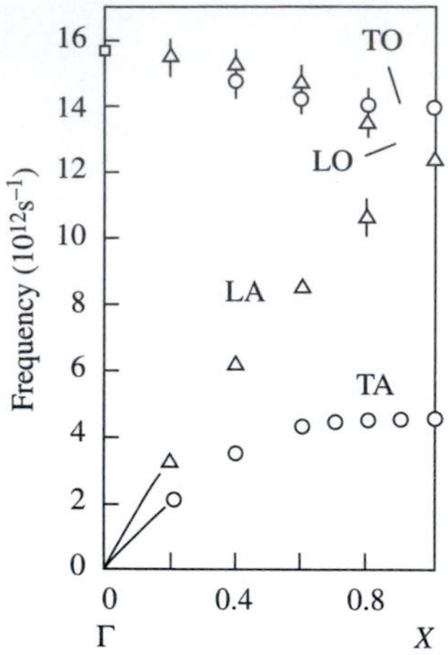

Fig. 7.7
Phonon spectrum for Si in the [100] direction (after Dolling 1962).

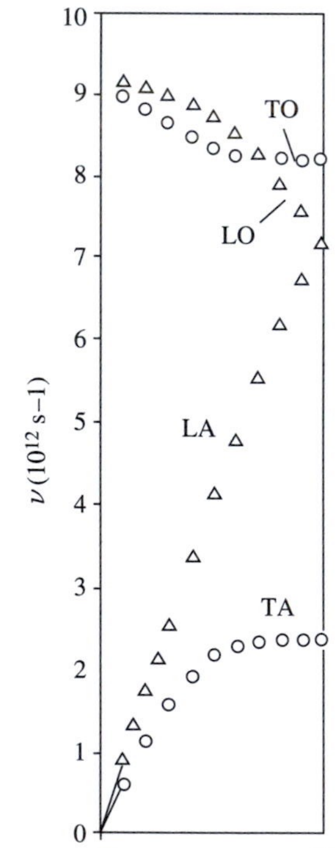

Fig. 7.8
Phonon spectrum for Ge in the [100] direction (after Nilsson and Nelin 1971).

7.5.1 Homopolar semiconductors

The inelastic neutron scattering technique has been employed to determine the phonon dispersion curves of Si (Dolling 1962) and Ge (Nilsson and Nelin 1971) which are displayed in Figs. 7.7 and 7.8, respectively. A striking feature of the results is the flatness of the transverse acoustic (TA) branch as it approaches the Brillouin zone boundary in the [100] and also the [111] directions. This flatness is difficult to reproduce with a model involving short range interatomic interactions between nearest neighbors and next-nearest neighbors. It has been shown (Herman 1959) that interactions out to and including fifth neighbors are required to give a reasonable fit to the experimental data for Ge.

The basic physical reason for the anomalous behavior of the TA branch in Si and Ge seems to be the high deformability of the electron charge distribution in these materials. This deformability was taken into account by the **shell model** (Dick and Overhauser 1958, Hanlon and Lawson 1959) which was first applied to alkali halides. The electron charge distribution is represented by a spherical shell about each nucleus. A given shell is coupled by Hooke's law forces to the nucleus inside the shell and to the shells about the nearest neighbor nuclei as shown in Fig. 7.9. In addition there are interactions between nearest neighbor nuclei.

The equations of motion for the shell model take the form

$$M_\kappa \frac{\partial^2 u_\alpha(\ell\kappa)}{\partial t^2} = \sum_{\ell'\kappa'\beta}\left\{\Phi^{nn}_{\alpha\beta}(\ell\kappa,\ell'\kappa')u_\beta(\ell'\kappa') + \Phi^{ns}_{\alpha\beta}(\ell\kappa,\ell'\kappa')r_\beta(\ell'\kappa')\right\} \tag{7.42a}$$

$$m \frac{\partial^2 r_\alpha(\ell\kappa)}{\partial t^2} = \sum_{\ell'\kappa'\beta}\left\{\tilde{\Phi}^{ns}_{\alpha\beta}(\ell\kappa,\ell'\kappa')u_\beta(\ell'\kappa')\Phi^{ss}_{\alpha\beta}(\ell\kappa,\ell'\kappa')r_\beta(\ell'\kappa')\right\}, \tag{7.42b}$$

where $r_\alpha(\ell\kappa)$ is the displacement component of the center of the shell about nucleus $\ell\kappa$, $\Phi^{nn}(\ell\kappa,\ell'\kappa')$, $\Phi^{ns}(\ell\kappa,\ell'\kappa')$, and $\Phi^{ss}(\ell\kappa,\ell'\kappa')$ are the force constant matrices for the nucleus–nucleus, shell–nucleus, and shell–shell interactions, and m is the electron mass. In practice, one invokes the adiabatic approximation and sets $m = 0$. Equation (7.42b) can then be used to eliminate the shell coordinates from Eq. (7.42a). The resulting equation can be written in matrix notation as

$$M_\kappa \frac{\partial^2 \boldsymbol{u}}{\partial t^2} = \left[\overleftrightarrow{\Phi}^{nn} + \overleftrightarrow{\Phi}^{ns} \cdot (\overleftrightarrow{\Phi}^{ss})^{-1} \cdot \overleftrightarrow{\tilde{\Phi}}^{ns}\right] \cdot \boldsymbol{u} \tag{7.43}$$

which can be solved in the manner developed in previous sections.

The shell model was applied to Si and Ge by Cochran (Cochran 1959). A suitable choice of the model parameters gave rather good agreement with the experimental phonon dispersion curves. In particular, the flat TA branches were reasonably well reproduced.

A variation on the shell model is the bond charge model (Weber 1977). Since the covalent bond leads to a pile-up of electron charge between nearest neighbor atoms, one can idealize this charge distribution by placing a negative point charge between each pair of nearest neighbor nuclei. The point charge is coupled by Hooke's law forces to these nuclei. Coupling is also introduced between nearest neighbor nuclei and between nearest neighbor point charges. The equations of motion are of the same general form as those for the shell model and may be solved in the same way using the adiabatic approximation. The results for the phonon dispersion curves of Si obtained by Weber are shown in Fig. 7.10. We see that the agreement between theory and experiment is excellent. Similar results have been obtained for Ge (Weber 1977).

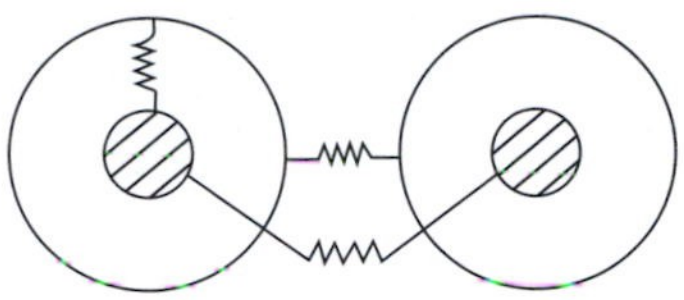

Fig. 7.9
Schematic representation of shell model interactions.

7.5.2 Heteropolar semiconductors

Phonon dispersion curves for a large number of heteropolar semiconductors have been obtained by inelastic neutron scattering. Both the shell model and the bond charge model have been applied to these materials. However, the theoretical analysis must include the partial ionic character of the electron-pair bonds and the effective charges on the ions. The situation is somewhat similar to that of ionic crystals such as alkali halides for which the **rigid ion model** (Kellerman 1941) and the **deformation dipole model** (Hardy 1962) have proved useful.

In the rigid ion model, the ions are regarded as nondeformable point charges interacting with each other through the Coulomb interaction. Short range repulsive interactions of the Born–Mayer type are also included. The Born–Mayer interaction has the form

$$V_{BM}(r) = Ae^{-\alpha r}, \tag{7.44}$$

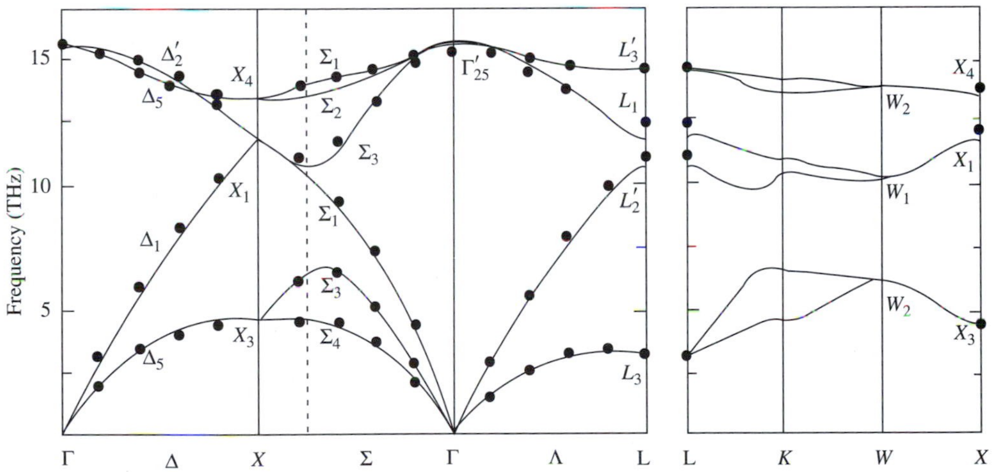

Fig. 7.10
Phonon dispersion curves for Si: solid circles, experimental data (after Dolling 1962, Nilsson and Nelin 1972); solid lines, calculated results (after Weber 1977).

where A and α are parameters which can, for example, be determined from the equilibrium lattice constant and the bulk modulus.

A more sophisticated model which takes into account the deformability of the electron charge distribution about an ion is the deformation dipole model (Hardy 1962, Karo and Hardy 1969, Kunc *et al.* 1975). This model is based on the adiabatic, harmonic, and linear dipole approximations. It describes the long range part of the interatomic forces, which is due to electrostatic interactions in the system of ion cores and loosely or tightly bound electrons. As a result of the displacements of the ions and the distortions of the electron distributions about the ions, an electric dipole moment $\boldsymbol{p}^d(\ell\kappa)$ is produced at lattice site $\ell\kappa$ whose components can be expanded in power series in the components of the ionic displacements $\boldsymbol{u}(\ell\kappa)$:

$$p_\alpha^d(\ell\kappa) = e^*(\kappa)u_\alpha(\ell\kappa) + \sum_{\ell'\kappa'\beta} m_{\alpha\beta}(\ell\kappa, \ell'\kappa')u_\beta(\ell'\kappa'). \tag{7.45}$$

Here $e^*(\kappa)$ is the effective charge on ions of type κ and $m_{\alpha\beta}(\ell\kappa, \ell'\kappa')$ is the deformation dipole matrix. The latter can typically be restricted to nearest neighbor interactions. An additional contribution to the dipole moment arises from the presence of an effective electric field $\boldsymbol{E}^{eff}$ which polarizes the ions. If $\alpha(\kappa)$ is the polarizability of ion κ, the total dipole moment can then be written as

$$p_\alpha(\ell\kappa) = e^*(\kappa)u_\alpha(\ell\kappa) + \sum_{\ell'\kappa'\beta'} m_{\alpha\beta}(\ell\kappa, \ell'\kappa')u_\beta(\ell'\kappa') + \alpha(\kappa)E_\alpha^{eff}(\ell\kappa). \tag{7.46}$$

The effective field is itself determined by the dipole moment and can be expressed as

$$E_\alpha^{eff}(\ell\kappa) = \sum_{\ell'\kappa'\beta} B_{\alpha\beta}(\ell\kappa, \ell'\kappa')p_\beta(\ell'\kappa'), \tag{7.47}$$

where $B_{\alpha\beta}(\ell\kappa, \ell'\kappa')$ is the appropriate Coulomb coefficient (Kunc 1973–74). Eliminating the effective field from Eq. (7.46) yields

$$\begin{aligned} p_\alpha\ell\kappa = e^*(\kappa)u_\alpha(\ell\kappa) &+ \sum_{\ell'\kappa'\beta} m_{\alpha\beta}(\ell\kappa, \ell'\kappa')u_\beta(\ell'\kappa) \\ &+ \alpha(\kappa)\sum_{\ell'\kappa'\beta} B_{\alpha\beta}(\ell\kappa, \ell'\kappa')p_\beta(\ell'\kappa'). \end{aligned} \tag{7.48}$$

The equations of motion for the displacements can be written as

$$\begin{aligned} M_\kappa\omega^2 u_\alpha(\ell\kappa) = &\sum_{\ell'\kappa'\beta} \Phi_{\alpha\beta}(\ell\kappa, \ell'\kappa')u_\beta(\ell'\kappa) \\ &- \sum_{\ell'\kappa'\beta} e^*(\kappa)B_{\alpha\beta}(\ell\kappa)p_\beta(\ell'\kappa') \\ &- \sum_{\ell'\kappa'\beta}\sum_{\ell''\kappa''\gamma} m_{\beta\alpha}(\ell'\kappa', \ell\kappa)B_{\beta\gamma}(\ell'\kappa', \ell''\kappa'')p_\gamma(\ell''\kappa''), \end{aligned} \tag{7.49}$$

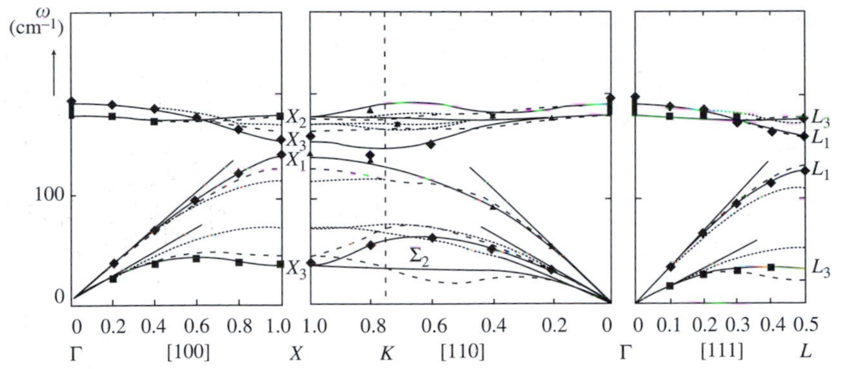

Fig. 7.11
Phonon dispersion curves for InSb: solid circles, experimental data (after Price *et al.* 1971); solid lines, 15-parameter deformation dipole model; dashed lines, second-neighbor central force model; dotted lines, first-neighbor central force model (after Kunc *et al.* 1975).

where $\Phi_{\alpha\beta}(\ell\kappa, \ell'\kappa')$ is the force constant matrix for short range interactions. Equations (7.48) and (7.49) are a set of simultaneous equations in the displacements and dipole moments. Solutions in the form of plane waves with wave vector $\boldsymbol{q}$ can be found that yield the frequency as a function of $\boldsymbol{q}$. Results for InSb are shown in Fig. 7.11 together with experimental data from inelastic neutron scattering. Parameters that appear in the force constant and deformation dipole matrices were chosen to give the best fit to the data.

A comment is desirable about the different behavior of the optical mode frequencies in homopolar and heteropolar semiconductors as $\boldsymbol{q} \to 0$. In homopolar semiconductors the transverse and longitudinal optical mode frequencies approach the same value, whereas in heteropolar semiconductors the transverse optical frequency ω_{TO} approaches a lower value than the longitudinal optical frequency ω_{LO}. This difference in limiting frequencies arises from the strong interaction of transverse optical modes with the transverse electromagnetic field at wave vectors on the order of ω_{TO}/c, where c is the speed of light. If the wave vector is strictly zero, however, the transverse and longitudinal optical mode frequencies in cubic heteropolar semiconductors have the same value by symmetry. When $q \cong \omega_{TO}/c$, the splitting of ω_{TO} and ω_{LO} becomes apparent. Since ω_{TO}/c is very small compared to wave vectors at the Brillouin zone boundary, the transition to identical values of ω_{TO} and ω_{LO} as $q \to 0$ cannot be seen on the scale of Fig. 7.11.

The interaction of a transverse optical phonon of small wave vector with an electromagnetic wave is conveniently expressed in terms of the modified relative displacement $\boldsymbol{w}$ given by

$$\boldsymbol{w} = (\bar{M}/\Omega_0)^{\frac{1}{2}}\boldsymbol{u}, \tag{7.50}$$

where $\boldsymbol{u}$ is the relative displacement of the two ions in a unit cell and $\bar{M}$ is their reduced mass. The equation of motion for $\boldsymbol{w}$ is

$$\frac{d^2\boldsymbol{w}}{dt^2} = b_{11}\boldsymbol{w} + b_{12}\boldsymbol{\mathcal{E}}, \tag{7.51}$$

in which b_{11} is a modified short-range force constant with dimensions of frequency squared, b_{12} is a modified electric charge, and $\boldsymbol{\mathcal{E}}$ is the macroscopic electric field set up by the vibrations. A second relation involving $\boldsymbol{w}$ and $\boldsymbol{\mathcal{E}}$ is provided by the **polarization** $\boldsymbol{P}$, or electric dipole moment per unit volume, given by

$$\boldsymbol{P} = b_{21}\boldsymbol{w} + b_{22}\boldsymbol{\mathcal{E}}. \tag{7.52}$$

The parameter b_{21} is a modified electric charge and is equal to b_{12} (Born and Huang 1954), while b_{22} is a modified polarizability.

Let us assume that $\boldsymbol{w}$, $\boldsymbol{\mathcal{E}}$, and $\boldsymbol{P}$ vary as $\exp(-i\omega t)$. We can then eliminate $\boldsymbol{w}$ from Eq. (7.52) using Eq. (7.51) and obtain

$$\boldsymbol{P} = \left(b_{22} - \frac{b_{12}^2}{b_{11} + \omega^2}\right)\boldsymbol{\mathcal{E}}. \tag{7.53}$$

Introducing this result into the expression for the electric displacement $\boldsymbol{\mathcal{D}}$ given by

$$\boldsymbol{\mathcal{D}} = \epsilon_0\boldsymbol{\mathcal{E}} + \boldsymbol{P} = \epsilon_0\epsilon(\omega)\boldsymbol{\mathcal{E}}, \tag{7.54}$$

where $\epsilon(\omega)$ is the dielectric function, we see that on eliminating $\boldsymbol{P}$,

$$\epsilon(\omega) = 1 + \frac{1}{\epsilon_0}\left(b_{22} - \frac{b_{12}^2}{b_{11} + \omega^2}\right). \tag{7.55}$$

Let us consider the limiting case $\omega \to \infty$ and $\epsilon(\omega) \to \epsilon_\infty$. Then

$$b_{22} = \epsilon_0(\epsilon_\infty - 1). \tag{7.56}$$

For transverse optical modes of long wavelength and frequency ω_{TO}, the macroscopic electric field $\boldsymbol{\mathcal{E}}$ vanishes, so from Eq. (7.51)

$$b_{11} = -\omega_{TO}^2. \tag{7.57}$$

Using this result and setting $\omega = 0$ in Eq. (7.55), we find that

$$b_{12}^2 = \epsilon_0(\epsilon_s - \epsilon_\infty)\omega_{TO}^2, \tag{7.58}$$

where $\epsilon_s = \epsilon(0)$. The dielectric function now takes the form

$$\epsilon(\omega) = \epsilon_\infty + \frac{(\epsilon_s - \epsilon_\infty)\omega_{TO}^2}{\omega_{TO}^2 - \omega^2}. \tag{7.59}$$

We now take the divergence of $\boldsymbol{\mathcal{D}}$ and focus on longitudinal optical modes of long wavelength and frequency ω_{LO}:

$$\nabla\cdot\boldsymbol{\mathcal{D}} = \epsilon_0\epsilon(\omega_{LO})\nabla\cdot\boldsymbol{\mathcal{E}}. \tag{7.60}$$

For longitudinal waves, $\nabla\cdot\boldsymbol{\mathcal{E}} \neq 0$, but if the semiconductor contains no space charge, $\nabla\cdot\boldsymbol{\mathcal{D}} = 0$. Consequently, $\epsilon(\omega_{LO}) = 0$, and

$$0 = \epsilon_\infty + (\epsilon_s - \epsilon_\infty)\omega_{TO}^2/(\omega_{TO}^2 - \omega_{LO}^2). \tag{7.61}$$

Rearranging this result gives the **Lyddane–Sachs–Teller relation**

$$\frac{\omega_{LO}^2}{\omega_{TO}^2} = \frac{\epsilon_s}{\epsilon_\infty}. \tag{7.62}$$

Since $\epsilon_s > \epsilon_\infty$, $\omega_{LO} > \omega_{TO}$. Thus, the limiting values of ω_{LO} and ω_{TO} as $\boldsymbol{q} \to 0$ are determined by the static and high-frequency dielectric constants.

7.5.3 First-principles methods

As pointed out in Chapter 2, the potential energy for the nuclear motion is the electronic ground state energy as a function of the nuclear coordinates plus the direct interaction between the nuclei. When the nuclei occupy their equilibrium positions, the electronic ground state energy can be calculated using Bloch's theorem and *ab initio* pseudopotentials (Bachelet *et al.* 1982) whose parameters are determined from atomic data only. When the nuclei are displaced from their equilibrium positions, the situation becomes more complicated and requires a more elaborate analysis.

7.5.3.1 Frozen phonon method

For certain normal modes of vibration, particularly those for $\boldsymbol{q} = 0$ and for $\boldsymbol{q}$ on the Brillouin zone boundary, the atomic displacement patterns are still periodic, but with a unit cell that may be the same size or larger than the primitive unit cell. One can therefore still use the Bloch theorem to calculate the electronic states. By varying the size of the atomic displacements and fitting a parabola to the results, one can obtain the force constants that control the motion in that particular mode. The calculation of the frequency of the mode is then straightforward (Kunc and Martin 1981).

7.5.3.2 Linear response method

A procedure that provides full phonon dispersion curves without fitting parameters to the experimental data is the **linear response method** (Baroni *et al.* 1987, Giannozzi *et al.* 1991, Quong and Klein 1992). The first step is to calculate Bloch functions $\psi_\nu(\boldsymbol{r})$ using the density functional method in the LDA as described in Section 3.4.4. One then expands the Hamiltonian in powers of the ionic displacements to yield a first-order change $H^{(1)}(\boldsymbol{r})$ of the form

$$H^{(1)}(\boldsymbol{r}) = V_b^{(1)}(\boldsymbol{r}) + V_h^{(1)}(\boldsymbol{r}) + V_{xc}^{(1)}(\boldsymbol{r}), \tag{7.63}$$

where $V_b^{(1)}(\boldsymbol{r})$, $V_h^{(1)}(\boldsymbol{r})$, and $V_{xc}^{(1)}(\boldsymbol{r})$ are the first-order changes in the electron–bare ion interaction, the electron–electron interaction, and the exchange–correlation interaction, respectively. The change in electron concentration $n^{(1)}(\boldsymbol{r})$ due to the first-order change in the Hamiltonian is given by

$$n^{(1)}(\boldsymbol{r}) = \sum_{\nu,\nu'} \frac{f_\nu - f_{\nu'}}{E_\nu - E_{\nu'}} \langle \nu' | H^{(1)} | \nu \rangle \psi_\nu^*(\boldsymbol{r}) \psi_{\nu'}(\boldsymbol{r}), \tag{7.64}$$

where f_ν is the Fermi–Dirac occupation number and E_ν is the Bloch state energy.

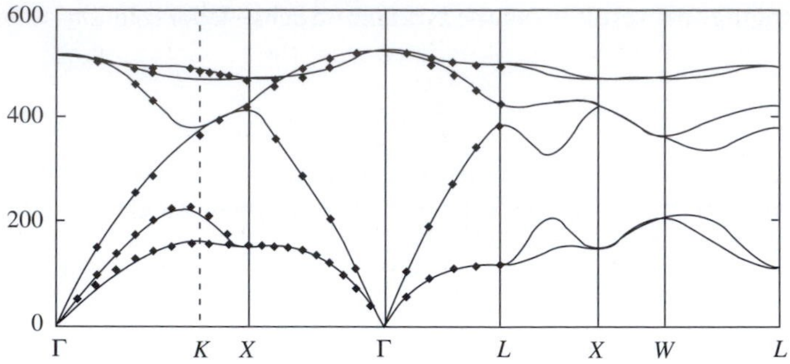

Fig. 7.12
Phonon dispersion curves for Si: solid circles, experimental data (after Dolling 1962, Nilsson and Nelin 1972); solid lines, calculated results (after Giannozzi *et al.* 1991).

Since $H^{(1)}(\boldsymbol{r})$ contains $n^{(1)}(\boldsymbol{r})$ through the terms $u_{xc}(n(\boldsymbol{r}))$ and $e^2 \int d^3r' n(r')/4\pi\epsilon_0|\boldsymbol{r} - \boldsymbol{r}'|$ in Eq. (3.48), one must solve Eqs. (7.63) and (7.64) self-consistently in order to obtain the converged values of $n^{(1)}(\boldsymbol{r})$ and $H^{(1)}(\boldsymbol{r})$. An important point is that this procedure handles nonlocal *ab initio* pseudopotentials without difficulty. The second-order change in the electronic energy of the system is specified by

$$E_{el}^{(2)} = \sum_{\nu} f_\nu \langle \nu | V_b^{(2)} | \nu \rangle + \frac{1}{2} \sum_{\nu, \nu'} \frac{f_\nu - f_{\nu'}}{E_\nu - E_{\nu'}} \langle \nu' | V_b^{(1)} | \nu \rangle \langle \nu | H^{(1)} | \nu' \rangle, \qquad (7.65)$$

where $V_b^{(2)}$ is the second-order electron-bare ion interaction. Adding the ion–ion interaction to $E_{el}^{(2)}$ and expanding in powers of the ion displacements to second order, one obtains the dynamical matrix from which phonon dispersion curves and other lattice dynamical properties can be calculated. Phonon dispersion curves for Si obtained by the linear response method are shown in Fig. 7.12.

7.6 Normal coordinate transformation

The basic characteristic of a normal mode of vibration is that all atoms vibrate with the same frequency when a particular mode is excited. In a harmonic crystal there is no transfer of energy from one normal mode to another. Each normal mode can be associated with a **normal coordinate** $Q(\boldsymbol{q}j)$ which is a linear combination of the physical displacements $u_\alpha(\ell\kappa)$. The **normal coordinate transformation** which carries the $u_\alpha(\ell\kappa)$ into the $Q(\boldsymbol{q}j)$ is given by

$$Q(\boldsymbol{q}j) = N^{-\frac{1}{2}} \sum_{\ell\kappa\alpha} M_\kappa^{\frac{1}{2}} e_{\alpha\kappa}^*(\boldsymbol{q}j) e^{-i\boldsymbol{q}\cdot\boldsymbol{R}(\ell)} u_\alpha(\ell\kappa), \qquad (7.66)$$

where the coefficients $e_{\alpha\kappa}(\boldsymbol{q}j)$ are the components of the eigenvectors of the dynamical matrix introduced in Section 7.1. Using the closure relation Eq. (7.19), the inverse transformation can be shown to be

$$u_\alpha(\ell\kappa) = (NM_\kappa)^{-\frac{1}{2}} \sum_{\boldsymbol{q}j} e_{\alpha\kappa}(\boldsymbol{q}j) e^{i\boldsymbol{q}\cdot\boldsymbol{R}(\ell)} Q(\boldsymbol{q}j). \qquad (7.67)$$

The normal coordinate transformation has the important property that it diagonalizes the harmonic Hamiltonian of the crystal given by

$$H = \frac{1}{2}\sum_{\ell\kappa\alpha} M_\alpha \dot{u}_\alpha^2(\ell\kappa) + \frac{1}{2}\sum_{\ell\kappa\alpha}\sum_{\ell'\kappa'\beta} \Phi_{\alpha\beta}(\ell\kappa;\ell'\kappa')u_\alpha(\ell\kappa)u_\beta(\ell'\kappa'), \quad (7.68)$$

the dot over the u denoting a first time derivative. Carrying out the normal coordinate transformation with the aid of Eqs. (7.18) and (7.67) yields the result

$$H = \frac{1}{2}\sum_{\boldsymbol{q}j}[\dot{Q}^2(\boldsymbol{q}j) + \omega_{\boldsymbol{q}j}^2 Q^2(\boldsymbol{q}j)]. \quad (7.69)$$

We see that the transformed Hamiltonian is a sum of independent harmonic oscillator terms. The Schrödinger equation for the nuclear motion, Eq. (2.4), can therefore be solved by separation of variables. The total vibrational energy E_v is the sum of the harmonic oscillator energies associated with the normal coordinates,

$$E_v = \sum_{\boldsymbol{q}j} \hbar\omega_{\boldsymbol{q}j}(n_{\boldsymbol{q}j} + \tfrac{1}{2}), \quad (7.70)$$

where the $n_{\boldsymbol{q}j}$ are the harmonic oscillator quantum numbers given by non-negative integers. The quantum of energy, $\hbar\omega_{\boldsymbol{q}j}$, of a normal mode of vibration is called a **phonon**.

A phonon can be viewed as a particle-like entity that serves as a carrier of vibrational energy in much the same way than an electron serves as a carrier of electrical charge. The group velocity of a phonon is given by

$$\boldsymbol{v}_g = \boldsymbol{\nabla}_{\boldsymbol{q}}\omega_{\boldsymbol{q}j}. \quad (7.71)$$

7.7 Vibrational specific heat

Using statistical mechanics and the energy given by Eq. (7.70) one can calculate thermodynamic properties, such as the specific heat and entropy, as functions of temperature. The first step is to compute the normal mode frequency as a function of wave vector. The specific heat can then be evaluated from the average energy $\langle E_v\rangle$ using the relation

$$C_\Omega = \left(\frac{\partial\langle E_v\rangle}{\partial T}\right)_\Omega, \quad (7.72)$$

where the temperature differentiation is at constant volume Ω. The average energy can be expressed in terms of the average oscillator quantum number $\langle n_{\boldsymbol{q}j}\rangle$ by

$$\langle E_v\rangle = \sum_{\boldsymbol{q}j} \hbar\omega_{\boldsymbol{q}j}(\langle n_{\boldsymbol{q}j}\rangle + \tfrac{1}{2}). \quad (7.73)$$

Averaging $n_{\boldsymbol{q}j}$ over a canonical ensemble, one obtains

$$\bar{n}_{\boldsymbol{q}j} = \frac{\sum_{n=0}^{\infty} n e^{-\beta\hbar\omega_{\boldsymbol{q}j}\left(n+\frac{1}{2}\right)}}{\sum_{n=0}^{\infty} e^{-\beta\hbar\omega_{\boldsymbol{q}j}\left(n+\frac{1}{2}\right)}} = \frac{1}{e^{\hbar\omega_{\boldsymbol{q}j}/k_B T} - 1}. \tag{7.74}$$

Combining Eqs. (7.72)–(7.74) yields the specific heat in the form

$$C_\Omega = k_B \sum_{\boldsymbol{q}j} \left(\frac{\hbar\omega_{\boldsymbol{q}j}}{k_B T}\right)^2 \frac{e^{\hbar\omega_{\boldsymbol{q}j}/k_B T}}{\left(e^{\hbar\omega_{\boldsymbol{q}j}/k_B T} - 1\right)^2}. \tag{7.75}$$

At an arbitrary temperature the calculation of C_Ω requires a numerical calculation based on a knowledge of the $\boldsymbol{q}$-dependence of $\omega_{\boldsymbol{q}j}$. The evaluation is much simplified at both high and low temperatures. At high temperatures where $k_B T > \hbar\omega_{\boldsymbol{q}j}$ for all $\boldsymbol{q}j$, the right hand side of Eq. (7.75) can be expanded in powers of $\hbar\omega_{\boldsymbol{q}j}/k_B T$ to give the limiting result

$$\begin{aligned} C_\Omega &= k_B \sum_{\boldsymbol{q}j} 1 \\ &= 3Nrk_B, \end{aligned} \tag{7.76}$$

where N is the number of unit cells and r is the number of atoms per unit cell. This result is the **Dulong–Petit** law.

At low temperatures where $k_B T \ll \hbar\omega_{max}$, only low frequency modes contribute significantly to C_Ω and elastic continuum theory can be used. We must recognize, however, that elastic continuum theory yields an infinite number of normal mode frequencies with no upper bound, whereas a finite crystal has a finite number of normal modes frequencies. Debye resolved this discrepancy by imposing a maximum frequency ω_D on the elastic continuum modes such that the total number of normal modes equals the total number of degrees of freedom:

$$\int_0^{\omega_D} F(\omega)d\omega = 3Nr. \tag{7.77}$$

Taking the isotropic approximation to an elastic continuum, we substitute Eq. (7.40) into this equation and solve for ω_D:

$$\omega_D = \left[18\pi^2 Nr \Big/ \left(\frac{2}{c_t^3} + \frac{1}{c_\ell^3}\right)\Omega\right]^{1/3}. \tag{7.78}$$

For diamond and zincblende structures the ratio Ω/Nr has the value $a^3/8$, where a is the lattice constant. The speeds of transverse and longitudinal sound waves can be calculated from the elastic moduli as described in Section 7.4 and averaged over various directions of propagation.

It is convenient in discussing the specific heat to introduce a characteristic temperature Θ known as the **Debye temperature** by the relation

$$\hbar\omega_D = k_B\Theta. \tag{7.79}$$

Table 7.4 Debye temperatures (after Böer 1990)

Semiconductor	Θ	Semiconductor	Θ	Semiconductor	Θ
Si	658	PbS	228	AlSb	263
Ge	366	PbTe	161	GaAs	345
ZnS	315	CdS	215	GaSb	269
ZnSe	273	CdSe	180	InAs	248
ZnTe	220	CdTe	162	InSb	200

Converting the sum in Eq. (7.75) to an integral by means of the phonon density-of-states, we can express the specific heat in the form

$$C_\Omega = 9Nrk_B\left(\frac{T}{\Theta}\right)^3 \int_0^{\frac{\Theta}{T}} \frac{x^4 e^x dx}{(e^x - 1)^2}. \quad (7.80)$$

As the temperature is lowered toward zero, the upper limit on the integral can be extended to infinity. The value of the integral is then $4\pi^4/15$ and

$$C_\Omega = \frac{12\pi^4}{5} Nrk_B\left(\frac{T}{\Theta}\right)^3. \quad (7.81)$$

This is the **Debye T^3-law**. It is well satisfied by solids at very low temperatures. Values of Θ for representative semiconductors are listed in Table 7.4. As the temperature increases, the specific heat starts to deviate significantly from the T^3 law. This behavior can be attributed to the excitation of higher frequency modes, including optical modes, whose frequencies are not well described by elastic continuum theory. The phonon density-of-states develops peaks due to Van Hove singularities. The specific heat curve flattens out toward the Dulong–Petit values as the Debye temperature is approached. Figure 7.13 shows this behavior for Si and Ge. Further increase in temperature to well beyond the Debye temperature leads to a renewed increase in the specific heat. This increase is due to anharmonic effects associated with cubic and higher order terms in the expansion of the potential energy in powers of the displacements.

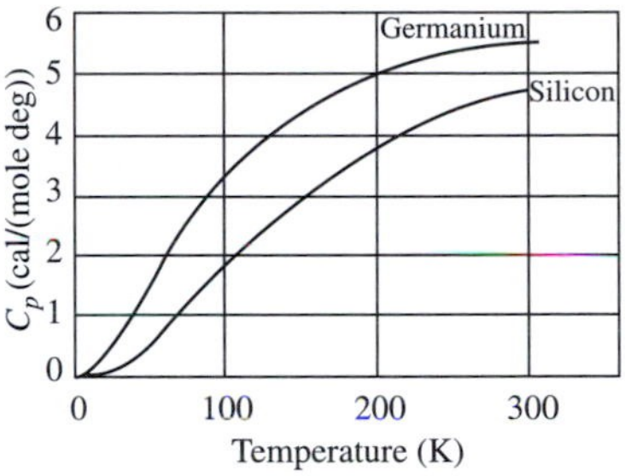

Fig. 7.13
Specific heat of Si and Ge versus absolute temperature (after Kittel 1986).

7.8 Anharmonic effects

7.8.1 Thermal expansion

If cubic anharmonic terms are included in the lattice potential energy, the interaction potential between two atoms becomes asymmetric with respect to the point of minimum potential. The potential rises more rapidly as the atomic separation decreases below the value at minimum potential and rises less rapidly as the separation increases above this value. A thermal distribution favors the larger values of separation. This behavior can be illustrated by considering a diatomic molecule with interaction potential

$$V_{int} = \beta(u - u_0)^2 - \gamma(u - u_0)^3, \quad (7.82)$$

where u_0 is the separation at the potential minimum. Letting $w = u - u_0$, we have for the average separation $\bar{u}$

$$\bar{u} = u_0 + \frac{\int_{-u_0}^{\infty} we^{-V_{int}/k_BT}dw}{\int_{-u_0}^{\infty} e^{-V_{int}/k_BT}dw}. \qquad (7.83)$$

To a good approximation we can extend the lower limits of the integrals to $-\infty$ and expand the integrals in powers of γ, retaining only the lowest-order nonvanishing terms. The result can be written as

$$\bar{u} = (1 + \alpha T)u_0, \qquad (7.84)$$

where α is the **thermal expansion coefficient** given by

$$\alpha \simeq \frac{3\gamma k_B}{4\beta^2 u_0}. \qquad (7.85)$$

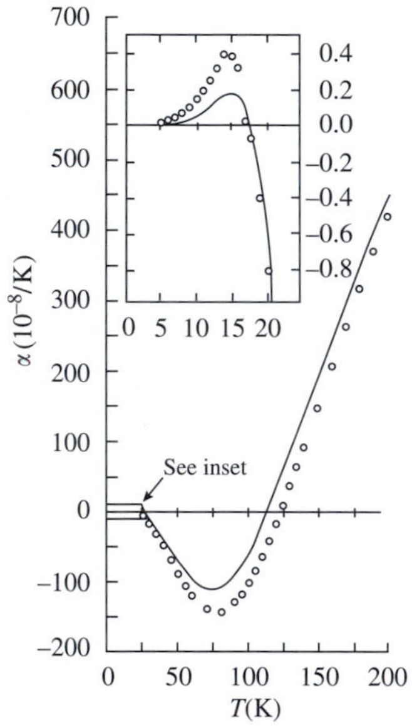

Fig. 7.14
Thermal expansion of Si as a function of temperature: open circles, experimental data (after Lyon *et al.*, 1977); solid curve, theoretical result (after Wanser 1982).

We see that α is proportional to the cubic anharmonic coefficient γ.

The foregoing treatment reveals no dependence of α on temperature. Real crystals, however, exhibit a decrease in α with decreasing T at low temperatures. This behavior is associated with the change in normal mode frequency due to change in volume that is described by the **Grüneisen parameter** Γ_i given by

$$\Gamma_i = -\left(\frac{\Omega}{\omega_i}\right)\frac{d\omega_i}{d\Omega}, \qquad (7.86)$$

where ω_i is the frequency of normal mode i. The thermal expansion is related to the Grüneisen parameters by (Mitra and Massa 1982)

$$\alpha = \sum_i \Gamma_i \frac{C_\Omega^{(i)}}{3B_0\Omega}, \qquad (7.87)$$

where $C_\Omega^{(i)}$ is the specific heat associated with normal mode i and B_0 is the bulk modulus. Anharmonicity enters α through the volume dependence of the normal mode frequencies and temperature through $C_\Omega^{(i)}$.

Semiconductors with the diamond or zincblende structure have negative thermal expansion coefficients over a range of low temperatures. In the case of Si, α becomes negative at ~ 120 K and becomes positive again at ~ 20 K as shown in Fig. 7.14. The Grüneisen parameters for certain low-frequency transverse acoustic modes are negative and produce the negative range of α.

7.8.2 Thermal conductivity

In semiconductors with very low carrier concentration the transport of heat is primarily associated with the flow of phonons. The anharmonic terms in the potential energy produce interactions between normal modes and collisions between phonons. Phonons therefore have a mean free path λ. Except under special circumstances, the flow of phonons and hence of heat is diffusive rather than ballistic in nature. The heat flux $\boldsymbol{Q}$ is related to the temperature gradient ∇T by **Fourier's law of heat conduction**

$$Q = -\kappa \nabla T, \tag{7.88}$$

where κ is the **thermal conductivity**. To derive an expression for κ, we use elementary kinetic theory in one-dimension. Let $u(x)$ be the energy density of phonons at point x and v_s the speed of sound. The net heat flux can be written as

$$Q = \frac{1}{6}[u(x-\lambda) - u(x+\lambda)]v_s. \tag{7.89}$$

Expanding $u(x \pm \lambda)$ in powers of λ and retaining only linear terms, we get

$$\begin{aligned} Q &= -\frac{1}{3}\frac{du}{dx}v_s\lambda = -\frac{1}{3}\frac{du}{dT}\frac{dT}{dx}v_s\lambda \\ &= -\frac{1}{3}C_\Omega v_s \lambda \frac{dT}{dx}. \end{aligned} \tag{7.90}$$

Comparing Eq. (7.88) with Eq. (7.90), we see that

$$\kappa = \frac{1}{3}C_\Omega v_s \lambda. \tag{7.91}$$

An appraisal of the temperature dependence of κ can be made by noting that at low T, $C_\Omega \sim T^3$, $v_s \simeq$ constant, and $\lambda \simeq$ constant, whereas at high T, $C_\Omega \simeq$ constant, $v_s \simeq$ constant, and $\lambda \sim 1/T$. The phonon mean free path λ at low T is determined by the sample dimensions or the impurity concentration, but at high T by the inverse of the mean number of phonons $\bar{n}_{ph} = 1/[\exp(\hbar\omega/k_B T) - 1] \sim T$. The thermal conductivity as a function of temperature is roughly a bell-shaped curve as shown in Fig. 7.15 for nearly isotopically pure Ge.

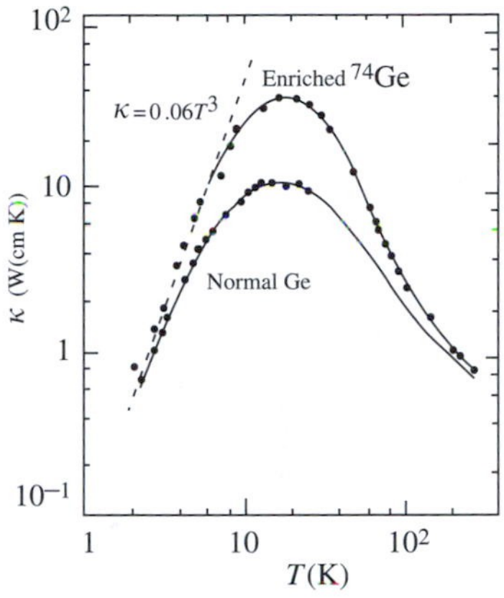

Fig. 7.15
Thermal conductivity of isotopically enriched Ge versus temperature (after Geballe and Hull 1958).

7.9 Impurity effects on lattice vibrations

The introduction of an impurity atom into a semiconductor can take place either substitutionally or interstitially. We shall focus on the substitutional process in which an atom of the pure crystal is replaced by an impurity atom. In general, both the mass of the atom and the force constants coupling it to its neighbors are changed by the substitution. In the case of isotopic substitution, only the mass changes. Under certain conditions a localized vibrational mode arises in which the amplitude of vibration is large at the impurity site and decreases exponentially going away from the site.

To illustrate the situation in a simple way, let us consider a monatomic linear chain with nearest neighbor interactions and a single isotopic impurity at site $\ell = 0$. The equations of motion are, assuming a time variation $\exp(i\omega t)$,

$$M\omega^2 u_\ell + \sigma(u_{\ell-1} + u_{\ell+1} - 2u_\ell) = 0, \quad \ell \neq 0 \tag{7.92a}$$

$$M'\omega^2 u_0 + \sigma(u_{-1} + u_{+1} - 2u_0) = 0, \tag{7.29b}$$

where M and M' are the host mass and impurity mass, respectively. We assume that there are $2N+1$ atoms in the chain and that fixed end boundary conditions are applied:

$$u_N = u_{-N} = 0. \tag{7.93}$$

Solutions to Eqs. (7.92) that satisfy Eq. (7.93) are given by

$$u_\ell = \begin{cases} A\sin(N-\ell)\phi & \ell \geq 0 \quad (7.94a) \\ B\sin(N+\ell)\phi. & \ell < 0. \quad (7.94b) \end{cases}$$

Substituting either expression for u_ℓ into Eq. (7.92a) gives, for $\ell > 0$ or $\ell < -1$,

$$M\omega^2 = 2\sigma(1-\cos\phi). \tag{7.95}$$

To determine A, B, and ϕ, we substitute Eqs. (7.94) into Eq. (7.92b) for $\ell = 0$ and into Eq. (7.92a) for $\ell = -1$. The resulting equations can be simplified with the aid of Eq. (7.95) to yield

$$A[2(Q-1)(1-\cos\phi)\sin N\phi - \sin(N+1)\phi] + B\sin(N-1)\phi = 0 \tag{7.96a}$$

$$(A-B)\sin N\phi = 0, \tag{7.96b}$$

where $Q = M'/M$. From the second of these equations we have

$$A = B \tag{7.97}$$

or

$$\sin N\phi = 0. \tag{7.98}$$

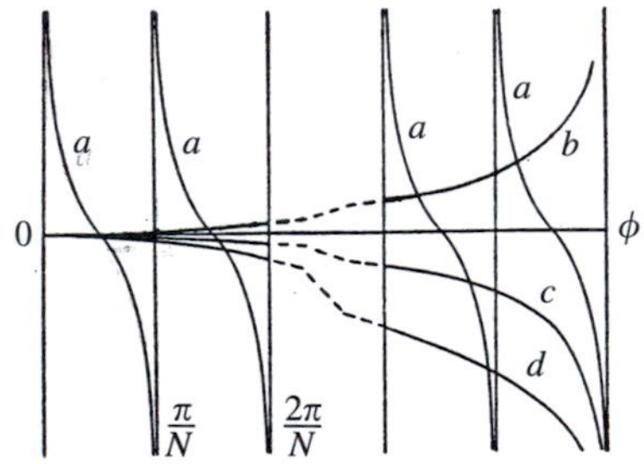

Fig 7.16
Graphical solution for localized vibrational modes. Curve a: $\cot N\phi$; curve b: $(Q-1)\tan\phi/2$, $Q>1$; curve c: $(Q-1)\tan\phi/2$, $Q<1$; curve d: $-\tan\phi/2$ (after Montroll and Potts 1955).

The case $A = B$ corresponds to even modes of vibration in which $u_{-\ell} = u_\ell$. From Eq. (7.96a) we have

$$(Q-1)(1-\cos\phi)\sin N\phi - \cos N\phi\sin\phi = 0$$

or

$$\cot N\phi = (Q-1)\tan(\phi/2). \tag{7.99}$$

This equation specifies the allowed values of ϕ for even modes. The graphical solution shown in Fig. 7.16 aids in understanding the nature of the modes. For $Q = 1$ one has the perfect lattice with fixed ends and $\cot N\phi = 0$. For $Q > 1$ one has a heavy impurity atom, the allowed values of ϕ and ω are downshifted by an amount of $O(1/N)$, and no modes are lost. For $Q < 1$, one has a light impurity atom, the allowed values of ϕ and ω are upshifted by an amount of $O(1/N)$, but the highest mode is "lost," i.e., the curves for $\cot(N\phi)$ and $(Q-1)\tan(\phi/2)$ never cross in the last segment to the right in Fig. 7.16. What has happened to the "lost" mode? The answer is that it has become a **localized vibrational mode** with a complex value of ϕ:

$$\phi = \pi + i\zeta. \tag{7.100}$$

Substituting this form for ϕ into Eq. (7.99), we obtain in the limit $N \to \infty$

$$(1 - Q)\coth(\zeta/2) = 1 \tag{7.101}$$

which determines the value of ζ and hence the value of the frequency via Eqs. (7.95) and (7.100):

$$M\omega^2 = 4\sigma \cosh^2(\zeta/2). \tag{7.102}$$

The localized mode frequency lies above the allowed band of the perfect lattice as shown in Fig. 7.17. The displacements of the localized mode are given by

$$u_\ell = \begin{cases} A(-1)^{N-\ell} \sinh(N - \ell)\zeta & \ell \geq 0 \quad (7.103a) \\ A(-1)^{N+\ell} \sinh(N + \ell)\zeta & \ell < 0 \quad (7.103b) \end{cases}$$

which describe an essentially exponential decay going away from the impurity atom as illustrated in Fig. 7.18.

The case $\sin N\phi = 0$ corresponds to odd modes with $B = -A$ and $u_{-\ell} = -u_\ell$. The values of ϕ are given by $\phi = \pi n/N$, where n is an integer, and are the same as for the perfect lattice with $M' = M$. The frequencies of the odd modes are unaffected by the impurity. This behavior is consistent with the fact that $u_0 = 0$, i.e., the impurity atom does not move in an odd mode.

For a heavy mass impurity, $M' > M$, the normal mode frequencies are lowered and no localized mode with frequency above the allowed band of the perfect lattice can appear in a monatomic crystal. However, a **resonance mode** can exist which is not a true normal mode, but which can persist for a considerable period of time. It has a large amplitude at the impurity and a frequency within the allowed band. In the case of a diatomic or polyatomic crystal, the replacement of one of the lighter constituent atoms by a heavier impurity can produce a localized mode with frequency in the forbidden gap between acoustic and optical branches. Replacement of a heavier constituent atom by a lighter impurity can also lead to a localized mode in the gap. A lighter constituent atom replaced by an even lighter impurity gives rise to a localized mode with frequency above the maximum frequency of the perfect lattice (Maradudin *et al.* 1971).

Localized impurity modes in three-dimensional lattices can be conveniently analyzed with the aid of Green's functions (Montroll and Potts 1955; Maradudin *et al.* 1971). The qualitative features are similar to those of one-dimensional lattices. In a diamond structure semiconductor such as Si, the replacement of a Si atom by a B atom leads to an impurity mode localized about the B atom. Such localized modes can be studied by optical techniques as discussed in Chapter 10.

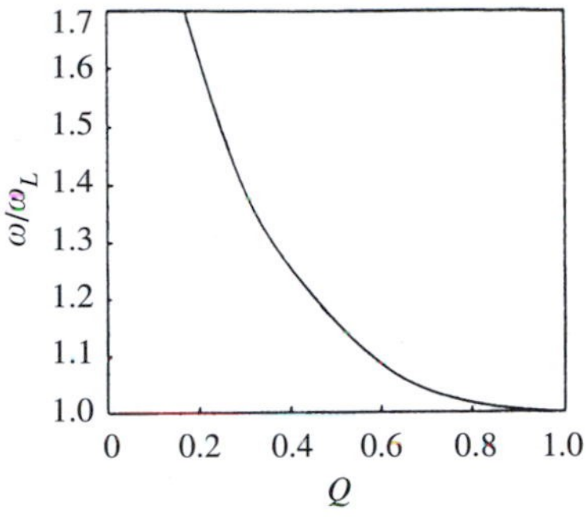

Fig. 7.17
Diagram of localized mode and perfect lattice frequencies for $M' < M$. ω_L is the maximum frequency of the perfect lattice and $Q = M'/M$.

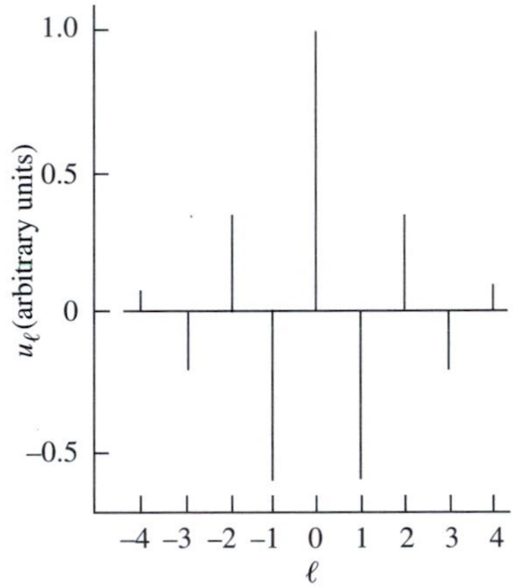

Fig. 7.18
Displacement pattern of a localized vibrational mode.

7.10 Piezoelectric effects

Some crystals have the property that if the crystal is strained, an electric dipole moment and associated electric field develop. This phenomenon is called **piezoelectricity.** Only crystals without a center of inversion can exhibit it. Under the inversion symmetry operation the crystal is transformed into itself and any physical property must be unchanged. The electric dipole moment $\boldsymbol{M}$ is an odd function of coordinates, however, so $\boldsymbol{M} = -\boldsymbol{M}$ under inversion. Hence $\boldsymbol{M} = 0$ and piezoelectric crystals cannot have a center of inversion. Thus, semiconductors of the zincblende structure such as GaAs and of the wurtzite structure such as CdS exhibit piezoelectricity, but diamond structure semiconductors such as Si do not.

The coupling between the elastic strain and the electric field is characterized by the third-rank **piezoelectric tensor** $e_{\alpha\beta\gamma}$ according to the relation

$$\sigma_{\alpha\beta} = \sum_{\gamma\delta} C_{\alpha\beta\gamma\delta}\frac{\partial u_\gamma}{\partial x_\delta} - \sum_{\gamma} e_{\gamma\alpha\beta}\mathcal{E}_\gamma \tag{7.104}$$

where $\sigma_{\alpha\beta}$ is the stress tensor, $C_{\alpha\beta\gamma\delta}$ is the fourth-rank elastic modulus tensor, u_γ is the particle displacement, and $\mathcal{E}_\gamma$ is the electric field. The equations of motion take the form

$$\begin{aligned}\rho\frac{\partial^2 u_\alpha}{\partial t^2} &= \sum_{\beta}\frac{\partial \sigma_{\alpha\beta}}{\partial x_\beta}\\ &= \sum_{\beta\gamma\delta} C_{\alpha\beta\gamma\delta}\frac{\partial^2 u_\gamma}{\partial x_\beta \partial x_\delta} - \sum_{\beta\gamma} e_{\gamma\alpha\beta}\frac{\partial \mathcal{E}_\gamma}{\partial x_\beta}.\end{aligned} \tag{7.105}$$

The constitutive relation involving the electric displacement $\mathcal{D}_\alpha$ and the dielectric tensor $\epsilon_{\alpha\beta}$ when piezoelectricity is present is

$$\mathcal{D}_\alpha = \sum_{\beta\gamma} e_{\alpha\beta\gamma}\frac{\partial u_\beta}{\partial x_\gamma} + \epsilon_0 \sum_{\beta} \epsilon_{\alpha\beta}\mathcal{E}_\beta. \tag{7.106}$$

Combining this relation with the Maxwell equation

$$\boldsymbol{\nabla}\cdot\mathcal{D} = 0 \tag{7.107}$$

yields the equation

$$\sum_{\alpha\beta\gamma} e_{\alpha\beta\gamma}\frac{\partial^2 u_\beta}{\partial x_\alpha \partial x_\gamma} + \sum_{\alpha\beta}\epsilon_0\epsilon_{\alpha\beta}\mathcal{E}_\beta = 0 \tag{7.108}$$

which is to be solved simultaneously with Eq. (7.105).

As a specific example, let us consider the wurtzite form of CdS with its sixfold axis parallel to the z-direction. For elastic waves polarized in the z-direction and propagating in the x-direction, there is a single element of the piezoelectric tensor to be considered, conventionally designated by e_{15}, and a single element of the dielectric tensor, ϵ_{xx}. The equations of motion and the Maxwell equation reduce to

$$\rho \frac{\partial^2 u_z}{\partial t^2} = C_{44} \frac{\partial^2 u_z}{\partial x^2} - e_{15} \frac{\partial \mathcal{E}_x}{\partial x} \tag{7.109}$$

$$0 = e_{15} \frac{\partial^2 u_z}{\partial x^2} + \epsilon_0 \epsilon_{xx} \frac{\partial \mathcal{E}_x}{\partial x}, \tag{7.110}$$

where C_{44} is the appropriate elastic modulus in the Voigt notation. Eliminating the electric field from the equation of motion, we obtain the wave equation

$$\rho \frac{\partial^2 u_z}{\partial t^2} = \bar{C}_{44} \frac{\partial^2 u_z}{\partial x^2}, \tag{7.111}$$

with $\bar{C}_{44}$ the effective elastic modulus specified by

$$\bar{C}_{44} = C_{44} + \frac{e_{15}^2}{\epsilon_0 \epsilon_{xx}}. \tag{7.112}$$

Taking the elastic wave to have the plane wave form

$$u_z(x, t) = U \cos(qx - \omega t), \tag{7.113}$$

the dispersion relation is found to be

$$\omega \sqrt{\frac{\bar{C}_{44}}{\rho}} q = c_t q, \tag{7.114}$$

where c_t is the speed of the wave. It is clear that the speed is enhanced by the piezoelectricity. Typically, the enhancement is on the order of 1–3 percent.

7.11 Effects of stress-induced atomic displacements

The application of an external stress to a semiconductor crystal produces a number of interesting effects. If the stress is uniform **pressure**, the crystal contracts with resulting changes in lattice constant, electronic energies, and phonon frequencies. The symmetry of the crystal is unchanged, so the degeneracies due to symmetry are unaffected. In cubic semiconductors the energy gap between conduction and valence bands at the Γ and L points increases with increasing pressure, whereas it decreases at the X point. In the case of phonons the frequencies of TA modes at X and L decrease with increasing pressure, whereas for TO modes at Γ the frequencies increase. LO modes show the same behavior as TO modes except for Si (Martinez 1980).

If a **uniaxial stress** is applied to a cubic crystal, the atoms typically displace in such a way as to reduce the symmetry and thereby removing degeneracies. In addition, displacements within a unit cell can occur that are referred to as **internal displacements** or **internal strains**.

Let a uniaxial stress be applied in the [111] direction of a diamond or zincblende structure crystal. A pair of nearest-neighbor atoms along that direction will undergo a change of separation which within the framework of classical elasticity theory would be Δd_1^e. However, it turns out that macroscopic elasticity theory is not adequate to completely specify Δd_1. Microscopic theory shows that Δd_1 is given by

$$\Delta d_1 = \Delta d_1^e(1 - \zeta), \tag{7.115}$$

where ζ is the **Kleinman internal displacement parameter** (Kleinman 1962). Using a lattice-dynamical model with nearest-neighbor central forces having force constant σ_c and tetrahedral angular forces having force constant σ_a, one can show that ζ is given by (Harrison 1980)

$$\zeta = \frac{\sigma_c - 4\sigma_a}{\sigma_c + 8\sigma_a}. \tag{7.116}$$

It is evident that $\zeta = 0$ ($\sigma_c = 4\sigma_a$) corresponds to no internal displacement and that Δd_1 is then fully described by macroscopic elasticity theory, whereas $\zeta = 1$ ($\sigma_a = 0$) corresponds to a full cancellation of the elastic displacement by the internal displacement.

X-ray diffraction can be employed to measure ζ. One finds that the values of ζ are 0.73 for Si (Cousins *et al.* 1982a) and 0.72 for Ge (Cousins *et al.* 1982b).

Application of a uniaxial compressive stress to Si in the [100] direction lowers the energy of the two [100] conduction band minima relative to that of the other four minima. Compressive stress in the [111] direction in Ge lowers the energy of the [111] conduction band minimum relative to that of the other three minima. In the case of the valence bands, uniaxial stress removes the degeneracy of the light and heavy hole bands (Pikus and Bir 1959). Degeneracies of impurity levels can also be removed by uniaxial stress. The three-fold degeneracy of the optical phonons at $\boldsymbol{q} = 0$ in Si and Ge is split by uniaxial stress as a result of inequivalent changes in the force constants in the three directions of polarization. One optical mode is polarized in the direction of the stress and the other two are polarized normal to the stress.

Problems

1. Consider a one-dimensional model of Si that has the same atomic arrangement as the [111] direction of the three-dimensional crystal. In the nearest-neighbor approximation two force constants σ_1 and σ_2 alternate along the chain corresponding to the two different nearest-neighbor distances.
   (a) Write down the equations of motion for the two atoms in the basis.
   (b) Using a plane-wave solution of the form $u(\ell\kappa; q, t) = M^{-\frac{1}{2}} \times W(\kappa) e^{i[qR(\ell) - \omega t]}$, $\kappa = 1, 2$, solve for the frequency ω as a function of the wave vector q. Plot ω versus q for $\sigma_1 = 100\,\mathrm{N/m}$ and $\sigma_2 = 0.5\,\sigma_1$.
   (c) Is there a gap in the frequency spectrum? If so, what is its value?
2. Repeat problem 1, but with the two atoms in the basis having different masses M_1 and M_2. Choose the masses to be those of GaSb.

3. Given that the energy of a monatomic chain with one atom per primitive unit cell is

$$E = \frac{1}{2} M \sum_{\ell} \left(\frac{du(\ell)}{dt} \right)^2 + \frac{1}{2}\sigma_1 \sum_{\ell} [u(\ell) - u(\ell+1)]^2 + \frac{1}{2}\sigma_2 \sum_{\ell} [u(\ell) - u(\ell+2)]^2,$$

derive the equations of motion. The quantities σ_1 and σ_2 are the first-neighbor and second-neighbor force constants. Using a plane wave solution, find the frequency versus wave vector relation and plot this relation for $\sigma_2 = 0.25\sigma_1$ and $\sigma_2 = 0.75\sigma_1$. Is the maximum frequency always at the Brillouin zone boundary?

4. Starting from the equations of motion of problem 3, pass to the continuum limit and obtain the elastic equation of motion. What is the speed of sound expressed in terms of σ_1, σ_2, M, and a?

5. An expression for the frequency distribution function in an s-dimensional system is

$$F_s(\omega) = \frac{\Omega_s}{(2\pi)^s} \sum_j \int d^s q \delta(\omega - \omega_{\boldsymbol{q}j}).$$

Calculate the frequency distribution function for the monatomic linear chain with nearest-neighbor interactions. The singularity in the result is an example of a Van Hove singularity.

6. Using a two-dimensional version of elasticity theory, derive an expression for the low-temperature specific heat and obtain an expression for the Debye temperature in terms of the speeds of longitudinal and transverse waves.

7. Consider a monatomic linear chain with nearest-neighbor interactions characterized by force constant σ and with a defect bond between a pair of nearest neighbors characterized by force constant σ'. Derive expressions that specify the frequency and displacements of a localized mode of vibration. For what values of σ'/σ does a localized mode exist?

References

G. B. Bachelet, D. R. Hamann, and M. Schlüter, *Phys. Rev.* **B26**, 4199 (1982).

S. Baroni, P. Giannozzi, and A. Testa, *Phys. Rev. Lett.* **58**, 1861 (1987).

K. W. Böer, *Survey of Semiconductor Physics* (Van Nostrand Reinhold, New York, 1990).

M. Born and K. Huang, *Dynamical Theory of Crystal Lattices* (Oxford University Press, Oxford, 1954).

W. Cochran, *Proc. Roy. Soc.* (London) **A253**, 260 (1959).

C. S. G. Cousins, L. Gerwald, J. S. Olsen, B. Selsmark, and B. J. Sheldon, *J. Appl. Crystallogr.* **15**, 154 (1982a).

C. S. G. Cousins, L. Gerwald, K. Nielsen, J. S. Olsen, B. Selsmark, B. J. Sheldon, and G. E. Webster, *J. Phys.* **C15**, L651 (1982b).

B. G. Dick and A. W. Overhauser, *Phys. Rev.* **112**, 90 (1958).

G. Dolling, in *Inelastic Scattering of Neutrons in Solids and Liquids* (International Atomic Energy Agency, Vienna, 1962), Vol. **II**, p. 37.

T. H. Geballe and G. W. Hull, *Phys. Rev.* **110**, 773 (1958).

P. Giannozzi, S. de Gironcoli, P. Pavone, and S. Baroni, *Phys. Rev.* **B43**, 7231 (1991).

J. E. Hanlon and A. W. Lawson, *Phys. Rev.* **113**, 472 (1959).

J. R. Hardy, *Phil. Mag.* **7**, 315 (1962).

W. A. Harrison, *Electronic Structure and the Properties of Solids* (W. H. Freeman, San Francisco, 1980).

F. Herman, *J. Phys. Chem. Solids* **8**, 405 (1959).
A. M. Karo and J. R. Hardy, *Phys. Rev.* **181**, 1272 (1969).
E. W. Kellerman, *Proc. Roy. Soc.* (London) **A178**, 17 (1941).
C. Kittel, *Introduction to Solid State Physics*, Sixth edition (John Wiley, New York, 1986).
L. Kleinman, *Phys. Rev.* **128**, 2614 (1962).
K. Kunc, *Ann. Phys.* (Paris) **8**, 319 (1973–74).
K. Kunc and R. M. Martin, *Phys. Rev.* **B24**, 2311 (1981).
K. Kunc, M. Balkanski and M. Nusimovici, *Phys. Rev.* **B12**, 4346 (1975).
K. G. Lyon, G. L. Salinger, C. A. Swenson, and G. K. White, *J. Appl. Phys.* **48**, 865 (1977).
A. A. Maradudin, E. W. Montroll, G. H. Weiss, and I. P. Ipatova, *Theory of Lattice Dynamics in the Harmonic Approximation*, Second edition (Academic Press, New York, 1971).
G. Martinez, in *Handbook on Semiconductors*, Vol. 2, eds. T. S. Moss and M. Balkanski (North-Holland, Amsterdam, 1980).
S. S. Mitra and M. E. Massa, in *Handbook on Semiconductors*, Vol 1, eds. T. S. Moss and W. Paul (North-Holland, Amsterdam, 1982).
E. W. Montroll and R. B. Potts, *Phys. Rev.* **100**, 525 (1955).
G. Nilsson and G. Nelin, *Phys. Rev.* **B3**, 364 (1971).
G. Nilsson and G. Nelin, *Phys. Rev.* **B6**, 3777 (1972).
G. E. Pikus and G. L. Bir, Fiz. Tverd. Tela **1**, 1642 (1959); *Sov. Phys. Solid State* **1**, 1502 (1959).
D. L. Price, J. M. Rowe, and R. M. Nicklow, *Phys. Rev.* **B3**, 1268 (1971).
A. A. Quong and B. M. Klein, *Phys. Rev.* **B46**, 10734 (1992).
H. B. Rosenstock and G. F. Newell, *J. Chem. Phys.* **21**, 1607 (1953).
R. Truell, C. Elbaum, and B. B. Chick, *Ultrasonic Methods in Solid State Physics* (Academic Press, New York, 1969).
L. van Hove, *Phys. Rev.* **89**, 1189 (1953).
K. H. Wanser, Thesis, University of California, Irvine, 1982.
W. Weber, *Phys. Rev.* **B15**, 4789 (1977).

Charge carrier scattering and transport properties

8

Key ideas

Charge carriers in semiconductors are characterized by their *mean free path*, *relaxation time*, and *mobility*.

The *Boltzmann equation* governs the behavior of the carrier *distribution function*.

The *mobility* of a carrier is proportional to the *average relaxation time*.

In general, the *relaxation time* of a carrier depends on its *energy* and on the nature of the *scatterers*.

Scattering mechanisms such as those due to *ionized impurities* and *phonons* contribute to the relaxation time.

The *electrical conductivity* is modified by an *external magnetic field*. The *Hall effect* enables one to measure the *carrier concentration*.

The presence of a *temperature gradient* gives rise to the *Seebeck effect*. An *electric current* can produce a *heat flux* through the *Peltier effect*.

Free carriers contribute to the *thermal conductivity* of a semiconductor.

Using *deep impurities*, *semi-insulating semiconductors* can be produced.

In *high electric fields*, free carriers have a *higher effective temperature* and a *lower mobility*. *Negative differential conductivity* can arise that produces *Gunn oscillations*. High-energy carriers can generate additional carriers by *impact ionization*.

In *disordered semiconductors* at low temperature, the electrical conductivity can have an $\exp(-BT^{-\frac{1}{4}})$ dependence due to *variable-range hopping* of carriers.

Transport properties

Transport properties such as electrical conductivity play a crucial role in the application of semiconductors to electronic devices. In this chapter we shall be concerned with the flow of a quantity such as electric charge in response to an externally applied force such as that due to an electric field. The carriers of electric charge of interest in semiconductors are primarily conduction electrons and holes.

8.1 Simple phenomenological introduction to transport in semiconductors

We start our discussion of transport with a simple phenomenological approach. When an external force is applied to a system of free charge carriers, the carriers are displaced under the influence of the field, and a current results. The external force may be due to an electric field producing an electric conduction current, to a concentration gradient giving rise to a diffusion current, or to a thermal gradient leading to a heat current associated with thermal conductivity.

8.1.1 Electric conduction current

To a first approximation a system of free carriers in an energy band can be considered as a gas of noninteracting charged particles. In the absence of an external force the charge carriers execute Brownian motion resulting from collisions due to the interaction of the carriers with impurities, lattice vibrations, and other perturbations to the periodic potential. During a collision, a carrier may undergo a sharp change in direction, but between collisions its motion is essentially rectilinear and characterized by its mean speed. For carriers obeying classical statistics, the mean speed is the thermal speed s_{th} specified by the law of equipartition,

$$\tfrac{1}{2}m^* s_{th}^2 = \tfrac{3}{2}k_B T,$$

or

$$s_{th} = \sqrt{\frac{3k_B T}{m^*}}, \tag{8.1}$$

where m^* is the carrier effective mass. In semiconductors, such as Si, where the carrier effective masses are on the order of the free electron mass m_0, the thermal speed is on the order of 10^7 cm/s at room temperature.

The average time between successive collisions is the **relaxation time** τ of the carrier. It depends on the purity and perfection of the material and on the temperature. Typical relaxation times lie in the range 10^{-13}–10^{-12} s. The mean distance a carrier travels between collisions is its **mean free path** Λ given by

$$\Lambda = s_{th}\tau \tag{8.2}$$

and typically lies in the range 100–1000 Å. In the case of **isotropic scattering**, the direction of a carrier immediately after a collision may be in any direction with equal probability. The **mean velocity** of the carrier is therefore zero.

When an electric field $\boldsymbol{\mathcal{E}}$ is applied at time t_0, a carrier of charge e_c is subject to a force $\boldsymbol{F} = e_c\boldsymbol{\mathcal{E}}$. For a carrier in a spherical parabolic band, the velocity component $v_\alpha(t)$ in the direction of the field is given by the equation of motion (see Eq. (4.19))

$$m^* \frac{dv_\alpha(t)}{dt} = e_c \mathcal{E}. \tag{8.3}$$

Integrating this equation from t_0 to $t_0 + t$, one gets

$$v_\alpha(t) = \frac{e_c \mathcal{E}}{m^*} t. \tag{8.4}$$

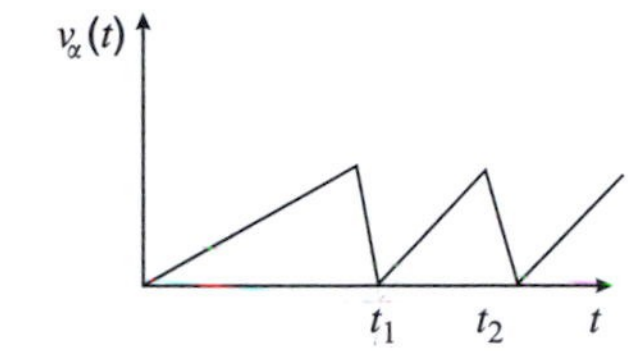

Fig. 8.1
Velocity component versus time for a carrier undergoing collisions.

This component increases linearly with time between two collisions and is returned to zero immediately after each collision as shown in Fig. 8.1.

Suppose that there are N carriers traveling with speed s in a given direction. The number of collisions made by the carriers in a small time interval dt is proportional to both N and dt. If $N(t)$ is the number of carriers which have **not** made a collision at time t, then

$$dN(t) = -\frac{N(t)}{\tau(s)} dt, \tag{8.5}$$

where $\tau(s)$ is the **relaxation time**. Integrating this equation, we have

$$N(t) = N_0 e^{-t/\tau(s)} \tag{8.6}$$

where $N_0 = N(0)$.

The probability $P(t)$ that a carrier has not made a collision at time t is given by

$$P(t) = \frac{1}{\tau(s)} e^{-t/\tau(s)}, \tag{8.7}$$

where $P(t)$ is normalized to unity:

$$\int_0^\infty P(t) dt = 1. \tag{8.8}$$

The mean time between collisions $\langle t \rangle$ is specified by

$$\langle t \rangle = \int_0^\infty t P(t) dt = \tau(s), \tag{8.9}$$

so $\tau(s)$ can be interpreted as the mean free time. Similarly, one obtains the mean value of the velocity component $v_\alpha(t)$ in the form

$$\begin{aligned} \langle v_\alpha \rangle &= \int_0^\infty v_\alpha(t) P(t) dt \\ &= \frac{e_c \mathcal{E}}{m^*} \int_0^\infty t P(t) dt = \frac{e_c \mathcal{E} \tau(s)}{m^*} \\ &= \pm \mu(s) \mathcal{E}, \end{aligned} \tag{8.10}$$

where $\mu(s)$ is the free carrier **mobility** given by

$$\mu(s) = \left|\frac{e_c\tau(s)}{m^*}\right| = \frac{e\tau(s)}{m^*}. \tag{8.11}$$

The units of μ are conventionally taken to be $\text{cm}^2/(\text{V s})$. In the **Drude model**, all carriers of a given mass have the same speed, so $\tau(s)$ and $\mu(s)$ are constants for such carriers.

The mobility is by definition positive and depends linearly on the relaxation time. For a given semiconductor the effective mass is a known fixed quantity. The mobility therefore varies from sample to sample through its dependence on τ which is a function of impurity content and temperature. The highest values of τ and hence of μ are obtained in very pure material at low temperature.

In semiconductors we can have positive charge carriers, holes, with mobility μ_h and velocity $\boldsymbol{v}_h = \mu_h\boldsymbol{\mathcal{E}}$, and negative charge carriers, conduction electrons, with mobility μ_e and velocity $\boldsymbol{v}_e = -\mu_e\boldsymbol{\mathcal{E}}$. The displacement of charge carriers under the influence of an applied electric field produces an electric current. The **current density** $\boldsymbol{j}$ is defined as the charge crossing unit surface area per unit time. For electrons and holes we have

$$\boldsymbol{j}_e = -ne\boldsymbol{v}_e = ne\mu_e\boldsymbol{\mathcal{E}} \tag{8.12a}$$

$$\boldsymbol{j}_h = pe\boldsymbol{v}_h = pe\mu_h\boldsymbol{\mathcal{E}}, \tag{8.12b}$$

where n and p are the concentrations of electrons and holes, respectively. If both electrons and holes are present, the total current $\boldsymbol{j}$ is the sum of the individual contributions, so that

$$\boldsymbol{j} = \boldsymbol{j}_e + \boldsymbol{j}_h = \sigma\boldsymbol{\mathcal{E}}, \tag{8.13}$$

where the **electrical conductivity** σ is given by

$$\sigma = ne\mu_e + pe\mu_h. \tag{8.14}$$

Using Eq. (8.11) we can re-express the conductivity as

$$\sigma = e^2\left(\frac{n\tau_e}{m_e^*} + \frac{p\tau_h}{m_h^*}\right). \tag{8.15}$$

Equation (8.13) is a microscopic statement of **Ohm's law**. For a homogeneous semiconductor slab of length ℓ and cross-sectional area S with a voltage V applied across the length, the conventional form of Ohm's law states that

$$V = IR, \tag{8.16}$$

where I is the current flowing and R is the resistance given by

$$R = \frac{\ell}{\sigma S}. \tag{8.17}$$

Table 8.1 Room temperature mobilities in $cm^2/(Vs)$ (after Kittel 1986)

Semiconductor	Si	Ge	GaAs	PbTe
Electron mobility μ_e	1350	3600	8000	2500
Hole mobility μ_h	480	1800	300	1000

It is a general observation that the mobility of electrons is significantly larger than that of holes. This can be attributed at least in part to the generally larger effective masses of holes compared to electrons and the inverse relationship of mobility to effective mass contained in Eq. (8.11). Mobilities tend to increase rapidly with decreasing temperature as a consequence of longer relaxation times at low temperature than at high temperature. The trapping of free carriers by ionized impurities and the decreased amplitude of atomic vibrations at low temperature both tend to decrease the effective collision rate and increase the relaxation time. Some representative mobilities at room temperature are shown in Table 8.1.

8.1.2 Conductivity effective mass

The expression we have obtained for the mobility, $\frac{e\tau}{m^*}$, contains the **conductivity effective mass** m^*. For a specific semiconductor it is necessary to determine what types of carriers are present and how their band structure effective masses are related to m^*. As a specific example we consider n-Si for which the conduction electrons are distributed among six equivalent valleys. Two valleys have their minima along each of the x-, y-, and z-axes at $\pm x_0$, $\pm y_0$, and $\pm z_0$. Each valley is characterized by a longitudinal effective mass m_ℓ and a transverse effective mass m_t due to the anisotropy of the conduction band constant energy surfaces. The total current density is the sum of the current densities associated with each valley,

$$\boldsymbol{j}_e = \boldsymbol{j}_{ex} + \boldsymbol{j}_{ey} + \boldsymbol{j}_{ez} \tag{8.18}$$

where $\boldsymbol{j}_{ei}$ is the current density from electrons in the two valleys with minima along the i-axis, $i = x, y, z$.

If the electric field is in the x-direction, the electrons in the two minima at $\pm x_0$ are characterized by the longitudinal effective mass m_ℓ, whereas the electrons in the four minima at $\pm y_0$ and $\pm z_0$ are characterized by the transverse effective mass m_t. We therefore have the following contributions to the current density using Eq. (8.12a):

$$j_{ex} = 2(n/6)e\mu_{ex}\mathcal{E} = \frac{ne^2\tau_e}{3m_\ell}\mathcal{E} \tag{8.19a}$$

$$j_{ey} = j_{ez} = \frac{ne^2\tau_e}{3m_t}\mathcal{E}. \tag{8.19b}$$

The total current density from Eq. (8.19) becomes

$$\boldsymbol{j}_e = \frac{1}{3}\left(\frac{1}{m_\ell} + \frac{2}{m_t}\right) ne^2\tau_e \boldsymbol{\mathcal{E}}. \tag{8.20}$$

Referring back to Eq. (8.12a) we see that

$$\mu_e = \frac{1}{3}\left(\frac{1}{m_\ell} + \frac{2}{m_t}\right) e\tau_e, \tag{8.21}$$

which upon comparison with Eq. (8.11) yields the inverse conductivity effective mass for the conduction band:

$$\frac{1}{m_c^*} = \frac{1}{3}\left(\frac{1}{m_\ell} + \frac{2}{m_t}\right). \tag{8.22}$$

For carriers in the valence band we have to consider the heavy hole and light hole bands which are degenerate at $\boldsymbol{k} = 0$ and will both contribute to the conduction process. The current density due to heavy holes with concentration p_h and mass m_{hh} is

$$\boldsymbol{j}_{hh} = p_h e \mu_{hh} \boldsymbol{\mathcal{E}} = \frac{p_h e^2 \tau_h}{m_{hh}} \boldsymbol{\mathcal{E}}, \tag{8.23}$$

while that for light holes with concentration p_ℓ and mass $m_{\ell h}$ is

$$\boldsymbol{j}_{\ell h} = \frac{p_\ell e^2 \tau_h}{m_{\ell h}} \boldsymbol{\mathcal{E}}, \tag{8.24}$$

where we have assumed that the relaxation times for light and heavy holes have the same value τ_h.

The total hole current density $\boldsymbol{j}_h$ is

$$\boldsymbol{j}_h = \boldsymbol{j}_{\ell h} + \boldsymbol{j}_{hh} = \left(\frac{p_h}{m_{hh}} + \frac{p_\ell}{m_{\ell h}}\right) e^2 \tau_h \boldsymbol{\mathcal{E}}, \tag{8.25}$$

which can be written in the alternative form

$$\boldsymbol{j}_h = \frac{p e^2 \tau_h}{m_v^*} \boldsymbol{\mathcal{E}}, \tag{8.26}$$

where m_v^* is the conductivity effective mass for valence band holes and $p = p_h + p_\ell$. Comparing Eqs. (8.25) and (8.26), we obtain

$$\frac{1}{m_v^*} = \frac{p_h}{p m_{hh}} + \frac{p_\ell}{p m_{\ell h}}.$$

In equilibrium the concentrations of light and heavy holes are proportional to $m_{\ell h}^{3/2}$ and $m_{hh}^{3/2}$, respectively, so

$$\frac{1}{m_v^*} = \frac{m_{hh}^{\frac{1}{2}} + m_{\ell h}^{\frac{1}{2}}}{m_{hh}^{3/2} + m_{\ell h}^{3/2}}. \tag{8.27}$$

8.1.3 Diffusion current

When the free carrier distribution is not uniform in space, the carriers tend to diffuse from regions of high concentration to regions of low concentration. Diffusion is governed by Fick's law, which states that the carrier flux J_n is proportional to the concentration gradient. The flux of a particular type of carrier i is the number of such carriers crossing unit area perpendicular to their direction of motion in unit time. In one dimension Fick's law takes the form

$$J_{nx}^{(i)} = -D_i \frac{dn_{ci}}{dx}. \tag{8.28}$$

where D_i is the **diffusion constant** for the *ith* type of carrier. In three dimensions, Fick's law becomes

$$\boldsymbol{J}_n^{(i)} = -D_i \boldsymbol{\nabla} n_{ci}. \tag{8.29}$$

Associated with a flux of charge carriers is an electric current called the **diffusion current** whose density $\boldsymbol{j}_i$ is $e_{ci}\boldsymbol{J}_n^{(i)}$. For conduction electrons and holes, we have

$$\boldsymbol{j}_e = eD_e \boldsymbol{\nabla} n \tag{8.30a}$$

and

$$\boldsymbol{j}_h = -eD_h \boldsymbol{\nabla} p. \tag{8.30b}$$

The total diffusion current is

$$\boldsymbol{j} = eD_e \boldsymbol{\nabla} n - eD_h \boldsymbol{\nabla} p. \tag{8.31}$$

When both an electric field $\boldsymbol{\mathcal{E}}$ and concentration gradients are present, the current densities satisfy

$$\boldsymbol{j}_e = ne\mu_e \boldsymbol{\mathcal{E}} + eD_e \boldsymbol{\nabla} n \tag{8.32a}$$

$$\boldsymbol{j}_h = pe\mu_h \boldsymbol{\mathcal{E}} - eD_h \boldsymbol{\nabla} p. \tag{8.32b}$$

The mobility μ is a measure of the ease with which a carrier responds to an electric field, whereas the diffusion constant D is a measure of the ease with which a carrier responds to a concentration gradient. From the form of Eq. (8.32) we see that the quantity $\boldsymbol{\nabla} n/n$ plays the role of an effective "field" which produces the diffusive motion of a carrier. It has been shown (Einstein 1905) that there is a direct connection, called the **Einstein relation**, between the mobility and the diffusion constant.

If we impose the condition of equilibrium on a semiconductor sample, the current density of each type of carrier must be zero. Taking electrons as an example, we have from Eq. (8.32a)

$$ne\mu_e \boldsymbol{\mathcal{E}} + eD_e \boldsymbol{\nabla} n = 0$$

or

$$n\mu_e \mathcal{E} = -D_e \nabla n. \tag{8.33}$$

The electric field $\mathcal{E}$ is related to the electric potential $V(\boldsymbol{r})$ by $\mathcal{E} = -\nabla V(\boldsymbol{r})$, so Eq. (8.33) becomes

$$ne\mu_e \nabla V = D_e \nabla n. \tag{8.34}$$

Furthermore, the conduction band edge becomes $E_c - eV(\boldsymbol{r})$, and under non-degenerate conditions, the conduction electron concentration at point $\boldsymbol{r}$ can be expressed by the following modification of Eq. (6.10):

$$n(\boldsymbol{r}) = \bar{N}_c e^{-\beta[E_c - eV(\boldsymbol{r}) - E_F]}. \tag{8.35}$$

Since the Fermi energy E_F is the chemical potential of the electrons, in equilibrium it must be a constant and independent of position. Taking the gradient of $n(\boldsymbol{r})$, we obtain

$$\nabla n(\boldsymbol{r}) = \beta e n(\boldsymbol{r}) \nabla V(\boldsymbol{r}). \tag{8.36}$$

Substituting Eq. (8.36) into Eq. (8.34) yields

$$D_e = \frac{\mu_e}{\beta e} = \frac{k_B T}{e} \mu_e, \tag{8.37}$$

which is the Einstein relation. The corresponding relation for holes is

$$D_h = \frac{k_B T}{e} \mu_h. \tag{8.38}$$

Taking into consideration the Einstein relation, we can rewrite the expressions for the electron and hole current densities given by Eqs. (8.32) in the form

$$\boldsymbol{j}_e = \mu_e (en\mathcal{E} + k_B T \nabla n) \tag{8.39a}$$

$$\boldsymbol{j}_h = \mu_h (ep\mathcal{E} - k_B T \nabla p). \tag{8.39b}$$

Thus, the current density is proportional to the mobility even when diffusion is taken into account.

8.1.4 Displacement current

Up to this point we have focused on electrical conductivity as a time-independent phenomenon. If a system of free carriers is subject to an electric field that oscillates in time of the form

$$\mathcal{E} = \mathcal{E}_0 e^{-i\omega t}, \tag{8.40}$$

a **displacement current** is set up whose density $\boldsymbol{j}_d$ is given by

$$\boldsymbol{j}_d = \frac{\partial \mathcal{D}}{\partial t}, \tag{8.41}$$

where $\mathcal{D}$ is the **electric displacement**. The electric displacement is related to the electric field by the equation

$$\mathcal{D} = \epsilon\epsilon_0\mathcal{E}, \tag{8.42}$$

in which ϵ is the dielectric constant and ϵ_0 is the permittivity of vacuum ($8.85 \times 10^{-12}\,\mathrm{C^2/Nm^2}$). Combining Eqs. (8.40)–(8.42), we obtain

$$j_d = -i\epsilon\epsilon_0\omega\mathcal{E}. \tag{8.43}$$

Adding this result to the static contribution to the current density given by Eq. (8.13) yields

$$j = (\sigma - i\epsilon\epsilon_0\omega)\mathcal{E}. \tag{8.44}$$

The displacement current contribution is important only at relatively high frequencies such as those found in certain electronic components and in optical phenomena.

8.2 The Boltzmann equation and its solution

In order to go beyond the simple treatment of Section 8.1, it is necessary to recognize that the relaxation time is in general a function of the energy of the free carrier. Furthermore, in the presence of an electric field, free carriers are not in equilibrium, and their distribution over energy is not specified by the equilibrium distribution function discussed in Chapter 6, but by a nonequilibrium distribution function that is determined by solving the **Boltzmann equation**.

The classical distribution function $f(\boldsymbol{v}, t)$ specifies the number of charge carriers having velocities in a unit range about the value $\boldsymbol{v}$ at time t. Since a carrier is characterized by its wave vector $\boldsymbol{k}$, we replace $\boldsymbol{v}$ by $\boldsymbol{k}$: $f(\boldsymbol{v}, t) \rightarrow f(\boldsymbol{k}, t)$. We shall deal only with spatially uniform systems, so there is no dependence of f on the spatial coordinate $\boldsymbol{r}$ of the carrier. For the moment it is assumed that only a single type of carrier is present.

A charge carrier moving in a perfect crystal with a periodic potential can be taken to be in a particular Bloch eigenstate of the periodic Hamiltonian. If the periodic potential is perturbed by the introduction of impurities or other flaws or by the vibrations of the nuclei away from their equilibrium positions, the carrier may undergo a scattering transition to another Bloch state.

The distribution function $f(\boldsymbol{k}, t)$ can change with time t as a result of collisions or as a result of an explicit dependence on t or an implicit dependence due to a dependence of $\boldsymbol{k}$ on t. The total derivative of f with respect to t can therefore be written as

$$\frac{df}{dt} = \frac{\partial f}{\partial t} + \sum_{\alpha=x,y,z} \frac{\partial f}{\partial k_\alpha}\frac{dk_\alpha}{dt} - \left.\frac{df}{dt}\right|_{coll}, \tag{8.45}$$

where $\frac{df}{dt}|_{coll}$ is the time rate of change of f due to collisions and the minus sign in front of this term reflects the fact that collisions tend to reduce the occupancy of state $\boldsymbol{k}$. If f is a function only of the carrier energy $E_{\boldsymbol{k}}$, as is frequently the case, then

$$\frac{\partial f}{\partial k_\alpha} = \frac{\partial f}{\partial E_{\boldsymbol{k}}} \frac{\partial E_{\boldsymbol{k}}}{\partial k_\alpha}, \tag{8.46}$$

$$\frac{\partial f}{\partial t} = 0, \tag{8.47}$$

and

$$\frac{df}{dt} = \frac{\partial f}{\partial E_{\boldsymbol{k}}} \left[\sum_\alpha \frac{\partial E_{\boldsymbol{k}}}{\partial k_\alpha} \frac{dk_\alpha}{dt} \right] - \left. \frac{df}{dt} \right|_{coll}. \tag{8.48}$$

From the basic equation of motion given by Eq. (4.15) we have

$$\hbar \frac{dk_\alpha}{dt} = F_\alpha, \tag{8.49}$$

where $\boldsymbol{F}$ is the external force acting on the carrier, so Eq. (8.48) becomes

$$\frac{df}{dt} = \frac{\partial f}{\hbar \partial E_{\boldsymbol{k}}} \boldsymbol{F} \cdot \nabla_{\boldsymbol{k}} E_{\boldsymbol{k}} - \left. \frac{df}{dt} \right|_{coll}. \tag{8.50}$$

But from Eq. (4.8), which specifies the group velocity of an electron wave packet,

$$\boldsymbol{v}_g = \frac{1}{\hbar} \nabla_{\boldsymbol{k}} E_{\boldsymbol{k}}. \tag{8.51}$$

Substituting this result into Eq. (8.50) yields

$$\frac{df}{dt} = \frac{\partial f}{\partial E_{\boldsymbol{k}}} \boldsymbol{F} \cdot \boldsymbol{v}_g - \left. \frac{df}{dt} \right|_{coll}. \tag{8.52}$$

In a steady state the two contributions to the time dependence on the right hand side cancel, so that

$$\frac{df}{dt} = 0, \tag{8.53}$$

and

$$\frac{\partial f}{\partial E_{\boldsymbol{k}}} \boldsymbol{F} \cdot \boldsymbol{v}_g = \left. \frac{df}{dt} \right|_{coll}. \tag{8.54}$$

For small deviations of f from the equilibrium distribution f_0, let us assume that the time rate of change of the distribution function due to collisions is proportional to the deviation of the distribution function from its equilibrium value:

$$\left. \frac{df}{dt} \right|_{coll} = -\frac{f - f_0}{\tau}. \tag{8.55}$$

The negative inverse of the proportionality constant is the relaxation time τ. Equation (8.54) can then be rewritten as

$$\frac{\partial f}{\partial E_{\boldsymbol{k}}}\boldsymbol{F}\cdot\boldsymbol{v}_g = -\frac{f-f_0}{\tau} \qquad (8.56)$$

which is the Boltzmann equation in the **relaxation time approximation**. It must be emphasized that a relaxation time cannot always be defined, particularly when the carriers are in anisotropic energy bands. A discussion of the requirements for the existence of a relaxation time can be found in the article by Roth (1992).

For the problem of electrical conductivity, the force $\boldsymbol{F}$ is

$$\boldsymbol{F} = e_c\boldsymbol{\mathcal{E}}, \qquad (8.57)$$

where e_c is the carrier charge and $\boldsymbol{\mathcal{E}}$ is the macroscopic electric field in the crystal. The Boltzmann equation becomes

$$\frac{\partial f}{\partial E_{\boldsymbol{k}}}e_c\boldsymbol{\mathcal{E}}\cdot\boldsymbol{v}_g = -\frac{f-f_0}{\tau}, \qquad (8.58)$$

which can be rearranged in the form

$$f = f_0 - e_c\tau\frac{\partial f}{\partial E_{\boldsymbol{k}}}\boldsymbol{\mathcal{E}}\cdot\boldsymbol{v}_g. \qquad (8.59)$$

If we assume that the perturbing electric field is weak, then to first order in $\boldsymbol{\mathcal{E}}$, we have

$$f \simeq f_0 - e_c\tau\frac{\partial f_0}{\partial E_{\boldsymbol{k}}}\boldsymbol{\mathcal{E}}\cdot\boldsymbol{v}_g, \qquad (8.60)$$

which is an explicit solution for the perturbed distribution function f. The deviation of f from its equilibrium value is proportional to both the electric field and the group velocity. This solution can be used to calculate the average values of physical quantities of interest such as $\boldsymbol{v}_g$.

8.3 Electrical conductivity and mobility

The total current density $\boldsymbol{j}$ is an average of the currents associated with the individual charge carriers and is given by

$$\boldsymbol{j} = n_c e_c\langle\boldsymbol{v}_g\rangle, \qquad (8.61)$$

where n_c is the concentration of charge carriers and the angular brackets denote a statistical mechanical average of v_g:

$$\langle\boldsymbol{v}_g\rangle = \frac{\int f\boldsymbol{v}_g\,d^3k}{\int f\,d^3k}. \qquad (8.62)$$

The carrier concentration can be expressed in terms of the distribution function. Combining the result for the group velocity (Eq. (8.51)) and the

Boltzmann equation (Eq. (8.60)), we can write

$$\int fd^3k = \int \left[f_0 - e_c\tau \frac{\partial f_0}{\partial E_{\boldsymbol{k}}} \boldsymbol{\mathcal{E}} \cdot \frac{1}{\hbar} \nabla_{\boldsymbol{k}} E_{\boldsymbol{k}} \right] d^3k. \tag{8.63}$$

The energy $E_{\boldsymbol{k}}$ is an even function of $\boldsymbol{k}$, but its gradient is an odd function. Since both the relaxation time τ and the equilibrium distribution function f_0 depend on $\boldsymbol{k}$ only through $E_{\boldsymbol{k}}$, the second term of the integrand in Eq. (8.63) is an odd function of $\boldsymbol{k}$, and its integral therefore vanishes. Hence, $\int fd^3k = \int f_0 d^3k$. The distribution function f_0 is the Fermi–Dirac distribution function. The carrier concentration is given by Eq. (6.3):

$$n_c = (1/4\pi^3) \int f_0 d^3k = (1/4\pi^3) \int fd^3k. \tag{8.64}$$

One can simplify the integral in the numerator of Eq. (8.62) by noting that $\boldsymbol{v}_g$ as expressed by Eq. (8.51) must be an odd function of $\boldsymbol{k}$, whereas f_0 is an even function. Hence,

$$\begin{aligned} \int f \boldsymbol{v}_g d^3k &= \int \left[f_0 - e_c\tau \frac{\partial f_0}{\partial E_{\boldsymbol{k}}} \boldsymbol{\mathcal{E}} \cdot \boldsymbol{v}_g \right] \boldsymbol{v}_g d^3k \\ &= -e_c \int \tau \frac{\partial f_0}{\partial E_{\boldsymbol{k}}} (\boldsymbol{\mathcal{E}} \cdot \boldsymbol{v}_g) \boldsymbol{v}_g d^3k. \end{aligned} \tag{8.65}$$

The average value of the group velocity, Eq. (8.62), now takes the form

$$\langle \boldsymbol{v}_g \rangle = -\frac{e_c}{4\pi^3 n_c} \int \tau \frac{\partial f_0}{\partial E_{\boldsymbol{k}}} (\boldsymbol{\mathcal{E}} \cdot \boldsymbol{v}_g) \boldsymbol{v}_g d^3k. \tag{8.66}$$

The quantity $\langle \boldsymbol{v}_g \rangle$ is called the **drift velocity** which is the average value of the group velocity of a scattered carrier in an electric field. By symmetry we see that if $\boldsymbol{k}$ enters τ only through the energy $E_{\boldsymbol{k}}$, the only nonvanishing component of $\boldsymbol{v}_g$ is that parallel to the electric field $\boldsymbol{\mathcal{E}}$. If $\boldsymbol{\mathcal{E}}$ is in the x-direction, Eq. (8.66) can be rewritten as

$$\langle \boldsymbol{v}_g \rangle = \langle v_{gx} \rangle = -\frac{e_c}{4\pi^3 n_c} \int \tau(E_{\boldsymbol{k}}) \frac{\partial f_0}{\partial E_{\boldsymbol{k}}} v_{gx}^2 \mathcal{E}_x d^3k. \tag{8.67}$$

The mobility μ is defined by the relation

$$\langle v_{gx} \rangle = (e_c/e)\mu \mathcal{E}_x. \tag{8.68}$$

Comparing Eqs. (8.67) and (8.68), we obtain

$$\mu = -\frac{e}{4\pi^3 n_c} \int \tau(E_{\boldsymbol{k}}) \frac{\partial f_0}{\partial E_{\boldsymbol{k}}} v_{gx}^2 d^3k. \tag{8.69}$$

The mobility, being proportional to the scattering time, is high when the time between collisions is long and low when the time is short. The DC electric conductivity σ is defined by Ohm's law,

$$j = \sigma \mathcal{E}, \tag{8.70}$$

which when combined with Eqs. (8.61) and (8.68) yields

$$\sigma = n_c e\mu. \tag{8.71}$$

The distribution function f_0 is the Fermi–Dirac distribution function,

$$f_0(E_{\boldsymbol{k}}) = \frac{1}{e^{\beta(E_{\boldsymbol{k}}-E_F)}+1} \tag{8.72}$$

which upon differentiation yields

$$\frac{\partial f_0(E_{\boldsymbol{k}})}{\partial E_{\boldsymbol{k}}} = -\frac{\beta e^{\beta(E_{\boldsymbol{k}}-E_F)}}{[e^{\beta(E_{\boldsymbol{k}}-E_F)}+1]^2} = -\beta f_0(1-f_0). \tag{8.73}$$

Substitution of Eqs. (8.51) and (8.73) into Eq. (8.69) gives the mobility in the form

$$\mu = \frac{e\beta}{4\pi^3 n_c \hbar^2}\int \tau(E_{\boldsymbol{k}}) f_0(1-f_0)(\boldsymbol{\nabla}_{\boldsymbol{k}} E_{\boldsymbol{k}})_x^2 d^3k \tag{8.74}$$

which is valid for both parabolic and nonparabolic bands.

For the special case of spherical parabolic bands with $E_{\boldsymbol{k}} = (\hbar^2/2m^*)k^2$,

$$(\boldsymbol{\nabla}_{\boldsymbol{k}} E_{\boldsymbol{k}})_x = \frac{\hbar^2}{m^*}k_x, \tag{8.75}$$

and

$$\mu = \frac{\hbar^2 e\beta}{4\pi^3 n_c m^{*2}}\int \tau(E_{\boldsymbol{k}}) f_0(1-f_0)k_x^2 d^3k. \tag{8.76}$$

Since k_x is a dummy integration variable, we can replace k_x by k_y and by k_z, add the three expressions and divide by 3 to give

$$\mu = \frac{\hbar^2 e\beta}{12\pi^3 n_c m^{*2}}\int \tau(E_{\boldsymbol{k}}) f_0(1-f_0)k^2 d^3k. \tag{8.77}$$

Introducing spherical coordinates, setting $E = (\hbar^2/2m^*)k^2$, and noting that the integrand in Eq. (8.77) depends only on E, we integrate over the angular variables and obtain in the classical limit

$$\mu = \frac{e\beta}{3\pi^2 n_c m^*}\left(\frac{2m^*}{\hbar^2}\right)^{3/2}\int \tau(E) f_0 E^{3/2} dE. \tag{8.78}$$

If we define the average value of $\tau(E)$ by

$$\langle\tau\rangle = \frac{\int \tau(E) f_0 E^{3/2} dE}{\int f_0 E^{3/2} dE}, \tag{8.79}$$

we can re-express Eq. (8.78) in the form

$$\mu = \frac{e\beta}{3\pi^2 n_c m^*}\left(\frac{2m^*}{\hbar^2}\right)^{3/2}\langle\tau\rangle\int f_0 E^{3/2} dE. \tag{8.80}$$

To treat the case of ellipsoidal parabolic bands, one "sphericalizes" the energy using the transformation in Eq. (6.21). The mobility tensor element $\mu_{\alpha\alpha}$ is obtained from Eq. (8.80) by replacing m^* by m^*_α and $\langle\tau\rangle$ by $\langle\tau_\alpha\rangle$, where $\alpha = t$ or ℓ.

In order to proceed further we must evaluate the integrals in Eqs. (8.79) and (8.80). The range of integration is from zero to E_m, the maximum energy of the band of interest. For not too large carrier concentrations and not too high temperatures, the exponential decrease of f_0 with increasing E enables us in good approximation to replace E_m by infinity. For semiconductors with temperatures, carrier concentrations, and effective masses satisfying the inequality (Tolman 1938)

$$n_c^{1/3} < \left(\frac{m^* k_B T}{2\pi}\right)^{\frac{1}{2}} / \hbar, \tag{8.81}$$

it is satisfactory to take f_0 to be the classical Boltzmann distribution given by

$$f_0(E) \simeq K e^{-\beta(E-E_F)}, \tag{8.82}$$

where K is the normalization constant determined by Eq. (8.64). When the inequality is not satisfied, the full Fermi–Dirac distribution must be used in Eq. (8.77) with the consequence that the integrals must be done numerically.

Restricting ourselves to the classical case, we find that

$$\int f_0 E^{3/2} dE = \frac{3\sqrt{\pi}}{4\beta^{5/2}} K e^{\beta E_F}, \tag{8.83a}$$

$$\begin{aligned} n_c &= \frac{1}{2\pi^2}\left(\frac{2m^*}{\hbar^2}\right)^{3/2} \int f_0 E^{\frac{1}{2}} dE \\ &= \frac{1}{4(\pi\beta)^{3/2}}\left(\frac{2m^*}{\hbar^2}\right)^{3/2} K e^{\beta E_F}. \end{aligned} \tag{8.83b}$$

Equation (8.80) for the mobility then becomes

$$\mu = \frac{e\langle\tau\rangle}{m^*}. \tag{8.84}$$

This result is a generalization of that for the **Drude model** in which all charge carriers are assumed to have the same τ regardless of energy. In the general case we must know $\tau(E)$ in order to evaluate $\langle\tau\rangle$. This problem is considered in the next section. We see that the mobility involves two physical parameters, the average relaxation time and the effective mass, which depend on the material. The effective mass is essentially a property of the pure material, but the average relaxation time is strongly affected by impurities and other imperfections.

8.4 Energy dependence of the relaxation time

In a scattering event a current carrier makes a transition from the Bloch state $|n\boldsymbol{k}\rangle$ to the Bloch state $|n'\boldsymbol{k}'\rangle$. For the present we restrict ourselves to **intraband scattering** for which $n' = n$. The scattering rate $W(\boldsymbol{k} \to \boldsymbol{k}')$ is determined quantum mechanically from Fermi's golden rule. The rate of change of the distribution function due to collisions is given by

$$\left.\frac{\partial f}{\partial t}\right|_{coll} = -\int \{W(\boldsymbol{k} \to \boldsymbol{k}')f(\boldsymbol{k})[1 - f(\boldsymbol{k}')] \\ - W(\boldsymbol{k}' \to \boldsymbol{k})f(\boldsymbol{k}')[1 - f(\boldsymbol{k})]\}d^3k', \qquad (8.85)$$

where the first and second terms on the right hand side correspond to scattering out of state $\boldsymbol{k}$ and scattering into state $\boldsymbol{k}$, respectively. Note that the Pauli principle has been taken into account. At equilibrium, the net collision rate vanishes, and we obtain the principle of detailed balance:

$$W(\boldsymbol{k} \to \boldsymbol{k}')f_0(\boldsymbol{k})[1 - f_0(\boldsymbol{k}')] - W(\boldsymbol{k}' \to \boldsymbol{k})f_0(\boldsymbol{k}')[1 - f_0(\boldsymbol{k})] = 0. \qquad (8.86)$$

A more convenient form of Eq. (8.85) can be obtained by using Eq. (8.60), which we rewrite as

$$f = f_0 - \varphi(\boldsymbol{k})\frac{\partial f_0}{\partial E_{\boldsymbol{k}}}, \qquad (8.87)$$

where

$$\varphi(\boldsymbol{k}) = e_c \tau \boldsymbol{\mathcal{E}} \cdot \boldsymbol{v}_g. \qquad (8.88)$$

Substituting Eq. (8.87) into Eq. (8.85) and exploiting the principle of detailed balance yields the result to terms linear in φ

$$\left.\frac{\partial f}{\partial t}\right|_{coll} = \int \Bigg\{ W(\boldsymbol{k} \to \boldsymbol{k}')\left[\varphi(\boldsymbol{k})\frac{\partial f_0(\boldsymbol{k})}{\partial E_{\boldsymbol{k}}}[1 - f_0(\boldsymbol{k}')] - \varphi(\boldsymbol{k}')\frac{\partial f_0(\boldsymbol{k}')}{\partial E_{\boldsymbol{k}'}}f_0(\boldsymbol{k})\right] \\ + W(\boldsymbol{k}' \to \boldsymbol{k})\Bigg[\varphi(\boldsymbol{k})\frac{\partial f_0(\boldsymbol{k})}{\partial E_{\boldsymbol{k}}}f_0(\boldsymbol{k}') \\ - \varphi(\boldsymbol{k}')\frac{\partial f_0(\boldsymbol{k}')}{\partial E_{\boldsymbol{k}'}}[1 - f_0(\boldsymbol{k})]\Bigg]\Bigg\}d^3k'. \qquad (8.89)$$

If we now use Eq. (8.73) and again take advantage of detailed balance, we obtain

$$\left.\frac{\partial f}{\partial t}\right|_{coll} = -\beta \int W(\boldsymbol{k} \to \boldsymbol{k}')[\varphi(\boldsymbol{k}) - \varphi(\boldsymbol{k}')]f_0(\boldsymbol{k})[1 - f_0(\boldsymbol{k}')]d^3k', \qquad (8.90)$$

which upon making the substitution

$$W(\boldsymbol{k} \to \boldsymbol{k}')[1 - f_0(\boldsymbol{k}')] = W(\boldsymbol{k}, \boldsymbol{k}')[1 - f_0(\boldsymbol{k})] \qquad (8.91)$$

becomes

$$\left.\frac{\partial f}{\partial t}\right|_{coll} = -\beta \int W(\boldsymbol{k},\boldsymbol{k}')[\varphi(\boldsymbol{k}) - \varphi(\boldsymbol{k}')] f_0(\boldsymbol{k})[1 - f_0(\boldsymbol{k})] d^3k'. \tag{8.92}$$

Again using Eq. (8.73), we can rewrite Eq. (8.92) as

$$\left.\frac{\partial f}{\partial t}\right|_{coll} = \frac{\partial f_0(\boldsymbol{k})}{\partial E_{\boldsymbol{k}}} \int W(\boldsymbol{k},\boldsymbol{k}')[\varphi(\boldsymbol{k}) - \varphi(\boldsymbol{k}')] d^3k'. \tag{8.93}$$

We consider two important cases for which a relaxation time can be defined (Roth 1992):

1. Elastic or inelastic scattering which is characterized by $W(\boldsymbol{k},\boldsymbol{k}')$, being an even function of $\boldsymbol{k}'$. This case is referred to as "$\boldsymbol{k}$-randomizing" (Herring and Vogt 1956).
2. Elastic scattering and isotropic energy bands with $W(\boldsymbol{k} \to \boldsymbol{k}')$ depending only on the angle between $\boldsymbol{k}$ and $\boldsymbol{k}'$.

For case (1) we note that since $W(\boldsymbol{k} \to \boldsymbol{k}')$ and $f_0(\boldsymbol{k}')$ are even functions of $\boldsymbol{k}'$, $W(\boldsymbol{k},\boldsymbol{k}')$ is also even. On the other hand $\varphi(\boldsymbol{k}')$ is an odd function of $\boldsymbol{k}'$ because $\boldsymbol{v}_g$ is an odd function of $\boldsymbol{k}'$ from its definition in Eq. (8.51). Consequently, the integral involving $\varphi(\boldsymbol{k}')$ in Eq. (8.93) vanishes, and

$$\left.\frac{\partial f}{\partial t}\right|_{coll} = \varphi(\boldsymbol{k}) \frac{\partial f_0(\boldsymbol{k})}{\partial E_{\boldsymbol{k}}} \int W(\boldsymbol{k},\boldsymbol{k}') d^3k'. \tag{8.94}$$

Use of Eq. (8.87) yields

$$\left.\frac{\partial f}{\partial t}\right|_{coll} = -(f - f_0) \int W(\boldsymbol{k},\boldsymbol{k}') d^3k', \tag{8.95}$$

which upon comparison with Eq. (8.55) gives the inverse relaxation time in the form

$$\frac{1}{\tau(E_{\boldsymbol{k}})} = \int W(\boldsymbol{k},\boldsymbol{k}') d^3k'. \tag{8.96}$$

Thus, for case (1) the inverse relaxation time depends only on the scattering rate $W(\boldsymbol{k},\boldsymbol{k}')$.

For case (2), we use the spherical parabolic band value for $\boldsymbol{v}_g, \hbar\boldsymbol{k}/m^*$, and write $\varphi(\boldsymbol{k})$ in the form

$$\varphi(\boldsymbol{k}) = \boldsymbol{\chi}(\boldsymbol{k}) \cdot \boldsymbol{v}_g = \boldsymbol{\chi}(\boldsymbol{k}) \cdot \hbar\boldsymbol{k}/m^*, \tag{8.97}$$

where

$$\boldsymbol{\chi}(\boldsymbol{k}) = e_c \tau(E_{\boldsymbol{k}}) \boldsymbol{\mathcal{E}}. \tag{8.98}$$

Substituting this expression for $\varphi(\boldsymbol{k})$ into Eq. (8.93) we obtain

$$\left.\frac{\partial f}{\partial t}\right|_{coll} = \frac{\hbar}{m^*} \frac{\partial f_0(\boldsymbol{k})}{\partial E_{\boldsymbol{k}}} \int W(\boldsymbol{k},\boldsymbol{k}')[\boldsymbol{\chi}(\boldsymbol{k}) \cdot (\boldsymbol{k} - \boldsymbol{k}')] d^3k', \tag{8.99}$$

where we have utilized the fact that for elastic scattering, $E_{k'} = E_k$ and $\tau(E_{k'}) = \tau(E_k)$. The term in square brackets can be written as

$$\chi(\boldsymbol{k})k(\cos\theta_{k,\chi} - \cos\theta_{k',\chi}), \tag{8.100}$$

where $\theta_{k,\chi}$ is the angle between $\boldsymbol{k}$ and $\boldsymbol{\chi}(\boldsymbol{k})$. The integral over $\boldsymbol{k}'$ is evaluated using spherical coordinates with the polar axis in the $\boldsymbol{k}$-direction. Then $\cos\theta_{k',\chi}$ becomes

$$\cos\theta_{k',\chi} = \cos\theta_{k,k'}\cos\theta_{k,\chi} - \sin\theta_{k,k'}\sin\theta_{k,\chi}\cos(\varphi_{k'} - \varphi_\chi). \tag{8.101}$$

The integral of the term containing $\cos(\varphi_{k'} - \varphi_\chi)$ vanishes because by assumption W depends only on $\theta_{k,k'}$ and not on $\varphi_{k'}$:

$$\left.\frac{\partial f}{\partial t}\right|_{coll} = \frac{\partial f_0(\boldsymbol{k})}{\partial E_k}\boldsymbol{\chi}(\boldsymbol{k})\cdot\frac{\hbar\boldsymbol{k}}{m^*}\int W(\boldsymbol{k},\boldsymbol{k}')[1 - \cos\theta_{k,k'}]d^3k'. \tag{8.102}$$

Use of Eqs. (8.97) and (8.87) yields

$$\left.\frac{\partial f}{\partial t}\right|_{coll} = -(f - f_0)\int W(\boldsymbol{k},\boldsymbol{k}')[1 - \cos\theta_{k,k'}]d^3k'. \tag{8.103}$$

Comparing this equation with Eq. (8.55) we see that

$$\frac{1}{\tau(E_k)} = \int W(\boldsymbol{k},\boldsymbol{k}')[1 - \cos\theta_{k,k'}]d^3k'. \tag{8.104}$$

Hence, for case (2) the inverse relaxation time depends not only on $W(\boldsymbol{k},\boldsymbol{k}')$ but also on the factor $1 - \cos\theta_{k,k'}$ which represents the fraction of forward momentum $\boldsymbol{k}$ that is randomized or lost in each collision. The relaxation time involved in low-field transport phenomena is the momentum relaxation time that is specified by Eq. (8.104).

In the classical limit to which we restrict ourselves, $W(\boldsymbol{k},\boldsymbol{k}')$ becomes equal to $W(\boldsymbol{k} \to \boldsymbol{k}')$ and can be calculated by the Fermi golden rule for the scattering mechanism of interest. We discuss the calculation of both the scattering rate $W(\boldsymbol{k},\boldsymbol{k}')$ and the relaxation time $\tau(E_k)$ for a variety of scattering mechanisms.

8.5 Relaxation times for specific scattering mechanisms

There are a variety of scattering processes of importance in semiconductors. They include scattering by ionized impurities, neutral impurities, other charge carriers, acoustic phonons, and optical phonons. Each scattering mechanism leads to a characteristic energy dependence of the relaxation time.

8.5.1 Ionized impurity scattering

If the weakly bound electron of a donor impurity such as P in Si is excited into the conduction band, the positive impurity ion that remains serves as a scattering center for free carriers. Since we typically deal with systems containing a finite concentration of free carriers, the Coulomb potential due to the positive impurity ion is screened by the free carriers. The interaction Hamiltonian is therefore of the form

$$H_{int}(\boldsymbol{r}) = \frac{-Ze^2}{4\pi\epsilon_0\epsilon|\boldsymbol{r}-\boldsymbol{R}|}e^{-q_0|\boldsymbol{r}-\boldsymbol{R}|}, \tag{8.105}$$

where $\boldsymbol{R}$ is the position vector of the ion, Ze is its charge, ϵ is the dielectric constant of the semiconductor, and q_0 is the inverse Debye screening length. If only a single type of free carrier with concentration n_c is present, q_0 is specified by (Debye and Hückel 1923)

$$q_0^2 = \frac{e^2 n_c}{\epsilon_0 \epsilon k_B T}. \tag{8.106}$$

Similar considerations apply to an acceptor impurity such as Al, but in this case the impurity ion is negatively charged and the free carrier is positively charged.

If we neglect multiple scattering and use the Born approximation, Fermi's golden rule (Fermi 1950) gives the scattering rate as

$$(\boldsymbol{k},\boldsymbol{k}') = \frac{2\pi}{\hbar}|\langle\boldsymbol{k}'|H_{int}(\boldsymbol{r})|\boldsymbol{k}\rangle|^2\delta(E_{\boldsymbol{k}'}-E_{\boldsymbol{k}})g(\boldsymbol{k}'), \tag{8.107}$$

where the δ-function expresses the fact that the scattering is elastic and the density-of-states $g(\boldsymbol{k}')$ is $\Omega/(2\pi)^3$. Taking the Bloch state $|\boldsymbol{k}\rangle$ as the product of a plane wave function and the periodic function $u_{\boldsymbol{k}}(\boldsymbol{r})$, the matrix element of H_{int} becomes

$$\langle\boldsymbol{k}'|H_{int}(\boldsymbol{r})|\boldsymbol{k}\rangle = \int e^{-i\boldsymbol{k}'\cdot\boldsymbol{r}}u_{\boldsymbol{k}'}^*(\boldsymbol{r})H_{int}(\boldsymbol{r})e^{i\boldsymbol{k}\cdot\boldsymbol{r}}u_{\boldsymbol{k}}(\boldsymbol{r})d^3r. \tag{8.108}$$

The product $u_{\boldsymbol{k}'}^*(\boldsymbol{r})u_{\boldsymbol{k}}(\boldsymbol{r})$ has the periodicity of the crystal and can be expanded in the Fourier series

$$u_{\boldsymbol{k}'}^*(\boldsymbol{r})u_{\boldsymbol{k}}(\boldsymbol{r}) = \sum_{\boldsymbol{G}} e^{i\boldsymbol{G}\cdot\boldsymbol{r}}B_{\boldsymbol{k}'\boldsymbol{k}}(\boldsymbol{G}). \tag{8.109}$$

In this equation

$$B_{\boldsymbol{k}'\boldsymbol{k}}(\boldsymbol{G}) = \frac{1}{\Omega}\int e^{-i\boldsymbol{G}\cdot\boldsymbol{r}}u_{\boldsymbol{k}'}^*(\boldsymbol{r})u_{\boldsymbol{k}}(\boldsymbol{r})d^3r \tag{8.110}$$

and $\boldsymbol{G}$ is a reciprocal lattice vector. Substituting Eq. (8.109) into Eq. (8.108) and evaluating the integral over the electronic coordinate yields

$$\langle\boldsymbol{k}'|H_{int}(\boldsymbol{r})|\boldsymbol{k}\rangle = \Omega\sum_{\boldsymbol{G}} B_{\boldsymbol{k}'\boldsymbol{k}}(\boldsymbol{G})\mathcal{H}(\boldsymbol{k}'-\boldsymbol{k}-\boldsymbol{G}), \tag{8.111}$$

where $\mathcal{H}(\boldsymbol{k}'-\boldsymbol{k}-\boldsymbol{G})$ is the Fourier transform of $H_{int}(\boldsymbol{r})$ given by

$$\mathcal{H}(\boldsymbol{k}'-\boldsymbol{k}-\boldsymbol{G}) = \frac{1}{\Omega}\int e^{-i(\boldsymbol{k}'-\boldsymbol{k}-\boldsymbol{G})\cdot\boldsymbol{r}}H_{int}(\boldsymbol{r})d^3r. \tag{8.112}$$

For a weak, slowly varying interaction potential the important Fourier components are those of small argument. We may therefore neglect the so-called umklapp terms with $\boldsymbol{G} \neq 0$. Using spherical coordinates with the polar axis parallel to $\boldsymbol{k} - \boldsymbol{k}'$ and placing the impurity ion at the origin, we find that $\mathcal{H}(\boldsymbol{k}' - \boldsymbol{k})$ is given by

$$\mathcal{H}(\boldsymbol{k}' - \boldsymbol{k}) = -\frac{Ze^2}{\epsilon_0 \epsilon \Omega} \left[\frac{1}{|\boldsymbol{k} - \boldsymbol{k}'|^2 + q_0^2} \right]. \tag{8.113}$$

In the absence of umklapp terms, we need only deal with the coefficient $B_{\boldsymbol{k}'\boldsymbol{k}}(0)$. The periodic function $u_{\boldsymbol{k}}(\boldsymbol{r})$ depends weakly on $\boldsymbol{k}$ for the states with high carrier occupancy near the band edge. Hence, we can approximate $u_{\boldsymbol{k}'}(\boldsymbol{r})$ by $u_{\boldsymbol{k}}(\boldsymbol{r})$ and from the normalization of $|\boldsymbol{k}\rangle$ obtain $B_{\boldsymbol{k}'\boldsymbol{k}}(0) \simeq 1/\Omega$. The matrix element of H_{int} simplifies to

$$\langle \boldsymbol{k}' | H_{int}(\boldsymbol{r}) | \boldsymbol{k} \rangle = \mathcal{H}(\boldsymbol{k}' - \boldsymbol{k}) = -\frac{Ze^2}{\epsilon_0 \epsilon \Omega} \left[\frac{1}{|\boldsymbol{k} - \boldsymbol{k}'|^2 + q_0^2} \right]. \tag{8.114}$$

The total scattering rate can now be obtained by substituting Eq. (8.114) into Eq. (8.107) and multiplying by the number of ionized impurities N_I:

$$W(\boldsymbol{k}, \boldsymbol{k}') = \frac{N_I}{4\pi^2 \hbar \Omega} \left[\frac{Ze^2/\epsilon_0 \epsilon}{|\boldsymbol{k} - \boldsymbol{k}'|^2 + q_0^2} \right]^2 \delta(E_{\boldsymbol{k}} - E_{\boldsymbol{k}'}). \tag{8.115}$$

If we assume spherical parabolic bands, we are dealing with the second of the two cases mentioned in Section 8.4. The relaxation time is therefore specified by Eq. (8.104) which, with the aid of Eq. (8.115) and setting $n_I = N_I/\Omega$, becomes

$$\frac{1}{\tau(E_{\boldsymbol{k}})} = \frac{n_I}{4\pi^2 \hbar} \int \left[\frac{Ze^2/\epsilon_0 \epsilon}{|\boldsymbol{k} - \boldsymbol{k}'|^2 + q_0^2} \right]^2 \delta(E_{\boldsymbol{k}} - E_{\boldsymbol{k}'})(1 - \cos\theta_{\boldsymbol{k},\boldsymbol{k}'}) d^3k'. \tag{8.116}$$

We introduce spherical coordinates with the polar axis parallel to $\boldsymbol{k}$ and note that since $E_{\boldsymbol{k}'} = E_{\boldsymbol{k}}$, we must have $k' = k$ and therefore

$$|\boldsymbol{k} - \boldsymbol{k}'|^2 = 2k^2(1 - \cos\theta). \tag{8.117}$$

The inverse relaxation time then takes the form

$$\frac{1}{\tau(E_{\boldsymbol{k}})} = \frac{n_I}{4\pi^2 \hbar} \int \left[\frac{Ze^2/\epsilon_0 \epsilon}{2k^2(1 - \cos\theta) + q_0^2} \right]^2 \times \delta(E_{\boldsymbol{k}} - E_{\boldsymbol{k}'})(1 - \cos\theta) k'^2 \, dk' \sin\theta \, d\theta \, d\varphi. \tag{8.118}$$

Eliminating k' in favor of $E_{k'}$ using $E_{k'} = \hbar^2 k'^2/2m^*$, and carrying out the integrals over $E_{k'}$ and φ gives

$$\frac{1}{\tau(E_k)} = \frac{n_I}{4\pi\hbar}\left(\frac{2m^*}{\hbar^2}\right)^{3/2}\int_{-1}^{+1}\left[\frac{Ze^2/\epsilon_0\epsilon}{2k^2(1-\cos\theta)+q_0^2}\right]^2 E_k^{\frac{1}{2}}(1-\cos\theta)\sin\theta\, d\theta. \tag{8.119}$$

The integral over $\cos\theta$ can be evaluated by elementary means to yield (Brooks 1955)

$$\frac{1}{\tau(E_k)} = \frac{\pi n_I Z^2 e^4}{\sqrt{2m^*}(4\pi\epsilon_0\epsilon)^2 E_k^{3/2}}\left[\log\left(1+\frac{8m^* E_k}{\hbar^2 q_0^2}\right) - \frac{1}{1+(\hbar^2 q_0^2/8m^* E_k)}\right]. \tag{8.120}$$

This result is the celebrated Brooks–Herring formula. For small q_0 (large screening length, low carrier concentration), the relaxation time is to a good approximation proportional to $E_k^{3/2}$. The average value of $\tau(E_k)$ can be evaluated starting from Eq. (8.79). The result is

$$\langle\tau\rangle = \frac{\sqrt{2m^*}(4\pi\epsilon_0\epsilon)^2}{\pi n_I Z^2 e^4} \times \frac{\int_0^\infty E^3[\log(1+8m^*E/\hbar^2 q_0^2) - (1/1+(\hbar^2 q_0^2/8m^*E))]^{-1} e^{-\beta E}\, dE}{\int_0^\infty E^{3/2} e^{-\beta E}\, dE}. \tag{8.121}$$

The factor in square brackets is a slowly varying function of E, whereas the function $h(E) = E^3\exp(-\beta E)$ is sharply peaked. We may therefore obtain a good approximation by replacing E in the square-bracketed factor by its value at the maximum of $h(E)$, i.e., $3k_BT$. The integrals that remain in Eq. (8.121) are proportional to gamma functions:

$$\int_0^\infty E^3 e^{-\beta E}\, dE = \beta^{-4}\Gamma(4) = (k_BT)^4\Gamma(4) \tag{8.122a}$$

$$\int_0^\infty E^{3/2} e^{-\beta E}\, dE = \beta^{-5/2}\Gamma\left(\frac{5}{2}\right) = (k_BT)^{5/2}\Gamma\left(\frac{5}{2}\right). \tag{8.122b}$$

Using the values $\Gamma(4) = 6$ and $\Gamma(5/2) = (3/4)\sqrt{\pi}$, we obtain

$$\langle\tau\rangle = \frac{4(4\pi\epsilon_0\epsilon)^2(2k_BT)^{3/2}\sqrt{m^*}}{\pi^{3/2} n_I Z^2 e^4}\left[\log\left(1+\frac{T^2}{T_0^2}\right) - \frac{1}{1+(T_0^2/T^2)}\right]^{-1}, \tag{8.123}$$

where $T_0^2 = \hbar^2 e^2 n_I/24m^*\epsilon_0\epsilon k_B^2$ and n_I replaces n_c in Eq. (8.106).

The mobility associated with screened ionized impurity scattering is obtained by multiplying Eq. (8.123) by e/m^*:

$$\mu = \frac{4(4\pi\epsilon_0\epsilon)^2(2k_BT)^{3/2}}{\pi^{3/2}\sqrt{m^*} n_I Z^2 e^3}\left[\log\left(1+\frac{T^2}{T_0^2}\right) - \frac{1}{1+(T_0^2/T^2)}\right]^{-1}. \tag{8.124}$$

The principal temperature dependence of the mobility is contained in the factor $T^{3/2}$. The increase in mobility with increasing temperature is associated with the fact that Coulomb scattering is more effective for slow carriers than for fast carriers. Higher temperatures produce faster carriers and therefore larger average relaxation time and higher mobility.

When $q_0 = 0$, the inverse relaxation time has a logarithmic divergence which is characteristic of the bare Coulomb interaction. An alternative means of eliminating the divergence is provided by the Conwell–Weisskopf method (Conwell and Weisskopf 1950). It is assumed that the Coulomb field of a particular ion ceases to be effective at a maximum impact parameter b_{max} (see Fig. 8.2) equal to one-half the mean distance between impurity ions:

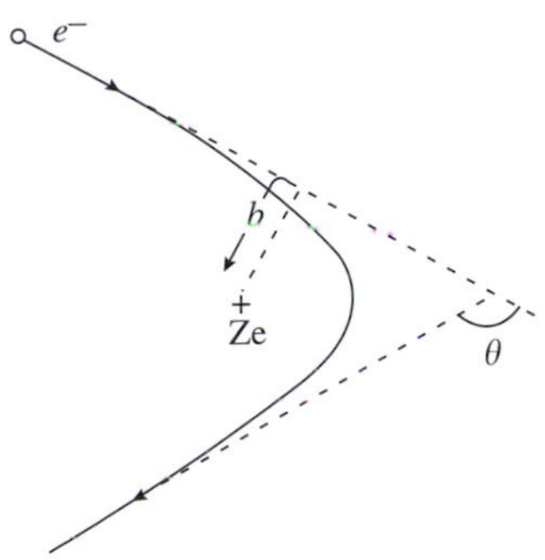

Fig. 8.2
Geometry for ionized impurity scattering.

$$b_{max} = \tfrac{1}{2} n_I^{1/3}. \tag{8.125}$$

Corresponding to b_{max} is a mimimum angle θ_{min} between $\boldsymbol{k}$ and $\boldsymbol{k}'$ specified by (Ridley 1988)

$$b_{max} = (Z/k)(\mathrm{Ry}^*/E_{\boldsymbol{k}})^{\frac{1}{2}} \cot(\theta_{min}/2), \tag{8.126}$$

where Ry^* is the effective Rydberg. The evaluation of the inverse relaxation time proceeds as in the derivation of the Brooks–Herring formula except that q_0 is set equal to zero and the range of the integration over θ is restricted to $\theta_{min} \le \theta \le \pi$. The result is

$$\frac{1}{\tau(E_{\boldsymbol{k}})} = \frac{\pi n_I}{(2m^* E_{\boldsymbol{k}}^3)^{\frac{1}{2}}} \left(\frac{Ze^2}{4\pi\epsilon_0\epsilon}\right)^2 \log\left[1 + \left(\frac{4\pi\epsilon_0\epsilon E_{\boldsymbol{k}}}{n_I^{1/3} Ze^2}\right)^2\right]. \tag{8.127}$$

For small impurity ion concentrations the dominant energy dependence of $\tau(E_{\boldsymbol{k}})$ is $E_{\boldsymbol{k}}^{3/2}$, just as with screened ionized impurity scattering.

The average value of $\tau(E_{\boldsymbol{k}})$ and the value of the mobility can be obtained in the same manner used in the Brooks–Herring treatment. The result for the mobility is

$$\mu = \frac{64\pi^{\frac{1}{2}}\epsilon_0^2\epsilon^2(2k_BT)^{3/2}}{\sqrt{m^*}n_I Z^2 e^3} \left\{ \log\left[1 + \left(\frac{12\pi\epsilon_0\epsilon k_BT}{n_I^{1/3}Ze^2}\right)^2\right]\right\}^{-1}. \tag{8.128}$$

The experimental verification of Eq. (8.124) or Eq. (8.128) is complicated by several factors. In materials such as n-Si and n-Ge, the energy bands are ellipsoidal, not spherical, and it is necessary to replace m_e by an appropriate density-of-states mass (Ham 1955). Of more importance is the fact that at high temperatures scattering by lattice vibrations dominates, whereas at very low temperatures most of the impurities are neutral rather than ionized, so neutral impurity scattering must be considered.

8.5.2 Neutral impurity scattering

If a charge carrier is scattered by a neutral impurity, there is no long range Coulomb interaction involved. The effective interaction potential is short range and can be modeled by a square well potential. It is known that a neutral hydrogen atom can bind a second electron with a binding energy E_T of 0.75 Ry*. By choosing the square well potential to reproduce the binding energy E_T, a reasonably realistic relaxation time can be obtained whose inverse has the form (Sclar 1956)

$$\frac{1}{\tau(E_{\boldsymbol{k}})} = \frac{4\pi n_I \hbar^2 E_{\boldsymbol{k}}^{\frac{1}{2}}}{\sqrt{2} m^{*3/2}(E_{\boldsymbol{k}} + E_T)}. \tag{8.129}$$

The corresponding mobility is

$$\mu = \frac{(2m^*/\pi)^{\frac{1}{2}} e E_T^{\frac{1}{2}}}{3\pi n_I \hbar^2} \left[2\left(\frac{k_B T}{E_T}\right)^{\frac{1}{2}} + \left(\frac{E_T}{k_B T}\right)^{\frac{1}{2}} \right]. \tag{8.130}$$

It has a weak temperature dependence with a minimum at $T = E_T/2k_B$. This weak dependence on temperature is in marked contrast to the strong dependence of ionized impurity scattering on temperature.

The values of the scattering rate are found to be small unless E_T is small. Consequently, neutral impurities that bind a carrier into a deep level will not act as strong scattering centers and will not be important in determining the mobility.

8.5.3 Lattice vibrational scattering

8.5.3.1 Acoustic phonon scattering

The vibrations of the nuclei in a semiconductor produce a deviation of the carrier potential from periodicity that scatters free charge carriers. If $\boldsymbol{R}$ and $\boldsymbol{R}_0$ denote the nuclear coordinates in the displaced and equilibrium configurations, respectively, we can expand the carrier potential in power series in the displacements $\boldsymbol{R} - \boldsymbol{R}_0$,

$$V(\boldsymbol{r}, \boldsymbol{R}) = V(\boldsymbol{r}, \boldsymbol{R}_0) + (\boldsymbol{R} - \boldsymbol{R}_0) \cdot \nabla_{\boldsymbol{R}} V(\boldsymbol{r}, \boldsymbol{R})|_{\boldsymbol{R}=\boldsymbol{R}_0} + \cdots, \tag{8.131}$$

and stop with the linear terms if the displacements are small. The first term on the right hand side of Eq. (8.131) is the periodic potential and the second term is the deviation from periodicity. Our task is to calculate the rate at which carriers are scattered out of a particular Bloch state by the deviation.

For long wavelength vibrations ($\lambda \gg a$) one can use elasticity theory and describe the vibrations in terms of the strain tensor $e_{\mu\nu}$ defined by

$$e_{\mu\nu} = \frac{1}{2}\left(\frac{\partial u_\mu}{\partial x_\nu} + \frac{\partial u_\nu}{\partial x_\mu}\right), \quad \mu, \nu = x, y, z \tag{8.132}$$

where

$$\boldsymbol{u} = \boldsymbol{R} - \boldsymbol{R}_0. \tag{8.133}$$

In the **deformation potential** procedure, the Bloch state energy $E_{\boldsymbol{k}}$ is expanded in powers of the $e_{\mu\nu}$:

$$E_{\boldsymbol{k}} = E_{\boldsymbol{k}0} + \sum_{\mu\nu} C_{\mu\nu} e_{\mu\nu} + \sum_{\mu\nu} C'_{\mu\nu} k_\mu k_\nu e_{\mu\nu} + \cdots. \tag{8.134}$$

For cubic crystals, one can simplify this expression to give

$$E_{\boldsymbol{k}} = E_{\boldsymbol{k}0} + C_1 \Delta + C_2 \left(\hat{k}_\mu \hat{k}_\nu e_{\mu\nu} - \tfrac{1}{3}\Delta \right) + \cdots \tag{8.135}$$

with Δ the dilation given by

$$\Delta = \sum_\mu e_{\mu\mu} = \frac{\partial u_x}{\partial x} + \frac{\partial u_y}{\partial y} + \frac{\partial u_z}{\partial z}. \tag{8.136}$$

For spherical bands, $C_2 = 0$.

Since the atomic vibrations of crystals are described in terms of normal modes, we introduce normal coordinates $Q(\boldsymbol{q}j)$ by means of the normal coordinate transformation

$$\boldsymbol{u}(\boldsymbol{r}) = \frac{1}{\sqrt{\rho\Omega}} \sum_{\boldsymbol{q}j} \boldsymbol{e}(\boldsymbol{q}j) e^{i\boldsymbol{q}\cdot\boldsymbol{r}} Q(\boldsymbol{q}j), \tag{8.137}$$

where $\boldsymbol{q}$ is the wave vector, j is the branch index restricted to acoustic branches, and $\boldsymbol{e}(\boldsymbol{q}j)$, is the polarization vector. Equation (8.137) is an adaptation of Eq. (7.46) to an elastic continuum of volume Ω and density ρ. In the limit $\boldsymbol{q} \to 0$, $\boldsymbol{e}(\boldsymbol{q}j)$ is related to $\boldsymbol{e}_\kappa(\boldsymbol{q}j)$ in Eq. (7.46) by $\boldsymbol{e}_\kappa(0j) = (M_\kappa / \sum_\kappa M_\kappa)^{\frac{1}{2}} \boldsymbol{e}(0j)$. As will be seen later, the important values of q are very small compared to the zone boundary value π/a, so we assume that this relation holds also for $\boldsymbol{q} \neq 0$.

In evaluating matrix elements involving phonon wave functions it is convenient to introduce creation and annihilation operators $a^+_{\boldsymbol{q}j}$ and $a_{\boldsymbol{q}j}$ by means of the relation

$$Q(\boldsymbol{q}j) = \left(\frac{\hbar}{2\omega_{\boldsymbol{q}j}} \right)^{\frac{1}{2}} \left(a_{\boldsymbol{q}j} + a^+_{-\boldsymbol{q}j} \right), \tag{8.138}$$

where $\omega_{\boldsymbol{q}j}$ is the normal mode frequency. Eliminating $Q(\boldsymbol{q}j)$ from Eq. (8.137) yields

$$\boldsymbol{u}(\boldsymbol{r}) = \left(\frac{\hbar}{2\rho\Omega} \right)^{\frac{1}{2}} \sum_{\boldsymbol{q}j} \left\{ \frac{\boldsymbol{e}(\boldsymbol{q}j)}{\sqrt{\omega_{\boldsymbol{q}j}}} e^{i\boldsymbol{q}\cdot\boldsymbol{r}} a_{\boldsymbol{q}j} + \frac{\boldsymbol{e}^*(\boldsymbol{q}j)}{\sqrt{\omega_{\boldsymbol{q}j}}} e^{-i\boldsymbol{q}\cdot\boldsymbol{r}} a^*_{\boldsymbol{q}j} \right\}, \tag{8.139}$$

where we have used the relations $\boldsymbol{e}(-\boldsymbol{q}j) = \boldsymbol{e}^*(\boldsymbol{q}j)$ and $\omega(-\boldsymbol{q}j) = \omega(\boldsymbol{q}j)$. The dilation specified by Eq. (8.136) then takes the form

$$\Delta(\boldsymbol{r}) - i \left(\frac{\hbar}{2\rho\Omega} \right)^{\frac{1}{2}} \sum_{\boldsymbol{q}j} \left\{ \frac{\boldsymbol{q}\cdot\boldsymbol{e}(\boldsymbol{q}j)}{\sqrt{\omega_{\boldsymbol{q}j}}} e^{i\boldsymbol{q}\cdot\boldsymbol{r}} a_{\boldsymbol{q}j} - \frac{\boldsymbol{q}\cdot\boldsymbol{e}^*(\boldsymbol{q}j)}{\sqrt{\omega_{\boldsymbol{q}j}}} e^{-i\boldsymbol{q}\cdot\boldsymbol{r}} a^+_{\boldsymbol{q}j} \right\}. \tag{8.140}$$

The Hamiltonian describing the interaction of a carrier in a spherical band with the acoustic modes of vibration can be taken as

$$H_{int} = C_1 \Delta(\boldsymbol{r}). \tag{8.141}$$

We now calculate the rate of scattering of a carrier from one Bloch state to another. We restrict ourselves to intraband scattering and suppress the band index. The full wave function $\Psi(\boldsymbol{r}, Q)$ for the crystal is a product of the Bloch function for the carrier and harmonic oscillator functions φ for the normal coordinates $Q(\boldsymbol{q}j)$

$$\Psi_{\boldsymbol{k}n}(\boldsymbol{r}, Q) = e^{i\boldsymbol{k}\cdot\boldsymbol{r}} u_{\boldsymbol{k}}(\boldsymbol{r}) \prod_{\boldsymbol{q}j} \varphi_{n_{\boldsymbol{q}j}}[Q(\boldsymbol{q}j)], \tag{8.142}$$

where n stands for the set of harmonic oscillator quantum numbers $n_{\boldsymbol{q}j}$ of the normal modes $\boldsymbol{q}j$ and Q stands for the set of normal coordinates. The scattering rate is given by Fermi's golden rule as

$$W(\boldsymbol{k}n \to \boldsymbol{k}'n') = \frac{2\pi}{\hbar} |\langle \boldsymbol{k}'n'|H_{int}|\boldsymbol{k}n\rangle|^2 \delta(E_{\boldsymbol{k}'n'} - E_{\boldsymbol{k}n}) g(\boldsymbol{k}'). \tag{8.143}$$

Since H_{int} involves the scalar product $\boldsymbol{q} \cdot \boldsymbol{e}(\boldsymbol{q}j)$, only the longitudinal component of a vibrational mode contributes to the scattering. We restrict ourselves to elastically isotropic systems for which the modes can be classified as either longitudinal or transverse. We then need only deal with the longitudinal acoustic modes and can drop the branch index j. For these modes the scalar product $\boldsymbol{q} \cdot \boldsymbol{e}(\boldsymbol{q})$ reduces to

$$\boldsymbol{q} \cdot \boldsymbol{e}(\boldsymbol{q}) = q. \tag{8.144}$$

The matrix element of H_{int} that appears in Eq. (8.143) involves the product of an integral over the electronic coordinate $\boldsymbol{r}$ and integrals over the normal coordinates $Q(\boldsymbol{q})$. The evaluation of the integrals over the normal coordinates is facilitated by exploiting the following properties of the creation operator $a^+_{\boldsymbol{q}}$ and destruction operator $a_{\boldsymbol{q}}$ (Maradudin *et al.* 1971):

$$a_{\boldsymbol{q}} \varphi_{n_{\boldsymbol{q}}}(Q(\boldsymbol{q})) = n_{\boldsymbol{q}}^{\frac{1}{2}} \varphi_{n_{\boldsymbol{q}}-1}(Q(\boldsymbol{q})) \tag{8.145}$$

$$a^+_{\boldsymbol{q}} \varphi_{n_{\boldsymbol{q}}}(Q(\boldsymbol{q})) = (n_{\boldsymbol{q}} + 1)^{\frac{1}{2}} \varphi_{n_{\boldsymbol{q}}+1}(Q(\boldsymbol{q})). \tag{8.146}$$

The matrix element of H_{int} can now be readily evaluated to give

$$\begin{aligned} \langle \boldsymbol{k}'n'|H_{int}|\boldsymbol{k}n\rangle = &-\frac{iC_1\hbar^{\frac{1}{2}}}{\sqrt{2\rho\Omega}} \sum_{\boldsymbol{q}} \frac{q}{\sqrt{\omega_{\boldsymbol{q}}}} \\ &\times \int e^{-i\boldsymbol{k}'\cdot\boldsymbol{r}} u^*_{\boldsymbol{k}'}(\boldsymbol{r}) \Big[e^{-i\boldsymbol{q}\cdot\boldsymbol{r}} (n_{\boldsymbol{q}} + 1)^{\frac{1}{2}} \delta_{n'_{\boldsymbol{q}}, n_{\boldsymbol{q}}+1} \\ &- e^{i\boldsymbol{q}\cdot\boldsymbol{r}} n_{\boldsymbol{q}}^{\frac{1}{2}} \delta_{n'_{\boldsymbol{q}}, n_{\boldsymbol{q}}-1} \Big] e^{i\boldsymbol{k}\cdot\boldsymbol{r}} u_{\boldsymbol{k}}(\boldsymbol{r}) d^3r. \end{aligned} \tag{8.147}$$

The integral over the electronic coordinate can be simplified by using the Fourier series for $u^*_{k'}(\boldsymbol{r})u_{\boldsymbol{k}}(\boldsymbol{r})$, Eq. (8.109), and neglecting umklapp terms ($\boldsymbol{G} \neq 0$). The result for the matrix element is

$$\langle \boldsymbol{k}'n'|\boldsymbol{H}_{int}|\boldsymbol{k}n\rangle = -\frac{iC_1\hbar^{\frac{1}{2}}}{\sqrt{2\rho\Omega}}\sum_{\boldsymbol{q}}\frac{q}{\sqrt{\omega_{\boldsymbol{q}}}}B_{\boldsymbol{k}'\boldsymbol{k}}(0)\Omega$$
$$\times\,[\delta_{\boldsymbol{k}',\boldsymbol{k}-\boldsymbol{q}}\delta_{n'_{\boldsymbol{q}},n_{\boldsymbol{q}}+1}(n_{\boldsymbol{q}}+1)^{\frac{1}{2}} - \delta_{\boldsymbol{k}',\boldsymbol{k}+\boldsymbol{q}}\delta_{n'_{\boldsymbol{q}},n_{\boldsymbol{q}}-1}n_{\boldsymbol{q}}^{\frac{1}{2}}], \tag{8.148}$$

where $B_{\boldsymbol{k}'\boldsymbol{k}}(0)$ can be approximated by $1/\Omega$.

The first term in the square brackets corresponds to **phonon emission** in the scattering process and the second term to **phonon absorption**. The matrix element vanishes unless the harmonic oscillator selection rule, $\Delta n = n'_{\boldsymbol{q}} - n_{\boldsymbol{q}} = \pm 1$, is satisfied. The normal mode $\boldsymbol{q}$ therefore gains or loses energy in the amount $\hbar\omega_{\boldsymbol{q}}$ and, by conservation of energy, the carrier must lose or gain an equal amount of energy. The scattering process is **inelastic**. However, the energy of the acoustic phonons of interest is typically so small compared to electronic energies that it can be neglected in the delta function in Eq. (8.143). Scattering by acoustic modes can therefore be treated as elastic and falls under case (2) in the discussion in Section 8.3 if the energy band of the carriers is isotropic.

The carrier part of the matrix element leads to a selection rule on the carrier wave vector, $\boldsymbol{k}' - \boldsymbol{k} = \pm\boldsymbol{q}$, where $+\boldsymbol{q}$ and $-\boldsymbol{q}$ are associated with $\Delta n = -1$ (phonon absorption) and $+1$ (phonon emission), respectively. This result is the **conservation of crystal momentum** in the scattering process.

The ingredients are now in hand to calculate the inverse relaxation time by combining Eqs. (8.104), (8.143), and (8.148). The scattering rate for phonon emission takes the form

$$W_e(\boldsymbol{k}n \rightarrow \boldsymbol{k}-\boldsymbol{q}n') = \frac{\Omega}{(2\pi)^3}\frac{\pi C_1^2 q}{\rho\Omega c_\ell}(n_{\boldsymbol{q}}+1)\delta_{n'_{\boldsymbol{q}},n_{\boldsymbol{q}}+1}\delta(E_{\boldsymbol{k}-\boldsymbol{q}} - E_{\boldsymbol{k}}), \tag{8.149}$$

where $\omega_{\boldsymbol{q}} = c_\ell q$ and c_ℓ is the longitudinal speed of sound. We now pass to the classical limit, so that $W_e(\boldsymbol{k}n \rightarrow \boldsymbol{k}-\boldsymbol{q}n') = W_e(\boldsymbol{k}n, \boldsymbol{k}-\boldsymbol{q}n')$. The inverse relaxation time is then obtained by starting from Eq. (8.104) and averaging over initial phonon states to give

$$\frac{1}{\tau_e} = \frac{C_1^2}{8\pi^2\rho c_\ell}\int q(\bar{n}_{\boldsymbol{q}}+1)[1-\cos\theta_{\boldsymbol{k},\boldsymbol{k}-\boldsymbol{q}}]\delta(E_{\boldsymbol{k}-\boldsymbol{q}} - E_{\boldsymbol{k}})d^3q, \tag{8.150}$$

where $\bar{n}_{\boldsymbol{q}}$ is the mean phonon occupation number introduced in Eq. (7.74).

The argument of the energy-conserving delta function is $(\hbar^2/2m^*)\times(q^2 - 2kq\cos\theta_{\boldsymbol{k},\boldsymbol{q}})$ for spherical parabolic bands. In order to have a non-zero contribution, it is necessary that

$$\cos\theta_{\boldsymbol{k},\boldsymbol{q}} - \frac{q}{2k} \tag{8.151}$$

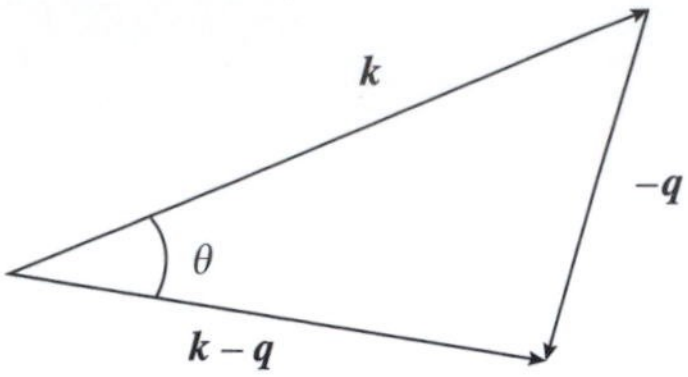

Fig. 8.3
Wave vector diagram for phonon emission.

Since $\cos\theta$ lies in the range $-1 \leq \cos\theta \leq +1$, the magnitude of the phonon wave vector must lie in the range $0 \leq q \leq q_{max} = 2k$. Furthermore, from the geometry shown in Fig. 8.3, we see that

$$q^2 = k^2 + |\boldsymbol{k} - \boldsymbol{q}|^2 - 2k|\boldsymbol{k} - \boldsymbol{q}| \cos\theta_{\boldsymbol{k},\boldsymbol{k}-\boldsymbol{q}}. \tag{8.152}$$

Applying energy conservation gives $k = |\boldsymbol{k} - \boldsymbol{q}|$ and hence

$$1 - \cos\theta_{\boldsymbol{k},\boldsymbol{k}-\boldsymbol{q}} = \frac{q^2}{2k^2}. \tag{8.153}$$

Utilizing the results just obtained, we find that Eq. (8.150) can be reduced to

$$\frac{1}{\tau_e} = \frac{C_1^2}{8\pi^2 \rho c_\ell} \int \frac{q^3}{2k^2} (\bar{n}_{\boldsymbol{q}} + 1)\delta(E_{\boldsymbol{k}-\boldsymbol{q}} - E_{\boldsymbol{k}}) d^3q. \tag{8.154}$$

At this point it is helpful to make an appraisal of the maximum energy, $\hbar\omega_{max} = \hbar c_\ell q_{max} = 2\hbar c_\ell k$, of phonons that are effective in scattering a carrier. As in Section 8.5.1, we assume that the carriers obey nondegenerate statistics and take k to be its value specified by equipartition $\hbar^2 k^2/2m^* = 3k_B T/2$:

$$\hbar k = (3m^* k_B T)^{\frac{1}{2}}. \tag{8.155}$$

The maximum energy is $2c_\ell (3m^* k_B T)^{\frac{1}{2}}$ and its ratio to $k_B T$ is

$$\frac{\hbar\omega_{max}}{k_B T} = 2c_\ell \left(\frac{3m^*}{k_B T}\right)^{\frac{1}{2}}. \tag{8.156}$$

Defining a characteristic temperature T_x by $T_x = 12m^* c_\ell^2/k_B$ and taking m^* and c_ℓ to have their values for Ge, we find that $T_x \simeq 3.9$ K. Consequently, for $T > 4$ K the ratio given by Eq. (8.156) is less than unity, and we may pass to the high-temperature limit:

$$\bar{n}_{\boldsymbol{q}} + 1 \simeq \bar{n}_{\boldsymbol{q}} = \frac{1}{\exp(\hbar\omega_{\boldsymbol{q}}(k_B T) - 1)} \simeq \frac{k_B T}{\hbar\omega_{\boldsymbol{q}}} > 1. \tag{8.157}$$

Substituting this result into Eq. (8.154) yields

$$\frac{1}{\tau_e} = \frac{C_1^2 k_B T}{16\pi^2 \hbar \rho c_\ell^2 k^2} \int q^2 \delta(E_{\boldsymbol{k}-\boldsymbol{q}} - E_{\boldsymbol{k}}) d^3q. \tag{8.158}$$

Introducing spherical coordinates with the polar axis parallel to $\boldsymbol{k}$ and integrating over the azimuthal angle φ, we get

$$\begin{aligned}
\frac{1}{\tau_e} &= \frac{C_1^2 k_B T}{8\pi\hbar\rho c_\ell^2 k^2} \int_0^{q_{max}} \int_{-1}^{+1} q^2 \delta\left[\left(\frac{\hbar^2 kq}{m^*}\right)\left(\frac{q}{2k} - \cos\theta\right)\right] q^2 dq\, d(\cos\theta) \\
&= \frac{C_1^2 k_B T}{8\pi\hbar\rho c_\ell^2 k^2} \left(\frac{m^*}{\hbar^2 k}\right) \int_0^{2k} q^3 dq \\
&= \frac{C_1^2 k_B T (2m^*)^{3/2}}{4\pi\hbar^4 \rho c_\ell^2} E_{\boldsymbol{k}}^{\frac{1}{2}}.
\end{aligned} \tag{8.159}$$

A similar analysis can be carried out for acoustic phonon absorption. The result corresponding to Eq. (8.154) is

$$\frac{1}{\tau_a} = \frac{C_1^2}{8\pi^2 \rho c_\ell} \int \frac{q^3}{2k^2} n_q \delta(E_{k+q} - E_k) d^3 q \tag{8.160}$$

which, in the high temperature limit, reduces to the same expression as for emission given by Eq. (8.159). The total inverse relaxation time is the sum of those for absorption and emission:

$$\frac{1}{\tau_{ac}} = \frac{C_1^2 k_B T (2m^*)^{3/2}}{2\pi \hbar^4 \rho c_\ell^2} E_k^{\frac{1}{2}}. \tag{8.161}$$

To determine the temperature dependence of the mobility, we must evaluate $\langle \tau_{ac} \rangle$. Using Eq. (8.79) in the nondegenerate statistics approximation, we obtain

$$\langle \tau_{ac} \rangle \simeq \frac{2\pi \hbar^4 \rho c_\ell^2}{C_1^2 k_B T (2m^*)^{3/2}} \frac{\int_0^\infty E\, e^{-\beta E} dE}{\int_0^\infty E^{3/2} e^{-\beta E} dE}, \tag{8.162}$$

which upon evaluation of the integrals becomes

$$\langle \tau_{ac} \rangle \simeq \frac{8\pi^{\frac{1}{2}} \hbar \rho c_\ell^2}{3C_1^2} \left(\frac{\hbar^2}{2m^* k_B T} \right)^{3/2}. \tag{8.163}$$

The mobility for spherical parabolic bands now follows from Eq. (8.84):

$$\mu_{ac} = \left(\frac{e}{m^*} \right) \frac{8\pi^{\frac{1}{2}} \hbar \rho c_\ell^2}{3C_1^2} \left(\frac{\hbar^2}{2m^* k_B T} \right)^{3/2}. \tag{8.164}$$

The $T^{-3/2}$ dependence follows from the simple consideration that the average relaxation time τ varies as $\Lambda/\langle v \rangle$, where Λ is the carrier mean free path and $\langle v \rangle$ is the mean carrier velocity. Since $\Lambda \sim 1/\bar{n}_q \sim 1/T$ and $\langle v \rangle \sim T^{1/2}$, $\langle \tau_{ac} \rangle \sim 1/T^{3/2}$ and $\mu_{ac} \sim 1/T^{3/2}$.

Our results for both the average relaxation time and the mobility exhibit an inverse three-halves power dependence on the temperature. A decrease in mobility as the temperature increases is to be expected physically. As the temperature increases, the mean amplitude of vibration of the atoms and the perturbation of the periodic potential caused by the vibrations increase. The rate of scattering of the carriers therefore increases with increasing temperature, whereas the average time between collisions and the mobility decrease.

The case of degenerate carrier statistics can be handled with the aid of Eqs. (8.72) and (8.77) and tabulations of Fermi–Dirac integrals. Spherical nonparabolic bands require that the energy dependence of the effective mass be taken into account.

8.5.3.2 Intervalley scattering

In the previous section we discussed the scattering of charge carriers between states associated with a single energy extremum or "valley" due to their interaction with acoustic phonons. We know from Chapter 3, however, that the conduction bands of semiconductors such as Si, Ge, and GaP are characterized by multiple valleys whose minima are at or near the Brillouin zone boundary. Phonon-assisted **intervalley scattering** is possible in such materials. For a transition from a state near the minimum of one valley to a state near the minimum of a different valley, the change in wave vector is relatively large and may be on the order of a nonzero reciprocal lattice vector. The phonon frequency corresponding to this wave vector is also large and only weakly dependent on the value of the wave vector.

The interaction giving rise to intervalley scattering between valleys a and b can be characterized by an interaction Hamiltonian of the form

$$H_{int} = \left(\frac{\bar{M}}{M_1 + M_2}\right)^{\frac{1}{2}} \boldsymbol{D}_{ab} \cdot \boldsymbol{u}(\boldsymbol{r}), \tag{8.165}$$

where M_1 and M_2 are the masses of the two atoms in the primitive unit cell and $\bar{M}$ is an effective oscillator mass. The displacement $\boldsymbol{u}(\boldsymbol{r})$ is an effective relative displacement of the two atoms in the unit cell and is related to the normal coordinates $Q(\boldsymbol{q}j)$ by

$$\boldsymbol{u}(\boldsymbol{r}) = \frac{1}{\sqrt{N\bar{M}}} \sum_{\boldsymbol{q}j} \boldsymbol{e}(\boldsymbol{q}j) e^{i\boldsymbol{q}\cdot\boldsymbol{r}} Q(\boldsymbol{q}j). \tag{8.166}$$

The deformation potential $\boldsymbol{D}_{ab}$ is in general a function of the wave vector $\boldsymbol{q}$ and can be expanded in powers of $\boldsymbol{q}$:

$$\boldsymbol{D}_{ab}(\boldsymbol{q}) = \boldsymbol{D}_{ab}^{(0)} + \boldsymbol{D}_{ab}^{(1)} \cdot \boldsymbol{q} + \cdots. \tag{8.167}$$

Both optical and acoustic phonons can be involved in intervalley scattering. Selection rules determine which, if any, phonons are allowed for a particular scattering process (Birman *et al.* 1966).

Zero-order intervalley scattering is determined by the term with the coefficient $\boldsymbol{D}_{ab}^{(0)}$ in Eq. (8.167). If $\boldsymbol{u}(\boldsymbol{r})$ is eliminated from Eq. (8.165) with the aid of Eq. (8.166) and the matrix elements of H_{int} evaluated with the wave functions given by Eq. (8.140), the result for zero-order scattering is

$$\langle \boldsymbol{k}'n'|H_{int}|\boldsymbol{k}n\rangle = \frac{1}{\sqrt{N(M_1 + M_2)}} \sum_{\boldsymbol{q}j} \boldsymbol{D}_{ab}^{(0)} \cdot \boldsymbol{e}(\boldsymbol{q}j)\langle n'|Q(\boldsymbol{q}j)|n\rangle \times \int e^{-(\boldsymbol{k}'-\boldsymbol{k}-\boldsymbol{q})\cdot\boldsymbol{r}} u_{\boldsymbol{k}'}^{*}(\boldsymbol{r}) u_{\boldsymbol{k}}(\boldsymbol{r}) d^3 r. \tag{8.168}$$

The integral over $\boldsymbol{r}$ can be evaluated with the aid of Eq. (8.109). Eliminating $Q(\boldsymbol{q}j)$ in favor of the phonon creation and annihilation operators and completing the evaluation of the matrix element, we obtain

$$\langle \boldsymbol{k}'n'|H_{int}|\boldsymbol{k}n\rangle = \sum_{\boldsymbol{q}j} \frac{\hbar^{\frac{1}{2}}\Xi_{ab}(\boldsymbol{q}j)}{\sqrt{2\rho\Omega\omega_{ab}}} \sum_{\boldsymbol{G}} [\delta_{\boldsymbol{k}',\boldsymbol{k}-\boldsymbol{q}+\boldsymbol{G}} \delta_{n'_{\boldsymbol{q}j},n_{\boldsymbol{q}j}+1}(n_{\boldsymbol{q}j}+1)^{\frac{1}{2}} + \delta_{\boldsymbol{k}',\boldsymbol{k}+\boldsymbol{q}+\boldsymbol{G}} \delta_{n'_{\boldsymbol{q}j},n_{\boldsymbol{q}j}-1}] B_{\boldsymbol{k}'\boldsymbol{k}}(\boldsymbol{G})\Omega, \tag{8.169}$$

where $\Xi_{ab}(\boldsymbol{q}j)$ is $\boldsymbol{D}_{ab}^{(0)} \cdot \boldsymbol{e}(\boldsymbol{q}j)$ and ω_{ab} is the intervalley phonon frequency. Intervalley phonons have wave vectors that are an appreciable fraction of the Brillouin zone boundary wave vector in the same direction. The dependence of ω_{ab} on $\boldsymbol{q}$ is weak and will be neglected.

Using the above result for $\langle \boldsymbol{k}'n'|H_{int}|\boldsymbol{k}n\rangle$, we find the scattering rate to be

$$W(\boldsymbol{k} \to \boldsymbol{k}') = \frac{\Omega}{(2\pi)^3} \frac{\pi \Xi_0^2}{\rho \Omega \omega_{ab}} \left[\bar{n}(\omega_{ab}) + \tfrac{1}{2} \mp \tfrac{1}{2}\right] \delta(E_{\boldsymbol{k}'} - E_{\boldsymbol{k}} \pm \hbar\omega_{ab}), \qquad (8.170)$$

where the upper (lower) sign refers to absorption (emission) of phonons and Ξ_0 is an average value of $\Xi_{ab}(\boldsymbol{q}j)$ multiplied by $B_{\boldsymbol{k}'\boldsymbol{k}}(\boldsymbol{G})\Omega$. Since $E_{\boldsymbol{k}'}$ and hence $W(\boldsymbol{k} \to \boldsymbol{k}')$ are even in $\boldsymbol{k}'$, zero-order intervalley scattering is $\boldsymbol{k}$-randomizing. The inverse relaxation time is therefore specified by Eq. (8.96). The result including both emission and absorption of intervalley phonons is (Roth 1992)

$$\frac{1}{\tau(E_{\boldsymbol{k}})} = \frac{(Z-1)\Xi_0^2}{4\pi\rho\omega_{ab}} \left(\frac{2m^*}{\hbar^2}\right)^{3/2} \left\{ [\bar{n}(\omega_{ab}) + 1](E_{\boldsymbol{k}} - \hbar\omega_{ab})^{\frac{1}{2}} \times \theta(E_{\boldsymbol{k}} - \hbar\omega_{ab}) + \bar{n}(\omega_{ab})(E_{\boldsymbol{k}} + \hbar\omega_{ab})^{\frac{1}{2}} \right\}, \qquad (8.171)$$

where $\theta(x)$ is the Heaviside step function and Z is the number of equivalent valleys.

In n-Si with conduction band minima along the principal axes but not at the Brillouin zone boundary, there are two types of intervalley scattering processes. Scattering between valleys along different axes is called f-scattering, and that between valleys along the same axis is called g-scattering. Umklapp processes involving nonzero reciprocal lattice vectors $\boldsymbol{G}$ are important because the transitions can cross the boundary of the first Brillouin zone. For example, g-scattering in Si involves $\boldsymbol{G}_{100}$, whereas f-scattering involves $\boldsymbol{G}_{111}$. It turns out that f-scattering with a phonon energy of 0.054 eV($\Theta_D = 630$ K) is the dominant process at temperatures above 200 K.

However, experimental mobility data can be fully understood only if first-order intervalley scattering associated with the second term on the right-hand side of Eq. (8.167) is included (Ferry 1976).

We shall not discuss first-order scattering in detail, but simply note that scattering by LA and TA phonons of energy 0.016 eV ($\Theta_D = 190$ K) has been shown by Ferry to contribute appreciably to the total scattering in n-Si. A plot of the mobility as a function of temperature is given in Fig. 8.4. The individual contributions of acoustic and intervalley scattering are plotted in Fig. 8.5. Acoustic phonon scattering dominates at temperatures below 200 K, while zero-order intervalley scattering dominates above 200 K.

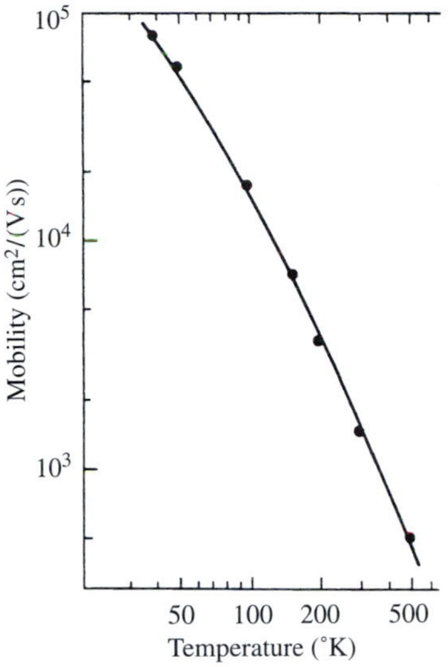

Fig. 8.4
Mobility versus temperature for n-Si (after Ferry 1976).

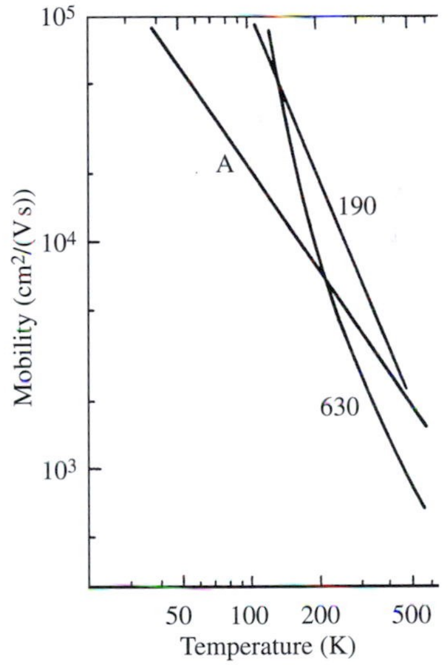

Fig. 8.5
Relative contributions to the mobility of n-Si as functions of temperature. A: acoustic intravalley contribution; 630: zero-order intervalley contribution; 190: first-order intervalley contribution (after Ferry 1976).

8.5.3.3 Polar optical phonon scattering

In Chapter 7 it was pointed out that the normal modes of vibration in semiconductors include optical as well as acoustic modes. An important characteristic of optical modes that distinguishes them from acoustic modes is that their frequencies do not approach zero as their wave vector approaches zero. Except at high temperatures, the carrier-optical phonon interaction must be treated as inelastic. Furthermore, in polar semiconductors the interaction is primarily Coulombian. The scattering depends on the angle between the initial and final wave vectors of the carrier and is not $\boldsymbol{k}$-randomizing. Neither case 1 nor case 2 of Section 8.3 applies, and the relaxation approximation time is not valid (Roth 1992). Under these circumstances the Boltzmann equation is customarily solved using a variational method (Howarth and Sondheimer 1953, Ehrenreich 1957). However, at sufficiently high temperatures the mean carrier energy greatly exceeds that of the phonons, and the phonon energy can be neglected in the energy conservation condition. The interaction then becomes elastic and a relaxation time can be introduced.

It is beyond the scope of this book to discuss the variational method in detail. We content ourselves with a treatment based on the relaxation time approximation that gives qualitatively reasonable results.

Compound semiconductors such as GaAs and InSb have polar character with the two atoms in the unit cell possessing effective charges $\pm e_L^*$. There is a strong interaction of charge carriers with the macroscopic electric field of longitudinal optical phonons of long wavelength. The interaction Hamiltonian can be written in the form (Ridley 1988)

$$H_{int}(\boldsymbol{r}) = -\frac{1}{\epsilon_0}\int \mathcal{E}_{vac}(\boldsymbol{r}-\boldsymbol{R})\cdot \boldsymbol{P}(\boldsymbol{R})d^3R, \tag{8.172}$$

where $\mathcal{E}_{vac}(\boldsymbol{r}-\boldsymbol{R})$ is the electric field in vacuum at $\boldsymbol{R}$ due to a carrier of charge e_c at $\boldsymbol{r}$ given by

$$\mathcal{E}_{vac}(\boldsymbol{r}-\boldsymbol{R}) = -\boldsymbol{\nabla}\left(\frac{e_c}{4\pi|\boldsymbol{r}-\boldsymbol{R}|}\right), \tag{8.173}$$

and $\boldsymbol{P}(\boldsymbol{R})$ is the polarization at $\boldsymbol{R}$ due to longitudinal optical phonons given by

$$\boldsymbol{P}(\boldsymbol{R}) = \frac{e_L^*\boldsymbol{u}(\boldsymbol{R})}{\Omega_0}. \tag{8.174}$$

The displacement $\boldsymbol{u}(\boldsymbol{R})$ is the relative displacement of the two atoms in the unit cell at $\boldsymbol{R}$ and can be expressed in terms of normal coordinates by

$$\boldsymbol{u}(\boldsymbol{R}) = (N\bar{M})^{-\frac{1}{2}}\sum_{\boldsymbol{q}j}\boldsymbol{e}_{op}(\boldsymbol{q}j)e^{i\boldsymbol{q}\cdot\boldsymbol{R}}Q(\boldsymbol{q}j), \tag{8.175}$$

where $\bar{M}$ is the reduced mass of the two atoms and j includes only optical modes. The vector $\boldsymbol{e}_{op}(\boldsymbol{q}j)$ is a unit vector that, for $\boldsymbol{q}=0$, is related to thepolarization vectors $\boldsymbol{e}_1(0j)$ and $\boldsymbol{e}_2(0j)$ by $\boldsymbol{e}_1(0j) = [M_2/(M_1+M_2)]^{\frac{1}{2}}$ $\boldsymbol{e}_{op}(0j)$ and $\boldsymbol{e}_2(0j) = -[M_1/(M_1+M_2)]^{\frac{1}{2}}\boldsymbol{e}_{op}(0j)$ (Maradudin *et al.* 1971). We assume that the same relations hold for $\boldsymbol{q}\neq 0$, because only small $\boldsymbol{q}$ lead to a significant interaction. Replacing $Q(\boldsymbol{q}j)$ by its expression in terms of creation and annihilation operators (Eq. (8.138)) yields

$$u(R) = \left(\frac{\hbar}{2N\bar{M}}\right)^{\frac{1}{2}} \sum_{qj} \frac{1}{\sqrt{\omega_{qj}}} \left[e_{op}(qj) e^{iq\cdot R} a_{qj} + e^{*}_{op}(qj) e^{-iq\cdot R} a^{+}_{qj} \right], \tag{8.176}$$

where q has been replaced by $-q$ in the terms involving a^{+}.

If Eqs. (8.172)–(8.174) are combined and an integration by parts is carried out, the result is

$$H_{int}(r) = -\frac{e^{*}_{L} e_{c}}{4\pi\epsilon_0 \Omega_0} \int \frac{1}{|r - R|} \nabla \cdot u(R) d^3R, \tag{8.177}$$

which, upon using Eq. (8.176), becomes

$$H_{int}(R) = -\frac{i e^{*}_{L} e_{c}}{4\pi\epsilon_0 \Omega_0} \left(\frac{\hbar}{2N\bar{M}}\right)^{\frac{1}{2}} \int \sum_{qj} \frac{1}{\sqrt{\omega_{qj}}|r - R|} \times \left[q \cdot e_{op}(qj) e^{iq\cdot R} a_{qj} - q \cdot e^{*}_{op}(qj) e^{-iq\cdot R} a^{+}_{qj} \right] d^3R. \tag{8.178}$$

Just as in the case of deformation potential scattering, only longitudinal modes contribute, although now we are dealing with optical modes. Taking $e_{op}(qL) = iq/q$, which satisfies $e^{*}_{op}(qL) = e_{op}(-qL)$, we obtain

$$H_{int}(r) = \frac{e^{*}_{L} e_{c}}{4\pi\epsilon_0 \Omega_0} \left(\frac{\hbar}{2N\bar{M}\omega_{LO}}\right)^{\frac{1}{2}} \times \int \sum_{q} \frac{q}{|r - R|} (e^{iq\cdot R} a_q + e^{-iq\cdot R} a^{+}_{q}) d^3R, \tag{8.179}$$

where ω_{LO} is the longitudinal optical phonon frequency assumed independent of q. The integral is simply the Fourier transform of the Coulomb potential, so

$$H_{int}(r) = \frac{e^{*}_{L} e_{c}}{\epsilon_0 \Omega_0} \left(\frac{\hbar}{2N\bar{M}\omega_{LO}}\right)^{\frac{1}{2}} \sum_{q} \frac{1}{q} \left(e^{iq\cdot r} a_q + e^{-iq\cdot r} a^{+}_{q} \right). \tag{8.180}$$

The analysis of long wavelength optical phonons at the end of Section 7.5.2 enables us to relate the effective charge e^{*}_{L} to the static and high-frequency dielectric constants. It was pointed out that $\epsilon(\omega_{LO}) = 0$, so from Eq. (7.54)

$$P = -\epsilon_0 \mathcal{E}. \tag{8.181}$$

Eliminating $\mathcal{E}$ from Eqs. (7.52) and (8.181) and utilizing Eq. (7.50) yields the relation

$$P = \frac{\epsilon_0 b_{21}}{\epsilon_0 + b_{22}} \left(\frac{\bar{M}}{\Omega_0}\right)^{\frac{1}{2}} u. \tag{8.182}$$

Making use of Eqs. (7.56) and (7.58), the fact that $b_{21} = b_{12}$, and the Lyddane–Sachs–Teller relation, we can express P in the form of

Eq. (8.174) with

$$e_L^* = \left[\epsilon_0 \bar{M} \Omega_0 \left(\frac{1}{\epsilon_\infty} - \frac{1}{\epsilon_s}\right)\right]^{\frac{1}{2}} \omega_{LO}. \tag{8.183}$$

Since e_L^* concerns interactions with longitudinal optical phonons, it is called the **longitudinal effective charge**. It is also known as the **Callen effective charge** after the person who introduced it (Callen 1949).

The scattering rate $W(\boldsymbol{k}n \to \boldsymbol{k}'n')$ can now be calculated in a manner similar to that used for deformation potential scattering. The result is

$$\begin{aligned} W(\boldsymbol{k}n \to \boldsymbol{k}'n') = & \frac{2(e_L^* e/4\pi\epsilon_0)^2}{\Omega_0 \bar{M} \omega_{LO} |\boldsymbol{k} - \boldsymbol{k}'|^2} \left[\bar{n}(\omega_{LO}) + \tfrac{1}{2} \pm \tfrac{1}{2}\right] \\ & \times \delta_{n'(\omega_{LO}), n(\omega_{LO}) \pm 1} \delta(E_{\boldsymbol{k}'} - E_{\boldsymbol{k}} \pm \hbar\omega_{LO}), \end{aligned} \tag{8.184}$$

where the upper (lower) sign refers to optical phonon emission (absorption). At this point we make the approximations that the scattering is elastic (valid in the high temperature limit) and that the energy band is isotropic. The terms $\pm\hbar\omega_{LO}$ in the energy-conserving delta functions are thereby neglected, and the situation falls under case (2) of Section 8.4. A treatment that does not neglect the $\pm\hbar\omega_{LO}$ terms is given by Ridley (1988). The inverse relaxation time in the approximation being considered is given by Eq. (8.104) and takes the form

$$\begin{aligned} \frac{1}{\tau(E_{\boldsymbol{k}})} &= \left(\frac{e_L^* e}{4\pi\epsilon_0}\right)^2 \frac{[2\bar{n}(\omega_{LO}) + 1 \pm 1]}{\Omega_0 \bar{M} \omega_{LO}} \int \frac{1 - \cos\theta_{\boldsymbol{k},\boldsymbol{k}'}}{|\boldsymbol{k} - \boldsymbol{k}'|^2} \delta(E_{\boldsymbol{k}'} - E_{\boldsymbol{k}}) d^3k' \\ &= \frac{e_L^{*2} e^2 (2m^*)^{\frac{1}{2}}}{8\pi\epsilon_0^2 \Omega_0 \bar{M} \hbar\omega_{LO} E_{\boldsymbol{k}}^{\frac{1}{2}}} \left[\bar{n}(\omega_{LO}) + \tfrac{1}{2} \pm \tfrac{1}{2}\right]. \end{aligned} \tag{8.185}$$

Adding the contributions from emission and absorption of optical phonons gives the total inverse relaxation time for polar optical phonon scattering:

$$\frac{1}{\tau(E_{\boldsymbol{k}})} = \frac{e_L^{*2} e^2 (2m^*)^{\frac{1}{2}}}{8\pi\epsilon_0^2 \Omega_0 \bar{M} \hbar\omega_{LO} E_{\boldsymbol{k}}^{\frac{1}{2}}} [2\bar{n}(\omega_{LO}) + 1]. \tag{8.186}$$

The value of $\langle\tau\rangle$ and the mobility can be calculated from Eqs. (8.79) and (8.80). Taking $\bar{n}(\omega_{LO}) \simeq k_B T/\hbar\omega_{LO}$ and using classical statistics for the carrier, we obtain for the mobility

$$\mu = \frac{e}{m^*} \cdot \frac{16\sqrt{2\pi}\epsilon_0^2 \Omega_0 \bar{M} (\hbar\omega_{LO})^2}{3 e_L^{*2} e^2 (m^* k_B T)^{\frac{1}{2}}}. \tag{8.187}$$

As in the case of deformation potential scattering, the mobility decreases as the temperature increases, but with a different power law dependence on temperature.

If the carrier concentration is sufficiently large, the field due to the ions that acts on a particular carrier will be screened by the other carriers. The bare Coulomb potential in Eq. (8.173) is then to be modified by multiplying e_c by $\exp[-q_0|\boldsymbol{r} - \boldsymbol{R}|]$, where q_0 is defined in Eq. (8.106). The calculation of the mobility is straightforward, but tedious, and will not be presented here.

A careful study of electron mobility in GaAs has been carried out by Rode (1970) whose theoretical results together with experimental data are shown in Fig. 8.6. At the higher temperatures the mobility is dominated by polar phonon scattering.

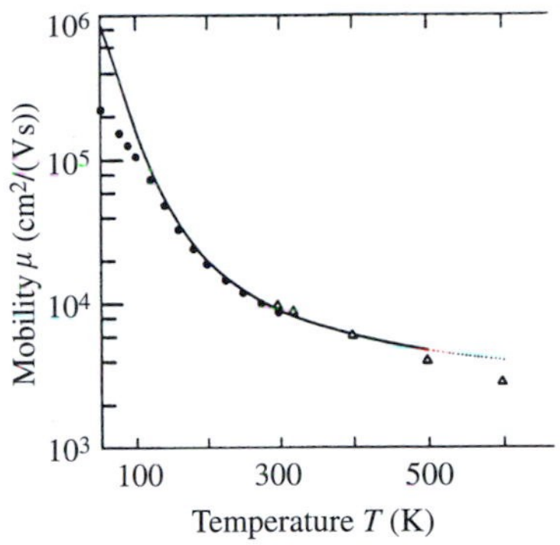

Fig. 8.6
Electron mobility versus temperature for GaAs (after Rode 1970).

8.5.3.4 Piezoelectric scattering

Compound semiconductors of the III–V and II–VI types lack a center of inversion symmetry and are consequently piezoelectric. The elastic strain associated with an acoustic mode is accompanied by an electric dipole moment or electric polarization. The electric field arising from the polarization interacts with charge carriers and produces scattering of the latter. If the concentration of carriers is significant, screening of the polarization field must be taken into account.

The piezoelectric effect exhibits a complicated dependence on the direction of propagation of the lattice vibrational wave. It is customary to carry out averages over direction separately for longitudinal and transverse waves. For zincblende structure materials the interaction Hamiltonian in the absence of screening can be expressed in the form (Ridley 1988, Roth 1992)

$$H_{int}(\boldsymbol{r}) = -i\sum_{\boldsymbol{q}j}{}' C_j\left[\frac{\hbar}{2(M_1+M_2)N\omega_{\boldsymbol{q}j}}\right]^{\frac{1}{2}} e^{i\boldsymbol{q}\cdot\boldsymbol{r}}(a_{\boldsymbol{q}j} + a^{+}_{-\boldsymbol{q}j}), \tag{8.188}$$

where the prime on the sum indicates that only acoustic modes are included. The coupling constants $C_{\ell(t)}$ for longitudinal (transverse) modes are given by

$$C_\ell = \left(\frac{12}{35}\right)^{\frac{1}{2}}\frac{e_c e_{14}}{\epsilon_0\epsilon} \tag{8.189a}$$

$$C_t = \left(\frac{16}{35}\right)^{\frac{1}{2}}\frac{e_c e_{14}}{\epsilon_0\epsilon}. \tag{8.189b}$$

Table 8.2 Piezoelectric coefficients e_{14} in C/m^2 for cubic III–V and II–VI semiconductors (after Ridley 1988)

III–V	e_{14}	II–VI	e_{14}
GaAs	0.160	ZnS	0.17
GaSb	0.126	ZnSe	0.045
InAs	0.045	ZnTe	0.027
InSb	0.071	CdTe	0.034

The quantity e_{14} is the single nonzero piezoelectric coefficient for the zincblende structure and ϵ is the dielectric constant. Values of e_{14} for some III–V and II–VI compounds are listed in Table 8.2.

From the interaction Hamiltonian the scattering rate can be calculated in the usual manner to give

$$W(\boldsymbol{k}n \to \boldsymbol{k}'n') = \pi\sum_{\boldsymbol{q}j}{}' \frac{C_j^2}{\rho\Omega\omega_{\boldsymbol{q}j}}\delta_{\boldsymbol{k}',\boldsymbol{k}\mp\boldsymbol{q}}\delta_{n'(\omega_{\boldsymbol{q}j}),n(\omega_{\boldsymbol{q}j})\pm1} \times \left[n(\omega_{\boldsymbol{q}j}) + \tfrac{1}{2} \pm \tfrac{1}{2}\right]\delta(E_{\boldsymbol{k}'} - E_{\boldsymbol{k}} \pm \hbar\omega_{\boldsymbol{q}j}), \tag{8.190}$$

where the upper (lower) sign refers to emission (absorption) of acoustic phonons. We make the approximation of elastic scattering by neglecting $\hbar\omega_{qj}$ in the energy-conserving delta function. Carrying out an average over initial phonon states and passing to the classical limit for the phonon statistics, we obtain

$$W(\boldsymbol{k} \to \boldsymbol{k}') = \frac{\pi k_B T}{\hbar\rho\Omega} \sum_j{}' \frac{C_j^2}{c_j^2 |\boldsymbol{k} - \boldsymbol{k}'|^2} \delta(E_{\boldsymbol{k}'} - E_{\boldsymbol{k}}), \tag{8.191}$$

where we have taken $\omega_{qj} = c_j q$. Noting that case 2 of Section 8.3 applies, we evaluate the inverse relaxation time with the aid of Eq. (8.104). The result after combining the contributions from emission and absorption of phonons is

$$\frac{1}{\tau_{E_k}} = \frac{e^2 (2m^*)^{\frac{1}{2}} k_B T K^2}{4\pi\epsilon_0\epsilon\hbar^2 E_{\boldsymbol{k}}^{\frac{1}{2}}} \tag{8.192}$$

with K a dimensionless measure of the piezoelectric interaction defined by

$$K^2 = \frac{\epsilon_0\epsilon}{e^2\rho} \sum_j{}' \frac{C_j^2}{c_j^2}. \tag{8.193}$$

We note that the energy dependence of $1/\tau$ is the same as that found for polar optical phonon scattering.

The calculation of $\langle\tau\rangle$ and the mobility follows previous procedures and yields the following result for the mobility:

$$\mu = \frac{e}{m^*} \cdot \frac{16\sqrt{2\pi}\hbar^2\epsilon_0\epsilon}{3e^2(m^* k_B T)^{\frac{1}{2}} K^2}. \tag{8.194}$$

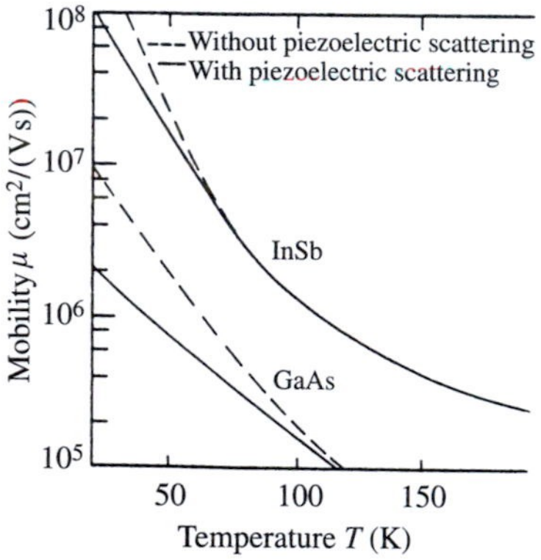

Fig. 8.7
Effect of piezoelectric scattering on electron mobility in GaAs and InSb. Polar optical and deformation potential acoustic scattering are included (after Rode 1970).

Although the mobilities for polar phonon scattering and piezoelectric scattering show the same temperature dependence in Eqs. (8.187) and (8.194), respectively, the latter equation remains valid to much lower temperatures than does the former.

In lightly doped n-GaAs and n-InSb, piezoelectric scattering dominates the scattering of thermal electrons at low temperatures. The influence of piezoelectric scattering on the mobility in these materials is illustrated in Fig. 8.7.

8.6 Magnetotransport properties

8.6.1 Magnetoresistance

In Chapter 4 it was shown that the application of an external magnetic field leads to phenomena such as cyclotron resonance which are very useful in determining basic parameters such as effective mass. We now analyze the effect of a magnetic field on the electrical conductivity.

The equation of motion for a carrier in a spherical parabolic band is

$$\frac{d\boldsymbol{v}}{dt} + \frac{1}{\tau}\boldsymbol{v} = \frac{e_c}{m^*}\boldsymbol{\mathcal{E}} + \frac{e_c}{m^*}(\boldsymbol{v} \times \boldsymbol{\mathcal{B}}), \tag{8.195}$$

where $\boldsymbol{\mathcal{E}}$ and $\boldsymbol{\mathcal{B}}$ are the applied electric and magnetic fields, respectively, and e_c, m^*, and τ are the carrier charge, mass, and relaxation time, respectively. Taking $\boldsymbol{\mathcal{B}}$ in the z-direction, the equations of motion in component form become

$$\frac{dv_x}{dt} + \frac{1}{\tau}v_x - \frac{e_c\mathcal{B}}{m^*}v_y = \frac{e_c}{m^*}\mathcal{E}_x \tag{8.196a}$$

$$\frac{dv_y}{dt} + \frac{1}{\tau}v_y + \frac{e_c\mathcal{B}}{m^*}v_x = \frac{e_c}{m^*}\mathcal{E}_y \tag{8.196b}$$

$$\frac{dv_z}{dt} + \frac{1}{\tau}v_z = \frac{e_c}{m^*}\mathcal{E}_z. \tag{8.196c}$$

We see from the last equation that the motion of the carrier parallel to the magnetic field is not affected by that field. This is a special result for the case where the inverse effective mass tensor is diagonal in a coordinate system containing the magnetic field direction as one axis.

Let us restrict our attention to stationary situations for which we can ignore the time derivative of $\boldsymbol{v}$. The equations of motion can then be solved to yield

$$v_x = \frac{e_c\tau}{m^*}\left(\frac{\mathcal{E}_x \pm \omega_c\tau\mathcal{E}_y}{1+\omega_c^2\tau^2}\right), \tag{8.197a}$$

$$v_y = \frac{e_c\tau}{m^*}\left(\frac{\mp\omega_c\tau\mathcal{E}_x + \mathcal{E}_y}{1+\omega_c^2\tau^2}\right), \tag{8.197b}$$

$$v_z = \frac{e_c\tau}{m^*}\mathcal{E}_z, \tag{8.197c}$$

where the upper and lower signs refer to holes and electrons, respectively, and ω_c is the cyclotron frequency $e\mathcal{B}/m^*$. The components of the current density are obtained by multiplying the velocity components by n_ce_c and may be expressed as $j_\alpha = \sum_\beta \sigma_{\alpha\beta}\mathcal{E}_\beta$, where the nonvanishing elements of the conductivity tensor $\sigma_{\alpha\beta}$ are given by

$$\sigma_{xx} = \sigma_{yy} = \frac{n_ce^2\tau}{m^*}\cdot\frac{1}{1+\omega_c^2\tau^2}, \tag{8.198a}$$

$$\sigma_{xy} = -\sigma_{yx} = \pm\frac{n_ce^2\tau}{m^*}\cdot\frac{\omega_c\tau}{1+\omega_c^2\tau^2}, \tag{8.198b}$$

$$\sigma_{zz} = \frac{n_ce^2\tau}{m^*} = \sigma_0, \tag{8.198c}$$

and n_c is the carrier concentration. Experimental data are usually presented in terms of magnetoresistance rather than magnetoconductance. The resistivity tensor is the inverse of the conductivity tensor and has elements

$$\rho_{xx} = \rho_{yy} = \frac{\sigma_{xx}}{\sigma_{xx}^2 + \sigma_{xy}^2}, \tag{8.199a}$$

$$\rho_{xy} = -\rho_{yx} = -\frac{\sigma_{xy}}{\sigma_{xx}^2 + \sigma_{xy}^2}, \tag{8.199b}$$

$$\rho_{zz} = \frac{1}{\sigma_{zz}} = \frac{1}{\sigma_0}. \tag{8.199c}$$

Magnetoresistance is expressed as $\Delta\rho/\rho$, where $\Delta\rho$ is the change in resistivity produced by the magnetic field. We see immediately from the last equation that the **longitudinal magnetoresistance** is zero for a spherical parabolic band.

The **transverse magnetoresistance** is given by

$$\frac{\Delta\rho}{\rho_0} = \frac{\rho_{xx} - \rho_0}{\rho_0}. \tag{8.200}$$

To simplify matters, let us first consider low magnetic fields for which we can expand quantities to second order in ω_c. Starting from Eqs. (8.198), we have

$$\sigma_{xx} \cong \frac{n_c e^2 \tau}{m^*}(1 - \omega_c^2 \tau^2) \tag{8.201a}$$

$$\sigma_{xy} = -\sigma_{yx} \cong \pm \frac{n_c e^2 \tau}{m^*} \omega_c \tau. \tag{8.201b}$$

Since τ is in general energy dependent, we must average the conductivity tensor elements over energy:

$$\langle \sigma_{xx} \rangle \cong \frac{n_c e^2}{m^*}(\langle \tau \rangle - \omega_c^2 \langle \tau^3 \rangle) \tag{8.202a}$$

$$\langle \sigma_{xy} \rangle \cong \pm \frac{n_c e^2 \omega_c}{m^*} \langle \tau^2 \rangle. \tag{8.202b}$$

Replacing σ_{xx} and σ_{xy} in Eq. (8.196a) by $\langle \sigma_{xx} \rangle$ and $\langle \sigma_{xy} \rangle$ yields

$$\langle \rho_{xx} \rangle = \rho_0 \left[1 + \omega_c^2 \frac{\langle \tau^3 \rangle}{\langle \tau \rangle} - \omega_c^2 \frac{\langle \tau^2 \rangle^2}{\langle \tau \rangle^2} \right],$$

where $\rho_0 = m^*/n_c e^2 \langle \tau \rangle$. From this result follows the transverse magnetoresistance

$$\frac{\Delta\rho}{\rho_0} = \left(\frac{e}{m^*}\right)^2 \frac{\langle \tau^2 \rangle^2}{\langle \tau \rangle^2}(A - 1)\mathcal{B}^2, \tag{8.203}$$

where

$$A = \frac{\langle\tau^3\rangle\langle\tau\rangle}{\langle\tau^2\rangle^2}. \tag{8.204}$$

For an energy-independent relaxation time, $A = 1$ and the transverse magnetoresistance vanishes. The energy dependences associated with the important scattering mechanisms lead to $A \neq 1$ and a nonvanishing transverse magnetoresistance. For τ varying as $(E/k_BT)^r$, $A = (3r + \frac{3}{2})!/[(2r + \frac{3}{2})!]^2$.

In the case of high magnetic fields ($\omega_c\tau \gg 1$), we see from Eqs. (8.198) that $\sigma_{xy} \gg \sigma_{xx}$, so

$$\langle\rho_{xx}\rangle \simeq \frac{\sigma_{xx}}{\sigma_{xy}^2} \simeq \left\langle\frac{n_ce^2\tau}{m^*}\frac{1}{\omega_c^2\tau^2}\right\rangle \Big/ \left\langle\frac{n_ce^2\tau}{m^*}\frac{1}{\omega_c\tau}\right\rangle^2 = \left\langle\frac{m^*}{n_ce^2\tau}\right\rangle. \tag{8.205}$$

The transverse magnetoresistance then becomes

$$\frac{\Delta\rho}{\rho_0} = \left\langle\frac{1}{\tau}\right\rangle\langle\tau\rangle - 1 \tag{8.206}$$

which corresponds to saturation as $\mathcal{B} \to \infty$.

Spherical nonparabolic bands can be handled by noting that the effective mass now depends on the carrier energy and that the combination of τ/m^* always occurs. One therefore averages this ratio over energy rather than τ alone.

The case of ellipsoidal parabolic bands requires the introduction of a tensor relaxation time (Roth 1992). One must also take into account the multiple valleys that typically accompany ellipsoidal bands. The full treatment is complicated, and the reader is referred to the article by Roth for details. The principal qualitative results are the following. There is a finite longitudinal magnetoresistance. The magnetoresistance is anisotropic and deviates from quadratic behavior at finite $\mathcal{B}$. At large fields it saturates. The anisotropy of the magnetoresistance of n-Si as measured by Pearson and Herring is shown in Fig. 8.8. The magnetoresistance of holes in p-type semiconductors is also anisotropic, primarily due to the warping of the heavy hole band.

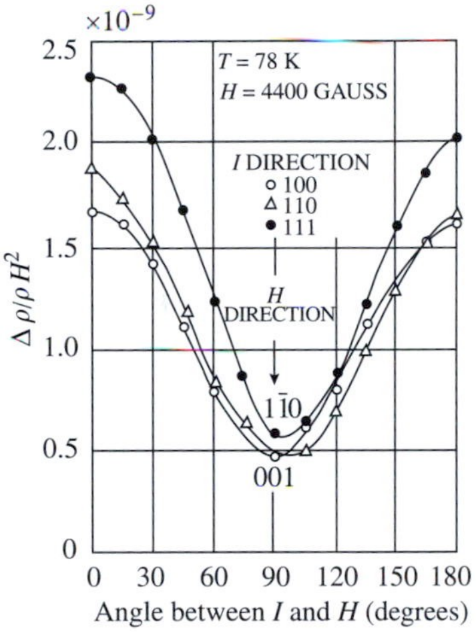

Fig. 8.8
Magnetoresistance in n-Si as a function of angle θ between the current and the magnetic field (after Pearson and Herring 1954).

8.6.2 Hall effect

The Hall effect was discussed from a simple point of view in Chapter 4. We now examine how the Hall coefficient is modified by complications such as an energy-dependent relaxation time. For small magnetic fields, Eqs. (8.197a) and (8.197b) for v_x and v_y can be reduced to

$$\langle v_x\rangle = \frac{e_c}{m^*}\left(\langle\tau\rangle\mathcal{E}_x \pm \omega_c\langle\tau^2\rangle\mathcal{E}_y\right) \tag{8.207a}$$

$$\langle v_y\rangle = \frac{e_c}{m^*}\left(\mp\omega_c\langle\tau^2\rangle\mathcal{E}_x + \langle\tau\rangle\mathcal{E}_y\right) \tag{8.207b}$$

after averaging over the carrier energy. In the Hall geometry with the applied electric field in the x-direction, $\langle v_y \rangle = 0$, so

$$\mathcal{E}_y = \pm\omega_c \frac{\langle \tau^2 \rangle}{\langle \tau \rangle} \mathcal{E}_x. \tag{8.208}$$

Eliminating $\mathcal{E}_x$ from Eq. (8.207a) and solving the resulting equation to lowest order in $\mathcal{B}$, we obtain

$$\mathcal{E}_y = R_H \mathcal{B} \langle j_x \rangle, \tag{8.209}$$

where $\langle j_x \rangle = n_c e_c \langle v_x \rangle$ and the Hall coefficient R_H is given by

$$R_H = \frac{1}{n_c e_c} \frac{\langle \tau^2 \rangle}{\langle \tau \rangle^2} = \frac{1}{n_c e_c} r_H. \tag{8.210}$$

We note that this result for the low field Hall coefficient differs from that for a constant relaxation time by the Hall factor $r_H = \langle \tau^2 \rangle / \langle \tau \rangle^2$. For an $(E/k_B T)^r$ dependence of τ on energy, $r_H = (2r + \frac{3}{2})!\frac{3}{2}!/[(r + \frac{3}{2})!]^2$. Since different scattering mechanisms dominate in different temperature ranges, the Hall coefficient can vary significantly with temperature due to the strong temperature dependence of r_H.

For a single type of carrier of concentration n_c, the conductivity σ defined by Eq. (8.14) is simply $n_c e \mu$. The Hall coefficient is related to σ by

$$|R_H| = \frac{\mu}{\sigma} r_H. \tag{8.211}$$

Instead of introducing r_H, one can define the **Hall mobility** μ_H by

$$|R_H| = \frac{\mu_H}{\sigma}, \tag{8.212}$$

where $\mu_H = \mu r_H$. Thus, μ_H is the product of two directly measurable quantities, whereas the **conductivity mobility** μ is not.

The Hall coefficient exhibits an interesting variation with carrier concentration when both electrons and holes are present. Since the two types of carriers conduct in parallel, we must add their conductivities to give the total conductivity and then invert the latter to give the resistivity. With the aid of Eq. (8.207b) for $\langle v_y \rangle$, we can express the y-component of the average current density as

$$\langle j_y \rangle = \frac{ne^2}{m_e^*} \left(\omega_{ce} \langle \tau^2 \rangle \mathcal{E}_x + \langle \tau \rangle \mathcal{E}_y \right) + \frac{pe^2}{m_h^*} \left(-\omega_{ch} \langle \tau^2 \rangle \mathcal{E}_x + \langle \tau \rangle \mathcal{E}_y \right), \tag{8.213}$$

where we have taken the average scattering time to be the same for both carriers. Imposing the Hall effect condition $\langle j_y \rangle = 0$ and solving for $\mathcal{E}_y$ gives

$$\mathcal{E}_y = \frac{\mathcal{B} r_H e}{\sigma_0} \left(p\mu_h^2 - n\mu_e^2 \right) \mathcal{E}_x, \tag{8.214}$$

where μ_e and μ_h are the mobilities of electrons and holes, respectively and $\sigma_0 = e(p\mu_h + n\mu_e)$. But from Eq. (8.209), $\mathcal{E}_y = R_H \mathcal{B}\langle j_x \rangle = R_H \mathcal{B}\sigma_0 \mathcal{E}_x$ to lowest order in $\mathcal{B}$. Comparing this expression with that of Eq. (8.214) yields the result

$$R_H = \frac{r_H e}{\sigma_0^2}(p\mu_h^2 - n\mu_e^2). \tag{8.215}$$

Eliminating σ_0, we obtain

$$R_H = \frac{r_H(p - nb^2)}{e(p + nb)^2} \tag{8.216}$$

with $b = \mu_e/\mu_h$. The Hall coefficient therefore changes sign at carrier concentrations that depend on the mobility ratio b.

If the constant energy surfaces are ellipsoidal, the Hall factor is modified to reflect the anisotropy. Introducing the longitudinal (transverse) scattering time $\tau_\ell(\tau_t)$ and effective mass $m_\ell^*(m_t^*)$, the anisotropy constant K is defined by

$$K = \frac{\tau_t m_\ell^*}{\tau_\ell m_t^*} \tag{8.217}$$

and the Hall factor is given by

$$r_H = \frac{\langle \tau_\ell^2 K(K+2)/3 \rangle}{\langle \tau_\ell(2K+1)/3 \rangle^2}. \tag{8.218}$$

If τ_ℓ and τ_t have the same energy dependence, K is a constant and the Hall factor reduces to

$$r_H = \frac{\langle \tau^2 \rangle 3K(K+2)}{\langle \tau \rangle^2 (2K+1)^2}. \tag{8.219}$$

A similar analysis can be applied to light and heavy holes in p-type semiconductors. Heavy holes have the higher concentration due to the higher density-of-states, but light holes have the higher mobility. The light holes contribute strongly to the low-field Hall effect.

The Hall coefficient exhibits a dependence on the magnetic field. The low-field limit for a single type of carrier is given by Eq. (8.210). The high-field limit can be obtained from Eq. (8.197a) by letting $\omega_c \to \infty$. Averaging over the carrier energy gives

$$\langle v_x \rangle \cong \frac{e}{m^* \omega_c}\mathcal{E}_y = \frac{1}{\mathcal{B}}\mathcal{E}_y. \tag{8.220}$$

The defining equation for the Hall coefficient becomes

$$\mathcal{E}_y = \frac{\mathcal{B}}{n_c e_c}\langle j_x \rangle, \tag{8.221}$$

so

$$R = \frac{1}{n_c e_c}. \tag{8.222}$$

8.7 Thermoelectric phenomena

If a crystal contains free charge carriers, an electric current is set up not only if an electric field or a carrier concentration gradient is present, but also if a temperature gradient is present. There are several electric phenomena associated with a temperature gradient. They are described by the basic equations

$$\boldsymbol{j} = \sigma[\boldsymbol{\mathcal{E}} - \alpha\boldsymbol{\nabla}T] \tag{8.223a}$$

$$\boldsymbol{Q} = (\sigma T\alpha)\boldsymbol{\mathcal{E}} - \kappa\boldsymbol{\nabla}T, \tag{8.223b}$$

where α is the **Seebeck coefficient** or **thermoelectric power**, $\boldsymbol{Q}$ is the heat flux, and κ is the thermal conductivity. Under open circuit conditions so that $\boldsymbol{j} = 0$, the **Seebeck effect** is the electromotive force developed by the presence of a temperature gradient in a material. It is the basis of thermocouples. If, on the other hand, there is no temperature gradient, the **Peltier effect** can occur in which a heat flux is produced by an electric current, $\boldsymbol{Q} = \pi_p\boldsymbol{j}$, where π_p is the **Peltier coefficient** given by the **Kelvin relation** $\pi_p = T\alpha$.

8.7.1 Thermoelectric power

We start our analysis with the Boltzmann equation which must now be generalized to include the spatial dependence of the distribution function: $f(\boldsymbol{k}) \to f(\boldsymbol{k}, \boldsymbol{r})$. The total time derivative of f becomes

$$\frac{df}{dt} = \frac{\partial f}{\partial t} + \boldsymbol{\nabla}_{\boldsymbol{k}} f \cdot \frac{d\boldsymbol{k}}{dt} + \boldsymbol{\nabla}_{\boldsymbol{r}} f \cdot \frac{d\boldsymbol{r}}{dt} - \left.\frac{\partial f}{\partial t}\right|_{coll}. \tag{8.224}$$

Now the terms involving the gradients of f are proportional to the perturbing forces, and therefore to first order we can replace f by f_0 in these terms. Proceeding as we did in Section 8.2 and restricting ourselves to steady states, we rewrite Eq. (8.224) in the form

$$\frac{\partial f_0}{\partial E_{\boldsymbol{k}}} \boldsymbol{F} \cdot \boldsymbol{v}_g + \boldsymbol{\nabla}_{\boldsymbol{r}} f_0 \cdot \boldsymbol{v} = \left.\frac{\partial f}{\partial t}\right|_{coll}, \tag{8.225}$$

where $\boldsymbol{v} = d\boldsymbol{r}/dt$. The velocity $\boldsymbol{v}$ is the carrier velocity and may be identified with $\boldsymbol{v}_g$.

The Fermi–Dirac distribution f_0 contains the Fermi energy E_F which in general is a function of $\boldsymbol{r}$. Operating on f_0 with $\boldsymbol{\nabla}_{\boldsymbol{r}}$ therefore yields

$$\boldsymbol{\nabla}_{\boldsymbol{r}} f_0 = \boldsymbol{\nabla}_{\boldsymbol{r}} \left[e^{(E_{\boldsymbol{k}}-E_F)/k_B T} + 1\right]^{-1} = -\frac{\partial f_0}{\partial E_{\boldsymbol{k}}} \left[\boldsymbol{\nabla}_{\boldsymbol{r}} E_F + \frac{(E_{\boldsymbol{k}} - E_F)}{T} \boldsymbol{\nabla}_{\boldsymbol{r}} T\right]. \tag{8.226}$$

Combining Eqs. (8.225) and (8.226) produces the generalized Boltzmann equation

$$\frac{\partial f_0}{\partial E_{\boldsymbol{k}}} \boldsymbol{v}_g \cdot \left[\boldsymbol{F} - \boldsymbol{\nabla}_{\boldsymbol{r}} E_F - \frac{(E_{\boldsymbol{k}} - E_F)}{T} \boldsymbol{\nabla}_{\boldsymbol{r}} T\right] = \left.\frac{\partial f}{\partial t}\right|_{coll}. \tag{8.227}$$

Let us assume that a relaxation time τ exists. Using Eq. (8.55) we can write the solution to the generalized Boltzmann equation as

$$f = f_0 - e_c \tau \frac{\partial f_0}{\partial E_{\boldsymbol{k}}} \boldsymbol{v}_g \cdot \boldsymbol{\mathcal{E}}_1, \tag{8.228}$$

where

$$\boldsymbol{\mathcal{E}}_1 = \boldsymbol{\mathcal{E}} - \frac{1}{e_c} \boldsymbol{\nabla}_{\boldsymbol{r}} E_F - \frac{E_{\boldsymbol{k}} - E_F}{e_c T} \nabla_{\boldsymbol{r}} T \tag{8.229}$$

and we have used $\boldsymbol{F} = e_c \boldsymbol{\mathcal{E}}$. The drift velocity $\langle v_g \rangle$ is calculated in the same fashion as in Section 8.3:

$$\langle \boldsymbol{v}_g \rangle = -\frac{e_c}{4\pi^3 n_c} \int \tau(E_{\boldsymbol{k}}) \frac{\partial f_0}{\partial E_{\boldsymbol{k}}} \boldsymbol{v}_g (\boldsymbol{v}_g \cdot \boldsymbol{\mathcal{E}}_1) d^3 k. \tag{8.230}$$

Eliminating $\partial f_0 / \partial E_k$ with the aid of Eq. (8.73) and restricting ourselves to spherical parabolic bands and the classical limit,

$$\langle \boldsymbol{v}_g \rangle = \frac{e_c \beta}{3\pi^2 n m^*} \left(\frac{2m^*}{\hbar^2} \right)^{3/2} \int \tau(E) f_0 E^{3/2} \boldsymbol{\mathcal{E}}_1 dE. \tag{8.231}$$

Introducing the expression for $\boldsymbol{\mathcal{E}}_1$ gives the result

$$\langle \boldsymbol{v}_g \rangle = \frac{e_c}{m^*} \left[\left(\boldsymbol{\mathcal{E}} - \frac{1}{e_c} \boldsymbol{\nabla}_{\boldsymbol{r}} E_F \right) \langle \tau \rangle - \left(\frac{\langle \tau E \rangle - E_F \langle \tau \rangle}{e_c T} \right) \boldsymbol{\nabla}_{\boldsymbol{r}} T \right], \tag{8.232}$$

where $\langle \tau \rangle$ is defined in Eq. (8.79) and

$$\langle \tau E \rangle = \frac{\int \tau(E) f_0 E^{5/2} dE}{\int f_0 E^{3/2} dE}. \tag{8.233}$$

Equation (8.232) can be rewritten in terms of the mobility μ as

$$\langle \boldsymbol{v}_g \rangle = \left(\frac{e_c}{e} \right) \mu \left[\boldsymbol{\mathcal{E}} - \frac{1}{e_c} \boldsymbol{\nabla}_{\boldsymbol{r}} E_F - \alpha \boldsymbol{\nabla}_{\boldsymbol{r}} T \right], \tag{8.234}$$

where α is the **thermoelectric power** given by

$$\alpha = \frac{1}{e_c T} \left(\frac{\langle \tau E \rangle}{\langle \tau \rangle} - E_F \right). \tag{8.235}$$

A physical interpretation of the thermoelectric power is provided by a sample in a temperature gradient under open-circuit conditions, so that the current density $\boldsymbol{j}$ and hence $\langle \boldsymbol{v}_g \rangle$ are zero. Under these conditions the sample remains in equilibrium; hence, E_F is uniform throughout the sample and $\nabla_r E_F = 0$. Equation (8.229) then yields

$$\alpha = \frac{\mathcal{E}}{\nabla_r T}. \tag{8.236}$$

In other words, the thermoelectric power is the electric field produced by unit temperature gradient when $\boldsymbol{j} = 0$.

An explicit expression for α can be obtained if $\tau(E) \sim E^{-s}$ and we assume that the carriers obey classical statistics. Evaluation of $\langle\tau\rangle$ and $\langle\tau E\rangle$ yields

$$\alpha = \frac{k_B}{e_c}\left(\frac{5}{2} - s - \frac{E_F}{k_B T}\right). \tag{8.237}$$

Note that the sign of α depends on the sign of the charge carriers. Measurements of thermoelectric power are therefore useful in determining the conductivity type (n or p) of a semiconductor sample. Having obtained the thermoelectric power, the Peltier coefficient can be calculated using the Kelvin relation.

8.7.2 Thermoelectric devices

A very familiar device based on the thermoelectric effect is the **thermocouple**, which consists of a pair of junctions involving two dissimilar metals. The junctions are maintained at different temperatures T_1 and T_2. As a result of the temperature gradient, a potential difference $\Delta\varphi$ exists between the two junctions given by $\Delta\varphi = (\alpha_2 - \alpha_1)(T_2 - T_1)$, where α_1 and α_2 are the thermoelectric powers of the two metals. By maintaining one junction at a fixed reference temperature, the temperature of the other junction can be measured.

In the case of semiconductors a thermoelectric device that has received considerable attention is the **thermoelectric refrigerator**. It is based on the Peltier effect. Under open circuit conditions we use Eq. (8.236) to eliminate $\mathcal{E}$ from Eq. (8.223b) and write

$$Q = \kappa(ZT - 1)\nabla_r T, \tag{8.238}$$

where $Z = \alpha^2/\kappa\rho$ and $\rho = 1/\sigma$. The quantity Z is the **thermoelectric figure of merit**. In a refrigerator heat is pumped from the cold region to the warm region, so one wants ZT large. Materials with small thermal conductivity and electrical resistivity, but large thermoelectric power, are desirable. At the present time these conditions are best fulfilled at room temperature by Bi_2Te_3/Sb_2Te_3 alloys with carrier concentration $\sim 10^{19}\ \text{cm}^{-3}$.

Thermoelectric refrigerators are highly reliable and offer the convenience of portability. They have found applications for cooling infrared detectors and central processors in computers as well as for such mundane things as beverage storage. Reversing the direction of the electric field or current converts the refrigerator into a heater.

8.8 Thermal conductivity

In Chapter 7 the conduction of heat by phonons was discussed. We now discuss the contribution of free carriers to heat conduction as described by Fourier's law. The heat flux associated with the motion of charge carriers of energy E is given by

$$\boldsymbol{Q} = \langle E n_c \boldsymbol{v}_g \rangle, \tag{8.239}$$

and is equal to the heat energy flowing in unit time through a unit area that is perpendicular to the temperature gradient. On comparing this result with that for $\langle \boldsymbol{v}_g \rangle$ in Eq. (8.232), we see that

$$\boldsymbol{Q} = \frac{n_c e_c}{m^*}\left[\left(\boldsymbol{\mathcal{E}} - \frac{1}{e_c}\nabla_r E_F\right)\langle \tau E \rangle - \left(\frac{\langle \tau E^2 \rangle - E_F \langle \tau E \rangle}{e_c T}\right)\boldsymbol{\nabla}_r T\right], \tag{8.240}$$

where

$$\langle \tau E^2 \rangle = \frac{\int \tau(E) f_0 (1 - f_0) E^{7/2} dE}{\int f_0 E^{3/2} dE}. \tag{8.241}$$

Thermal conductivity is generally measured under conditions of zero current and hence zero $\langle \boldsymbol{v}_g \rangle$. Setting $\langle v_q \rangle = 0$ in Eq. (8.232) and eliminating $\boldsymbol{\mathcal{E}}$ from Eq. (8.240) yields the result

$$\boldsymbol{Q} = \frac{n_c}{m^*}\left[\frac{\langle \tau E \rangle^2 - \langle \tau \rangle \langle \tau E^2 \rangle}{\langle \tau \rangle T}\right]\boldsymbol{\nabla}_r T. \tag{8.242}$$

Comparing this result with Fourier's law, Eq. (7.75), we obtain the carrier contribution to the thermal conductivity

$$\kappa = \frac{n_c[\langle \tau \rangle \langle \tau E^2 \rangle - \langle \tau E \rangle^2]}{m^* \langle \tau \rangle T}. \tag{8.243}$$

Recalling that the electrical conductivity σ is $n_c e^2 \langle \tau \rangle / m^*$, we can re-express κ as

$$\kappa = (k_B/e)^2 \mathcal{L} \sigma T, \tag{8.244}$$

where $\mathcal{L}$ is a dimensionless quantity known as the **Lorentz number** given by

$$\mathcal{L} = \frac{\langle \tau \rangle \langle \tau E^2 \rangle - \langle \tau E \rangle^2}{(k_B T)^2 \langle \tau \rangle^2} \tag{8.245}$$

For a nondegenerate semiconductor with $\tau(E) = aE^{-s}$,

$$\langle \iota E^n \rangle = a(k_B T)^n \Gamma\left(\tfrac{5}{2} + n - s\right)/\Gamma\left(\tfrac{5}{2}\right). \tag{8.246}$$

Since $\Gamma(x+1) = x\Gamma(x)$, we find that $\mathcal{L} = \frac{5}{2} - s$.

To the electronic contribution must be added the lattice vibrational contribution to the thermal conductivity. Except at low temperatures, the lattice contribution varies as $1/T^x$ with $1 < x < 2$ (Ashcroft and Mermin 1976). It typically dominates the electronic contribution except when the carrier concentration is sufficiently high due to high doping levels or high temperatures.

8.9 Semi-insulating semiconductors

In the modern microelectronics industry considerable use is made of devices consisting of a thin layer of semiconductor grown epitaxially on a substrate. The substrate is generally the same basic material as the thin layer. However, in order for the components grown on the substrate not to be short circuited, the substrate must have a very high resistivity, i.e., it must be semi-insulating. The industrial problem is to produce the semi-insulating substrate in the least expensive way with the easiest technology. Let us consider a possible procedure for attaining this objective with GaAs.

The electrical resistivity of a semiconductor is the reciprocal of the electrical conductivity and is given by

$$\rho = (ne\mu_e + pe\mu_h)^{-1}, \tag{8.247}$$

where μ_e is the mobility of electrons with concentration n and μ_h is the mobility of holes with concentration p. For GaAs the mobilities at room temperature are (Table 8.1): $\mu_e = 8000\,\mathrm{cm^2/(V\,s)}$ and $\mu_h = 300\,\mathrm{cm^2/(V\,s)}$. We now consider several scenarios that may lead to high resistivity.

8.9.1 Pure GaAs

Pure GaAs is a possible candidate for a high resistivity material because it contains no donor or acceptor impurities that can augment the carrier concentrations beyond the intrinsic value n_i. The latter is specified by Eq. (6.17). For GaAs the energy gap is 1.43 eV at 300 K and the effective masses are 0.07 m for electrons (Table 4.2) and 0.7 m for heavy holes (Table 4.3). To simplify the calculation we neglect the contribution of light holes, since their concentration is small compared to that of heavy holes (cf. Eqs. (6.30)–(6.34)). The intrinsic carrier concentration is then found from Eq. (6.17) to be $n_i = 2.6 \times 10^6\,\mathrm{cm^{-3}}$. The resistivity is then given by

$$\rho = [n_i e(\mu_e + \mu_h)]^{-1} \tag{8.248}$$

which yields the result $\rho = 3.2 \times 10^8\,\Omega\,\mathrm{cm}$. This is a very high resistivity.

8.9.2 Impure GaAs: shallow impurities

Technically it is impossible to produce GaAs with an impurity concentration lower than $10^{14}\,\mathrm{cm^{-3}}$. One can assume that the residual impurities are shallow donors and that the impurity potential is Coulomb-like. The impurity ionization energy E_d is given by the effective Rydberg (Eq. (5.10)) with $Z = 1$. For GaAs with dielectric constant $\epsilon = 11$ one finds that

$E_d \simeq 8 \times 10^{-3}$ eV which is smaller than the value 26×10^{-3} eV of $k_B T$ at 300 K. The characteristic concentration n_c defined by Eqs. (6.27) and (6.57) is found to be 1.7×10^{17} cm^{-3}, which far exceeds the residual impurity concentration n_d of 10^{14} cm^{-3}. Essentially all the impurities are ionized in accordance with Eq. (6.56), and so $n \simeq n_d$. The resistivity is then given by

$$\rho = \frac{1}{n_d e \mu_e}, \tag{8.249}$$

so $\rho = 7.8\,\Omega$ cm. This value is far lower than that of pure GaAs and raises the question of how to reduce the effect of the residual impurities in order to obtain semi-insulating material.

8.9.3 Impure GaAs: deep impurities

A possible procedure for increasing the resistivity of GaAs containing residual shallow impurities is to introduce deep impurities that will bring the Fermi energy down to near mid-gap. Such a deep impurity for GaAs is chromium which has a doubly degenerate energy level situated 0.7 eV below the conduction band edge. It is therefore almost precisely in the middle of the forbidden gap. Electrons in shallow donor states will drop down to the Cr level and thereby lower the Fermi energy to the vicinity of that level. The concentration of conduction electrons therefore decreases and the resistivity increases.

The conduction electrons originate either from the valence band, the Cr levels of concentration n_{Cr}, or the shallow donors of concentration n_d. The concentration of electrons trapped on the Cr levels is $P_1^{Cr} n_{Cr}$, where P_1^{Cr} is given by Eq. (6.46). Similarly, the concentration of electrons trapped on the shallow donors is $P_1^{sd} n_d$. The condition of charge neutrality can be expressed as

$$p + (1 - P_1^{Cr}) n_{Cr} + (1 - P_1^{sd}) n_d = n + P_1^{Cr} n_{Cr} + P_1^{sd} n_d. \tag{8.250}$$

Since the Cr concentration is the highest of all, $n_{Cr} \simeq 10^{17}$ cm^{-3} $> n_d, n, p$, this equation can be simplified to

$$(1 - P_1^{Cr}) = P_1^{Cr} \tag{8.251}$$

or $P_1^{Cr} = 0.5$. Replacing P_1 in Eq. (6.46) by this result and E_I^d by E_{Cr}, the energy of the Cr level, we see that $E_{Cr} - E_F \simeq 1/\beta = k_B T$. At room temperature $k_B T$ is small compared to E_g, so $E_F \simeq E_{Cr}$. One says that the Fermi energy is **pinned** at E_{Cr}, which is close to the center of the gap.

The free electron and hole concentrations given by Eqs. (6.10) and (6.14) can be re-expressed as

$$n = \left(\frac{m_c^*}{m}\right)^{3/2} \bar{N}_0 e^{\beta(E_{Cr} - E_c)} \tag{8.252}$$

$$p = \left(\frac{m_v^*}{m}\right)^{3/2} \bar{N}_0 e^{\beta(E_V - E_{Cr})}, \tag{8.253}$$

where

$$\bar{N}_0 = 2(mk_BT/2\pi\hbar^2)^{3/2}. \tag{8.254}$$

Substituting the values of the parameters for GaAs and taking $T = 300\,\text{K}$, we find that $n \simeq 2.5 \times 10^5\,\text{cm}^{-3}$ and $p \simeq 5 \times 10^6\,\text{cm}^{-3}$, so $\rho = 1.5 \times 10^5\,\Omega\,\text{cm}$. This value of ρ is four orders of magnitude larger than that for GaAs with residual shallow impurities. Thus, the addition of chromium makes possible the production of high resistivity material from a semiconductor that necessarily contains impurities and has a relatively low resistivity. The high-resistivity material is a **semi-insulating semiconductor.** It can be used as a substrate upon which can be grown epitaxial layers of good crystalline quality whose electrical properties can be controlled by appropriate doping with shallow impurities.

8.10 Hot carrier phenomena

In our discussion of electrical conductivity we have so far assumed that the applied electric field is weak in the sense that the Boltzmann distribution of the carriers is perturbed only to first order in the field. The drift velocity of a carrier is proportional to the field and the mobility is independent of the field. As the applied field is increased, the quasi-equilibrium of the carriers with the lattice cannot be maintained. Consequently, the carriers acquire a higher effective temperature than the lattice temperature. The scattering of the carriers by phonons increases due to their higher temperature. Their mobility therefore decreases.

8.10.1 Distribution function in high electric fields

To develop a quantitative treatment of high electric field effects we turn to the Boltzmann equation whose first-order solution is given by Eq. (8.60). A solution valid for higher fields can be expressed as a power series in $\mathcal{E}_x$:

$$f = f_0 - \frac{e_c}{e}\bar{\mu}\mathcal{E}_x\frac{\partial f_0}{\partial v_x} + \bar{\mu}^2\mathcal{E}_x^2\frac{\partial^2 f_0}{\partial v_x^2} - \frac{e_c}{e}\bar{\mu}^3\mathcal{E}_x^3\frac{\partial^3 f_0}{\partial v_x^3} + \cdots \tag{8.255}$$

where $\bar{\mu} = e\tau/m^*$. The drift velocity v_d is given by

$$\begin{aligned} v_d &\equiv \langle v_{gx}\rangle \\ &= \left(\int v_x[f_0 - (e_c/e)\bar{\mu}\mathcal{E}_x(\partial f_0/\partial v_x) + \bar{\mu}^2\mathcal{E}_x^2(\partial^2 f_0/\partial v_x^2) \right. \\ &\quad \left. - (e_c/c)\bar{\mu}^3\mathcal{E}_x^3(\partial^3 f_0/\partial v_x^3) + \cdots]d^3v\right)\Big/\int fd^3v. \end{aligned} \tag{8.256}$$

Noting that $f_0(v)$ is an even function of v_x, we obtain

$$v_d = \frac{e_c}{e}\mu_0(1 - \beta_2\bar{\mu}^2\mathcal{E}_x^2 + \cdots)\mathcal{E}_x, \tag{8.257}$$

where μ_0 is the low field mobility and

$$\beta_2 = \frac{\int(\partial^2 f_0/\partial v_x^2)d^3v}{\int f_0 d^3v} - \frac{\int v_x(\partial^3 f_0/\partial v_x^3)d^3v}{\int v_x(\partial f_0/\partial v_x)d^3v} \tag{8.258}$$

with corresponding definitions for higher β_n. Only terms even in $\mathcal{E}_x$ appear inside the parentheses in Eq. (8.257). Writing $v_d = \mu(\mathcal{E}_x)\mathcal{E}_x$, where $\mu(\mathcal{E}_x)$ is the field-dependent mobility, we see that

$$\mu(\mathcal{E}_x) = \mu_0\left(1 - \beta_2\bar{\mu}^2\mathcal{E}_x^2 + \cdots\right). \tag{8.259}$$

Since the coefficient β_2 is positive, the mobility decreases with increasing electric field, as shown in Fig. 8.9 for n-Ge.

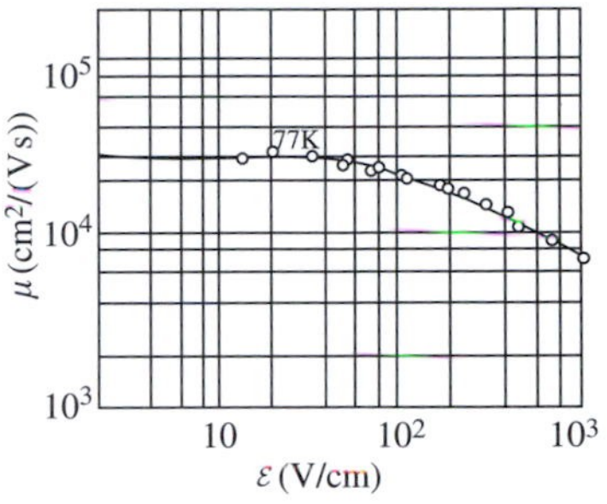

Fig. 8.9
Mobility as a function of applied electric field for n-Ge (after Conwell 1953).

An alternative procedure replaces the perturbed Boltzmann distribution by an equilibrium distribution with a carrier temperature T_e higher than the lattice temperature T. The difference between T_e and T can be related to the increase in drift velocity Δv_d in the direction of the field by

$$\Delta E \simeq m^* v_{d0}\Delta v_d = e\tau\mu_0\left(\beta_2\mathcal{E}_x^4 + \cdots\right) = \tfrac{3}{2}k_B(T_e - T), \tag{8.260}$$

where $v_{d0} = \mu_0\mathcal{E}_x$, $\Delta v_d = v_d - v_{d0}$, and the quadratic term in Δv_d has been neglected. If the only correction term needed involves β_2, the carriers are warm, whereas if terms with higher-order coefficients are needed, the carriers are hot. The field-dependent mobility can be written in the form

$$\mu(\mathcal{E}_x) = \mu_0[1 + (T_e - T)\zeta(T) + \cdots], \tag{8.261}$$

where $\zeta(T)$ involves only the lattice temperature and the scattering mechanism.

8.10.2 Gunn effect

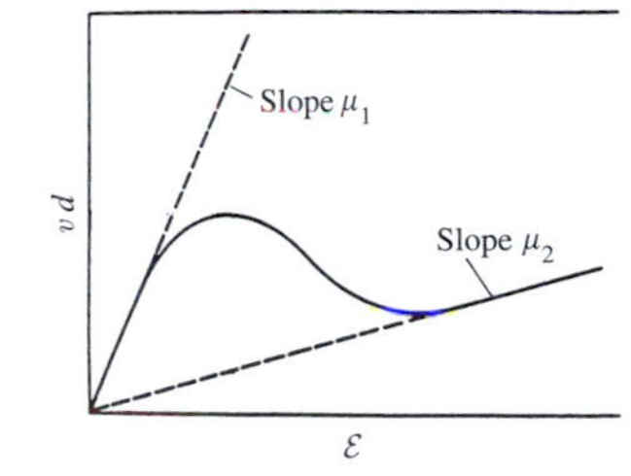

Fig. 8.10
Drift velocity versus electric field.

In certain semiconductors such as GaAs, the conduction band minimum at the Γ-point may be lower by only a small amount of energy than that of a subsidiary minimum at the L-point. In GaAs this energy difference is 0.36 eV. A conduction electron accelerated by a high applied electric field can acquire sufficient energy to make a transition from the vicinity of the Γ-point to the vicinity of the L-point. The effective mass is higher near the L-point, and the mobility is lower. In fact a field range can exist in which the drift velocity **decreases** with increasing field as shown in Fig. 8.10. The differential conductivity is negative, an instability can arise, and current oscillations can occur. These oscillations are called **Gunn oscillations** after the person who first observed them (Gunn 1963).

The physical origin of Gunn oscillations can be described as follows. Associated with an instability is a fluctuation in electric field as shown in Fig. 8.11a. This high-field **domain** drifts with the current carriers through the crystal. Most of the carriers inside the domain are in the L-point valley, while those outside the domain are primarily in the Γ-point valley. The moving leading edge of the domain overtakes light Γ-electrons which, because of the high electric field, are converted to heavy L-electrons. The latter have low mobility and accumulate at the trailing edge of the domain. The electron concentration profile is shown in Fig. 8.11b. Eventually the domain disappears into the anode and a new domain is generated at the cathode. The periodic appearance and disappearance of domains constitute Gunn oscillations. If v_d is the drift velocity and L is the sample

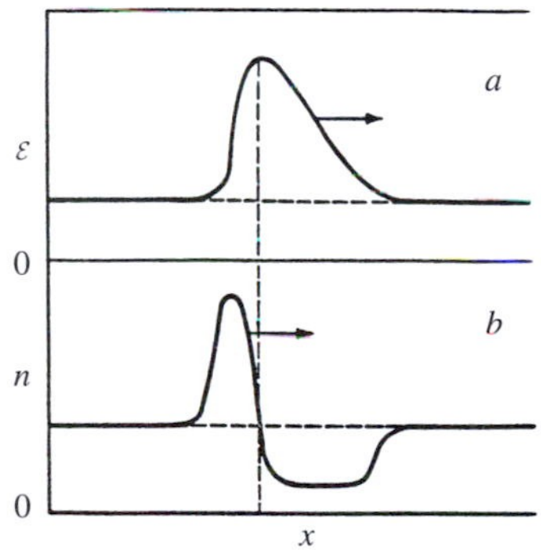

Fig. 8.11
Electric field and electron concentration versus position for a domain moving to the right (after Butcher *et al.* 1966).

length, the frequency ν_G of Gunn oscillations is v_d/L. For $v_d \simeq 10^5$ m/s and $L = 10\,\mu$m, $\nu_G \simeq 10$ GHz which lies in the microwave region. Microwave generators based on the Gunn effect are known as **Gunn diodes** and have found applications in radar and telecommunications.

8.10.3 Field ionization

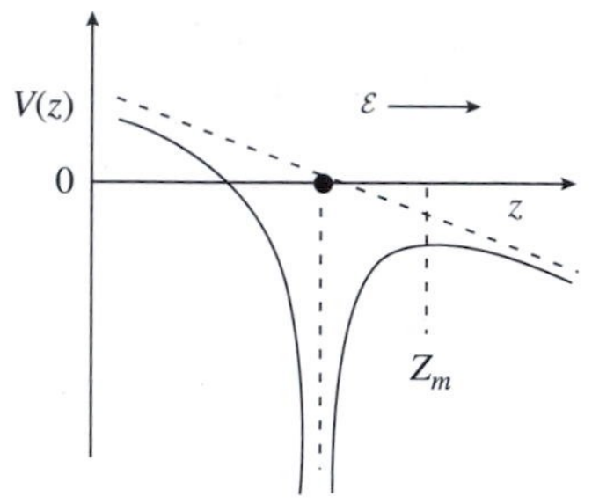

Fig. 8.12
Impurity potential versus position in the presence of an electric field.

The application of an external electric field to a semiconductor containing donor or acceptor impurities tilts the energy bands and leads to a lower ionization energy of the impurities. For a donor impurity with a pure Coulomb potential placed in a constant external field, the potential energy as a function of position is given by

$$V(\mathbf{r}) = -\frac{e^2 Z}{4\pi\epsilon_s\epsilon_0 r} - e\mathcal{E}z, \tag{8.262}$$

where the field $\mathcal{E}$ is in the z-direction. The potential energy along the z-axis is plotted in Fig. 8.12. In the positive z-direction the potential starts from $-\infty$ at $z = 0$, arises to a maximum and then decreases as z increases. The maximum is determined by $dV/dz = 0$ and occurs at z_m given by

$$z_m = \left(\frac{eZ}{4\pi\epsilon_s\epsilon_0\mathcal{E}}\right)^{\frac{1}{2}}. \tag{8.263}$$

The potential energy at the maximum is

$$V_m = -e\left(\frac{eZ\mathcal{E}}{\pi\epsilon_s\epsilon_0}\right)^{\frac{1}{2}}. \tag{8.264}$$

In zero external field the maximum potential energy is zero at $z = \infty$. In the presence of the field the height of the barrier that traps the electron is lowered by the amount

$$\delta E = |V_m| = e\left(\frac{eZ\mathcal{E}}{\pi\epsilon_s\epsilon_0}\right)^{\frac{1}{2}}, \tag{8.265}$$

and the probability of thermal ionization is enhanced by the factor $\exp(\beta\delta E)$.

> **Exercise.** A critical field $\mathcal{E}_{cr}$ for thermal ionization can be defined by setting $\delta E = k_B T$. Calculate $\mathcal{E}_{cr}$ for P-doped Si at 300 K if $\epsilon_s = 12$.
> **Answer.** 1.3×10^4 V/cm.

8.10.4 Impact ionization

If an external electric field is applied to a semiconductor containing free carriers, the carriers are accelerated and gain energy. If the energy gained exceeds the ionization energy of an impurity atom and the free carrier passes sufficiently close to the impurity, energy can be transferred from the free carrier to a carrier trapped at the impurity, thereby exciting the trapped carrier to the conduction or valence band. The new free carrier thus produced can itself be accelerated and cause the ionization of yet another impurity. The number of free carriers multiplies, thus leading to **avalanche** formation associated with a sharply increasing current. At very high current densities dielectric breakdown can occur, which results in structural changes in the material. In narrow-gap semiconductors it is possible for an accelerated carrier to acquire enough energy to excite an electron from the valence band to the conduction band with the creation of a pair of free carriers.

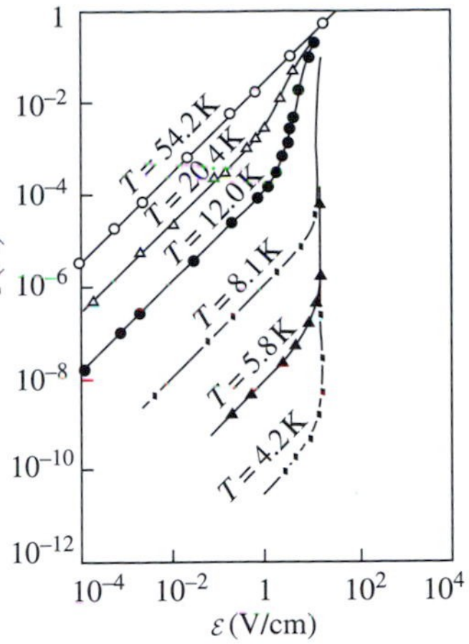

Fig. 8.13
Current versus electric field for n-Ge (after Lautz 1961).

8.10.4.1 Impact ionization of shallow impurities

The observation of impact ionization is facilitated by operating at low temperatures where the impurity states are well occupied, the mobility is high, and the mean free path is long. In n-Ge at temperatures below 10 K in fields on the order of 10 V/cm, impact ionization occurs as shown in Fig. 8.13. We note that as the temperature rises, the low-field part of the curve rises as a result of thermal ionization of the impurities.

8.10.4.2 Impact ionization by interband transitions

In the absence of phonon interactions an accelerated carrier must have an energy at least equal to the band gap to produce electron–hole pairs by impact ionization. If the details of the conservation of energy and momentum are considered, however, one finds that the threshold energy can be larger than the band gap.

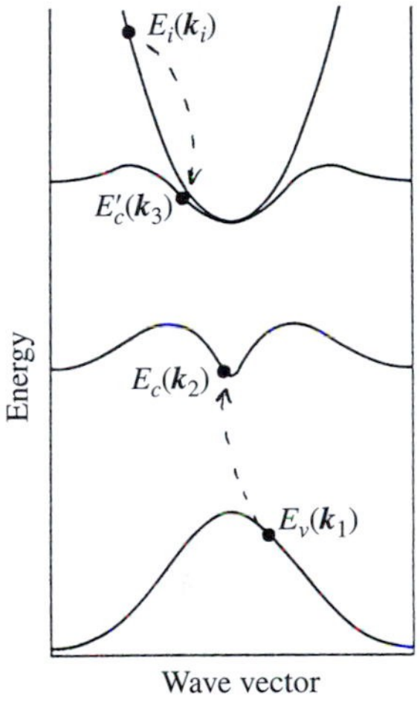

Fig. 8.14
Diagram for an impact-ionization process (after Anderson and Crowell 1972).

Let us consider the situation shown in Fig. 8.14. In the initial state there is an initiating electron in conduction band i with wave vector $\boldsymbol{k}_i$ and an electron in valence band v with wave vector $\boldsymbol{k}_1$. In the final state there is an electron in conduction band c with wave vector $\boldsymbol{k}_2$, an electron in conduction band c' with wave vector $\boldsymbol{k}_3$, and a hole replacing the electron in the valence band. The basic problem in obtaining the threshold energy is to find the minimum energy of the system consistent with given values of the energy $E_i(\boldsymbol{k}_i)$ and wave vector $\boldsymbol{k}_i$ of the initiating electron.

Conservation of energy and momentum give the relations

$$E_i(\boldsymbol{k}_i) = E_c(\boldsymbol{k}_2) + E_{c'}(\boldsymbol{k}_3) - E_v(\boldsymbol{k}_1) \tag{8.266a}$$

$$\boldsymbol{k}_i = \boldsymbol{k}_2 + \boldsymbol{k}_3 - \boldsymbol{k}_1. \tag{8.266b}$$

To minimize the energy of the final particles we take the variation of Eqs. (8.266) and obtain

$$0 = d\boldsymbol{k}_2 \cdot \boldsymbol{v}_2 + d\boldsymbol{k}_3 \cdot \boldsymbol{v}_3 - d\boldsymbol{k}_1 \cdot \boldsymbol{v}_1 \tag{8.267a}$$

$$0 = d\boldsymbol{k}_2 + d\boldsymbol{k}_3 - d\boldsymbol{k}_1, \tag{8.267b}$$

where we have used the definition of the group velocity given by Eq. (8.51). Eliminating $d\boldsymbol{k}_1$ from Eq. (8.267a) yields

$$0 = d\boldsymbol{k}_2 \cdot (\boldsymbol{v}_2 - \boldsymbol{v}_1) + d\boldsymbol{k}_3 \cdot (\boldsymbol{v}_3 - \boldsymbol{v}_1). \tag{8.268}$$

Since $d\boldsymbol{k}_2$ and $d\boldsymbol{k}_3$ are linearly independent, we must have

$$\boldsymbol{v}_1 = \boldsymbol{v}_2 = \boldsymbol{v}_3. \tag{8.269}$$

Therefore, a necessary condition that an initiating particle have minimum energy consistent with production of electron–hole pairs is that the final particles all have the same group velocity (Anderson and Crowell 1972).

The calculation of the threshold energy for a realistic band structure must be done numerically. However, for the case of spherical parabolic conduction and valence bands, a simple treatment is possible. The group velocity is then $\hbar k/m^*$, so the equality of the group velocities gives

$$\frac{k_1}{m_v^*} = \frac{k_2}{m_c^*} = \frac{k_3}{m_c^*} \tag{8.270}$$

and

$$\boldsymbol{k}_i = (2+\gamma)\boldsymbol{k}_2, \tag{8.271}$$

where $\gamma = m_v^*/m_c^*$ and m_v^* has been taken to be positive. Using Eq. (8.266a) the energy of the initiating electron takes the form

$$E_i(\boldsymbol{k}_i) = \frac{\hbar^2 k_2^2}{2m_c^*}(2+\gamma) + E_g, \tag{8.272}$$

where the zero of energy has been taken to be the bottom of the conduction band. Alternatively, we can write $E_i(k_i)$ as

$$E_i(k_i) = \frac{\hbar^2 k_i^2}{2m_c^*} = \frac{\hbar^2 k_2^2}{2m_c^*}(2+\gamma)^2. \tag{8.273}$$

Eliminating $\hbar^2 k_2^2/2m_c^*$ from Eqs. (8.272) and (8.273) yields the threshold energy $E_{th,e}$:

$$E_{th,e} = E_i(k_i) = \frac{(2+\gamma)}{(1+\gamma)}E_g. \tag{8.274}$$

If the effective masses of the conduction and valence bands are equal, $\gamma = 1$ and one obtains the "$\frac{3}{2}$-band gap" rule $E_{th,e} = \frac{3}{2}E_g$.

When the initiating particle is a hole one can show that the threshold energy $E_{th,h}$ is given by

$$E_{th,h} = \frac{(1+2\gamma)}{(1+\gamma)}E_g, \tag{8.275}$$

which also reduces to the $\frac{3}{2}$-band gap rule if $\gamma = 1$.

To obtain reliable results for real semiconductors it is necessary to take into account the detailed band structure as well as phonon contributions and umklapp processes. In a great many cases the factor multiplying E_g exceeds $\frac{3}{2}$ (Anderson and Crowell 1972).

8.11 Variable-range hopping conductivity

As was pointed out in the last section of Chapter 5, impurities or other traps for carriers in a semiconductor form a disordered system in which both localized and delocalized states appear that are separated by the mobility edge. If the Fermi energy of the electrons lies below the mobility edge in a three-dimensional, n-type semiconductor, the conductivity at low temperatures occurs by **variable-range hopping** (Mott 1969). The rate of hopping of an electron between two traps a distance R apart is proportional to the wave function overlap factor $\exp(-2\alpha R)$, where α is an inverse effective Bohr radius. In addition there is a temperature-dependent factor arising from an activation energy. The latter can be estimated by noting that if the electron jumps a distance no greater than R, the number of states in the range dE is $(4\pi R^3/3)N(E)dE$, so that the average spacing between energies is $W = 3/4\pi R^3 N(E)$. The activation energy is taken to be W. The hopping frequency ν_{hop} is then given by

$$\nu_{hop} = \nu_0 e^{-(2\alpha R + W/k_B T)}. \tag{8.276}$$

The maximum frequency ν^m_{hop} is obtained by varying R to give $2\alpha = (9/4\pi)/R^4 N(E)k_B T$. Eliminating R from Eq. (8.276) yields ν^m_{hop}. The conductivity, which is proportional to ν^m_{hop}, takes the form

$$\sigma(T) = A\exp(-B/T^{\frac{1}{4}}). \tag{8.277}$$

This behavior is exhibited by amorphous Ge at low temperatures.

Problems

1. The inelastic character of acoustic phonon scattering can be taken into account by replacing the energy conserving delta function in Eq. (8.149) by $\delta(E_{\boldsymbol{k}-\boldsymbol{q}} - E_{\boldsymbol{k}} + \hbar\omega_{\boldsymbol{q}})$, where $\omega_{\boldsymbol{q}}$ is the frequency of the emitted acoustic phonon. Derive an expression for the inverse scattering time associated with the emission of acoustic phonons. Repeat the derivation for the case of acoustic phonon absorption.
2. Carry out a derivation analogous to that of problem 1 for the case of polar optical phonon scattering.
3. Homopolar semiconductors have contributions to the inverse scattering time that arise from a deformation potential-type interaction with optical phonons. The interaction Hamiltonian has the form

$$H_{int} = \boldsymbol{D}_0 \cdot \boldsymbol{u},$$

   where $\boldsymbol{D}_0$ is the interaction coefficient and $\boldsymbol{u}$ is the relative displacement of the two atoms in the unit cell. Derive expressions for the inverse scattering time for optical phonon emission and absorption both for the elastic scattering approximation and for the inelastic case.
4. For an intrinsic semiconductor, show that the thermoelectric power is given by the expression

$$\alpha = \frac{\sigma_n\alpha_n + \sigma_p\alpha_p}{\sigma_n + \sigma_p},$$

   where the subscripts n and p refer to conduction electrons and holes, respectively.

References

C. L. Anderson and C. R. Crowell, *Phys. Rev.* B**5**, 2267 (1972).
N. W. Ashcroft and N. D. Mermin, *Solid State Physics* (Holt, Rinehart and Winston, New York, 1976).
J. Birman, M. Lax, and R. Loudon, *Phys. Rev.* **145**, 620 (1966).
H. Brooks, in *Advances in Electronics and Electron Physics*, Vol. 7, ed. L. Marton (Academic Press, New York, 1955).
P. N. Butcher, W. Fawcett, and C. Hilsum, *Brit. J. Appl. Phys.* **17**, 841 (1966).
H. B. Callen, *Phys. Rev.* **76**, 1394 (1949).
E. M. Conwell, *Phys. Rev.* **90**, 769 (1953).
E. Conwell and V. Weisskopf, *Phys. Rev.* **77**, 388 (1950).
P. Debye and E. Hückel, *Physik. Z.* **24**, 305 (1923).
H. Ehrenreich, *J. Phys. Chem. Solids* **2**, 131 (1957).
A. Einstein, *Ann. Physik* **17**, 549 (1905).
E. Fermi, *Nuclear Physics* (Univ. Chicago Press, Chicago, 1950).
D. K. Ferry, *Phys. Rev.* B**14**, 1605 (1976).
J. B. Gunn, *Solid State Commun.* **1**, 88 (1963).
F. S. Ham, *Phys. Rev.* **100**, 1251 (1955).
C. Herring and E. Vogt, *Phys. Rev.* **101**, 944 (1956).
D. Howarth and E. Sondheimer, *Proc. Roy. Soc.* (London) A**219**, 53 (1953).
C. Kittel, *Introduction to Solid State Physics*, Sixth Edition (John Wiley, New York, 1986).
G. Lautz, in *Halbleiterprobleme*, Vol. 6, ed. F. Sauter (Fried. Vieweg & Sohn, Braunschweig, 1961).
A. A. Maradudin, E. W. Montroll, G. H. Weiss, and I. P. Ipatova, *Theory of Lattice Dynamics in the Harmonic Approximation* (Academic Press, New York, 1971).
N. F. Mott, *Phil. Mag.* **19**, 835 (1969).
G. L. Pearson and C. Herring, *Physica* **20**, 975 (1954).
B. K. Ridley, *Quantum Processes in Semiconductors*, Second edition (Clarendon Press, Oxford, 1988).
D. L. Rode, *Phys. Rev.* B**2**, 1012 (1970).
L. M. Roth, in *Handbook on Semiconductors*, Second edition, Vol. 1, ed. P. T. Landsberg (North-Holland, Amsterdam, 1992).
N. Sclar, *Phys. Rev.* **104**, 1548, 1559 (1956).
R. C. Tolman, *Principles of Statistical Mechanics* (Oxford University Press, Oxford, 1938).

Surface properties of semiconductors

9

Key ideas

The *wave function amplitude* of *electronic surface states* decreases strongly from the surface to the interior of the crystal.

The *wave function* of a *surface state* in the *nearly free electron approximation* is constructed from Bloch functions with *complex wave vectors*. The *energy eigenvalue* lies in the *gap* between the energy bands of the bulk crystal.

In the *tight binding method* the surface state wave function is a *linear combination of atomic orbitals* with coefficients that *decay exponentially* toward the interior of the crystal.

Surface elastic waves have displacement amplitudes that *decay exponentially* from the surface to the interior. Their *velocity* is *less* than that of *bulk waves*.

In diatomic semiconductors *surface modes of vibration* can occur with frequencies in the gap between acoustic and optical branches.

The *shell model*, the *bond charge model* and *ab initio methods* can be used to calculate the properties of surface modes in real semiconductors.

Surface vibrational modes can be studied experimentally using *Brillouin* scattering, *Raman* scattering, *electron* scattering, *helium atom* scattering, and *infrared spectroscopy*.

The *recombination* of electrons and holes can be *enhanced* at a surface or interface of a semiconductor.

Surface Properties

The physical boundaries of a semiconductor are found to produce significant effects on its electronic and vibrational properties. If the boundary separates the semiconductor from vacuum or a gas, it is referred to as a **surface**. If the semiconductor is separated from a liquid or a solid, the boundary is called an **interface**. Many of the technological applications of semiconductors are based on surface and interface effects. In the present chapter we focus on surface effects and defer our consideration of interface effects to Chapter 15.

9.1 Surface effects on electronic states

As discussed in Chapter 2, the allowed states of an electron moving in a perfect, periodic crystal lie in energy bands with forbidden energy gaps between bands. The qualitative effect of creating a free surface is to introduce **surface states** (Tamm 1932) whose energies lie in the forbidden gaps of the perfect crystal. The wave functions associated with surface states have amplitudes that are large at or near the surface and decay to essentially zero in the interior of the crystal. In two- and three-dimensional crystals, the surface states may form **surface energy bands.**

The nature of the atomic arrangement at and near a surface can vary widely from one type of crystal to another. In the simplest situation, the spacing between adjacent atomic layers retains its bulk value right up to the surface layer, and the atomic arrangement within the surface layer is the same as that in corresponding interior layers. More generally, the interlayer spacing may change as the surface is approached, giving rise to **surface relaxation**. Furthermore, the atomic arrangement within the surface layer may differ from that of a corresponding layer in the bulk as a result of **surface reconstruction**. Both surface relaxation and surface reconstruction occur rather frequently in semiconductor surfaces and can lead to significant modifications of the surface energy bands.

9.1.1 Nearly free electron approximation

A number of the qualitative aspects of surface electronic states can be developed with the aid of the nearly free electron (NFE) approximation and the two-band model. We first consider a one-dimensional semiconductor that has energy eigenvalues specified by Eq. (2.56) with E_k having some real value in the gap between the two allowed bands, but k being complex: $k = \zeta \pm i\alpha$. Solving the resulting equation for the wave vector k, we find that $\zeta = G_0/2$ and hence that

$$k = \tfrac{1}{2} G_0 \pm i\alpha, \tag{9.1}$$

where α is real. The resulting expression for E_k is

$$E_k = V_0 + \frac{\hbar^2}{2m}\left(\frac{G_0^2}{4} - \alpha^2\right) \pm \left[|V_{G_0}|^2 - \left(\frac{\hbar^2 \alpha G_0}{2m}\right)^2\right]^{\frac{1}{2}}. \tag{9.2}$$

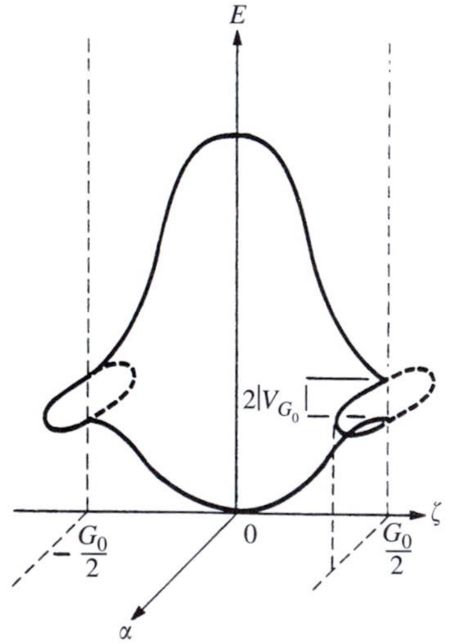

Fig. 9.1
Energy versus complex wave vector for the one-dimensional two-band model.

Exercise. Consider a one-dimensional system with E_k having the value $V_0 + (\hbar^2/2m)(G_0/2)^2$ corresponding to the middle of the gap at $G_0/2$. Calculate the value of α.
Answer. $\alpha = [\tfrac{1}{2}(G_0^4 + 16m^2|V_{G_0}|^2/\hbar^4)^{\frac{1}{2}} - \tfrac{1}{2}G_0^2]^{\frac{1}{2}}$

The behavior of the energy E_k as a function of the complex wave vector k is shown in Fig. 9.1 for the one-dimensional case of the two-band model. The edges of the two bands at the Brillouin zone boundary are connected together by loops oriented along the imaginary axis of the complex wave vector. The factor $\exp(ikx)$ in the Bloch function therefore acquires the factor $\exp(\pm\alpha x)$, which, with proper choice of the algebraic sign, makes possible a surface state.

To illustrate the calculation of the wave function and energy of a surface electronic state, we consider a one-dimensional monatomic crystal with lattice constant a. The potential energy $V(x)$ is assumed to have the form

$$V(x) = V_0 + 2V_{G_0}\cos(G_0 x), \tag{9.3}$$

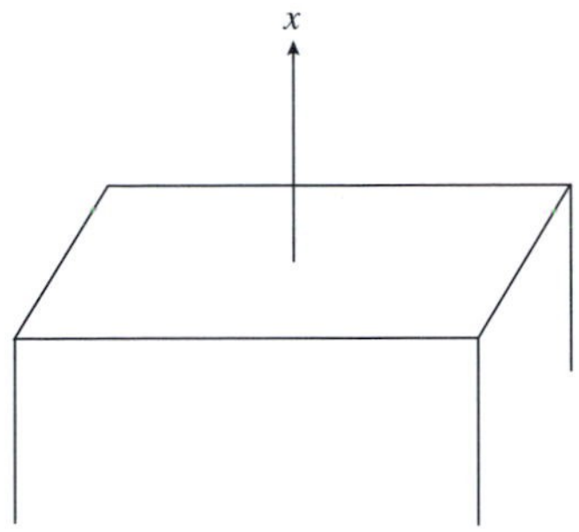

Fig. 9.2
Geometry of a crystal with a surface.

where both V_0 and V_{G_0} are real and negative. The crystal occupies the region $x \le 0$, and vacuum occupies the region $x > 0$, as shown in Fig. 9.2. The vacuum level is taken to be the zero of energy.

We take the wave function of the electron within the crystal to be a linear combination of Bloch functions with complex wave vector such that the wave function decays exponentially in the direction away from the vacuum into the crystal. Within the two-band model the appropriate linear combination follows from Eq. (2.26) by retaining only the terms having $G = 0$ and $G = G_0$,

$$\psi(x) = e^{\alpha x + i(G_0/2)x}[C(k) + C(k - G_0)e^{-iG_0 x}], \tag{9.4}$$

with $k = \frac{1}{2}G_0 - i\alpha$ and $x < 0$. The wave function in the vacuum region is also localized at the interface with the crystal and can be written as

$$\psi(x) = \psi_0 e^{-\alpha_0 x}, \quad x > 0. \tag{9.5}$$

The quantities α and α_0 are the decay constants describing the decay of the wave function from the surface into the crystal and into the vacuum, respectively.

The boundary conditions at the interface require that $\psi(x)$ and its first derivative $\psi'(x)$ be continuous. Using Eqs. (9.4) and (9.5) leads to the equations

$$\psi_0 = C(k) + C(k - G_0) \tag{9.6a}$$

$$-\alpha_0\psi_0 = \left(\alpha + i\frac{G_0}{2}\right)C(k) + \left(\alpha - i\frac{G_0}{2}\right)C(k - G_0). \tag{9.6b}$$

Eliminating ψ_0 yields the equation

$$\left(\alpha + \alpha_0 + i\frac{G_0}{2}\right)C(k) + \left(\alpha + \alpha_0 - i\frac{G_0}{2}\right)C(k - G_0) = 0. \tag{9.7}$$

The complex conjugate of the last result is

$$\left(\alpha + \alpha_0 - i\frac{G_0}{2}\right)C^*(k) + \left(\alpha + \alpha_0 + i\frac{G_0}{2}\right)C^*(k - G_0) = 0, \tag{9.8}$$

which, upon comparison with Eq. (9.7), shows that one can take

$$C^*(k) = C(k - G_0). \tag{9.9}$$

Expressions satisfying this condition are

$$C(k) = \bar{C}e^{i\delta} \tag{9.10a}$$

$$C(k - G_0) = \bar{C}e^{-i\delta}, \tag{9.10b}$$

where $\bar{C}$ is a real normalization constant and δ is to be determined. Substitution of Eqs. (9.10) into Eq. (9.7) gives

$$\alpha + \alpha_0 = \frac{G_0}{2}\tan\delta. \tag{9.11}$$

The quantity δ is a measure of the difference in phase of the wave function components with amplitudes $C(k)$ and $C(k - G_0)$.

Another relation between α and δ can be obtained by substituting Eqs. (9.10) into Eqs. (2.53) and eliminating $E_{\boldsymbol{k}} - V_0$. The result is

$$\alpha = -\frac{2mV_{G_0}}{\hbar^2 G_0}\sin 2\delta, \tag{9.12}$$

where V_{G_0} has been taken to be real in accordance with Eq. (9.3).

To determine δ we take the boundary condition expressed by Eq. (9.11) and rewrite it as

$$-\frac{2mE}{\hbar^2} = \left(\alpha - \frac{G_0}{2}\tan\delta\right)^2 \tag{9.13}$$

where we have used the fact that in the vacuum region E and α_0 are related by

$$E = -\frac{\hbar^2\alpha_0^2}{2m}. \tag{9.14}$$

An additional relation is provided by substituting Eqs. (9.10) into Eq. (2.53a) and eliminating E_k using Eq. (9.2):

$$\left\{E_k^{(0)} - \frac{\hbar^2}{2m}\left(\frac{G_0^2}{4} - \alpha^2\right) \mp \left[|V_{G_0}|^2 - \left(\frac{\hbar^2}{2m}\right)^2\right]^{\frac{1}{2}}\right\}e^{i\delta} + V_{G_0}e^{-i\delta} = 0. \tag{9.15}$$

The unperturbed energy $E_k^{(0)}$ for k given by $\frac{1}{2}G_0 - i\alpha$ is

$$E_k^{(0)} = \frac{\hbar^2}{2m}\left(\frac{G_0}{2} - i\alpha\right)^2. \tag{9.16}$$

Substituting into Eq. (9.15) and taking the real part of the resulting equation yields

$$\alpha G_0 \tan\delta = \frac{2m}{\hbar^2}\left\{-V_{G_0} \pm \left[|V_{G_0}|^2 - \left(\frac{\hbar^2\alpha G_0}{2m}\right)^2\right]^{\frac{1}{2}}\right\}. \quad (9.17)$$

If we take Eq. (9.13) and eliminate both E and $\alpha G_0 \tan\delta$, the result is

$$\sec^2\delta = -\frac{8m}{\hbar^2 G_0^2}(V_0 + V_{G_0}), \quad (9.18)$$

which specifies δ in terms of the potential energy coefficients and the zone-boundary wave vector. The decay constants α and α_0 can now be calculated from Eqs. (9.11) and (9.12) and the surface state energy eigenvalue from Eq. (9.14).

The energy eigenvalue E_s as a function of the coefficient V_{G_0} is plotted in Fig. 9.3 for $V_0 = -1$ in units of $\hbar^2 G_0^2/8m$. The boundaries of the forbidden gap are also shown. The surface state energy lies in the lower part of the gap and moves farther from the midpoint as V_{G_0} increases. A plot of the decay constants α and α_0 versus V_{G_0} is given in Fig. 9.4.

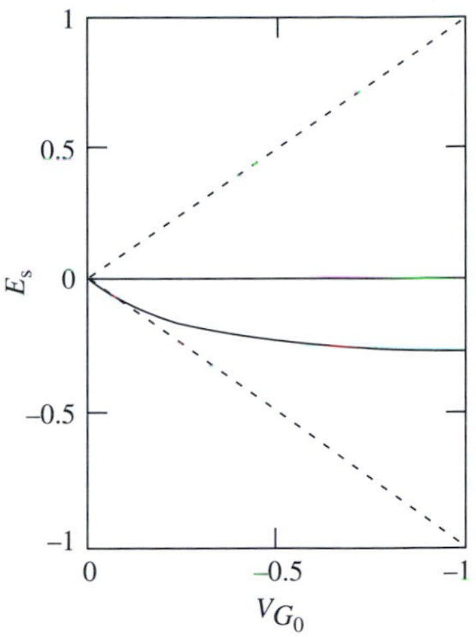

Fig. 9.3
Surface state energy versus Fourier coefficient of the potential energy for $V_0 = -1$. Dashed lines: boundaries of the forbidden gap.

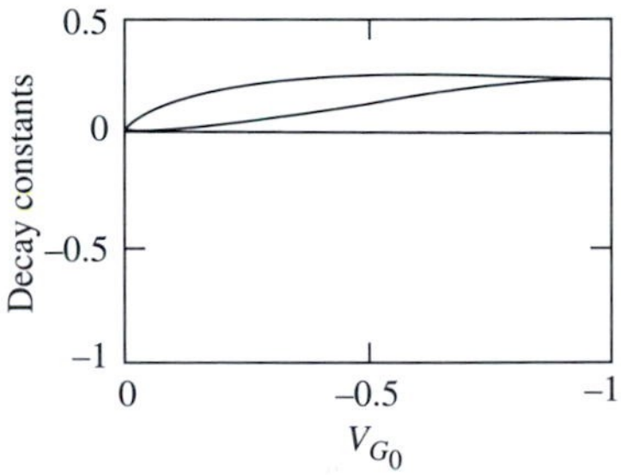

Fig. 9.4
Decay constants versus Fourier coefficient of the potential energy for $V_0 = -1$. Upper curve: α_0/V_{G_0}; lower curve: α/V_{G_0}.

The one-dimensional case just considered can readily be extended to the three-dimensional case (Goodwin 1939). If the surface coincides with a principal lattice plane and is normal to the reciprocal lattice vector $\boldsymbol{G}_0$ the key equations are simple generalizations of those for one dimension. We introduce a wave vector $\boldsymbol{k}$ given by

$$\boldsymbol{k} = \boldsymbol{k}_\perp + \boldsymbol{k}_\parallel, \quad (9.19)$$

where $\boldsymbol{k}_\perp$ is perpendicular to the surface and $\boldsymbol{k}_\parallel$ is a two-dimensional vector parallel to the surface. Similarly, we introduce an electron position vector $\boldsymbol{r}$ by

$$\boldsymbol{r} = \boldsymbol{r}_\perp + \boldsymbol{r}_\parallel. \quad (9.20)$$

The expression for the wave function in Eq. (9.4) is replaced by

$$\psi(\boldsymbol{r}) = e^{i\boldsymbol{k}\cdot\boldsymbol{r}}\left[C(\boldsymbol{k}) + C(\boldsymbol{k} - \boldsymbol{G}_0)e^{i\boldsymbol{G}_0\cdot\boldsymbol{r}}\right] \quad (9.21)$$

with $\boldsymbol{k}_\perp$ specified by Eq. (9.1). Similarly, Eq. (9.5) is replaced by

$$\psi(\boldsymbol{r}) = \psi_0 e^{-\alpha_0 r_\perp + i\boldsymbol{k}_\parallel\cdot\boldsymbol{r}_\parallel}. \quad (9.22)$$

Key equations (9.11), (9.12), and (9.18) are unchanged, but the term $\hbar^2 k_\parallel^2/2m$ must be added to the right hand side of Eqs. (9.2), (9.14) and (9.16). Thus, when $\boldsymbol{G}_0$ is normal to the surface, the net effect is simply to superpose the translational kinetic energy parallel to the surface on the energy of the motion normal to the surface. When $\boldsymbol{G}_0$ is not normal to the surface, the situation is more complicated, but can be handled by the methods just described.

9.1.2 Tight binding method

Fig. 9.5
One-dimensional model of silicon.

We have seen in Chapter 2 that the tight binding method provides a simple, qualitative picture of how energy bands vary as the lattice constant is changed. It also proves useful in the discussion of surface states (Davison and Steslicka 1992).

Let us consider a one-dimensional model of silicon (or other diamond structure material) along the [111] direction as shown in Fig. 9.5. There are two different separations between adjacent atoms that alternate along the chain. We assume that a plane normal to [111] cuts the chain midway between two adjacent atoms in neighboring unit cells. The orbital at a lattice site is taken to be an sp hybrid of the form

$$\varphi_{\pm}(\mathbf{r}) = \frac{1}{\sqrt{2}}(|s\rangle \pm |p_z\rangle), \tag{9.23}$$

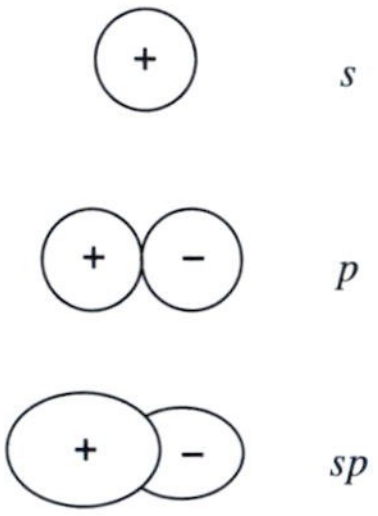

Fig. 9.6
Hybrid orbitals in a one-dimensional model of silicon.

where the $+$ and $-$ signs alternate from one site to the next and the chain is aligned along the z-direction. The orientation of the orbitals along the chain is indicated in Fig. 9.6.

The Hamiltonian has the tight-binding form

$$H = -\frac{\hbar^2\nabla^2}{2m} + \sum_{\ell\kappa} v(\mathbf{r} - \mathbf{R}_{\ell\kappa}) \tag{9.24}$$

where $v(\mathbf{r} - \mathbf{R}_{\ell\kappa})$ is the atomic potential energy at site $\ell\kappa$. Since the system under study is not periodic, the Bloch form for the wave function in Eq. (2.74) is no longer appropriate. We write a solution to the Schrödinger equation

$$H\psi(\mathbf{r}) = E\psi(\mathbf{r}) \tag{9.25}$$

as an expansion

$$\psi(\mathbf{r}) = \sum_{\ell\kappa} c_{\ell\kappa}\varphi_{\kappa}(\mathbf{r} - \mathbf{R}(\ell\kappa)), \tag{9.26}$$

where the coefficients $c_{\ell\kappa}$ are to be determined. The parameter κ takes on the values $+$ and $-$. We choose the origin of coordinates such that $\mathbf{R}(\ell\kappa) = 0$ for $\ell = 0$ and $\kappa = +$.

Substituting the expansion for $\psi(\mathbf{r})$ into the Schrödinger equation, multiplying the result by $\varphi_+^*(\mathbf{r})$, and integrating over $\mathbf{r}$, we obtain

$$\sum_{\ell\kappa} c_{\ell\kappa} \int \varphi_+^*(\mathbf{r}) H \varphi_{\kappa}(\mathbf{r} - \mathbf{R}(\ell\kappa)) d^3r$$
$$= E \sum_{\ell\kappa} c_{\ell\kappa} \int \varphi_+^*(\mathbf{r}) \varphi_{\kappa}(\mathbf{r} - \mathbf{R}(\ell\kappa)) d^3r \tag{9.27}$$

It is convenient to rewrite the Hamiltonian as

$$H = H_0 + \sum_{\ell'\kappa' \neq 0+} v(\mathbf{r} - \mathbf{R}(\ell'\kappa')), \tag{9.28}$$

where

$$H_0 = -\frac{\hbar^2\nabla^2}{2m} + v(\boldsymbol{r}), \tag{9.29}$$

$$H_0\varphi_\pm(\boldsymbol{r}) = \frac{1}{\sqrt{2}}\left[\epsilon_s|s\rangle \pm \epsilon_p|p\rangle\right], \tag{9.30}$$

and ϵ_s and ϵ_p are the energy eigenvalues of the s- and p-orbitals, respectively. With these results Eq. (9.27) becomes

$$\begin{aligned}\sum_{\ell\kappa} c_{\ell\kappa} \int \frac{1}{\sqrt{2}}[\epsilon_s\langle s| + \epsilon_p\langle p|]\varphi_\kappa(\boldsymbol{r} - \boldsymbol{R}(\ell\kappa))d^3r \\ + \sum_{\ell\kappa}\sum_{\ell'\kappa'\neq 0+} c_{\ell\kappa}\int \varphi_+^*(\boldsymbol{r})v(\boldsymbol{r} - \boldsymbol{R}(\ell'\kappa'))\varphi_\kappa(\boldsymbol{r} - \boldsymbol{R}(\ell\kappa))d^3r \\ = E\sum_{\ell\kappa} c_{\ell\kappa}\int \varphi_+^*(\boldsymbol{r})\varphi_\kappa(\boldsymbol{r} - \boldsymbol{R}(\ell\kappa))d^3r.\end{aligned} \tag{9.31}$$

This equation can be simplified by exploiting the localized character of the atomic orbitals $\varphi_\kappa(\boldsymbol{r})$ and the atomic potentials $v(\boldsymbol{r})$. We make the approximations

$$\int \frac{1}{\sqrt{2}}[\epsilon_s\langle s| + \epsilon_p\langle p|]\varphi_\kappa(\boldsymbol{r} - \boldsymbol{R}(\ell\kappa))d^3r = \tfrac{1}{2}(\epsilon_s + \epsilon_p)\delta_{\ell 0}\delta_{\kappa +} \tag{9.32a}$$

$$\int \varphi_\kappa^*(\boldsymbol{r} - \boldsymbol{R}(\ell\kappa))\varphi_{\kappa'}(\boldsymbol{r} - \boldsymbol{R}(\ell'\kappa'))d^3r = \delta_{\ell\ell'}\delta_{\kappa\kappa'} \tag{9.32b}$$

and retain only those terms containing $v(\boldsymbol{r})$ that involve a single atomic site or two adjacent atomic sites.

Equation (9.31) then reduces to

$$\left[E - \tfrac{1}{2}(\epsilon_s + \epsilon_p) - J - J'\right]c_{0+} = Kc_{0-} + K'c_{-1,-}, \tag{9.33}$$

where

$$J = \int \varphi_+^*(\boldsymbol{r})v(\boldsymbol{r} - \boldsymbol{R}(-))\varphi_+(\boldsymbol{r})d^3r \tag{9.34a}$$

$$J' = \int \varphi_+^*(\boldsymbol{r})v(\boldsymbol{r} + \boldsymbol{a} - \boldsymbol{R}(-))\varphi_+(\boldsymbol{r})d^3r \tag{9.34b}$$

$$K = \int \varphi_+^*(\boldsymbol{r})v(\boldsymbol{r} - \boldsymbol{R}(-))\varphi_-(\boldsymbol{r} - \boldsymbol{R}(-))d^3r \tag{9.34c}$$

$$K' = \int \varphi_+^*(\boldsymbol{r})v(\boldsymbol{r} + \boldsymbol{a} - \boldsymbol{R}(-))\varphi_-(\boldsymbol{r} + \boldsymbol{a} - \boldsymbol{R}(-))d^3r \tag{9.34d}$$

and $\boldsymbol{a}$ is the primitive translation vector. Similarly, for an arbitrary value of the lattice site index ℓ, we have

$$\left[E - \tfrac{1}{2}(\epsilon_s + \epsilon_p) - J - J'\right]c_{\ell+} = Kc_{\ell-} + K'c_{\ell-1,-}. \tag{9.35}$$

The integrals J and J' are called **Coulomb integrals** because they represent the Coulomb energies of a charge distribution $\rho(\boldsymbol{r}) = |\phi_+(r)|^2$ if $v(\boldsymbol{r})$ is the potential of a point charge. The integrals K and K', on the other hand, are **exchange integrals** because the electron is "exchanged" between two different atomic sites.

A second equation can be obtained by multiplying the Schrödinger equation by $\varphi^*_-(\boldsymbol{r} - \boldsymbol{R}(-))$ and repeating the same steps as before. The result for arbitrary ℓ is

$$\left[E - \tfrac{1}{2}(\epsilon_s + \epsilon_p) - J - J'\right]c_{\ell-} = K' c_{\ell+1,+} + K c_{\ell+}. \tag{9.36}$$

Introducing

$$X = \left[E - \tfrac{1}{2}(\epsilon_s + \epsilon_p) - J - J'\right]/K \tag{9.37a}$$

$$\eta = K'/K, \tag{9.37b}$$

leads to the set of equations

$$X c_{\ell+} = c_{\ell-} + \eta c_{\ell-1,-} \tag{9.38a}$$

$$X c_{\ell-} = \eta c_{\ell+1,+} + c_{\ell+}. \tag{9.38b}$$

Equations (9.38) describe a periodic crystal if ℓ takes on all positive and negative integer values. A solution can be found by setting

$$c_{\ell+} = u^{\ell} \tag{9.39a}$$

$$c_{\ell-} = A u^{\ell} \tag{9.39b}$$

and substituting into Eqs. (9.38) to yield

$$X u^{\ell} = A(u^{\ell} + \eta u^{\ell-1}) \tag{9.40a}$$

$$A X u^{\ell} = \eta u^{\ell+1} + u^{\ell}. \tag{9.40b}$$

Eliminating A from these equations, we find that

$$u = \frac{1}{2\eta}\left\{X^2 - (1+\eta^2) \pm [((1+\eta^2) - X^2)^2 - 4\eta^2]^{\frac{1}{2}}\right\}. \tag{9.41}$$

Since we expect to get Bloch states for the wave function of the infinite crystal, we let

$$u = e^{i\theta}, \tag{9.42}$$

where $\theta = ka$ and k is the wave vector. Inverting Eq. (9.41) gives

$$X = \pm\left[1 + \eta^2 + 2\eta\cos\theta\right]^{\frac{1}{2}}, \tag{9.43}$$

or with the aid of Eq. (9.37a),

$$E = \tfrac{1}{2}(\epsilon_s + \epsilon_p) + J + J' \pm K[1 + \eta^2 + 2\eta\cos\theta]^{\frac{1}{2}}. \tag{9.44}$$

The energy eigenvalues are thus seen to lie in two bands corresponding to the + and − signs with θ in the range $0 \le \theta \le \pi$. A plot of X versus θ is shown in Fig. 9.7 for $\eta = \frac{1}{9}$.

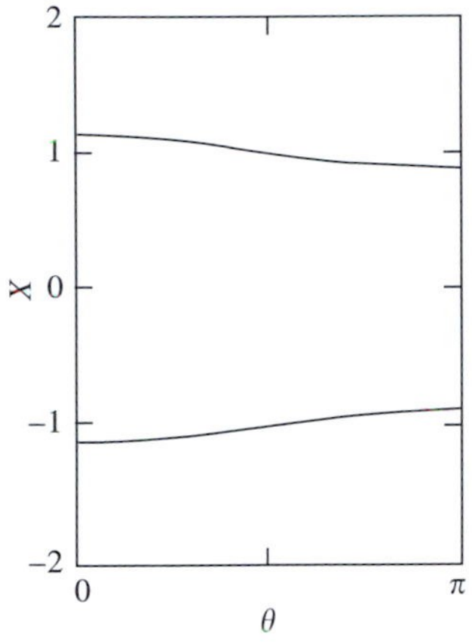

Fig. 9.7
Energy parameter X versus reduced wave vector θ for $\eta = \frac{1}{9}$.

We now turn to the question of surface electronic states of an atomic chain with *sp*-hybrid orbitals. As mentioned earlier, the "surface" is created in our one-dimensional model by setting to zero the parameters that couple two adjacent atomic sites. Let these sites be 0+ and −1, −. Equations (9.38) for $\ell = 0$ take the form

$$X' c_{0+} = c_{0-} \tag{9.45a}$$

$$X c_{0-} = \eta c_{1+} + c_{0+}, \tag{9.45b}$$

where

$$X' = X + \xi \tag{9.46a}$$

$$\xi = J'/K. \tag{9.46b}$$

Eliminating c_{0-} from Eqs. (9.45) yields

$$X^2 - 1 + X\xi = \eta \frac{c_{1+}}{c_{0+}}. \tag{9.47}$$

For a surface state it is necessary that the amplitudes $c_{\ell+}$ and $c_{\ell-}$ tend to zero as $\ell \to \infty$. Such behavior can be achieved if we take θ in Eq. (9.42) to have the form $\theta = \pi + i\alpha$. Then

$$u = e^{i(\pi + i\alpha)} = -e^{-\alpha}, \tag{9.48}$$

where α is the **decay constant**. From Eq. (9.39a) we obtain

$$c_{\ell+} = (-1)^\ell e^{-\ell\alpha}, \tag{9.49}$$

which has the desired limit for large ℓ. Equations (9.43) and (9.47) then become

$$X = \pm[1 + \eta^2 - 2\eta\cosh\alpha]^{\frac{1}{2}} \tag{9.50}$$

$$X^2 - 1 + X\xi = -\eta e^{-\alpha}. \tag{9.51}$$

Eliminating X from these equations gives

$$\eta - e^\alpha \pm [1 + \eta^2 - 2\eta\cosh\alpha]^{\frac{1}{2}} \frac{\xi}{\eta} = 0, \tag{9.52}$$

which is the **surface boundary condition** that specifies α for given values of ξ and η. To have a surface state, α must be real and positive or complex with positive real part. Equation (9.52) is a cubic equation in e^{α} whose solution can be expressed analytically. After α is determined, the amplitudes $c_{\ell+}$ and energy parameter X can be calculated from Eqs. (9.49) and (9.50), respectively.

An expression for the amplitudes $c_{\ell-}$ follows from Eqs. (9.39b) and (9.48):

$$c_{\ell-} = A(-1)^{\ell} e^{-\ell\alpha}. \tag{9.53}$$

The constant A is determined by substituting Eqs. (9.49) and (9.53) into Eq. (9.38b) to give the result

$$A = (1 - \eta e^{-\alpha})/X. \tag{9.54}$$

A special case arises if $\xi = 0$. From Eq. (9.52) we see that $\eta = e^{\alpha}$ and from Eq. (9.50) that $X = 0$. Equations (9.38a) and (9.45a) then lead to the conclusion that all $c_{\ell-}$ are zero. The surface state for this case is therefore associated with $X = 0$ and $\eta > 1$.

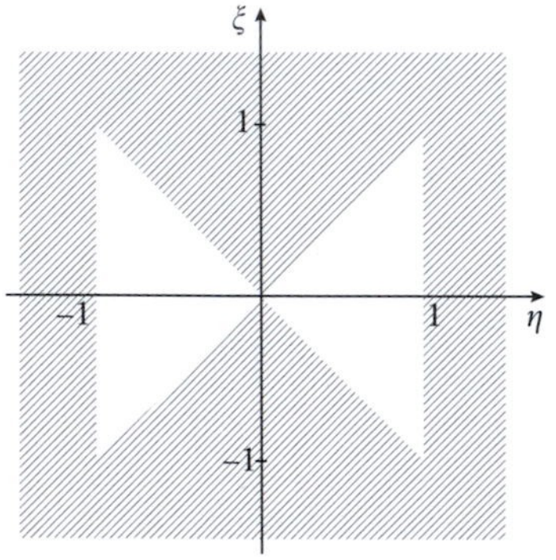

Fig. 9.8
Ranges of ξ and η (shaded region) for a surface state. The shaded region extends indefinitely beyond that shown.

Only certain ranges of ξ and η correspond to values of α that are consistent with a surface state if $|\eta| < 1$. These ranges are shown in Fig. 9.8. The energy parameter X_s for the surface state is plotted versus η in Fig. 9.9 for $\xi = 2/3$ and $\eta < 1$.

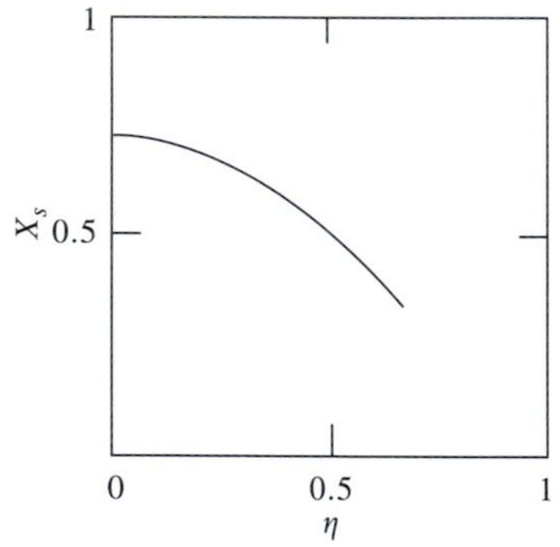

Fig. 9.9
Energy parameter X_s versus η for $\xi = 2/3$. The curve ends at $\eta = 2/3$ where $\alpha = 0$. For $\eta > 1$ there is an additional branch of the curve with $X_s = 0$.

In treating real semiconductors one must take into account the three-dimensional character of these materials. The chains of atoms parallel to the [111] direction are coupled together as a result of the valence bonds coupling atoms in adjacent chains. In the absence of coupling, the surface states of the various chains are degenerate. Introduction of the coupling splits the degeneracy and leads to **surface state bands**. Their theoretical analysis is conveniently carried out with the aid of Green's functions (Garcia-Moliner and Velasco 1994). Photoemission is a powerful tool for investigating surface states experimentally and is discussed in Chapter 10.

As a result of the rupture of chemical bonds that occurs in the creation of a semiconductor surface, the surface atoms have **dangling bonds** containing a single electron each. There is a tendency for a pair of adjacent dangling bands to pair up their lone electrons to form electron-pair bonds. Consequently, forces arise that act on the surface atoms to distort the surface geometry and produce **surface reconstruction**. A reconstructed surface is shown in Fig. 9.10 for Si(100) and is characterized by a surface unit cell that is larger than that of the unreconstructed surface. If the latter is Si(100) and is subjected to suitable heat treatment, the Si(100) 2×1 reconstructed surface arises as illustrated in Fig. 9.10. The pattern exhibits Si dimers that result from the pairing up of dangling bonds.

Experimentally, a valuable technique for determining surface structure is **low energy electron diffraction** (LEED). Electrons of low energies (20–200 eV) have de Broglie wavelengths on the order of typical lattice spacings and penetrate only a few atomic layers into the crystal. They exhibit diffraction patterns that are characteristic of the surface geometry.

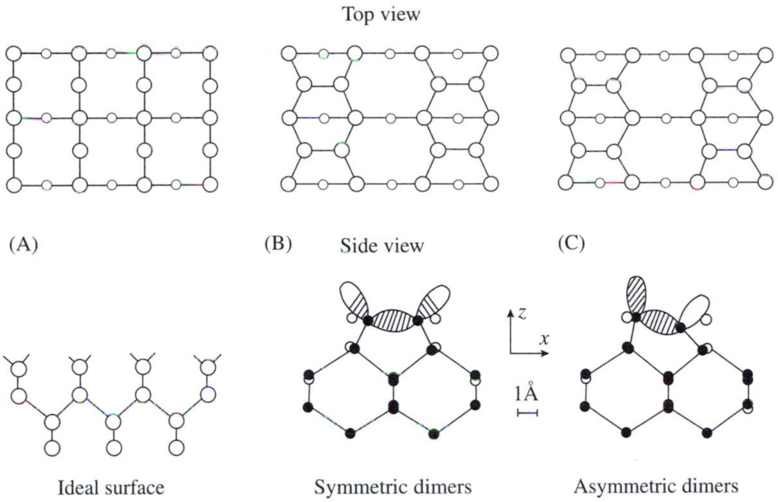

Fig. 9.10
Top and side views of Si(100) surfaces. (A) Ideal 1×1; (B) symmetric dimers (unstable); and (C) asymmetric dimers. Small circles (top view) are third row positions. Open circles (side view) are positions that would be occupied by the atoms if they were in the bulk. Hatching indicates degree of filling of surface bonds (after Le Lay 1987).

9.2 Surface effects on lattice vibrations

As pointed out in Chapter 7, typical semiconductors have both acoustic and optical modes of vibration as a consequence of having more than one atom per unit cell. If a surface is present, surface modes can arise from both acoustic and optical branches. In the following we treat each of these types of modes.

9.2.1 Surface acoustic modes

Surface modes derived from the acoustic branch can occur with wavelengths varying from much greater than the lattice constant to being comparable to the lattice constant. We start by considering the long wavelength case which can be handled by the rather simple but general procedures of elastic continuum theory.

We restrict our attention to cubic crystals, the class to which the more common semiconductors belong. Denoting the displacement components of a material point at (x, y, z) by (u, v, w), we can write the equations of motion as

$$\rho\frac{\partial^2 u}{\partial t^2} = \frac{\partial}{\partial x}\left(C_{11}\frac{\partial u}{\partial x} + C_{12}\frac{\partial v}{\partial y} + C_{12}\frac{\partial w}{\partial z}\right) + C_{44}\left[\frac{\partial}{\partial y}\left(\frac{\partial u}{\partial y} + \frac{\partial v}{\partial x}\right) + \frac{\partial}{\partial z}\left(\frac{\partial u}{\partial z} + \frac{\partial w}{\partial x}\right)\right] \tag{9.55a}$$

$$\rho\frac{\partial^2 v}{\partial t^2} = C_{44}\frac{\partial}{\partial x}\left(\frac{\partial u}{\partial y} + \frac{\partial v}{\partial x}\right) + \frac{\partial}{\partial y}\left(C_{12}\frac{\partial u}{\partial x} + C_{11}\frac{\partial v}{\partial y} + C_{12}\frac{\partial w}{\partial z}\right) + C_{44}\frac{\partial}{\partial z}\left(\frac{\partial v}{\partial z} + \frac{\partial w}{\partial y}\right) \tag{9.55b}$$

$$\rho\frac{\partial^2 w}{\partial t^2} = C_{44}\left[\frac{\partial}{\partial x}\left(\frac{\partial u}{\partial z}+\frac{\partial w}{\partial x}\right)+\frac{\partial}{\partial y}\left(\frac{\partial v}{\partial z}+\frac{\partial w}{\partial y}\right)\right]$$
$$+\frac{\partial}{\partial z}\left(C_{12}\frac{\partial u}{\partial x}+C_{12}\frac{\partial v}{\partial y}+C_{11}\frac{\partial w}{\partial z}\right), \qquad (9.55c)$$

where ρ is the density and C_{11}, C_{12}, C_{44} are the cubic elastic moduli in the Voigt notation. It should be noted that the elastic moduli actually form a fourth-rank tensor that leads to anisotropy of the elastic properties. The velocity of propagation of an acoustic wave therefore depends on the direction of propagation relative to the crystallographic axes.

Let us consider surface waves associated with a free (001) surface at $z = 0$ and seek solutions to the equations of motion of the form

$$(u, v, w) = (U, V, W)e^{-q\alpha z+iq(\ell x+my-ct)}, \qquad (9.56)$$

where U, V, W are amplitudes associated with the decay constant α and q is the wave vector. The wave front in the plane $z = 0$ makes an angle θ with the x-axis specified by $\ell = \cos\theta$, $m = \sin\theta$. Substituting Eq. (9.56) into Eqs. (9.55), one obtains a set of linear homogeneous algebraic equations for the amplitudes U, V, W whose nontrivial solution requires that

$$\begin{vmatrix} g_1\ell^2+m^2-p^2-\alpha^2 & \ell m(g_2+1) & \ell\alpha(g_2+1) \\ \ell m(g_2+1) & \ell^2+g_1m^2-p^2-\alpha^2 & m\alpha(g_2+1) \\ \ell\alpha(g_2+1) & m\alpha(g_2+1) & p^2+g_1\alpha^2-1 \end{vmatrix} = 0, \qquad (9.57)$$

where $g_1 = C_{11}/C_{44}$, $g_2 = C_{12}/C_{44}$, $p^2 = \rho c^2/C_{44}$. The quantity p is the ratio of the surface wave velocity to that of transverse bulk waves propagating in the [100] direction.

Equation (9.57) is a bicubic in α and p. To a given value of p correspond three values of α^2 that we denote by α_j^2, $j = 1, 2, 3$. The displacement components u, v, w given by Eq. (9.56) decrease toward zero as z increases provided the constants α_j are positive real numbers or complex numbers with positive real parts. For a given α_j, the corresponding amplitudes U_j, V_j, iW_j are specified by

$$\frac{U_j}{\xi_j} = \frac{V_j}{\eta_j} = \frac{iW_j}{\zeta_j} = K_j, \quad j = 1, 2, 3 \qquad (9.58)$$

where

$$\xi_j = (\ell^2+g_1m^2-p^2-\alpha_j^2)(p^2+g_1\alpha_j^2-1)-m^2\alpha_j^2(g_2+1)^2 \qquad (9.59a)$$

$$\eta_j = \ell m(g_2+1)[\alpha_j^2(g_2+1-g_1)+1-p^2] \qquad (9.59b)$$

$$\zeta_j = \ell\alpha_j(g_2+1)[m^2(g_2+1-g_1)-\ell^2+p^2+\alpha_j^2] \qquad (9.59c)$$

and the K_j are constants to be determined by the boundary conditions. The general solution for the displacement components is

$$(u, v, iw) = \sum_{j=1,2,3} (\xi_j, \eta_j, \zeta_j) K_j e^{-q\alpha_j z + iq(\ell x + my - ct)}. \tag{9.60}$$

The boundary conditions to be imposed correspond to the vanishing of the three components of stress at the surface $z = 0$. They can be written as (Stoneley 1955)

$$\frac{\partial w}{\partial x} + \frac{\partial u}{\partial z} = 0, \quad \frac{\partial w}{\partial y} + \frac{\partial v}{\partial z} = 0, \tag{9.61a}$$

$$C_{11}\frac{\partial w}{\partial z} + C_{12}\left(\frac{\partial u}{\partial x} + \frac{\partial v}{\partial y}\right) = 0, \tag{9.61b}$$

at $z = 0$. Substitution of Eq. (9.60) into Eqs. (9.61) leads to a set of three linear homogeneous equations in the K_j. Setting the determinant of the coefficients of the K_j to zero yields

$$D(p) \equiv |f_{ij}| = 0, \quad i, j = 1, 2, 3 \tag{9.62}$$

where $f_{1j} = \ell\zeta_j - \alpha_j\xi_j$, $f_{2j} = m\zeta_j - \alpha_j\eta_j$, and $f_{3j} = \ell\xi_j + m\eta_j + (C_{11}/C_{12})\zeta_j$. Equation (9.62) determines the velocity parameter p for given values of the elastic moduli and direction of propagation. The actual calculation of p for a general direction of propagation proceeds as follows. An assumed value of p is used in Eq. (9.57) to obtain the three α_j. The latter are then substituted into Eqs. (9.59) and thence into Eq. (9.62) to obtain a new value of p. This process is repeated until convergence in the value of p is achieved. The resulting solution is a surface wave only if the final values of the α_j have positive real parts.

For illustrative purposes it is worth while to analyze the special case of propagation in the [100] direction on a (001) surface. Then $m = 0$, all $\eta_j = 0$, and $v = 0$. There is no dependence of u and w on y. The displacement vector at a material point lies in the sagittal plane defined by the propagation direction and the surface normal. Equation (9.57) reduces to a 2×2 determinantal equation which upon expansion takes the form

$$(g_1 - p^2 - \alpha^2)(1 - p^2 - g_1\alpha^2) + \alpha^2(g_2 + 1)^2 = 0. \tag{9.63}$$

The two solutions of this equation, α_1^2 and α_2^2, form the basis for the general solution

$$(u, iw) = \sum_{j=1,2} (\xi_j', \zeta_j') K_j' e^{-q\alpha_j z + iq(x - ct)}, \tag{9.64}$$

where

$$\xi_j' = (1 - p^2 - \alpha_j^2)(p^2 + g_1\alpha_j^2 - 1) \tag{9.65a}$$

$$\zeta_j' = \alpha_j(g_2 + 1)(p^2 + \alpha_j^2 - 1). \tag{9.65b}$$

Satisfaction of the two nontrivial boundary condition equations requires that

$$g_2(g_2 + p^2)(1 - p^2) + g_1^2\alpha_1^2\alpha_2^2 + g_1g_2(1 - p^2)(\alpha_1^2 + \alpha_2^2 + \alpha_1\alpha_2) \\ - g_1(g_2 + p^2)\alpha_1\alpha_2 = 0. \tag{9.66}$$

The quantities $\alpha_1^2 + \alpha_2^2$ and $\alpha_1^2\alpha_2^2$ can be taken directly from Eq. (9.63). They are given by

$$\alpha_1^2 + \alpha_2^2 = [g_1(g_1 - p^2) + 1 - p^2 - (g_2 + 1)^2]/g_1 \tag{9.67a}$$

$$\alpha_1^2\alpha_2^2 = (g_1 - p^2)(1 - p^2)/g_1. \tag{9.67b}$$

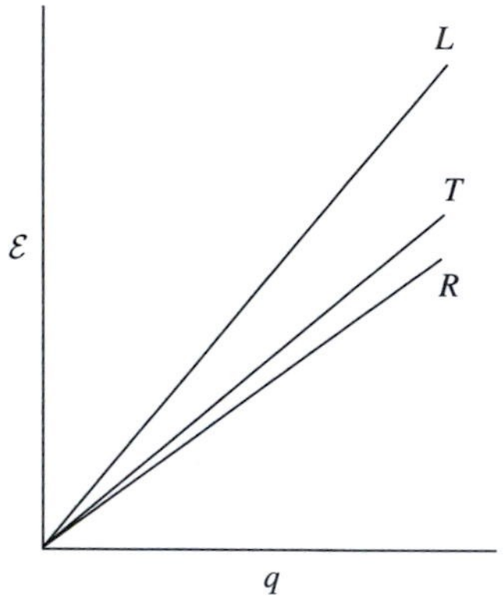

Fig. 9.11
Long wavelength dispersion relations for longitudinal (L) and transverse (T) bulk waves and Rayleigh (R) surface waves for an isotropic elastic solid. The velocities of the waves are given by the slopes of the lines.

Eliminating these quantities from Eq. (9.66) then yields the **surface wave velocity equation**

$$(1 - p^2)(g_1^2 - g_2^2 - g_1p^2)^2 = g_1p^4(g_1 - p^2). \tag{9.68}$$

Only one of the solutions for p^2, when entered into Eqs. (9.67), produces values of q_1 and q_2 that are consistent with surface waves.

In similar fashion one can obtain the equation that specifies the velocity of surface waves propagating in the [110] direction on a (001) surface:

$$(1 - p^2)(g_1g_3 - g_2^2 - g_1p^2)^2 = g_1p^4(g_3 - p^2), \tag{9.69}$$

where $g_3 = \frac{1}{2}(g_1 + g_2 + 2)$.

An important special case is that of an **isotropic elastic solid** for which $C_{11} = C_{12} + 2C_{44}$, or $g_1 = g_2 + 2$. The surface wave velocity equation then reduces to that for **Rayleigh surface waves** (Lord Rayleigh 1887). The velocity of Rayleigh waves is always less than that of both longitudinal and transverse bulk waves as shown in Fig. 9.11. Isotropy implies that the velocities of these three types of waves do not vary with change in the direction of propagation. The two decay constants of the Rayleigh wave, q_1 and q_2 are always real and positive.

In the case of **anisotropic elastic solids**, which includes most real crystals, the qualitative behavior of the surface waves is similar to that of the isotropic case if the elastic moduli lie in a certain range. For surface waves propagating in the [100] direction on a (001) surface, the criterion for Rayleigh-like behavior is to a good approximation (Gazis *et al.* 1960):

$$C_{11} \geq C_{12} + 2C_{44}. \tag{9.70}$$

On the other hand, if

$$C_{11} < C_{12} + 2C_{44}, \tag{9.71}$$

the decay constants q_1 and q_2 form a complex conjugate pair. The imaginary parts of q_1 and q_2 cause the displacement components to decay in an oscillatory fashion going away from the surface instead of monotonically as in the Rayleigh wave case. We refer to the oscillatory decaying surface wave as a **generalized Rayleigh wave**. The boundary between the ordinary and

Table 9.1 Surface wave velocity ratio p for propagation in the [100] and [110] directions on the (001) surface of various semiconductors

	Si	Ge	GaAs	GaSb	InSb	PbS
[100]	0.841	0.826	0.820	0.818	0.804	0.975
[110]	0.867	0.859	0.862	0.858	0.855	0.968

generalized Rayleigh wave regions is very close to the line of isotropy in the space of g_1 and g_2.

Surface waves propagating in the [110] direction on a (001) surface can be investigated in a similar manner (Stoneley 1955, Gazis *et al.* 1960). Both ordinary Rayleigh waves and generalized Rayleigh waves can exist with criteria for their existence very similar to those in Eqs. (9.70) and (9.71).

Semiconductors of the diamond and zincblende structures such as Si, Ge, and GaAs exhibit generalized Rayleigh surface waves. The tetrahedral bonds have high resistance to deformation of the bond angle, thus causing the crystal to have high resistance to shear and a relatively large value of C_{44}. The elastic anisotropy produces a variation of surface wave velocity with direction of propagation. This is illustrated in Table 9.1 which lists the surface wave velocities in the [100] and [110] directions on a (001) surface for several semiconductors. For the diamond and zincblende structure semiconductors considered the surface wave velocity in the [110] direction is larger than that in the [100] direction, whereas the reverse is true in the rock salt structure semiconductor PbS. As the wavelength of a surface wave decreases, the group velocity decreases and typically becomes zero at the surface Brillouin zone boundary. Lattice dynamics rather than elastic continuum theory is required to properly treat this effect. Another result from a lattice dynamical treatment is the existence of gaps in the frequency spectrum within which surface modes of a different character than Rayleigh waves can appear. These gap surface modes are usually highly localized at the surface with the amplitude decaying to negligible values within a few atomic layers of the surface.

9.2.2 Surface optical modes

A monatomic semiconductor such as Si has two atoms per primitive unit cell and consequently both acoustic and optical modes of vibration. There is no absolute gap between the acoustic and optical branches, since the longitudinal acoustic and longitudinal optical branches meet at the L-point. In a diatomic semiconductor such as InP, however, there is an absolute gap as a consequence of the different masses of the two atoms in the primitive unit cell. If a surface is created, the optical mode at the bottom of the optical branch can drop down into the gap and become a **surface optical mode** (Wallis 1957). We demonstrate the existence of such a mode with the aid of a diatomic linear chain of atoms with masses M_1 and M_2, nearest-neighbor harmonic interactions with force constant σ, and a free end as shown in Fig. 9.12. This model describes one-dimensional NaCl along the [111] direction.

Fig. 9.12 Linear diatomic chain of atoms with a free end.

The equations of motion of the atoms are given by

$$M_1\ddot{u}(01) = \sigma[u(02) - u(01)] \tag{9.72a}$$

$$M_2\ddot{u}(02) = \sigma[u(11) - u(02)] + \sigma[u(01) - u(02)] \tag{9.72b}$$

$$M_1\ddot{u}(\ell 1) = \sigma[u(\ell 2) - u(\ell 1)] + \sigma[u(\ell - 1, 2) - u(\ell 1)], \quad \ell \geq 1 \tag{9.72c}$$

$$M_2\ddot{u}(\ell 2) = \sigma[u(\ell + 1, 1) - u(\ell 2)] + \sigma[u(\ell 1) - u(\ell 2)], \quad \ell \geq 1, \tag{9.72d}$$

where $u(\ell\kappa)$ is the displacement of the κth atom in the ℓth unit cell. We assume that $M_1 < M_2$, so the atom at the end of the chain has the lighter mass. Since a surface mode can be expected to have displacements that decrease exponentially from the free end, we take

$$u(\ell 1) = e^{i\ell(\pi + i\alpha) - i\omega t} \tag{9.73a}$$

$$u(\ell 2) = Ae^{i\ell(\pi + i\alpha) - i\omega t}, \tag{9.73b}$$

where α is the decay constant. The constant A is determined from the free end boundary condition that can be expressed as the difference between Eq. (9.72a) and Eq. (9.72c) with $\ell = 0$:

$$u(-1, 2) = u(01). \tag{9.74}$$

Substituting Eqs. (9.73) into the last equation yields

$$A = -e^{-\alpha}. \tag{9.75}$$

Finally, α can be related to the masses M_1 and M_2 by substituting Eqs. (9.73) and (9.75) into Eqs. (9.72c) and (9.72d) and taking the ratio of the resulting equations:

$$e^{-\alpha} = \frac{M_1}{M_2}. \tag{9.76}$$

Since we are assuming $M_1 < M_2$, α is real and positive. The displacements can now be expressed as

$$u(\ell 1) = (-1)^{\ell} \left(\frac{M_1}{M_2}\right)^{\ell} e^{-i\omega t} \tag{9.77a}$$

$$u(\ell 2) = (-1)^{\ell+1} \left(\frac{M_1}{M_2}\right)^{\ell+1} e^{-i\omega t}. \tag{9.77b}$$

A plot of the displacements versus ℓ for $M_2 = 2M_1$ is shown in Fig. 9.13. The surface character of the mode is evident. Note that the mode can be viewed as that of an array of diatomic molecules with their phases chosen to yield no net force acting between molecules.

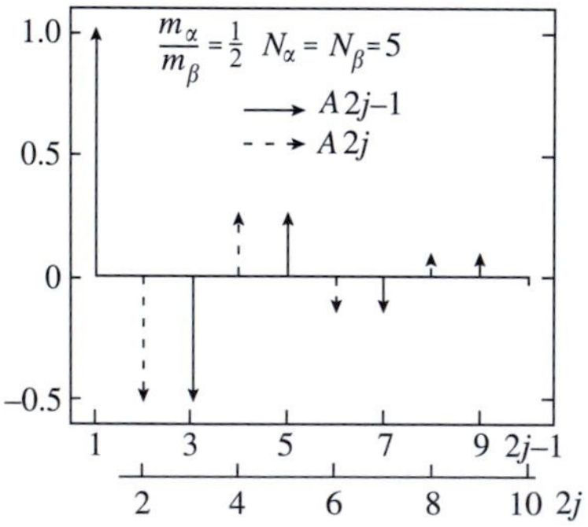

Fig. 9.13
Maximum atomic displacement versus atomic position in the lattice for the surface mode of a diatomic chain.

The surface optical mode frequency ω_s can be obtained by substituting Eqs. (9.77) into any of Eqs. (9.72):

$$\omega_s^2 = \sigma\left(\frac{1}{M_1} + \frac{1}{M_2}\right) = \frac{\sigma}{\bar{M}}. \tag{9.78}$$

Comparison of this result with the frequencies at the edges of the forbidden gap given in Table 7.1 shows that in terms of frequency squared, ω_s^2 lies at the center of the forbidden gap. That the surface mode can have its frequency in the forbidden gap is due to the surface breaking the periodicity of the system.

9.2.3 Surface vibrational modes in real semiconductors

In order to develop a comprehensive picture of surface vibrational modes in real semiconductors, it is necessary to go beyond elasticity theory and one-dimensional models and treat the lattice dynamics problem in three dimensions. For homopolar semiconductors such as Si and Ge, reasonable results can be obtained with the shell model or the bond-charge model. In the case of heteropolar semiconductors such as GaAs, it is necessary to include the Coulomb interaction between ions. On a more fundamental level *ab initio* calculations based on density-functional theory in the local density approximation provide the force constant tensor from which normal mode frequencies and eigenvectors are obtained. Specific results from such approaches will be presented in the following section in connection with experimental data.

9.2.4 Experimental observation of surface vibrational modes

9.2.4.1 Brillouin scattering

Brillouin scattering is the inelastic scattering of light by elementary excitations such as acoustic phonons. Both surface and bulk phonons can be studied. Monochromatic light of frequency ω_i and wave vector $\boldsymbol{k}_i$ is incident on the crystal surface. As a result of the modulation of the dielectric constant by acoustic vibrations, scattered light appears. By analyzing the latter with a monochromator, one can observe peaks in the scattered intensity which are associated with surface acoustic modes of frequency ω and parallel wave vector $\boldsymbol{q}$. By conservation of energy and parallel momentum,

$$\hbar\omega_i = \hbar\omega_s \pm \hbar\omega \tag{9.79a}$$

$$\hbar\boldsymbol{k}_{i\|} = \hbar\boldsymbol{k}_{s\|} \pm \hbar\boldsymbol{q}, \tag{9.79b}$$

where ω_s and $\boldsymbol{k}_{s\|}$ are the frequency and the wave vector parallel to the surface, respectively, of the scattered radiation.

The plus signs refer to **Stokes scattering**, in which a phonon is excited by the radiation, and the minus signs refer to **anti-Stokes scattering**, in which an already excited phonon gives up energy to the radiation. From measured values of ω_i, ω_s, $\boldsymbol{k}_{i\|}$, and $\boldsymbol{k}_{s\|}$, one can determine ω and $\boldsymbol{q}$ of the surface mode.

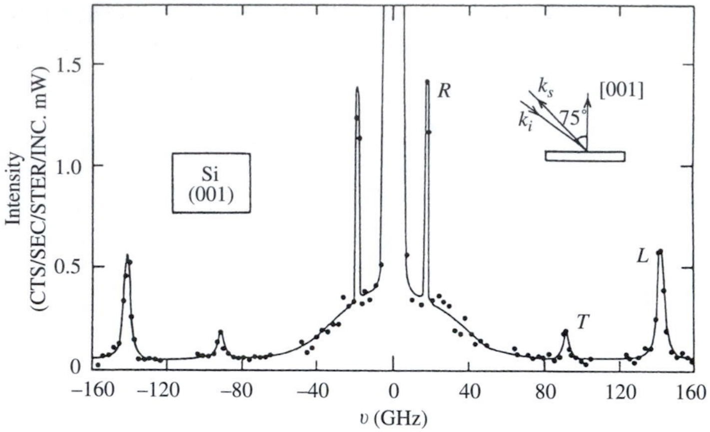

Fig. 9.14
Brillouin scattering from an Si(001) surface with oblique incidence light (insert). R refers to the Rayleigh phonon, T and L to transverse and longitudinal bulk phonon scattering, respectively. The scattered intensity is expressed in units of counts per second per steradian per unit incident power (after Sandercock 1978).

An example of a Brillouin scattering spectrum is shown in Fig. 9.14 for Si(001). The surface acoustic mode (R) as well as bulk transverse (T) and longitudinal (L) acoustic modes exhibit well-defined peaks.

9.2.4.2 Raman scattering

Raman scattering is the inelastic scattering of light by elementary excitations such as optical phonons. The main application of this technique to surface optical phonons has been to polar semiconductors in which the optical phonons couple strongly to electromagnetic radiation to give **surface polaritons**. This subject will be discussed in Chapter 10.

9.2.4.3 Electron energy loss spectroscopy

In **electron energy loss spectroscopy (EELS)**, relatively low energy electrons ($E \lesssim 400$ eV) are scattered inelastically from a crystal surface with the concomitant excitation of a surface phonon. This technique has been used more for metals than semiconductors. However, in an early example of the EELS technique, a surface phonon on the Si(111) 2×1 reconstructed surface was observed at 55 meV (Ibach 1971).

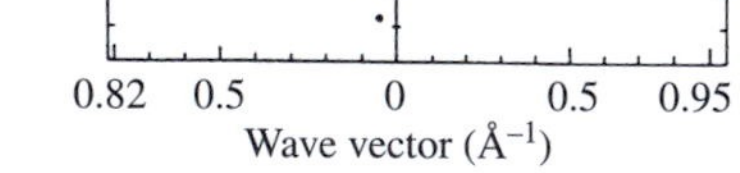

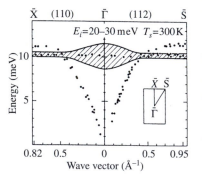

Fig. 9.15
Surface phonon dispersion curves for Si(111) 2×1 at 300 K. Circles: experimental points; shaded region: full width at half-maximum for the measured peaks due to the 10.5 meV mode (after Harten *et al.* 1986).

9.2.4.4 Helium atom energy loss spectroscopy

Helium atoms have the advantage for surface studies that they are scattered primarily by the surface atoms of a crystal, provided their energy is sufficiently low ($E \lesssim 20$ meV). Highly monoenergetic helium atom beams can be produced whose inelastic scattering from a crystal surface enables one to determine surface phonon dispersion curves over the entire surface Brillouin zone.

Inelastic helium atom scattering has been applied to the study of low-frequency surface phonons on Si(111) 2×1 (Harten *et al.* 1986). A nearly dispersionless surface phonon was observed with an energy of 10.5 meV as shown in Fig. 9.15. In addition, surface acoustic modes with energies ranging from 1 meV to 12 meV were found.

A variety of theoretical investigations have been carried out to elucidate the nature of the surface modes at 10.5 and 56 meV. The results support the

π-bonded chain model (Pandey 1981) for the 2×1 reconstructed surface. The 10.5 meV mode is attributed to vibrations of the chains perpendicular to the surface, whereas the 55 meV mode involves longitudinal optical vibrations of the chains parallel to the surface (Ancilotto *et al.* 1990). The calculations of Ancilotto *et al.* reveal that there is a large dynamic effective charge associated with the 55 meV mode that leads to strong coupling to electrons and the mode's large cross section for inelastic electron scattering.

9.2.4.5 Infrared spectroscopy

Another optical technique that is useful in studying surface vibrations is infrared (IR) spectroscopy. In Chapter 10 it is shown that this technique is very valuable for the experimental investigation of long-wavelength optical phonons in bulk semiconductors. The application of IR spectroscopy to the case of surface vibrations, however, encounters difficulties associated with the typically small surface-to-volume ratio. These difficulties can be overcome by using multiple internal reflections in thin layers which substantially enhance the sensitivity of the method (Chabal 1994). Of particular interest are infrared-active modes associated with chemisorbed atoms such as H on Si surfaces.

9.3 Surface recombination

Under certain conditions a surface or interface of a semiconductor may be a very effective site for the recombination of electrons and holes. Indeed, in semiconductor devices excessively high surface recombination may be a problem that must be avoided.

Let us focus on the rate of recombination of holes whose concentration in the bulk is p. We need to calculate the rates at which holes flow toward and away from the surface. In the classical regime the Maxwell–Boltzmann distribution of velocities can be used to calculate these rates. For the rate of flow toward the surface, one obtains the result $\frac{1}{4}v_t p$ per unit area, where v_t is the average thermal velocity. In considering the flow away from the surface we note that incident holes may be reflected back toward the bulk with a probability r and that holes may be generated at the surface at a rate S per unit area. The total rate of flow from the surface is therefore $\frac{1}{4}rv_t p + S$. At equilibrium one must have

$$\tfrac{1}{4}v_t p_0 = \tfrac{1}{4}rv_t p_0 + S_0, \tag{9.80}$$

where the subscript zero indicates equilibrium values. Solving for S_0 yields

$$S_0 = \tfrac{1}{4}(1 - r)v_t p_0. \tag{9.81}$$

Away from equilibrium the net rate of trapping of holes $\bar{S}$ is given by

$$\bar{S} = \tfrac{1}{4}v_t p - \tfrac{1}{4}rv_t p - S_0 = \tfrac{1}{4}(1 - r)v_t(p - p_0). \tag{9.82}$$

The quantity $s = \frac{1}{4}(1 - r)v_t$ is called the **surface recombination velocity**. The net rate $\bar{S}$ takes the simple form

$$\bar{S} = s\Delta p, \tag{9.83}$$

where $\Delta p = p - p_0$. Under steady state conditions the net rate of trapping of holes must be equal to the rate at which the holes recombine with electrons.

In the bulk the net rate of recombination $\mathcal{R}$ is proportional to $pn - n_i^2 = (n_0 + \Delta n)(p_0 + \Delta p) - n_i^2 = \Delta p(n_0 + p_0 + \Delta p)$,

$$\mathcal{R} = A\Delta p(n_0 + p_0 + \Delta p), \tag{9.84}$$

where $\Delta n = \Delta p$ and A is a constant.

The dependence of s on the carrier concentrations can be obtained if we assume that the recombination rate given by Eq. (9.84) applies to the surface case. Comparing Eq. (9.84) to Eq. (9.83) gives

$$s = A(n_0 + p_0 + \Delta p). \tag{9.85}$$

Experimental data on Ge (Schultz 1954) are consistent with this relation. Values of s vary from 10^2 to 10^6 cm/s, the smaller value being typical of etched surfaces and the larger value of sand-blasted surfaces. In the latter case essentially all holes that reach the surface are trapped, so $s \simeq \frac{1}{4}v_t \simeq 2 \times 10^6$ cm/s.

The variation of the excess hole concentration near the surface is described by the diffusion equation

$$D_h \frac{d^2\Delta p}{dx^2} = \frac{\Delta p}{\tau_p} - \mathcal{R}_e, \tag{9.86}$$

where D_h is the diffusion constant for holes, τ_p is the hole lifetime, and $\mathcal{R}_e$ is the uniform generation rate of excess carriers throughout the sample. At the surface $x = 0$, the net rate of diffusion of holes must equal the net rate of trapping of holes

$$D_h \frac{d\Delta p}{dx} = s\Delta p, \tag{9.87}$$

while in the bulk $x \to \infty$, the excess hole concentration is uniform, diffusion is negligible, and

$$\Delta p = \mathcal{R}_e \tau_p. \tag{9.88}$$

The solution of the diffusion equation subject to the boundary condition at $x \to \infty$ is

$$\Delta p(x) = \Delta p_1 e^{-x/L_h} + \mathcal{R}_e \tau_p, \tag{9.89}$$

where $L_h = (D_h \tau_p)^{\frac{1}{2}}$ is the diffusion length of holes and Δp_1 is determined by the boundary condition at $x = 0$,

$$-(D_h/L_h)\Delta p_1 = s\Delta p(0) = s(\Delta p_1 + \mathcal{R}_e \tau_p). \tag{9.90}$$

Solving for Δp_1 and eliminating D_h, we find that

$$\Delta p_1 = -\frac{s\mathcal{R}_e\tau_p^2}{L_h + s\tau_p} \tag{9.91}$$

and

$$\Delta p(x) = \mathcal{R}_e\tau_p\left[1 - \frac{s\tau_p}{L_h + s\tau_p}e^{-x/L_p}\right]. \tag{9.92}$$

We see that the excess hole concentration is reduced in the surface region due to surface recombination at the surface,

$$\Delta p(0) = \frac{\mathcal{R}_e\tau_p L_h}{L_h + s\tau_p}. \tag{9.93}$$

The region of reduced Δp extends over a distance on the order of L_h and the magnitude of the reduction depends on the ratio s/v_p, where $v_p = L_h/\tau_p$. For Ge typical values are $L_h = 0.05\,\text{cm}$, $\tau_p = 100\,\mu s$, and hence $v_p = 500\,\text{cm/s}$. If s is much larger than v_p, the reduction in Δp at the surface is large.

Problems

1. Carry out a tight-binding treatment of electronic surface states for one-dimensional [111] Si in which the "surface" is created by cutting the bond between atoms 0+ and 0−, i.e., atoms in the same unit cell. Compare the results with those for the cut between atoms 0+ and −1, − discussed in the text.
2. Set up a tight-binding treatment of surface states for one-dimensional [111] GaAs. Note that there are now two atomic potentials, $v_{Ga}(\boldsymbol{r})$ and $v_{As}(\boldsymbol{r})$, of atoms in the primitive unit cell. Consider the two cases mentioned in problem 1.
3. Show that the surface wave velocity equations for the [100] and [110] directions reduce to the same equation if the solid is isotropic. Calculate the reduced velocity p as a function of the ratio $C_{11}/C_{44} = g_1$ and plot p versus g_1.
4. Investigate surface modes of vibration in one-dimensional [111] ZnS. Calculate the force constants that couple a given atom to its neighbors on each side by fitting the experimental LO and LA bulk mode frequencies at the L-point given by 338 and 200 cm^{-1}, respectively. Determine the frequencies of surface modes, if they exist, for the two possibilities of cutting bonds to create a "surface."

References

F. Ancilotto, W. Andreoni, A. Selloni, R. Car, and M. Parrinello, *Phys. Rev. Lett.* **65**, 3148 (1990).

Y. J. Chabal, in *Handbook on Semiconductors*, Vol. 2, ed. M. Balkanski (North-Holland, Amsterdam, 1994).

S. G. Davison and M. Steslicka, *Basic Theory of Surface States* (Oxford University Press, Oxford, 1992).

F. Garcia-Moliner and V. R. Velasco, *Surf. Sci.* **299**/**300**, 332 (1994).

D. C. Gazis, R. Herman, and R. F. Wallis, *Phys. Rev.* **119**, 533 (1960).
E. T. Goodwin, *Proc. Camb. Phil. Soc.* **35**, 205 (1939).
U. Harten, J. P. Toennies, and Ch. Wöll, *Phys. Rev. Lett.* **57**, 2947 (1986).
H. Ibach, *Phys. Rev. Lett.* **27**, 253 (1971).
G. Le Lay, in *Semiconductor Interfaces: Formation and Properties*, eds. G. Le Lay, J. Derrier, and N. Boccara (Springer-Verlag, Berlin, 1987).
K. C. Pandey, *Phys. Rev. Lett.* **47**, 1913 (1981).
Lord Rayleigh, *Proc. London Math. Soc.* **17**, 4 (1887).
J. R. Sandercock, *Solid State Commun.* **26**, 547 (1978).
B. H. Schultz, *Physica* **20**, 1031 (1954).
R. Stoneley, *Proc. Roy. Soc.* (London) A**232**, 447 (1955).
I. Tamm, *Physik. Z. Sowj.* **1**, 733 (1932).
R. F. Wallis, *Phys. Rev.* **105**, 540 (1957).

Optical properties of semiconductors

10

Key ideas

The *dielectric tensor* $\overleftrightarrow{\epsilon}$ characterizes the interaction of an electrically polarizable medium with the radiation field. The dependence of $\overleftrightarrow{\epsilon}$ on wave vector $\boldsymbol{k}$ is known as *spatial dispersion*.

The propagation of an *electromagnetic wave* in a medium is characterized by the *refractive index* $N(\omega) = \sqrt{\epsilon(\omega)}$. The *phase velocity* of the wave is $c/N(\omega)$.

The *refractive index* is a complex quantity: $N(\omega) = n(\omega) + iK(\omega)$. The imaginary part $K(\omega)$ is the *extinction coefficient*. Two measurable quantities are the *reflectivity* and *transmissivity*. The latter is determined by the *absorption coefficient*.

The coupling of different types of *elementary excitations* to the radiation field gives rise to characteristic forms of the *dielectric function*.

The *intrinsic absorption coefficient* increases sharply for photon energies above the *energy gap*. In *direct transitions* the wave vector change of the excited electron is zero. In *indirect transitions* it is nonzero.

Free carrier absorption of radiation varies with wavelength λ as λ^s. *Free carrier reflectivity* is large below the *plasma frequency* ω_p.

Impurities can give rise to optical absorption through the *excitation* of either electrons or phonons.

Lattice vibration absorption is sharply peaked at the *transverse optical phonon frequency* ω_{TO}. Lattice vibration reflectivity is large between ω_{TO} and the *longitudinal optical phonon frequency* ω_{LO}.

Radiative recombination of electrons and holes can be *direct* or *indirect*. It can be *spontaneous* or *induced*.

Coupled electron–photon and optical phonon–photon excitations localized at a surface are *surface polaritons*.

Raman and *Brillouin scattering* provide energies and symmetries of elementary excitations.

Photoemission provides experimental information about energy band structure.

Optical Properties

The optical properties of matter, whether in the form of a solid, a liquid, or a gas, are associated with the absorption, dispersion, and scattering of electromagnetic radiation. These processes arise as a result of the perturbation of the material system by the electromagnetic field. The perturbation can involve the electronic states or the vibrational states or both and leads to transitions between states. From calculations of the transition rate one can evaluate the parameters that characterize the optical property under consideration.

In this chapter we first present a general formulation of the electromagnetic response of matter. We then focus on semiconductors and discuss a number of specific processes that lead to important optical phenomena. These phenomena include intrinsic interband absorption, free carrier absorption, impurity absorption, lattice vibration absorption and Raman scattering, and electronic Raman scattering. The subjects of radiative recombination, surface polaritons, and photoemission are also discussed.

10.1 Fundamentals of electromagnetic response

10.1.1 Maxwell's equations

The basic equations of electrodynamics in the presence of matter can be formulated within the framework of Maxwell's classical electromagnetic theory. Maxwell's equations for a collection of charges can be written, using MKS units, in the form

$$\nabla \cdot \mathcal{E} = \frac{\rho}{\epsilon_0} \tag{10.1a}$$

$$\nabla \times \mathcal{E} = -\mu_0 \frac{\partial \mathcal{H}}{\partial t} \tag{10.1b}$$

$$\nabla \cdot \mathcal{H} = 0 \tag{10.1c}$$

$$\nabla \times \mathcal{H} = \epsilon_0 \frac{\partial \mathcal{E}}{\partial t} + \boldsymbol{J}, \tag{10.1d}$$

where $\mathcal{E}$ is the electric field, $\mathcal{H}$ is the magnetic field, ρ is the electric charge density, and $\boldsymbol{J}$ is the electric current density. The quantity ϵ_0 is the permittivity constant whose experimental value can be expressed as $1/4\pi\epsilon_0 \simeq 9 \times 10^9$ newton meter^2/coulomb^2. The permeability constant μ_0 is defined to be $4\pi \times 10^{-7}$ weber/ampere meter.

In the case of magnetic materials, Maxwell's equations (10.1b) and (10.1c) should be written in the form

$$\nabla \times \mathcal{E} = -\frac{\partial \mathcal{B}}{\partial t} \tag{10.2a}$$

$$\nabla \cdot \mathcal{B} = 0, \tag{10.2b}$$

where $\mathcal{B}$ is the magnetic induction. Let us introduce the magnetization $\boldsymbol{M}$, i.e., the magnetic dipole moment per unit volume, and the magnetic permeability tensor $\overleftrightarrow{\mu}$. The magnetic induction is expressed in terms of these quantities by

$$\mathcal{B} = \mu_0(\mathcal{H} + \boldsymbol{M}) = \mu_0 \overleftrightarrow{\mu} \cdot \mathcal{H}. \tag{10.3}$$

Furthermore, the magnetization and magnetic field are related by

$$\boldsymbol{M} = \overleftrightarrow{\chi}_m \cdot \mathcal{H}, \tag{10.4}$$

where $\overleftrightarrow{\chi}_m$ is the magnetic susceptibility tensor. From Eqs. (10.3) and (10.4) we see that

$$\overleftrightarrow{\mu} = I + \overleftrightarrow{\chi}_m \tag{10.5}$$

with I the unit tensor. For nonmagnetic materials, $\overleftrightarrow{\chi}_m = 0$ and $\mathcal{B} = \mu_0 \mathcal{H}$.

To establish the relations which characterize the interaction of the radiation field with an electrically polarizable medium, it is convenient to introduce the polarization vector $\boldsymbol{P}$, defined as the electric dipole moment per unit volume, and the electric displacement vector $\mathcal{D}$, which is related to $\mathcal{E}$ and $\boldsymbol{P}$ by

$$\mathcal{D} = \epsilon_0 \mathcal{E} + \boldsymbol{P}. \tag{10.6}$$

When a crystal is placed in an electric field $\mathcal{E}$, it acquires an electric dipole moment and hence a polarization which is specified in the limit of linear response by

$$\boldsymbol{P} = \epsilon_0 \overleftrightarrow{\chi} \cdot \mathcal{E}, \tag{10.7}$$

where $\overleftrightarrow{\chi}$ is the dielectric susceptibility tensor. Substituting Eq. (10.7) into Eq. (10.6) yields

$$\mathcal{D} = \epsilon_0(I + \overleftrightarrow{\chi}) \cdot \mathcal{E}. \tag{10.8}$$

Introducing the dielectric tensor $\overleftrightarrow{\epsilon}$ by the relation

$$\overleftrightarrow{\epsilon} = I + \overleftrightarrow{\chi}, \tag{10.9}$$

we obtain

$$\mathcal{D} = \epsilon_0 \overleftrightarrow{\epsilon} \cdot \mathcal{E}. \tag{10.10}$$

The dielectric tensor characterizes a dielectric medium and contains the information on how the medium responds to an electromagnetic field. In a cubic crystal the tensors, $\overleftrightarrow{\chi}$ and $\overleftrightarrow{\epsilon}$, are scalar multiples of the unit tensor,

$$\overleftrightarrow{\chi} = \chi I \tag{10.11a}$$

$$\overleftrightarrow{\epsilon} = \epsilon I, \tag{10.11b}$$

and the medium can now be characterized by the scalar quantities χ and ϵ.

In general, the field quantities $\mathcal{E}$, $\mathcal{H}$, $\mathcal{D}$, and $\mathcal{B}$ are functions of both position $\boldsymbol{r}$ and time t. If the field quantities vary in time with a frequency ω, the material parameters $\overleftrightarrow{\chi}$, $\overleftrightarrow{\epsilon}$ and $\overleftrightarrow{\mu}$ are in general functions of ω. In writing Eqs. (10.6), (10.7), and (10.8) we have assumed local response, by which we mean that the field quantities on each side of the respective equations are evaluated at the same position $\boldsymbol{r}$. We do not deal with non-local response in this volume.

10.1.2 Propagation of an electromagnetic wave in a conducting medium

In a conducting medium the total electric current density $\boldsymbol{J}_t$ is related to the electric field $\boldsymbol{\mathcal{E}}$ in the linear response approximation by

$$\boldsymbol{J}_t = \tilde{\sigma}\boldsymbol{\mathcal{E}}, \tag{10.12}$$

where $\tilde{\sigma}$ is the complex electrical conductivity. The total electric current density contains contributions from a steady part due to the motion of unbound charges and from a time-dependent part due to the motion of bound charges. The contribution of the latter to the current density is $\partial \boldsymbol{P}/\partial t$, so we may write the sum of these contributions as

$$\boldsymbol{J} = \boldsymbol{J}_{cond} + \frac{\partial \boldsymbol{P}}{\partial t}, \tag{10.13}$$

where $\boldsymbol{J}_{cond}$ is the steady contribution to the current density from unbound charges given by $\boldsymbol{J}_{cond} = \sigma\boldsymbol{\mathcal{E}}$. Substitution of Eq. (10.13) into Eq. (10.1d) and use of Eq. (10.6) yields

$$\nabla \times \boldsymbol{\mathcal{H}} = \frac{\partial \boldsymbol{\mathcal{D}}}{\partial t} + \boldsymbol{J}_{cond} = \boldsymbol{J}_t. \tag{10.14}$$

In considering the optical properties of semiconductors one is dealing with time-dependent currents. Any steady current $\boldsymbol{J}_{cond}$ associated with free carriers can be disregarded in first approximation. One can then associate $\boldsymbol{J}$ entirely with $\partial \boldsymbol{P}/\partial t$,

$$\boldsymbol{J} = \frac{\partial \boldsymbol{P}}{\partial t}, \tag{10.15}$$

and write Eq. (10.14) as

$$\nabla \times \boldsymbol{\mathcal{H}} = \frac{\partial \boldsymbol{\mathcal{D}}}{\partial t}. \tag{10.16}$$

In what immediately follows, we shall use Eq. (10.16) rather than Eqs. (10.1d) and (10.14).

Another relationship that applies to an electrically neutral solid can be derived by taking the divergence of Eq. (10.6) and using Eq. (10.1a) to give

$$\nabla \cdot \boldsymbol{\mathcal{D}} = \rho + \nabla \cdot \boldsymbol{P}. \tag{10.17}$$

If Eq. (10.15) is substituted into the equation of continuity,

$$\frac{\partial \rho}{\partial t} + \nabla \cdot \boldsymbol{J} = 0, \tag{10.18}$$

one finds that

$$\rho = -\nabla \cdot \boldsymbol{P}. \tag{10.19}$$

Substituting Eq. (10.19) into Eq. (10.17) gives

$$\nabla \cdot \mathcal{D} = 0. \tag{10.20}$$

This completes the derivation of the electrodynamic equations for an electrically neutral polarizable medium with $\boldsymbol{J}_{cond}$ neglected.

Let us now turn our attention to the derivation of the wave equation for a nonmagnetic medium. If we take the curl of Eq. (10.1b) and use Eq. (10.16), we obtain the equation

$$\nabla \times \nabla \times \mathcal{E} = \mu_0 \frac{\partial^2 \mathcal{D}}{\partial t^2}. \tag{10.21}$$

At this point we restrict our attention to cubic crystals, so that

$$\mathcal{D} = \epsilon_0 \epsilon \mathcal{E} \tag{10.22}$$

with ϵ a scalar dielectric constant. Then Eq. (10.21) becomes

$$\nabla \times \nabla \times \mathcal{E} = -\frac{\epsilon}{c^2} \frac{\partial^2 \mathcal{E}}{\partial t^2}, \tag{10.23}$$

where $c = (\mu_0 \epsilon_0)^{-\frac{1}{2}}$ is the speed of light in vacuum. With the aid of an identity from vector analysis, Eq. (10.23) becomes

$$\nabla(\nabla \cdot \mathcal{E}) - \nabla^2 \mathcal{E} = -\frac{\epsilon}{c^2} \frac{\partial^2 \mathcal{E}}{\partial t^2}. \tag{10.24}$$

Using Eqs. (10.20) and (10.22) we can reduce Eq. (10.24) to the **wave equation**

$$\nabla^2 \mathcal{E} = \frac{\epsilon}{c^2} \frac{\partial^2 \mathcal{E}}{\partial t^2}. \tag{10.25}$$

Let us seek a plane wave solution to the wave equation of the form

$$\mathcal{E}(\boldsymbol{r}, t) = \mathcal{E}_0 \exp[i(\boldsymbol{k} \cdot \boldsymbol{r} - \omega t)], \tag{10.26}$$

where $\mathcal{E}_0$ is the electric field amplitude, $\boldsymbol{k}$ is the wave vector, and ω is the frequency. Substitution of Eq. (10.26) into Eq. (10.25) yields the following condition for a nontrivial solution:

$$c^2 k^2 = \omega^2 \epsilon(\omega). \tag{10.27}$$

Equation (10.27) is the dispersion relation for electromagnetic waves propagating in the dielectric medium. Its explicit evaluation requires a knowledge of the dielectric function $\epsilon(\omega)$, whose determination we discuss in later sections of this chapter. Using Eq. (10.27) we can rewrite Eq. (10.26) in the form

$$\boldsymbol{\mathcal{E}}(\mathbf{r}, t) = \boldsymbol{\mathcal{E}}_0 \exp\left[i\omega\left(\frac{N(\omega)}{c}\hat{k} \cdot \boldsymbol{r} - t\right)\right], \tag{10.28}$$

where $\hat{k}$ is a unit vector in the direction of $\boldsymbol{k}$ and the refractive index $N(\omega)$ is given by

$$N(\omega) = [\epsilon(\omega)]^{\frac{1}{2}}. \tag{10.29}$$

The phase velocity of the wave is $c/N(\omega)$.

Equations (10.25) and (10.27) are valid only for crystals whose symmetry is at least cubic. For noncubic crystals, one must use Eqs. (10.8) and (10.21).

In the foregoing discussion we have associated the current density $\boldsymbol{J}$ with the time rate of change of the polarization and have thus been able to characterize the propagation of the electromagnetic wave in terms of the dielectric constant ϵ. Now we need to take the conduction current density into account. Doing so leads to expressions for a complex conductivity $\tilde{\sigma}$ and a complex dielectric constant $\tilde{\epsilon}$ which can be obtained by equating $\boldsymbol{J}_t$ in Eq. (10.12) to its expression in Eq. (10.14) and using Eq. (10.10) with the assumption that $\boldsymbol{\mathcal{D}} \sim \exp(-i\omega t)$. The result is

$$\tilde{\sigma} = \sigma - i\omega\epsilon_0\epsilon. \tag{10.30}$$

We can rearrange the expression for $\boldsymbol{J}_t$ to give $-i\omega\epsilon_0\tilde{\epsilon}\boldsymbol{\mathcal{E}}$ where

$$\tilde{\epsilon} = \epsilon + (i/\epsilon_0\omega)\sigma = (i/\omega\epsilon_0)\tilde{\sigma}. \tag{10.31}$$

10.1.3 Optical constants

We have seen in Section 10.1.2 that the propagation characteristics of an electromagnetic wave in a nonmagnetic cubic crystal are determined by the **dielectric function** $\epsilon(\omega)$ and by the **refractive index** $N(\omega) = [\epsilon(\omega)]^{1/2}$. It has also been seen that $\epsilon(\omega)$ must be treated in general as a complex quantity, and hence $N(\omega)$ is also a complex quantity. We write

$$\tilde{\epsilon} = \epsilon(\omega) = \epsilon'(\omega) + i\epsilon''(\omega) \tag{10.32}$$

and

$$N(\omega) = n(\omega) + iK(\omega), \tag{10.33}$$

where $K(\omega)$, the imaginary part of the refractive index, is the so-called **extinction coefficient**. Substituting Eqs. (10.32) and (10.33) into the squared form of Eq. (10.29), we obtain the relations

$$\epsilon'(\omega) = n^2 - K^2 \tag{10.34a}$$

$$\epsilon''(\omega) = 2nK. \tag{10.34b}$$

None of the foregoing quantities is directly measurable in an optical experiment. Two quantities which are measurable, however, are the **reflectivity** and **transmissivity**. They are defined as the fractions of the incident electromagnetic energy that are reflected and transmitted by the crystal, respectively. The transmissivity is determined in part by a quantity known as the **absorption coefficient**. The latter is defined as the inverse of the distance an electromagnetic wave must travel in a crystal in order for its intensity to decrease by a factor $1/e$ of its original intensity. Since the intensity is proportional to the square of the magnitude of the electric field, we can readily obtain an expression for the absorption coefficient, $\alpha(\omega)$ by taking the square of the magnitude of Eq. (10.28) and using Eq. (10.33). The result is

$$\alpha(\omega) = \frac{2\omega K(\omega)}{c}. \tag{10.35}$$

Alternatively, $\alpha(\omega)$ can be expressed in terms of $\epsilon''(\omega)$ with the aid of Eq. (10.34b):

$$\alpha(\omega) = \frac{\omega\epsilon''(\omega)}{cn(\omega)}. \tag{10.36}$$

Frequently $\epsilon''(\omega)$ possesses structure in a frequency region where $n(\omega)$ is slowly varying. The structure in $\epsilon''(\omega)$ is then revealed directly by $\alpha(\omega)$.

The reflectivity at the surface of a semi-infinite crystal can be calculated by considering incident and reflected waves in the vacuum and a transmitted wave in the crystal. By applying electromagnetic boundary conditions at the crystal surface (continuity of the tangential components of $\mathcal{E}$ and $\mathcal{H}$ and the normal components of $\mathcal{D}$ and $\mathcal{B}$) one can obtain an expression for the reflectivity. This expression is somewhat complicated for an arbitrary angle of incidence and can be found in Born and Wolf (1965). We content ourselves here with the case of normal incidence. The amplitude of the electric field in the reflected wave, $\mathcal{E}_{ref}$, is then related to that in the incident wave, $\mathcal{E}_{inc}$, by the equation

$$\mathcal{E}_{ref} = r(\omega)\mathcal{E}_{inc}, \tag{10.37}$$

where $r(\omega)$ is given by

$$r(\omega) = \rho(\omega)\exp[i\theta(\omega)] = \frac{n + iK - 1}{n + iK + 1}, \tag{10.38}$$

$\rho(\omega)$ is the modulus of $r(\omega)$, and $\theta(\omega)$ is the phase difference between the electric fields of the reflected and incident waves.

The experimentally measurable reflectivity, $R(\omega)$, relates the incident and reflected intensities, I_{inc} and I_{ref},

$$I_{ref} = R(\omega)I_{inc}, \tag{10.39}$$

where

$$R(\omega) = |r(\omega)|^2 = \rho^2(\omega) = \frac{(n-1)^2 + K^2}{(n+1)^2 + K^2}. \tag{10.40}$$

In order to determine both n and K we need a second piece of experimental information. The two pieces of information can be obtained by measuring the reflectivity at two different angles of incidence or with two different polarizations at non-normal incidence or by measuring both the reflectivity and transmissivity at normal incidence. We assume that n and K have been determined by one of these methods and refer the reader to Born and Wolf (1965) for a more general discussion.

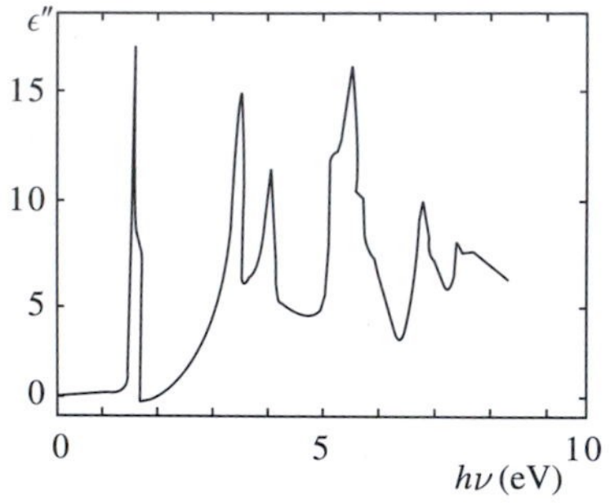

Fig. 10.1
Imaginary part of the dielectric function versus frequency for electronic transitions in CdTe (after Balkanski 1972).

10.1.4 Dielectric function of a crystal

In a polar dielectric medium an electromagnetic wave is coupled at low photon energies with optical phonons and with free current carriers if the latter are present. As the photon energy is raised, the electromagnetic wave will eventually start to couple to electronic transitions between valence and conduction bands. At very high energies in the far ultraviolet and X-ray regions, coupling will occur to atomic core levels. Each type of coupling makes a contribution to the dielectric function of the crystal. The situation is illustrated in Fig. 10.1 where the imaginary part of the dielectric function as determined from reflectivity data is plotted against frequency for CdTe. All peaks shown there correspond to interband electronic transitions. At much lower energies the same procedure yields the dielectric function for transitions due to phonons. Figure 10.2 shows the dielectric function variation due to phonon transitions as a function of the frequency for CdS.

In cases where two elementary excitations have nearly the same frequency, strong coupling between the two excitations can occur. This may lead to significant modification of the optical properties and thus provide a means of obtaining experimental information about the coupled excitations.

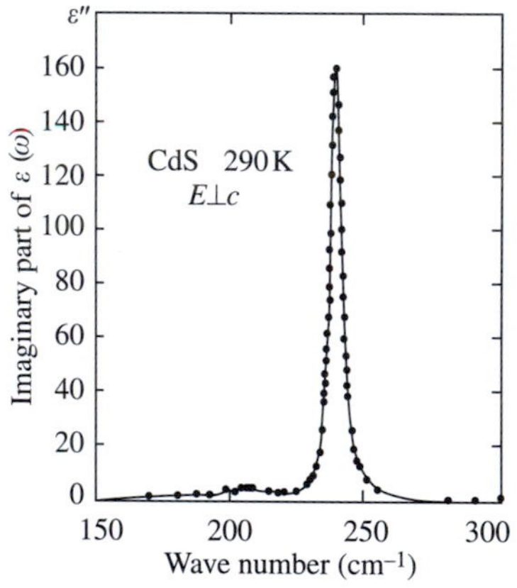

Fig. 10.2
Imaginary part of the dielectric function versus frequency for phonon transitions in CdS (after Balkanski 1972).

In Fig. 10.3 we show the modification of the reflectivity spectrum as the plasma frequency is shifted into the region of the normal mode frequencies by displacement of the Fermi energy as a function of temperature in HgTe. When the plasmon frequency is close to the LO-phonon frequency, the distinction between the pure plasmon and the pure longitudinal optical phonon is no longer possible.

10.1.5 Optical spectroscopies

The optical properties of a solid are manifestations of the interaction of the electrons and nuclei of the solid with the radiation field. When an effect of this interaction is presented as a function of the electromagnetic wave frequency, the procedure is termed **spectroscopy**. Spectroscopy provides an excellent tool for probing the microscopic character of matter. For example, a solid may absorb a photon from the electromagnetic field and be raised to an excited energy state. Analysis of the absorption coefficient as a function of frequency yields information about the energy separation of the initial and final states. In this case we are dealing with **absorption spectroscopy**, which is also a **resonance spectroscopy**, i.e., light is absorbed when its photon energy coincides with the energy change in an allowed transition of the solid.

Excited states can be associated with so-called **elementary excitations**. In the case of electrons the elementary excitations include single-particle excitations, collective excitations such as plasmons, and electron–hole pair excitations such as excitons. In the case of nuclei the elementary excitations are phonons, and in the case of magnetic systems, where the excited states are spin waves, they are magnons.

An elementary excitation can decay with the emission of a photon. In this case the associated spectroscopy is **emission spectroscopy** which is also a resonance spectroscopy. It is particularly useful in providing information about the lifetimes of excited states and forms the basis of **recombination spectroscopy** discussed in Section 10.6.

When the radiation field is scattered by an elementary excitation, one deals with both incident and scattered light whose frequency difference is characteristic of the elementary excitation. This spectroscopy is **light scattering spectroscopy**.

We now discuss the various types of spectroscopies and the information they can provide about semiconductors.

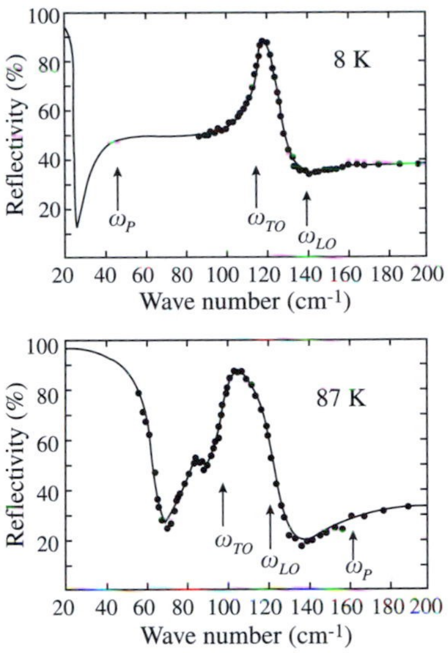

Fig. 10.3
Reflectivity spectrum of HgTe in the region of plasmon–phonon interaction (after Grynberg *et al.* 1974).

10.2 Intrinsic interband absorption

When photons with sufficient energy are incident on a pure semiconductor crystal, absorption of photons can take place with the simultaneous creation of electron–hole pairs, i.e., the excitation of an electron from the valence band to the conduction band. This process is known as **intrinsic interband absorption**. The threshold photon energy required is related to the fundamental band gap. The absorption coefficient increases rapidly above threshold.

10.2.1 Absorption coefficient

Since electronic energy bands are essentially quantum mechanical in nature, it is convenient to employ quantum mechanics to calculate the absorption coefficient. We consider transitions of an electron between states of the same or different energy bands. The rate of such transitions, W, can be calculated by perturbation theory using Fermi's golden rule. We now consider the relation between the transition rate and the absorption coefficient $\alpha(\omega)$.

Consider a beam of electromagnetic radiation of intensity I incident on a sample of thickness dx. The absorption coefficient is defined by the equation

$$dI = -I\alpha(\omega)dx, \tag{10.41}$$

where dI is the change in intensity of the beam after passing through the sample. If A is the cross-sectional area of the sample, then $-AdI$ is the rate of energy absorption in the sample:

$$\begin{aligned}\frac{dE}{dt} &= -AdI \\ &= I\alpha(\omega)Adx.\end{aligned} \tag{10.42}$$

But the rate of energy absorption is also given by

$$\frac{dE}{dt} = \hbar\omega W. \tag{10.43}$$

Equating these two results yields the expression

$$\alpha(\omega) = \frac{\hbar\omega W}{I\Omega}, \tag{10.44}$$

where $\Omega = Adx$ is the volume of the sample.

A convenient measure of the intensity is the mean value of the Poynting vector $\boldsymbol{S} = \boldsymbol{\mathcal{E}} \times \boldsymbol{\mathcal{H}}$. The electric field vector $\boldsymbol{\mathcal{E}}$ and the magnetic field vector $\boldsymbol{\mathcal{H}}$ can be expressed in terms of the vector potential $\boldsymbol{\mathcal{A}}$. Using the Coulomb gauge, one has

$$\boldsymbol{\mathcal{E}} = -\frac{\partial \boldsymbol{\mathcal{A}}}{\partial t} \tag{10.45}$$

$$\mu_0 \boldsymbol{\mathcal{H}} = \boldsymbol{\nabla} \times \boldsymbol{\mathcal{A}}. \tag{10.46}$$

Taking $\boldsymbol{\mathcal{A}}$ to have the form of a standing wave

$$\boldsymbol{\mathcal{A}} = \boldsymbol{\mathcal{A}}_0 \cos(\boldsymbol{q} \cdot \boldsymbol{r} - \omega t), \tag{10.47}$$

we have

$$\boldsymbol{\mathcal{E}} = -\omega \boldsymbol{\mathcal{A}}_0 \sin(\boldsymbol{q} \cdot \boldsymbol{r} - \omega t) \tag{10.48}$$

$$\mu_0 \boldsymbol{\mathcal{H}} = -\boldsymbol{q} \times \boldsymbol{\mathcal{A}}_0 \sin(\boldsymbol{q} \cdot \boldsymbol{r} - \omega t). \tag{10.49}$$

The Poynting vector now takes the form

$$\boldsymbol{S} = \frac{\omega}{\mu_0} \boldsymbol{\mathcal{A}}_0 \times (\boldsymbol{q} \times \boldsymbol{\mathcal{A}}_0 \sin^2(\boldsymbol{q} \cdot \boldsymbol{r} - \omega t)). \tag{10.50}$$

Using the identity from vector analysis $A \times (\boldsymbol{B} \times \boldsymbol{C}) = (A \cdot \boldsymbol{C})B - (A \cdot B)\boldsymbol{C}$ yields

$$\boldsymbol{S} = \frac{\omega}{\mu_0} |\boldsymbol{\mathcal{A}}_0|^2 \boldsymbol{q} \sin^2(\boldsymbol{q} \cdot \boldsymbol{r} - \omega t), \tag{10.51}$$

where use has been made of the fact that $\boldsymbol{q}$ and $\boldsymbol{\mathcal{A}}_0$ are orthogonal. Neglecting the imaginary part of the refractive index, the dispersion relation in Eq. (10.27) reduces to $q = (\omega/c)n(\omega)$. Taking the time average of the magnitude of $\boldsymbol{S}$, we obtain

$$\langle S \rangle = \frac{\omega^2 n(\omega)}{2\mu_0 c} |\boldsymbol{\mathcal{A}}_0|^2. \tag{10.52}$$

Identifying $\langle S \rangle$ with I in Eq. (10.44), the expression for the absorption coefficient becomes

$$\alpha(\omega) = \frac{2\mu_0 c \hbar W}{\omega n(\omega) |\boldsymbol{\mathcal{A}}_0|^2 \Omega}. \tag{10.53}$$

The actual calculation of the absorption coefficient requires the evaluation of the transition probability W.

10.2.2 Transition probability

We assume that the intensity of the electromagnetic radiation is sufficiently low that the interaction between an electron and the radiation can be treated by perturbation theory. In the semiclassical approach the Hamiltonian of an electron moving in the radiation field can be written as

$$\begin{aligned} H &= \frac{1}{2m}(\boldsymbol{p}+e\boldsymbol{\mathcal{A}})^2 + V(\boldsymbol{r}) \\ &= \frac{p^2}{2m} + \frac{e}{2m}(\boldsymbol{p}\cdot\boldsymbol{\mathcal{A}}+\boldsymbol{\mathcal{A}}\cdot\boldsymbol{p}) + \frac{e^2}{2m}\boldsymbol{\mathcal{A}}^2 + V(\boldsymbol{r}), \end{aligned} \tag{10.54}$$

where $\boldsymbol{p}$ is the electron momentum, $\boldsymbol{\mathcal{A}}$ is the vector potential of the radiation, and $V(\boldsymbol{r})$ is the potential energy of the electron. At low intensities we can neglect the term in $\boldsymbol{\mathcal{A}}^2$ and take the interaction Hamiltonian to be

$$H_{int} = \frac{e}{2m}(\boldsymbol{p}\cdot\boldsymbol{\mathcal{A}}+\boldsymbol{\mathcal{A}}\cdot\boldsymbol{p}). \tag{10.55}$$

The wave vector $\boldsymbol{q}$ of the radiation of interest is very small compared to typical electron wave vectors. We can therefore neglect the term arising from the operation of $\boldsymbol{p}$ on $\boldsymbol{\mathcal{A}}$ and rewrite H_{int} as

$$H_{int} = \frac{e}{m}\boldsymbol{\mathcal{A}}\cdot\boldsymbol{p}. \tag{10.56}$$

Using time-dependent perturbation theory the transition probability of an electron from a valence band Bloch state $|\boldsymbol{k}v\rangle$ to a conduction band Bloch state $|\boldsymbol{k}'c\rangle$ is given by Fermi's golden rule as (Fermi 1950)

$$W(\boldsymbol{k}v \to \boldsymbol{k}'c) = \frac{2\pi}{\hbar}|\langle \boldsymbol{k}'c|H_{int}|\boldsymbol{k}v\rangle|^2\delta(E_{\boldsymbol{k}'c} - E_{\boldsymbol{k}v} - \hbar\omega). \tag{10.57}$$

We take $\boldsymbol{\mathcal{A}}$ to be given by Eq. (10.47). The interband matrix element of H_{int} for absorption processes can then be expressed as

$$\langle \boldsymbol{k}'c|H_{int}|\boldsymbol{k}v\rangle = (e/2m)\langle \boldsymbol{k}'c|\boldsymbol{\mathcal{A}}_0\cdot\boldsymbol{p}|\boldsymbol{k}v\rangle, \tag{10.58}$$

where we have exploited the smallness of $\boldsymbol{q}$ by setting

$$\boldsymbol{k}+\boldsymbol{q} \simeq \boldsymbol{k}. \tag{10.59}$$

From these results we obtain

$$W(\boldsymbol{k}v \to \boldsymbol{k}'c) = \frac{\pi e^2}{2\hbar m^2}|\boldsymbol{\mathcal{A}}_0|^2|\langle \boldsymbol{k}'c|p_{\mathcal{A}}|\boldsymbol{k}v\rangle|^2\delta(E_{\boldsymbol{k}'c} - E_{\boldsymbol{k}v} - \hbar\omega), \tag{10.60}$$

where $p_{\mathcal{A}}$ is the component of $\boldsymbol{p}$ in the direction of $\boldsymbol{\mathcal{A}}$.

To evaluate the momentum matrix element we utilize the Bloch form given by Eq. (2.38) for the eigenstates $|\boldsymbol{k}v\rangle$ and $|\boldsymbol{k}'c\rangle$:

$$\langle \boldsymbol{k}'c|p_{\mathcal{A}}|\boldsymbol{k}v\rangle = \int_{crystal} e^{-i\boldsymbol{k}'\cdot\boldsymbol{r}} u^*_{c\boldsymbol{k}'}(\boldsymbol{r}) p_{\mathcal{A}} e^{i\boldsymbol{k}\cdot\boldsymbol{r}} u_{v\boldsymbol{k}}(\boldsymbol{r}) d^3r. \tag{10.61}$$

Dividing the crystal into unit cells leads to the alternative expression

$$\langle \boldsymbol{k}'c|p_{\mathcal{A}}|\boldsymbol{k}v\rangle = \sum_{\boldsymbol{\ell}} \int_{cell\boldsymbol{\ell}} e^{i(\boldsymbol{k}-\boldsymbol{k}')\cdot\boldsymbol{r}} u^*_{c\boldsymbol{k}'}(\boldsymbol{r})(ik_{\mathcal{A}} + p_{\mathcal{A}}) u_{v\boldsymbol{k}}(\boldsymbol{r}) d^3r. \tag{10.62}$$

Exploiting the periodicity of $u_{n\boldsymbol{k}}(\boldsymbol{r})$ and $p_{\mathcal{A}}$, we let

$$\boldsymbol{r} = \boldsymbol{R}(\boldsymbol{\ell}) + \boldsymbol{\rho}, \tag{10.63}$$

so that

$$u_{n\boldsymbol{k}}(\boldsymbol{r}) = u_{n\boldsymbol{k}}(\boldsymbol{R}(\boldsymbol{\ell}) + \boldsymbol{\rho}) = u_{n\boldsymbol{k}}(\boldsymbol{\rho}). \tag{10.64}$$

Then

$$\langle \boldsymbol{k}'c|p_{\mathcal{A}}|\boldsymbol{k}v\rangle = \sum_{\boldsymbol{\ell}} e^{i(\boldsymbol{k}-\boldsymbol{k}')\cdot\boldsymbol{R}(\boldsymbol{\ell})} \int_{cell0} e^{i(\boldsymbol{k}-\boldsymbol{k}')\cdot\boldsymbol{\rho}} \times u^*_{c\boldsymbol{k}'}(\boldsymbol{\rho})(ik_{\mathcal{A}} + p_{\mathcal{A}}) u_{v\boldsymbol{k}}(\boldsymbol{\rho}) d^3\rho. \tag{10.65}$$

Using Eqs. (1.1), (2.14) and (2.28), the lattice sum is found to be $N\delta_{\boldsymbol{k},\boldsymbol{k}'}$, where N is the number of unit cells. Noting that $u_{c\boldsymbol{k}}(\boldsymbol{\rho})$ and $u_{v\boldsymbol{k}}(\boldsymbol{\rho})$ are orthogonal, we obtain

$$\langle \boldsymbol{k}'c|p_{\mathcal{A}}|\boldsymbol{k}v\rangle = N\delta_{\boldsymbol{k},\boldsymbol{k}'} \int_{cell0} u^*_{c\boldsymbol{k}}(\boldsymbol{\rho}) p_{\mathcal{A}} u_{v\boldsymbol{k}}(\boldsymbol{\rho}) d^3\rho. \tag{10.66}$$

In typical semiconductors the integral over the unit cell depends only weakly on $\boldsymbol{k}$, and can be approximated by a constant. Introducing

$$P = N \int_{cell0} u^*_{c\boldsymbol{k}}(\boldsymbol{\rho}) p_{\mathcal{A}} u_{v\boldsymbol{k}}(\boldsymbol{\rho}) d^3\rho, \tag{10.67}$$

the expression for $W(\boldsymbol{k}v \to \boldsymbol{k}'c)$ becomes

$$W(\boldsymbol{k}v \to \boldsymbol{k}'c) = \frac{\pi e^2}{2\hbar m^2} |\mathcal{A}_0|^2 P^2 \delta(E_{\boldsymbol{k}'c} - E_{\boldsymbol{k}v} - \hbar\omega)\delta_{\boldsymbol{k},\boldsymbol{k}'}. \tag{10.68}$$

We note the following. The Kronecker delta gives the selection rule $\boldsymbol{k} = \boldsymbol{k}'$ corresponding to **direct** or **vertical transitions** from the valence band to the conduction band as illustrated in Fig. 10.4. If the crystal possesses a center of inversion, the functions $u_{v\boldsymbol{k}}(\boldsymbol{\rho})$ and $u_{c\boldsymbol{k}}(\boldsymbol{\rho})$ must have opposite parity for an allowed transition.

The total probability of transitions from the valence band to the conduction band is obtained by summing $W(\boldsymbol{k}v \to \boldsymbol{k}'c)$ over $\boldsymbol{k}$ and $\boldsymbol{k}'$, taking

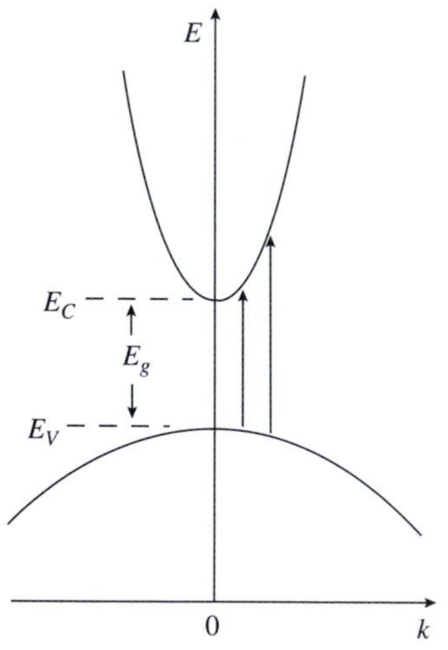

Fig. 10.4
Direct transitions from the valence band to the conduction band.

into account the Pauli principle. The result is

$$W_{vc} = 2\sum_{k}\sum_{k'} W(kv \to k'c) f_{kv}(1 - f_{k'c})$$

$$= \frac{e^2\Omega}{8\pi^2\hbar m^2}|\mathcal{A}_0|^2 P^2 \int \delta(E_{kc} - E_{kv} - \hbar\omega) f_{kv}(1 - f_{kc}) d^3k, \qquad (10.69)$$

where f is the Fermi–Dirac distribution function (Eq. (6.1)), a factor of two has been introduced for spin, and the sum over $\boldsymbol{k}$ has been converted to an integral. If the spin–orbit interaction is important, as it is in many semiconductors, the eigenstates $\boldsymbol{k}v$ and $\boldsymbol{k}'c$ must be labeled with spin indices, and the factor of two must be replaced by sums over these indices. Furthermore, the operator $\boldsymbol{p}$ is to be replaced by the operator $\boldsymbol{\pi}$ defined in Section 3.5.2.

A particularly simple case to treat is that involving zero temperature and spherical, parabolic energy bands with the spin–orbit interaction and the degeneracy of the valence band neglected. In this **two-band model** we have $f_{kv} = 1, f_{kc} = 0$, and

$$E_{kc} - E_{kv} = E_g + \frac{\hbar^2k^2}{2m_c^*} + \frac{\hbar^2k^2}{2m_v^*}$$

$$= E_g + \frac{\hbar^2k^2}{2\bar{m}^*}, \qquad (10.70)$$

where $\bar{m}^*$ is the reduced effective mass of electrons and holes. The integral over $\boldsymbol{k}$ in Eq. (10.69) can be evaluated in spherical coordinates to give

$$\int \delta(E_{kc} - E_{kv} - \hbar\omega) d^3k = 4\pi \int_0^\infty k^2 \delta\left(E_g - \hbar\omega + \frac{\hbar^2k^2}{2\bar{m}^*}\right) dk$$

$$= 2\pi\left(\frac{2\bar{m}^*}{\hbar^2}\right)^{3/2} \sqrt{\hbar\omega - E_g}, \quad \hbar\omega \geq E_g \qquad (10.71a)$$

$$= 0, \quad \hbar\omega < E_g. \qquad (10.71b)$$

The total transition probability and the absorption coefficient can now be obtained explicitly. The result for the absorption coefficient is

$$\alpha(\omega) = \frac{e^2}{2\pi\epsilon_0 cm^2\omega n(\omega)}\left(\frac{2\bar{m}^*}{\hbar^2}\right)^{3/2} P^2\sqrt{\hbar\omega - E_g}, \quad \hbar\omega \geq E_g \qquad (10.72a)$$

$$= 0, \quad \hbar\omega < E_g. \qquad (10.72b)$$

This result justifies the statement at the beginning of this section that radiation is not absorbed due to interband transitions if the photon energy is less than the band gap and the Coulomb interaction between electrons and holes is neglected. A qualitative plot of $\alpha(\omega)$ versus $\hbar\omega$ is shown in Fig. 10.5.

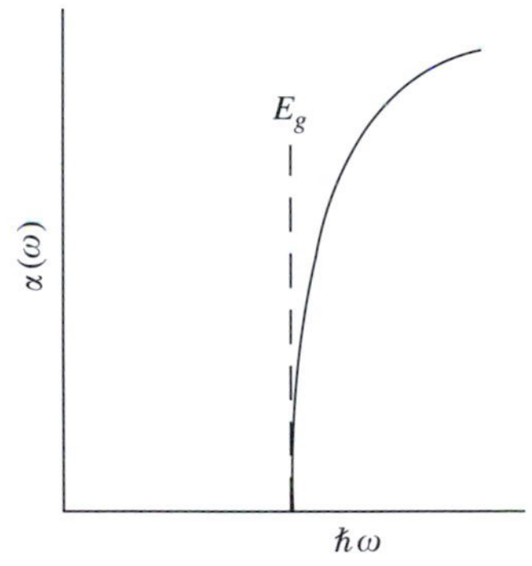

Fig. 10.5
Absorption coefficient versus frequency for direct interband transitions near the band gap.

10.2.3 Oscillator strength

It is convenient when discussing the magnitude of the absorption coefficient to introduce a dimensionless quantity called the **oscillator strength** which is denoted by f_{vc} and defined by

$$f_{vc} = \frac{2|\langle \boldsymbol{k}c|p_A|\boldsymbol{k}v\rangle|^2}{m\hbar\omega} \simeq \frac{2P^2}{m(E_{\boldsymbol{k}c} - E_{\boldsymbol{k}v})}. \tag{10.73}$$

In terms of the oscillator strength the absorption coefficient takes the form

$$\alpha(\omega) = \frac{e^2(2\bar{m}^*)^{3/2}}{4\pi\epsilon_0 cmn(\omega)\hbar^2} f_{vc}\sqrt{\hbar\omega - E_g}, \quad \hbar\omega \geq E_g. \tag{10.74}$$

An order of magnitude estimate for f_{vc} can be obtained from the f-sum rule (Bardeen *et al.* 1956), which shows that if we sum the oscillator strengths $f_{vj}(\boldsymbol{k})$ over transitions from the state $|\boldsymbol{k}v\rangle$ to states j with the same value of $\boldsymbol{k}$ in all other possible bands, we get

$$\begin{aligned} \sum_{j\neq v} f_{vj}(\boldsymbol{k}) &= 1 - \frac{m}{\hbar^2}\frac{\partial^2 E_{\boldsymbol{k}v}}{\partial k^2} \\ &= 1 + \frac{m}{m_h^*}. \end{aligned} \tag{10.75}$$

Since the smallest value of $|E_{\boldsymbol{k}j} - E_{\boldsymbol{k}v}|$ typically occurs when v is the highest valence band and j is the lowest conduction band c, the largest value of $f_{vj}(\boldsymbol{k})$ is f_{vc} and Eq. (10.75) can be approximated by

$$f_{vc} \cong 1 + \frac{m}{m_h^*}. \tag{10.76}$$

The hole mass m_h^* can generally be approximated by m, so that f_{vc} is of the order of magnitude two. With this result we can estimate the magnitude of the absorption coefficient.

Example 10.1: Direct interband absorption in Ge
Calculate the absorption coefficient for direct interband transitions in Ge for $\hbar\omega - E_g = 0.01$ eV.
Solution. The values of the pertinent parameters are $n(\omega) = 4$, $f_{vc} = 2$, and $2\bar{m}^*/m \cong 1$. Substituting into Eq. (10.74), we find that

$$\alpha(\omega) = \frac{e^2 m^{\frac{1}{2}}}{8\pi\epsilon_0 c\hbar^2}\sqrt{\hbar\omega - E_g}.$$

For $h\nu - E_g = 0.01$ eV, $\alpha = 1.3 \times 10^4\,\text{cm}^{-1}$. This is a strong absorption coefficient that is associated with direct transitions from a p-like valence band to an s-like conduction band. Direct transitions between other bands produce additional absorption at higher frequencies.

10.2.4 Excitons

Until now we have neglected the Coulomb interaction between an electron in the conduction band and a hole in the valence band. If this interaction is taken into account, bound states of an electron–hole pair arise which are analogous to the bound states of a hydrogen atom. A bound electron–hole pair is called an **exciton**. A photon with energy slightly smaller than the minimum energy gap can be absorbed with the simultaneous creation of an exciton.

Let us consider the case of conduction and valence bands that are spherical and parabolic with extrema at $\boldsymbol{k} = 0$. In the presence of the Coulomb interaction, the wave function $\psi(\boldsymbol{r}_e, \boldsymbol{r}_h)$ of an electron–hole pair can be written as (Dimmock 1967)

$$\psi(\boldsymbol{r}_e, \boldsymbol{r}_h) = \Phi(\boldsymbol{r}_e, \boldsymbol{r}_h) u_{c0}(\boldsymbol{r}_e) u_{v0}(\boldsymbol{r}_h), \tag{10.77}$$

where $\boldsymbol{r}_e$ and $\boldsymbol{r}_h$ are the position vectors of electrons and holes, respectively, and the functions $u_{c0}(\boldsymbol{r}_e)$, $u_{v0}(\boldsymbol{r}_h)$ are the periodic parts of the Bloch functions at $\boldsymbol{k} = 0$. The function $\Phi(\boldsymbol{r}_e, \boldsymbol{r}_h)$ is the envelope function and is a slowly varying function of its arguments. It has the form

$$\Phi(\boldsymbol{r}_e, \boldsymbol{r}_h) = \varphi_i(\boldsymbol{r}) e^{i\boldsymbol{K}\cdot\boldsymbol{R}}, \tag{10.78}$$

where $\boldsymbol{R}$ and $\boldsymbol{r}$ are the center-of-mass coordinate and relative coordinate, respectively, of the electron–hole pair and $\boldsymbol{K}$ is the wave vector of the center of mass. The function $\varphi_i(\boldsymbol{r})$ is the localized, hydrogen-like wave function that satisfies the Schrödinger equation

$$\left(\frac{p^2}{2\bar{m}^*} - \frac{e^2}{4\pi\epsilon_0\epsilon|\boldsymbol{r}|}\right)\varphi_i(\boldsymbol{r}) = E_i\varphi_i(\boldsymbol{r}). \tag{10.79}$$

The energy eigenvalues E_i correspond to hydrogenic energy levels with the effective Rydberg involving the electron–hole reduced mass $\bar{m}^*$. The total energy of the electron–hole pair, E_{eh}, is given by

$$E_{eh} = E_g + \frac{\hbar^2 K^2}{2(m_e^* + m_h^*)} + E_i. \tag{10.80}$$

Figure 10.6 displays excitonic energy levels for $\boldsymbol{K} = 0$.

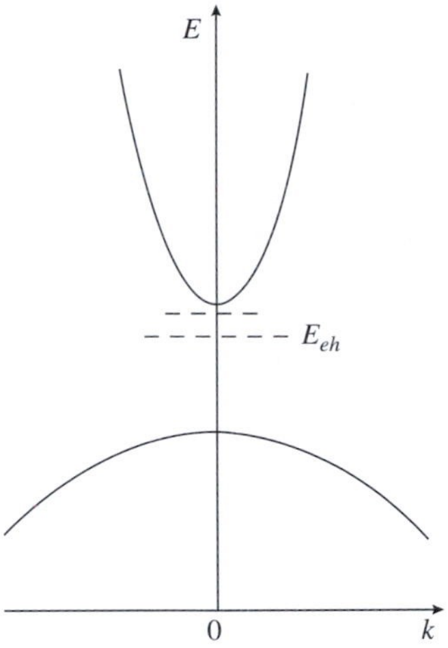

Fig. 10.6
Excitonic energy levels for $\boldsymbol{K} = 0$.

Experimentally, evidence for excitonic effects is provided by an absorption peak just below the band gap, as shown in Fig. 10.7 for GaAs at 21 K. As the temperature is increased, the exciton peak broadens and becomes difficult to discern at room temperature.

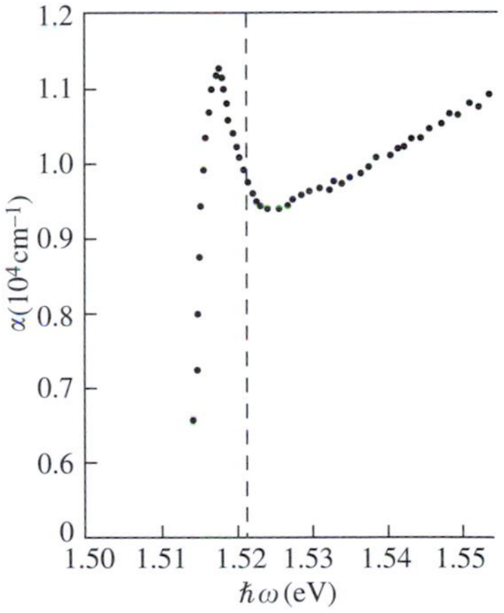

Fig. 10.7
Optical absorption spectrum near the band gap (dashed line) for GaAs at 21 K (after Sturge 1962).

10.2.5 Burstein–Moss effect

Another complication that may modify direct interband absorption arises from free carriers due to donor or acceptor impurities. At high doping levels with small effective masses and low temperatures, the carriers may be degenerate and the Fermi energy may lie within the conduction band

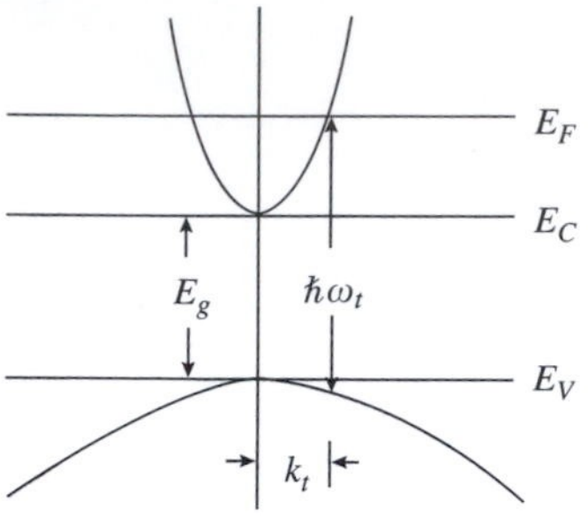

Fig. 10.8
Energy level diagram for high donor impurity concentration.

(donors) or the valence band (acceptors). Turning to the diagram for the donor impurity case shown in Fig. 10.8, we see that the Pauli principle blocks vertical transitions for wave vectors lying in the range $0 \le |\boldsymbol{k}| \le k_t$. The threshold wave vector k_t is given in the degenerate limit by $\hbar^2 k_t^2/2m_c^* = E_F - E_C$. The photon energy at the threshold for allowed transitions is given by

$$\begin{aligned}\hbar\omega_t &= E_F - E_V + \frac{\hbar^2 k_t^2}{2m_v^*} \\ &= E_g + (E_F - E_C)\left(1 + \frac{m_c^*}{m_v^*}\right).\end{aligned} \tag{10.81}$$

The rise of $\hbar\omega_t$ above E_g with increasing E_F and carrier concentration is the **Burstein–Moss effect** as shown in Fig. 10.9 in n-InSb.

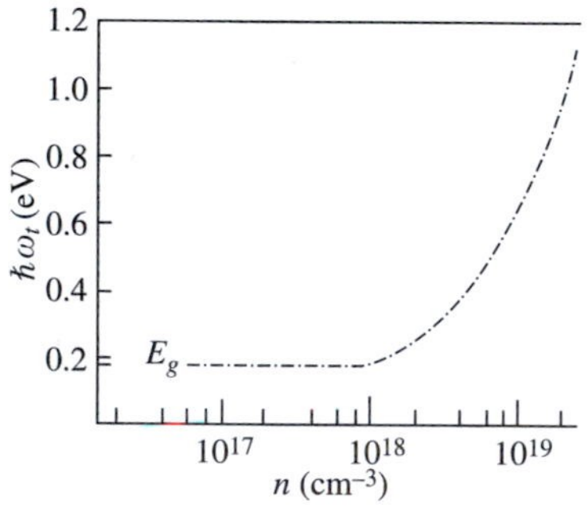

Fig. 10.9
Burstein–Moss effect in n-InSb (after Burstein 1954).

10.2.6 Indirect interband absorption

The threshold for interband optical absorption is associated with direct transitions if the conduction band minimum and valence band maximum occur at the same wave vector. To a good approximation this situation applies to a number of III–V semiconductors such as GaAs and InSb and II–VI semiconductors such as CdS and ZnS. However, in other semiconductors such as Si, Ge, and GaP, the band extrema do not occur at the same point in the Brillouin zone, as we have seen in Chapter 3.

In the latter materials direct optical transitions do not occur between the band extrema. However, there can be thermal excitation of electrons from states near the maximum of the valence band to states near the minimum of the conduction band. The energy gap between the extrema can be determined by measuring the temperature dependence of the electrical conductivity and is referred to as the **thermal gap** or **indirect gap**. A diagram of the situation is given in Fig. 10.10.

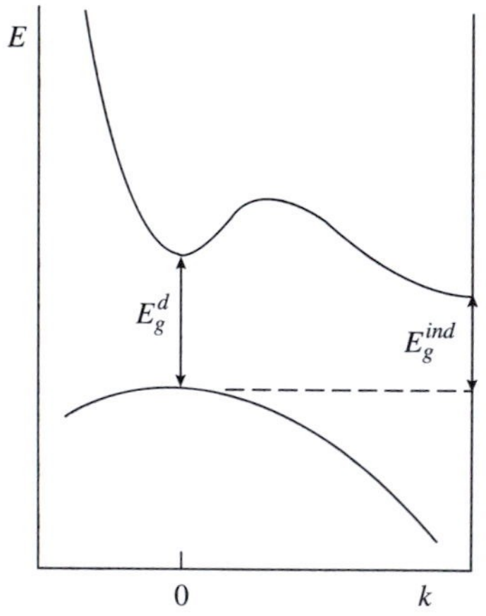

Fig. 10.10
Direct and indirect gaps in the energy band structure of a semiconductor.

Weak optical absorption is associated with **indirect transitions** between states near the band extrema. In order to conserve crystal momentum or wave vector it is necessary that phonons participate in the absorption process. Wave vector conservation can then be expressed as

$$\boldsymbol{k}_c = \boldsymbol{k}_v \pm \boldsymbol{q}, \tag{10.82}$$

where $\boldsymbol{q}$ is the phonon wave vector. Conservation of energy takes the form

$$\hbar\omega = E_c(\boldsymbol{k}_c) - E_v(\boldsymbol{k}_v) \pm \hbar\omega_{\boldsymbol{q}j}, \tag{10.83}$$

where $\hbar\omega_{\boldsymbol{q}j}$ is the energy of the phonon that is absorbed (lower sign) or emitted (upper sign) in the process. Two transitions via an intermediate state are involved: a vertical transition due to the electron–radiation interaction and a nonvertical transition due to the electron–phonon interaction, as shown in Fig. 10.11. There are two possibilities for the intermediate state: one in the conduction band and one in the valence band. To simplify matters we take all pertinent phonons to have the same average frequency $\bar{\omega}_{ph}$.

The analysis of the absorption coefficient for indirect interband absorption can be carried out by starting with a modification of Eq. (10.69) for the total transition probability. A factor involving the square of the magnitude of the appropriate electron–phonon interaction matrix element must be inserted to the left of the integral. The argument of the delta function must be changed to $E_c(\boldsymbol{k}_c) - E_v(\boldsymbol{k}_v) \pm \hbar\bar{\omega}_{ph} - \hbar\omega$. Furthermore, a factor involving the mean number of phonons $\bar{n}_{ph} = 1/[\exp(\beta\hbar\bar{\omega}_{ph}) - 1]$ must be inserted before the integral. The absorption coefficient then has two contributions $\alpha_\pm(\omega)$ which can be written as

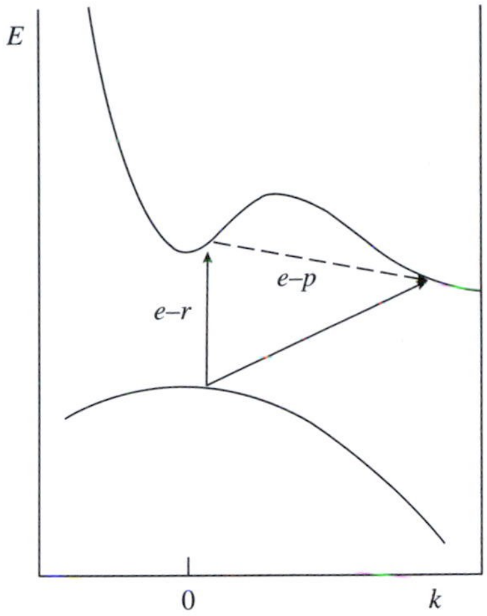

Fig. 10.11
Vertical and nonvertical transitions in crossing an indirect gap.

$$\alpha_\pm(\omega) = (K/\omega)|P_{va}|^2|H_{ac}|^2\left(\bar{n}_{ph} + \tfrac{1}{2} \pm \tfrac{1}{2}\right) \times \int d^3k_v \int d^3k_c \delta[E_c(\boldsymbol{k}_c) - E_v(\boldsymbol{k}_v) \pm \hbar\bar{\omega}_{ph} - \hbar\omega], \tag{10.84}$$

where H_{ac} is an effective electron–phonon interaction matrix element, K contains universal constants and P_{va} is the momentum matrix element. The factor containing $\bar{n}_{ph}$ imparts a temperature dependence to the absorption coefficient.

We take the zero of energy to be the valence band maximum and set

$$E_c(\boldsymbol{k}_c) = E_g + E, \tag{10.85}$$

where E is given by

$$E = \frac{\hbar^2|\boldsymbol{k}_c - \boldsymbol{k}_{co}|^2}{2m_c^*} \tag{10.86}$$

and $\boldsymbol{k}_{co}$ is the wave vector at the conduction band minimum. The valence band energy E_v is

$$E_v(\boldsymbol{k}_v) = -\frac{\hbar^2 k_v^2}{2m_v^*} = -E'. \tag{10.87}$$

Introducing the conduction band density-of-states $N_c(E)$ given by

$$N_c(E) = N_{co}E^{\frac{1}{2}} \tag{10.88}$$

and the corresponding valence band density-of-states $N_v(E')$, the expression for the absorption coefficient becomes

$$\alpha_\pm(\omega) = (K/\omega)|P_{va}|^2|H_{ac}|^2 N_{co}N_{vo}\left(\bar{n}_{ph} + \tfrac{1}{2} \pm \tfrac{1}{2}\right) \times \int dE'E'^{\frac{1}{2}} \int dEE^{\frac{1}{2}}\delta(E_g + E + E' \pm \hbar\bar{\omega}_{ph} - \hbar\omega). \tag{10.89}$$

Evaluating the integral over E gives

$$\alpha_\pm(\omega) = (K/\omega)|P_{va}|^2|H_{ac}|^2 N_{co}N_{vo}\left(\bar{n}_{ph} + \tfrac{1}{2} \pm \tfrac{1}{2}\right) \times \int_0^{E'_m} dE'E'^{\frac{1}{2}}[\hbar\omega \mp \hbar\bar{\omega}_{ph} - E_g - E']^{\frac{1}{2}}, \tag{10.90}$$

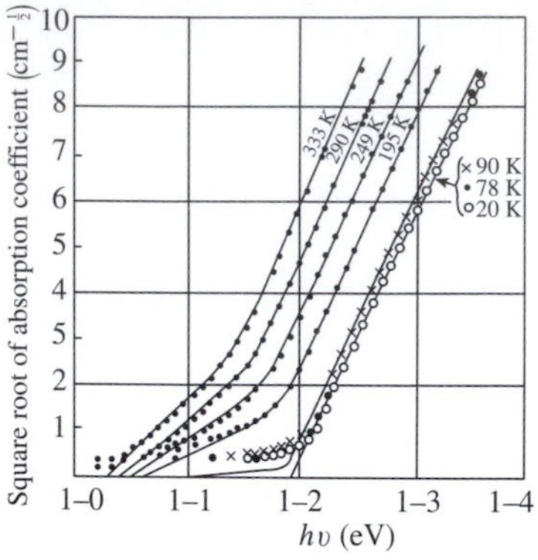

Fig. 10.12
Indirect interband absorption in Si (after MacFarlane and Roberts 1955).

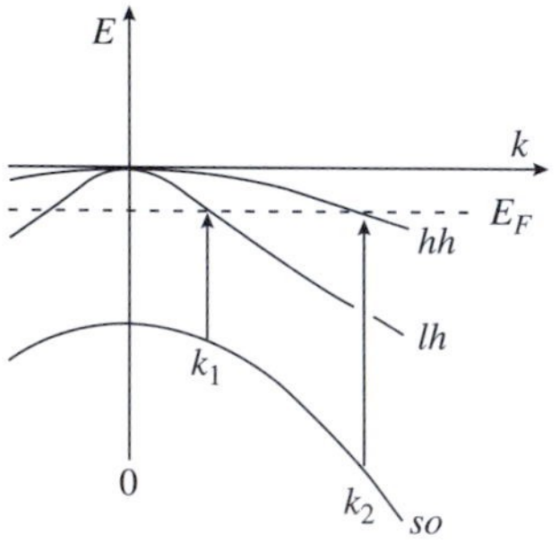

Fig. 10.13
Intervalence band transitions in a p-type semiconductor.

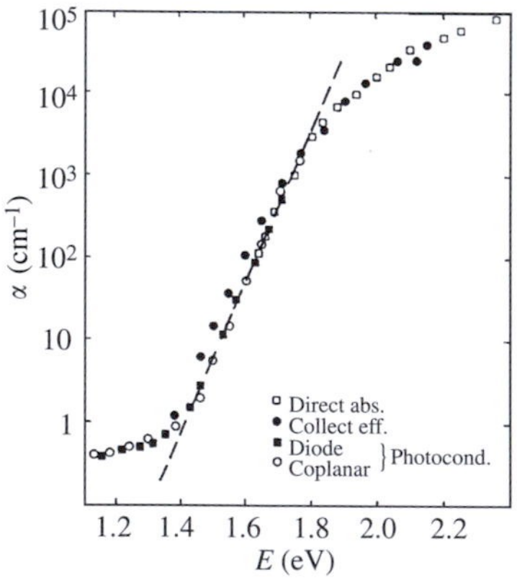

Fig. 10.14
Absorption coefficient versus photon energy in α-SiH$_x$ (after Abeles *et al.* 1980).

where E'_m is the largest value of E' that makes the integrand real:

$$E'_m = \hbar\omega - E_g \mp \hbar\bar{\omega}_{ph}. \tag{10.91}$$

The integral over E' can be evaluated by elementary means. The result for the total absorption coefficient $\alpha_+(\omega) + \alpha_-(\omega)$ is

$$\begin{aligned}\alpha(\omega) = {} & (\pi K/8\omega)|P_{va}|^2|H_{ac}|^2 N_{co}N_{vo} \\ & \times [\bar{n}_{ph}x^2\theta(x) + (\bar{n}_{ph}+1)y^2\theta(y)],\end{aligned} \tag{10.92}$$

where $\theta(x)$ is the Heaviside step function, $x = \hbar\omega - E_g + \hbar\bar{\omega}_{ph}$, and $y = \hbar\omega - E_g - \hbar\bar{\omega}_{ph}$. A plot of $\alpha(\omega)^{1/2}$ versus ω should be nearly a straight line as is shown in Fig. 10.12 for Si.

10.2.7 Extrinsic interband absorption

In Chapter 3 it was pointed out that the valence bands in typical semiconductors have a complex structure involving light hole, heavy hole, and split-off valence bands. At elevated temperatures or in heavily doped p-type material having the Fermi energy below the valence band edge, intervalence band transitions are possible from the light hole band to the heavy hole band and from the split-off band to the light and heavy hold bands, as shown in Fig. 10.13 for degenerate p-type material. If the bands are spherical and $T = 0$ K, the light to heavy hole transitions are restricted to the wave vector range from k_1 to k_2. The split-off to light hole band and split-off to heavy hole band transitions occur in the wave vector ranges 0 to k_1 and 0 to k_2, respectively. The intervalence band transitions are forbidden at $\boldsymbol{k} = 0$, because the valence bands at this point are p-like. Direct (vertical) transitions become possible for $\boldsymbol{k} \neq 0$. We speak of these transitions as **direct forbidden transitions**. The momentum matrix element for these transitions is proportional to k. They give rise to broad absorption bands which, in the case of spherical bands, have high- and low-frequency cutoffs. If the valence bands are nonspherical, the sharp absorption edges at the cutoff frequencies are smeared out.

10.2.8 Interband absorption in amorphous semiconductors

As pointed out in Chapter 3, band edges in amorphous semiconductors are smeared out and band gaps are not clearly defined. These features are manifested in the optical absorption spectrum due to interband transitions. Instead of having a low-frequency cutoff associated with a band gap, the absorption has a long tail called the **Urbach tail** that extends well into the gap. The absorption coefficient in the Urbach tail varies exponentially with photon energy as shown in Fig. 10.14 for hydrogenated α-Si (α − SiH$_x$). The exponential behavior is well described by the empirical relation $\alpha(\omega) = \alpha_0 \exp[(\hbar\omega - E_g)/E_0]$ over four decades of the absorption coefficient. The quantity E_0 is the **Urbach parameter** that specifies the slope of a $\log[\alpha(\omega)]$ versus $\hbar\omega$ plot.

10.3 Optical properties of free carriers

10.3.1 Free carrier absorption

If free carriers are present in a semiconductor as a result of interband excitation or ionization of impurities, absorption of electromagnetic radiation can take place as a result of transitions of the carriers between states of the same energy band. To get some insight into these **intraband optical transitions**, let us consider a carrier with wave vector $\boldsymbol{k}$ in a spherical parabolic band. If the carrier makes a transition to a state of wave vector $\boldsymbol{k}'$ in the same band with the simultaneous absorption of a photon of energy $\hbar\omega$ and wave vector $\boldsymbol{k}_0$, both energy and wave vector must be conserved. The conservation of energy condition is

$$\hbar\omega = E' - E = \frac{\hbar^2}{2m^*}(k'^2 - k^2). \tag{10.93}$$

The conservation of momentum condition is

$$\boldsymbol{k}' = \boldsymbol{k} + \boldsymbol{k}_0, \tag{10.94}$$

where $\boldsymbol{k}_0 = (\omega n(\omega)/c)\hat{e}$ and $\hat{e}$ is a unit vector in the direction of propagation of the photon. Introducing the angle θ between $\boldsymbol{k}$ and $\boldsymbol{k}_0$, we can re-express the wave vector conservation condition as

$$k'^2 = k^2 + k_0^2 - 2kk_0\cos\theta. \tag{10.95}$$

Eliminating k' from Eqs. (10.93) and (10.95), we obtain

$$\hbar\omega = 2\left[m^*\left(\frac{c}{n(\omega)}\right)^2 + \hbar k\left(\frac{c}{n(\omega)}\right)\cos\theta\right]. \tag{10.96}$$

Taking $m^* = 0.1\,m$, $n(\omega) = 3$, and $k = 0$, we find that $m^*(c/n(\omega))^2 \simeq 10^4$ eV. The photon energy is therefore on the order of 10^4 eV, which is much larger than typical widths of energy bands near the fundamental gap. Radiative transitions of the type just discussed cannot account for intraband free carrier absorption.

In order to achieve intraband absorption, it is necessary to consider second order processes which include the scattering of free carriers by phonons, impurities, and other defects. In Chapter 8 expressions for the mean relaxation time associated with a number of scattering processes were derived. These relaxation times can be incorporated into a classical treatment of the electromagnetic response of free carriers using the Drude model.

The classical treatment gives a good approximation to the main features of this process. We shall first deduce the dielectric response of an electron gas and then calculate the absorption coefficient. The usefulness of reflectivity spectra in determining semiconductor characteristics such as free carrier concentration and effective mass will be discussed.

10.3.1.1 Dielectric response of an electron gas

A very simple model of an electron gas in a solid is obtained by considering only forces on the electrons arising from the macroscopic electric field and from damping processes such as collisions. The classical equation of motion of the ith electron can then be written as

$$m^* \ddot{\boldsymbol{x}}_i + m^* \gamma \dot{\boldsymbol{x}}_i = -e\boldsymbol{\mathcal{E}}, \tag{10.97}$$

where $\boldsymbol{x}_i$, m^* and γ are the position vector, effective mass, and damping constant, respectively, of the ith electron, and we have assumed that the effective mass is isotropic. If the time dependences of $\boldsymbol{\mathcal{E}}$ and $\boldsymbol{x}_i$ are taken to be $\boldsymbol{\mathcal{E}}(t) = \boldsymbol{\mathcal{E}}_0 \exp(-i\omega t)$ and $\boldsymbol{x}_i(t) = \boldsymbol{x}_{i0} \exp(-i\omega t)$, then Eq. (10.97) can be solved to yield

$$\boldsymbol{x}_i = \frac{e\boldsymbol{\mathcal{E}}}{m^* \omega(\omega + i\gamma)}. \tag{10.98}$$

The polarization or electric dipole moment per unit volume is given by

$$\boldsymbol{P} = -e \sum_i \boldsymbol{x}_i / \Omega = -\frac{ne^2 \boldsymbol{\mathcal{E}}}{m^* \omega(\omega + i\gamma)}, \tag{10.99}$$

where n is the electron concentration. The dielectric susceptibility and dielectric function can now be obtained using Eqs. (10.7) and (10.10). The result for the dielectric function is

$$\epsilon(\omega) = 1 - \frac{ne^2}{\epsilon_0 m^* \omega(\omega + i\gamma)}. \tag{10.100}$$

In an electrically neutral solid with free electrons one has, in fact, a plasma with equal concentrations of positive and negative charges, the negative charges being mobile. At a displaced position $\boldsymbol{\xi}$, the electron is subject to a restoring force proportional to the displacement. This leads to a harmonic oscillation whose frequency, the plasma frequency ω_p, can be evaluated from the equation of motion neglecting damping,

$$m^* \ddot{\boldsymbol{\xi}} = -e\boldsymbol{\mathcal{E}}, \tag{10.101}$$

and the equation of continuity,

$$\dot{n} + \boldsymbol{\nabla} \cdot (n\dot{\boldsymbol{\xi}}) = 0. \tag{10.102}$$

Let us write

$$n = \bar{n} + n_1, \tag{10.103}$$

where $\bar{n}$ is the mean electron density and $e\bar{n}$ is equal to the mean positive charge density. Treating both n_1 and $\boldsymbol{\xi}$ as small quantities, Eq. (10.102) becomes, in first-order approximation,

$$\dot{n}_1 + \bar{n} \boldsymbol{\nabla} \cdot \dot{\boldsymbol{\xi}} = 0. \tag{10.104}$$

Assuming that all time-dependent quantities vary as $\exp(-i\omega t)$, we obtain from Eq. (10.104)

$$n_1 = -\bar{n}\,\boldsymbol{\nabla}\cdot\boldsymbol{\xi}. \tag{10.105}$$

Let us now take the divergence of Eq. (10.101) and use the Maxwell equation

$$\nabla\cdot\boldsymbol{\mathcal{E}} = -\frac{en_1}{\epsilon_0}. \tag{10.106}$$

We then obtain

$$-m^*\omega^2\boldsymbol{\nabla}\cdot\boldsymbol{\xi} = -e\nabla\cdot\boldsymbol{\mathcal{E}} = \frac{e^2 n_1}{\epsilon_o} \tag{10.107}$$

which, with the aid of Eq. (10.105) becomes

$$-m^*\omega^2\boldsymbol{\nabla}\cdot\boldsymbol{\xi} = -\frac{e^2\bar{n}\,\boldsymbol{\nabla}\cdot\boldsymbol{\xi}}{\epsilon_0} \tag{10.108}$$

or

$$\omega^2 = \omega_p^2 = \frac{e^2\bar{n}}{\epsilon_0 m^*}. \tag{10.109}$$

If n in Eq. (10.100) is interpreted as $\bar{n}$, we can then write the dielectric function for the free electron gas as

$$\epsilon(\omega) = 1 - \frac{\omega_p^2}{\omega(\omega + i\gamma)}. \tag{10.110}$$

In the case of a solid such as a semiconductor, it is necessary to take into account interband electronic transitions. If ω is less than typical interband transition frequencies, one can replace Eq. (10.110) by

$$\epsilon(\omega) = \epsilon_\infty - \frac{\omega_p^2}{\omega(\omega + i\gamma)}, \tag{10.111}$$

where ϵ_∞ represents the contribution of interband transitions to the dielectric constant. It is convenient to redefine the plasma frequency by

$$\omega_p^2 = \frac{e^2\bar{n}}{\epsilon_\infty\epsilon_0 m^*}. \tag{10.112}$$

Then we can rewrite Eq. (10.111) as

$$\epsilon(\omega) = \epsilon_\infty\left[1 - \frac{\omega_p^2}{\omega(\omega + i\gamma)}\right]. \tag{10.113}$$

The plasma frequency as defined by Eq. (10.112) is the frequency at which the dielectric function becomes zero in the absence of damping. The plasma

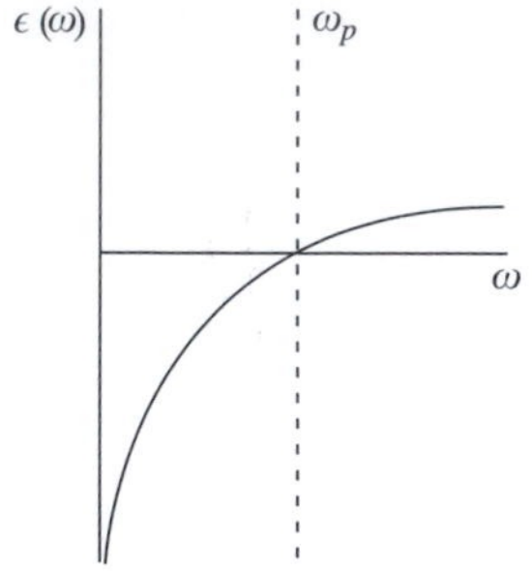

Fig. 10.15
Dielectric function versus frequency for a free electron gas without damping.

frequency is therefore analogous to ω_{LO} in the optical phonon case. One can show, in fact, that plasma oscillations are longitudinal oscillations. The analogue of ω_{TO} is zero, since the free electron gas cannot support a shear stress.

A plot of $\epsilon(\omega)$ for the free electron gas without damping is given in Fig. 10.15. We note that the dielectric function is negative for $\omega < \omega_p$. The dispersion relation describing wave propagation for $\omega > \omega_p$ can be obtained by substituting Eq. (10.113) into Eq. (10.27) and taking $\gamma = 0$. The result is

$$\omega^2 = \omega_p^2 + \frac{c^2k^2}{\epsilon_\infty}. \tag{10.114}$$

This relation applies to transverse electromagnetic waves in a solid such as a semiconductor that contains a plasma.

10.3.1.2 Drude model for electrical conductivity and optical absorption

Highly doped or degenerate semiconductors are characterized by a relatively high electrical conductivity due to the presence of a high concentration of free carriers. The conductivity is in general frequency-dependent and has a nonzero value as $\omega \to 0$. The Drude model based upon the equation of motion given by Eq. (10.97) provides a simple phenomenological picture of the conductivity due to conduction electrons in a parabolic band. If one differentiates Eq. (10.98) with respect to time, one obtains the velocity of an electron in the form

$$\dot{x}_i = -\frac{ie\mathcal{E}}{m^*(\omega + i\gamma)}. \tag{10.115}$$

Multiplying the velocity by the electron charge, $-e$, and by the electron concentration n, gives the conduction current density

$$J_{cond} = \frac{ine^2\mathcal{E}}{m^*(\omega + i\gamma)}. \tag{10.116}$$

Comparing this result with the equation $J_{cond} = \sigma(\omega)\mathcal{E}$ we see that the conductivity is given by

$$\sigma(\omega) = \frac{ine^2}{m^*(\omega + i\gamma)}. \tag{10.117}$$

In the limit of zero frequency, the real part of the conductivity reduces to the standard expression for the DC conductivity

$$\sigma'(0) = \frac{ne^2}{m^*\gamma} = \frac{ne^2\tau}{m^*}, \tag{10.118}$$

where $\tau = 1/\gamma$ is the scattering time.

For the discussion of optical properties we need the real and imaginary parts of the dielectric function which follow from Eq. (10.113):

$$\epsilon'(\omega) = \epsilon_\infty\left(1 - \frac{\omega_p^2}{\omega^2 + \gamma^2}\right) \tag{10.119}$$

$$\epsilon''(\omega) = \frac{\epsilon_\infty \omega_p^2 \gamma}{\omega(\omega^2 + \gamma^2)}. \tag{10.120}$$

A limit of interest is that where $\gamma \ll \omega$. Then

$$\epsilon''(\omega) \cong \frac{\epsilon_\infty \omega_p^2 \gamma}{\omega^3}. \tag{10.121}$$

If one introduces the absorption coefficient $\alpha(\omega)$ with the aid of Eq. (10.36), one finds that

$$\alpha(\omega) = \frac{\epsilon_\infty \omega_p^2 \gamma}{cn(\omega)\omega^2}. \tag{10.122}$$

Consider a semiconductor at frequencies such that $\omega \gg \omega_p$. Then from Eqs. (10.34) and (10.113) we find that the refractive index $n(\omega) \cong \epsilon_\infty^{1/2}$. The absorption coefficient should therefore vary as ω^{-2} or as λ^2 where λ is the wavelength. Many materials exhibit such a square law dependence on wavelength, but there are also materials in which the absorption coefficient varies as λ^s, where $s > 2$. In the latter case this behavior may be attributed, at least in part, to the fact that the damping constant γ and its inverse, the scattering time τ, are frequency-dependent. The development of a proper theory of this dependence requires the application of techniques (Holstein 1964, Wallis and Balkanski 1986) that are beyond the scope of this book.

10.3.2 Free carrier reflectivity

It was seen in Section 10.1.3 that the normal incidence reflectivity can be expressed in terms of the real and imaginary parts of the refractive index by Eq. (10.40). The latter quantities in turn are related to the real and imaginary parts of the dielectric function by Eq. (10.34). Using the expressions for $\epsilon'(\omega)$ and $\epsilon''(\omega)$ given by Eqs. (10.119) and (10.120) we can determine all of the quantities necessary to calculate the normal incidence reflectivity.

The simplest case to consider neglects scattering of the free carriers, so that the relaxation time is infinite and the damping constant γ is zero. The dielectric function is then real, $\epsilon''(\omega) = 0$ and $K = 0$. If $\omega > \omega_p$, $\epsilon' > 0$, $n = \sqrt{\epsilon'} > 0$, and $0 < R(\omega) < 1$. On the other hand, if $\omega < \omega_p$, $\epsilon' < 0$. The real part of the refractive index n is then zero, since $\sqrt{\epsilon'}$ is pure imaginary and n is by definition real. The imaginary part of the refractive index K is given by $K = \sqrt{-\epsilon'}$, and the normal incidence reflectivity is unity.

If scattering of the carriers is now taken into account, $\gamma > 0$ and $\epsilon'' > 0$. The optical constants are specified by

$$n^2 = \tfrac{1}{2}[\epsilon' + (\epsilon'^2 + \epsilon''^2)^{\frac{1}{2}}] \quad (10.123a)$$

$$K^2 = \tfrac{1}{2}[-\epsilon' + (\epsilon'^2 + \epsilon''^2)^{\frac{1}{2}}]. \quad (10.123b)$$

Both n^2 and K^2 are positive regardless of the sign of ϵ' and $R(\omega) < 1$ except at $\omega = 0$ where $R(\omega) = 1$.

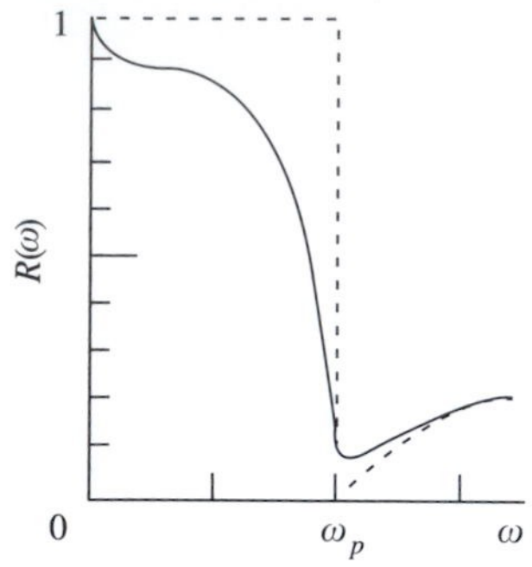

Fig. 10.16
Normal incidence reflectivity versus frequency for free carriers for $\gamma = 0$ (dashed lines) and $\gamma = \omega_p/5$ (solid curve).

A plot of the normal incidence reflectivity as a function of frequency is shown in Fig. 10.16 for both $\gamma = 0$ and $\gamma > 0$. In both cases $R(\omega)$ is relatively small for ω above ω_p. For $\omega < \omega_p$, $R(\omega)$ is relatively large. It is close to unity if $\gamma \ll \omega_p$. The frequency dependence of the reflectivity provides a useful means of determining parameters such as ω_p from experimental measurements of reflectivity. By varying ω_p and γ to give the best fit of the theoretical curve for $R(\omega)$ to experimental data, experimental values of ω_p and γ can be obtained provided ϵ_∞ is known. If the carrier effective mass is also known, the carrier concentration can be determined from Eq. (10.112).

10.4 Absorption due to electronic transitions of impurities

It was pointed out in Chapter 5 that there are localized electronic states associated with impurities in semiconductors. The energy levels of these states lie in the forbidden energy gap between the valence and conduction bands. The energy levels of donor impurities typically lie near the conduction band edge, whereas those of acceptor impurities typically lie near the valence band edge.

Effective mass theory shows that there is a correspondence between the localized impurity states in a semiconductor and bound states of an isolated atom. In certain cases such as P in Si, the localized states correspond to those of a hydrogen atom and can be labeled as 1*s*, 2*p*, 3*d*, etc. The lower symmetry of a crystal compared to that of an isolated atom, however, can lead to a splitting of impurity levels that would be degenerate in an atom.

Optical transitions can occur between localized impurity levels with selection rules similar to those that apply to isolated atoms. Because the binding energies of impurity carriers are typically much smaller than those of electrons in atoms, the frequencies of photons absorbed or emitted in transitions between impurity levels typically lie in the infrared rather than the visible or ultraviolet. Optical transitions can take place not only from one localized state to another localized state, but also from a localized state to an unbound state lying above the band edge. Such a transition corresponds to photoionization of the impurity carrier and causes an increase in the electrical conductivity of the material. A variety of infrared detectors are based on this effect.

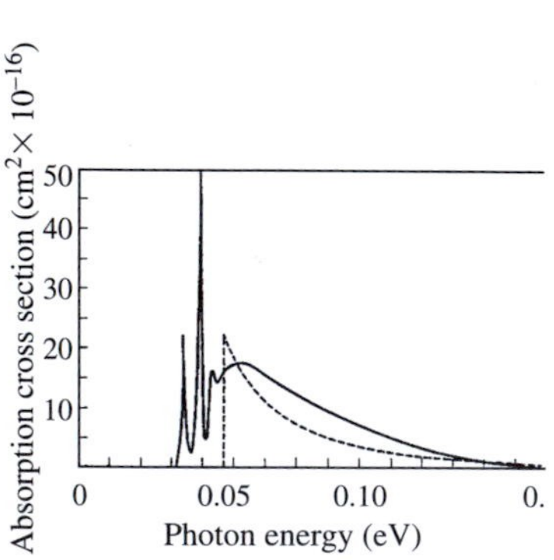

Fig. 10.17
Absorption coefficient versus photon energy for boron-doped silicon (after Burstein *et al.* 1956).

The experimental absorption coefficient for boron-doped silicon is shown in Fig. 10.17. The sharp peaks correspond to transitions between localized states, and the broad tail at higher photon energies corresponds to transitions to unbound states.

10.5 Optical properties due to lattice vibrations

In Chapter 7 we have seen that the equations of motion of the atoms in a perfect lattice with periodic boundary conditions have nontrivial solutions in which the frequency ω is specified by the dispersion relation as a continuous function of the wave vector $\boldsymbol{q}$ within the first Brillouin zone. For crystals with two atoms per primitive unit cell the solutions fall into two sets of branches separated by a range of forbidden frequencies. The three low-frequency branches constitute the acoustic branch. Two of the branches may be classified as transverse and the third as longitudinal if the symmetry is sufficiently high. The three high-frequency branches constitute the optical branch with transverse and longitudinal modes as in the acoustic case. In polar crystals the TO modes of long wavelength interact strongly with the radiation field giving rise to characteristic absorption and reflection spectra. Combining experimental spectra with a theoretical analysis enables one to determine the frequencies and other characteristics of the optical modes. Light scattering spectra are also of interest for determining these quantities.

10.5.1 Dielectric response of polar lattice vibrations

Let us consider an insulating polar crystal where the lattice is made up of positively and negatively charged ions. For simplicity we restrict our attention to cubic crystals with two atoms per unit cell, one positively charged and the other negatively charged. In an optical mode of vibration, the two ions in a unit cell vibrate against one another and produce an electric dipole moment. Particularly important are the optical modes of very long wavelength where the dipole moments in the various unit cells are in phase and add up to give a large macroscopic polarization. This polarization can be written in the form

$$\boldsymbol{P} = Ne_T^*(\boldsymbol{u}_1 - \boldsymbol{u}_2)/\Omega + \epsilon_0(\epsilon_\infty - 1)\boldsymbol{\mathcal{E}}, \tag{10.124}$$

where $\boldsymbol{u}_1$ and $\boldsymbol{u}_2$ are the displacements of the two ions in a unit cell, N is the number of unit cells in the crystal, Ω is the volume of the crystal, e_T^* is the **transverse effective charge**, and ϵ_∞ is the dielectric constant at frequencies high compared to the optical phonon frequencies. The contribution proportional to $\boldsymbol{\mathcal{E}}$ on the right hand side of Eq. (10.124) is due to interband electronic transitions.

In order to obtain the dielectric function we must calculate the displacement difference $\boldsymbol{u} \equiv \boldsymbol{u}_1 - \boldsymbol{u}_2$ as a function of $\boldsymbol{\mathcal{E}}$. We extend the treatment given in Section 7.4 by including phonon damping and write the equations of motion in the form (Born and Huang 1954)

$$M_1(\ddot{\boldsymbol{u}}_1 + \Gamma\dot{\boldsymbol{u}}_1) + 2\bar{\sigma}(\boldsymbol{u}_1 - \boldsymbol{u}_2) = e_T^*\boldsymbol{\mathcal{E}} \tag{10.125a}$$

$$M_2(\ddot{\boldsymbol{u}}_2 + \Gamma\dot{\boldsymbol{u}}_2) + 2\bar{\sigma}(\boldsymbol{u}_2 - \boldsymbol{u}_1) = -e_T^*\boldsymbol{\mathcal{E}}, \tag{10.125b}$$

where M_1 and M_2 are the masses of the two ions in the unit cell, Γ is a phenomenological damping constant, and $\bar{\sigma}$ is the effective Hooke's law force constant for the interaction between nearest neighbor ions.

Multiplying the first equation by M_2, the second by M_1, subtracting the second of the resulting equations from the first, and dividing by $M_1 + M_2$ yields the equation of motion for the relative displacement $\boldsymbol{u} = \boldsymbol{u}_1 - \boldsymbol{u}_2$,

$$\bar{M}(\ddot{\boldsymbol{u}} + \Gamma\dot{\boldsymbol{u}}) + 2\bar{\sigma}\boldsymbol{u} = e_T^* \boldsymbol{\mathcal{E}}, \tag{10.126}$$

where $\bar{M}$ is the reduced mass of the positive and negative ions. Dividing Eq. (10.126) by $\bar{M}$ converts it to

$$\ddot{\boldsymbol{u}} + \Gamma\dot{\boldsymbol{u}} + \omega_{TO}^2\boldsymbol{u} = (e_T^*/\bar{M})\boldsymbol{\mathcal{E}}, \tag{10.127}$$

where $\omega_{TO} = \sqrt{2\bar{\sigma}/\bar{M}}$ is the frequency of the optical phonon of long wavelength that is excited by the coupling to the external radiation field. This optical phonon is transverse because the external radiation field is transverse. To solve Eq. (10.127), let

$$\boldsymbol{u} = \boldsymbol{u}_0 \exp(-i\omega t) \tag{10.128}$$

$$\boldsymbol{\mathcal{E}} = \boldsymbol{\mathcal{E}}_0 \exp(-i\omega t). \tag{10.129}$$

Substituting into Eq. (10.127) we find that

$$\boldsymbol{u} = \frac{(e_T^*/\bar{M})\boldsymbol{\mathcal{E}}}{\omega_{TO}^2 - \omega^2 - i\omega\Gamma}. \tag{10.130}$$

Combining Eqs. (10.124) and (10.130) and using Eq. (10.7), we obtain from Eq. (10.10) the following result for $\epsilon(\omega)$:

$$\epsilon(\omega) = \epsilon_\infty + \frac{e_T^{*2}/\bar{M}\Omega_0\epsilon_0}{\omega_{TO}^2 - \omega^2 - i\omega\Gamma}, \tag{10.131}$$

where Ω_0 is the volume of a unit cell. Setting $\omega = 0$ in Eq. (10.131) yields the static dielectric constant

$$\epsilon_s \equiv \epsilon(0) = \epsilon_\infty + \frac{e_T^{*2}/\bar{M}\Omega_0\epsilon_0}{\omega_{TO}^2}. \tag{10.132}$$

The transverse effective charge can then be expressed as

$$e_T^* = [\bar{M}\Omega_0\epsilon_0(\epsilon_s - \epsilon_\infty)]^{\frac{1}{2}}\omega_{TO} \tag{10.133}$$

which involves experimentally measurable quantities.

Eliminating the quantity $e_T^{*2}/\bar{M}\Omega_0\epsilon_0$ from Eqs. (10.131) and (10.132) gives

$$\epsilon(\omega) = \epsilon_\infty + \frac{\omega_{TO}^2(\epsilon_s - \epsilon_\infty)}{\omega_{TO}^2 - \omega^2 - i\omega\Gamma}. \tag{10.134}$$

The real and imaginary parts of $\epsilon(\omega)$ are

$$\epsilon'(\omega) = \epsilon_\infty + \frac{\omega_{TO}^2(\omega_{TO}^2 - \omega^2)(\epsilon_s - \epsilon_\infty)}{(\omega_{TO}^2 - \omega^2)^2 + \omega^2\Gamma^2} \qquad (10.135a)$$

$$\epsilon''(\omega) = \frac{\omega_{TO}^2\omega\Gamma(\epsilon_s - \epsilon_\infty)}{(\omega_{TO}^2 - \omega^2)^2 + \omega^2\Gamma^2}. \qquad (10.135b)$$

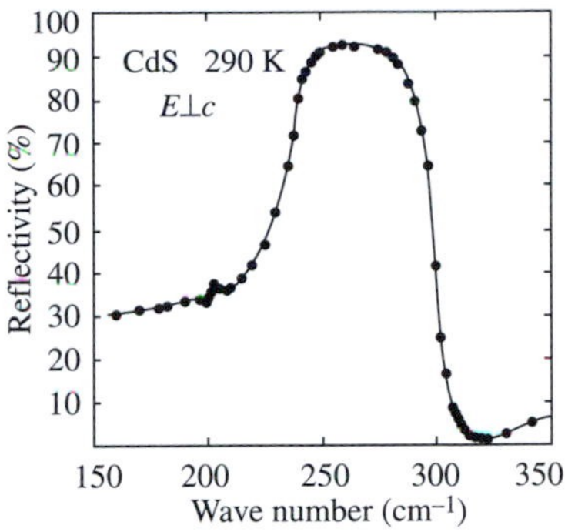

Fig. 10.18
Reflectivity versus wave number for CdS at 290 K (after Balkanski 1972).

By choosing values for the four parameters $\epsilon_s, \epsilon_\infty, \omega_{TO}$, and Γ, one can calculate $\epsilon'(\omega)$ and $\epsilon''(\omega)$ as functions of frequency. Using Eqs. (10.34) and (10.40) one can then calculate $n(\omega)$, $K(\omega)$, and $R(\omega)$. By adjusting $\epsilon_s, \epsilon_\infty, \omega_{TO}$ and Γ to give the best fit of the calculated reflectivity curve to an experimental curve, one can determine the values of these four parameters for a particular material. An example of such a procedure is shown for CdS in Fig. 10.18. For further details see Balkanski (1972).

It is frequently not possible to obtain an entirely satisfactory fit between calculated and experimental reflectivity curves. One reason for this is that the damping parameter Γ is not really a constant, but is a function of frequency. Anharmonic interactions provide a mechanism whereby Γ becomes frequency-dependent (Wallis and Balkanski 1986).

If Eq. (10.134) is substituted into Eq. (10.27) we obtain the dispersion relation for electromagnetic waves propagating in a polar cubic crystal in the frequency range of optical phonons. Let us for the moment ignore damping by setting $\Gamma = 0$. Then the dispersion relation becomes

$$\frac{c^2k^2}{\omega^2} = \epsilon_\infty + \frac{\omega_{TO}^2(\epsilon_s - \epsilon_\infty)}{\omega_{TO}^2 - \omega^2}, \qquad (10.136)$$

where k is the magnitude of the wave vector. This equation may be simplified by introducing the frequency of longitudinal optical phonons of long wavelength, ω_{LO}, through the Lyddane–Sachs–Teller relation given by Eq. (7.62). One then obtains

$$\frac{c^2k^2}{\omega^2} = \frac{\epsilon_\infty(\omega_{LO}^2 - \omega^2)}{\omega_{TO}^2 - \omega^2} = \epsilon(\omega)_{\Gamma=0}. \qquad (10.137)$$

Using values of ϵ_s, ϵ_∞, and ω_{TO} determined from reflectivity measurements, one can calculate the value of ω_{LO}.

The longitudinal optical mode whose frequency appears in Eq. (10.137) is characterized by an oscillating macroscopic electric field even in the absence of an external radiation field. If the latter field is zero, then the electric displacement $\mathcal{D}$ is zero. Since $\mathcal{D} = \epsilon_0\mathcal{E} + \boldsymbol{P}$ and $\boldsymbol{P}$ is given by Eq. (10.124), setting $\mathcal{D}$ equal to zero gives the result

$$\mathcal{E} = -\frac{e_T^*\boldsymbol{u}}{\epsilon_0\epsilon_\infty\Omega_0}. \qquad (10.138)$$

Eliminating $\mathcal{E}$ from Eq. (10.127), we obtain the equation of motion for longitudinal optical phonons of long wavelength:

$$\ddot{\boldsymbol{u}} + \Gamma\dot{\boldsymbol{u}} + \omega_{LO}^2\boldsymbol{u} = 0, \qquad (10.139)$$

Table 10.1 Optical mode parameters of heteropolar semiconductors (after Burstein *et al.* 1971).

Semiconductor	ϵ_∞	ϵ_s	ω_{TO} cm^{-1}	ω_{LO} cm^{-1}	e_T^*/e
AlP	7.56	9.83	440	501	2.28
AlAs	9	11	361	404	2.3
AlSb	10.2	11.6	319	340	1.93
GaP	8.5	10.7	367	403	2.04
GaAs	10.9	12.9	269	292	2.16
GaSb	14.4	16.1	230	243	2.15
InP	9.6	12.4	304	345	2.55
InAs	12.3	14.9	219	241	2.53
InSb	15.6	17.7	185	197	2.42
ZnS	5.1	8.7	271	352	2.15
ZnSe	5.9	8.8	207	253	2.03
ZnTe	7.3	9.9	177	206	2.00
CdTe	7.3	10.8	140	170	2.35
HgTe	14.0	20.1	116	139	2.96

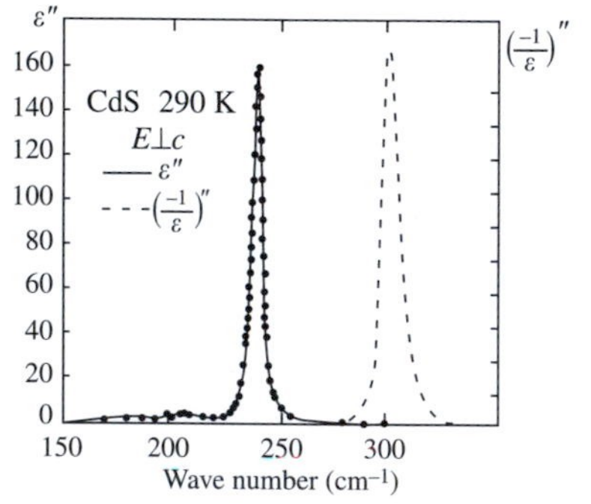

Fig. 10.19
$\epsilon''(\omega)$ and $(-1/\epsilon(\omega))''$ versus wave number for CdS at 290 K (after Balkanski 1972).

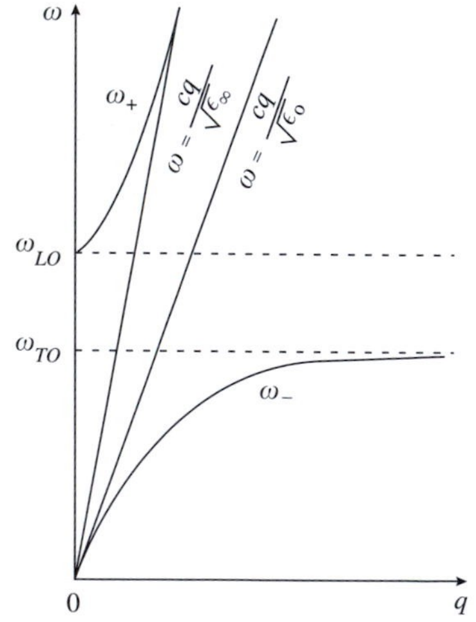

Fig. 10.20
Dispersion curves for mixed photon-optical phonon modes.

where

$$\omega_{LO}^2 = \omega_{TO}^2 + \frac{e_T^{*2}}{\epsilon_0 \epsilon_\infty \bar{M} \Omega_0}. \tag{10.140}$$

This result for ω_{LO}^2 also arises if ϵ_s is eliminated from Eq. (10.132) using the Lyddane–Sachs–Teller relation. For heteropolar semiconductors such as GaAs, $e_T^* \neq 0$, and $\omega_{LO} > \omega_{TO}$, whereas for homopolar semiconductors such as Si, $e_T^* = 0$ and $\omega_{LO} = \omega_{TO}$. Optical mode parameters of heteropolar semiconductors are listed in Table 10.1.

We note from Eq. (10.135b) that $\epsilon''(\omega)$ has its maximum at $\omega = \omega_{TO}$. If e_T^{*2} is eliminated from Eq. (10.131) with the aid of Eq. (10.140), one can obtain the quantity $(-1/\epsilon(\omega))''$ in the form

$$\left(-\frac{1}{\epsilon(\omega)}\right)'' = \frac{\omega\Gamma(\omega_{LO}^2 - \omega_{TO}^2)}{\epsilon_\infty[(\omega_{LO}^2 - \omega^2)^2 + \omega^2\Gamma^2]} \tag{10.141}$$

which peaks at $\omega = \omega_{LO}$. The behavior of $\epsilon''(\omega)$ and $(-1/\epsilon(\omega))''$ is shown in Fig. 10.19, for CdS (Balkanski 1972).

Equation (10.137) yields two solutions, ω_+ and ω_-, which are plotted as functions of k in Fig. 10.20. We see that ω_+ lies in the region $\omega > \omega_{LO}$ and ω_- in the region $\omega < \omega_{TO}$. The curves for ω_+ and ω_- are typical of those arising from interacting systems. In this case the interacting systems are photons and transverse optical phonons. The coupled modes are known as **polaritons.**

When $|\boldsymbol{k}| \gg \omega_{TO}(\epsilon_\infty^{1/2}/c)$, the solutions to Eq. (10.137) are $\omega_+ \approx c|\boldsymbol{k}|/\epsilon_\infty^{1/2}$ and $\omega_- \approx \omega_{TO}$. The first solution corresponds to a polariton having the character of a photon with velocity $c/\epsilon_\infty^{1/2}$. The second solution corresponds to a polariton having the character of an optical phonon of frequency ω_{TO}. For small values of $|\boldsymbol{k}|$, the energy of the polariton is a mixture of mechanical energy and electromagnetic energy (Huang 1951). The fraction τ of mechanical energy contained in the total energy of a polariton of frequency ω is given by

$$\tau = \frac{E_{\text{mec}}}{E_{\text{polariton}}} = \frac{(\epsilon_s - \epsilon_\infty)[1 + (\omega/\omega_{TO})^2]}{2\{\epsilon_\infty[1 - (\omega/\omega_{TO})^2]^2 + (\epsilon_s - \epsilon_\infty)\}}. \tag{10.142}$$

For each of the two branches ω_+ and ω_-, we have plotted in Fig. 10.21 the fraction τ as a function of $|\boldsymbol{k}|$.

If we refer back to Fig. 10.20 we note that no electromagnetic wave propagates in the crystal in the frequency interval between ω_{TO} and ω_{LO} when damping is neglected. The crystal is therefore totally reflecting in this interval. If infrared radiation of broad bandwidth is reflected repeatedly from the surface of a heteropolar crystal, only the frequency range between ω_{LO} and ω_{TO} remains in the reflected beam. This radiation constitutes the **residual rays** or **reststrahlen.**

The absence of propagation in the reststrahlen region can be understood in terms of the dielectric function, which is plotted in Fig. 10.22 as a function of frequency when $\Gamma = 0$. We see that between ω_{TO} and ω_{LO} the dielectric function is negative and hence the refractive index from Eq. (10.29) is imaginary. Consequently, from Eq. (10.28), an electromagnetic wave incident on the crystal will decay exponentially into the crystal and will not propagate. Therefore, the crystal is 100% reflecting in the reststrahlen region in the absence of damping. Experimentally the measured reflectivity is not 100% due to the presence of damping and peaks at a frequency between ω_{TO} and ω_{LO}. It is worth noting that ω_{LO} is the frequency at which the dielectric function in the absence of damping goes to zero and ω_{TO} is the frequency at which it goes to infinity, as may be seen from Eq. (10.137). The reststrahlen region is evident in the reflectivity data shown in Fig. 10.18 for CdS with $\omega_{TO} = 240\,\text{cm}^{-1}$ and $\omega_{LO} = 300\,\text{cm}^{-1}$.

When damping is included in the dielectric function, the dispersion relation cannot be solved to give a unique plot of the real part of ω versus the real part of k. One can take ω real and solve for complex k or take k real and solve for complex ω, for example. The dispersion curves obtained with ω real and k complex for a zincblende crystal are shown in Fig. 10.23. We note that the dispersion curve exhibits backbending in the vicinity of ω_{TO} and that it passes continuously through the reststrahlen region. The backbending is not found if one solves for complex ω with real k. Whether or not backbending is observed depends on the experimental situation.

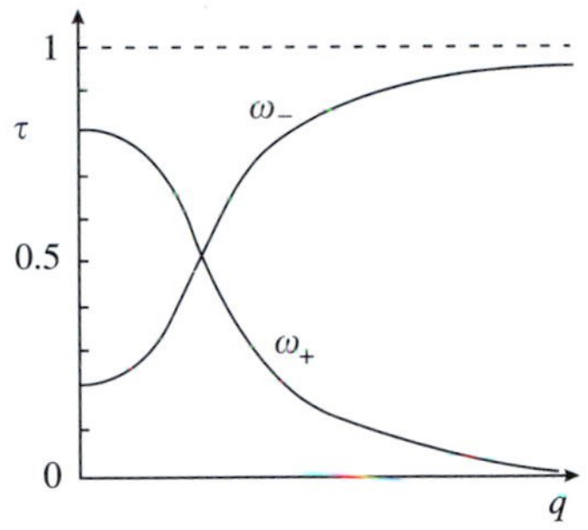

Fig. 10.21
Fraction of the mechanical energy in the total energy of a polariton versus wave vector.

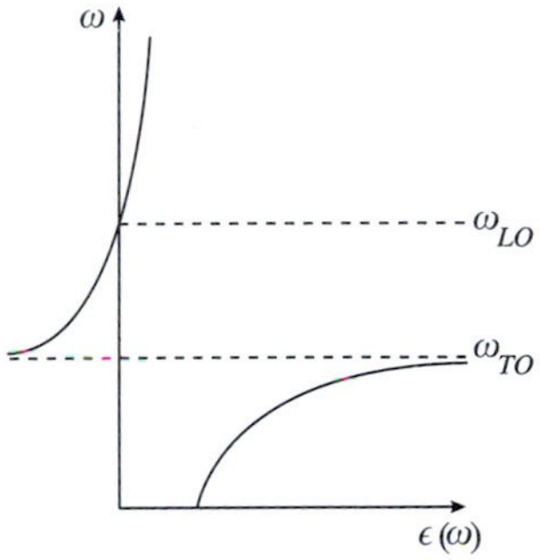

Fig. 10.22
Dielectric function versus frequency in the optical phonon region for $\Gamma = 0$.

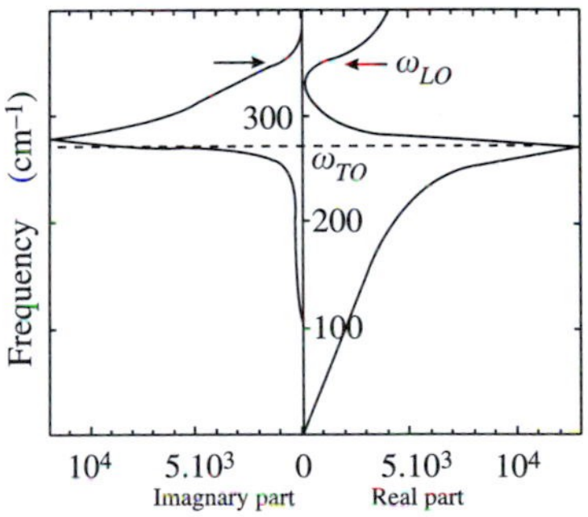

Fig. 10.23
Real and imaginary parts of the wave vector versus frequency for $\Gamma \neq 0$ (after Le Toullec 1968).

10.5.2 Lattice vibration absorption

Strong optical absorption is observed at the frequency of the transverse optical mode ω_{TO}. The power absorbed is proportional to the real part of the electrical conductivity $\sigma'(\omega)$. The latter can be calculated from the expression for $\boldsymbol{u}$ given by Eq. (10.130). Since $\dot{\boldsymbol{u}} = -i\omega\boldsymbol{u}$, we can write the current density $\boldsymbol{J}$ as

$$\boldsymbol{J} = \frac{e_T^* \dot{\boldsymbol{u}}}{\Omega_0} = \left(\frac{e_T^{*2}}{\bar{M}\Omega_0}\right)\left(\frac{-i\omega}{\omega_{TO}^2 - \omega^2 - i\omega\Gamma}\right)\boldsymbol{\mathcal{E}}. \tag{10.143}$$

The conductivity defined by $\jmath = \sigma(\omega)\mathcal{E}$ is

$$\sigma(\omega) = -\frac{ie_T^{*2}[\omega(\omega_{TO}^2 - \omega^2 + i\omega\Gamma)]}{\bar{M}\Omega_0[(\omega_{TO}^2 - \omega^2)^2 + \omega^2\Gamma^2]}. \tag{10.144}$$

The power absorbed is proportional to

$$\sigma'(\omega) = \frac{e_T^{*2}\omega^2\Gamma}{\bar{M}\Omega_0[(\omega_{TO}^2 - \omega^2)^2 + \omega^2\Gamma^2]} \tag{10.145}$$

which has a Lorentzian shape with maximum at $\omega = \omega_{TO}$.

Although homopolar semiconductors have no first-order infrared absorption, they do have lattice vibration absorption arising from second-order terms in the expansion of the polarization in powers of the atomic displacements (Lax and Burstein 1955). This process enables the two atoms in the unit cell to develop effective charges which then couple to the radiation field and produce absorption over a broad range of frequency.

If the semiconductor contains impurities, localized vibrational modes can occur as discussed in Chapter 7. Since an impurity atom can acquire a charge through ionization of a charge carrier, absorption of radiation can occur as a result of the interaction of the vibrating impurity ion with the electromagnetic field (Lax and Burstein 1955, Maradudin and Wallis 1960).

Experimental observations of localized impurity mode absorption in semiconductors have been made by Balkanski and Nazarewicz (1964, 1966) and by Spitzer and Waldron (1965), who studied boron-doped silicon compensated with lithium. Compensation is necessary in order to prevent the free carrier absorption from concealing the impurity-mode absorption. The B^- ions and Li^+ ions tend to aggregate to form B^-Li^+ pairs. The absorption spectrum can have lines associated with impurity modes due to isolated B^- and Li^+ ions and to B^-Li^+ pairs. Experimental data are shown in Fig. 10.24.

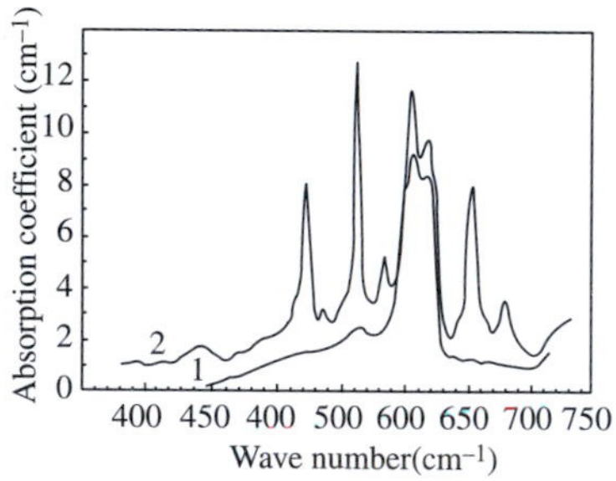

Fig. 10.24
Absorption spectrum of silicon doped with boron and lithium: curve 1, pure silicon; curve 2, B and Li in natural isotopic abundances (after Balkanski and Nazarewicz 1966).

10.6 Radiative recombination

10.6.1 Internal quantum efficiency

The interband optical absorption process leads to the production of free carriers in the conduction and valence bands. These free carriers, constituting excited states, have relatively short lifetimes and tend to recombine with the emission of photons. This process is called radiative recombination. One can make the same distinctions in radiative recombination as in optical absorption according to the band structure of the material. A radiative recombination transition can be direct or indirect depending on whether or not it is phonon-assisted. In a direct gap material where the conduction and valence band extrema coincide in $\boldsymbol{k}$ space, radiative recombination occurs without phonon participation, and the process is direct. The frequency of the emitted light corresponds to the direct (optical) gap energy.

In an indirect gap material the excited carriers thermalize in the lowest conduction band minimum, which typically is at $k \neq 0$. The radiative recombination process is accompanied by phonon emission or absorption, necessary to account for the momentum change of the carrier. Because momentum must be conserved in a band-to-band electron–hole recombination, and since the momentum of the emitted photon is nearly zero, the sum of the recombining carrier momenta must be matched by the momentum of the phonon participating in the process. The probability of band-to-band radiative recombination is much lower for the indirect gap than for the direct gap.

According to the method of producing the free carriers, the luminescence processes are given different names:

- *Photoluminescence*. When the excitation is caused by a light beam producing the free electron–hole population.
- *Cathodoluminescence*. When the excitation is produced by electron bombardment.
- *Injection luminescence*. When the radiative recombination occurs as a consequence of carrier injection in a p–n junction.

The excess population created by any of these is an excess electron concentration Δn, equal to the excess hole concentration Δp. The time dependence of the excess carrier concentration is governed by the rate equation

$$\frac{d\Delta n}{dt} = -\frac{\Delta n}{\tau}, \tag{10.146}$$

where τ is the carrier **lifetime**. Integration of this equation shows that the excess carrier concentration varies exponentially with time:

$$\Delta n(t) = \Delta n(0)e^{-t/\tau}. \tag{10.147}$$

We shall find it convenient to introduce the recombination rate R defined by

$$R = -\frac{d\Delta n}{dt} = \frac{\Delta n}{\tau}. \tag{10.148}$$

The carrier lifetime is determined by the fundamental properties of the material, as well as by the defects present. It is an essential parameter in radiative recombination and determines the main features of an electro-optic device. The lifetime τ represents the average time a carrier spends in a given band before recombining. It should be clearly distinguished from the relaxation time or characteristic time in transport, corresponding to intraband scattering of a carrier. The relaxation time is the time between two collisions in which the carrier changes state. Generally, lifetime is much greater than relaxation time.

Free carrier recombination can occur at the surface, giving rise to **surface recombination**, or in the bulk, giving rise to **bulk recombination**. Surface recombination processes can be more important than bulk recombination processes.

The thermalization of carriers is not exclusively radiative. Nonradiative recombination may occur through the successive emission of phonons, producing heat, or through Auger processes where the energy produced by recombination is transferred to a third carrier, which is elevated into the conduction band if it is an electron or lowered into the valence band if it is a hole.

Let the radiative recombination rate in a material per unit volume be R_r and the nonradiative be R_{nr}. The total recombination rate for the spontaneous process is

$$R_{sr} = R_r + R_{nr}. \tag{10.149}$$

The internal quantum efficiency is just the ratio of the radiative recombination rate to the total recombination rate

$$\eta_i = \frac{R_r}{R_r + R_{nr}}. \tag{10.150}$$

Assuming an exponential decay, the lifetime for a radiative recombination process is $\tau_r = \Delta n / R_r$, and for a nonradiative recombination process is $\tau_{nr} = \Delta n / R_{nr}$. The internal quantum efficiency can therefore be given in terms of the carrier lifetimes as

$$\eta_i = \frac{\tau_r^{-1}}{\tau_r^{-1} + \tau_{nr}^{-1}} = \left(1 + \frac{\tau_r}{\tau_{nr}}\right)^{-1}. \tag{10.151}$$

For good quantum efficiency the ratio τ_r / τ_{nr} should be kept as small as possible, i.e., the nonradiative processes should be minimized.

10.6.2 Carrier lifetime limited by band-to-band recombination

10.6.2.1 Dependence of radiative lifetime on carrier concentration

For a nondegenerate material, i.e., when the doping is not too high, it will be possible to use the measured absorption coefficient to calculate the radiative carrier lifetime when the recombination is direct band-to-band. This is because the absorption and radiative processes are related by the principle of detailed balance (Van Roosbroek and Shockley 1954).

The spontaneous recombination rate is directly proportional to the product of the free carrier concentrations, and it depends on the fundamental properties of the material expressed by a characteristic parameter $B_r[\mathrm{cm}^3/\mathrm{s}]$:

$$R_{sr} = B_r np. \tag{10.152}$$

At thermal equilibrium the hole and electron concentrations p_0 and n_0 are related by $p_0 n_0 = n_i^2$ where n_i denotes the intrinsic carrier concentration. Out of equilibrium where Δn and Δp are produced by external perturbation, the spontaneous recombination rate becomes

$$R_r = B_r(n_0 + \Delta n)(p_0 + \Delta p). \tag{10.153}$$

For optical excitation across the gap, $\Delta n = \Delta p$. Letting R_{sr}^{exc} be the recombination rate of the injected excess carriers, we have

$$R_{sr} = R_{sr}^{0} + R_{sr}^{exc}, \tag{10.154}$$

where R_{sr}^{0} is the spontaneous recombination rate at thermal equilibrium. Hence,

$$R_{sr} = B_r[n_0p_0 + \Delta n(p_0 + n_0) + (\Delta n)^2]. \tag{10.155}$$

Since

$$R_{sr}^{0} = B_r n_0 p_0 = B_r n_i^2 \tag{10.156}$$

and

$$R_{sr}^{exc} = R_{sr} - R_{sr}^{0} = B_r \Delta n(p_0 + n_0 + \Delta n), \tag{10.157}$$

then from Eq. (10.148)

$$\tau_r = [B_r(n_0 + p_0 + \Delta n)]^{-1}. \tag{10.158}$$

Depending on the relative importance of the injected and equilibrium carrier concentrations, one can distinguish two limits for the radiative lifetime:

1. *High injection rate*:

$$\Delta n > n_0 \text{ or } p_0$$
$$\tau_r = [B_r(\Delta n)]^{-1}.$$

This is the bimolecular recombination regime.

2. *Low injection rate*:

$$\Delta n < n_0 \text{ or } p_0$$
$$\tau_r \cong [B_r(n_0 + p_0)]^{-1}.$$

For p-type material: $p_0 \gg n_0$, p_0 will be the determining factor.
For n-type material: $n_0 \gg p_0$, n_0 will be the determining factor.
For any type material we can use the relation $n_0p_0 = n_i^2$ to re-express τ_r as

$$\tau_r = \frac{n_0}{B_r(n_0^2 + n_i^2)}. \tag{10.159}$$

A plot of τ_r versus n_0 is presented in Fig. 10.25. The longest possible radiative lifetime corresponds to $n_0 = n_i$, i.e., a pure intrinsic material:

$$\tau_r(max) = (2B_r n_i)^{-1}. \tag{10.160}$$

If the material is made n- or p-type, the radiative lifetime is reduced.

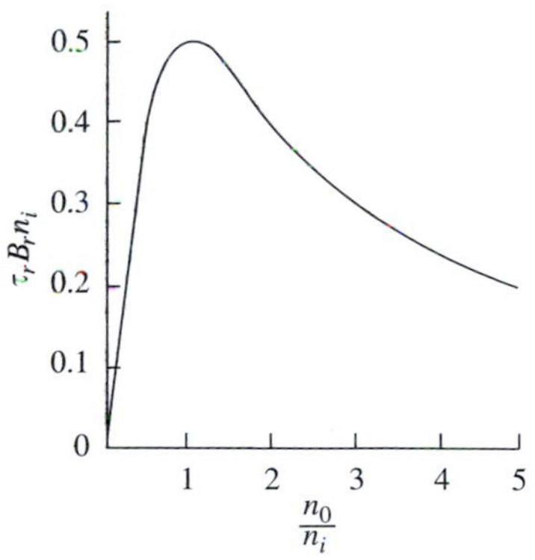

Fig. 10.25
Radiative recombination time versus conduction electron concentration.

B_r depends on the band structure of the material. Direct gap semiconductors have a much larger B_r and consequently a much smaller τ_r, than do indirect gap semiconductors. B_r can be deduced from the measured absorption coefficient (Van Roosbroeck and Shockley 1954) or by microscopic calculation (Dumke 1957).

10.6.2.2 Radiative lifetime from measured absorption coefficient

We follow the treatment of Van Roosbroeck and Shockley (1954). If a semiconductor is in equilibrium the principle of detailed balance (Tolman 1938) states that the rate of photoexcitation of carriers across the gap in a frequency interval $d\nu$ must be equal to the rate of generation of photons in $d\nu$ by electron–hole recombination. The rate of photoexcitation R^0_{pe} is given by

$$R^0_{pe} = \int c'\rho(\nu)\alpha(\nu)d\nu, \tag{10.161}$$

where c' is the velocity of light in the material, $\rho(\nu)$ is the photon density of the radiation, and $\alpha(\nu)$ is the absorption coefficient. Since $\rho(\nu)$ increases rapidly with wavelength, the main contribution to the integral comes from the vicinity of the absorption edge where the absorption is relatively weak and the dispersion is small. It is a good approximation to ignore dispersion and take the refractive index n to be a constant. One then has $c' = c/n$ and

$$\rho(\nu) = \frac{8\pi\nu^2 n^3}{c^3(e^{h\nu/k_B T} - 1)}. \tag{10.162}$$

The Planck function for surface emission $D(\nu)$ is related to $\rho(\nu)$ by

$$\rho(\nu) = 4\nu^{-2}n^3 D(\nu). \tag{10.163}$$

Eliminating $\rho(\nu)$ in favor of $D(\nu)$ in Eq. (10.161) and replacing the integration variable ν by the wavelength $\lambda = c/\nu$, we obtain

$$R^0_{pe} = 4n^2 \int_0^\infty D(\lambda)\alpha(\lambda)d\lambda. \tag{10.164}$$

Invoking the principle of detailed balance,

$$R^0_{sr} = R^0_{pe}. \tag{10.165}$$

Equation (10.152) then gives the result

$$B_r = \frac{R^0_{pe}}{n_i^2} \tag{10.166}$$

which upon substituting into Eq. (10.160) gives

$$\tau_r(\max) = \frac{n_i}{2R^0_{pe}}. \tag{10.167}$$

10.6.2.3 Radiative lifetime from microscopic calculations

The microscopic evaluation of the radiative lifetime (Dumke 1957) involves the calculation of the rate of recombination of electrons and holes by a procedure similar to that employed in Section 10.2 for direct and indirect interband absorption. In Ge, electrons excited to the lowest conduction band minima at the L points and the higher minimum at the Γ point can recombine with holes via indirect and direct transitions, respectively. The corresponding radiative lifetimes τ_{r1} and τ_{r2} calculated by Dumke for intrinsic Ge at room temperature are 0.75 s and 0.29 s, respectively. The total radiative lifetime τ specified by

$$\frac{1}{\tau} = \frac{1}{\tau_1} + \frac{1}{\tau_2} \tag{10.168}$$

is 0.21 s, which can be compared to the value 0.25 s determined from experimental absorption data by Van Roosbroeck and Shockley (1954) as corrected by Haynes (1955).

10.6.2.4 Recombination via traps (recombination centers) in the energy gap

When deep energy levels are present in the material, having the possibility to capture electrons or holes, the radiative recombination rate will depend on the capture cross-section of these centers and on their concentration. The theory for recombination via traps was originally worked out by Shockley and Read (1952).

The general expression for recombination through traps is

$$R_{trap} = \frac{\sigma_e \sigma_h v_{th} n_t (np - n_i^2)}{\sigma_e \left[n + \bar{N}_c \exp\left(\frac{E_t - E_C}{k_B T}\right)\right] + \sigma_h \left[p + \bar{N}_v \exp\left(\frac{E_V - E_t}{k_B T}\right)\right]}, \tag{10.169}$$

where σ_e and σ_h are the electron and hole trapping cross-sections respectively; v_{th} is the thermal velocity of the electrons, n_t is the trap concentration, and E_t is the energy of the center (which is often near midgap $\sim \frac{E_G}{2}$). Recombination via traps often competes with the indirect gap recombination and will therefore be important mainly in indirect gap materials.

10.7 Surface polaritons

In Section 10.5.1 the result of coupling a photon to an elementary excitation such as an optical phonon to give a polariton was discussed for the case of bulk materials. If the system of interest has a surface or an interface between two materials, a **surface polariton** or an **interface polariton** can arise in which the field amplitudes are localized at the surface or interface.

Let us consider a system with a planar interface between two materials A and B that are characterized by dielectric functions $\epsilon_A(\omega)$ and $\epsilon_B(\omega)$, respectively. The interface is taken to be the plane $z = 0$ with material A occupying the region $z > 0$ and Material B occupying the region $z < 0$. The electric vector of the radiation field must satisfy the wave equation in each material:

$$\nabla^2 \boldsymbol{\mathcal{E}} = \frac{\epsilon_{A,B}(\omega)}{c^2}\frac{\partial^2 \boldsymbol{\mathcal{E}}}{\partial t^2}. \tag{10.170}$$

We seek solutions to the wave equation that are localized at the interface and that propagate in the x-direction parallel to the interface. Such solutions can be written as

$$\boldsymbol{\mathcal{E}}_A = (\mathcal{E}_{Ax0}, 0, \mathcal{E}_{Az0})e^{-\alpha_A z}e^{ikx-i\omega t} \tag{10.171a}$$

$$\boldsymbol{\mathcal{E}}_B = (\mathcal{E}_{Bx0}, 0, \mathcal{E}_{Bz0})e^{\alpha_B z}e^{ikx-i\omega t}, \tag{10.171b}$$

where α_A and α_B are the decay constants for materials A and B, respectively, k is the wave vector, and ω is the frequency. Substituting Eqs. (10.171) into the wave equation, one finds for nontrivial solutions that

$$\alpha_A^2 = k^2 - \epsilon_A(\omega)\frac{\omega^2}{c^2} \tag{10.172a}$$

$$\alpha_B^2 = k^2 - \epsilon_B(\omega)\frac{\omega^2}{c^2}. \tag{10.172b}$$

In order to obtain the dispersion relation for surface or interface polaritons, we must consider the boundary conditions at $z = 0$ which we take to be the continuity of tangential $\boldsymbol{\mathcal{E}}$ and normal $\boldsymbol{\mathcal{D}}$:

$$\mathcal{E}_{Ax0} = \mathcal{E}_{Bx0} \tag{10.173a}$$

$$\epsilon_A(\omega)\mathcal{E}_{Az0} = \epsilon_B(\omega)\mathcal{E}_{Bz0}. \tag{10.173b}$$

To complete the specification of the field components we invoke Maxwell's equation $\boldsymbol{\nabla} \cdot \boldsymbol{\mathcal{D}} = 0$ for zero space charge and obtain

$$ik\mathcal{E}_{Ax0} - \alpha_A\mathcal{E}_{Az0} = 0 \tag{10.174a}$$

$$ik\mathcal{E}_{Bx0} + \alpha_B\mathcal{E}_{Bz0} = 0. \tag{10.174b}$$

The condition for a nontrivial solution to Eqs. (10.173) and (10.174) is

$$\frac{\epsilon_A(\omega)}{\epsilon_B(\omega)} = -\frac{\alpha_A}{\alpha_B}. \tag{10.175}$$

Eliminating α_A and α_B with the aid of Eqs. (10.172) yields the dispersion relation for surface or interface polaritons

$$\frac{\epsilon_A(\omega)}{\epsilon_B(\omega)} = -\frac{\sqrt{k^2 - \epsilon_A(\omega)\omega^2/c^2}}{\sqrt{k^2 - \epsilon_B(\omega)\omega^2/c^2}}, \tag{10.176}$$

from which ω as a function of k can be calculated for specified $\epsilon_A(\omega)$ and $\epsilon_B(\omega)$. If material B is taken to be vacuum, one is dealing with surface polaritons associated with material A. Note that $\epsilon_A(\omega)$ and $\epsilon_B(\omega)$ must have different signs at the frequency of the polariton for given k. We now consider specific examples of surface polaritons.

10.7.1 Surface plasmon polaritons

Free carriers in a semiconductor produce a contribution to the dielectric function given by Eq. (10.113). For frequencies less than the plasma frequency the dielectric function is negative, so we can expect to have surface polaritons in this range. Taking $\epsilon_B(\omega) = 1$ for vacuum and solving Eq. (10.176) gives the dispersion relation in the form

$$k^2 = \frac{\omega^2}{c^2} \cdot \frac{\epsilon_A(\omega)}{\epsilon_A(\omega) + 1}. \tag{10.177}$$

When $\epsilon_A(\omega) = -1$, $k = \infty$ corresponding to a nonretarded surface plasmon with frequency ω_{sp}^{∞} given by

$$\omega_{sp}^{\infty} = \frac{\omega_p}{\sqrt{1 + \frac{1}{\epsilon_\infty}}}. \tag{10.178}$$

Since ω_{sp}^{∞} is less than ω_p, it lies in a region of high reflectivity.

A second limiting case is $\omega \ll \omega_p$. Then $\omega \simeq kc$, and the dispersion curve is very close to the light line. A plot of the dispersion curve for n-InSb is shown in Fig. 10.26.

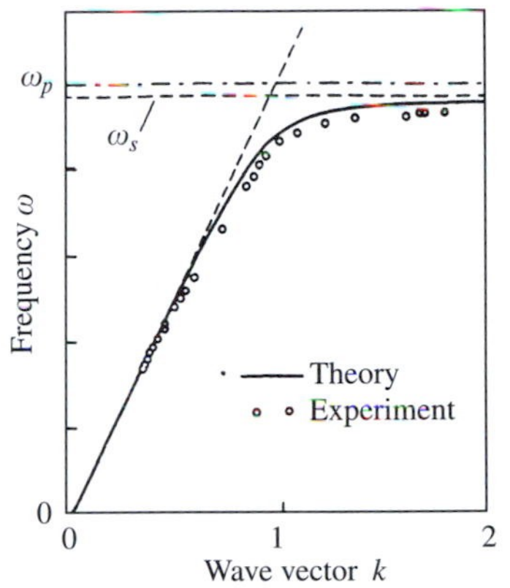

Fig. 10.26
Surface plasmon polariton dispersion curve for n-InSb Circles: experimental data; solid curve: theoretical result. ω_s is ω_{sp}^{∞} (after Marschall *et al.* 1971).

10.7.2 Surface optical phonon polaritons

The dielectric function associated with optical phonons is given by Eq. (10.137). Substituting this expression into Eq. (10.176) yields the dispersion relation for surface optical phonon polaritons. Considering the limit $k \to \infty$ with $\epsilon_A(\omega) = -1$, we obtain the nonretarded surface optical phonon frequency ω_{sop}^{∞}:

$$\omega_{sop}^{\infty} = \left[\frac{\epsilon_\infty \omega_{LO}^2 + \omega_{TO}^2}{\epsilon_\infty + 1}\right]^{\frac{1}{2}}. \tag{10.179}$$

It is evident that ω_{sop}^{∞} is between ω_{TO} and ω_{LO}, i.e., it is in the reststrahlen region of high reflectivity. As k decreases from large values, the frequency of the surface optical phonon polariton decreases until it reaches the value ω_{TO} at the light line where $k = \omega_{TO}/c$. For smaller values of k, Eq. (10.177) has no solutions corresponding to real ω. A plot of the dispersion curve for surface optical phonon polaritons in GaAs on a sapphire substrate is given in Fig. 10.27.

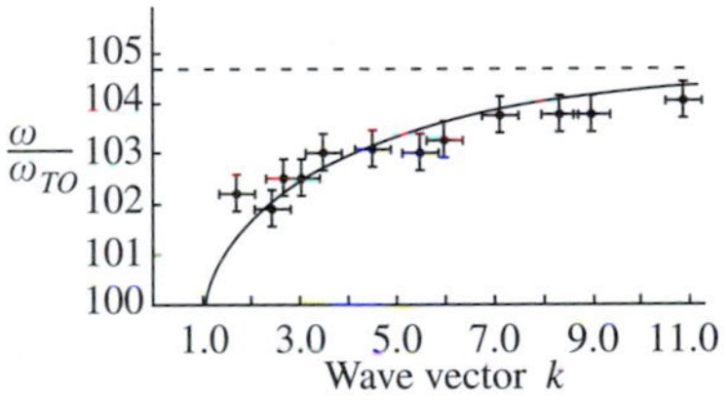

Fig. 10.27
Surface optical phonon polariton dispersion curve for GaAs on a sapphire substrate of effective dielectric constant ϵ_m. Crossed dots: experimental data; solid curve: theoretical result (after Evans *et al.* 1973).

10.7.3 Experimental observation of surface polaritons

We have noted above that the surface polariton frequency lies in a region of high reflectivity of the material being studied. However, if the frequency and wave vector of radiation incident on a crystal satisfy the dispersion relation for a surface polariton, the conditions exist for a transfer of energy from the radiation field to the crystal and the creation of a surface polariton. Dissipative processes can cause this energy to be retained by the crystal and not reflected back into the radiation field. In other words one has **attenuated total reflection** (ATR) associated with the creation of surface polaritons that is characterized by a dip in reflectivity when the wave vector and frequency satisfy the surface polariton dispersion relation. By varying the wave vector and noting the frequency at which the dip occurs, one can determine the dispersion curve. The ATR method was used to obtain the experimental points in Fig. 10.26 and Raman scattering, to be discussed in the next section, was used to obtain the data points in Fig. 10.27.

10.8 Light scattering

When a monochromatic light beam passes through a transparent medium, a small part of the light is scattered out of the incident direction with a change in frequency. The scattering processes are usually classified into three groups.

1. In **Brillouin scattering** the light is scattered with a small frequency shift that varies continuously with scattering angle.
2. In **Raman scattering** the light is scattered with a relatively large frequency shift that is independent of scattering angle. The possibility of observing a given transition, however, depends on the orientation of the crystal relative to the polarization of the incident light.
3. In **Rayleigh scattering** the light is scattered without frequency shift. In this case the scattering is elastic with the incident and scattered frequencies equal: $\omega_i = \omega_s$. Rayleigh scattering is particularly useful in the study of critical phenomena or aspects related to the size and polarizability of particles.

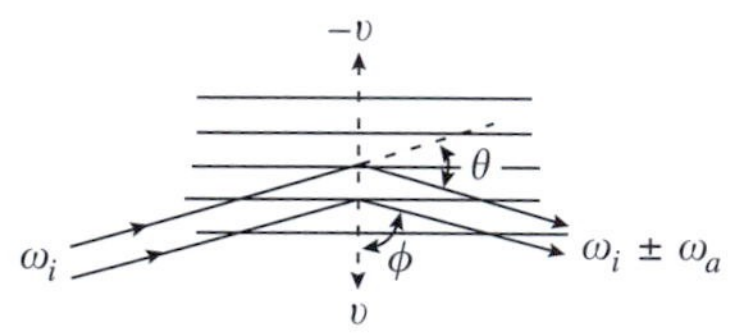

Fig. 10.28
Diagram showing light scattering at angle ϕ and Bragg reflection at θ by moving thermal waves.

10.8.1 Brillouin scattering

Brillouin scattering is caused by the interaction of light with the acoustic modes of vibration of the crystal. Let us consider an acoustic wave of frequency ω_a propagating with velocity $\pm v$ as shown in Fig. 10.28. Light with incident frequency ω_i interacts strongly with acoustic waves satisfying the Bragg condition $\omega_a = 2\omega_i(v/c)n\sin(\theta/2)$, where n is the refractive index of the material and θ is the angle of deviation of the scattered beam. Since the acoustic waves are moving with velocity $\pm v$, the scattered light suffers a Doppler shift in frequency and exhibits a frequency doublet at the angle θ given by the **Brillouin equation** (Brillouin 1922)

$$\omega_i \pm \omega_a = \omega_i \pm 2\omega_i(v/c)n\sin(\theta/2). \qquad (10.180)$$

In terms of a quantum picture, Brillouin scattering corresponds to scattering of incident photons of frequency ω_i and wave vector $\boldsymbol{k}_i$ into scattered photons of frequency ω_s and wave vector $\boldsymbol{k}_s$ with the emission or absorption of an acoustic phonon of frequency ω_a and wave vector $\boldsymbol{q}$.

A diagram illustrating the phonon emission case is given in Fig. 10.29. The predicted Brillouin spectrum is shown in Fig. 10.30. It consists of **Stokes** ($\omega_s = \omega_i - \omega_a$) and **anti-Stokes** ($\omega_s = \omega_i + \omega_a$) components corresponding to phonon emission and absorption, respectively.

The frequency shift $|\omega_i - \omega_s|$ for a well-defined scattering angle θ directly gives the acoustic phonon frequency and velocity. The width Γ of the Brillouin component of the spectrum is a measure of the damping or attenuation of the wave.

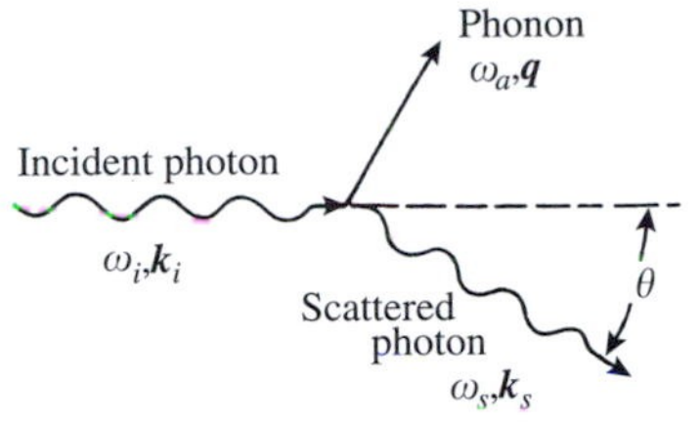

Fig. 10.29
Schematic representation of light scattering by acoustic phonons.

10.8.2 Raman scattering

The essential difference between Brillouin and Raman scattering is that in Raman scattering the incident light beam is scattered with relatively large frequency shift independent of the scattering angle. The same basic considerations apply to Raman scattering that apply to Brillouin scattering. The Raman spectrum has Stokes and anti-Stokes branches corresponding to the emission and absorption, respectively, of an elementary excitation. A variety of elementary excitations are important. They include optical phonons and, in the case of magnetic materials, magnons. Also of interest are electronic excitations such as intraband single-particle excitations, interband excitations, and collective excitations (plasmons).

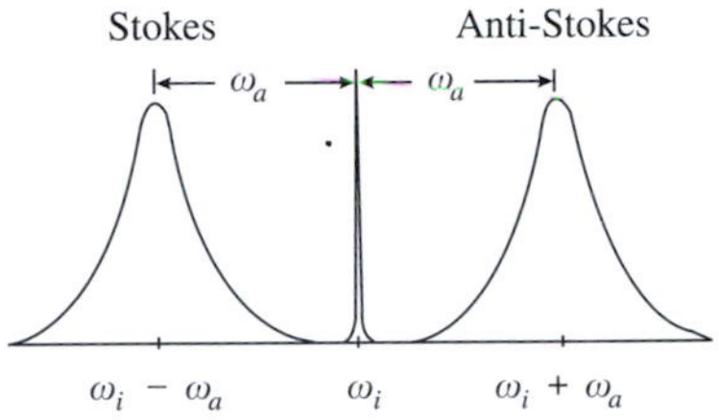

Fig. 10.30
Schematic diagram of a Brillouin spectrum.

10.8.2.1 Theory of Raman scattering by phonons

Qualitative considerations

The basic mechanism of Raman scattering is the modulation of the dielectric susceptibility tensor $\overleftrightarrow{\chi}(\omega_i)$ by fluctuations associated with an elementary excitation. We start by treating optical phonons as the elementary excitation. For a particular phonon the atomic displacement $\boldsymbol{u}$ can be expressed as

$$\boldsymbol{u} = \boldsymbol{u}_0 \cos(\boldsymbol{q} \cdot \boldsymbol{r} - \omega_{\boldsymbol{q}} t), \tag{10.181}$$

where $\boldsymbol{q}$ is the phonon wave vector and $\omega_{\boldsymbol{q}}$ is its frequency. The susceptibility tensor is a function of $\boldsymbol{u}$ and can be expanded in power series

$$\overleftrightarrow{\chi}(\omega_i, \boldsymbol{u}) = \overleftrightarrow{\chi}_0(\omega_i) + (\boldsymbol{\nabla} \overleftrightarrow{\chi}) \cdot \boldsymbol{u} + \cdots, \tag{10.182}$$

where $\boldsymbol{\nabla} \overleftrightarrow{\chi}$ is a third-rank tensor with elements

$$\chi_{\alpha\beta\gamma} = \left.\frac{\partial \chi_{\alpha\beta}}{\partial u_\gamma}\right|_{\boldsymbol{u}=0}. \tag{10.183}$$

The incident radiation has an electric field that can be expressed as

$$\boldsymbol{\mathcal{E}} = \boldsymbol{\mathcal{E}}_0 \cos(\boldsymbol{k}_i \cdot \boldsymbol{r} - \omega_i t). \tag{10.184}$$

This field gives rise to an electric polarization given by

$$\begin{aligned} \boldsymbol{P} &= \epsilon_0 \overleftrightarrow{\chi}(\omega_i, \boldsymbol{u}) \cdot \boldsymbol{\mathcal{E}} \\ &\simeq \epsilon_0 \overleftrightarrow{\chi}_0(\omega_i) \cdot \boldsymbol{\mathcal{E}} + \epsilon_0 \overleftrightarrow{\chi}_1(\boldsymbol{u}) \cdot \boldsymbol{\mathcal{E}} + \cdots, \end{aligned} \tag{10.185}$$

where $\overleftrightarrow{\chi}_1(\boldsymbol{u}) = (\boldsymbol{\nabla} \overleftrightarrow{\chi}) \cdot \boldsymbol{u}$.

The polarization represents an oscillating electric dipole that radiates electromagnetic waves with frequency ω_s. The first term on the right hand side of Eq. (10.185) corresponds to Rayleigh scattering with $\omega_s = \omega_i$. The second term corresponds to first-order Raman scattering and will be labeled $\boldsymbol{P}_1$. In view of Eq. (10.181) we can write $\overleftrightarrow{\chi}_1(\boldsymbol{u})$ as

$$\overleftrightarrow{\chi}_1(\boldsymbol{u}) = (\boldsymbol{\nabla}\overleftrightarrow{\chi}) \cdot \boldsymbol{u}_0 \cos(\boldsymbol{q}\cdot\boldsymbol{r} - \omega_{\boldsymbol{q}} t). \tag{10.186}$$

Combining Eqs. (10.184)–(10.186) yields the following result for P_1:

$$\begin{aligned}\boldsymbol{P}_1 &= \tfrac{1}{2}\epsilon_0(\boldsymbol{\nabla}\overleftrightarrow{\chi}) \cdot \boldsymbol{u}_0 \cdot \boldsymbol{\mathcal{E}}_0\{\cos[(\boldsymbol{k}_i - \boldsymbol{q})\cdot\boldsymbol{r} - (\omega_i - \omega_{\boldsymbol{q}})t] \\ &\quad + \cos[(\boldsymbol{k}_i + \boldsymbol{q})\cdot\boldsymbol{r} - (\omega_i + \omega_{\boldsymbol{q}})t]\}.\end{aligned} \tag{10.187}$$

The wave vector $\boldsymbol{k}_s$ and frequency ω_s that characterize $\boldsymbol{P}_1$ satisfy the conditions $\boldsymbol{k}_s = \boldsymbol{k}_i - \boldsymbol{q}$, $\omega_s = \omega_i - \omega_0$ for the Stokes branch and $\boldsymbol{k}_s = \boldsymbol{k}_i + \boldsymbol{q}$, $\omega_s = \omega_i + \omega_0$ for the anti-Stokes branch. A phonon is emitted in the Stokes branch and absorbed in the anti-Stokes branch. The higher-order terms in Eq. (10.185) correspond to the emission or absorption of two or more phonons and describe higher-order Raman scattering. Since the wave vectors $\boldsymbol{k}_i$ and $\boldsymbol{k}_s$ in a typical experiment are very small compared to those at the Brillouin zone boundary, first-order Raman scattering yields information only about optical phonons near the zone center. This restriction is relaxed in higher-order Raman scattering.

Let us introduce a unit vector $\hat{e}$ in the direction of the displacement $\boldsymbol{u}$. The quantity $\boldsymbol{\nabla}\overleftrightarrow{\chi} \cdot \hat{e}$ is a second-rank tensor $\mathsf{R}(\hat{e})$ called the **Raman tensor** that is associated with the phonon specified by $\boldsymbol{u}$. The Raman tensor relates the polarization of the scattered radiation to that of the incident radiation. Its nonvanishing elements are determined by the symmetry of the crystal.

The analysis of second-order Raman scattering can be carried out in an analogous manner except that the energy and momentum conservation conditions involve both participating phonons. The presence of impurities in a crystal gives rise to localized vibrational modes whose Raman spectrum allows the identification of the impurity and a precise description of its environment.

Semiclassical theory of Raman scattering

The modification of the susceptibility associated with Raman scattering gives rise to a quantum mechanical transition between initial and final vibrational states $|v\rangle$ and $|v'\rangle$ characterized by a matrix element $\langle v|(\boldsymbol{\nabla}\overleftrightarrow{\chi})\cdot\boldsymbol{u}|v'\rangle$. The energy scattered per unit time per unit area per unit solid angle at a large distance R from the scattering region is given by (Born and Huang 1954, Hayes and Loudon 1978)

$$S_s = \frac{V\bar{V}\omega_s^4\epsilon_0}{8\pi^2R^2c^3}|\langle v|\hat{\eta}_s\cdot(\boldsymbol{\nabla}\overleftrightarrow{\chi})\cdot\boldsymbol{u}|v'\rangle\cdot\boldsymbol{\mathcal{E}}_0|^2, \tag{10.188}$$

where an average over a vibrational period has been taken, $\hat{\eta}_s$ is the unit polarization vector of the scattered radiation, V is the crystal volume, and $\bar{V}$ is the volume of the illuminated part of the crystal.

The energy flux of the incident radiation averaged over a vibrational period is given by

$$S_i = 2c\epsilon_0|\boldsymbol{\mathcal{E}}_0|^2. \tag{10.189}$$

The differential scattering cross section can be obtained by considering two different but equivalent expressions for the energy scattered per unit time into solid angle $d\Omega$ when flux S_i is incident on area $d\sigma$:

$$\frac{dE_s}{dt} = S_s R^2 d\Omega = S_i d\sigma. \tag{10.190}$$

In writing down this equation we have made the simplifying approximation $\omega_i/\omega_s \simeq 1$. This is a good approximation under typical situations in which ω_i and ω_s are in the visible region ($\hbar\omega \sim 4\,\text{eV}$) and ω_0 is in the far infrared ($\hbar\omega_0 \sim 0.03\,\text{eV}$). Using the above expressions for S_s and S_i, we obtain the differential scattering cross section in the form

$$\frac{d\sigma}{d\Omega} = \frac{V\bar{V}}{(4\pi)^2}\left(\frac{\omega_s}{c}\right)^4 |\langle v|\hat{\eta}_s \cdot [(\boldsymbol{\nabla}\overleftrightarrow{\chi}) \cdot \boldsymbol{u}] \cdot \hat{\eta}_i|v'\rangle|^2, \tag{10.191}$$

where $\hat{\eta}_i$ is the unit polarization vector of the incident radiation. For the $\boldsymbol{q} \simeq 0$ optical phonons of importance for first-order Raman scattering, the displacement $\boldsymbol{u}$ is the relative displacement of the two atoms in the unit cell $\boldsymbol{u}_1 - \boldsymbol{u}_2$. Introducing creation and annihilation operators as in Eq. (8.176), evaluating the vibrational matrix element, and averaging over a canonical ensemble, we obtain for the differential scattering cross section

$$\frac{d\sigma}{d\Omega} = \frac{\hbar V\bar{V}\omega_s^4}{32\pi^2\bar{M}Nc^4\omega_0}|\hat{\eta}_s \cdot [(\boldsymbol{\nabla}\overleftrightarrow{\chi}) \cdot \hat{e}_0] \cdot \hat{\eta}_i|^2 \left\{ \begin{array}{c} \bar{n}_0 + 1 \\ \bar{n}_0 \end{array} \right\}, \tag{10.192}$$

where $\bar{M}$ is the reduced mass, $\hat{e}_0$ and ω_0 are the unit polarization vector and frequency of the optical phonon, the factors $\bar{n}_0 + 1$ and $\bar{n}_0$ refer to the Stokes and anti-Stokes processes, respectively, and $\bar{n}_0$ is the phonon population factor given by

$$\bar{n}_0 = \left[e^{\hbar\omega_0/k_BT} - 1\right]^{-1}. \tag{10.193}$$

The ratio of the intensities of Stokes and anti-Stokes scattering is

$$\frac{I_s}{I_{as}} = e^{\hbar\omega_0/k_BT}. \tag{10.194}$$

This expression shows that the Stokes scattering is always more intense than the anti-Stokes scattering. The asymmetry increases as T decreases, until finally the anti-Stokes line vanishes at $T = 0$.

Quantum theory of Raman scattering

In developing the quantum theory of Raman scattering, one goes beyond the phenomenological approach of the preceding section and treats the system on a microscopic level. An inelastic light scattering event involves the destruction of a photon of frequency ω_i incident from a light source, the creation of a scattered photon of frequency ω_s, and the creation or destruction of an optical phonon of frequency ω_0. The first two processes arise from the electron–radiation interaction given by Eq. (10.56). It is convenient in evaluating the required matrix elements to express the vector potential in Fourier series

$$\mathcal{A}(r) = \sum_{k} \left(\frac{\hbar}{2\epsilon_0 [n(\omega)]^2 V \omega_k} \right)^{\frac{1}{2}} \hat{\eta}_k \left(a_k e^{ik\cdot r} + a_k^{+} e^{-ik\cdot r} \right), \tag{10.195}$$

where a_k and a_k^{+} are photon destruction and creation operators, respectively. The last process arises from the electron–phonon interaction that we treat using an optical deformation potential specified by

$$H_{ep} = \Xi \boldsymbol{u}/a, \tag{10.196}$$

where a is the lattice constant and Ξ has the dimensions of energy and is a function of the electronic coordinates. We shall consider in this discussion the Stokes component of the scattering with $\omega_i = \omega_s + \omega_0$. The properties of the anti-Stokes component can be obtained by making appropriate changes.

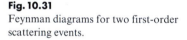

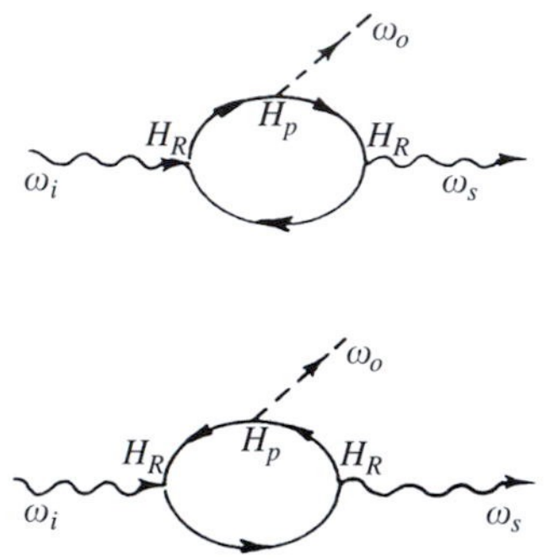

Fig. 10.31
Feynman diagrams for two first-order scattering events.

Figure 10.31 shows the Feynman diagrams for two typical first-order scattering events in a perfect semiconductor crystal. Four additional diagrams arise by interchanging pairs of vertices. These diagrams correspond to third-order terms in time-dependent perturbation theory which have the form

$$R^{\gamma}_{\alpha\beta}(-\omega_i, \omega_s, \omega_0) = \frac{1}{V} \sum_{a,b} \left\{ \frac{p^{\beta}_{0b} \Xi^{\gamma}_{ba} p^{\alpha}_{a0}}{(\omega_b + \omega_0 - \omega_i)(\omega_a - \omega_i)} + \frac{p^{\alpha}_{0b} \Xi^{\gamma}_{ba} p^{\beta}_{a0}}{(\omega_b + \omega_0 + \omega_s)(\omega_a + \omega_s)} + \text{four similar terms} \right\}, \tag{10.197}$$

where the subscripts a, b on the matrix elements of the electronic momentum operator $\boldsymbol{p}$ and the deformation potential parameter Ξ refer to electron–hole pair states with energies $\hbar\omega_a$ and $\hbar\omega_b$, and the subscript 0 refers to the electronic ground state. The subscripts α and β on R are the polarization directions of the incident and scattered photons, respectively, and the superscript γ is the polarization direction of the phonon. The matrix elements p^{α}_{0a} and p^{β}_{0b} originate from the matrix elements of the electron–radiation interaction Hamiltonian.

The quantity $R^{\gamma}_{\alpha\beta}(-\omega_i, \omega_s, \omega_0)$ is the **Raman amplitude** and is related to $\chi_{\alpha\beta\gamma}$ defined in Eq. (10.183). Under circumstances such that $\omega_0 \ll \omega_i, \omega_s$ one has (Hayes and Loudon 1978)

$$\chi_{\alpha\beta\gamma} \simeq -\frac{e^2}{\epsilon_0 m^2 \hbar^2 \omega_i^2 a} R^{\gamma}_{\beta\alpha}(-\omega_i, \omega_s, \omega_0). \tag{10.198}$$

Equation (10.197) indicates that the Raman amplitude has the symmetry $R^{\gamma}_{\alpha\beta}(-\omega_i, \omega_s, \omega_0) = R^{\gamma}_{\beta\alpha}(\omega_s, -\omega_i, \omega_0)$. Using the symmetries $\boldsymbol{p}_{\alpha\beta} = -\boldsymbol{p}_{\beta\alpha}$, $\Xi^{\gamma}_{\alpha\beta} = \Xi^{\gamma}_{\beta\alpha}$, and energy conservation $\omega_i = \omega_0 + \omega_s$, one can show that $R^{\gamma}_{\alpha\beta}(-\omega_i, \omega_s, \omega_0) = R^{\gamma}_{\alpha\beta}(\omega_i, -\omega_s, -\omega_0)$. Combining these two results and passing to the limit $\omega_0 \to 0$ shows that the Raman amplitude is symmetric under the interchange of the photon polarizations $\alpha, \beta : R^{\gamma}_{\alpha\beta}(-\omega_i, \omega_s, 0) = R^{\gamma}_{\beta\alpha}(-\omega_i, \omega_s, 0)$. The differential scattering cross section expressed in terms of the Raman amplitude follows from Eqs. (10.192) and (10.198). The result for Stokes scattering is

$$\frac{d\sigma}{d\Omega} \simeq \frac{V\bar{V}(n_0 + 1)}{(4\pi\epsilon_0)^2 2\hbar^3 \bar{M} N \omega_0 a^2} \left(\frac{e}{mc}\right)^4 \times |\hat{\eta}_s \cdot [\vec{\boldsymbol{R}}(-\omega_i, \omega_s, \omega_0) \cdot \hat{e}_0] \cdot \hat{n}_i|^2, \tag{10.199}$$

where letter in bold italic with the symbol $\rightarrow$ above it denotes a third-rank tensor.

The scattering efficiency S is related to the scattering cross section by

$$\mathcal{S} = \frac{1}{A}\frac{d\sigma}{d\Omega}, \tag{10.200}$$

where A is the area illuminated. Using Eq. (10.199) we obtain

$$\mathcal{S} = \frac{VL(n_0 + 1)}{32\pi^2 \bar{M} N \omega_0 a^2 \hbar^3} \left(\frac{e}{mc}\right)^4 \times |\hat{\eta}_s \cdot [\vec{\boldsymbol{R}}(-\omega_i, \omega_s, \omega_0) \cdot \hat{e}_0] \cdot \hat{\eta}_i|^2, \tag{10.201}$$

L being the length of the illuminated path. Loudon has made numerical estimates that indicate Raman scattering efficiencies are typically of order 10^{-6} or 10^{-7}.

The result given by Eq. (10.201) applies to both TO and LO modes in homopolar semiconductors. It also applies to TO modes in polar semiconductors provided $cq \gg \omega_{TO}$, but must be modified for LO modes whose macroscopic electric field leads to an additional contribution to the scattering cross section. The macroscopic field is a consequence of the Coulomb field of the ionic charges and gives rise to an electron–phonon interaction whose Hamiltonian is specified by Eq. (8.180).

Using third-order perturbation theory we can write the differential scattering cross section for the Stokes component of polar mode scattering as

$$\frac{d\sigma}{d\Omega} \simeq \frac{V\bar{V}(n_{LO} + 1)}{(4\pi\epsilon_0)^2 2\hbar^3 \bar{M} N \omega_{LO}} \left(\frac{e}{mc}\right)^4 \left| \frac{i}{a}\hat{\eta}_s \cdot [\vec{\boldsymbol{R}}(-\omega_i, \omega_s, \omega_{LO}) \cdot \hat{e}_{LO}] \cdot \hat{\eta}_i + \frac{ee^*_L}{\epsilon_0 m \Omega_0}\hat{\eta}_s \cdot [\vec{\boldsymbol{P}}(-\omega_i, \omega_s, \omega_{LO}) \cdot \hat{e}_{LO}] \cdot \hat{\eta}_i \right|^2, \tag{10.202}$$

where $\vec{\boldsymbol{P}}(-\omega_i, \omega_s, \omega)$ is obtained from $\vec{\boldsymbol{R}}(-\omega_i, \omega_s, \omega)$ by replacing Ξ^{γ}_{ba} by $p^{\gamma}_{ba}/(\omega_b - \omega_a)$ with similar replacements in the other four terms. Due to the presence of the macroscopic field term $\vec{\boldsymbol{P}}$ in the scattering cross section for LO phonons, there is no simple relationship between the scattering intensities for TO and LO phonons in polar semiconductors.

For the regime $cq \simeq \omega_{TO}$ in polar semiconductors it is necessary to take into account the polariton nature of the coupled photon-transverse optical phonon modes and their associated macroscopic electric field (Mills and Burstein 1974). The electric field is related to the relative displacement $\boldsymbol{u}$ by Eq. (10.130) in which ω is the polariton frequency. The susceptibility $\overleftrightarrow{\chi}$ is modulated by $\boldsymbol{\mathcal{E}}$ as a result of the electro-optic effect. The expansion that appears in Eq. (10.182) must be generalized to include powers of $\boldsymbol{\mathcal{E}}$:

$$\chi_{\alpha\beta}(\omega_i, \boldsymbol{u}, \boldsymbol{\mathcal{E}}) - \chi_{0\alpha\beta}(\omega_i) = \sum_{\gamma}\left[\frac{\partial\chi_{\alpha\beta}}{\partial u_\gamma}u_\gamma + \frac{\partial\chi_{\alpha\beta}}{\partial\mathcal{E}_\gamma}\mathcal{E}_\gamma\right]$$
$$= \sum_{\gamma}[a_{\alpha\beta\gamma}u_\gamma + b_{\alpha\beta\gamma}\mathcal{E}_\gamma], \qquad (10.203)$$

where $a_{\alpha\beta\gamma}$ specifies the atomic displacement contribution and $b_{\alpha\beta\gamma}$ specifies the electro-optic contribution. Making use of Eq. (10.130) with $\Gamma = 0$, we can eliminate $\boldsymbol{\mathcal{E}}$ in favor of $\boldsymbol{u}$ and obtain

$$\chi_{\alpha\beta}(\omega_i, \boldsymbol{u}, \boldsymbol{\mathcal{E}}) - \chi_{0\alpha\beta}(\omega_i) = \sum_{\gamma}\left[a_{\alpha\beta\gamma} + \frac{\bar{M}}{e_T^*}(\omega_{TO}^2 - \omega^2)b_{\alpha\beta\gamma}\right]u_\gamma. \quad (10.204)$$

From here on, the derivation of the differential scattering cross section proceeds as before, but it is now evident that the Raman amplitude has contributions from both mechanical and electro-optic origins. An exception occurs for the lower polariton branch with $q \gg \omega_{TO}/c$ and $\omega \simeq \omega_{TO}$. Then the electro-optic contribution is very small compared to the mechanical contribution. If $\omega > \omega_{TO}$ as in the upper polariton branch, destructive interference between the mechanical and electro-optic contributions can occur.

10.8.2.2 Selection rules in Raman scattering

Not all elementary excitations in semiconductors scatter light. The Raman active modes are determined by **selection rules** established using group-theoretical methods. The various normal modes in a given crystal correspond to various symmetries of the vibrations of the atoms in the crystal and are characterized by the irreducible representations of the space group of the crystal lattice. One can show that a normal mode can participate in a first-order Raman transition if and only if its irreducible representation is the same as one of the irreducible representations that occur in the reduction of the representation of the Raman tensor.

An important result of group theory is the **rule of mutual exclusion** which states that, in crystals with a center of inversion, excitations that are active in the first-order infrared spectrum are inactive in the first-order Raman spectrum, and conversely, excitations that are active in the first-order Raman spectrum are inactive in the first-order infrared spectrum. In particular, the first-order spectra due to optical phonons in NaCl are infrared active, but Raman inactive, whereas in Si they are infrared inactive, but Raman active. This difference is related to the fact that each atomic site in NaCl is a center of inversion and the active optical phonons have odd parity whereas each midpoint between two nearest-neighbor atomic sites in Si is a center of inversion and the active optical phonons have even parity.

In a crystal with the zincblende structure, there is no center of inversion, and the crystal is both infrared and Raman active. The only nonvanishing elements of the Raman tensor $\vec{R}(\hat{e})$ with $\hat{e}\|\hat{z}$ are the xy and yx elements. To observe the Raman effect of an LO phonon propagating in the z-direction, one can arrange the polarization of the incident light parallel to the x-axis and observe the scattered light with its polarization parallel to the y-axis or vice versa. An analogous set of constraints applies to TO phonons. A typical spectrum is illustrated in Fig. 10.32 for GaAs.

For higher-order Raman processes in which several phonons participate, it is the product of the irreducible representations of the phonons involved that must be the same as an irreducible representation of the Raman tensor. Tables of products of space group representations can be found in the literature (Birman 1974).

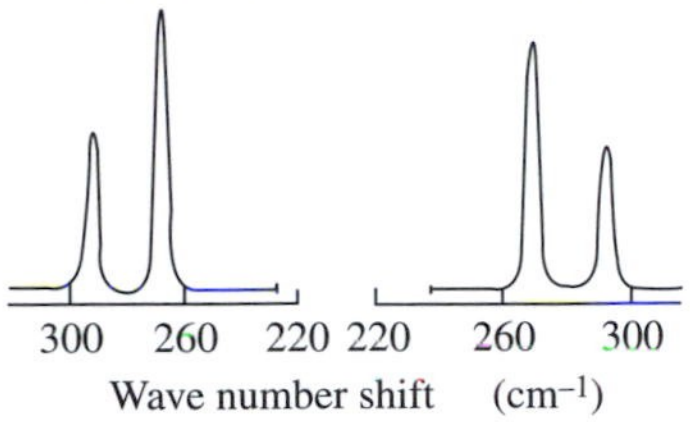

Fig. 10.32
Raman spectrum for optical phonons in GaAs (after Mooradian and Wright 1966).

10.8.2.3 Geometrical aspects of first-order Raman scattering

The observation of light scattering by optical phonons in transparent crystals is usually done in a geometry in which the linearly polarized incident light beam is directed along, say, the x-axis and the scattered beam is observed along the y-axis (see Fig. 10.33). When the crystal is not transparent, as is often the case in semiconductors, the observation is made in the back scattering geometry in which the scattered beam is observed in the opposite direction to the incident beam.

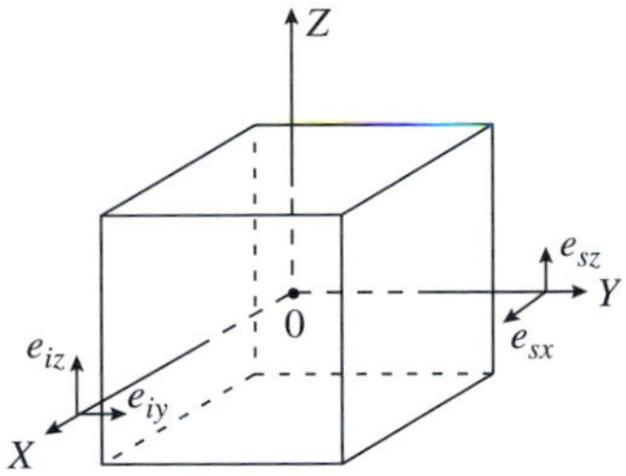

Fig. 10.33
Geometry for light scattering experiments in transparent crystals.

The scattering geometry affects the range of phonon wave vectors that is accessible in first-order Raman scattering. The conservation of wave vector condition is

$$\boldsymbol{k}_i - \boldsymbol{k}_s = \pm \boldsymbol{q}, \tag{10.205}$$

where $\boldsymbol{q}$ is the phonon wave vector and the plus (minus) sign refers to the Stokes (anti-Stokes) process. The Stokes geometry is shown in Fig. 10.34 and has the following relation satisfied:

$$q^2 = k_i^2 + k_s^2 - 2k_ik_s\cos\theta. \tag{10.206}$$

Forward scattering is characterized by $\theta = 0$ and a minimum value of q given, for isotropic media, by

$$q_{min} = [n(\omega_i)\omega_i - n(\omega_s)\omega_s]/c, \tag{10.207}$$

where $n(\omega_i)$ and $n(\omega_s)$ are the refractive indices of the crystal for the incident and scattered light, respectively, and $k_{i(s)} = n(\omega_{i(s)})\omega_{i(s)}/c$. An alternative expression for q can be obtained from Eq. (10.205) by writing k_s as

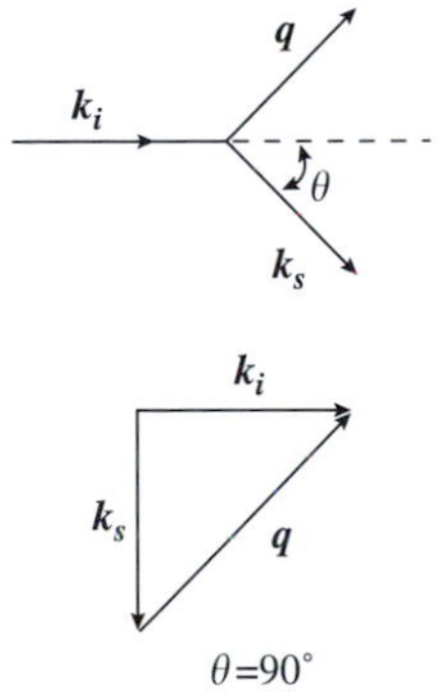

Fig. 10.34
Diagram for Stokes processes.

$$k_s(\omega_s) \simeq k_i(\omega_i) + \left.\frac{\partial k}{\partial \omega}\right|_{\omega_i}(\omega_s - \omega_i). \tag{10.208}$$

For small θ, Eq. (10.206) becomes

$$\begin{aligned} q^2 &\simeq (k_i - k_s)^2 + k_ik_s\theta^2 \\ &\simeq \left(\left.\frac{\partial k}{\partial \omega}\right|_{w_i}\omega_{\boldsymbol{q}}\right)^2 + k_ik_s\theta^2, \end{aligned} \tag{10.209}$$

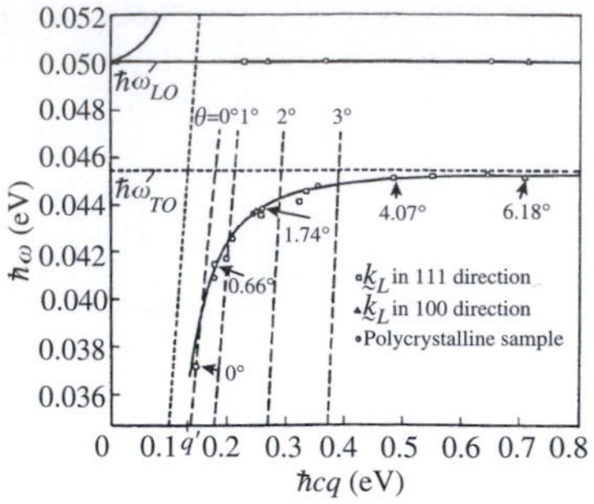

Fig. 10.35
Phonon–polariton dispersion curves in GaP. Solid lines are theoretical curves and dashed lines are uncoupled photons and phonons. Experimental data denoted by □, △, ○ (after Henry and Hopfield 1965).

where we have used the energy conservation condition for Stokes scattering $\omega_i = \omega_s + \omega_{\boldsymbol{q}}$. For $\theta = 0$,

$$q \simeq \left.\frac{\partial k}{\partial \omega}\right|_{\omega_i} \omega_{\boldsymbol{q}}, \tag{10.210}$$

which specifies a straight line when $\omega_{\boldsymbol{q}}$ is plotted against q. This line has the same slope as the dispersion curve of the incident light and intersects the lower branch of the phonon–polariton dispersion curve as shown in Fig. 10.35 for GaP. As θ is increased from zero, the lines from Eq. (10.209) appear to the right of the $\theta = 0$ line and also intersect the polariton dispersion curve. These intersections correspond to the conditions for the experimental observation of polaritons. It is evident that the polariton dispersion curve to the right of the $\theta = 0$ intersection can be determined by Raman scattering.

Back scattering is characterized by the maximum value of q when $\theta = 180°$ and is given by

$$q_{max} = [n(\omega_i)\omega_i + n(\omega_s)\omega_s]/c. \tag{10.211}$$

For typical light scattering experiments in the visible region the range of the incident wave vector is $0 < k_i < 10^6\ \mathrm{cm}^{-1}$. This implies that for first-order scattering processes the accessible range of q under conditions of wave vector conservation is small compared to a nonzero reciprocal wave vector. Light scattering experiments yield the frequencies of optical modes at essentially the center of the Brillouin zone.

The energy and momentum conservation rules have to be modified when the lifetime of the crystal excitations are strongly limited by their decay into other crystal excitations. Momentum conservation breaks down in imperfect crystals, in solids lacking translational symmetry like amorphous materials, and in crystals which are opaque to incident and scattered light.

In those cases where the incident and scattered waves are damped inside the scattering volume, such as occur in small gap semiconductors that are opaque at the light frequencies involved, $\boldsymbol{k}_1$ and $\boldsymbol{k}_2$ are complex. The inelastic scattering is due to excitations having a range of wave vector

$$\Delta q = |\mathrm{Im}\ \boldsymbol{k}_1| + |\mathrm{Im}\ \boldsymbol{k}_2| \tag{10.212}$$

about $\boldsymbol{q} = \mathrm{Re}(\boldsymbol{k}_1 - \boldsymbol{k}_2)$. Effects associated with such a wave vector uncertainty have been reported in Raman scattering spectra of III–V semiconductor compounds (Buchner and Burstein 1974).

10.8.2.4 Second-order Raman scattering

Second-order Raman scattering involves two phonons rather than a single phonon as in first-order scattering. The possible second-order processes are the following: two phonons may be created, giving a Stokes component in the scattered light, or one is created and the other destroyed, giving a Stokes or anti-Stokes component, or both may be destroyed giving an anti-Stokes component. For each of these cases there are the possibilities of a pair of first-order electron–phonon interactions or a single second-order electron–phonon interaction.

In the case of second-order Raman scattering the wave vector conservation condition is $\boldsymbol{q}_1 \pm \boldsymbol{q}_2 \simeq 0$, where $\boldsymbol{q}_1$ and $\boldsymbol{q}_2$ are the wave vectors of the phonons involved. This condition places no restriction on the magnitudes of the individual wave vectors, other than $|\boldsymbol{q}_1| \simeq |\boldsymbol{q}_2|$, which is in contrast to first-order scattering where $q \simeq 0$. The possible phonon frequency pairings associated with second-order processes are $\omega_1 + \omega_2$ (combination band), $\omega_1 - \omega_2$ (difference band), and $2\omega_1$ (overtone band). The second-order scattering spectrum covers a broad range of frequencies. The overtone spectrum provides a measure of the phonon density-of-states.

10.8.2.5 Resonant light scattering

General formulation

Resonant Raman scattering occurs when the incident or scattered photon energy is close to the energy of an intermediate electronic state relative to the ground electronic state. Certain terms in the Raman amplitude given by Eq. (10.197) then diverge leading to a very large scattering cross section. We see that divergences occur in the Stokes spectrum when $\hbar\omega_i = \hbar\omega_a$ and $\hbar\omega_s = \hbar\omega_b$. The photon energies **resonate** with the excitation energies of the intermediate states a or b. If a and b are the same state, resonance occurs nearly simultaneously for both ω_i and ω_s and gives rise to a particularly strong enhancement of the scattering efficiency. It is thus clear that the resonance phenomenon is specific to the nature of the intermediate state, and its investigation leads to basic information concerning the electronic states of the system. We shall now examine the resonance behavior in several cases where the intermediate states are carrier Bloch states, free excitons, or bound excitons.

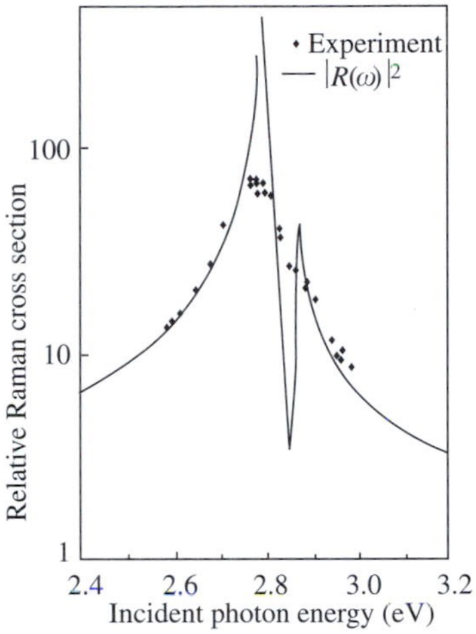

Fig. 10.36
Raman cross section as a function of incident photon energy for allowed first-order TO-phonon scattering in GaP at room temperature. The experimental results (crosses) have been adjusted to agree with theory at 2.64 eV (Bell *et al.* 1973).

Resonance at the fundamental band gap

The divergent behavior of the Raman amplitude is associated with the factor $(\hbar\omega_i - E_g)^{-1}$, where E_g is the lowest direct band gap of the material. Resonance at this gap in GaP has been observed experimentally (Bell *et al.* 1973). The scattering cross section as a function of incident photon energy is shown in Fig. 10.36.

Resonance at free exciton states

The Raman intensity as a function of the incident photon energy has a Lorentzian line shape in the range of a single intermediate state. It is centered at the resonance frequency and has a width determined by the lifetime of the intermediate state. The distinction between scattering at resonance and scattering off resonance in the wings of the Lorentzian is that at resonance the intensity is determined by the exciton lifetime, whereas off resonance it is determined by the frequency separation from resonance. Since in many instances excitons have very long lifetimes, they can lead to a very large Raman cross section at resonance.

Resonance at bound exciton states

Scattering induced by the presence of impurities involving bound exciton states has the following essential features: (a) observation of sharp resonances at the energies of the bound exciton which form discrete levels below the free exciton resonance; (b) dependence upon impurity concentration; (c) involvement of LO phonons of wave vector of the order of the inverse of the impurity state radius. Resonance at bound exciton energies has been observed in impurity induced resonant Raman scattering by LO phonons in CdS (Damen and Shah 1971). The resonance is centered at the absorption peak for the impurity state and has approximately Lorentzian shape. If the impurity gives rise to a localized vibrational mode, bound exciton resonance in Raman scattering due to the local mode can occur. An example of this resonance has been found in CdS containing Cl impurities (Yu *et al.* 1978).

10.8.3 Anharmonic effects on Raman spectra

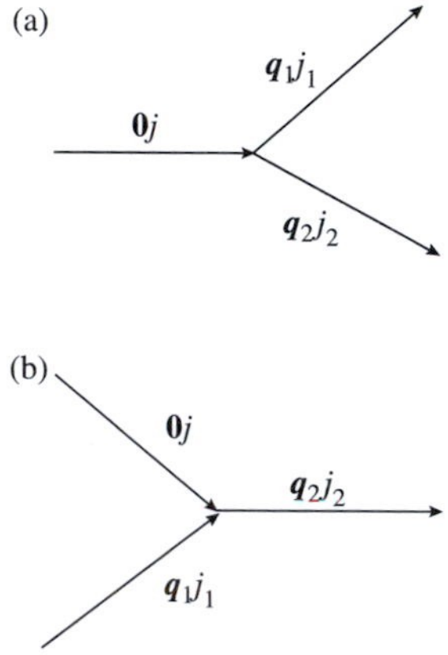

Fig. 10.37
Diagrams representing three-phonon anharmonic processes contributing to the decay of the Raman active optical mode.

In a perfect crystal whose vibrations are harmonic and which is transparent to the incident and scattered light, the Raman line shape is a delta function centered on the optical phonon frequency. Experimentally one finds that even in crystals of very high quality, the Raman line is broadened into a roughly Lorentzian shape with a width that increases with increasing temperature. This width can be attributed to anharmonic terms in the vibrational Hamiltonian.

Let us consider the cubic anharmonic terms in the expansion of the nuclear potential energy given by Eq. (7.1). If normal coordinates are introduced using Eq. (7.66) and then creation and destruction operators via Eq. (8.138), we obtain a number of terms, one of which contains the product $a_{\boldsymbol{0}j}a^{+}_{\boldsymbol{q}_1j_1}a^{+}_{\boldsymbol{q}_2j_2}$. This term gives rise to a Stokes light scattering process involving the absorption of a photon $\hbar\omega_i$, the emission of a photon $\hbar\omega_s$, and the creation of an optical phonon $\boldsymbol{0}j$ which then decays via cubic anharmonicity into two phonons $\boldsymbol{q}_1 j_1$ and $\boldsymbol{q}_2 j_2$. Energy and wave vector conservation give the relations $\omega_i - \omega_s = \omega_{\boldsymbol{q}_1j_1} + \omega_{\boldsymbol{q}_2j_2}$ and $\boldsymbol{k}_i - \boldsymbol{k}_s = \boldsymbol{q}_1 + \boldsymbol{q}_2$, respectively. A diagram representing this process is shown in Fig. 10.37a.

A second process of interest arises from the product $a_{\boldsymbol{0}j}a_{\boldsymbol{q}_1j_1}a^{+}_{\boldsymbol{q}_2j_2}$ and is represented diagrammatically in Fig. 10.37b. In this process the phonon $\boldsymbol{q}_1j_1$ is destroyed rather than being created. The conservation conditions are $\omega_i - \omega_s = -\omega_{\boldsymbol{q}_1j_1} + \omega_{\boldsymbol{q}_2j_2}$ and $\boldsymbol{k}_i - \boldsymbol{k}_s = -\boldsymbol{q}_1 + \boldsymbol{q}_2$.

It should be noted that the extreme smallness of the photon wave vectors $\boldsymbol{k}_i$ and $\boldsymbol{k}_s$ leads to the approximate wave vector conservation conditions $\boldsymbol{q}_2 \simeq -\boldsymbol{q}_1$ and $\boldsymbol{q}_2 \simeq \boldsymbol{q}_1$, but no restriction on the magnitudes of $\boldsymbol{q}_1$ and $\boldsymbol{q}_2$ other than $q_1 \simeq q_2$. Consequently, the frequency shift $\omega_i - \omega_s$ can vary over a range about $\omega_{\boldsymbol{0}j}$ and thereby impart a nonzero width to the Raman line. By treating the anharmonic terms as a perturbation, one can show that the half-width at half-maximum associated with cubic anharmonicity and long wavelength optical phonons is given by

$$\Gamma(\boldsymbol{0}j,\omega) = \frac{18\pi}{\hbar^2}\sum_{\boldsymbol{q}}\sum_{j_1j_2}|V(\boldsymbol{0}j,\boldsymbol{q}j_1,-\boldsymbol{q}j_2)|^2\{(n_1+n_2+1)$$
$$\times\,\delta(\omega-\omega_1-\omega_2) - 2(n_1-n_2)\delta(\omega-\omega_1+\omega_2)\}, \qquad (10.213)$$

where the subscripts 1 and 2 on n and ω have replaced $\boldsymbol{q}j_1$ and $-\boldsymbol{q}j_2$, respectively, and

$$V(\boldsymbol{0}j, \boldsymbol{q}j_1, -\boldsymbol{q}j_2) = \frac{1}{12}\left(\frac{\hbar^3}{2N\omega_{0j}\omega_1\omega_2}\right)^{\frac{1}{2}} \times \sum_{\kappa\alpha}\sum_{\ell'\kappa'\beta}\sum_{\ell''\kappa''\gamma} \Phi_{\alpha\beta\gamma}(0\kappa; \ell'\kappa'; \ell''\kappa'') e_{\alpha\kappa}(\boldsymbol{0}j) e_{\beta\kappa'}(\boldsymbol{q}j_1) \times e_{\gamma\kappa''}(-\boldsymbol{q}j_2)(M_\kappa M_{\kappa'} M_{\kappa''})^{-\frac{1}{2}} e^{i\boldsymbol{q}\cdot[\boldsymbol{R}(\ell')-\boldsymbol{R}(\ell'')]}. \tag{10.214}$$

The temperature dependence of the line width is determined by the phonon occupation factors n_1 and n_2. As the temperature approaches zero, n_1 and n_2 approach zero, and the width approaches the value due to the zero-point motion of the nuclei. At temperatures above the Debye temperature, the width due to cubic anharmonicity becomes proportional to T. Quartic anharmonicity imparts a T^2 dependence to the width which becomes significant at very high temperatures. The experimental data in Fig. 10.38 show evidence of both contributions.

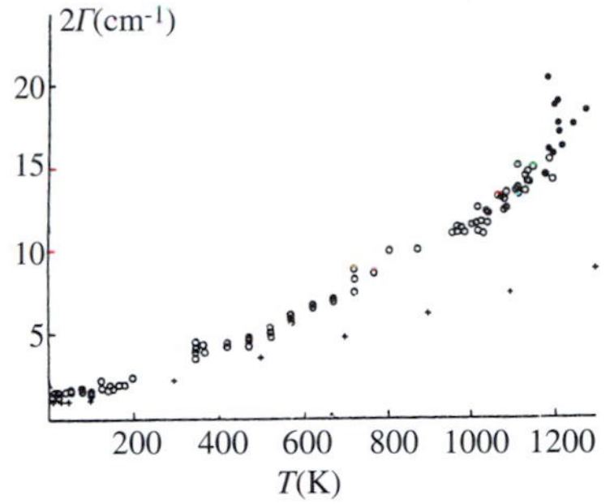

Fig. 10.38
The quantity $2\Gamma(0j, \omega)$ versus absolute temperature: theory based on cubic anharmonicity (crosses), experiment (open and solid circles) (after Haro *et al.* 1986).

In addition to line broadening, anharmonicity causes a shift of the frequency of peak intensity to lower values as the temperature increases. Contributions to the shift proportional to T and T^2 arise in the high-temperature regime from cubic and quartic anharmonicity, respectively.

10.8.4 Light scattering due to electronic excitations

10.8.4.1 Light scattering by plasmons

Plasmons are similar to LO phonons in that they have a macroscopic electric field associated with them. The field modulates the electric susceptibility and gives rise to Raman scattering. The difference in frequency of the incident and scattered light is equal to the plasma frequency ω_p specified by $\omega_p^2 = e^2 n/\epsilon_0\epsilon_\infty m^*$, where n is the electron concentration, m^* is their effective mass, and ϵ_∞ is the high-frequency dielectric constant.

To calculate the differential scattering cross section one must deal with the Coulomb interaction between electrons. This can be done using standard many-body techniques. The result is (Wallis and Balkanski 1986)

$$\frac{d^2\sigma}{d\omega d\Omega} = \frac{\hbar r_0^2}{\Phi}\left(\frac{\omega_s}{\omega_i}\right)(\hat{e}_i \cdot \hat{e}_s)^2[n(\omega) + 1]\delta[\mathrm{Re}\,\epsilon(\boldsymbol{q}, \omega)], \tag{10.215}$$

where r_0 is the classical electron radius $e^2/4\pi\epsilon_0 mc^2$, $\Phi = e^2/\epsilon_0\epsilon_\infty Vq^2$, V is the volume, $n(\omega)$ is the Bose factor $[e^{\hbar\omega/k_BT} - 1]^{-1}$, and $\epsilon(\boldsymbol{q}, \omega)$ is the Lindhard dielectric function for the interacting electron gas. In the regime of small q, $\mathrm{Re}\,\epsilon(\boldsymbol{q}, \omega)$ can be approximated by

$$\epsilon(0, \omega) = \epsilon_\infty\left(1 - \frac{\omega_p^2}{\omega^2}\right). \tag{10.216}$$

Thus, the scattering cross section has a delta-function peak at $\omega = \omega_p$. This is the well-known plasma resonance in the scattering cross section.

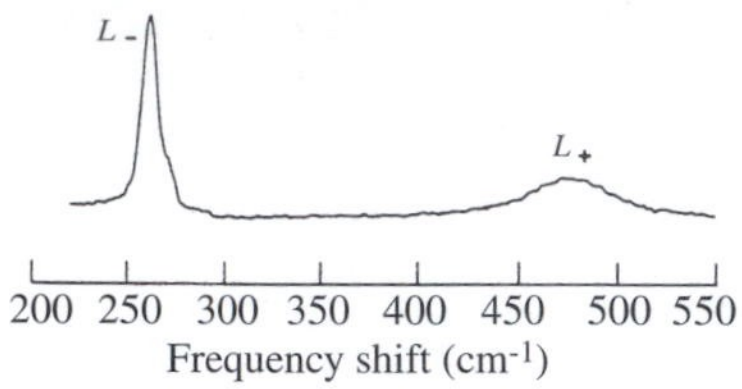

Fig. 10.39
Raman spectrum for n-GaAs (after Mooradian and McWhorter 1967).

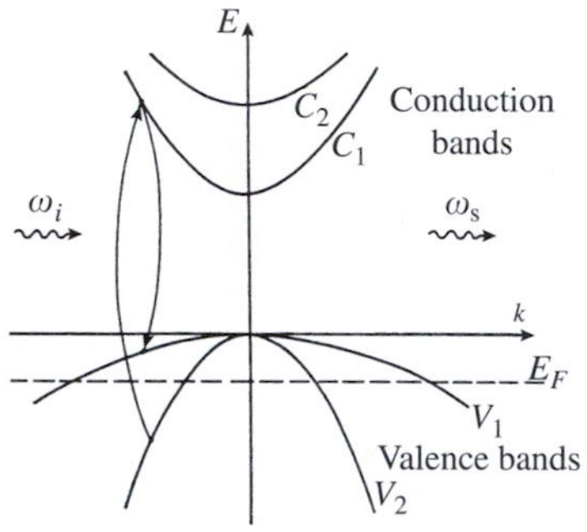

Fig. 10.40
Electronic transitions occurring in intervalence band Raman scattering.

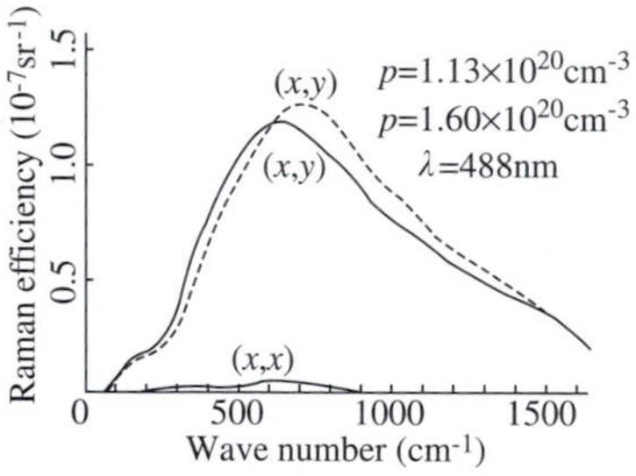

Fig. 10.41
Intervalence band Raman spectrum for anisotropic bands at 0 K (after Kanehisa *et al.* 1982).

To obtain a non-zero width of the plasma resonance line in the present approximation, it is necessary to include interactions of the free carriers with impurities or phonons (Platzman 1965).

When the plasma frequency is close to the LO phonon frequency, there is a plasmon–phonon interaction via the macroscopic electric fields that leads to coupled modes and forces the frequencies apart. This behavior is evident in Fig. 10.39 for n-GaAs.

10.8.4.2 Light scattering associated with interband transitions

The plasmons dealt with in the preceding section are an **intraband** type of excitation. Also of interest are **interband** transitions in which a carrier undergoes a transition from one band to a different band during the scattering process. An example is provided by p-type silicon in which an electron makes a transition from an occupied state in the light hole band to an unoccupied state in the heavy hole band via the conduction band as shown in Fig. 10.40. For the case of isotropic bands the scattering efficiency spectrum at 0 K has sharp edges. If the anisotropy of the valence bands is taken into account, the sharp edges of the spectrum are smoothed out, as can be seen in Fig. 10.41.

The experimental Raman spectrum for p-Si exhibits the broad continuum predicted by theory but the comparison between theory and experiment is complicated by the presence of discrete Raman lines due optical phonons. The resulting distortion of the spectrum is associated with **Fano interference** that occurs between the discrete phonon lines and the electronic continuum.

10.9 Photoemission

10.9.1 Direct photoemission

The photoelectric effect is a well-known phenomenon in metals. The corresponding effect in semiconductors is known as **photoemission**. If a photon incident on a semiconductor has a sufficiently high energy, it can be absorbed with the simultaneous ejection of an electron from a valence band into the vacuum outside. By analyzing the energies and momenta of the photon and the electron, information about the energy–wave vector relation of the valence band under consideration can be obtained.

Let the photon have energy $\hbar\omega$ and wave vector $\boldsymbol{k}_{ph}$, and let the semiconductor be intrinsic. Conservation of energy then leads to the equation

$$\hbar\omega + E_{v\boldsymbol{k}} = E_g + \chi + E_{KE}, \tag{10.217}$$

where $E_{v\boldsymbol{k}}$ is the energy of the valence band state from which the electron is ejected, χ is the electron affinity, and E_{KE} is the kinetic energy of the ejected electron. A diagram showing these energies is given in Fig. 10.42. Since the zero of energy is taken to be the valence band edge, $E_{v\boldsymbol{k}}$ is negative for $\boldsymbol{k} \neq \boldsymbol{0}$.

The minimum or **threshold photon energy**, $\hbar\omega_t$, required for photo emission corresponds to $E_{KE} = 0$ and $E_{v\boldsymbol{k}} = 0$. Equation (10.217) then yields

$$\hbar\omega_t = E_g + \chi. \tag{10.218}$$

From a measured value of $\hbar\omega_t$ and a value of E_g derived from interband optical absorption, one obtains the electron affinity χ.

Conservation of momentum gives the relation

$$\hbar \boldsymbol{k}_{ph} + \hbar \boldsymbol{k} = m\boldsymbol{v} \tag{10.219}$$

with $\boldsymbol{v}$ the velocity of the ejected electron. The photon wave vector is specified by the direction of the incident light beam and by its magnitude given by $k_{ph} = \omega/c$. the value of v can be obtained from the measured value of $E_{KE} = \frac{1}{2}mv^2$. The direction of $\boldsymbol{v}$ is determined by experiment. The information is now in hand to deduce the electron wave vector $\boldsymbol{k}$ from Eq. (10.219).

By carrying out experiments with values of $\hbar\omega$ above threshold and using the corresponding measured values of E_{KE}, one obtains values of $E_{v\boldsymbol{k}}$ from Eq. (10.217). Thus, both $\boldsymbol{k}$ and $E_{v\boldsymbol{k}}$ can be determined by photoemission. An example of valence band structure obtained in this way is given in Fig. 10.43 for Ge.

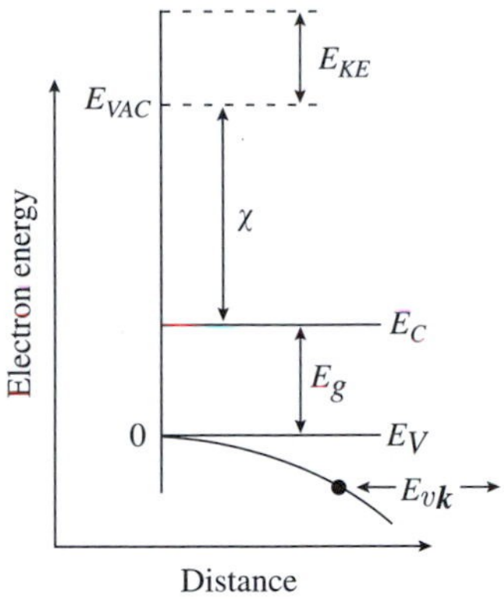

Fig. 10.42
Band diagram showing the pertinent energies for photoemission.

If the semiconductor is heavily doped with impurities so that the highest occupied electron state is no longer the valence band edge E_V, Eq. (10.218) for the threshold photon energy must be modified. In the case of donor impurities with the Fermi energy lying above the conduction band edge by an amount Δ_n, $\hbar\omega_t$ is given by

$$\hbar\omega_t = \chi - \Delta_n \tag{10.220}$$

as shown in Fig. 10.44. For acceptor impurities, on the other hand, with the Fermi energy an amount Δ_p below the valence band edge, one has

$$\hbar\omega_t = E_g + \chi + \Delta_p \tag{10.221}$$

as shown in Fig. 10.45.

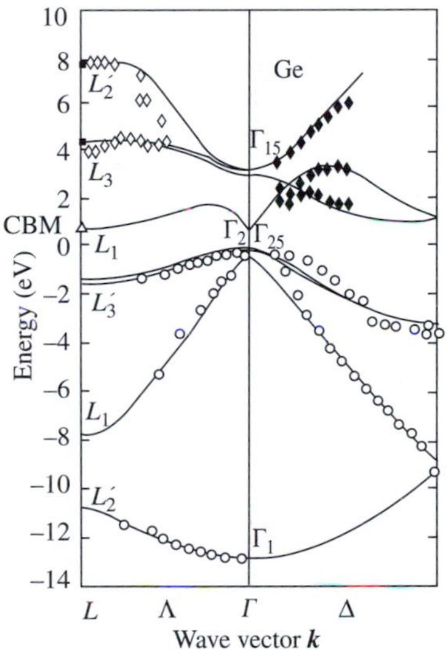

Fig. 10.43
Energy bands of Ge. Solid lines are theoretical results. Experimental results are denoted by ○, △, ◇, □ (after Ortega and Himpsel, 1993).

10.9.2 Inverse photoemission

So far we have discussed the use of photoemission in the determination of energies of band states that are occupied by electrons. We now turn to the inverse process which enables one to determine the energies of unoccupied states. **Inverse photoemission** consists of injecting an electron of known energy and wave vector into a solid and measuring the frequency of photons emitted in a given direction. The injected electron is trapped in an empty state which is typically a conduction band state. If the latter has energy $\Delta_{n\boldsymbol{k}}$ above the conduction band edge, conservation of energy leads to the relation

$$\hbar\omega + \Delta_{n\boldsymbol{k}} = \chi + E_{KE} \tag{10.222}$$

where E_{KE} is now the kinetic energy of an incident electron and $\hbar\omega$ is the energy of an emitted photon. From a previously determined value of χ and measured values of E_{KE} and $\hbar\omega$, one obtains the value of $\Delta_{n\boldsymbol{k}}$. Conservation of momentum yields the value of $\boldsymbol{k}$. Experimental results for conduction band energies in Ge are presented in Fig. 10.43.

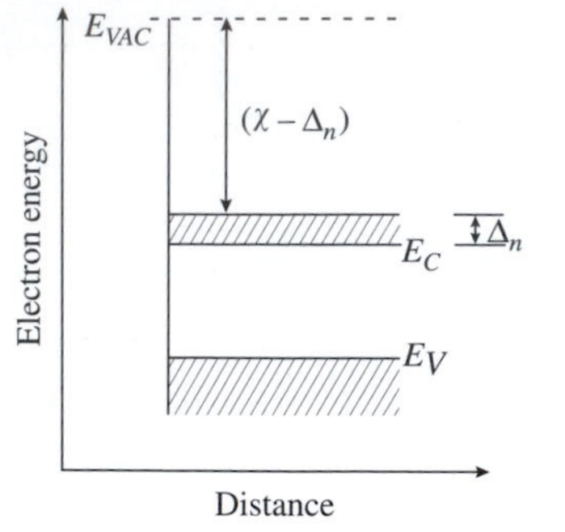

Fig. 10.44
Band diagram for n-type material.

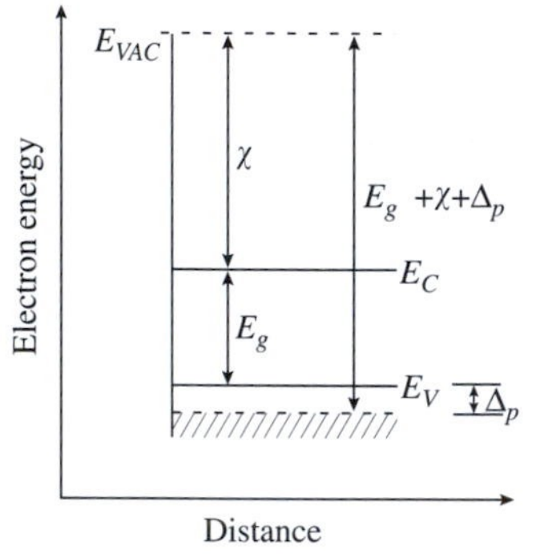

Fig. 10.45
Band diagram for p-type material.

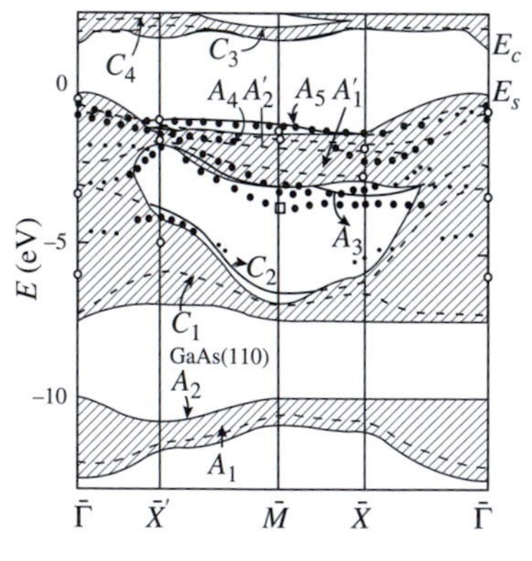

Fig. 10.46
Surface states of clean GaAs(110). Solid and dashed lines are theoretical results. Experimental data are denoted by dots (after Sébenne 1994).

10.9.3 Surface state energies

Photoemission experiments involving electrons in the 10–40 eV range are well suited to the study of electronic surface states because such electrons have mean free paths of only a few atomic layers. Conservation of momentum now involves only the component of wave vector parallel to the surface. An example of surface states is given in Fig. 10.46 for clean cleaved GaAs(110). The fact that the surface state band closest to the valence band edge lies inside the valence band is due to surface reconstruction. Without reconstruction it would lie in the gap and be associated with the dangling bonds of the As surface atoms.

A very interesting extension of the photoemission technique is to surfaces with adsorbed atoms (Sébenne 1994, Himpsel 1994). The large number of adsorbate species available makes possible a rich variety of systems and phenomena to be explored. There are two general effects due to adsorbed atoms. The first is the elimination of dangling bonds and surface states associated with them. The second is the generation of new surface states that arise from the presence of the adsorbate.

An example of the latter situation is provided by the alkali metal Na adsorbed on Si(111). The alkali s-electron can be viewed as pairing with the single electron of the dangling bond of a surface Si atom to give a lone-pair-like occupied surface state labeled by S in Fig. 10.47 and an unoccupied alkali state labeled by U_1' in Fig. 10.47. These states have been explored experimentally by photoemission and inverse photoemission, respectively, with results indicated by the crosses in Fig. 10.47 (Reihl *et al.* 1992).

Problems

1. Discuss the temperature dependence of indirect interband absorption in terms of the emission and absorption of phonons. Derive an expression for the absorption coefficient that gives its explicit temperature dependence.
2. In the semiconductor Cu_2O the conduction band minimum and valence band maximum occur at $\boldsymbol{k} = 0$, but at this point both bands have even parity and an electric dipole transition between them is forbidden. For $\boldsymbol{k} \neq 0$ odd parity states are admixed into the wave functions and the interband matrix element of the momentum operator takes the approximate form (Bardeen *et al.* 1956) $\langle c|\boldsymbol{p}|v\rangle \simeq (m/m_T^*)\hbar\boldsymbol{k}$, where m_T^* is an effective mass for the transition. Derive an expression for the interband absorption coefficient $\alpha(\omega)$ that reveals its frequency dependence explicitly. What function involving $\alpha(\omega)$ gives a straight line when plotted against ω?
3. The first-order polarization given by Eq. (10.124) has no vibrational contribution in homopolar semiconductors such as Si and Ge because $e_T^* = 0$. However, a second-order polarization does exist of the form
$$P^{(2)} = \sum_{\ell\kappa\alpha}\sum_{\ell'\kappa'\beta} P^{(2)}_{\ell\kappa\alpha,\,\ell'\kappa'\beta} u_\alpha(\ell\kappa) u_\beta(\ell'\kappa').$$
By introducing normal coordinates and thence creation and destruction operators, evaluate the matrix element of $P^{(2)}$ with respect to vibrational wave functions consisting of products of harmonic oscillator wave functions for the normal coordinates. Using Fermi's golden rule develop expressions for the transition rate and for the conservative laws of wave vector and energy. Obtain an expression for the absorption coefficient using a one-dimensional model of

Si and discuss the principal features of the absorption spectrum (Reference: Lax and Burstein 1955).

4. Consider a semiconductor such as n-InSb with a free surface. Derive expressions for the frequencies of coupled surface plasmon-surface optical phonon excitations. Calculate the frequencies for electron concentrations such that the range from $\omega_p \ll \omega_{LO}$ to $\omega_p \gg \omega_{LO}$ is covered. Plot the frequencies versus electron concentration.
5. Derive an expression that describes the Burstein-Moss effect when the nonparabolicity of the conduction band is taken into account. The band energy can be expressed as $E_{c\mathbf{k}} - E_C = y(1 - E_4 y)$, where $y = (\hbar^2 k^2/2m_c^*)$ and E_4 is the nonparabolicity parameter. For non parabolic bands and complete degeneracy of the carriers, derive a relation between the threshold photon energy and the carrier concentration.
6. PbTe crystallizes in the NaCl structure. Discuss its phonon Raman spectrum in terms of first-order and second-order processes. Derive an expression for the differential scattering cross section associated with second-order processes and develop a qualitative picture of the Raman spectrum. (Reference: Born and Huang 1954).

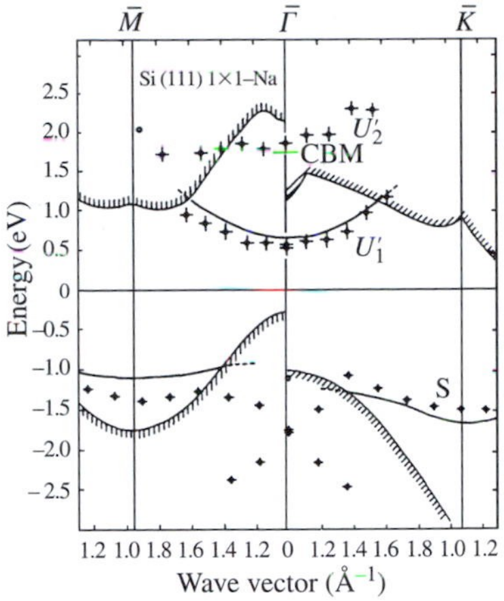

Fig. 10.47
Surface states of Si(111)-Na. Solid lines are theoretical results. Experimental data are denoted by crosses (after Himpsel 1994).

References

B. Abeles, C. R. Wronski, T. Tiedje, and G. D. Cody, *Solid State Commun.* **36**, 537 (1980).

M. Balkanski, in *Optical Properties of Solids*, ed. F. Abeles (North-Holland, Amsterdam, 1972) p. 533.

M. Balkanski and W. Nazarewicz, *J. Phys. Chem. Solids* **25**, 437 (1964).

M. Balkanski and W. Nazarewicz, *J. Phys. Chem. Solids* **27**, 671 (1966).

J. Bardeen, F. J. Blatt, and L. H. Hall, in *Proceedings of the Photoconductivity Conference held at Atlantic City*, eds. R. G. Breckenridge, B. R. Russell, and E. E. Hahn (John Wiley, New York, 1956), p. 146.

M. I. Bell, R. N. Tyte, and M. Cardona, *Solid State Commun.* **13**, 1833 (1973).

J. L. Birman, *Encyclopedia of Physics*, Vol. XXV 12b (Springer-Verlag, Berlin, 1974).

M. Born and K. Huang, *Dynamical Theory of Crystal Lattices* (Oxford University Press, Oxford, 1954).

M. Born and E. Wolf, *Principles of Optics*, Third edition (Pergamon Press, Oxford, 1965).

L. Brillouin, *Ann. Physique* **17**, 88 (1922).

S. Buchner and E. Burstein, *Phys. Rev. Lett.* **33**, 908 (1974).

E. Burstein, *Phys. Rev.* **93**, 632 (1954).

E. Burstein, G. Picus, and N. Sclar, in *Proceedings of the Photoconductivity Conference held at Atlantic City*, eds. R. G. Breckenridge, B. R. Russell, and E. E. Hahn (John Wiley, New York, 1956), p. 353.

E. Burstein, A. Pinczuk, and R. F. Wallis, in *The Physics of Semimetals and Narrow Gap Semiconductors*, eds. D. L. Carter and R. Bate (Pergamon Press, Oxford, 1971) p. 251.

T. C. Damen and J. Shah, *Phys. Rev. Lett.* **27**, 1506 (1971).

J. O. Dimmock, in *Semiconductors and Semimetals*, Vol. III, eds. R. K. Willardson and A. C. Beer (Academic Press, New York, 1967), p. 259.

W. Dumke, *Phys. Rev.* **105**, 139 (1957).

D. J. Evans, S. Ushioda, and J. D. McMullen, *Phys. Rev. Lett.* **31**, 369 (1973).

E. Fermi, *Nuclear Physics* (University of Chicago Press, Chicago, 1950), p. 142.

M. Grynberg, R. Le Toullec, and M. Balkanski, *Phys. Rev. B* **9**, 517 (1974).

E. Haro, M. Balkanski, R. F. Wallis, and K. H. Wanser, *Phys. Rev.* B**34**, 5358 (1986).

W. Hayes and R. Loudon, *Scattering of Light by Crystals* (John Wiley, New York, 1978).

J. R. Haynes, *Phys. Rev.* **98**, 1866 (1955).

C. H. Henry and J. J. Hopfield, *Phys. Rev. Lett.* **15**, 964 (1965).
F. J. Himpsel, in *Handbook on Semiconductors*, Second edition, Vol. 2, ed. M. Balkanski (North-Holland, Amsterdam, 1994).
T. Holstein, *Ann. Phys.* **29**, 410 (1964).
K. Huang, *Proc. Roy. Soc.* (London) A**208**, 352 (1951).
M. A. Kanehisa, R. F. Wallis, and M. Balkanski, *Phys. Rev. B* **25**, 7619 (1982),
M. Lax and E. Burstein, *Phys. Rev.* **97**, 39 (1955).
R. Le Toullec, Thesis, Paris (1968).
G. G. MacFarlane and V. Roberts, *Phys. Rev.* **98**, 1865 (1955).
A. A. Maradudin and R. F. Wallis, *Prog. Theor. Phys.* **24**, 1055 (1960).
N. Marschall, B. Fischer, and H. J. Queisser, *Phys. Rev. Lett.* **27**, 95 (1971).
D. L. Mills and E. Burstein, *Rep. Prog. Phys.* **37**, 817 (1974).
A. Mooradian and A. L. McWhorter, *Phys. Rev. Lett.* **19**, 849 (1967).
A. Mooradian and G. B. Wright, *Solid State Commun.* **4**, 431 (1966).
J. E. Ortega and F. J. Himpsel, *Phys. Rev. B* **47**, 2130 (1993).
P. M. Platzman, *Phys. Rev.* A**139**, 379 (1965).
B. Reihl, R. Dudde, L. S. O. Johansson, K. O. Magnusson, S. L. Sorensen, and S. Wiklund, *Appl. Surf. Sci.* **56**/58, 123 (1992).
C. A. Sébenne, in *Handbook on Semiconductors*, Second Edition, Vol 2, ed. M. Balkanski (North-Holland, Amsterdam, 1994).
F. Seitz, *Modern Theory of Solids* (McGraw-Hill, New York, 1940).
W. Shockley and W. T. Read, *Phys. Rev.* **87**, 835 (1952).
W. Spitzer and M. Waldron, *Phys. Rev. Lett.* **14**, 223 (1965).
M. D. Sturge, *Phys. Rev.* **127**, 768 (1962).
R. C. Tolman, *Principles of Statistical Mechanics* (Oxford University Press, Oxford, 1938).
W. van Roosbroeck and W. Shockley, *Phys. Rev.* **94**, 1558 (1954).
R. F. Wallis and M. Balkanski, *Many-Body Aspects of Solid State Spectroscopy* (North-Holland, Amsterdam, 1986).
P. Y. Yu, M. H. Pilkhun, and F. Evangelisti, *Solid State Commun.* **25**, 371 (1978).

Magneto-optical and electro-optical phenomena

11

Key ideas

The elements of the *magnetoconductivity tensor* $\overleftrightarrow{\sigma}(\omega)$ satisfy the general relation $\sigma_{ij}(-\omega) = \sigma_{ij}^*(\omega)$ which results from the reality of the electric field.

The *dispersion relation* for longitudinal propagation is $c^2k_\pm^2 = \omega^2\epsilon_\pm$. ϵ_+ and ϵ_- are the effective dielectric constants for right circularly polarized (RCP) and left circularly polarized (LCP) light, respectively.

The effective dielectric constants for transverse propagation are the perpendicular or Voigt dielectric constant $\epsilon_\perp$ and the parallel dielectric constant $\epsilon_\|$.

The *Faraday effect* is the rotation of the plane of polarization of a plane-polarized electromagnetic wave as it propagates parallel to an external magnetic field. Faraday rotation is expressed in terms of the angle of rotation per unit sample thickness. The extinction coefficients for the two circularly polarized waves are different: an initially plane-polarized light beam becomes elliptically polarized corresponding to *magnetodichroism*. Faraday rotation is a nonresonant phenomenon and concerns wave propagation. *Magneto-absorption* is a resonant phenomenon and can have intraband and interband contributions.

The *Voigt effect* is the phase difference introduced by a magnetic field between the parallel and perpendicular components of a plane-polarized wave propagating perpendicular to $\mathcal{B}_0$.

Faraday rotation is a convenient method to investigate effective masses and band structures at moderate magnetic fields in the infrared region of the spectrum.

In *parabolic bands* the effective mass is independent of the free carrier concentration and is equal to the effective mass at the band edge. In *nonparabolic bands* the effective mass is a measure of the band curvature at the Fermi energy. By varying the carrier concentration one can explore the energy band.

For electrons LCP radiation is associated with σ'_- and gives resonant absorption at the cyclotron frequency ω_c. For holes it is RCP radiation that gives resonant absorption at ω_c.

In the presence of a magnetic field the energy eigenvalues of free carriers are quantized into subbands or *Landau levels*. For parabolic bands the

Effects of magnetic and electric fields

Landau level energies are parabolic functions of the wave vector component $k_{\|}$ parallel to the field.

11.6 Quantum mechanical theory of cyclotron resonance

When the degeneracy of the valence bands at $\boldsymbol{k} = 0$ and the spin–orbit interaction are included in the presence of a magnetic field, four sets of Landau levels degenerate at $k_z = 0$ arise which are called *ladders*. The spacing of the levels in a given ladder is not uniform near the valence band edge as a result of *quantum effects*.

11.7 Interband magneto-absorption

Transitions between Landau levels in the valence band and those in the conduction band lead to *interband magneto-absorption*. The absorption coefficient shows singularities when the transitions involve Landau level edges.

11.8 Electro-optical effects

The *Pockels effect* arises from changes in the refractive index that are linear in the applied field. For plane polarized light the material becomes *birefringent*. Semiconductors with a center of inversion symmetry do not exhibit the Pockels effect.

The *Kerr effect* arises from third-order nonlinearities and is quadratic in the applied electric field.

The *Franz–Keldysh effect* is the shift of the interband absorption edge to lower photon energies under the influence of an applied electric field.

11.9 Modulation spectroscopy

Application of an external electric field modifies the optical reflectivity above the absorption edge. The relative changes in reflectivity show behavior that can be used to give information about *energy band structure*.

This chapter is divided into three major parts. In the first part magneto-optical phenomena are discussed from a classical point of view. Optical absorption as exemplified by cyclotron resonance and dispersion as exemplified by the Faraday and Voigt effects are analyzed. In the second part a quantum mechanical treatment is presented. The theory of the Landau levels of an electron moving in a uniform magnetic field is developed and applied to optical properties. In the third part the effects of an external electric field on optical properties are discussed.

11.1 Frequency-dependent conductivity tensor

The presence of a magnetic field introduces an anistropy into a conducting medium. In the case of a cubic lattice with a uniform magnetic field parallel to the z-axis, the conductivity tensor $\overleftrightarrow{\sigma}$ is given by

$$\overleftrightarrow{\sigma}(\omega) = \begin{pmatrix} \sigma_{xx} & \sigma_{xy} & 0 \\ \sigma_{yx} & \sigma_{yy} & 0 \\ 0 & 0 & \sigma_{zz} \end{pmatrix}, \qquad (11.1)$$

$$\sigma_{xx} = \sigma_{yy}, \quad \sigma_{yx} = -\sigma_{xy}.$$

The elements of $\overleftrightarrow{\sigma}(\omega)$ satisfy a number of general relations which we now discuss.

When studying the optical properties of semiconductors, the applied radiation field is not in general strictly monochromatic. The electric field $\mathcal{E}(t)$ can be represented in the form

$$\mathcal{E}(t) = \frac{1}{2\pi}\int_{-\infty}^{+\infty} \mathcal{E}(\omega)e^{-i\omega t}\,d\omega. \tag{11.2}$$

The current density is then

$$\boldsymbol{j}(t) = \frac{1}{2\pi}\int_{-\infty}^{+\infty} \overleftrightarrow{\sigma}(\omega)\cdot\mathcal{E}(\omega)e^{-i\omega t}\,d\omega. \tag{11.3}$$

The reality of the electric field $\boldsymbol{\mathcal{E}}(t)$ and the current density $j(t)$ leads to the property

$$\sigma_{ij}(-\omega) = \sigma^{*}_{ij}(\omega) \tag{11.4}$$

for all the elements of the tensor $\overleftrightarrow{\sigma}$.

Explicit expressions for the elements of the magnetoconductivity tensor associated with free carriers in a spherical energy band can be derived from the classical equation of motion

$$m^{*}\frac{d\boldsymbol{v}}{dt} + \frac{m^{*}}{\tau}\boldsymbol{v} = e_c[\boldsymbol{\mathcal{E}} + \boldsymbol{v}\times\boldsymbol{\mathcal{B}}_0], \tag{11.5}$$

where v is the carrier velocity, $\boldsymbol{\mathcal{E}}$ is the applied electric field, and $\boldsymbol{\mathcal{B}}_0$ is the applied (static) magnetic field in the z-direction. We re-express Eq. (11.5) in component form as

$$m^{*}\frac{dv_x}{dt} + \frac{m^{*}}{\tau}v_x = e_c\mathcal{E}_x + e_c\mathcal{B}_0 v_y \tag{11.6a}$$

$$m^{*}\frac{dv_y}{dt} + \frac{m^{*}}{\tau}v_y = e_c\mathcal{E}_y - e_c\mathcal{B}_0 v_x \tag{11.6b}$$

$$m^{*}\frac{dv_z}{dt} + \frac{m^{*}}{\tau}v_z = e_c\mathcal{E}_z. \tag{11.6c}$$

Assuming that $\boldsymbol{v}$ and $\boldsymbol{\mathcal{E}}$ vary with time as $\exp(-i\omega t)$, we now solve Eqs. (11.6) for the components of $\boldsymbol{v}$ in terms of the components of $\boldsymbol{\mathcal{E}}$. Multiplying the components of $\boldsymbol{v}$ by the carrier concentration n_c and carrier charge e_c yields the components of the current density $\boldsymbol{j}$. Comparing these results with the defining equation for the conductivity tensor $\overleftrightarrow{\sigma}$,

$$\boldsymbol{j} = \overleftrightarrow{\sigma}\cdot\boldsymbol{\mathcal{E}}, \tag{11.7}$$

we obtain the components of the conductivity tensor,

$$\sigma_{xx} = \sigma_{yy} = \frac{\sigma_0(1 - i\omega\tau)}{1 - (\omega^2 - \omega_c^2)\tau^2 - 2i\omega\tau} \tag{11.8a}$$

$$\sigma_{zz} = \frac{\sigma_0}{1 - i\omega\tau} \tag{11.8b}$$

$$\sigma_{xy} = -\sigma_{yx} = \frac{\sigma_0\omega_c\tau}{1 - (\omega^2 - \omega_c^2)\tau^2 - 2i\omega\tau} \tag{11.8c}$$

$$\sigma_{xz} = \sigma_{zx} = \sigma_{yz} = \sigma_{zy} = 0, \tag{11.8d}$$

where $\omega_c = e_c\mathcal{B}_0/m^*$ and $\sigma_0 = n_c e_c^2\tau/m^*$. It should be noted that w_c is now positive for holes and negative for conduction electrons. The elements of the dielectric tensor $\overleftrightarrow{\epsilon}$ follow from Eqs. (11.8) using the relation

$$\overleftrightarrow{\epsilon} = \overleftrightarrow{\mathrm{I}} + (i/\epsilon_0\omega)\,\overleftrightarrow{\sigma}\,. \tag{11.9}$$

11.2 Propagation of an electromagnetic wave in the presence of a magnetic field

11.2.1 Longitudinal propagation, $\boldsymbol{k}\|\mathcal{B}_0$

In the case of propagation of an electromagnetic wave with wave vector k parallel to the magnetic field $\mathcal{B}_0$, substitution of the plane wave solution given by Eq. (10.26) into the wave equations specified by Eq. (10.25) generalized to tensor $\overleftrightarrow{\epsilon}$ yields the coupled equations

$$c^2k^2\mathcal{E}_x = \omega^2(\epsilon_{xx}\mathcal{E}_x + \epsilon_{xy}\mathcal{E}_y) \tag{11.10a}$$

$$c^2k^2\mathcal{E}_y = \omega^2(-\epsilon_{xy}\mathcal{E}_x + \epsilon_{xx}\mathcal{E}_y). \tag{11.10b}$$

where we have used the relations $\epsilon_{yx} = -\epsilon_{xy}$, $\epsilon_{xx} = \epsilon_{yy}$, $\epsilon_{xz} = \epsilon_{yz} = 0$, and $\boldsymbol{k} = (0, 0, k)$. Multiplying Eq. (11.10b) by $\pm i$ and adding term by term to Eq. (11.10a) we find that

$$c^2k^2\mathcal{E}_\pm = \omega^2\epsilon_\pm\mathcal{E}_\pm, \tag{11.11}$$

where we see appearing the radiation fields

$$\mathcal{E}_\pm = \mathcal{E}_x \pm i\mathcal{E}_y \tag{11.12}$$

and the associated dielectric constants

$$\epsilon_\pm = \epsilon_{xx} \mp i\epsilon_{xy} \tag{11.13}$$

which characterize circularly polarized light. Cancelling out the common factor $\mathcal{E}_\pm$ in Eq. (11.11) we obtain the dispersion relation for longitudinal propagation in the form

$$c^2k_\pm^2 = \omega^2\epsilon_\pm. \tag{11.14}$$

We shall follow the traditional convention (Born and Wolf 1965) and define right circularly polarized (RCP) light to have an electric vector which rotates *clockwise* when viewed by an observer looking *back* in the direction from which the light is coming. For left circularly polarized (LCP) light, the electric vector rotates *counterclockwise*. The electric field components specified by Eq. (10.26) satisfy the relations

$$\text{RCP}: \begin{cases} \mathcal{E}_x = \mathcal{E}_{0x}e^{i(\boldsymbol{k}\cdot\boldsymbol{r}-\omega t)} & (11.15\text{a}) \\ \mathcal{E}_y = -i\mathcal{E}_x = -i\mathcal{E}_{0x}e^{i(\boldsymbol{k}\cdot\boldsymbol{r}-\omega t)} & (11.15\text{b}) \end{cases}$$

$$\text{LCP}: \begin{cases} \mathcal{E}_x = \mathcal{E}_{0x} e^{i(\boldsymbol{k}\cdot\boldsymbol{r}-\omega t)} & (11.16a) \\ \mathcal{E}_y = i\mathcal{E}_x = i\mathcal{E}_{0x} e^{(i(\boldsymbol{k}\cdot\boldsymbol{r}-\omega t)}. & (11.16b) \end{cases}$$

It is frequently useful to deal with real rather than complex field components. Taking the real part of each of these equations we obtain

$$\text{RCP}: \begin{cases} \text{Re } \mathcal{E}_x = \mathcal{E}_{0x} \cos(\boldsymbol{k}\cdot\boldsymbol{r} - \omega t) & (11.17a) \\ \text{Re } \mathcal{E}_y = \mathcal{E}_{0x} \sin(\boldsymbol{k}\cdot\boldsymbol{r} - \omega t) & (11.17b) \end{cases}$$

$$\text{LCP}: \begin{cases} \text{Re } \mathcal{E}_x = \mathcal{E}_{0x} \cos(\boldsymbol{k}\cdot\boldsymbol{r} - \omega t) & (11.18a) \\ \text{Re } \mathcal{E}_y = -\mathcal{E}_{0x} \sin(\boldsymbol{k}\cdot\boldsymbol{r} - \omega t). & (11.18b) \end{cases}$$

If we substitute Eqs. (11.15) and (11.16) into Eqs. (11.10) and refer to Eq. (11.13), we see that ϵ_+ and ϵ_- are associated with RCP and LCP radiation, respectively.

11.2.2 Transverse propagation, $\boldsymbol{k} \perp \boldsymbol{\mathcal{B}}_0$

In this case the propagation vector $\boldsymbol{k}$ is in the x–y plane perpendicular to $\boldsymbol{\mathcal{B}}_0$ which is always taken along the z-direction. Since $\epsilon_{xx} = \epsilon_{yy}$ the dielectric constant is isotropic in the x–y plane and $\boldsymbol{k}$ can be taken in the y-direction, for example. Then Eqs. (11.10) are replaced by the set of equations

$$c^2k^2\mathcal{E}_x = \omega^2(\epsilon_{xx}\mathcal{E}_x + \epsilon_{xy}\mathcal{E}_y) \tag{11.19a}$$

$$-c^2k^2\mathcal{E}_y + c^2k^2\mathcal{E}_y = 0 = \omega^2(-\epsilon_{xy}\mathcal{E}_x + \epsilon_{xx}\mathcal{E}_y) \tag{11.19b}$$

$$c^2k^2\mathcal{E}_z = \omega^2\epsilon_{zz}\mathcal{E}_z. \tag{11.19c}$$

Due to the polarization of the medium in this case, an electric field component appears along the direction of propagation, given by

$$\mathcal{E}_y = \frac{\epsilon_{xy}}{\epsilon_{xx}}\mathcal{E}_x, \tag{11.20}$$

It is necessary to distinguish two cases depending on whether $\mathcal{E}$ is perpendicular or parallel to $\boldsymbol{\mathcal{B}}_0$. In the first case, $\boldsymbol{\mathcal{E}} \perp \boldsymbol{\mathcal{B}}_0$, one has $\mathcal{E}_z = 0$. Substituting Eq. (11.20) into Eq. (11.19a) gives

$$c^2k_\perp^2 = \omega^2\left(\epsilon_{xx} + \frac{\epsilon_{xy}^2}{\epsilon_{xx}}\right) = \omega^2\epsilon_\perp. \tag{11.21}$$

For $\boldsymbol{\mathcal{E}} \parallel \boldsymbol{\mathcal{B}}_0$, one has from Eq. (11.19c),

$$c^2k_\parallel^2 = \omega^2\epsilon_{zz}. \tag{11.22}$$

Equations (11.21) and (11.22) are characteristic of transverse propagation. The effective dielectric constants are the perpendicular or Voigt dielectric constant $\epsilon_\perp$ given by

$$\epsilon_\perp = \epsilon_{xx} + \frac{\epsilon_{xy}^2}{\epsilon_{xx}} \tag{11.23}$$

and the parallel dielectric constant $\epsilon_\parallel$ given by ϵ_{zz}. For spherical bands $\epsilon_\parallel$ is independent of $\mathcal{B}_0$ and equal to the zero field dielectric constant $\epsilon(\omega)$.

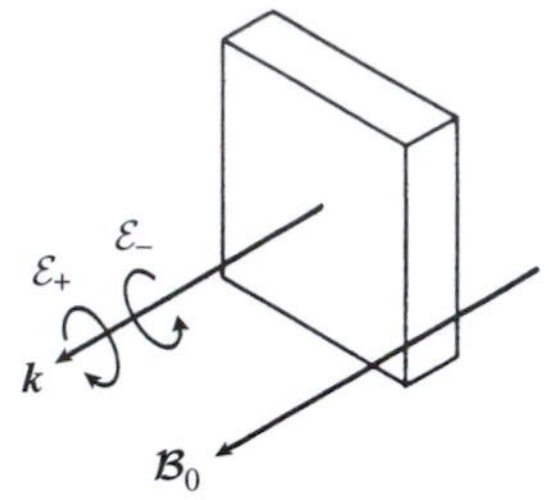

Fig. 11.1
Diagram of longitudinal propagation in the Faraday configuration.

11.3 Macroscopic expressions for magnetodispersion and magneto-absorption

11.3.1 Longitudinal propagation: Faraday configuration

The Faraday configuration is represented in Fig. 11.1. Introducing the optical constants $n(\omega)$, refractive index, and $K(\omega)$, extinction coefficient, which have been defined in Section 10.1, Eq. (11.14) can be rewritten as

$$c^2 k_\pm^2 = \omega^2 \epsilon_\pm(\omega) = \omega^2 (n_\pm + iK_\pm)^2. \tag{11.24}$$

The **Faraday effect** is the rotation of the plane of polarization of a plane-polarized electromagnetic wave as it propagates parallel to an external magnetic field. It arises because RCP and LCP waves propagate with different phase velocities, $v_\pm = c/n_\pm$, and a phase difference results between the two waves. A plane-polarized (PP) wave can be represented as the superposition of RCP and LCP waves with equal amplitudes. From Eqs. (11.17) and (11.18) we have

$$\mathrm{PP}: \begin{cases} \operatorname{Re}\mathcal{E}_x = \mathcal{E}_{0x}[\cos(\boldsymbol{k}_+ \cdot \boldsymbol{r} - \omega t) + \cos(\boldsymbol{k}_- \cdot \boldsymbol{r} - \omega t)] \\ \operatorname{Re}\mathcal{E}_y = \mathcal{E}_{0x}[\sin(\boldsymbol{k}_+ \cdot \boldsymbol{r} - \omega t) - \sin(\boldsymbol{k}_- \cdot \boldsymbol{r} - \omega t)], \end{cases} \tag{11.25}$$

where the magnitude of $\boldsymbol{k}_\pm$ is specified by Eq. (11.24) and $\boldsymbol{k}_\pm = (0, 0, k_\pm)$. For typical situations of interest the absorption is weak, so we can ignore the extinction coefficient $K_\pm$. Then $k_\pm = (\omega/c)n_\pm$.

At $z = 0$ and $t = 0$, the PP wave is polarized parallel to the x-axis. At $z = L$, the plane of polarization has been rotated by an angle Θ that can be determined as follows. Using the trigonometric identities

$$\cos u + \cos v = 2\cos\tfrac{1}{2}(u+v)\cos\tfrac{1}{2}(u-v) \tag{11.26a}$$

$$\sin u - \sin v = 2\cos\tfrac{1}{2}(u+v)\sin\tfrac{1}{2}(u-v), \tag{11.26b}$$

we can write $\tan\Theta$ as

$$\begin{aligned} \tan\Theta &= \frac{\operatorname{Re}\mathcal{E}_y}{\operatorname{Re}\mathcal{E}_x} = \frac{\sin\frac{1}{2}(k_+ - k_-)L}{\cos\frac{1}{2}(k_+ - k_-)L} \\ &= \tan\tfrac{1}{2}(k_+ - k_-)L. \end{aligned} \tag{11.27}$$

The angle Θ is therefore

$$\Theta = \tfrac{1}{2}(k_+ - k_-)L = \frac{\omega}{2c}(n_+ - n_-)L. \tag{11.28}$$

It is customary to express Faraday rotation in terms of the angle of rotation per unit thickness defined by

$$\theta = \frac{\Theta}{L} = \frac{\omega}{2c}(n_+ - n_-). \tag{11.29}$$

Positive θ corresponds to a clockwise rotation for an observer looking back toward the approaching wave.

Equation (11.24) also shows that the extinction coefficients for the two circularly polarized waves are different, as a consequence of which an initially linearly polarized wave becomes elliptically polarized. This effect is known as **magnetodichroism** and is characterized by the ellipticity ξ:

$$\xi = \tanh\frac{\omega}{2c}(K_+ - K_-) \simeq \frac{\omega}{2c}(K_+ - K_-). \tag{11.30}$$

Experimentally, Faraday rotation is measured mainly under non-resonant conditions, i.e., for frequencies far from resonance, where the absorption is weak. In this case the angle of rotation reduces to

$$\theta = \frac{\omega}{2cn(\omega)}\epsilon''_{xy}. \tag{11.31}$$

In terms of the conductivity tensor this angle is given by

$$\theta = \frac{\sigma'_{xy}}{2\epsilon_0 cn(\omega)}. \tag{11.32}$$

The prime and double prime refer to real and imaginary parts.

The extinction coefficient can be determined from absorption measurements. Knowing that the absorption coefficient is given by

$$\alpha(\omega) = \frac{2\omega K(\omega)}{c} \tag{11.33}$$

and that $K(\omega) = \epsilon''(\omega)/2n(\omega)$ from Eq. (10.34b), we see that it is possible to express the absorption for RCP and LCP waves as

$$\alpha_\pm = \frac{\omega}{cn_\pm}\epsilon''_\pm = \frac{\omega}{cn_\pm}[\epsilon''_{xx} \mp \epsilon'_{xy}]. \tag{11.34}$$

The magneto-absorption described by this equation is a resonant phenomenon involving strong absorption at frequencies equal to the resonance frequencies of the system. This is in contrast to Faraday rotation which is a nonresonant phenomenon and concerns wave propagation. We can therefore consider two different aspects of the interaction of the radiation field with matter when a magnetic field is present: on the one hand the magnetodispersion characteristic of nonresonant propagation in the medium gives a measure of the intensity of the effect far from resonance; on the other hand, the magneto-absorption is characteristic of the resonances of the radiation field with electronic levels and leads to the possibility of a spectral analysis of the resonance frequencies.

We now examine the relation between Faraday rotation and magneto-absorption. It is important to notice that the relation concerning the elements of the conductivity tensor stated in Eq. (11.4) has an analogue for the real and imaginary parts of ϵ_+ and ϵ_-. For the latter quantities one finds in fact that

$$\epsilon_+(-\omega) = \epsilon_-^*(\omega). \tag{11.35}$$

From Eqs. (11.31) and (11.34) one obtains the results

$$\epsilon''_{xy}(\omega) = \frac{2cn(\omega)}{\omega}\theta \tag{11.36}$$

$$\epsilon'_{xy}(\omega) = -\frac{c}{2\omega}(n_+\alpha_+ - n_-\alpha_-). \tag{11.37}$$

Using the Kramers–Kronig relation between $\epsilon'_{xy}(\omega)$ and $\epsilon''_{xy}(\omega)$ (Wallis and Balkanski 1986) we find that

$$\theta(\omega_0) = \frac{\omega_0^2}{2\pi n(\omega_0)}\mathcal{P}\int_0^\infty \frac{n_+\alpha_+ - n_-\alpha_-}{\omega(\omega^2 - \omega_0^2)}d\omega. \tag{11.38}$$

This equation gives the relation between magneto-absorption and Faraday rotation. It can be used to calculate the Faraday rotation when the magneto-absorption is known for a broad frequency range, in principle from 0 to ∞. Generally in semiconductors one distinguishes two types of transitions that contribute to the rotation:

(i) *Intraband transitions*. These transitions are due to excitation of free carriers within a band and determine the contribution to the rotation at low frequencies.

(ii) *Interband transitions*. These transitions are due to excitation of electrons across the band gap and determine the contribution to the rotation near the absorption edge.

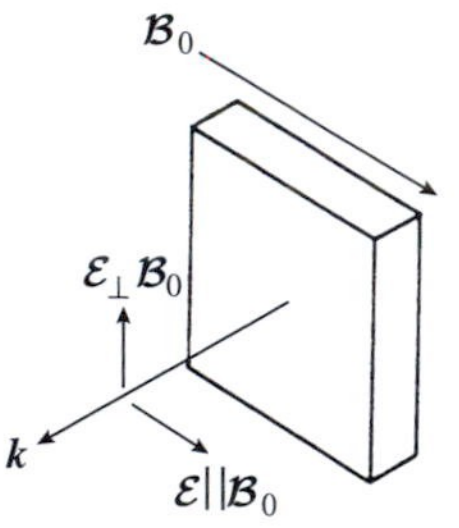

Fig. 11.2
Diagram of transverse propagation in the Voigt configuration.

11.3.2 Transverse propagation: Voigt configuration

The Voigt configuration is shown in Fig. 11.2. Equations (11.21) and (11.22) can be transformed as follows:

$$c^2k_\perp^2 = \omega^2\epsilon_\perp = \omega^2(n_\perp + iK_\perp)^2 \quad \text{for } \mathcal{E} \perp \mathcal{B}_0 \tag{11.39}$$

$$c^2k_\parallel^2 = \omega^2\epsilon_\parallel = \omega^2(n_\parallel + iK_\parallel)^2 \quad \text{for } \mathcal{E} \parallel \mathcal{B}_0. \tag{11.40}$$

A linearly polarized wave making an angle of 45° with $\mathcal{B}_0$ will have its parallel and perpendicular components with respect to $\mathcal{B}_0$ passing through the medium with different phase velocities $c/n_\parallel$ and $c/n_\perp$. The wave thereby becomes elliptically polarized.

The **Voigt effect** is the phase difference introduced by the magnetic field between the two components. This phase difference is given by (Teitler *et al.* 1961)

$$\delta = \frac{\omega}{c}(n_\parallel - n_\perp)L \tag{11.41}$$

for a medium of thickness L. Experimentally this effect is investigated by measuring the ellipticity Δ defined by

$$\Delta^2 = \frac{1-\cos\delta}{1+\cos\delta} = \tan^2\left(\frac{\delta}{2}\right). \tag{11.42}$$

The difference in the extinction coefficients $K_\parallel$ and $K_\perp$ leads to a rotation of the ellipsoid by an angle φ given by (Palik and Wright 1967)

$$\tan 2\varphi = \frac{2f\cos\delta}{f^2-1} \tag{11.43}$$

where $f = \exp[(\omega L/c)(K_\parallel - K_\perp)]$. The effect of the magnetic field in this configuration can also be studied using the absorption coefficient. From Eqs. (11.33), (11.39), and (11.40) one finds

$$\alpha_\perp = \frac{\omega}{cn_\perp}\,\mathrm{Im}\,\epsilon_\perp \quad \text{for } \mathcal{E} \perp \mathcal{B}_0, \tag{11.44}$$

$$\alpha_\parallel = \frac{\omega}{cn_\parallel}\,\mathrm{Im}\,\epsilon_\parallel \quad \text{for } \mathcal{E} \parallel \mathcal{B}_0. \tag{11.45}$$

11.4 Faraday rotation due to intraband transitions

11.4.1 Classical model for Faraday rotation due to free carriers

We utilize the expressions for the conductivity and dielectric tensors given in Section 11.1. For a magnetic field $\mathcal{B}_0$ in the z-direction one can calculate the current densities $\boldsymbol{j}_\pm = \boldsymbol{j}_x \pm i\boldsymbol{j}_y$ relative to the RCP and LCP waves associated with $\mathcal{E}_\pm = \mathcal{E}_x \pm i\mathcal{E}_y$. The corresponding conductivities $\sigma_\pm$ defined by $j_\pm = \sigma_\pm \mathcal{E}_\pm$ are found to have the form

$$\sigma_\pm = \frac{n_c e^2}{m^*}\frac{1}{\frac{1}{\tau} - i(\omega \mp \omega_c)}, \tag{11.46}$$

where n_c is the carrier concentration.

In the approximation of weak fields, $\omega_c \ll \omega$, and high frequencies, $\omega\tau \gg 1$, Eq. (11.46) can be written as

$$\sigma_\pm = \frac{in_c e^2}{m^*\omega}\left[1 \pm \frac{\omega_c}{\omega} + \frac{1}{i\omega\tau}\right]. \tag{11.47}$$

Since $\sigma_{xy} = (1/2i)(\sigma_- - \sigma_+)$, the macroscopic expression for the Faraday rotation given by Eq. (11.32) now becomes for conduction electrons

$$\theta = -\frac{e^3 n_c \mathcal{B}_0}{2\epsilon_0 n(\omega) c m^{*2}\omega^2} = -\frac{e^3 n_c \mathcal{B}_0 \lambda^2}{8\pi^2\epsilon_0 n(\omega) c^3 m^{*2}}. \tag{11.48}$$

On substituting for the values of the constants one finds that

$$\theta = -0.15 \times 10^{-22}\frac{n_c \mathcal{B}_0 \lambda^2}{n(\omega)(m^*/m)^2}\,[^\circ/\mathrm{cm}] \tag{11.49}$$

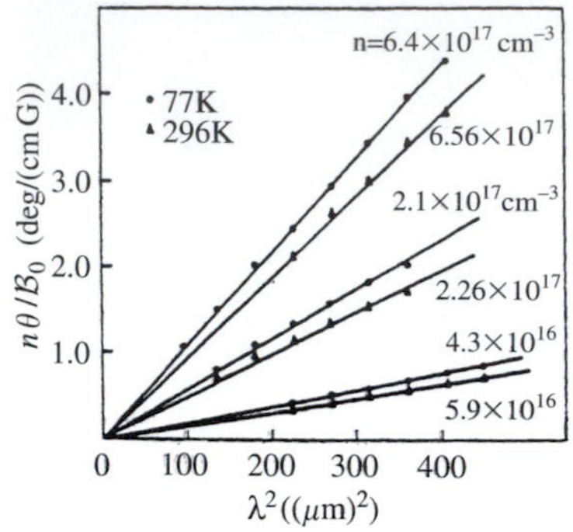

Fig. 11.3
Faraday rotation magnitude versus wavelength squared for several samples of n-InSb of various carrier concentrations (after Pidgeon 1962).

with B_0 in gauss, n_c in cm^{-3} and wavelength λ in micrometers. The rotation is negative for electrons and positive for holes in accordance with our choice of sign for ω_c. Experimental data of θ versus λ^2 are shown in Fig. 11.3 for n-InSb.

It can be seen from Eq. (11.48) that the rotation does not depend on $\omega_c\tau$ and can be measured at moderate magnetic fields in the infrared region of the spectrum where a cyclotron resonance measurement would require high magnetic fields and large relaxation times. This is therefore a convenient method for the investigation of effective masses and band structure of solids.

The analysis of the effective mass determined by this method in terms of the band structure depends on the statistics used to determine the free carrier distribution. The foundations of this analysis have been laid by Stephen and Lidiard (1958).

11.4.2 Conductivity tensor deduced from the Boltzmann equation

The distribution function f in the case of a homogeneous solid at constant temperature satisfies the Boltzmann equation

$$\frac{\partial f}{\partial t} + \frac{1}{\hbar}\boldsymbol{F}\cdot\nabla_k f = \left(\frac{\partial f}{\partial t}\right)_{\text{collisions}}. \tag{11.50}$$

In the presence of an electric field $\mathcal{E}$ and a magnetic field $\boldsymbol{\mathcal{B}}$ the second term on the left is given by

$$\frac{1}{\hbar}\boldsymbol{F}\cdot\nabla_k f = -\frac{e}{\hbar}\left[\mathcal{E} + \frac{1}{c}\boldsymbol{v}\times\boldsymbol{\mathcal{B}}_0\right]\cdot\nabla_k f. \tag{11.51}$$

Equation (11.50) can be linearized by introducing $f = f_0 + g$, where g is a perturbation to first order in $\mathcal{E}$, and f_0 is the equilibrium distribution function

$$f_0 = \frac{1}{1 + \exp[(E - E_F)/k_B T]}. \tag{11.52}$$

Introducing a relaxation time τ, we can rewrite the collision term as

$$\left(\frac{\partial f}{\partial t}\right)_{\text{collisions}} = -\frac{g}{\tau}. \tag{11.53}$$

With $\mathcal{E} = \mathcal{E}_0 \exp(-i\omega t)$ and $g = g_o \exp(-i\omega t)$, we have $\partial f/\partial t = -i\omega g$. Using a relaxation time τ^{prime} defined by $\tau^{prime-1} = \tau^{-1} - i\omega$, we can write Eq. (11.50) as

$$\frac{e}{\hbar}\left[\mathcal{E} + \frac{1}{c}\boldsymbol{v}\times\boldsymbol{\mathcal{B}}_0\right]\cdot\nabla_k f = \frac{g}{\tau'}. \tag{11.54}$$

In first approximation we can take $f = f_0$ and get

$$g = e\tau'(\mathcal{E}\cdot\boldsymbol{v})\frac{\partial f_0}{\partial E}, \tag{11.55}$$

where we have used the results

$$\nabla_k f = \frac{\partial f_0}{\partial E}\nabla_k E, \quad \boldsymbol{v} = \frac{1}{\hbar}\nabla_k E. \tag{11.56}$$

The term $\boldsymbol{v} \times \boldsymbol{\mathcal{B}}_0$ does not contribute because $\boldsymbol{v} \times \boldsymbol{\mathcal{B}}_0 \cdot \boldsymbol{v} = 0$. Substituting Eq. (11.55) into Eq. (11.54) one obtains to first order in $\mathcal{E}$:

$$g = e\tau'(\mathcal{E} \cdot \boldsymbol{v})\frac{\partial f_0}{\partial E} + \frac{e^2\tau'^2}{c\hbar^2}\frac{\partial f_0}{\partial E}(\boldsymbol{v} \times \boldsymbol{\mathcal{B}}_0) \cdot \nabla_k(\mathcal{E} \cdot \nabla_k E). \tag{11.57}$$

The current density is given by the expression

$$\boldsymbol{j} = -\frac{2}{(2\pi)^3}\int e\boldsymbol{v}g(\boldsymbol{k})d^3k. \tag{11.58}$$

It is now sufficient to replace g by the value given by Eq. (11.57) and execute the different vectorial operations in order to obtain the elements of the conductivity tensor σ'_{xy} for $\omega\tau \gg 1$:

$$\sigma'_{xy} = \frac{e^3\mathcal{B}_0}{4\pi^3\hbar^4 c\omega^2}\int \frac{\partial f_0}{\partial E}\frac{\partial E}{\partial k_x}\left[\frac{\partial E}{\partial k_y}\frac{\partial^2 E}{\partial k_x \partial k_y} - \frac{\partial E}{\partial k_x}\frac{\partial^2 E}{\partial k_y^2}\right]d^3k. \tag{11.59}$$

For spherical energy surfaces this result can be written as

$$\sigma'_{xy} = \frac{e^3\mathcal{B}_0}{3\pi^2\hbar^4 c\omega^2}\int_0^\infty f_0\left[\frac{dE}{dk} + 2k\frac{d^2E}{dk^2}\right]dE. \tag{11.60}$$

With this expression we can now carry out the analysis of the effective mass obtained by Faraday rotation.

11.4.3 Analysis of the effective mass obtained by Faraday rotation

Substituting Eq. (11.60) into the expression for the rotation, Eq. (11.32), and comparing the result with the free carrier result, Eq. (11.48), we obtain

$$\frac{n_c}{m^{*2}} = \frac{1}{3\pi^2\hbar^4}\int_0^\infty f_0\left[\frac{dE}{dk} + 2k\frac{d^2E}{dk^2}\right]dE, \tag{11.61}$$

where the free carrier concentration n_c is given by

$$n_c = \frac{1}{4\pi^3}\int f_0 d^3k = \frac{1}{\pi^2}\int_0^\infty k^2 f_0 dk. \tag{11.62}$$

Equations (11.61) and (11.62) offer the possibility of treating the following simple cases:

1. *Parabolic bands*. This is the case where $E = \hbar^2k^2/2m_0^*$, so that $m^* = m_0^*$. The effective mass determined by Faraday rotation is independent of the free carrier concentration and is equal to m_0^*, the effective mass at the bottom of the conduction band.

2. *Nonparabolic bands.* For a degenerate distribution the free carrier concentration is given by

$$n_c = \frac{2}{(2\pi)^3} \frac{4\pi}{3} k_F^3 \tag{11.63}$$

where k_F is the wave vector at the Fermi surface. In this case the effective mass is given by

$$\frac{1}{m^*} = \frac{1}{\hbar^2 k_F} \left(\frac{dE}{dk}\right)_F. \tag{11.64}$$

The effective mass obtained by experiment will then measure the slope of $E(\boldsymbol{k})$ at the Fermi surface. By varying the carrier density, which means shifting the Fermi energy, one can explore the energy band, determining m^* experimentally as a function of $\boldsymbol{k}_F$. A detailed account of Faraday rotation due to free carriers is given in a review article by Balkanski and Amzallag (1968) where effective masses m^* for a nonparabolic band obtained by different experimental techniques are compared and discussed.

11.4.4 Cyclotron resonance absorption

In the preceding sections we have presented a discussion of Faraday rotation, which is a dispersive phenomenon. We now turn our attention to cyclotron resonance, which is the corresponding absorptive phenomenon.

Since the optical absorption coefficient is determined by the imaginary part of the dielectric constant (cf. Eq. (10.36)), which in turn is determined by the real part of the conductivity (cf. Eq. (10.31)), let us take the real part of $\sigma_\pm$ specified by Eq. (11.47),

$$\sigma'_\pm(\omega) = \frac{n_c e^2}{m^*} \cdot \frac{\Gamma}{(\omega \mp \omega_c)^2 + \Gamma^2}, \tag{11.65}$$

where

$$\Gamma = \frac{1}{\tau}. \tag{11.66}$$

It is evident from Eq. (11.65) that for electrons ($\omega_c < 0$), RCP radiation associated with σ'_+ gives nonresonant absorption, whereas LCP radiation associated with σ'_- gives resonant absorption at the cyclotron frequency ω_c. This is the situation for electrons, which are negatively charged. For positively charged current carriers such as holes ($\omega_c > 0$), it is RCP radiation which gives the resonant absorption at ω_c. The resonant absorption is known as cyclotron resonance absorption.

The quantity Γ is a measure of the width of the cyclotron resonance absorption line. If we let $\Gamma \to 0$, the expression for $\sigma'_\pm$ reduces to a delta function:

$$\sigma'_\pm(\omega) \to \frac{\pi n_c e^2}{m^*} \delta(\omega \mp \omega_c). \tag{11.67}$$

Mechanisms which can contribute to the line width are scattering of the current carriers by impurities and by lattice vibrations. In the following sections we shall develop a theory of the line width due to lattice vibrations. Another mechanism of line broadening results from nonparabolicity of the energy band occupied by the carriers (Wallis 1958). In a nonparabolic band, the effective mass of the carriers is not constant, but depends on the energy of a carrier. The cyclotron frequency accordingly is also not constant, and the cyclotron resonance line is inhomogeneously broadened.

Figure 11.4 shows an experimental cyclotron resonance absorption spectrum for conduction electrons in n-type InAs (Palik and Wallis 1961). The peak in absorption at the magnetic field where the cyclotron frequency is equal to the frequency of the radiation (111 cm^{-1}) is clearly evident. The increase in line width in going from 80 K to 300 K may be attributed to the combined effects of lattice vibration scattering and nonparabolicity.

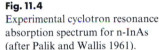

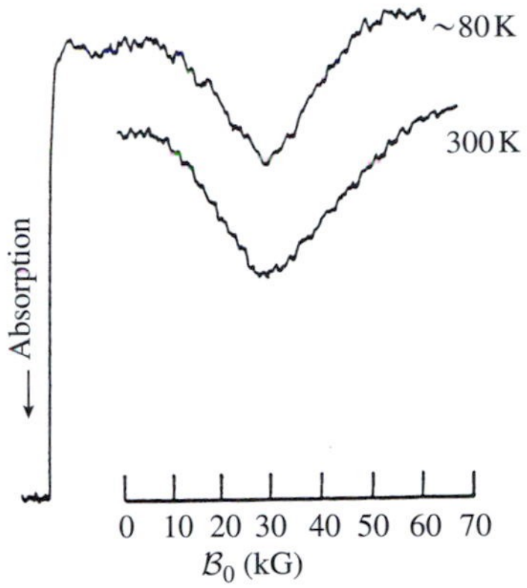

Fig. 11.4
Experimental cyclotron resonance absorption spectrum for n-InAs (after Palik and Wallis 1961).

11.5 Electronic eigenstates in a constant magnetic field

The development so far has been based on classical considerations. We now turn to the quantum theory of magneto-optical phenomena. The starting point is the set of eigenfunctions and energy eigenvalues of a free carrier moving in a constant magnetic field. We shall work within the effective mass approximation (Luttinger and Kohn 1955) that has been applied to the problem of impurity states in Chapter 5. We restrict our attention to carriers in a simple parabolic band with effective mass m^*. The effective mass Hamiltonian can be written as

$$H = \frac{1}{2m^*}[\boldsymbol{p} - e_c\boldsymbol{\mathcal{A}}(\boldsymbol{r})]^2, \tag{11.68}$$

where $\boldsymbol{p}$ is the momentum operator of the carrier, e_c is its charge, and $\mathcal{A}(\boldsymbol{r})$ is the vector potential giving rise to the magnetic field $\boldsymbol{\mathcal{B}}(\boldsymbol{r})$.

For a given magnetic field, the vector potential is not unique, since the two quantities are related by

$$\boldsymbol{\mathcal{B}}(\boldsymbol{r}) = \boldsymbol{\nabla} \times \boldsymbol{\mathcal{A}}(\boldsymbol{r}). \tag{11.69}$$

The vector potential can be expressed in terms of various **gauges**, all of which give the same results for $\boldsymbol{\mathcal{B}}(\boldsymbol{r})$. For a constant magnetic field $\boldsymbol{\mathcal{B}}_0$ in the z-direction, the **Landau gauge** specifies $\boldsymbol{\mathcal{A}}(\boldsymbol{r})$ by the relation

$$\boldsymbol{\mathcal{A}}(\boldsymbol{r}) = (-y\mathcal{B}_0, 0, 0). \tag{11.70}$$

The Schrödinger equation obtained by combining Eqs. (11.68) and (11.69) takes the form

$$\frac{1}{2m^*}[(p_x + e_c y\mathcal{B}_0)^2 + p_y^2 + p_z^2]F(\boldsymbol{r}) = EF(\boldsymbol{r}), \tag{11.71}$$

where $F(\mathbf{r})$ is the effective mass wave function. In the coordinate representation, this equation becomes

$$-\frac{\hbar^2}{2m^*}\left[\frac{\partial^2}{\partial x^2}-\frac{2}{i}\left(\frac{e_c\mathcal{B}_0}{\hbar}\right)y\frac{\partial}{\partial x}-\left(\frac{e_c\mathcal{B}_0}{\hbar}\right)^2y^2+\frac{\partial^2}{\partial y^2}+\frac{\partial^2}{\partial z^2}\right]F(\mathbf{r})=EF(\mathbf{r}). \tag{11.72}$$

At this point we specialize to the case of electrons with $e_c = -e$ and set

$$s=\frac{e\mathcal{B}_0}{\hbar}. \tag{11.73}$$

The quantity s has the dimensions of inverse length squared. Its inverse square root is a measure of the cyclotron radius in the ground state of the carrier. Eliminating $\mathcal{B}_0$ from Eq. (11.72) in favor of s yields

$$-\frac{\hbar^2}{2m^*}\left[\frac{\partial^2}{\partial x^2}+\frac{2}{i}sy\frac{\partial}{\partial x}-s^2y^2+\frac{\partial^2}{\partial y^2}+\frac{\partial^2}{\partial z^2}\right]F(\mathbf{r})=EF(\mathbf{r}). \tag{11.74}$$

The presence of the coordinates x and z only through their respective derivatives suggests that the eigenfunction has plane wave character in the x- and z-directions. We therefore substitute the trial function

$$F(\mathbf{r})=(L_xL_z)^{-\frac{1}{2}}\varphi(y)\exp(ik_xx+ik_zz) \tag{11.75}$$

into the Schrödinger equation and obtain

$$-\frac{\hbar^2}{2m^*}\left[\frac{d^2}{dy^2}-(sy-k_x)^2\right]\varphi(y)=\left(E-\frac{\hbar^2k_z^2}{2m^*}\right)\varphi(y). \tag{11.76}$$

The quantities L_x and L_z in Eq. (11.75) are the dimensions of the sample in the x- and z-directions and $(L_xL_z)^{-\frac{1}{2}}$ is a normalization factor. Equation (11.76) is seen to be the Schrödinger equation for a displaced harmonic oscillator. By making the transformation

$$\eta=y-k_x/s \tag{11.77}$$

we reduce Eq. (11.76) to the form

$$-\frac{\hbar^2}{2m^*}\frac{d^2\varphi}{d\eta^2}+\frac{1}{2}m^*\omega_c^2\eta^2\varphi=\left(E-\frac{\hbar^2k_z^2}{2m^*}\right)\varphi, \tag{11.78}$$

which is the Schrödinger equation for an ordinary harmonic oscillator of frequency ω_c. For simplicity we return to treating the cyclotron frequency as a positive quantity and write

$$\omega_c=\frac{e\mathcal{B}_0}{m^*}. \tag{11.79}$$

In terms of the variable $\xi = \sqrt{s}\eta$, the eigenfunctions of Eq. (11.78) have the form

$$\varphi_\ell(\xi) = \left(\frac{\sqrt{s}}{2^\ell \ell! \sqrt{\pi}}\right)^{\frac{1}{2}} e^{-\xi^2/2} H_\ell(\xi) \tag{11.80}$$

where $H_\ell(\xi)$ is the Hermite polynomial of order ℓ with ℓ a nonnegative integer. The energy eigenvalues and corresponding effective mass functions are

$$E(\ell, k_z) = \hbar\omega_c(\ell + \tfrac{1}{2}) + \frac{\hbar^2 k_z^2}{2m^*} \tag{11.81a}$$

and

$$F_\ell(\boldsymbol{r}) = (L_x L_z)^{-\frac{1}{2}} \varphi_\ell(\xi) \exp(ik_x x + ik_z z). \tag{11.81b}$$

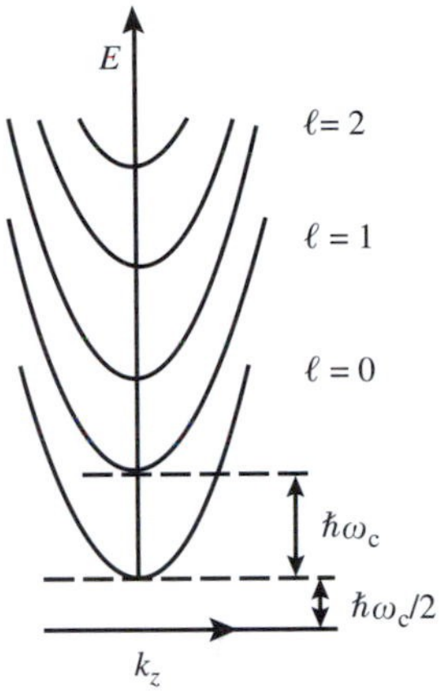

Fig. 11.5. Diagram of Landau levels for a simple parabolic band.

In the absence of the magnetic field, the energy eigenvalues of free carriers form a continuum; in the presence of the magnetic field, they are quantized into subbands or **Landau levels** characterized by the harmonic oscillator quantum number ℓ. Each Landau level forms a parabola as a function of k_z. A schematic representation of the Landau levels for a simple band is given in Fig. 11.5. It should be noted that the energy eigenvalues are independent of k_x; hence, all states of the same ℓ and k_z but different k_x are degenerate. The degeneracy can be specified by requiring that the center of the oscillator, which is determined by k_x, be within the confines of the sample of length L_y in the y-direction. One then obtains for the degeneracy N_B:

$$N_B = \frac{e\mathcal{B}_0}{h} L_x L_y. \tag{11.82}$$

We see that the degeneracy is proportional to the magnetic field.

11.6 Quantum mechanical theory of cyclotron resonance

The phenomenon of cyclotron resonance can now be interpreted quantum mechanically. If we include the vector potential of the radiation field, $\mathcal{A}_R(\boldsymbol{r})$, the total vector potential can be written as

$$\mathcal{A}(\boldsymbol{r}) = \mathcal{A}_0(\boldsymbol{r}) + \mathcal{A}_R(\boldsymbol{r}), \tag{11.83}$$

where $\mathcal{A}_0(\boldsymbol{r})$ specifies the external magnetic field $\mathcal{B}_0$. The electron–radiation interaction Hamiltonian to first order in $\mathcal{A}_R(\boldsymbol{r})$ follows from Eq. (11.68) with $e_c = -e$:

$$H_{ER} = \frac{e}{2m^*}[(p + e\mathcal{A}_0) \cdot \mathcal{A}_R(\boldsymbol{r}) + \mathcal{A}_R(\boldsymbol{r}) \cdot (p + e\mathcal{A}_0(\boldsymbol{r}))]. \tag{11.84}$$

Furthermore, in the electric dipole approximation we can neglect the r-dependence of $\mathcal{A}_R(r)$ and write

$$H_{ER} = \frac{e}{m^*}\mathcal{A}_R \cdot [p + e\mathcal{A}_0(r)]. \tag{11.85}$$

Using semiclassical perturbation theory the matrix element for an optical transition between an initial state i and a final state f has the form

$$\langle f|H_{ER}|i\rangle = \int \psi_f^*(r) H_{ER} \psi_i(r) d^3r, \tag{11.86}$$

where to lowest order in effective mass theory,

$$\psi_{i,f}(r) = F_\ell^{i,f}(r) u_0^{i,f}(r), \tag{11.87}$$

and $u_0^{i,f}(r)$ is the Bloch function at $k = 0$ for the initial or final state. The integral required in the evaluation of $\langle f|H_{ER}|i\rangle$ is then

$$I_1 = \int F_\ell^{f*}(r) u_0^{f*}(r)[p + e\mathcal{A}_0] F_{\ell'}^i(r) u_0^i(r) d^3r. \tag{11.88}$$

By exploiting the fact that $u_0^{f*}(r)u_0^i(r)$ and $u_0^{f*}(r)pu_0^i(r)$ have the periodicity of the crystal lattice, one can show that

$$I_1 = \int F_\ell^{f*}(r)(p + e\mathcal{A}_0)F_{\ell'}^i(r) d^3r \int u_0^{f*}(r) u_0^i(r) d^3r + \int F_\ell^{f*}(r) F_{\ell'}^i(r) d^3r \int u_0^{f*}(r) p u_0^i(r) d^3r. \tag{11.89}$$

Excercise. Discuss the result in Eq. (11.89).
Answer. The orthonormality of the Bloch functions $u_0(r)$ means that the first term on the right-hand side of Eq. (11.89) is zero unless the initial and final state bands are the same. This term therefore describes intraband transitions, i.e., cyclotron resonance. On the other hand, the second term describes interband magneto-optic effects, which will be taken up in Section 11.7.

The case of cyclotron resonance requires the evaluation of the integral

$$I_\alpha = \int F_\ell^{f*}(r)(p + e\mathcal{A}(r))_\alpha F_{\ell'}^i(r) d\tau, \tag{11.90}$$

where $\alpha = x$ or y. The function $F_\ell(r)$ is given by Eq. (11.81b). Carrying out the evaluation, one obtains

$$I_\alpha = u\hbar s^{\frac{1}{2}} \left[\pm \left(\frac{\ell}{2}\right)^{\frac{1}{2}} \delta_{\ell',\ell-1} + \left(\frac{\ell+1}{2}\right)^{\frac{1}{2}} \delta_{\ell',\ell+1} \right], \tag{11.91}$$

where the upper sign refers to $\alpha = x$, the lower sign refers to $\alpha = y$, and the quantity $u = 1(i)$ for $\alpha = x(y)$.

Equation (11.91) specifies the selection rules for cyclotron resonance transitions. The harmonic oscillator quantum number ℓ changes by ± 1 and there is no change in k_x and k_z. This means that cyclotron resonance transitions can be represented in Fig. 11.5 by vertical arrows connecting a given point on one Landau level to the points directly above or below in the adjacent Landau levels. Since the vertical energy separation of adjacent Landau levels is $\hbar\omega_c$, as can be seen from Eq. (11.81a), the frequency of adsorbed or emitted radiation in cyclotron resonance is ω_c, just as in the classical treatment.

The foregoing development can be readily generalized to include electron spin. The interaction of the spin with the external magnetic field contributes a term in the Hamiltonian of the form

$$H_{SB} = g\mu_B \boldsymbol{S} \cdot \boldsymbol{\mathcal{B}}_0, \tag{11.92}$$

where g is the Landé g-factor, μ_B is the Bohr magneton given by $e\hbar/2mc$, and $\boldsymbol{S}$ is the spin operator. If the magnetic field is in the z-direction, H_{SB} becomes

$$H_{SB} = g\mu_B S_z \mathcal{B}_0. \tag{11.93}$$

For electrons in solids, one finds that it is necessary to use an effective g-factor, g^*, which can deviate significantly from the value 2 for electrons in vacuum. Since the eigenvalues of S_z are $\pm\frac{1}{2}$, the eigenvalues of H_{SB} are

$$\pm\tfrac{1}{2}g^*\mu_B\mathcal{B}_0, \tag{11.94}$$

and the expression for the energy eigenvalues of an electron in a magnetic field given by Eq. (11.81a) must be generalized to the form

$$\epsilon(\ell, k_z, \sigma) = \hbar\omega_c(\ell + \tfrac{1}{2}) + \frac{\hbar^2 k_z^2}{2m^*} \pm \tfrac{1}{2}g^*\mu_B\mathcal{B}_0. \tag{11.95}$$

The effective g-factor depends on the energy band just as the effective mass does. In semiconductors, g^* can be large in magnitude and negative in sign, e.g., conduction electrons in InSb have $g^* \simeq -50$. The large value of g^* is a consequence of the large spin–orbit splitting of the valence band in InSb, as is revealed by the formula due to Roth *et al.* (1959),

$$g^* = 2\left[1 - \left(\frac{m}{m^*} - 1\right)\frac{\Delta}{3E_g + 2\Delta}\right], \tag{11.96}$$

where Δ is the spin–orbit splitting of the valence bands at $\boldsymbol{k} = 0$. A schematic representation of spin-split Landau levels is shown in Fig. 11.6.

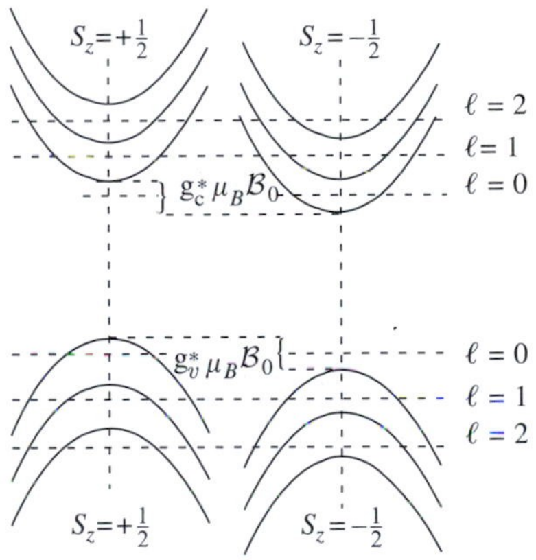

Fig. 11.6
Diagram of spin-split Landau levels for simple parabolic bands.

The analysis of Landau levels presented above assumes that the current carrier occupies a parabolic energy band that is nondegenerate aside from spin. The complications introduced by degenerate bands, such as the valence bands of group IV and III–V semiconductors, were adequately treated for the first time by Luttinger and Kohn (1955) and by Luttinger (1956). When the degeneracy of the valence bands at $\boldsymbol{k} = 0$ and the spin–orbit interaction are included in the theory, the following picture emerges.

The uppermost valence bands corresponding to light and heavy holes have a fourfold degeneracy at $\boldsymbol{k}=0$ when spin is included. In the presence of a magnetic field four sets of Landau levels arise which are called **ladders**. The spacing of the levels in a given ladder is not uniform near the valence band edge as a result of **quantum effects**. The latter lead to a complex structure in the cyclotron resonance spectra of holes at low temperature in materials such as Ge.

The effect of nonparabolicity, which is particularly striking in the conduction band of InSb, is to reduce the separation between adjacent Landau levels as the energy of a conduction electron increases. Consequently, a broadening of the cyclotron resonance peak appears on the low-frequency side that increases as the temperature increases. Furthermore, the frequency of the peak for the $n \rightarrow n+1$ transition decreases as n increases, in contrast to the parabolic case where it is independent of n.

11.7 Interband magneto-absorption

In the preceding section, we noted that the matrix element for optical transitions contains a term that corresponds to interband transitions. This term arises from the second term on the right hand side of Eq. (11.89), since the integral

$$\int u_0^{f*}(\boldsymbol{r})\boldsymbol{p}u_0^i(\boldsymbol{r})d^3r \tag{11.97}$$

vanishes if $u_0^f(\boldsymbol{r})$ and $u_0^i(\boldsymbol{r})$ are associated with the same band and is nonvanishing only if the initial and final bands are different. In a crystal with a center of inversion such as Ge, the two bands must have opposite parity to give a nonvanishing result. The highest valence bands and lowest conduction band at $\boldsymbol{k}=0$ satisfy this condition, and hence one expects interband absorption associated with transitions between these bands.

In addition to the above integral, the second term on the righthand side of Eq. (11.89) contains a factor consisting of the overlap integral of the effective mass functions $F_\ell^f(\boldsymbol{r})$ and $F_{\ell'}^i(\boldsymbol{r})$. We note that these two functions both involve the function $\varphi_n(\xi)$, which are defined by Eq. (11.80) and which form an orthonormal set. Using the expression for $F_\ell(\boldsymbol{r})$ contained in Eq. (11.81b), we obtain the result

$$\int F_\ell^{f*}(\boldsymbol{r})F_{\ell'}^i(\boldsymbol{r})d^3r = \delta_{\ell\ell'}\delta_{k_xk_{x'}}\delta_{k_zk_{z'}} \tag{11.98}$$

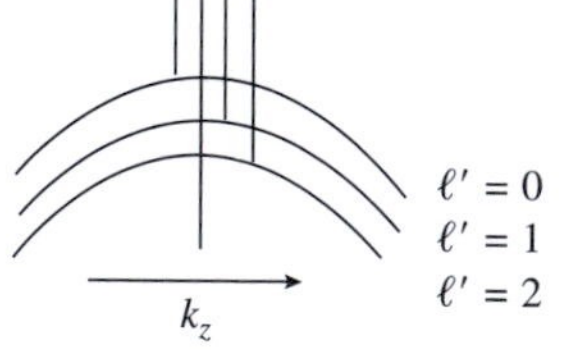

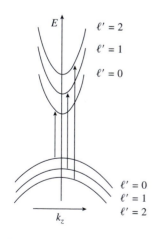

Fig. 11.7
Direct allowed interband transitions for simple parabolic bands in a magnet field.

which specifies the selection rules for interband transitions: $\Delta\ell=0, \Delta k_x=0, \Delta k_z=0$. Typical allowed interband transitions are shown in Fig. 11.7. The transition energies $\Delta E(\ell, k_z)$ follow from Eq. (11.92) and are given by

$$\Delta E(\ell, k_z) = E_g + \frac{\hbar e\mathcal{B}_0}{\bar{m}c}(\ell+\tfrac{1}{2}) + \frac{\hbar^2k_z^2}{2\bar{m}} \pm \tfrac{1}{2}(g_c^* + g_v^*)\mu_B\mathcal{B}_0, \tag{11.99}$$

where $\bar{m}$ is the reduced mass of electrons and holes, g_c^* and g_v^* are the effective g-factors for electrons and holes and μ_B is the Bohr magneton.

Since shifts away from the zero field band edge occur in both the valence band the conduction band, there is an increase in the optical band gap due to an applied magnetic field (Burstein *et al.* 1956). The field-dependent gap $E_g(\mathcal{B}_0)$ is obtained from Eq. (11.95) by setting $\ell = k_z = 0$:

$$E_g(\mathcal{B}_0) = E_g + \frac{\hbar e \mathcal{B}_0}{2\bar{m}c} - \tfrac{1}{2}|g_c^* + g_v^*|\mu_B \mathcal{B}_0. \qquad (11.100)$$

From the expression for the Landau levels given by Eq. (11.95), we see that a carrier in a magnetic field behaves as a free carrier only in the direction parallel to the field. In the directions transverse to the field, the carrier behaves as a two-dimensional bound oscillator. This one-dimensional character of free carrier motion in a magnetic field produces a dramatic effect on the transition density-of-states $N(\Delta E)$, which varies as

$$N(\Delta E) \sim \frac{dk_z}{d\Delta E}. \qquad (11.101)$$

Using Eq. (11.99) we find that

$$N(\Delta E) \sim \frac{(\bar{m}/2)^{\frac{1}{2}}}{\hbar[\Delta E - E_g - (\hbar e\mathcal{B}_0/\bar{m}c)(\ell + \frac{1}{2}) \mp \frac{1}{2}(g_c^* + g_v^*)\mu_B \mathcal{B}_0]^{\frac{1}{2}}}. \qquad (11.102)$$

Equation (11.102) tells us that the transition density-of-states diverges when the transition energy equals the energy separation of the extrema of a Landau level in the conduction band and the Landau level in the valence band with the same n value.

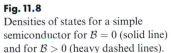

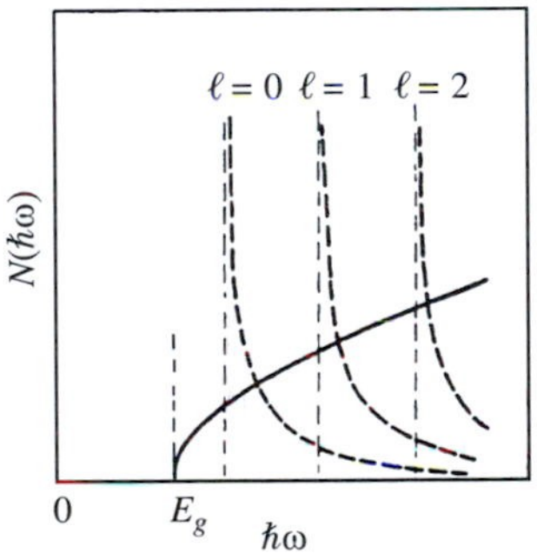

Fig. 11.8
Densities of states for a simple semiconductor for $\mathcal{B} = 0$ (solid line) and for $\mathcal{B} > 0$ (heavy dashed lines).

In Chapter 10 it was shown that transitions between valence and conduction bands in the absence of a magnetic field lead to optical absorption that increases with increasing photon energy when the latter is above the minimum energy gap. The dependence of the absorption coefficient on photon energy closely follows the energy dependence of the transition density-of-states: $N(\Delta E) \sim (\Delta E - E_g)^{\frac{1}{2}}$. The monotonically increasing behavior exhibited here contrasts sharply with the singular behavior exhibited by the magnetic field case in Eq. (11.102).

Aside from a factor that varies slowly with photon frequency ω, the absorption coefficient is determined by $N(\hbar\omega)$ which is plotted in Fig. 11.8 for $\mathcal{B}_0 = 0$ and $\mathcal{B}_0 \neq 0$. The singularities in the magnetic field case are evident. If damping processes are included in the theory, the singularities are eliminated and replaced by rounded peaks.

When one deals with real semiconductors such as Ge and InSb, one must take into account the complications arising from the multiple ladders of Landau levels in the valence band. These complications cause additional structure to appear in the interband magneto-absorption spectrum.

Experimental studies of interband magneto-absorption have been carried out for a number of semiconductors including Ge (Zwerdling and Lax 1957, Burstein *et al.* 1959) and InSb (Burstein *et al.* 1956, Zwerdling *et al.* 1957). The results give excellent confirmation of the effective mass theory of magnetic field effects on interband transitions and provide precise values of energy band parameters. A particularly noteworthy study is that of Pidgeon and Brown (1966) on InSb.

Complementary to interband magneto-absorption is the dispersive effect known as interband Faraday rotation. The discrete nature of the Landau levels shows up in the oscillatory behavior of the Faraday rotation that is observed experimentally in Ge (Mitchell and Wallis 1963). The behavior of the rotation near the zero-field band gap reveals effects due to excitons arising from the Coulomb interaction of electrons and holes.

11.8 Electro-optical effects

The application of a constant electric field to an intrinsic semiconductor leads to a number of effects that arise from changes in the real and imaginary parts of the dielectric tensor. We assume that the temperature is low enough that the concentration of free carriers is negligible.

11.8.1 Pockels effect

The Pockels effect arises from a change in the refractive index that is linear in the applied electric field $\mathcal{E}_0$. We restrict our attention to the transparent region of the semiconductor well below the fundamental absorption edge where the effects can be attributed to a single oscillator of frequency ω_0 which represents the interband transitions. The equation of motion takes the form

$$m_0\left(\frac{d^2}{dt^2} + \omega_0^2 u + \beta_2 u^2\right) = -e\mathcal{E}_0, \tag{11.103}$$

where the term in β_2 is a nonlinear contribution that is essential for the Pockels effect and m_0 is the effective mass of the oscillator. The electric field causes a static displacement of the oscillator given by

$$\bar{u} = -\frac{e\mathcal{E}_0}{m_0\omega_0^2}, \tag{11.104}$$

where we have neglected the nonlinear term in u. Inclusion of the latter leads to a shift in the resonance frequency that can be obtained by linearizing the nonlinear term according to

$$u^2 = uu \simeq \bar{u}u + u\bar{u} = 2u\bar{u}. \tag{11.105}$$

Substituting this result into Eq. (11.103), we see that the effective oscillator frequency squared is

$$\omega_{eff}^2 = \omega_0^2 + 2\beta_2\bar{u} = \omega_0^2 + \Delta(\omega_0^2), \tag{11.106}$$

where

$$\Delta(\omega_0^2) = -\frac{2\beta_2 e\mathcal{E}_0}{m_0\omega_0^2}. \tag{11.107}$$

The change in the refractive index due to the applied field can now be calculated using the results for the real part of the dielectric constant due to a harmonic oscillator presented in Eq. (10.135a). Replacing ω_{TO}^2 by $\omega_0^2 + \Delta(\omega_0^2)$ and ϵ_∞ by 1 and then expanding to first order in $\Delta(\omega_0^2)$, we obtain for the real part of the dielectric constant in the absence of damping

$$\epsilon'(\mathcal{E}_0) = \epsilon'\left[1 - \frac{\Delta(\omega_0^2)(\epsilon' - 1)\omega^2}{\epsilon'(\omega_0^2 - \omega^2)\omega_0^2}\right], \tag{11.108}$$

where $\epsilon' = \epsilon'(0)$. The refractive index $n(\mathcal{E}_0)$ is the square root of $\epsilon'(\mathcal{E}_0)$. The change in n is given to first order in $\mathcal{E}_0$ by

$$\Delta n = -\frac{(n^2 - 1)\omega^2\Delta(\omega_0^2)}{2n\omega_0^2(\omega_0^2 - \omega^2)} = \frac{(n^2 - 1)\omega^2\beta_2 e\mathcal{E}_0}{nm_0\omega_0^4(\omega_0^2 - \omega^2)}. \tag{11.109}$$

In the experimental arrangement for observing the Pockels effect, the light beam propagates perpendicular to the applied electric field $\boldsymbol{\mathcal{E}}_0$. If the light is plane polarized parallel to $\boldsymbol{\mathcal{E}}_0$, its velocity is determined by the refractive index $n + \Delta n$, whereas if it is polarized perpendicular to $\boldsymbol{\mathcal{E}}_0$, its velocity is determined by n. The material has become **birefringent**.

Semiconductors with a center of symmetry such as Si and Ge do not exhibit the Pockels effect because Eq. (11.103) does not properly transform under the inversion operation $u \to -u$, $\mathcal{E}_0 \to -\mathcal{E}_0$ unless $\beta_2 = 0$. This symmetry operation does not apply to semiconductors without a center of symmetry such as GaAs, so the Pockels effect is not excluded in these materials.

11.8.2 Kerr effect

The Kerr effect arises if a third-order nonlinear term $m_0\beta_3 u^3$ is inserted into Eq. (11.103). The change in ω_0^2 due to this term is

$$\Delta(\omega_0^2) = \frac{3\beta_3 e^2\omega^2\mathcal{E}_0^2}{m_0^2\omega_0^4} \tag{11.110}$$

and the change in refractive index is

$$\Delta n = -\frac{3\beta_3(n^2 - 1)e^2\omega^2\mathcal{E}_0^2}{2nm_0^2\omega_0^6(\omega_0^2 - \omega^2)}. \tag{11.111}$$

The leading term in the Kerr effect is thus quadratic in the applied field. It can be observed in Si and Ge because the equation of motion involving the u^3 term transforms into a multiple of itself under inversion and therefore behaves properly under this operation. Semiconductors without inversion symmetry also exhibit the Kerr effect.

11.8.3 Franz–Keldysh effect

So far we have focused on changes in the refractive index in our discussion of electro-optic effects. Now we take up an important effect involving the absorption coefficient, i.e., the shift in the intrinsic absorption edge under an applied electric field (Franz 1958, Keldysh 1958). This is the **Franz–Keldysh effect.**

We restrict our attention to direct allowed transitions between spherical parabolic valence and conduction bands. The analysis parallels that for the corresponding case of magneto-absorption. The absorption coefficient can be written in the form

$$\alpha(\omega) = \frac{\pi e^2}{\epsilon_0 ncm^2\omega}\sum_i P^2|\phi(0)|^2\delta(E_f - E_i - \hbar\omega), \tag{11.112}$$

where P is the interband momentum matrix element and $\phi(\boldsymbol{r})$ satisfies the effective mass equation

$$-\left[\frac{\hbar^2\nabla^2}{2\mu^*} + e\boldsymbol{\mathcal{E}}_0\cdot\boldsymbol{r}\right]\phi(\boldsymbol{r}) = E\phi(\boldsymbol{r}). \tag{11.113}$$

The wave function $\phi(\boldsymbol{r})$ is that for the relative motion of electron and hole, $\boldsymbol{\mathcal{E}}_0$ is the external electric field, and μ^* is the reduced effective mass. We take $\boldsymbol{\mathcal{E}}_0$ to be in the z-direction. Aside from the plane wave solutions in the x- and y-directions, the solution in the z-direction has the form (Aspnes 1966)

$$\phi(z) = C\,Ai(\xi), \tag{11.114}$$

where $Ai(\xi)$ is the Airy function, C is a normalization constant,

$$\xi = \left(\frac{2\mu^*}{\hbar^2}\right)^{\frac{1}{3}}\left[\frac{E_z}{(e\mathcal{E}_0)^{2/3}} - (e\mathcal{E}_0)^{\frac{1}{3}}z\right], \tag{11.115}$$

and E_z is the energy eigenvalue for the z-direction.

The absorption coefficient is obtained by substituting Eq. (11.114) into Eq. (11.112) and replacing the sum over i by an integral over energy. The result is

$$\alpha(\omega, \mathcal{E}_0) \sim \int Ai^2\left[\left(\frac{2\mu^*}{\hbar^2}\right)^{\frac{1}{3}}\frac{E_z}{(e\mathcal{E}_0)^{2/3}}\right]\delta(E_g - \hbar\omega - E_z)dE_z. \tag{11.116}$$

Evaluating the integral with the aid of the asymptotic expression

$$Ai^2(x) \sim \exp(-\tfrac{4}{3}x^{3/2}) \tag{11.117}$$

yields

$$\alpha(\omega, \mathcal{E}_0) \sim \exp\left[-\frac{4}{3}\left(\frac{2\mu^*}{\hbar^2}\right)^{\frac{1}{2}}(E_g - \hbar\omega)^{3/2}/e\mathcal{E}_0\right]. \tag{11.118}$$

For $\hbar\omega < E_g$ the absorption coefficient increases with increasing applied electric field. This result corresponds to a shift of the absorption edge to lower photon energies.

At photon energies above the band gap it is the reflectivity that is typically measured. Using an asymptotic expansion for the Airy function that is appropriate for $\hbar\omega > E_g$, one obtains the relative change in reflectivity in the form

$$\frac{\Delta R}{R} \sim (\hbar\omega - E_g)^{-1} \cos\left[\frac{4}{3}\left(\frac{2\mu^*}{\hbar^2}\right)^{\frac{1}{2}} \frac{(\hbar\omega - E_g)^{3/2}}{e\mathcal{E}_0} - \frac{\pi}{2}\right]. \qquad (11.119)$$

This expression describes decaying oscillations in $\Delta R/R$ that are known as **Franz–Keldysh oscillations**. An example is shown in Fig. 11.9 for GaAs. The transitions involved are from the upper valence bands of Γ_7 and Γ_8 symmetry to higher conduction bands that are also of Γ_7 and Γ_8 symmetry (Aspnes and Studna 1973).

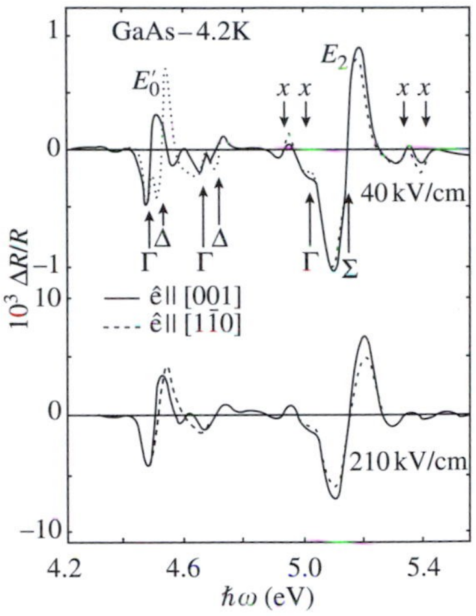

Fig. 11.9
Franz–Keldysh oscillations in the relative reflectivity change in GaAs at 4.2 K (after Aspnes and Studna 1973).

11.9 Modulation spectroscopy

We have seen in the previous section that the application of an external electric field produces a shift of the optical absorption edge toward lower photon energies and modifies the reflectivity above the absorption edge. It was pointed out by Seraphin (Seraphin 1964, Seraphin and Hess 1965) that modulation of the reflectivity provides a useful technique for studying interband transitions well above the fundamental absorption edge. Since the absorption coefficient becomes very large and difficult to measure at photon energies in this region, it is advantageous to work with the reflectivity which is relatively easy to measure.

The reflectivity is given in terms of the real and imaginary parts of the refractive index by Eq. (10.40). The latter are related to the real and imaginary parts of the dielectric function by Eqs. (10.34). The relative change of the reflectivity can then be expressed as

$$\frac{dR}{Rd\mathcal{E}} = \alpha_S \frac{d\epsilon'}{d\mathcal{E}} + \beta_S \frac{d\epsilon''}{d\mathcal{E}}, \qquad (11.120)$$

where α_S and β_S are the **Seraphin coefficients** defined by

$$\alpha_S = \frac{\partial \log R}{\partial \epsilon'} \qquad (11.121a)$$

$$\beta_S = \frac{\partial \log R}{\partial \epsilon''} \qquad (11.121b)$$

and $\mathcal{E}$ is the modulating electric field. By using an AC electric field or chopping the incident light beam one can take advantage of lock-in amplifiers to obtain very precise electroreflectivity spectra.

Optical parameters such as the absorption coefficient, reflectivity, and dielectric function exhibit Van Hove singularities at various photon energies at and above the fundamental absorption edge. The dielectric function, for example, can be written near a singular point at E_i in the form

$$\epsilon(\omega) = A(\hbar\omega - E_i)^n, \qquad (11.122)$$

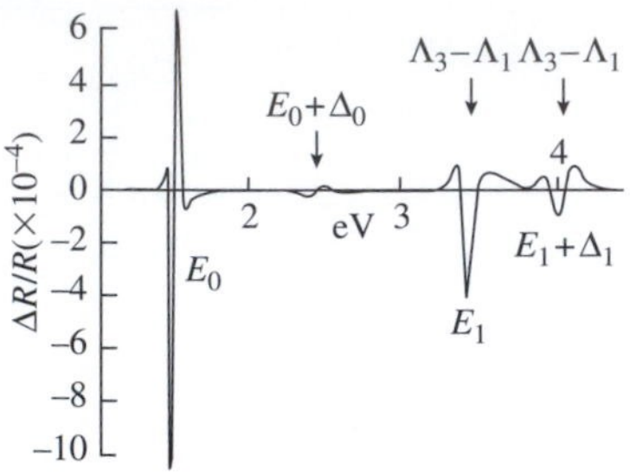

Fig. 11.10
Electroreflectivity spectrum of CdTe at 300 K. There are oscillations associated with two interband transitions E_0 and E_1 (after Ludeke and Paul 1967).

where A is a constant and n takes the values $-\frac{1}{2}$, 0 (logarithmic), or $\frac{1}{2}$ for one-, two-, or three-dimensional critical points. For the three-dimensional case one has

$$\epsilon(\omega) = A(\hbar\omega - E_i)^{\frac{1}{2}} \tag{11.123}$$

and

$$\frac{d\epsilon(\omega)}{d\mathcal{E}} = \tfrac{1}{2}A(\hbar\omega - E_i)^{-\frac{1}{2}}\frac{dE_i}{d\mathcal{E}}. \tag{11.124}$$

Thus, the modulated reflectivity spectrum characterized by $d\epsilon/d\mathcal{E}$ has a much more dramatic singularity than the reflectivity itself, which is characterized by ϵ. An example of an electroreflectivity spectrum is given in Fig. 11.10 for CdTe. The sharpness of the singularities is evident.

A number of other physical parameters can be varied to yield modulated spectra. They include temperature, stress, and frequency.

Problems

1. Starting from Eqs. (11.9) and (11.46), derive expressions for the dielectric constants ϵ_+ and ϵ_- in terms of ω and ω_c. Using these results, the dispersion relation given by Eq. (11.14), and the high field limit $\omega_c\tau > 1$, show that under the conditions $\omega_c > \omega$ and $\omega_p > \omega$, only one type of circularly polarized wave propagates, i.e., k is real. This low-frequency propagating wave is called a **helicon**.
2. Consider the Voigt configuration with the radiation propagating perpendicular to the external magnetic field. Using the classical equations of motion for the carriers, derive expressions for $\epsilon_\parallel$ and $\epsilon_\perp$. Taking $\tau = \infty$ obtain the refractive indices $n_\parallel$ and $n_\perp$ and then the Voigt phase shift δ in the small field limit $\omega_c \ll \omega$. What is the dependence of δ on $\mathcal{B}_0$?
3. Investigate the reflectivity due to free carriers in an external magnetic field in the Faraday configuration. Calculate the real and imaginary parts of the refractive index for n-InSb taking the carrier concentration $n_c = 10^{18}\,\text{cm}^{-3}$, $\tau = 3\times 10^{-13}\,\text{s}$, $m^*/m = 0.035$ and $\mathcal{B}_0 = 0$ and 25 kG. Then calculate the reflectivity for RCP and LCP radiation for frequencies in the vicinity of the plasma frequency and make a plot of reflectivity versus frequency for each case. What physical quantities can be determined from experimental results on the magnetoreflectivity of free carriers?
4. Generalize the result in Eq. (3.84) for the energy of a conduction electron in a nonparabolic band to include an external magnetic field. Ignore the term $\hbar^2k^2/2m$ and replace the quantity $[(\hbar\mathbf{k}/m)\cdot|\langle c0|\mathbf{p}|v0\rangle|]^2$ by $E_g(\ell+\frac{1}{2})\hbar\omega_c$, where ℓ is the Landau quantum number. For InSb calculate the energy separations for the transitions $\ell=0 \rightarrow \ell=1$, $\ell=1\rightarrow\ell=2$, and $\ell=2\rightarrow\ell=3$ and compare the results to those for the parabolic case. Take $E_g = 0.23$ eV, $m^*/m = 0.014$, and $\mathcal{B}_0 = 10$ kG.

References

D. E. Aspnes, *Phys. Rev.* **B147**, 554 (1966).
D. E. Aspnes and A. A. Studna, *Phys. Rev.* **B7**, 4605 (1973).
M. Balkanski and E. Amzallag, *Phys. Stat. Solidi* **30**, 407 (1968).
M. Born and E. Wolf, *Principles of Optics*, Third edition (Pergamon Press, Oxford, 1965).

E. Burstein, G. S. Picus, H. A. Gebbie, and F. Blatt, *Phys. Rev.* **103**, 826 (1956).
E. Burstein, G. S. Picus, R. F. Wallis, and F. Blatt, *Phys. Rev.* **113**, 15 (1959).
W. Franz, *Z. Naturforsch.* **13a**, 484 (1958).
L. V. Keldysh, *Sov. Phys. — JETP* **7**, 778 (1958).
R. Ludeke and W. Paul, in *II–VI Semiconductor Compounds*, ed. D. G. Thomas (Benjamin, New York, 1967).
J. M. Luttinger, *Phys. Rev.* **102**, 1030 (1956).
J. M. Luttinger and W. Kohn, *Phys. Rev.* **97**, 869 (1955).
D. L. Mitchell and R. F. Wallis, *Phys. Rev.* B**131**, 1965 (1963).
E. D. Palik and R. F. Wallis, *Phys. Rev.* **123**, 131 (1961).
E. D. Palik and G. B. Wright, in *Semiconductors and Semimetals*, Vol. 3, eds. R. K. Willardson and A. C. Beer (Academic, New York, 1967).
C. R. Pidgeon, *Thesis* (Reading 1962).
C. R. Pidgeon and R. N. Brown, *Phys. Rev.* **146**, 575 (1966).
L. M. Roth, B. Lax, and S. Zwerdling, *Phys. Rev.* **114**, 90 (1959).
B. O. Seraphin, in *Proceedings of the 7th International Conference on the Physics of Semiconductor*, ed. M. Hulin (Dunod, Paris, 1964).
B. O. Seraphin and R. B. Hess, *Phys. Rev. Lett.* **14**, 138 (1965).
M. J. Stephen and A.B. Lidiard, *J. Phys. Chem. Solids* **9**, 43 (1958).
S. Teitler, E. D. Palik, and R. F. Wallis, *Phys. Rev.* **123**, 1631 (1961).
R. F. Wallis, *J. Phys. Chem. Solids* **4**, 101 (1958).
R. F. Wallis and M. Balkanski, *Many-Body Aspects of Solid State Spectroscopy* (North-Holland, Amsterdam, 1986).
S. Zwerdling and B. Lax, *Phys. Rev.* **106**, 51 (1957).
S. Zwerdling, B. Lax, and L. M. Roth, *Phys. Rev.* **108**, 1402 (1957).

P–N junctions in semiconductors

12

Key ideas

The *p–n junction* represents the interface between two regions, one of which is p-type and the other n-type.

On both sides of the interface electrons and holes diffuse and recombine leaving uncompensated charged ions that constitute a *space charge region* or *depletion layer*. From the region of positive ions on the n-type side toward the region of negative ions on the p-type side exists a *built-in electric field*.

The charge density is *positive* on the *n-type side* and *negative* on the *p-type side* of the space charge region.

The contact potential or *diffusion potential* is the potential difference across the junction, i.e., the difference in potential of the neutral n-type region and the neutral p-type region.

The band gap width remains constant throughout the space charge region. The extrema of the conduction and valence bands at a particular point are shifted by an energy equal to the product of the electron charge and the potential at that point.

The higher the ionized impurity concentration on a given side of the junction, the smaller the penetration of the space charge region on that side and the total width of the space charge region.

The concentration of a carrier does not drop abruptly to zero in passing from the neutral region into the adjoining space charge region. A measure of the penetration depth of the carrier is the *Debye length*.

Under an applied external voltage called the *bias voltage* the diffusion current is no longer balanced by the drift current.

The bias voltage changes the width of the space charge region and modifies the electric field.

Under forward bias the width of the space charge region is decreased. The potential difference also decreases and the energy band diagram is modified.

The electric field arising from a reverse bias voltage is directed from the n-type side to the p-type side and is parallel to the built-in field. The width of the space charge region is increased.

The *diffusion current* of electrons consists of majority carrier electrons that diffuse from the n-type side toward the p-type side. The *generation*

P–N junctions

12.1 Abrupt junction in thermal equilibrium

12.2 P–N junction under an applied voltage

current of electrons arises from minority electrons on the p-type side generated by thermal excitation across the band gap. Holes contribute analogous currents. At equilibrium the diffusion and generation currents are equal in magnitude and opposite in direction. Under forward bias the diffusion current increases exponentially with bias voltage and the generation current remains constant. Under reverse bias the diffusion current decreases and the generation current is constant. The total current reaches a saturation value as the reverse bias voltage increases.

12.3 Graded p–n junction

The impurity distribution near a graded p–n junction is not uniform.

12.4 P–N junction capacitance

A p–n junction has *capacitance* which can be easily varied.

Storage capacitance is due to a rearrangement of minority carriers as the voltage is changed.

Transition capacitance arises from majority carriers flowing in response to a change in reverse bias potential.

12.5 Avalanche breakdown and Zener breakdown

At sufficiently high fields impact ionization and carrier multiplication take place resulting in *avalanche breakdown* or in *Zener breakdown*.

The preceding chapters have treated the basic physics of semiconductors. At this point we turn to the applications of semiconductors and the devices that arise therefrom. Perhaps the simplest semiconductor device is the **p–n junction** which is useful as a rectifier. It also serves as the basis for the **transistor** which has important applications as an amplifier.

12.1 Abrupt junction at thermal equilibrium

We have seen in Chapter 5 that impurities of different types can be introduced into bulk intrinsic semiconductors. If the impurity is the source of free charge carriers that are negatively charged electrons, the material is designated n-type, whereas if the impurity is the source of positively charged holes, the material is designated p-type. In n-type material the principal impurities are donors, and the **majority carriers** are conduction electrons arising primarily from the ionization of donor atoms. In p-type material, the principal impurities are acceptors, and the majority carriers are holes created by the transfer of electrons from the valence band to acceptor atoms leaving positively charged holes in the valence band. In both types of material at nonzero temperature, conduction electrons and holes arising from the excitation of electrons across the energy gap are also present. The thermally excited holes in n-type material and conduction electrons in p-type material are called **minority carriers**.

It is possible to spatially control the distribution of impurities in bulk material and create systems that contain both n- and p-type regions. Let us consider a rectangular slab of a pure undoped semiconductor, such as Si or Ge, whose length is large compared to its lateral dimensions as shown in

Fig. 12.1. The slab can be selectively doped by diffusion of impurity atoms deposited on the end faces. We discuss the particular case in which a certain quantity of an acceptor impurity such as Ga is deposited on the left-hand end face at position A and a donor impurity such as As is deposited on the right-hand end face at position B. If we now heat the slab by putting it into a furnace, the Ga and As atoms will diffuse into the interior of the slab, each with its own diffusion front. The two diffusion fronts will meet toward the middle of the slab at position 0. If the slab is now removed from the furnace, one finds that the semiconductor changes from p-type to the left of position 0 to n-type to the right of position 0. This internal interface between p- and n-type regions is called a **p–n junction**.

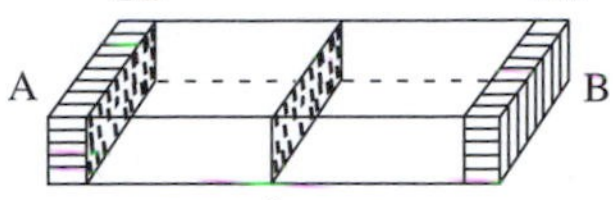

Fig. 12.1
Fabrication of a p–n junction.

In an idealized situation called an **abrupt junction** the concentration of doping atoms changes discontinuously from a uniform value n_a on the p-type side to another uniform value n_d on the n-type side. On the other hand, in a **graded junction** the concentration of doping atoms varies continuously from one side of the junction to the other. In this section we discuss the abrupt junction.

The p–n junction forms the basis for many of the devices employed in modern electronics. We shall therefore go into some detail in describing its basic properties.

12.1.1 Space charge region

We shall assume that the left hand side of the junction shown in Fig. 12.1 contains shallow acceptors with a small ionization energy E_a^i on the order of k_BT for room temperature. Then practically all the acceptors are ionized at room temperature. This means that the concentration of holes p in the valence band is very nearly equal to the concentration of acceptor impurities $n_a$introduced into the sample.

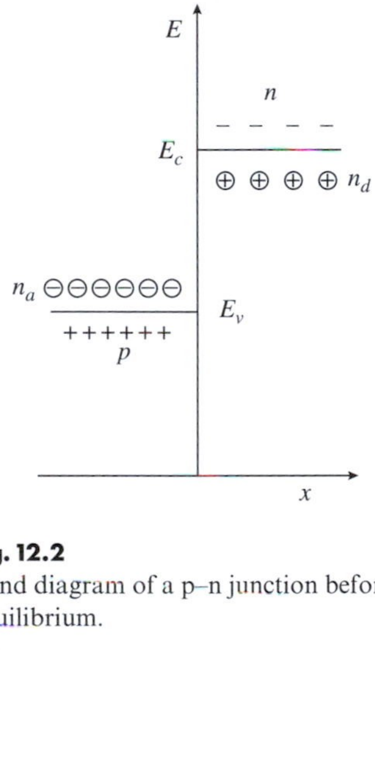

Fig. 12.2
Band diagram of a p–n junction before equilibrium.

Similarly, if the donor impurities on the right hand side of the junction are shallow, they are practically all ionized at room temperature. The free electron concentration n is then essentially the same as the donor atom concentration n_d. We therefore have free negative carriers (electrons) in the conduction band of the n-type region and free positive carriers (holes) in the valence band of the p-type region as shown in Fig. 12.2. The ionized donors and acceptors are indicated by encircled + and − signs.

It must be emphasized that the p–n junction represented by Fig. 12.2 is not in thermal equilibrium. Very large concentration gradients of electrons and holes exist at the interface between the n-type and p-type regions. Consequently, electrons will diffuse from the n-type region into the p-type region, and holes will diffuse from the p-type region into the n-type region. Diffusion will continue until the basic criterion for thermal equilibrium is attained, i.e. the Fermi energy is constant throughout the system. As the electrons and holes intermix on each side of the junction, they recombine, thereby annihilating each other and leaving a region of vastly reduced free carrier concentration. The n-type side of the region contains ionized donors (positively charged), but is greatly depleted of compensating free electrons. The p-type side of the region contains ionized acceptors (negatively charged), but is greatly depleted of compensating free holes.

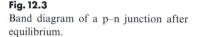

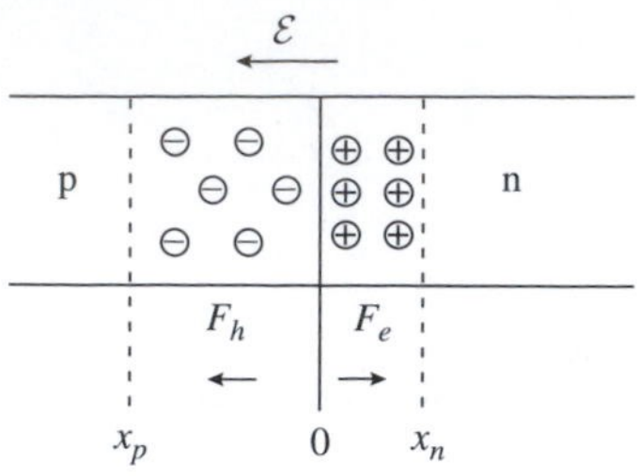

Fig. 12.3
Band diagram of a p–n junction after equilibrium.

The uncompensated charged ions are referred to as **space charge** and the region in which they are located is the **space charge region** or **depletion layer** shown in Fig. 12.3. The space charge region extends from x_p to x_n.

It is clear from Fig. 12.3 that the positive and negative ions in the space charge region form separate layers much like the plates of a charged condenser. Therefore, an electric field $\mathcal{E}$ exists in this region. Its direction is to the left from the region of positive ions toward the region of negative ions. This field is called the **built-in electric field** and is shown in Fig. 12.3. In Fig. 12.3 are also shown the forces $\boldsymbol{F}_e$ and $\boldsymbol{F}_h$ which the electric field exerts on electrons and holes, respectively. These forces exerted by the built-in electric field oppose the diffusion of electrons out of the n-type region and of holes out of the p-type region. Equilibrium is established when

$$\boldsymbol{J}_e = en\mu_e\mathcal{E} + eD_e\nabla n = 0 \qquad (12.1a)$$

$$\boldsymbol{J}_h = ep\mu_h\mathcal{E} - eD_h\nabla p = 0, \qquad (12.1b)$$

where D_e and D_h are the diffusion coefficients for electrons and holes, respectively, μ_e and μ_h are their mobilities, and $\boldsymbol{J}_e$ and $\boldsymbol{J}_h$ are their current densities. Equilibrium requires that the current densities be zero.

12.1.2 Charge density variation

The charge density $\rho(x)$ at any point x in the junction can be written as

$$\rho(x) = e[n_d(x) - n_a(x) + p(x) - n(x)]. \qquad (12.2)$$

The discussion of Section 12.1.1 leads us to the following results. In the n-type neutral region $x_n \leq x$, the conduction electron concentration n_n is given by $n_n \simeq n_d$, $n_a \cong 0$, and the hole concentration p_n is given by the law of mass action (Eq. (6.16))

$$p_n = \frac{n_i^2}{n_n} \simeq \frac{n_i^2}{n_d}. \qquad (12.3)$$

At 300 K, $n_i \simeq 1.5 \times 10^{10}\,\text{cm}^{-3}$ for Si. If $n_d = 10^{15}\,\text{cm}^{-3}$, $p_n \simeq 2.2 \times 10^5\,\text{cm}^{-3}$, which is very small compared to n_n. Hence, the electrons are the **majority carriers** and the holes are the **minority carriers** in the n-type neutral region.

On the n-type side of the space charge region $0 \leq x \leq x_n$, we have $n \cong p \cong n_a \cong 0$ and $n_d \neq 0$. On the p-type side of the space charge region $x_p \leq x \leq 0$, we have $n \cong p \cong n_d \cong 0$ and $n_a \neq 0$. In the p-type neutral region $x \leq x_p$, we have $p_p \simeq n_a$, $n_d \cong 0$, and n_p is given by

$$n_p = \frac{n_i^2}{p_p} \simeq \frac{n_i^2}{n_a}. \qquad (12.4)$$

The holes are now the majority carriers and the electrons are the minority carriers.

The charge density as specified by Eq. (12.2) can be expressed as

$$\rho(x) = 0 \qquad \text{for } x \leq x_p \qquad \text{and } x \geq x_n \tag{12.5a}$$

$$\rho(x) = -en_a \quad \text{for } x_p \leq x \leq 0 \tag{12.5b}$$

$$\rho(x) = en_d \quad \text{for } 0 \leq x \leq x_n. \tag{12.5c}$$

A plot of $\rho(x)$ versus x for an abrupt p–n junction is shown in Fig. 12.4.

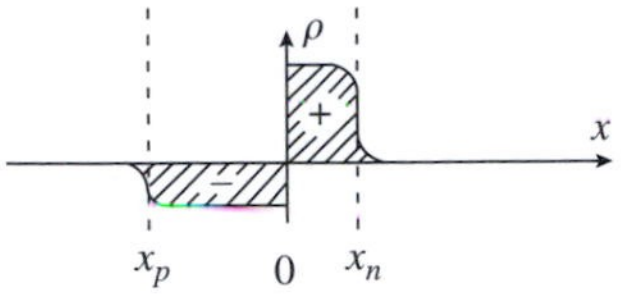

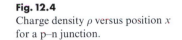

Fig. 12.4
Charge density ρ versus position x for a p–n junction.

12.1.3 Diffusion potential

The particular charge density distribution that exists in the space charge region creates the built-in electric field $\mathcal{E}$. This electric field corresponds to the negative gradient of the electrostatic potential $V(\boldsymbol{r})$ within the p–n junction:

$$\mathcal{E} = -\nabla V(\boldsymbol{r}). \tag{12.6}$$

The potential difference across the junction is the difference in the potentials of the neutral n-type region V_n and the neutral p-type region V_p and is called the **contact potential** or the **diffusion potential** V_d:

$$V_d = V_n - V_p. \tag{12.7}$$

The potentials in the various regions of a p–n junction are shown schematically in Fig. 12.5.

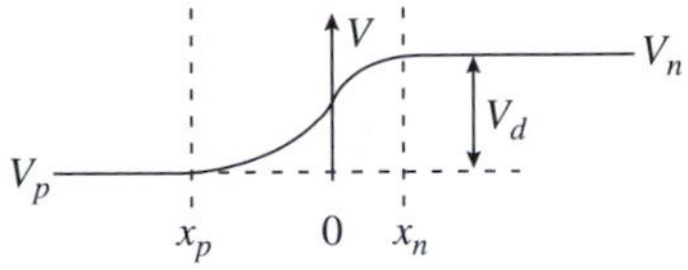

Fig. 12.5
Potential V versus position x for a p–n junction.

We now consider an ideal abrupt junction in which the built-in electric field is confined to the space charge region. The electrostatic potential is therefore constant in the regions outside the space charge region. The diffusion potential can be calculated by taking into account the fact that the system is in equilibrium. There are two ways to specify this condition:

1. require that the Fermi energy be constant throughout the junction, or
2. require that the electron and hole currents be zero.

Starting with procedure (1), we note from Eq. (6.10) that the electron concentration in the n-type neutral region can be written as

$$n_n = \bar{N}_c e^{\beta(E_F - E_{Cn})} \tag{12.8}$$

and in the p-type neutral region as

$$n_p = \bar{N}_c e^{\beta(E_F - E_{Cp})}, \tag{12.9}$$

where $\bar{N}_c$ is the effective density of conduction band states and the quantities E_{Cn} and E_{Cp} are the energies of the conduction band minima in the n-type and p-type neutral regions.

The identity of E_F on both sides of the p–n junction means that we can eliminate E_F from Eqs. (12.8) and (12.9) to yield

$$E_{Cp} - E_{Cn} = \frac{1}{\beta}\log\left(\frac{n_n}{n_p}\right) = k_B T \log\left(\frac{n_n}{n_p}\right). \tag{12.10}$$

Taking $n_n = n_d$ and n_p from Eq. (12.4), we obtain

$$E_{Cp} - E_{Cn} = k_B T \log\left(\frac{n_d n_a}{n_i^2}\right). \tag{12.11}$$

The energies E_{Cn} and E_{Cp} are related to the potentials V_n and V_p by

$$E_{Cn} = -eV_n \tag{12.12}$$

$$E_{Cp} = -eV_p. \tag{12.13}$$

The diffusion potential is then given by

$$V_d = V_n - V_p = \frac{1}{e}(E_{Cp} - E_{Cn}), \tag{12.14}$$

and from Eq. (12.11) follows the result

$$V_d = \frac{k_B T}{e} \log\left(\frac{n_d n_a}{n_i^2}\right). \tag{12.15}$$

For Si at 300 K and $n_d = n_a = 10^{15}$ cm^{-3}, V_d is found to be ~ 0.6 volt.

Turning now to procedure (2) we start from Eqs. (12.1) and use the Einstein relation (Eqs. (8.37) and (3.38)) to eliminate the diffusion constants:

$$\boldsymbol{J}_e = \mu_e[en\boldsymbol{\mathcal{E}} + k_B T\, \boldsymbol{\nabla} n] = 0 \tag{12.16a}$$

$$\boldsymbol{J}_h = \mu_h[ep\boldsymbol{\mathcal{E}} - k_B T\, \boldsymbol{\nabla} p] = 0. \tag{12.16b}$$

For the one-dimensional case under consideration, these equations can be rewritten as

$$\frac{dn}{n} = -\frac{e}{k_B T}\mathcal{E}dx = \frac{e}{k_B T}dV \tag{12.17a}$$

$$\frac{dp}{p} = \frac{e}{k_B T}\mathcal{E}dx = -\frac{e}{k_B T}dV. \tag{12.17b}$$

Integrating Eq. (12.17a) from x_p to x_n across the whole space charge region yields

$$\log\frac{n(x_n)}{n(x_p)} = \frac{e}{k_B T}(V_n - V_p) = \frac{e}{k_B T}V_d. \tag{12.18}$$

Noting that $n(x_n) = n_d$ and $n(x_p) = n_i^2/n_a$, we get

$$V_d = \frac{k_B T}{e} \log\left(\frac{n_d n_a}{n_i^2}\right) \tag{12.19}$$

which agrees with the result from procedure (1).

12.1.4 Electric field in the space charge region

The potential in the space charge region can be obtained by solving Poisson's equation with the charge density given by Eq. (12.2). The electric field is then obtained by taking the negative gradient of the potential.

At any point x, Poisson's equation is

$$\frac{d^2V}{dx^2} = -\frac{\rho(x)}{\epsilon_0\epsilon}, \tag{12.20}$$

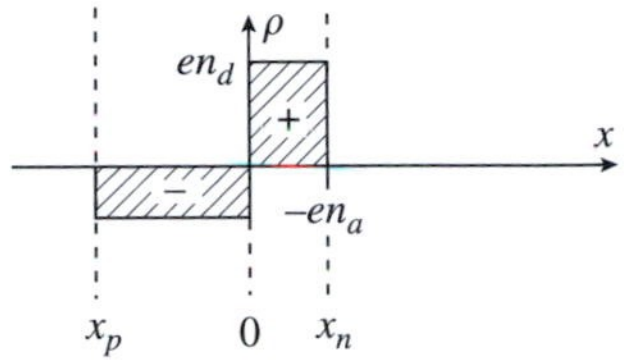

Fig. 12.6
Diagram showing charge neutrality.

where ϵ is the dielectric constant of the junction. The charge density $\rho(x)$ is illustrated in Fig. 12.6. Overall charge neutrality requires that

$$n_d x_n = -n_a x_p, \tag{12.21}$$

which means that the areas of the two rectangles on the left and right hand sides of the p–n junction in Fig. 12.6 are equal.

We first consider the region $x_p \leq x \leq 0$ where Poisson's equation has the form

$$\frac{d^2V}{dx^2} = -\frac{d\mathcal{E}}{dx} = \frac{en_a}{\epsilon_0\epsilon}. \tag{12.22}$$

Integration gives

$$\frac{dV}{dx} = -\mathcal{E}(x) = \frac{en_a x}{\epsilon_0\epsilon} + c_1, \tag{12.23}$$

where c_1 is a constant of integration determined by the boundary condition that $\mathcal{E}(x) = 0$ at $x = x_p$. The value of c_1 is

$$c_1 = -\frac{en_a x_p}{\epsilon_0\epsilon}, \tag{12.24}$$

and hence

$$\mathcal{E}(x) = -\frac{en_a(x - x_p)}{\epsilon_0\epsilon}, \quad x_p \leq x \leq 0. \tag{12.25}$$

On the n-type side of the junction where the space charge is positive, Poisson's equation is

$$\frac{d^2V}{dx^2} = -\frac{d\mathcal{E}}{dx} = -\frac{en_d}{\epsilon_0\epsilon}, \quad 0 \leq x \leq x_n. \tag{12.26}$$

Integration yields the result

$$\frac{dV}{dx} = -\mathcal{E}(x) = -\frac{en_d x}{\epsilon_0\epsilon} + c_2. \tag{12.27}$$

The integration constant c_2 is determined by the boundary condition that $\mathcal{E}(x) = 0$ at $x = x_n$ and is found to be

$$c_2 = \frac{en_d x_n}{\epsilon_0 \epsilon}. \tag{12.28}$$

The built-in electric field is then

$$\mathcal{E}(x) = \frac{en_d(x - x_n)}{\epsilon_0 \epsilon}, \quad 0 \leq x \leq x_n. \tag{12.29}$$

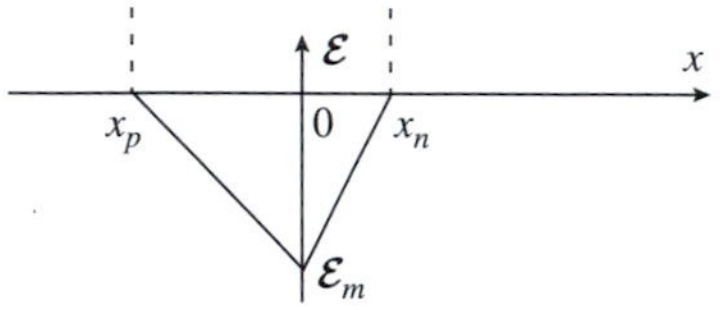

Fig. 12.7
Built-in electric field $\mathcal{E}$ versus position x.

The electroneutrality condition, Eq. (12.21), shows that $\mathcal{E}(x)$ is continuous at $x = 0$:

$$\mathcal{E}(0) = \frac{en_a x_p}{\epsilon_0 \epsilon} = -\frac{en_d x_n}{\epsilon_0 \epsilon}. \tag{12.30}$$

$\mathcal{E}(0)$ is the maximum field $\mathcal{E}_m$ in the depletion region. A plot of the built-in electric field as a function of x is shown in Fig. 12.7. We note that $\mathcal{E}(x)$ is negative over the entire space charge region $x_p \leq x \leq x_n$, in agreement with Eqs. (12.25) and (12.29).

12.1.5 Electronic energy bands in the space charge region

If the potential $V(x)$ is multiplied by $-e$, the result is the potential energy of an electron in the junction. The electron energy is therefore changed by the amount $-eV(x)$; in particular, if the valence band edge E_V is taken to be zero in the p-type neutral region, it has the value

$$E_V(x) = -eV(x) \tag{12.31}$$

in the space charge region. An expression for the conduction band edge E_C can be obtained by noting that the energy gap $E_g = E_C - E_V$ is a constant in real space; hence,

$$E_C(x) = E_g - eV(x). \tag{12.32}$$

The variation of the band structure is thus specified at each point of the space charge region.

To evaluate $V(x)$ we integrate the equation

$$\frac{dV}{dx} = -\mathcal{E}(x)$$

over the region $x_p \leq x \leq 0$ using Eq. (12.25) and obtain

$$V(x) = \frac{en_a(x - x_p)^2}{2\epsilon_0 \epsilon} + V_p, \tag{12.33}$$

where we have used the boundary condition that in the neutral p-type region the electrostatic potential is constant and equal to V_p: $V(x_p) = V_p$. For the region $0 \leq x \leq x_n$ we use Eq. (12.29) and the boundary condition

$V(x_n) = V_n$ and get

$$V(x) = -\frac{en_d(x - x_n)^2}{2\epsilon_0\epsilon} + V_n. \quad (12.34)$$

The behavior of the valence and conduction band edges throughout the p–n junction is obtained by substituting Eqs. (12.33) and (12.34) into Eqs. (12.31) and (12.32), respectively. In Fig. 12.8 are presented the variations of the valence and conduction band edges $E_V(x)$ and $E_C(x)$ as functions of x. As noted previously the band gap E_g is independent of x. In the space charge region the "tilting" of the band edges by the built-in electric field is evident. Since we are considering the p–n junction at thermal equilibrium, the Fermi energy is constant throughout the structure. If E_{Fn} and E_{Fp} are the **quasi-Fermi energies** in the neutral n- and p-type regions before equilibrium is established, with $E_{Fn} > E_{Fp}$, then after equilibrium is established,

$$E_{Fp} = E_{Fn} - eV_d = E_F. \quad (12.35)$$

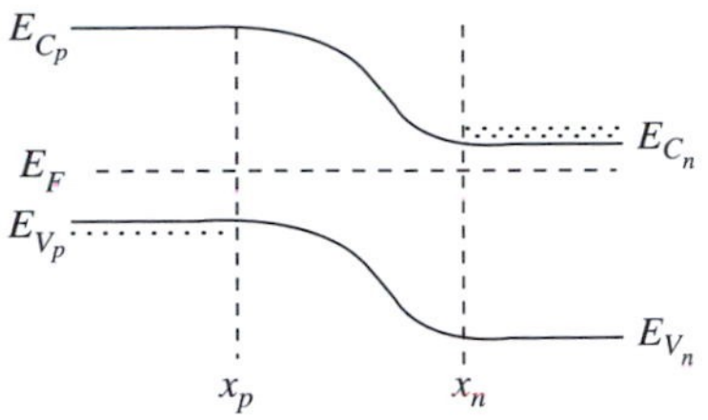

Fig. 12.8
Conduction and valence band edges versus x.

12.1.6 Width of the space charge region

The condition of charge neutrality as expressed by Eq. (12.21) can be rewritten in terms of $W_p = |x_p| = -x_p$ and $W_n = |x_n| = x_n$:

$$n_a W_p = n_d W_n. \quad (12.36)$$

This result means that the space charge is wider spread in the less doped region. The total width W of the space charge region is

$$W = W_n + W_p. \quad (12.37)$$

To calculate W_n and W_p individually, we need to supplement Eq. (12.37) with a second relation that is obtained from the condition requiring continuity of $V(x)$ at $x = 0$. Using Eqs. (12.33) and (12.34), this condition takes the form

$$\frac{en_a x_p^2}{2\epsilon_0\epsilon} + V_p = -\frac{en_d x_n^2}{2\epsilon_0\epsilon} + V_n, \quad (12.38)$$

or

$$V_d = V_n - V_p = \frac{e}{2\epsilon_0\epsilon}(n_d W_n^2 + n_a W_p^2). \quad (12.39)$$

Taking into consideration Eq. (12.36) we get

$$V_d = \frac{en_d W_n^2}{2\epsilon_0\epsilon}\left(1 + \frac{n_d}{n_a}\right) = \frac{en_a W_p^2}{2\epsilon_0\epsilon}\left(1 + \frac{n_a}{n_d}\right). \quad (12.40)$$

Solving for W_n and W_p separately:

$$W_n^2 = \frac{2\epsilon_0\epsilon V_d}{en_d(1+(n_d/n_a))} \tag{12.41a}$$

$$W_p^2 = \frac{2\epsilon_0\epsilon V_d}{en_a(1+(n_a/n_d))}. \tag{12.41b}$$

With V_d given by Eq. (12.19), we obtain

$$W_n = L_{Dn}\left[\frac{1}{(1+(n_d/n_a))}\log\frac{n_a n_d}{n_i^2}\right]^{\frac{1}{2}} \tag{12.42a}$$

$$W_p = L_{Dp}\left[\frac{1}{(1+(n_a/n_d))}\log\frac{n_a n_d}{n_i^2}\right]^{\frac{1}{2}}, \tag{12.42b}$$

where

$$L_{Dn} = \left(\frac{2\epsilon_0\epsilon k_B T}{e^2 n_d}\right)^{\frac{1}{2}} \tag{12.43a}$$

$$L_{Dp} = \left(\frac{2\epsilon_0\epsilon k_B T}{e^2 n_a}\right)^{\frac{1}{2}}. \tag{12.43b}$$

L_{Dn} and L_{Dp} are referred to as the **Debye lengths** in the n- and p-type regions, respectively.

The total width W of the space charge region is

$$W = \left[\frac{2\epsilon_0\epsilon k_B T}{e^2}\left(\frac{n_a+n_d}{n_a n_d}\right)\log\frac{n_a n_d}{n_i^2}\right]^{\frac{1}{2}}. \tag{12.44}$$

For Si with $\epsilon = 11.7$ and equal doping on both sides of the junction, $n_a = n_d = 10^{15}\,\text{cm}^{-3}$, one finds that W is about 1.3×10^{-4} cm at 300 K. This is a typical value for a p–n junction in Si.

One can obtain simple expressions for the widths of the space charge regions W_n and W_p on the n and p sides of the junction by utilizing the overall neutrality condition $n_a W_p = n_d W_n$. The total width of the space charge region can then be written as

$$W = W_p + W_n = W_p + \frac{n_a}{n_d}W_p = W_p\left(1+\frac{n_a}{n_d}\right),$$

from which we get

$$W_p = \frac{W}{1+(n_a/n_d)}. \tag{12.45}$$

We can express W_n as

$$W_n = W - W_p = \frac{W}{1 + (n_d/n_a)}. \tag{12.46}$$

These results indicate that the larger the ionized impurity concentration on a given side of the junction, the smaller the penetration of the space charge region into that side.

If we consider the case $n_a \gg n_d$, the n-type region will be the wider region. Its width as specified by Eq. (12.42a) is

$$W_n \cong L_{Dn}\left[\log \frac{n_a n_d}{n_i^2}\right]^{\frac{1}{2}}. \tag{12.47}$$

For Si at 300 K with $n_a = 10^{18}$ cm^{-3} and $n_d = 10^{17}$ cm^{-3}, the extent of the space charge region into the n-type side is roughly

$$W_n \cong 6L_{Dn},$$

or 0.1μ.

12.1.7 Physical interpretation of the Debye length

The concentration of electrons does not drop abruptly to zero in passing from the neutral n-type region into the space charge region at x_n. Some electrons diffuse into the space charge region, and their concentration is not zero, but depends on position x: $n = n(x)$.

To determine $n(x)$ we generalize Eq. (12.8) and write the electron concentration at any point x as

$$n(x) = \bar{N}_c e^{\beta[E_C(x) - E_F]}, \tag{12.48}$$

where $E_C(x)$ is the conduction band edge at x given by

$$E_C(x) = E_{Cn} - eV(x) \tag{12.49}$$

and $V(x)$ is the electrostatic potential given by Eq. (12.34).

In the neutral part of the n-type region, $V(x) = V_n$ which we take to be the zero of potential. Then we recover Eq. (12.8), which for shallow donors at room temperature becomes

$$n(x) = n_n = \bar{N}_c e^{-\beta(E_{Cn} - E_F)} \simeq n_d. \tag{12.50}$$

In the depleted region the electron concentration is

$$n(x) = \bar{N}_c e^{-\beta[E_{Cn} - eV(x) - E_F]}. \tag{12.51}$$

Rearranging this equation with the aid of Eq. (12.50) yields

$$n(x) = n_d e^{\beta eV(x)}. \tag{12.52}$$

Replacing $V(x)$ by the expression given in Eq. (12.34) with $V_n = 0$, we obtain

$$n(x) = n_d \exp\left[-\left(\frac{x - x_n}{L_{Dn}}\right)^2\right]. \tag{12.53}$$

This result provides a physical interpretation of the Debye length L_{Dn}. It plays the role of a diffusion length and is a measure of the penetration depth of electrons into the space charge region.

Example 12.1: Diffusion and drift currents in a p–n junction at thermal equilibrium
Calculate the diffusion and drift currents of electrons and holes in a silicon p–n junction at thermal equilibrium.
Solution. In the neutral n- and p-type regions the net charge density is zero, the electric field is zero and the electron and hole currents are separately zero. At the p–n interface, $x=0$, the electric field has its maximum magnitude, and the drift current of each type of free charge carrier can be large even if their densities are small. The drift current of a particular type of carrier is canceled by that carrier's diffusion current which can also be large because the concentration gradient is large.
A simple estimate of the drift current can be made by considering a symmetric junction for which $n_a = n_d$ and the carrier masses are equal. At $x=0$, E_F is exactly one-half of the energy gap: $E_F = \frac{1}{2}E_g$. The free carrier concentrations at this point are the intrinsic concentrations. The drift current density is $J = n_i e\mu\mathcal{E}$. For Si at 300 K with $n_i = 10^{10}$ cm^{-3}, $\mu \simeq 1000$ cm^2 V^{-1}s^{-1}, and $n_a = n_d = 10^{15}$ cm^{-3}, one finds from Eq. (12.39) that $\mathcal{E} \simeq V_d/W \simeq 10^4$ Vcm^{-1} if the space charge width is $\sim 1\mu$m. The current density is then ~ 0.01 A/cm^2.

12.2 P–N junction under an applied voltage

12.2.1 Qualitative effects of an applied voltage

When an external voltage is applied across a p–n junction, the equilibrium situation no longer holds, and an electric current flows through the junction. This means that the diffusion current of a given type of carrier is no longer balanced by the drift current. An external voltage is referred to as a **bias voltage**.

A bias voltage changes the difference in potential of the n- and p-type neutral regions and modifies the electric field in the space charge region. Furthermore, a bias voltage produces a change in the width of the space charge region. If the bias voltage is such that the n-type region becomes more negative, the junction is said to have a **forward bias**. If the n-type region becomes more positive, the junction has a **reverse bias**.

12.2.2 Forward bias: n-type region biased negatively

12.2.2.1 Effect on the built-in electric field

For simplicity we assume that the space charge region is completely depleted of free carriers and hence is a region of much higher electrical resistivity than the neutral n- and p-type regions. Therefore, the bias voltage V_a appears entirely across the space charge region of width W', where W' is the width under forward bias. Recalling that the built-in electric field $\mathcal{E}$ in the junction at equilibrium is directed from the n-type side to the p-type side of the space charge region, we see that making the n-type region more negative (forward bias) will tend to neutralize some of the positive space charge and reduce the built-in electric field by the amount $\mathcal{E}_a = V_a/W'$. In this case the magnitude of the resulting electric field in the space charge region is equal to $\mathcal{E} - \mathcal{E}_a$. The vectors $\mathcal{E}$ and $\mathcal{E}_a$ are antiparallel as represented in Fig. 12.9.

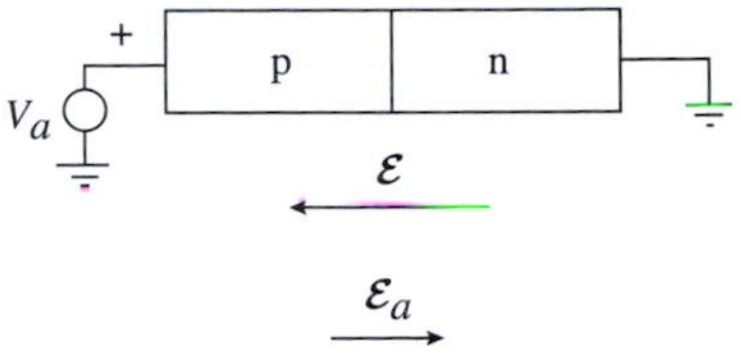

Fig. 12.9
Diagram of a forward bias voltage V_a applied to a p–n junction.

12.2.2.2 Effect on the width of the space charge region

Under forward bias the effect of an applied voltage V_a is to reduce the width of the space charge region to the value $W' = W'_n + W'_p$. This follows from the reduction of the electric field, since a smaller field must arise from fewer uncompensated ionized impurity atoms in the space charge region. Consequently, the space charge region becomes narrower, as we can see from Eqs. (12.41) in which V_d is to be replaced by $V_d - V_a$:

$$W'_n = \left[\frac{2\epsilon_0\epsilon(V_d - V_a)n_a}{en_d(n_a + n_d)}\right]^{\frac{1}{2}} \tag{12.54a}$$

$$W'_p = \left[\frac{2\epsilon_0\epsilon(V_d - V_a)n_d}{en_a(n_a + n_d)}\right]^{\frac{1}{2}}. \tag{12.54b}$$

12.2.2.3 Effect on the difference in electrostatic potential between the n- and p-type regions of the junction

The difference in potential between the neutral n- and p-type regions can be written as $W'\mathcal{E}_{ave}$, where $\mathcal{E}_{ave}$ is the average value of the electric field in the space charge region. Since both the average field and the width of the space charge region decrease under forward bias, the potential difference also decreases. The decrease in potential in fact is the bias voltage V_a, a positive quantity. The potential difference $V_n - V_p$, which at equilibrium is the diffusion potential V_d, is now

$$V_n - V_p = V_d - V_a. \tag{12.55}$$

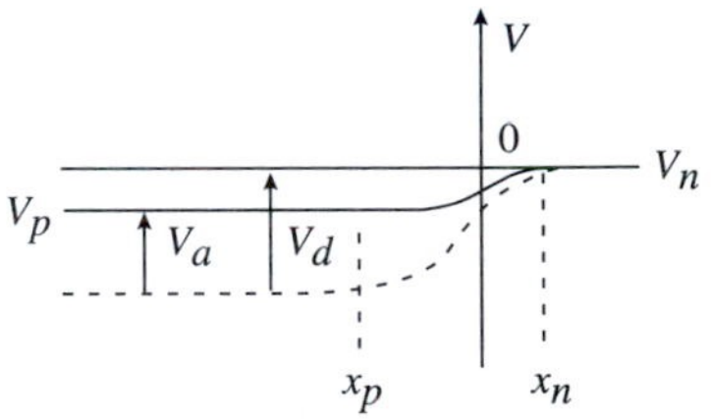

Fig. 12.10
Potential versus position for a p–n junction with (solid curve) and without (dashed curve) an applied forward bias.

The electrostatic potential $V(x)$ at point x in the junction can be calculated by generalizing the treatment in Section 12.1.5. The behavior of $V(x)$ both at equilibrium and under forward bias is illustrated in Fig. 12.10.

12.2.2.4 Effect of a forward bias voltage on the electron potential energy $-eV(x)$ and the energy band edges

The conduction and valence band edges for an abrupt p–n junction at equilibrium are shown in Fig. 12.11 as dotted lines. The energy barrier between the n- and p-sides is eV_d. The application of a forward bias has the effect of reducing the energy barrier to the value $e(V_d - V_a)$ and thus leads to a modification of the energy band diagram, as shown by full lines in Fig. 12.11.

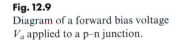

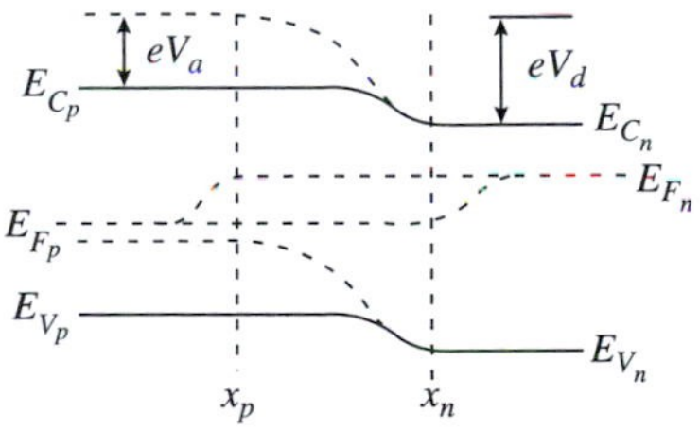

Fig. 12.11
Conduction and valence band edges versus position with (solid curve) and without (dashed curve) an applied forward bias.

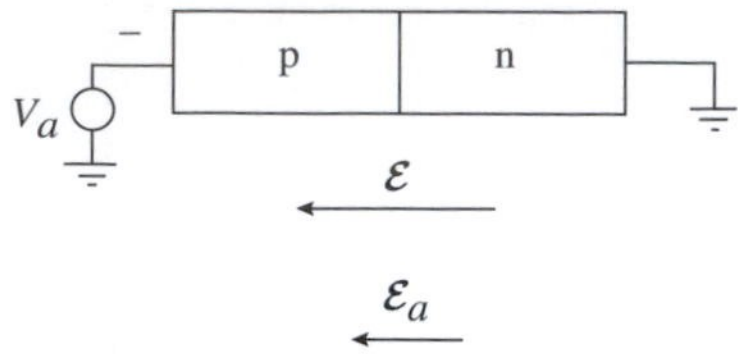

Fig. 12.12
Diagram of a reverse bias voltage V_a applied to a p–n junction.

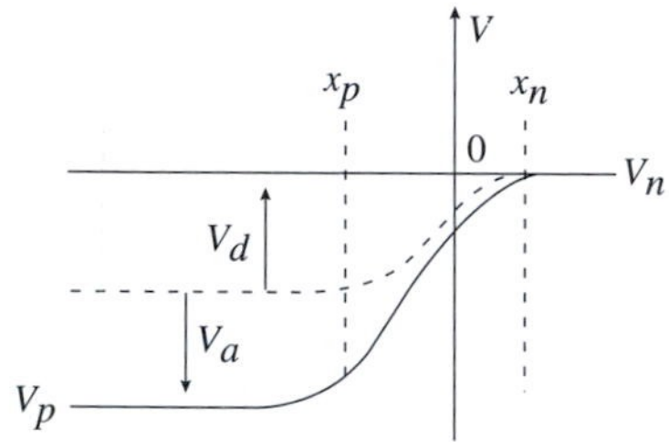

Fig. 12.13
Potential versus position for a p–n junction with (solid curve) and without (dashed curve) an applied reverse bias.

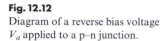

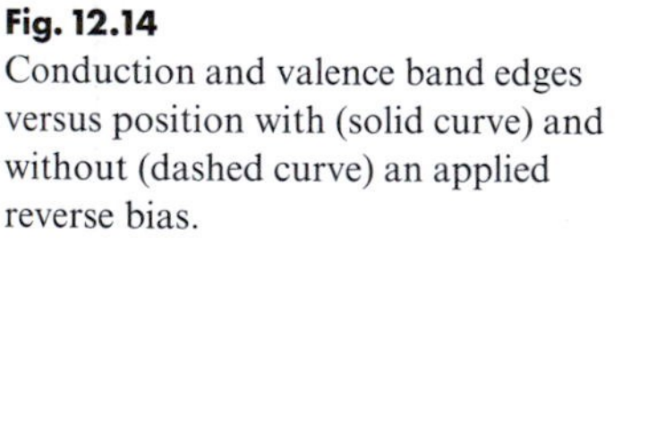

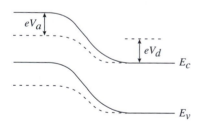

Fig. 12.14
Conduction and valence band edges versus position with (solid curve) and without (dashed curve) an applied reverse bias.

12.2.3 Reverse bias: n-type region biased positively

12.2.3.1 Effect on the built-in electric field

Under reverse bias the p–n junction is polarized with the n-type side connected to the positive terminal of the generator as shown in Fig. 12.12. The electric field $\mathcal{E}_a$ arising from the bias voltage is now directed from the n-type side to the p-type side of the space charge region and is parallel to the built-in field $\mathcal{E}$. Consequently, the total electric field in the space charge region has the magnitude $\mathcal{E} + \mathcal{E}_a$.

12.2.3.2 Effect on the width of the space charge region

There are increased numbers of positively charged ionized impurities on the n-type side and of negatively charged impurities on the p-type side and therefore an increased width of the space charge region. The widths of the n- and p-type regions are obtained by replacing V_a by $-V_a$ in Eqs. (12.54).

12.2.3.3 Effect on the electrostatic potential between the n- and p-type regions of the junction

Since the electrostatic potential difference is $W'\mathcal{E}_{ave}$ and since both W' and $\mathcal{E}_{ave}$ increase under reverse bias, the potential difference must also increase. The bias potential V_a is a positive quantity which adds to the diffusion potential V_d to yield the total potential difference between the n- and p-type regions:

$$V_n - V_p = V_d + V_a. \tag{12.56}$$

This situation is shown in Fig. 12.13.

12.2.3.4 Effect on the electron potential energy and the energy band edges

The potential energy barrier between the n- and p-type regions, which has the magnitude $e(V_n - V_p) = eV_d$ in equilibrium, is increased under reverse bias to $e(V_d + V_a)$. This increase in barrier height produces a change in the energy band edges, as shown in Fig. 12.14, and reduces the diffusion currents of both electrons and holes.

12.2.4 Qualitative description of current flow in a biased junction

To describe current flow in a biased junction we need to identify more precisely the nature of the currents discussed in Section 12.1.1. Considering first electrons, we have introduced the diffusion current density $eD_e\nabla_n = J_{e1}$, which consists of majority carrier electrons that diffuse from the n-type side toward the p-type side and that have sufficient energy to overcome the barrier due to the diffusion potential. Only a small fraction of majority electrons participate in the diffusion current. A second current introduced is a drift current with current density $J_{e2} = en\mu_e\mathcal{E}$. It consists of minority carrier electrons from the p-type side that slide down the energy barrier under the influence of the built-in electric field $\mathcal{E}$. The minority carriers are generated by thermal excitation across the band gap. For this reason the term drift current will be abandoned and replaced by the term **generation current**. Since the minority carrier electrons participating in this process originate within a Debye length L_{Dn} on the p-type side of the space charge region and since essentially all of them are swept into the n-type side, the generation current is independent of the barrier height.

A diagram illustrating the diffusion and generation currents is presented in Fig. 12.15. Similar considerations apply to the current due to holes.

At equilibrium the diffusion and generation currents are equal in magnitude and opposite in direction for both electrons and holes. The net current is therefore zero. Under the application of a forward bias, the energy barrier is lowered and a larger fraction of the majority carrier electrons have energy sufficient to surmount the barrier and flow from the n-type region to the p-type region. The diffusion current j_{e1} therefore increases exponentially with bias voltage V_a. Since the generation current j_{e2} is independent of barrier height and hence of V_a, it remains constant. The net current due to electrons, $j_{e1} + j_{e2}$, consequently increases. A similar analysis applies to holes. The total net current due to both electrons and holes increases exponentially as the forward bias voltage increases.

Reverse bias increases the energy barrier and reduces the fraction of majority carrier electrons that can cross it. The diffusion current j_{e1} from the n-type side to the p-type side decreases as the reverse bias increases. The minority carrier current j_{e2} from the p-type side to the n-type side remains practically constant. Eventually a reverse bias voltage is reached such that the diffusion current due to majority carriers is insignificant compared to the generation current due to the minority carriers. Since the latter current is constant, the net electron current becomes constant. The same arguments hold for the case of holes. Combining the currents from electrons and holes yields the total current, which reaches a saturation value at large reverse bias voltage. The saturation value is the sum of the generation currents of electrons and holes.

From these considerations one can conclude that under forward bias the total current in the junction increases continuously with increasing bias voltage, whereas under reverse bias the current reverses direction and reaches a saturation value. This qualitative appraisal is shown graphically in Fig. 12.16.

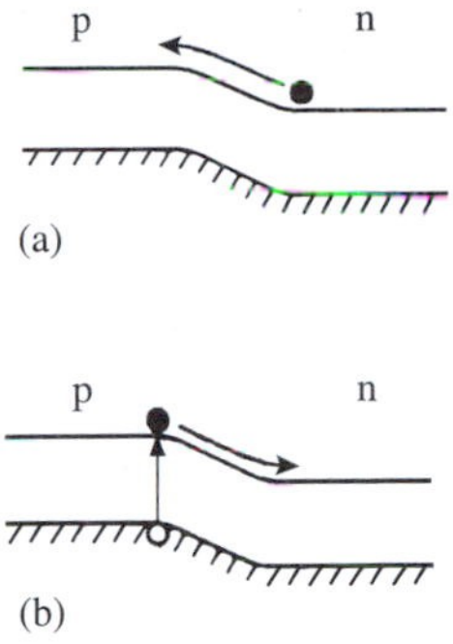

Fig. 12.15
Diagrams of (a) diffusion current j_{e1} and (b) generation current j_{e2} for electrons in a p–n junction (after Dalven 1990).

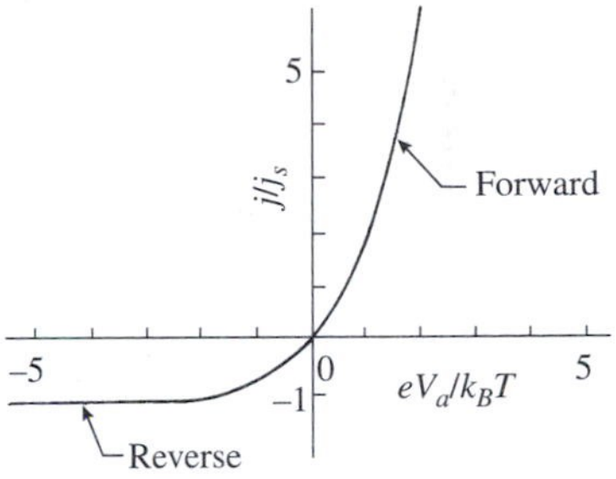

Fig. 12.16
Total current versus bias voltage in a p–n junction.

12.2.5 Quantitative treatment of current flow in a biased junction

12.2.5.1 Minority carrier injection

At thermal equilibrium, the electron and hole currents flowing through the junction are both zero as a result of the balance between the diffusion and drift currents embodied in Eqs. (12.1). When an external voltage is applied, the electric field in the space charge region changes, decreasing under forward bias, and increasing under reverse bias. The carrier concentrations also change.

The electron and hole current densities j_e and j_h are given by a generalization of Eqs. (12.16),

$$j_e = \mu_e \left[en\mathcal{E}' + k_B T \frac{dn}{dx} \right] \quad (12.57a)$$

$$j_h = \mu_h \left[ep\mathcal{E}' - k_B T \frac{dp}{dx} \right], \quad (12.57b)$$

where $\mathcal{E}'$ is the total electric field equal to the sum of the built-in and applied electric fields. Of particular importance in determining the net current is the process known as **minority carrier injection.** In the weak injection regime to be discussed next, the net electron and hole currents are much smaller than their individual components due to diffusion and drift.

One can make a simple estimate of the diffusion current density in the space charge region. The gradient of the hole concentration can be written as

$$\frac{dp}{dx} \simeq \frac{\Delta p}{\Delta x} \simeq \frac{p_p - p_n}{W} \simeq \frac{p_p}{W}, \tag{12.58}$$

since the concentration of minority holes is much less than that of majority holes. The hole diffusion current density is then given by

$$\begin{aligned} J_{hd} &= -\mu_p k_B T \frac{dp}{dx} = -\mu_p k_B T \frac{p_p}{W} \\ &= -\frac{k_B T \sigma_p}{eW}, \end{aligned} \tag{12.59}$$

where σ_p is the conductivity of the p-type region of the junction.

Example 12.2: Diffusion current in a p–n junction
Calculate the diffusion current density of a junction with $\sigma_p = 100\,\Omega^{-1}\,\mathrm{cm}^{-1}$ and $W = 1\,\mu\mathrm{m} = 1 \times 10^{-4}\,\mathrm{cm}$.
Solution. At a temperature of 300 K, $k_{BT} = 4.14 \times 10^{-14}\,\mathrm{erg} = 4.14 \times 10^{-21}\,\mathrm{J}$. Taking $e = 1.6 \times 10^{-19}\,\mathrm{C}$ and using Eq. (12.59), one obtains $|j_{hd}| \simeq 10^4\,\mathrm{A/cm^2}$. This current density is equivalent to 1 A per $10^4\,\mu\mathrm{m}^2$ of cross-sectional area.

The current density of electrons due to diffusion is found to have approximately the same magnitude as that for holes. Both current densities are much larger than the net current densities specified by Eqs. (12.57). It is therefore reasonable to set J_e and J_h equal to zero in these equations and obtain the relations

$$\frac{dn}{n} = -\frac{e}{k_B T}\mathcal{E}' dx = \frac{e}{k_B T} dV \tag{12.60a}$$

$$\frac{dp}{p} = \frac{e}{k_B T}\mathcal{E}' dx = -\frac{e}{k_B T} dV. \tag{12.60b}$$

Equations (12.60) are analogous to Eqs. (12.17) that apply to the case without bias and that for electrons have the solution

$$\log \frac{n(x_n)}{n(x_p)} = \frac{e}{k_B T} V_d. \tag{12.61}$$

For the electron concentrations we have $n(x_n) = n_n$ and $n(x_p) = n_p$, where n_n and n_p are the majority and minority electron concentrations at equilibrium in the neutral n- and p-type regions which terminate at x_n and x_p, respectively. Thus

$$\log \frac{n_n}{n_p} = \frac{e}{k_B T} V_d. \tag{12.62}$$

Under forward bias, integration of Eq. (12.60a) yields

$$\log \frac{n'(x_n)}{n'(x_p)} = \frac{e}{k_B T}(V_n - V_p) = \frac{e}{k_B T}(V_d - V_a), \tag{12.63}$$

where use has been made of Eq. (12.55). The majority electron concentration is relatively large and is little changed by the application of the bias. To a good approximation $n'(x_n) \simeq n_n$. The minority electron concentration, however, undergoes significant change, and we set $n'(x_p) = n'_p$. Equation (12.63) becomes

$$\log \frac{n_n}{n'_p} = \frac{e}{k_B T}(V_d - V_a). \tag{12.64}$$

Taking the differences of each side of Eqs. (12.62) and (12.64), we get

$$\log \frac{n'_p}{n_p} = \frac{e}{k_B T} V_a, \tag{12.65}$$

or

$$n'_p = n_p e^{eV_a/k_B T}. \tag{12.66}$$

A similar treatment can be carried out for the analogous minority hole concentrations p'_n and p_n. It yields

$$p'_n = p_n e^{eV_a/k_B T}. \tag{12.67}$$

We see that the minority carrier concentrations increase exponentially with increasing forward bias voltage. This behavior forms the basis for the statement that electrons are **injected** into the p-type region and holes are injected into the n-type region by a forward bias. However, under the weak injection conditions being discussed, the majority carrier concentrations are essentially unchanged by a forward bias.

The equilibrium minority carrier concentrations can be eliminated from Eqs. (12.66) and (12.67) with the aid of Eqs. (12.3) and (12.4) to give

$$n'_p = \frac{n_i^2}{p'_p} e^{eV_a/k_B T} \tag{12.68a}$$

$$p'_n = \frac{n_i^2}{n'_n} e^{eV_a/k_B T} \tag{12.68b}$$

or

$$n'_p p'_p = n'_n p'_n = n_i^2 e^{eV_a/k_B T}. \tag{12.69}$$

This result expresses the mass action law under forward bias.

In the case of reverse bias the sign of V_a must be reversed in accordance with Eq. (12.56). The minority carrier concentrations then **decrease** with increasing reverse bias voltage and become less than their equilibrium concentrations: $n'_p < n_p, p'_n < p_n$. Electrons are therefore **extracted** from the p-type region and holes are extracted from the n-type region.

12.2.5.2 Minority carrier distribution in the neutral regions

The concentrations of majority carriers in the neutral regions are uniform over the regions and do not vary with bias voltage. The minority carriers, on the other hand, are not uniformly distributed over the neutral regions. Their distribution can be calculated with the aid of the continuity equation taking into account carrier diffusion and carrier recombination.

For electrons in the p-type neutral region the equation of continuity is

$$\frac{\partial n}{\partial t} = \frac{1}{e}\frac{\partial j_{ed}(x)}{\partial x} - \frac{n - n_p}{\tau_n}, \tag{12.70}$$

where n_p and n are the equilibrium and nonequilibrium concentrations of electrons, j_{ed} is the diffusion current density, and τ_n is the minority electron lifetime. Since the electric field is negligible in the neutral regions, we can neglect the drift current. The diffusion current is (see Eq. (12.1))

$$j_{ed}(x) = eD_e\frac{\partial n}{\partial x}. \tag{12.71}$$

Substituting this result into the equation of continuity gives

$$\frac{\partial n}{\partial t} = D_e\frac{\partial^2 n}{\partial x^2} - \frac{n - n_p}{\tau_n}. \tag{12.72}$$

With a constant bias voltage a stationary state is established for which $\partial n/\partial t = 0$. Hence,

$$\frac{\partial^2 n}{\partial x^2} - \frac{n - n_p}{L_e^2} = 0, \tag{12.73}$$

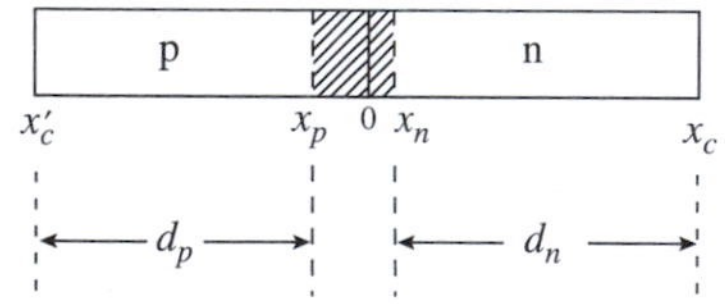

Fig. 12.17
Diagram showing the boundaries of the space charge and neutral regions of a p–n junction.

where $L_e = (D_e\tau_n)^{\frac{1}{2}}$ is the **diffusion length** of electrons.

The general solution to this differential equation is

$$n(x) - n_p = Ae^{x/L_e} + Be^{-x/L_e}, \tag{12.74}$$

where A and B are constants to be determined by the boundary conditions. The latter can be visualized with the aid of Fig. 12.17. At x_p, the boundary of the space charge region on the p-type side, the value of n is given by Eq. (12.66): $n(x_p) = n'_p$. As the minority electrons diffuse away from the space charge region into the neutral p-type region, they recombine with majority holes and their concentration decreases toward its equilibrium value n_p. We assume that $n(x)$ reaches n_p at $x = x'_c$ where a metallic contact accelerates the recombination process. The boundary conditions are therefore

$$n(x) = n'_p \quad \text{at} \quad x = x_p \tag{12.75a}$$

$$= n_p \quad \text{at} \quad x = x'_c \tag{12.75b}$$

and the solution for $n(x) - n_p$ is

$$n(x) - n_p = \frac{(n'_p - n_p)\sinh[(x - x'_c)/L_e]}{\sinh(d_p/L_e)}, \tag{12.76}$$

where $d_p = x_p - x'_c$ is the width of the p-type neutral region.

A similar analysis can be made for the concentration of minority holes in the n-type neutral region. The result is

$$p(x) - p_n = \frac{(p'_n - p_n)\sinh[(x_c - x)/L_h]}{\sinh(d_n/L_h)}, \tag{12.77}$$

where $d_n = x_c - x_n$, $L_h = (D_h\tau_p)^{\frac{1}{2}}$, and τ_p is the minority hole lifetime.

Alternative expressions for the minority carrier concentrations can be obtained by eliminating n'_p and p'_n using Eqs. (12.66) and (12.67):

$$n(x) - n_p = \frac{n_p}{\sinh(d_p/L_e)} \times (e^{eV_a/k_BT} - 1)\sinh[(x - x'_c)/L_e] \tag{12.78a}$$

$$p(x) - p_n = \frac{p_n}{\sinh(d_n/L_h)} \times (e^{eV_a/k_BT} - 1)\sinh[(x_c - x)/L_h]. \tag{12.78b}$$

If the diffusion lengths L_e and L_h are much smaller than the extents of the neutral regions d_p and d_n, respectively, and if the point x is not too far away from the space charge region, the minority carrier distributions take the simple forms

$$n(x) - n_p \simeq n_p(e^{eV_a/k_BT} - 1)e^{(x-x_p)/L_e} \tag{12.79a}$$

$$p(x) - p_n \simeq p_n(e^{eV_a/k_BT} - 1)e^{(x_n-x)/L_h}, \tag{12.79b}$$

which exhibit an exponential decrease of concentration with increasing distance from the space charge region. On the other hand, if $L_e \gg d_p$ and $L_h \gg d_n$, the hyperbolic sines can be replaced by their arguments to give

$$n(x) - n_p \simeq n_p(e^{eV_a/k_BT} - 1)(x - x'_c)/d_p \tag{12.80a}$$

$$p(x) - p_n \simeq p_n(e^{eV_a/k_BT} - 1)(x_c - x)/d_n \tag{12.80b}$$

The various results for the minority carrier concentrations in the neutral regions of a p–n junction are represented diagrammatically in Fig. 12.18.

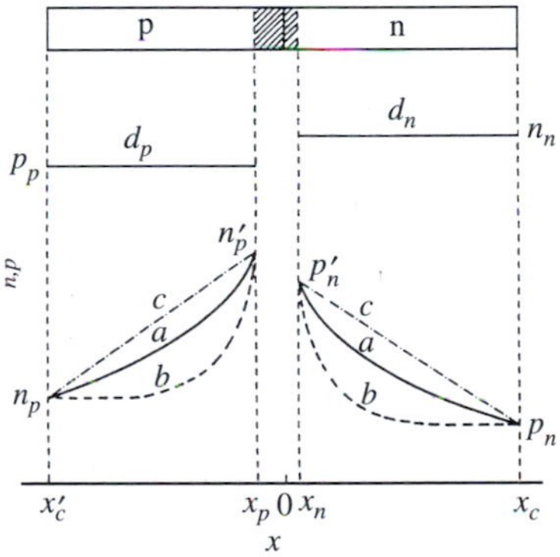

Fig. 12.18
Concentration profiles of minority electrons and minority holes in the p- and n-type regions, respectively: (a) arbitrary values of d_n, d_p; (b) $d_p \gg L_e$, $d_n \gg L_h$; (c) $d_p \ll L_e$, $d_n \ll L_h$ (after Mathieu 1987).

12.2.5.3 Minority carrier currents in the neutral regions

In calculating the current through the junction we continue to make the assumption that the electric field in the neutral regions is zero and that the current is due to minority carrier diffusion:

$$j_e(x) = eD_e\frac{dn(x)}{dx} \tag{12.81a}$$

$$j_h(x) = -eD_h\frac{dp(x)}{dx}. \tag{12.81b}$$

Substituting the equilibrium values $n_p = n_i^2/n_a$ and $p_n = n_i^2/n_d$ into Eqs. (12.78)–(12.80) and carrying out the differentiation with respect to x, we obtain the following results for the current densities:

1. *arbitrary values of d_p/L_e and d_n/L_h*

$$j_e(x) = \frac{en_i^2 D_e}{n_a L_e \sinh(d_p/L_e)} \times \left(e^{eV_a/k_B T} - 1\right) \cosh[(x - x_c')/L_e] \quad (12.82a)$$

$$j_h(x) = \frac{en_i^2 D_h}{n_d L_h \sinh(d_n/L_h)} \times \left(e^{eV_a/k_B T} - 1\right) \cosh[(x_c - x)/L_h] \quad (12.82b)$$

2. $d_p \gg L_e$, $d_n \gg L_h$

$$j_e(x) = \frac{en_i^2 D_e}{n_a L_e} \left(e^{eV_a/k_B T} - 1\right) e^{-(x_p - x)/L_e} \quad (12.83a)$$

$$j_h(x) = \frac{en_i^2 D_h}{n_d L_h} \left(e^{eV_a/k_B T} - 1\right) e^{-(x - x_n)/L_h} \quad (12.83b)$$

3. $d_p \ll L_e$, $d_n \ll L_h$

$$j_e = \frac{en_i^2 D_e}{n_a d_p} \left(e^{eV_a/k_B T} - 1\right) \quad (12.84a)$$

$$j_h = \frac{en_i^2 D_h}{n_d d_n} \left(e^{eV_a/k_B T} - 1\right). \quad (12.84b)$$

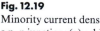

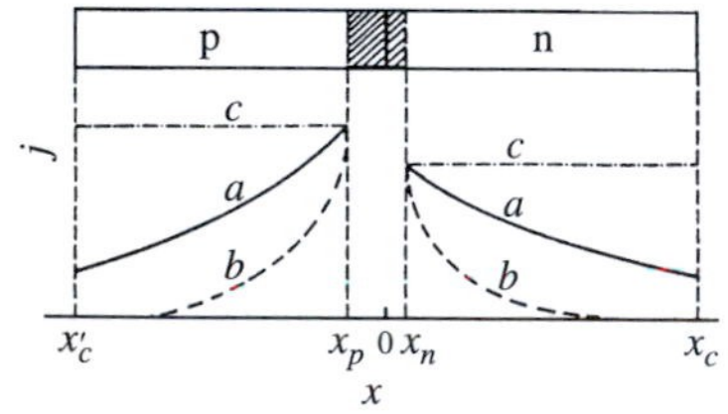

Fig. 12.19
Minority current density profiles for a p–n junction: (a) arbitrary values of d_n, d_p; (b) $d_p \gg L_e$, $d_n \gg L_h$; (c) $d_p \ll L_e$, $d_n \ll L_h$ (after Mathieu 1987).

The minority currents for these various cases are shown graphically in Fig.12.19.

12.2.5.4 Current-voltage characteristic of a p–n junction

The calculation of the total current across a p–n junction can be simplified by assuming that the transit time for carriers across the space charge region is so short that negligible recombination occurs within that region. The currents of electrons and of holes must each be constant throughout the space charge region in the steady state. One can therefore evaluate the currents at a particular point that is convenient computationally. The results apply at every point of the space charge region, because each of the currents is independent of x.

Two points that are particularly convenient are the boundary points x_n and x_p between the space charge region and the neutral regions. Our assumption that the electric field vanishes in the neutral regions requires that the drift currents associated with both majority and minority carriers at $x = x_n, x_p$ be zero. Furthermore, since we assume that the majority carrier concentrations are uniform in the neutral regions, the diffusion currents of majority carriers are zero at x_n, x_p. The only currents we need consider are the minority carrier diffusion currents. They satisfy the relations

$$j_e(x_n) = j_e(x_p) \tag{12.85a}$$

$$j_h(x_n) = j_h(x_p). \tag{12.85b}$$

Let us focus on the point x_n where the total current density is

$$j = j_e(x_n) + j_h(x_n). \tag{12.86}$$

The hole current density $j_h(x_n)$ can be obtained directly from Eq. (12.82b) or from the approximate expressions in Eqs. (12.83b) and (12.84b). The electron current density j_e is conveniently evaluated at x_p using Eq. (12.82a) or the approximate expressions in Eqs. (12.83a) and (12.84a). Equation (12.85a) is then used to obtain $j_e(x_n)$. With these results the total current density takes the form of the **Shockley diode equation**

$$j = j_s\left(e^{eV_a/k_BT} - 1\right), \tag{12.87}$$

where j_s is the **saturation current density**. The expressions for j_s corresponding to the various relative values of d_p, L_e, d_n and L_h are:

1. *arbitrary values of* d_p/L_e *and* d_n/L_h:

$$j_s = \frac{en_i^2 D_e}{n_a L_e \tanh(d_p/L_e)} + \frac{en_i^2 D_h}{n_d L_h \tanh(d_n/L_h)} \tag{12.88a}$$

2. $d_p \gg L_e$, $d_n \gg L_h$:

$$j_s = en_i^2\left(\frac{D_e}{n_a L_e} + \frac{D_h}{n_d L_h}\right) \tag{12.88b}$$

3. $d_p \ll L_e$, $d_n \ll L_h$:

$$j_s = en_i^2\left(\frac{D_e}{n_a d_p} + \frac{D_h}{n_d d_n}\right). \tag{12.88c}$$

The behavior of the total current as specified by Shockley's diode equation contrasts strongly between forward and reverse bias conditions:

1. *Forward bias,* $V_a > 0$. The importance of the potential barrier decreases under forward bias, and the current increases exponentially with V_a when $V_a \gg k_BT/e$:

$$j \simeq j_s e^{eV_a/k_BT}. \tag{12.89}$$

   The resistance of the junction itself becomes negligible compared to the series resistance of the n- and p-type regions.
2. *Reverse bias,* $V_a < 0$. When the magnitude of V_a under reverse bias reaches a few k_BT/e, the exponential term can be neglected compared to unity, and we have

$$j \simeq -j_s. \tag{12.90}$$

The current density has the saturation value and is independent of V_a. The saturation current is in fact the sum of the generation currents of electrons and holes that were discussed in Section 12.2.4 and is also referred to as the generation current. The variation of the total current as a function of the applied voltage is shown in Fig. 12.16.

Our calculation of the current is based on minority carriers and gives the electron and hole currents in the space charge region. These currents together with their sum, the total current, are constant over this region. Unlike the minority carrier currents, however, the total current is constant over the entire device including the neutral regions. The majority carrier currents in each of the neutral regions can therefore be obtained by subtracting the minority carrier current from the total current. We focus on the case $d_p \gg L_e$, $d_n \gg L_h$ which involves a simple result and yet is physically realistic. The minority carrier current density in the n-type region is given by Eq. (12.83b), which can be rewritten as

$$j_h(x) = j_h(x_n)e^{-(x-x_n)/L_h}. \tag{12.91}$$

Subtracting this result from the total current density specified by Eq. (12.86) yields the majority electron current density in the n-type region

$$\begin{aligned} j_e(x) &= j_e(x_n) + j_h(x_n) - j_h(x_n)e^{-(x-x_n)/L_h} \\ &= j_e(x_n) + j_h(x_n)[1 - e^{-(x-x_n)/L_h}], \end{aligned} \tag{12.92}$$

where from Eqs. (12.83)

$$j_e(x_n) = j_e(x_p) = \frac{en_i^2 D_e}{n_a L_e}\left(e^{eV_a/k_BT} - 1\right) \tag{12.93a}$$

$$j_h(x_n) = \frac{en_i^2 D_h}{n_d L_h}\left(e^{eV_a/k_BT} - 1\right). \tag{12.93b}$$

A similar analysis can be carried out for the majority hole current density in the p-type region. It can be written as

$$\begin{aligned} j_h(x) &= j_h(x_n) + j_e(x_n)[1 - e^{-(x_p-x)/L_e}] \\ &= j_h(x_p) + j_e(x_p)[1 - e^{-(x_p-x)/L_e}] \end{aligned} \tag{12.94}$$

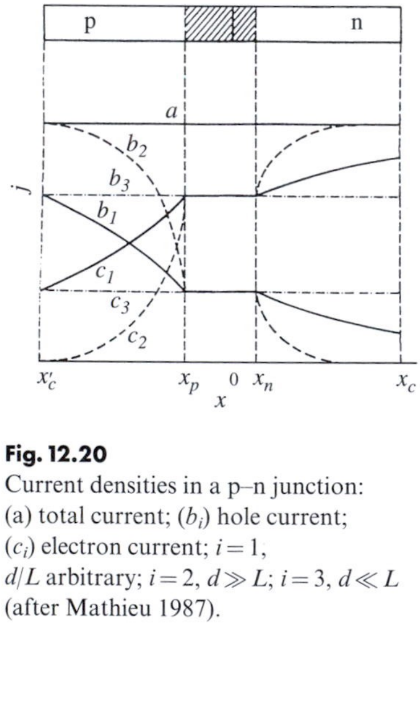

Fig. 12.20
Current densities in a p–n junction: (a) total current; (b_i) hole current; (c_i) electron current; $i=1$, d/L arbitrary; $i=2$, $d \gg L$; $i=3$, $d \ll L$ (after Mathieu 1987).

In Fig. 12.20 are shown the total current density $j = j_e(x) + j_h(x)$ through the entire junction and its electron and hole components $j_e(x)$ and $j_h(x)$, respectively, as functions of position x. It is assumed that there is no generation or recombination of electrons and holes in the space charge region. At large distances from this region the current is composed entirely of majority carriers: electrons on the n-type side and holes on the p-type side. Both majority carrier currents are drift currents produced by the small electric field that exists in the neutral regions, but which has been neglected in the derivation of the currents. One can regard the majority carrier currents as a result of the injection of minority carriers. In order to maintain electroneutrality majority carriers must flow from the metallic electrodes toward the space charge region from which the minority carriers are injected. Upon reaching the boundary of the neutral region the majority carriers enter the space charge region, pass across it, and are injected into the neutral region on the opposite side, thereby becoming minority carriers.

12.2.5.5 Summarizing remarks on the abrupt p–n junction

The difference in Fermi energies between the n- and p-type regions leads to the creation of a space charge region and a difference in electrostatic potential between the two regions. The energy barrier thus produced can be modulated by an applied voltage, thereby causing the currents through the junction to be modulated. The exponential increase in current under increasing forward bias voltage and the saturation current arising from increasing reverse bias voltage lead to a nonlinear current–voltage characteristic. This nonlinear characteristic is responsible for the useful rectifying properties of a p–n junction.

12.3 Graded p–n junction

In the discussion developed in Sections 12.1 and 12.2 the impurity distribution was considered to be uniform on each side of the junction. We now consider the more general case in which the impurity distribution near the junction is not uniform, and the change from p-type to n-type is not abrupt at $x=0$ but occurs progressively. In order to present a clear description and a simple calculation of the relevant parameters, we shall treat the particular situation where the n- and p-type neutral regions have widths that are small compared to the diffusion lengths of the free carriers in these regions: $d_p \ll L_e$, $d_n \ll L_h$.

We have seen in Section 12.2.5.2 that if $d_p \ll L_e$ and $d_n \ll L_h$, the minority carrier distributions vary linearly with position x as indicated in Eqs. (12.80) and the minority current densities are independent of x as indicated in Eqs. (12.84). Let us now calculate the current voltage characteristic and the saturation current j_s of the junction.

We focus first on the n-type region where the donor impurity concentration is $n_d(x)$. The injection rate of holes is assumed to be low, so that the majority electron concentration satisfies $n(x) \simeq n_d(x)$ and the minority hole concentration satisfies $p(x) \ll n(x)$. The current densities in this region can be written as

$$j_h(x) = eD_h\left[p(x)\frac{e}{k_BT}\mathcal{E}(x) - \frac{dp(x)}{dx}\right] \tag{12.95a}$$

$$j_e(x) = eD_e\left[n(x)\frac{e}{k_BT}\mathcal{E}(x) + \frac{dn(x)}{dx}\right] \tag{12.95b}$$

Replacing $n(x)$ by the donor impurity distribution $n_d(x)$ and eliminating $\mathcal{E}(x)$ from the two equations, we get

$$\frac{j_h(x)}{eD_h} - \frac{p(x)}{n_d(x)}\frac{j_e(x)}{eD_e} = -\frac{dp(x)}{dx} - \frac{p(x)}{n_d(x)}\frac{dn_d(x)}{dx}. \tag{12.96}$$

In the n-type region the minority carrier concentration $p(x)$ is much lower than $n_d(x)$. Therefore we can neglect the second term on the left-hand side and write

$$\frac{dp(x)}{dx}+\frac{d\log n_d(x)}{dx}p(x)+\frac{j_h(x)}{eD_h}=0. \tag{12.97}$$

If we consider the limit of an abrupt junction with homogeneous n- and p-type regions, $n_d(x)$ is a constant, its logarithmic derivative vanishes, and we recover Eq. (12.81b). The hole current in the n-type region is a purely diffusive current. In the case of a graded junction we note that the term involving the gradient of $n_d(x)$ in Eq. (12.97) is proportional to $p(x)$ and may be interpreted as contributing a drift current associated with the electric field resulting from the nonuniform distribution of the doping ions $n_d(x)$.

The solution of Eq. (12.97) is facilitated by the fact that when the widths of the neutral regions are small compared to the respective diffusion lengths, the minority carrier current densities are constant. Consequently, $j_h(x)$ is independent of x: $j_h(x)=j_h$. Equation (12.97) is a first-order, linear differential equation that can be written in the standard form

$$y'(x)+a(x)y(x)=b, \tag{12.98}$$

where $y(x)=p(x)$, $b=-j_h/eD_h$, and

$$a(x)=\frac{d\log n_d(x)}{dx}. \tag{12.99}$$

The solution is

$$y(x)=e^{-\int a(x)dx}\left[C+b\int e^{\int a(x)dx}dx\right], \tag{12.100}$$

where C is the constant of integration. The integral of $a(x)$ is trivial and yields

$$e^{\int a(x)dx}=n_d(x). \tag{12.101}$$

The minority hole concentration can then be expressed as

$$p(x)=\frac{1}{n_d(x)}\left[C-\frac{j_h}{eD_h}\int_{x_n}^{x}n_d(x')dx'\right]. \tag{12.102}$$

The constant of integration C is determined by the boundary condition at $x=x_n$. We have

$$p(x_n)=p_n'=\frac{C}{n_d(x_n)}, \tag{12.103}$$

so

$$C=n_d(x_n)p_n' \tag{12.104}$$

and

$$p(x) = \frac{1}{n_d(x)}\left[n_d(x_n)p_n' - \frac{j_h}{eD_h}\int_{x_n}^{x} n_d(x')dx'\right]. \tag{12.105}$$

An expression for the minority hole current density j_h can be obtained by evaluating this equation at $x = x_c$ where the n-type region terminates at an ohmic contact. Letting $p(x_c) = p_c$ and solving for j_h, we obtain

$$j_h = \frac{eD_h[p_n' n_c(x_n) - p_c n_d(x_d)]}{\int_{x_n}^{x_c} n_d(x)dx}. \tag{12.106}$$

Using the expression for the injected hole concentration p_n' given by Eq. (12.67) and assuming as before that the hole concentration at the ohmic contact has its equilibrium value $n_i^2/n_d(x_c)$, we can write j_h as

$$j_h = \frac{en_i^2 D_h}{\int_{x_n}^{x_c} n_d(x)dx}\left(e^{eV_a/k_BT} - 1\right). \tag{12.107}$$

Introducing the average value of $n_d(x)$ over the interval $x_n \leq x \leq x_c$ by

$$\bar{n}_d = \frac{1}{d_n}\int_{x_n}^{x_c} n_d(x)\,dx$$

we get

$$j_h = \frac{en_i^2 D_h}{\bar{n}_d d_n}(e^{eV_a/k_BT} - 1). \tag{12.108}$$

The same considerations apply to the minority electron current in the p-type region. Analogous calculations yield the following result for the electron current density j_e:

$$j_e = \frac{en_i^2 D_e}{\bar{n}_a d_p}(e^{eV_a/k_BT} - 1). \tag{12.109}$$

The quantity $\bar{n}_a$ is the average value of the acceptor impurity concentration in the p-type region defined by

$$\bar{n}_a = \frac{1}{d_p}\int_{x_c'}^{x_p} n_a(x)dx. \tag{12.110}$$

It is evident that the current densities j_h and j_e specified by Eqs. (12.107) and (12.109) are independent of x within their respective neutral regions. By continuity the hole and electron current densities in the space charge region must be equal to these values of j_h and j_e at the boundaries of the space charge region with the respective neutral regions. In the absence of generation and recombination these currents are constant over the whole space charge region, and their sum is the total current density j in the junction. Hence, under the conditions $d_n \ll L_h$ and $d_p \ll L_e$,

$$j = j_s(e^{eV_a/k_BT} - 1), \tag{12.111}$$

where the saturation current density j_s is given by

$$j_s = en_i^2\left(\frac{D_h}{\bar{n}_d d_n} + \frac{D_e}{\bar{n}_a d_p}\right). \tag{12.112}$$

This result reduces to Eq. (12.88c) in the limit of uniform doping of the neutral regions.

Example 12.3: Saturation current for an arbitrary junction with nonuniform doping of the neutral regions
Develop an expression for j_s for a nonuniformly doped junction with arbitrary values of d_n/L_h and d_p/L_e.
Solution. Consider the denominator of Eq. (12.107) for j_h. Using the relation $x_c = x_n + d_n$ it becomes

$$\int_{x_n}^{x_n+d_n} n_d(x)d_x.$$

For uniform dopage, the denominator reduces to $n_d d_n$. We note, however, that with arbitrary d_n/L_h and uniform dopage, the denominator is $n_d L_h \tanh(d_n/L_h)$ from Eq. (12.88a). This suggests that we replace the denominator in Eq. (12.107) for nonuniform dopage by

$$\int_{x_n}^{x_n+L_h\tanh(d_n/L_h)} n_d(x)dx.$$

Similar considerations for the p-type region lead us to replace the denominator in Eq. (12.109) by

$$\int_{x_p-L_e\tanh(d_p/L_e)}^{x_p} n_a(x)dx.$$

The saturation current for an arbitrary junction with nonuniform doping is then

$$j_s = en_i^2\left[\frac{D_h}{\int_{x_n}^{x_n+L_h\tanh(d_n/L_h)} n_d(x)dx} + \frac{D_e}{\int_{x_p-L_e\tanh(d_p/L_e)}^{x_p} n_a(x)dx}\right]. \tag{12.113}$$

In the limits $L_h \gg d_n$ and $L_e \gg d_p$, Eq. (12.113) reduces to Eq. (12.112).

12.4 P–N junction capacitance

During a change in voltage across a p–n junction the current is not equal solely to that due to the instantaneous voltage, but has an additional component which is proportional to the time rate of change of voltage. The second component is a capacitive current and corresponds to a capacitance of the p–n junction.

A p–n junction has two basic types of capacitance. The first type is the **differential storage capacitance** or **diffusion capacitance** which arises from a rearrangement of the minority carrier distribution due to diffusion into the bulk semiconductor as the voltage is changed. It is dominant under forward bias conditions. The second type of junction capacitance is the **transition capacitance** or **depletion-layer capacitance.** It is due to the variation of space charge with voltage in the depletion region and is dominant under reverse bias conditions.

12.4.1 Storage capacitance

We need to calculate the variation of the number of minority carriers in the neutral regions with applied voltage. For electrons the concentration is given by Eq. (12.79a) for the case of diffusion length small compared to extent x_p of the neutral p-side. Integrating this equation from $-\infty$ to x_p and multiplying by the electron charge e yields the stored electron charge per unit area.:

$$Q_e = en_pL_e\left(e^{eV_a/k_BT} - 1\right). \tag{12.114}$$

By differentiating this result with respect to V_a we obtain the differential storage capacitance per unit area for electrons in the form

$$C_e = \frac{e^2 n_p L_e}{k_BT} e^{eV_a/k_BT}. \tag{12.115}$$

Starting from Eq. (12.79b) a similar treatment can be carried out for holes on the neutral n-side to give the differential storage capacitance per unit area for holes:

$$C_h = \frac{e^2 p_n L_h}{k_BT} e^{eV_a/k_BT}. \tag{12.116}$$

The total storage capacitance per unit area is the sum of C_e and C_h:

$$C_{stor} = \frac{e^2}{k_BT}\left(n_pL_e + p_nL_h\right)e^{eV_a/k_BT}. \tag{12.117}$$

Clearly, the storage capacitance increases exponentially with increasing forward bias V_a. It is a differential capacitance, because it is a small change in charge due to a small change in voltage and not simply the total charge divided by the voltage. The alternative term diffusion capacitance indicates the importance of diffusion in the process producing the capacitance.

12.4.2 Transition capacitance

Under reverse bias the storage capacitance decreases rapidly toward zero and is replaced as the dominant capacitance by the transition capacitance. The later is a result of the flow of majority carriers that changes the widths W_n and W_p of the n- and p-type parts of the space charge region in response to a change in the applied reverse bias potential V_a.

We start by combining Eqs. (12.39) and (12.56) to give

$$V_d + V_a = \frac{e}{2\epsilon_0\epsilon}(n_d W_n^2 + n_a W_p^2). \tag{12.118}$$

A relatively simple result can be obtained if we consider a very asymmetric junction with, for example, $n_a \gg n_d$. Making use of the condition of electroneutrality, Eq. (12.36), we can rewrite Eq. (12.118) as

$$V_d + V_a = \frac{en_d W_n^2}{2\epsilon_0\epsilon}\left(1 + \frac{n_d}{n_a}\right) \simeq \frac{en_d W_n^2}{2\epsilon_0\epsilon}. \tag{12.119}$$

Solving for W_n yields

$$W_n = \left[\frac{2\epsilon_0\epsilon}{en_d}(V_d + V_a)\right]^{\frac{1}{2}}. \tag{12.120}$$

If the applied bias is varied, a variation in W_n arises according to

$$dW_n = \frac{1}{2}\left(\frac{2\epsilon_0\epsilon}{en_d}\right)^{\frac{1}{2}}(V_d + V_a)^{-\frac{1}{2}}dV_a. \tag{12.121}$$

The corresponding variation in space charge per unit area dQ is given by

$$dQ = en_d dW_n. \tag{12.122}$$

Eliminating dW_n from Eqs. (12.121) and (12.122) results in

$$dQ = \left(\frac{\epsilon_0\epsilon e n_d}{2}\right)^{\frac{1}{2}}(V_d + V_a)^{-\frac{1}{2}}dV_a \tag{12.123}$$

from which follows the transition capacitance per unit area

$$C_t = \left[\frac{\epsilon_0\epsilon e n_d}{2(V_d + V_a)}\right]^{\frac{1}{2}}. \tag{12.124}$$

Under the condition $V_a \gg V_d$, the transition capacitance varies as $V_a^{-\frac{1}{2}}$.

12.4.3 Applications of p–n junction capacitance

The dependence of junction capacitance on bias voltage has been exploited in devices such as the **varactor** which is an abbreviation for variable reactor. For example, the bulky variable plate capacitor in the tuning circuit of a radio receiver can be replaced by a much smaller varactor which has the added benefit of greater dependability. Varactors are also employed in parametric amplifiers and harmonic generation.

12.5 Avalanche breakdown and Zener breakdown

12.5.1 Avalanche breakdown

In the discussion of high electric field effects on carrier transport in Chapter 8 it was pointed out that at sufficiently high fields impact ionization and carrier multiplication take place resulting in avalanche breakdown. Since very high electric fields can arise in the depletion region of a p–n junction, avalanche breakdown is an important phenomenon to be considered. It typically occurs under reverse bias.

The parameters that characterize avalanche breakdown are the **ionization rates** α_e and α_h defined by

$$\alpha_e = \frac{1}{n}\frac{dn}{dx} \tag{12.125a}$$

$$\alpha_h = \frac{1}{p}\frac{dp}{dx}, \tag{12.125b}$$

which are the relative increases in electron and hole concentrations per unit length, respectively. We assume that the avalanche process is initiated by holes with a current I_{h0} incident on the depletion region from the p-type side. The **multiplication factor** of holes M_h is defined by

$$M_h = \frac{I_h(W)}{I_{h0}} \tag{12.126}$$

and is related to α_h and α_e by (Sze 1981)

$$1 - \frac{1}{M_h} = \int_0^W \alpha_h \exp\left[-\int_0^x (\alpha_h - \alpha_e)dx'\right]dx. \tag{12.127}$$

Breakdown occurs at a voltage where M_h becomes infinite, i.e.,

$$\int_0^W \alpha_h \exp\left[-\int_0^x (\alpha_h - \alpha_e)dx'\right]dx = 1. \tag{12.128}$$

For semiconductors such as GaP and GaAs which have equal ionization rates, $\alpha_h = \alpha_e = \alpha$, the criterion for breakdown reduces to

$$\int_0^W \alpha dx = 1. \tag{12.129}$$

The voltage V_B required to initiate breakdown can be derived by solving Poisson's equation with appropriate boundary conditions. For an abrupt junction with lightly doped and heavily doped sides, the result is

$$V_B = \frac{\epsilon_0 \epsilon_s \mathcal{E}_m^2}{2en_B}, \tag{12.130}$$

where $\mathcal{E}_m$ is the maximum electric field in the depletion region and n_B is the ionized background impurity concentration on the lightly doped side.

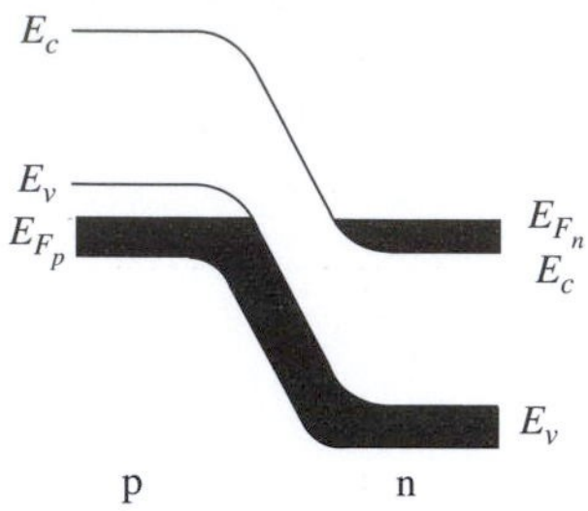

Fig. 12.21
Energy band diagram for a p–n junction with degenerate p- and n-regions (after Streetman 1990).

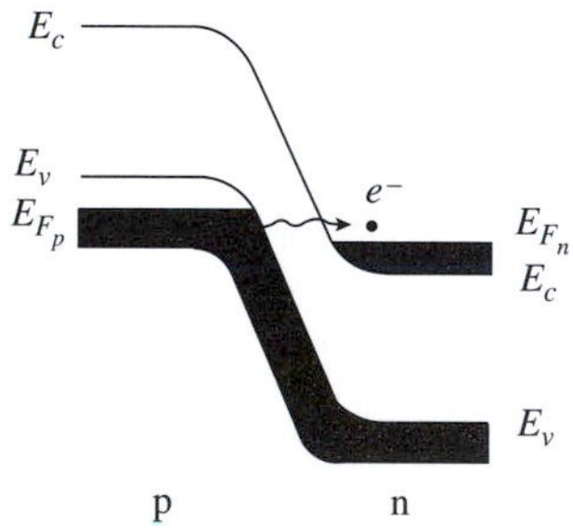

Fig. 12.22
Energy band diagram for a heavily doped p–n junction under reverse bias (after Streetman 1990).

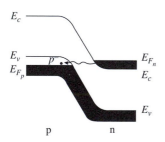

Fig. 12.23
Energy band diagram for a heavily doped p–n junction under forward bias (after Streetman 1990).

12.5.2 Zener breakdown

If the electric field in the depletion region exceeds a critical value $\mathcal{E}_{cr}$, the conduction and valence band edges are tilted to the extent that the two bands have the same energy at spatially separated points in the junction as shown in Fig. 12.21. Under these conditions a carrier can tunnel across the forbidden gap between the two bands. In Si junctions the critical field is $\sim 10^6$ V/cm and is typically achieved only if the doping level is high on the order of 10^{18} cm^{-3}. The physical separation of the valence and conduction band edges at the same energy is ~ 100 Å. At thermal equilibrium the Fermi energy is the same throughout the junction as indicated in Fig. 12.21, but no net current flows because the tunneling currents in each direction balance each other.

Let us now consider the effect of a reverse bias. The energy level diagram takes the form illustrated in Fig. 12.22. Electrons can tunnel from the valence band on the p-type side to empty states in the conduction band on the n-type side resulting in a net current that increases sharply with increasing bias voltage. This behavior constitutes **Zener breakdown**. On the other hand, a forward bias initially produces the situation shown in Fig. 12.23. Electrons in the conduction band on the n-type side can tunnel into empty states in the valence band on the p-type side, and a sharp rise in forward current appears that corresponds to Zener breakdown. However, as the bias voltage increases further, the filled conduction band states on the n-type side become adjacent to states in the forbidden gap on the p-type side, and the tunnel current drops to very low values. Eventually the bias voltage reaches values where the ordinary p–n junction current takes over. The behavior of the current for these various regimes of bias voltage is presented in Fig. 12.24. In typical Si junctions the first breakdown process to appear with increasing reverse bias voltage is Zener breakdown. Above 4 V avalanche breakdown starts to appear and becomes dominant above 8 V.

Problems

1. Calculate the width of the space charge region in an ideal abrupt silicon junction at 300 K with 10^{16} cm^{-3} ionized impurities on both the n- and p-type sides. Take the electron, light hole, and heavy hole masses from Table 4.1, the indirect gap from Table 3.4, and the dielectric constant $\epsilon = 12$.
2. For the situation in problem 1, calculate the hole diffusion current density in the space charge region taking the hole mobility from Table 8.1.
3. Calculate the diffusion potential of an ideal abrupt silicon junction for acceptor concentrations in the range 10^{15} cm$^{-3} \leq n_a \leq 10^{17}$ cm^{-3} for $T = 300$ K and $n_d = 10^{16}$ cm^{-3}. Plot V_d versus n_a.
4. Calculate the saturation current density for an ideal abrupt silicon junction at 300 K for $n_a = n_d = 10^{16}$ cm^{-3} and widths of the neutral regions very large compared to the diffusion lengths. Take the lifetimes of both carriers to be 0.1 μs. The required diffusion constants can be obtained from the mobilities using Eqs. (8.37) and (8.38). See Grove (1967) for experimental values.
5. For the situation in problem 4, calculate the current density under a forward bias of 0.2 V.

6. Consider an ideal abrupt Si p–n junction with $n_a = 10^{17}\,\text{cm}^{-3}$ on one side and $n_d = 10^{16}\,\text{cm}^{-3}$ on the other side. Calculate the quasi-Fermi energies at 300 K in the n- and p-type regions. Use these results to calculate the diffusion potential and compare with the value obtained in problem 3.
7. Derive an expression for the transition capacitance that involves only ϵ_0, ϵ, and W_n if $n_a \gg n_d$. Compare the result with that for a simple parallel plate capacitor.
8. For the case of avalanche breakdown in an abrupt Si junction with $n_a \gg n_d$, derive an expression that relates the breakdown voltage V_B to the width W of the space charge region. If $n_d = 10^{15}\,\text{cm}^{-3}$ and the breakdown voltage is 300 V, what is W?

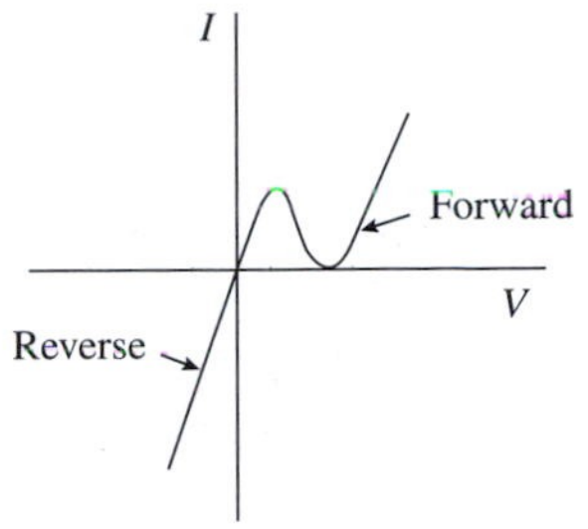

Fig. 12.24
Current versus bias voltage for a heavily doped p–n junction.

References

R. Dalven, *Introduction to Applied Solid State Physics*, Second edition (Plenum Press, New York, 1990).

D. A. Fraser, *The Physics of Semiconductor Devices*, Fourth edition (Oxford University Press, Oxford, 1986).

A. S. Grove, *Physics and Technology of Semiconductoir Devices* (John Wiley, New York, 1967).

H. Mathieu, *Physique des Semiconducteurs et des Composants Electroniques* (Masson, Paris, 1987).

J. L. Moll, *Physics of Semiconductors* (McGraw-Hill, New York, 1964).

W. Shockley, *Electrons and Holes in Semiconductors* (Van Nostrand, New York, 1950).

B.G. Streetman, *Solid State Electronic Devices*, Third edition (Prentice Hall, Englewood Cliffs, 1990).

S. M. Sze, *Physics of Semiconductor Devices*, Second edition (John Wiley, New York, 1981).

Bipolar junction transistor

13

Key ideas

The *transistor* is a three-terminal semiconductor device that can amplify an electrical signal. In the *bipolar junction transistor* (BJT) a thin layer of p-type semiconductor, the *base*, is sandwiched between two layers of an n-type semiconductor, the *emitter* and the *collector*. This is an n–p–n BJT. The n- and p-type materials can be interchanged to give a p–n–p BJT. Typically, the emitter has a higher dopant concentration than the collector, e.g., the p^+–n–p BJT.

Forward bias at the p^+–n junction and reverse bias at the n–p junction constitute the *forward active bias mode* or *amplification mode*. Amplification is possible because a small fluctuation in the majority carrier current can cause a large fluctuation in the minority carrier current.

Modulation of the base current by an impressed signal leads to a modulation of the collector current. In an operation in the amplification mode the modulation of the collector current can exceed the modulation of the base current, thus giving rise to amplification or gain. The base-to-collector *current gain* is the ratio of the collector current modulation to the base current modulation. The emitter-to-collector *current amplification factor* is the ratio of the collector current modulation to the emitter current modulation. The *emitter efficiency* is the ratio of the hole current from the emitter to the total emitter current. The *base transport factor* is the ratio of the hole current reaching the collector to the hole current injected by the emitter. Long minority hole lifetime favors high gain. Indirect gap semiconductors are preferred for the fabrication of BJTs.

The signal is small if the output current or voltage response is a linear function of the input signal voltage.

Transistors

In Chapter 12 the p–n junction and the metal–semiconductor junction were discussed. These devices are two-terminal devices which are useful as rectifiers, but cannot be used as amplifiers because they contain no mechanism for modulating the electric current in such a way as to produce gain. What is needed is a third terminal to provide the necessary modulation.

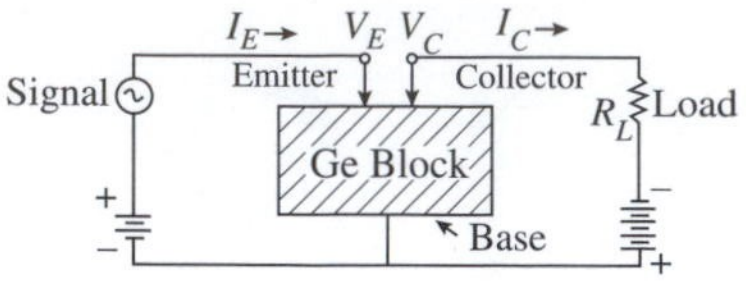

Fig. 13.1
Diagram of the point contact transistor (after Bardeen and Brattain 1948).

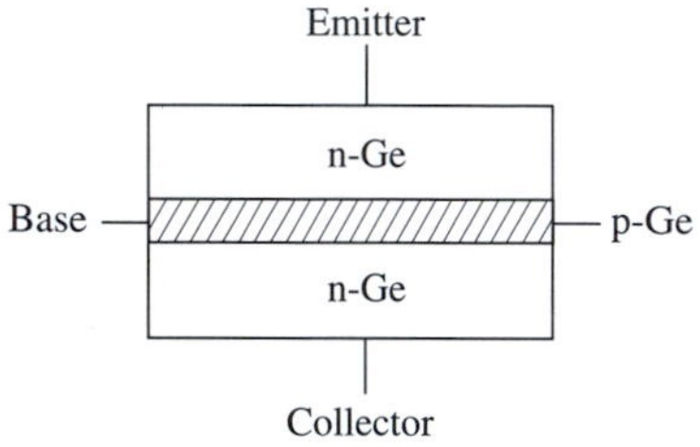

Fig. 13.2
Diagram of a bipolar junction transistor based on Ge.

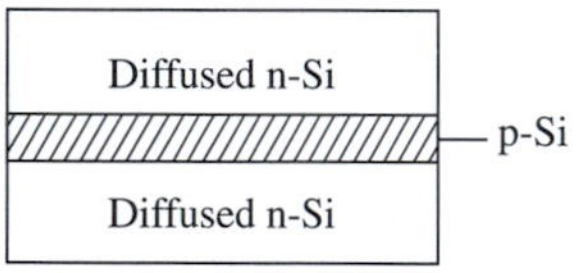

Fig. 13.3
Diagram of a double-diffused silicon transistor.

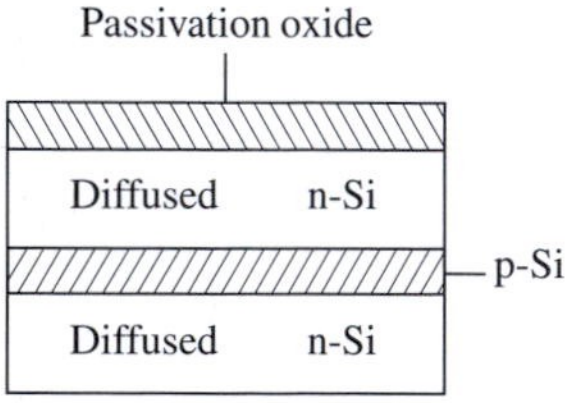

Fig. 13.4
Diagram of a double-diffused oxide passivated silicon planar transistor.

13.1 Fabrication of transistors

The **transistor** is a three-terminal semiconductor device that can amplify an electrical signal. The first transistor to be manufactured successfully was developed by Bardeen and Brattain in 1947. It consisted of two sharp metal electrodes making contact with an n-type Ge single crystal at two closely spaced points on the surface as shown in Fig. 13.1. One point contact was forward biased and the other was reverse biased. Negative ions present on the surface repel the electrons in the n-type Ge and create a p-type surface inversion layer. The forward-biased contact (emitter) injects holes into the inversion layer. The holes drift along the inversion layer until they enter the space charge layer of the reverse-biased contact (collector). The high electric field of the space charge layer accelerates the holes which then create numerous electron–hole pairs by interband impact excitation. The result is that an increasing collector voltage produces a onsaturating collector current.

The next step in the evolution of the transistor was the placement of the metal electrodes on opposite faces of a thin Ge disk by Shive (1948) to give the **coaxial transistor**. Soon thereafter, Shockley (1949) proposed the **bipolar junction transistor** (BJT) in which a thin layer of p-Ge is sandwiched between two layers of n-Ge. Electrodes are attached to each of the three layers as indicated in Fig. 13.2. One electrode is attached to an n-type layer (the **emitter**), the second is attached to the other n-type layer (the **collector**), and the third is attached to the p-type layer (the **base**).

The technology of fabricating BJTs advanced in several stages. In the **alloy junction bipolar transistor** two metal alloy dots containing an acceptor impurity such as indium are alloyed into the two faces of a thin disk of n-Ge. The In is diffused into the disk, creating thereby two p–n junctions. The metal alloy dots form the emitter and collector electrodes. The **grown junction bipolar transistor** is fabricated by changing the dopant impurity type during crystal growth from n-type to p-type to n-type.

In none of the BJTs discussed so far is the base layer thin enough to permit high-frequency amplification. Two advances contributed to overcoming this problem. The first was the development of solid state diffusion technology in which a semiconductor surface is exposed at high temperature to a gaseous impurity which diffuses into the semiconductor. The second was the switch from Ge to Si as the basic material, the latter with its higher melting point and larger energy band gaps allowing the use of higher temperatures than the former. In Si very tight control of the thickness of diffused layers can be achieved, making possible the production of base layers with thickness less than one micron. These advances culminated in the **double-diffused silicon transistor** depicted in Fig. 13.3. However, the performance of this device was adversely affected by interaction of the exposed junction surfaces with air. This problem was solved by the **surface passivation technique** which protects the junction surfaces with a thermally grown oxide layer that is shown in Fig. 13.4. At the same time the more efficient planar stripe geometry was adopted resulting in the **double-diffused oxide passivated silicon planar transistor**. In the modern fabrication process some of the impurity introduction has been accomplished using ion implantation rather than diffusion.

Silicon BJTs can operate in the microwave region with switching speeds on the order of nanoseconds. Placing many of them coupled together on a single Si chip creates an **integrated circuit**.

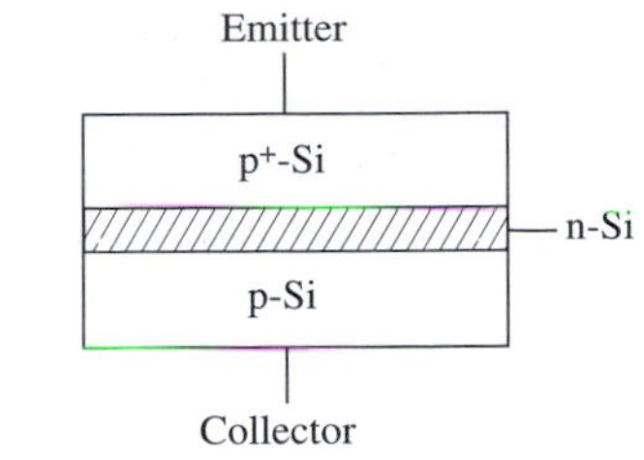

Fig. 13.5
Diagram of a p^+–n–p BJT.

13.2 Physical basis of the BJT

The BJT can be fabricated in either the n–p–n or the p–n–p configuration. In a typical operating transistor the configuration is not symmetric in the sense that the dopant concentrations of the emitter and collector are the same. Instead, the emitter has the higher concentration as indicated in Fig. 13.5 for the p^+–n–p BJT. For this case the concentrations n_a and n_d of ionized acceptors and donors in the emitter and base, respectively, satisfy the inequality

$$n_a \gg n_d. \tag{13.1}$$

In the absence of applied bias, the law of mass action (Eqs. (12.3) and (12.4)) requires that

$$n_p = \frac{n_i^2}{p_p} \simeq \frac{n_i^2}{n_a} \tag{13.2a}$$

$$p_e = \frac{n_i^2}{n_n} \simeq \frac{n_i^2}{n_d}, \tag{13.2b}$$

where n_p and p_n are the concentrations of minority electrons in the emitter and minority holes in the base, respectively.

We therefore conclude from Eqs. (13.1) and (13.2) that

$$p_n \gg n_p. \tag{13.3}$$

In other words, the concentration of minority holes in the base is much larger than that of minority electrons in the emitter. From Eq. (12.88a) the saturation current density, when the neutral regions are much wider than the respective diffusion lengths, is given by

$$j_s = \frac{eD_e n_p}{L_e \tanh(d_p/L_e)} + \frac{eD_h p_n}{L_h \tanh(d_n/L_h)} \simeq \frac{eD_h p_n}{L_h \tanh(d_n/L_h)}. \tag{13.4}$$

Under applied bias V_E between emitter and base, the density of the current injected from the emitter into the base is (cf. Eq. (12.87))

$$j \simeq \frac{eD_h p_n}{L_h \tanh(d_n/L_h)} (e^{eV_E/k_BT} - 1), \tag{13.5}$$

thus establishing that most of the injection current is carried by holes.

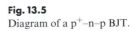

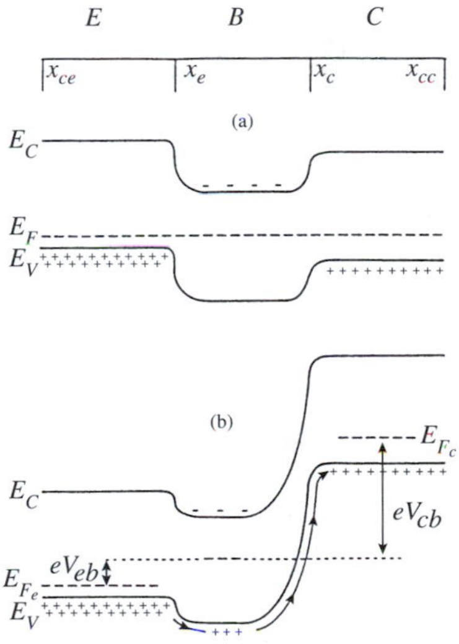

Fig. 13.6
Energy band diagram for a p^+–n–p BJT: (a) zero bias; (b) nonzero bias (after Mathieu 1987).

The energy band diagram of the p^+–n–p BJT with zero bias is shown in Fig. 13.6a. When the p^+–n emitter junction is forward biased and the n–p collector junction is reverse biased, the energy band diagram is modified as

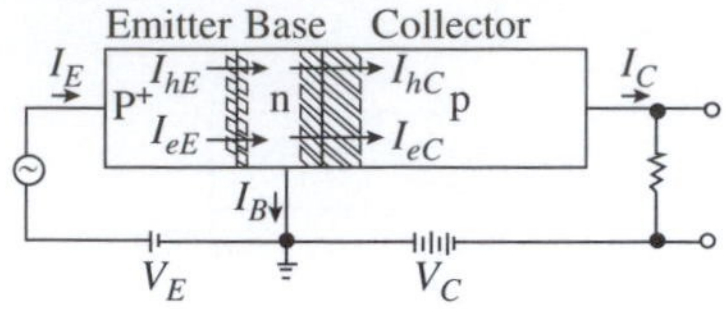

Fig. 13.7
Diagram of individual current flows in a p^+–n–p BJT (after Sze 1981).

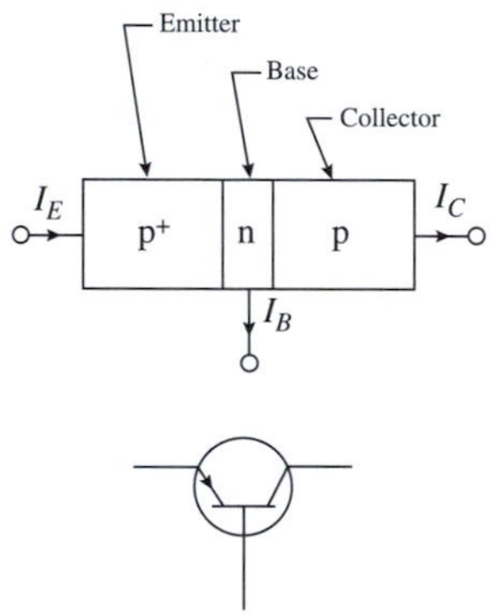

Fig. 13.8
Diagram of overall current flows in a p^+–n–p BJT (after Sze 1981).

in Fig. 13.6b. The energy barrier for passage of holes from the emitter into the base has been lowered, leading to the hole current I_{hE} across the p^+–n junction depicted in Fig. 13.7. A certain fraction of the injected holes diffuse across the base and reach the n–p junction. At the n–p junction a strong electric field exists that sweeps the holes down the potential hill and into the collector, producing hole current I_{hC}. However, some of the holes injected into the base recombine with majority electrons. The electrons thus lost are replenished by electron current I_{eB} that flows into the base from the external current. A similar replenishment takes care of electrons that flow from the base into the emitter, constituting electron current I_{eE}. Finally, minority electrons in the collector pass across the n–p junction into the base, giving rise to the electron current I_{eC}. In connection with Fig. 13.7 it must be emphasized that a hole current is in the same direction as hole flow, whereas an electron current is in the opposite direction to electron flow.

A diagram showing the overall current flows in a p^+–n–p BJT is given in Fig. 13.8. The arrows indicate the directions of flow under so-called normal operating conditions, i.e., forward bias at the p^+–n junction and reverse bias at the n–p junction. These conditions constitute the **forward active bias mode** or **amplification mode**. Also shown in Fig. 13.8 is the conventional symbol for this case. The p^+–n–p BJT illustrated in Figs. 13.7 and 13.8 has the **common-base configuration** with emitter and collector bias voltages corresponding to the amplification mode. Diagrams for the n^+–p–n BJT can be obtained by reversing the polarities of the bias voltages and the directions of the currents.

The basic principles of operation of a BJT can be summarized as follows. Minority carriers are injected into the base from the emitter by the application of a forward bias across the emitter base junction. An impressed low-input signal voltage modulates the flow of majority carriers into the base from the external circuit. Electroneutrality requires that the minority carrier current in the base, and thence the collector current, be modulated also. The fact that a small fluctuation of the majority carrier current can cause a large fluctuation in the minority carrier current leads to the possibility of amplification. To understand in detail how these effects occur, it is necessary to derive the DC current–voltage characteristics of the BJT.

13.3 DC characteristics of the BJT

13.3.1 Injected minority carrier concentrations

We consider a p^+–n–p BJT with abrupt junctions in the common base configuration as shown in Fig. 13.7. The BJT is assumed to have a uniform cross sectional area A, so the calculation of the currents reduces to a one-dimensional problem of determining the current densities. The neutral regions of the emitter, base, and collector are assumed to have zero electric field, with all of the potential drops occurring across the depletion regions of the junctions. The motion of the carriers in these regions is therefore diffusive, and no drift currents are involved.

Our basic procedure is to calculate the carrier concentrations as functions of position and obtain the currents from the concentration gradients.

The steady-state motion of minority holes in the neutral region of the base is governed by the hole analogue of the diffusion equation given by Eq. (12.73),

$$\frac{\partial^2 p}{\partial x^2} - \frac{p - p_0}{L_h^2} = 0, \tag{13.6}$$

where p_0 and p are the equilibrium and nonequilibrium concentrations of holes, $L_h = (D_h \tau_p)^{\frac{1}{2}}$, D_h is the hole diffusion constant, and τ_p is the minority hole lifetime. The general solution of this differential equation is the hole analogue of Eq. (12.74):

$$\Delta p(x) = p(x) - p_0 = Ae^{x/L_h} + Be^{-x/L_h}. \tag{13.7}$$

The constants A and B are determined by the boundary conditions at the edges of the neutral region of the base, $x = 0$ and $x = W_B$, as shown in Fig. 13.8. At $x = 0$, the hole concentration is that of minority holes in a forward-biased p–n junction given by Eq. (12.67). Converting to our present notation, we have

$$p(0) = p_0 e^{eV_E/k_B T} \tag{13.8}$$

or

$$\Delta p(0) = p(0) - p_0 = p_0(e^{eV_E/k_B T} - 1) = A + B. \tag{13.9}$$

At $x = W_B$, the hole concentration is that of minority holes in a reverse biased n–p junction

$$p(W_B) = p_0 e^{eV_C/k_B T}, \tag{13.10}$$

where V_C is negative for reverse bias. Subtracting p_0 from both sides of Eq. (13.10) and using Eq. (13.7), we obtain

$$\begin{aligned} \Delta p(W_B) &= p(W_B) - p_0 = p_0(e^{eV_C/k_B T} - 1) \\ &= Ae^{W_B/L_h} + Be^{-W_B/L_h}. \end{aligned} \tag{13.11}$$

Equations (13.9) and (13.11) can be solved for A and B to yield

$$A = \frac{\Delta p(W_B) - \Delta p(0)e^{-W_B/L_h}}{2\sinh(W_B/L_h)} \tag{13.12a}$$

$$B = -\frac{\Delta p(W_B) - \Delta p(0)e^{W_B/L_h}}{2\sinh(W_B/L_h)}, \tag{13.12b}$$

from which, with the aid of Eq. (13.7), we obtain $\Delta p(x)$ in the form

$$\Delta p(x) = \frac{\Delta p(W_B)\sinh(x/L_h) + \Delta p(0)\sinh[(W_B - x)/L_h]}{\sinh(W_B/L_h)}. \tag{13.13}$$

The minority electron concentrations in the neutral regions of the emitter and collector can be derived in similar fashion, starting from the diffusion equation for electrons, Eq. (12.73), and the general solution, Eq. (12.74). As in our discussion of the p–n junction in Chapter 12, we assume that the metallic contacts to the emitter and collector accelerate the recombination of the minority electrons, so that $n(x_E) = n_{0E}$ and $n(x_C) = n_{0C}$. At the boundary between the neutral region of the emitter and the space charge region of the p$^+$–n junction, the following condition must be satisfied:

$$\Delta n(x_{EB}) = n(x_{EB}) - n_{0E} = n_{0E}(e^{eV_E/k_BT} - 1). \tag{13.14}$$

The corresponding boundary condition for the collector is

$$\Delta n(x_{CB}) = n(x_{CB}) - n_{0C} = n_{0C}(e^{eV_C/k_BT} - 1). \tag{13.15}$$

The solutions to the diffusion equation subject to these boundary conditions can be taken directly from that in Eq. (12.76) by making the appropriate changes in the notation for the coordinates of the various boundaries. For the emitter the result is

$$\Delta n_E(x) = n_{0E}(e^{eV_E/k_BT} - 1)\frac{\sinh[(x - x_E)/L_E]}{\sinh(W_E/L_E)}, \tag{13.16}$$

where W_E is the width of the neutral region of the emitter and L_E is the diffusion length for minority electrons in the emitter. For the collector,

$$\Delta n_C(x) = n_{0C}(e^{eV_C/k_BT} - 1)\frac{\sinh[(x_C - x)/L_C]}{\sinh(W_C/L_C)}, \tag{13.17}$$

where W_C and L_C are the corresponding quantities for the collector. If the diffusion lengths are small compared to the widths of the neutral regions, approximate expressions analogous to that in Eq. (12.79a) can be employed.

13.3.2 Currents in the BJT

In our simple model of a BJT, the electric field is zero in the neutral regions, there is no drift of carriers, and the motion of the carriers is entirely diffusive. The current density due to carrier motion is therefore proportional to the concentration gradient of the carrier as given by Eqs. (12.81a) and (12.81b) for minority electrons and holes, respectively. The current density due to holes in the base is

$$j_h(x) = -eD_h\frac{d\Delta p(x)}{dx}. \tag{13.18}$$

Making use of the expression for $\Delta p(x)$ in Eq. (13.13), we obtain

$$j_h(x) = -\frac{eD_h}{L_h}\left\{\frac{\Delta p(W_B)\cosh(x/L_h) - \Delta p(0)\cosh[(W_B - x)/L_h]}{\sinh(W_B/L_h)}\right\}. \tag{13.19}$$

The current density due to electrons in the emitter or the collector is

$$j_e(x) = eD_e \frac{d\Delta n(x)}{dx}. \tag{13.20}$$

Substitution of Eqs. (13.16) and (13.17) into Eq. (13.20) yields the emitter and collector current densities $j_{eE}(x)$ and $j_{eC}(x)$,

$$j_{eE}(x) = \frac{eD_E \Delta n(x_E)}{L_E} \frac{\cosh[(x - x_E)/L_E]}{\sinh(W_E/L_E)} \tag{13.21a}$$

$$j_{eC}(x) = -\frac{eD_C \Delta n(x_C)}{L_C} \frac{\cosh[(x_C - x)/L_C]}{\sinh(W_C/L_C)}, \tag{13.21b}$$

where D_E and D_C are the electron diffusion constants in the emitter and collector, respectively.

We are now in position to evaluate the total emitter and collector current densities. The total emitter current density j_E is given by

$$j_E = j_h(0) + j_{eE}(x_{EB}), \tag{13.22}$$

if generation–recombination currents in the emitter–base space charge layer are neglected. From Eqs. (13.19) and (13.21a),

$$\begin{aligned} j_E = {} & \frac{eD_h}{L_h}\left\{\Delta p(0)\coth\left(\frac{W_B}{L_h}\right) - \Delta p(W_B)\operatorname{csch}\left(\frac{W_B}{L_h}\right)\right\} \\ & + \frac{eD_E}{L_E}\Delta n(x_E)\coth\left(\frac{W_E}{L_E}\right). \end{aligned} \tag{13.23}$$

Similarly, Eqs. (13.19) and (13.21b) yield the total collector current density j_C:

$$\begin{aligned} j_C &= j_h(W_B) + j_{eC}(x_{CB}) \\ &= \frac{eD_h}{L_h}\left\{\Delta p(0)\operatorname{csch}\left(\frac{W_B}{L_h}\right) - \Delta p(W_B)\coth\left(\frac{W_B}{L_h}\right)\right\} \\ &\quad - \frac{eD_C}{L_C}\Delta n(x_C)\coth\left(\frac{W_C}{L_C}\right). \end{aligned} \tag{13.24}$$

If the widths of the emitter and collector are large compared to the respective diffusion lengths, the approximation $\coth(W_E/L_E) \simeq \coth(W_C/L_C) \simeq 1$ can be made. However, since a significant fraction of minority holes must be able to diffuse across the base, satisfaction of the inequality $W_B \gg L_h$ would be detrimental. In fact, $L_h \gg W_B$ is desirable.

Conservation of charge requires that the base current density j_B be given by

$$j_B = j_E - j_C. \tag{13.25}$$

Making use of identities involving hyperbolic functions together with Eqs. (13.23) and (13.24), we write the base current as

$$j_B = \frac{eD_h}{L_h}[\Delta p(0) + \Delta p(W_B)] \tanh\left(\frac{W_B}{2L_h}\right) + \frac{eD_E}{L_E}\Delta n(x_{EB}) \times \coth\left(\frac{W_E}{L_E}\right) + \frac{eD_C}{L_C}\Delta n(x_{CB}) \coth\left(\frac{W_C}{L_C}\right). \tag{13.26}$$

The explicit dependence of j_E, j_C, and j_B on V_E and V_C is revealed by replacing $\Delta p(0)$, $\Delta p(W_B)$, $\Delta n(x_{EB})$, and $\Delta n(x_{CB})$ by their expressions contained in Eqs. (13.9), (13.11), (13.14), and (13.15):

$$j_E = \frac{eD_h p_0}{L_h}\left\{(e^{eV_E/k_BT} - 1)\coth\left(\frac{W_B}{L_h}\right) - (e^{eV_C/k_BT} - 1)\operatorname{csch}\left(\frac{W_B}{L_h}\right)\right\} + \frac{eD_E n_{0E}}{L_E}(e^{eV_E/k_BT} - 1)\coth\left(\frac{W_E}{L_E}\right) \tag{13.27a}$$

$$j_C = \frac{eD_h p_0}{L_h}\left\{(e^{eV_E/k_BT} - 1)\operatorname{csch}\left(\frac{W_B}{L_h}\right) - (e^{eV_C/k_BT} - 1)\coth\left(\frac{W_B}{L_h}\right)\right\} - \frac{eD_C n_{0C}}{L_E}(e^{eV_C/k_BT} - 1)\coth\left(\frac{W_C}{L_C}\right) \tag{13.27b}$$

$$j_B = \frac{eD_h p_0}{L_h}\left[(e^{eV_E/k_BT} + e^{eV_C/k_BT} - 2)\tanh\left(\frac{W_B}{2L_h}\right)\right] + \frac{eD_E n_{0E}}{L_E}(e^{eV_E/k_BT} - 1)\coth\left(\frac{W_E}{L_E}\right) + \frac{eD_C n_{0C}}{L_C}(e^{eV_C/k_BT} - 1)\coth\left(\frac{W_C}{L_C}\right). \tag{13.27c}$$

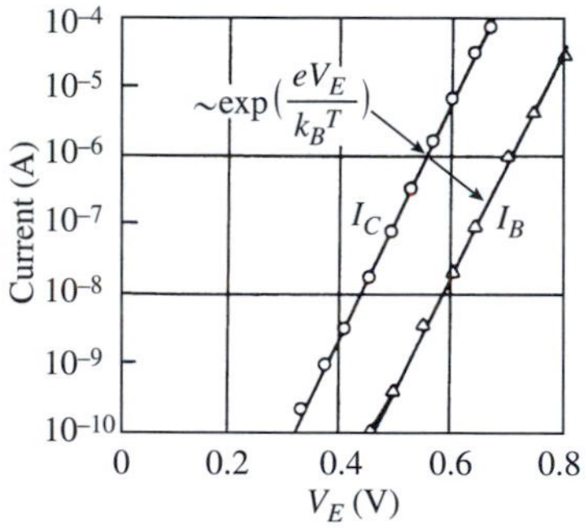

Fig. 13.9
Collector and base currents versus emitter voltage for theory (solid lines) and experiment (circles and triangles) with $V_C = 0$ (after Jespers 1977).

Equations (13.27) constitute the **Shockley bipolar junction transistor equations** (SBJT equations). They give a good account of the DC current–voltage behavior of the BJT under high forward current operating conditions, as seen in Fig. 13.9. However, Sah *et al.* (1957) have shown that under low forward or reverse current operating conditions, it is necessary to include the contributions of generation–recombination currents in the space charge layers, which are neglected in the SBJT equations. The generation–recombination current densities, j_{EB} and j_{CB}, in the emitter–base and collector–base space charge layers are given by (Sah 1991)

$$j_{EB} = (en_i W_{EB}/\tau_{EB})(e^{eV_E/2k_BT} - 1) \tag{13.28a}$$

and

$$j_{CB} = (en_i W_{CB}/\tau_{CB})(e^{eV_C/2k_BT} - 1), \tag{13.28b}$$

where n_i is the intrinsic carrier concentration, W_{EB} and W_{CB} are the widths, and τ_{EB} and τ_{CB} are the effective lifetimes of carriers in the respective space charge layers.

Example 13.1: Diffusion and generation–recombination currents
Demonstrate that under large forward bias the generation–recombination current is small compared to the diffusion current.
Solution. Under normal operating conditions with forward bias at the p^+–n junction, the diffusion and generation recombination currents are dominated by the exponential terms $\exp(eV_E/k_BT)$ and $\exp(eV_E/2k_BT)$, respectively. The term of 1/2 in the exponent of the second term makes the latter much smaller than the first term at large V_E. Consequently, the generation–recombination current is small compared to the diffusion current under large forward bias.

13.3.3 Current gain in the BJT

We have noted earlier that the base current I_B influences the collector current I_C. If the BJT is operated in the amplification mode, the output current I_C can exceed the input current I_B, thus giving rise to amplification or gain. The base-to-collector **current gain** β is defined by

$$\beta = \frac{I_C}{I_B} = \frac{j_C}{j_B} \tag{13.29}$$

where the second form applies if the cross sectional area of the BJT is uniform. The emitter-to-collector **current amplification factor** α is defined by

$$\alpha = \frac{I_C}{I_E} = \frac{j_C}{j_E}. \tag{13.30}$$

From the relation $j_B = j_E - j_C$ follows the result

$$\frac{1}{\beta} = \frac{1}{\alpha} - 1$$

or

$$\beta = \frac{\alpha}{1-\alpha}. \tag{13.31}$$

The parameter α is always less than unity, but in a well-designed BJT it is very close to unity, e.g., ~ 0.99. The value of β is therefore high, on the order of 100.

To proceed further, we make some approximations that are reasonable for a typical BJT in the amplification mode. Since the base–collector junction is reverse biased with $V_C \ll 0$, we neglect the current contributions in Eqs. (13.27) that are associated with V_C:

$$j_E \simeq \left[\frac{eD_h p_0}{L_h}\coth\left(\frac{W_B}{L_h}\right) + \frac{eD_E n_{0E}}{L_E}\coth\left(\frac{W_E}{L_E}\right)\right]\left(e^{eV_E/k_BT} - 1\right) \tag{13.32a}$$

$$j_C \simeq \left[\frac{eD_h p_0}{L_h}\operatorname{csch}\left(\frac{W_B}{L_h}\right)\right]\left(e^{eV_E/k_BT} - 1\right). \tag{13.32b}$$

The **emitter injection efficiency** γ is defined as the ratio of the hole current from the emitter $I_{hE}(0)$ to the total emitter current I_E:

$$\gamma = \frac{I_{hE}(0)}{I_E} = \frac{j_{hE}(0)}{j_E}. \tag{13.33}$$

The current density $j_{hE}(0)$ corresponds to the first term in Eq. (13.32a). We find that

$$\gamma = \frac{1}{1 + (D_E n_{0E} L_h / D_h p_0 L_E)\tanh(W_B/L_h)\coth(W_E/L_E)}. \tag{13.34}$$

Another parameter of interest is the **base transport factor** B defined as the ratio of the hole current reaching the collector $I_{hE}(W_B)$ to the hole current injected by the emitter $I_{hE}(0)$. The current $I_{hE}(W_B)$ is $Aj_{hE}(W_B) = Aj_C$ with j_C given by Eq. (13.32b). Hence,

$$B = \frac{I_{hE}(W_B)}{I_{hE}(0)} = \frac{j_C}{j_{hE}(0)} = \operatorname{sech}\left(\frac{W_B}{L_h}\right). \tag{13.35}$$

Physically, B is the fraction of minority holes that avoid recombination while diffusing across the base from emitter to collector. Like the current amplification factor α, the parameters γ and B are always less than unity. For the desirable situation $L_h \gg W_B$, γ and B can be written in the approximate forms

$$\gamma \simeq 1 - \frac{D_E n_{0E} W_B}{D_h p_0 L_E} \tag{13.36a}$$

$$B \simeq 1 - \frac{W_B^2}{2L_h^2}. \tag{13.36b}$$

We see that B is only slightly less than unity. If $L_h = 10W_B$, $B \simeq 0.995$.

An expression for the current amplification factor α can be obtained with our simple model by using Eqs. (13.32):

$$\begin{aligned}\alpha &= \frac{I_C}{I_E} \\ &= \frac{1}{\cosh(W_B/L_h) + (D_E n_{0E} L_h / D_h p_0 L_E)\sinh(W_B/L_h)\coth(W_E/L_E)}.\end{aligned} \tag{13.37}$$

Use of Eqs. (13.34) and (13.35) yields the relation

$$\alpha = \gamma B < 1. \tag{13.38}$$

Substituting this result into Eq. (13.31) gives the current gain in the form

$$\beta = \frac{\gamma B}{1 - \gamma B}. \tag{13.39}$$

With the aid of Eqs. (13.34) and (13.37), an explicit expression for β can be obtained,

$$\beta = \frac{1}{\sinh(W_B/L_h)[\tanh(W_B/2L_h) + (D_E n_{0E} L_h/D_h p_0 L_E)\coth(W_E/L_E)]}, \tag{13.40}$$

where the relation $\cosh x - 1 = \sinh x \tanh(x/2)$ has been used. In a typical BJT the emitter width is much greater than the emitter diffusion length, so one can set $\coth(W_E/L_E) = 1$.

The law of mass action stated in Eqs. (13.2) enables one to eliminate the minority carrier concentrations, noting that the intrinsic carrier concentration n_i depends only on the temperature and the energy band parameters of the transistor material and is therefore uniform throughout the device. Hence, $n_{0E}/p_0 = n_{dB}/n_{aE}$ and

$$\beta = \frac{1}{\sinh(W_B/L_h)[\tanh(W_B/2L_h) + (D_E n_{dB} L_h/D_h n_{aE} L_E)]}. \tag{13.41}$$

This result shows us how to choose the various parameters to maximize the gain. It is clearly favorable to minimize the ratio W_B/L_h by minimizing the width W_B and maximizing the diffusion length L_h of the neutral region of the base. From the relation

$$L_h^2 = D_h \tau_p, \tag{13.42}$$

we see that L_h can be increased by increasing both D_h and τ_p. This can be achieved by minimizing the base doping n_{dB}. Light doping serves to increase the mobility of the carriers in the base, thus increasing the diffusion constant D_h, and to increase the recombination time τ_p by lowering the concentration of recombination centers.

It is also evident from Eq. (13.41) that heavily doping the emitter to make n_{aE} large will enhance the gain. The combination of light doping of the base and heavy doping of the emitter causes the space charge region at the p^+–n junction to extend fairly far into the base, thereby reducing W_B. Note also that the mass action law enables Eq. (13.34) to be rewritten as (with $\coth(W_E/L_E) = 1$)

$$\gamma = \frac{1}{1 + (D_E n_{dB} L_h/D_h n_{aE} L_E)\tanh(W_B/L_h)}, \tag{13.43}$$

so reducing n_{dB}/n_{aE} brings the emitter efficiency γ closer to unity. The inverse relationship between doping level and diffusion constant suggests that the ratio D_E/D_h should be small, thereby causing γ to be even closer to unity and β to be even higher.

We have seen that a long minority hole lifetime τ_p favors high gain. For this reason indirect gap semiconductors are preferred for the fabrication of BJTs. We also note that increasing τ_p, and thereby L_h, brings the base transport factor B closer to unity, in accordance with Eq. (13.35). A larger fraction of the injected holes are able to cross the base and reach the collector.

13.4 Small-signal characteristics of the BJT

The BJT is frequently used as an amplifier under small-signal conditions in either wide band or tuned circuits. A signal consists of a time-dependent voltage applied to a pair of terminals. The signal is small if the output current or voltage response is a linear function of the input signal voltage. The currents and voltages of the input and output can be Fourier analyzed in time. Under the small-signal condition, the output reproduces a given Fourier component of the input with only a phase delay and without distortion of the amplitude or generation of harmonics.

The major nonlinearity arises from the exponential dependence on voltage arising from the Boltzmann factors in the SBJT equations. Since in the amplification mode, the collector–base junction is reverse-biased, the term $\exp(eV_C/k_BT)$ is negligible and can be omitted. The term $\exp(eV_E/k_BT)$ associated with the forward-biased emitter–base junction is large, however. It is handled by writing V_E as the sum of a DC part $V_E^{(0)}$ and a time-dependent fluctuation $\delta V_E(t)$, expanding in powers of $\delta V_E(t)$,

$$e^{eV_E/k_BT} \simeq e^{eV_E^{(0)}/k_BT}\left(1 + \frac{e\delta V_E(t)}{k_BT} + \cdots\right), \tag{13.44}$$

and neglecting higher powers of $\delta V_E(t)$. Each of the various current densities and carrier concentrations is written as the DC value plus a time-dependent fluctuation. Use of these expressions together with the time-dependent diffusion equation given by Eq. (12.72) gives rise to a set of time-dependent equations that are linear in the fluctuations.

To illustrate the procedure, we set

$$p = p^{(0)} + \delta p(t) \tag{13.45}$$

and substitute into the time-dependent diffusion equation:

$$\frac{\partial \delta p(t)}{\partial t} = D_h \frac{\partial^2 [p^{(0)} + \delta p(t)]}{\partial x^2} - \frac{p^{(0)} + \delta p(t) - p_0}{\tau_p}. \tag{13.46}$$

For a Fourier component varying as $\exp(i\omega t)$,

$$D_h \frac{\partial^2 \delta p(t)}{\partial x^2} - \left(\frac{1}{\tau_p} + i\omega\right)\delta p(t) = 0$$

or

$$D_h \frac{\partial^2 \delta p(t)}{\partial x^2} - \frac{\delta p(t)}{\bar{\tau}_p} = 0, \tag{13.47}$$

where $\bar{\tau}_p = \tau_p/(1 + i\omega\tau_p)$ is an effective recombination time. Setting $\bar{L}_h = (D_h\bar{\tau}_p)^{\frac{1}{2}}$ leads to the equation

$$\frac{\partial^2 \delta p(t)}{\partial x^2} - \frac{\delta p(t)}{\bar{L}_h^2} = 0 \tag{13.48}$$

which has the same form as that satisfied by $\Delta p(x) = p(x) - p_0$ for the DC case:

$$\frac{\partial^2 \Delta p}{\partial x^2} - \frac{\Delta p}{L_h^2} = 0. \tag{13.49}$$

The solutions for the latter can therefore be carried over to the time-dependent case by making the replacement $L_h \to \bar{L}_h$.

One can define small-signal parameters analogous to those for the DC case. For example, the small-signal current gain β_s is given by

$$\beta_s = \frac{\delta I_C}{\delta I_B}. \tag{13.50}$$

Starting from the definition of β in Eq. (13.29), the following relation between β_s and β can be derived:

$$\beta_s = \frac{\beta}{1 - (I_C/\beta)(d\beta/dI_C)}. \tag{13.51}$$

If the current gain β is independent of the collector current I_C, then $\beta_s = \beta$. When recombination in the emitter–base space charge region is taken into account, β does depend on I_C as shown in Fig. 13.10.

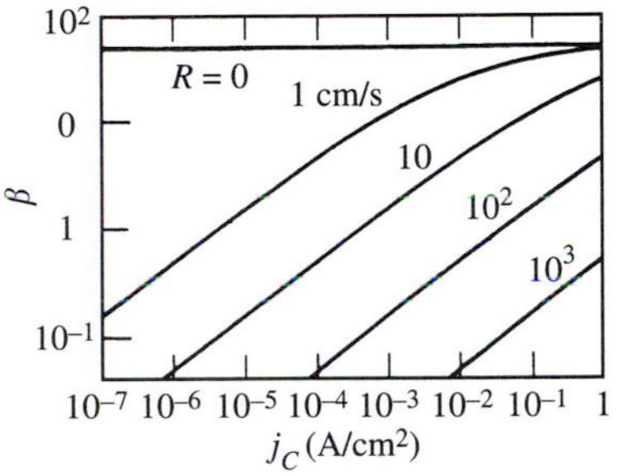

Fig. 13.10
Current gain β versus collector current density j_C for various space charge region recombination rates R for an Si transistor (after Grove 1967).

Problems

1. Starting from Eqs. (13.41) and (13.42) develop an approximate expression for the current gain β that is valid if $W_B \ll L_h$ and $n_{dB} \ll n_{aE}$. The result can be written as $\beta = \tau_p/\tau_{tr}$, where τ_{tr} is the **transit time** required for holes to diffuse from the emitter to the collector. Write down an expression for τ_{tr} and calculate its value for a p$^+$–n–p silicon transistor at 300 K with $W_B = 10^{-4}$ cm.
2. Under the influence of an electric field the transit time of a minority carrier through the bias region due to the field only is the **drift transit time**. What potential difference across the base region would be necessary so that the drift transit time is equal to the diffusion transit time?
3. For a p$^+$–n–p$^+$ silicon transistor with $n_a = 10^{17}$ cm^{-3}, $n_d = 10^{15}$ cm^{-3}, $\tau_n = 0.1\ \mu s$, $\tau_p = 10\ \mu s$, $\mu_e = 700$ cm^2/(V s), $\mu_h = 440$ cm^2/(V s), $W_B = 10^{-4}$ cm, and $W_E \gg L_E$, calculate:
   (a) the diffusion lengths L_h and L_E
   (b) the emitter injection efficiency γ.
4. Calculate the current amplification factor α for the transistor of problem 3. How much of the deviation of α from unity arises from incomplete injection and how much from recombination losses?
5. For a p$^+$–n–p$^+$ bipolar transistor derive an expression for the change of the neutral base width due to the collector–base voltage V_C if V_C is much larger than the diffusion voltage V_d of the collector–base junction.
6. A p$^+$–n–p$^+$ silicon transistor has base and collector doping concentrations of 1×10^{17} cm^{-3} and 1×10^{18} cm^{-3}, respectively. Given that the base width is 1×10^{-4} cm for $V_C = 0$, estimate the value of V_C which causes the base width to change by 10%.
7. For the transistor of problem 6, at 300 K calculate the capacitance per unit area of the base–collector junction with $V_C = 0$.

References

J. Bardeen and W. H. Brattain, *Phys. Rev.* **74**, 230 (1948).

R. Dalven, *Introduction to Applied Solid State Physics*, Second edition (Plenum Press, New York, 1990).

A. S. Grove, *Physics and Technology of Semiconductor Devices* (John Wiley, New York, 1967).

P. G. A. Jespers, in *Process and Device Modeling for Integrated Circuit Design*, eds. F. Van de Wiele, W. L. Engl, and P. G. Jespers (Noordhoff, Leyden, 1977).

H. Mathieu, *Physique des Semiconducteurs et des Composants Electroniques* (Masson, Paris, 1987).

C. T. Sah, *Fundamentals of Solid-State Electronics* (World Scientific, Singapore, 1991).

C. T. Sah, R. N. Noyce, and W. Shockley, *Proc. IRE* **45**, 1228 (1957).

J. N. Shive, *Phys. Rev.* **75**, 689 (1949).

W. Shockley, *Bell System Tech. J.* **28**, 453 (1949).

S. M. Sze, *Physics of Semiconductor Devices*, Second edition (John Wiley, New York, 1981).

Semiconductor lasers and photodevices

14

Key ideas

*L*ight *A*mplification by the *S*timulated *E*mission of *R*adiation = LASER.

A system in an excited state is *stimulated* by the radiation field to emit photons. The rate of stimulated emission is proportional to the energy density of the radiation. A larger population of excited states than ground states means *population inversion*. When population inversion exists, stimulated emission can produce an increase in energy density or *optical gain*. If the optical gain exceeds the optical loss due to absorption, *amplification* results. Stimulated emission is a *coherent* process and leads to *spectral narrowing*. The coherence is due to the emitted radiation having its phase related to that of the stimulating field.

A p–n junction is a convenient system for establishing population inversion of energy levels. *Direct gap* materials are most appropriate for semiconductor lasers.

Optical gain can occur if the emitted photon energy $h\nu$ lies between the band gap energy E_g and the difference of the *quasi-Fermi energies* ΔE_F. Optical gain is equivalent to a negative absorption coefficient. Gain means enhancement whereas absorption means attenuation of a light beam.

At current densities above the *transparency current density*, the optical gain is positive.

The gain increases linearly with nominal current density at a rate that is inversely proportional to the temperature.

The threshold current density for lasing decreases with increasing separation of the mirrors of the cavity.

Radiative recombinaton at p–n junctions in semiconductors provides light sources known as *light-emitting diodes*.

Free carriers created by absorption of light produce an increase in the electrical conductivity of a semiconductor that is called *photoconductivity*.

High photoconductive gain is favored by long recombination time and high mobility.

The *responsivity* is the ratio of the change in current to the incident radiation power. The *detectivity* is the normalized signal-to-noise ratio.

Lasers and photodevices

It is enhanced by a large decay time, small carrier concentration, and small semiconductor thickness.

14.4 Solar cells

Solar cells convert electromagnetic energy into electric energy by means of the *photovoltaic effect*.

The term **laser** is an acronym for **Light Amplification by the Stimulated Emission of Radiation**. The concept of **stimulated emission** is due to Einstein, whose studies of the statistical mechanics of electromagnetic radiation indicated that the usual treatment based on induced absorption and spontaneous emission violates the principle of detailed balance. In order to correct this deficiency, Einstein introduced the notion that a system in an excited state is stimulated by the radiation field to emit a photon and drop to a lower energy state. The rate of stimulated emission is proportional to the energy density u_ν of the radiation, whereas the rate of spontaneous emission is independent of u_ν.

14.1 General features of stimulated emission

Let us consider a system of atoms of which N_1 are in the ground state of energy E_1 and N_2 are in the excited state of energy E_2. We assume that these states are coupled by electric dipole radiation. Under ordinary conditions of thermal equilibrium, $N_1 > N_2$. If a radiation field of frequency $\nu_{12} = (E_2 - E_1)/h$ is switched on, a net absorption of energy from the field takes place. However, if by some means the condition $N_2 > N_1$ is achieved, a net emission of radiation takes place. This condition is called **population inversion**. A light beam of frequency ν_{12} incident on the system will produce additional photons of the same frequency by stimulated emission of even more photons. Consequently, the energy density increases in time. This phenomenon is called **optical gain** and is characterized by a negative absorption coefficient. If the latter is larger in magnitude than the ordinary absorption coefficient that describes optical loss, **amplification** of the light beam results. In order to achieve amplification, the exciting light beam must have an intensity exceeding a critical value necessary for the stimulated emission to overcome the losses.

An important aspect of stimulated emission is that a photon thus generated has the same frequency, polarization, direction of propagation, and phase as the incident photon. In other words the process is **coherent**. However, if the incident light beam is composed of incoherent photons, the resulting stimulated emission will also be composed of incoherent photons. The stimulated emission process favors the production of photons at the peak of the energy density distribution of the incident beam, so **spectral narrowing** occurs.

The establishment and maintainance of a population inversion is facilitated by the reabsorption of an emitted photon by an atom in the ground state to recreate an atom in the excited state. This reabsorption can be enhanced by placing the system in a cavity with highly reflecting walls.

An emitted photon that would otherwise escape through the boundary is reflected back into the system and thus has a higher probability of being reabsorbed. Another benefit is that it favors the growth of a single mode, which has a single frequency and a single phase. The output beam is therefore highly coherent with high directionality and very narrow spectral linewidth. This coherence, which is a characteristic feature of laser emission, arises from the fact that the emitted radiation has its phase related to that of the stimulating field.

14.2 Physical basis of semiconductor lasers

14.2.1 Qualitative aspects

A semiconductor p–n junction is a convenient system for establishing population inversion of energy levels. In a heavily doped junction at equilibrium the conduction band edge on the n-type side is typically below the valence band edge on the p-type side as shown in Fig. 14.1. If a forward bias V_a is applied, where $V_a \simeq E_g/e$, the conduction and valence bands are shifted with respect to each other as indicated in Fig. 14.2. Each side of the junction in this nonequilibrium situation is characterized by its **quasi-Fermi energy** indicated by a horizontal dashed line. In a narrow region about the midpoint of the junction, both conduction and valence bands are occupied by free carriers that are able to recombine. The applied bias enables the electrons and holes to be continuously supplied, so that the population inversion can be maintained even though recombination is occurring.

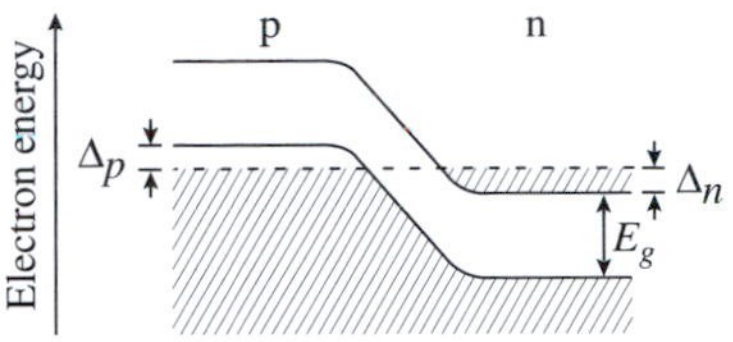

Fig. 14.1
Band edge diagram for a degenerate p–n junction at equilibrium. The Fermi energy is indicated by the horizontal dashed line (after Dalven 1990).

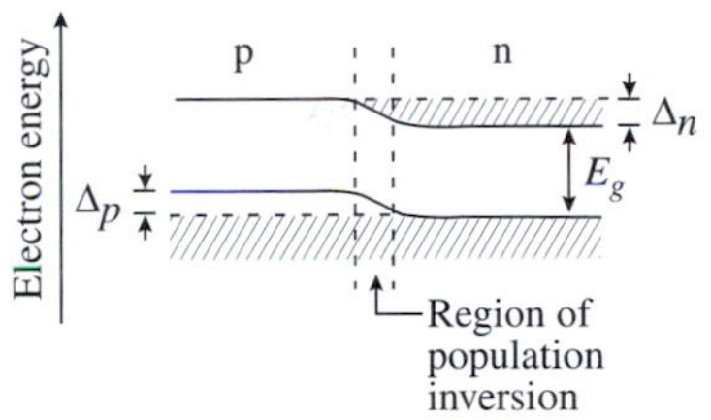

Fig. 14.2
Band edge diagram for a degenerate p–n junction under forward bias $V_a \simeq E_g/e$ (after Dalven 1990).

In terms of an energy level diagram the nonequilibrium situation can be represented as shown in Fig. 14.3. We restrict our attention to direct gap materials such as GaAs which have a high transition probability for electron–hole recombination. In indirect gap materials, electron–hole recombination near the absorption edge involves phonon-assisted transitions that have a relatively low transition probability. These materials are therefore less attractive for laser applications than those with direct gaps.

It is clear from Fig. 14.3 that radiative recombination can occur in the photon energy range $E_g \leq h\nu \leq \Delta E_F$, where $\Delta E_F = E_{Fc} - E_{Fv}$, E_{Fc} and E_{Fv} are the quasi-Fermi energies of the conduction and valence bands, respectively, and the temperature is assumed low enough to give sharp Fermi surfaces in the conduction and valence bands. Optical gain is possible only in this range. This means that the output photon energies of semiconductor lasers are restricted to the regions near the band gaps of appropriate materials such as III–V compounds and their alloys. Lasers of this type are possible in the visible and near infrared.

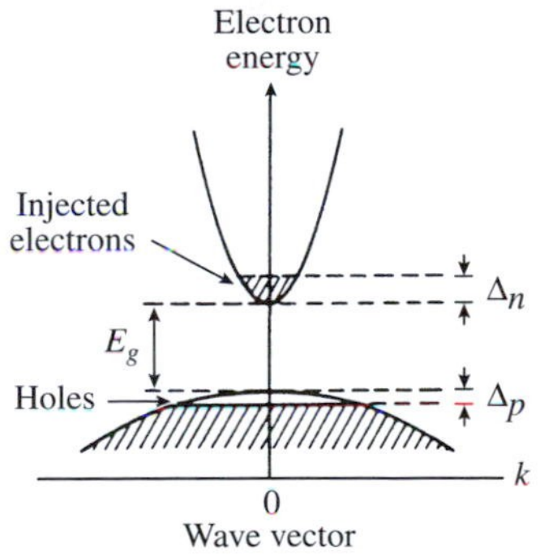

Fig. 14.3
Energy level diagram for the region of population inversion (after Dalven 1990).

14.2.2 Optical gain in direct gap semiconductors

14.2.2.1 Transition rates

To analyze the performance of semiconductor lasers, it is necessary to develop expressions for the rates of stimulated emission, spontaneous emission, and absorption. We assume initially that the impurity concentration of the semiconductor is not so high as to invalidate the vertical transition selection rule for interband transitions. The rate r_{st} for stimulated emission per unit volume associated with the transition from the conduction band state of energy E_c' to the valence band state of energy E_v' can be written as

$$r_{st} = B_{cv} f_c(E_c')[1 - f_v(E_v')]\rho_{red}(h\nu)\phi(h\nu), \tag{14.1}$$

where $\rho_{red}(\hbar\omega)$ is the reduced density-of-states for direct interband transitions, $\phi(h\nu)$ is the photon density for the mode of frequency ν, f_c and f_v are the distribution functions for conduction electrons and holes, respectively, B_{cv} is the stimulated emission rate constant that contains the interband matrix element, and $h\nu = E_c' - E_v'$. The rate r_{sp} for spontaneous emission per unit volume is given by

$$r_{sp} = A_{cv} f_c(E_c')[1 - f_v(E_v')]\rho_{red}(h\nu), \tag{14.2}$$

where A_{cv} is the rate constant for spontaneous emission. Finally, the rate r_{ab} for absorptive transitions takes the form

$$r_{ab} = B_{vc} f_v(E_v')[1 - f_c(E_c')]\rho_{red}(h\nu)\phi(h\nu), \tag{14.3}$$

where B_{vc} is the rate constant for absorption.

A relation between the rate constants can be obtained by considering the condition of equilibrium. Then detailed balance requires that the rate of absorption be equal to the sum of the rates of stimulated and spontaneous emission:

$$r_{ab} = r_{st} + r_{sp}. \tag{14.4}$$

The distributions $f_c(E_c')$ and $f_v(E_v')$ take the Fermi–Dirac forms

$$f_c(E_c') = \frac{1}{1 + e^{\beta(E_c' - E_{Fc})}} \tag{14.5a}$$

$$f_v(E_v') = \frac{1}{1 + e^{\beta(E_v' - E_{Fv})}}, \tag{14.5b}$$

and in equilibrium $E_{Fc} = E_{Fv}$. One can then show that

$$f_v(E_v')[1 - f_c(E_c')] = f_c(E_c')[1 - f_v(E_v')]e^{\beta(E_c' - E_v')}. \tag{14.6}$$

Substituting Eqs. (14.1)–(14.3) into Eq. (14.4) and using Eq. (14.6),

$$B_{vc}\phi(h\nu)e^{\beta(E_c' - E_v')} = A_{cv} + B_{cv}\phi(h\nu). \tag{14.7}$$

Solving for $\phi(h\nu)$,

$$\phi(h\nu) = \frac{A_{cv}}{B_{vc}e^{\beta h\nu} - B_{cv}}, \tag{14.8}$$

where the relation $E_c' - E_v' = h\nu$ has been used. Comparing Eq. (14.8) to the known equilibrium result for photons (Thompson 1980),

$$\phi(h\nu) = \frac{1}{\Omega(e^{\beta h\nu} - 1)}, \tag{14.9}$$

where Ω is the volume of the system, we see that

$$B_{vc} = B_{cv} \tag{14.10a}$$

$$A_{cv} = B_{cv}/\Omega. \tag{14.10b}$$

Comment. It is worth noting the dimensions of the various quantities introduced above. In terms of volume Ω, time t, and energy E, the rates r_{st}, r_{sp}, and r_{ab} have dimensions $\Omega^{-1}E^{-1}t^{-1}$, while $\rho_{red}(h\nu) \sim \Omega^{-1}E^{-1}$, $\phi(h\nu) \sim \Omega^{-1}$, $B_{cv} \sim \Omega t^{-1}$, and $A_{cv} \sim t^{-1}$. If the radiation spectrum is continuous rather than a single mode, the photon density $P(h\nu)$ per unit energy interval is given by

$$P(h\nu) = \frac{8\pi n^3 (h\nu)^2}{(hc)^3} \cdot \frac{1}{e^{\beta h\nu} - 1}, \tag{14.11}$$

where n is the refractive lindex and $P(h\nu)$ has dimensions $\Omega^{-1}E^{-1}$. The rate constant A_{cv} still has dimensions t^{-1} and satisfies the relation

$$A_{cv} = \frac{8\pi n^3 (h\nu)^2}{(hc)^3} \cdot B_{cv}. \tag{14.12}$$

However, B_{cv} now has dimensions $\Omega E t^{-1}$.

With the relationships between A_{cv}, B_{cv}, and B_{vc} established, Eqs. (14.1)–(14.3) for the transition rates can be applied to the calculation of the net rate of stimulated emission and the gain.

14.2.2.2 Net stimulated emission rate

The net rate of stimulated emission r'_{st} is obtained by taking the difference of the rates of stimulated emission and absorption:

$$r'_{st} = r_{st} - r_{ab}. \tag{14.13}$$

Using Eqs. (14.1), (14.3), and (14.10a), we have

$$r'_{st} = B_{cv}[f_c(E'_c) - f_v(E'_v)]\rho_{red}(h\nu)\phi(h\nu). \tag{14.14}$$

This result can be cast into an especially revealing form if we eliminate $B_{cv}\rho_{red}(h\nu)$ with the aid of Eqs. (14.2) and (14.10b)

$$r'_{st} = \Omega r_{sp}\phi(h\nu)[1 - e^{\beta(h\nu - \Delta E_F)}], \tag{14.15}$$

where the relation

$$\frac{f_c(E'_c) - f_v(E'_v)}{f_c(E'_c)[1 - f_v(E'_v)]} = 1 - e^{\beta(E'_c - E'_v - E_{Fc} + E_{Fv})} \tag{14.16}$$

has been used.

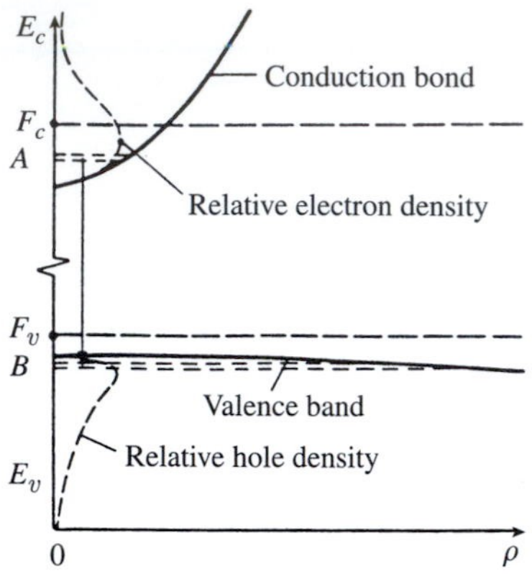

Fig. 14.4
Energy band diagram showing the conditions for net stimulated emission (after Thompson 1980).

From Eq. (14.15) one sees that net stimulated emission can occur only if the quantity in square brackets is positive and this can be true only if

$$h\nu < \Delta E_F = E_{Fc} - E_{Fv}. \tag{14.17}$$

Equation (14.17) is the **Bernard–Duraffourg condition** (Bernard and Duraffourg 1961). Physically, it states that to have net stimulated emission, the photon energy of the emitted radiation must be less than the difference of the quasi-Fermi energies. Furthermore, as stated earlier, it is necessary that $h\nu > E_g$, i.e., the photon energy must exceed the energy gap so that an allowed transition can take place between conduction and valence bands. Thus, we have the augmented condition

$$E_g < h\nu < \Delta E_F \tag{14.18}$$

for net stimulated emission to be possible. Both quasi-Fermi energies cannot therefore lie inside the energy gap; at least one of them must be outside. A diagram illustrating the conditions for net stimulated emission is given in Fig. 14.4.

An exception to the condition $h\nu > E_g$ can occur if the junction is heavily doped and a hole impurity band exists in the gap above the valence band edge. Then transitions are possible between the conduction band and hole states in the impurity band with transition energies less than E_g.

14.2.2.3 Evaluation of the rate constants

It has been established that the three rate constants B_{vc}, B_{cv}, and A_{cv} can be expressed in terms of just one of them through Eqs. (14.10). In order to evaluate them, a microscopic treatment is necessary such as that for interband absorption due to direct allowed transitions given in Chapter 10, Section 10.2. This treatment concerns an intrinsic direct gap semiconductor with negligible impurity content for which the wave vector selection rule $\boldsymbol{k} = \boldsymbol{k}_c$ applies to the initial and final wave vectors $\boldsymbol{k}_v$ and $\boldsymbol{k}_c$ of the transition. Under these conditions, the rate constant $B = B_{cv} = B_{vc}$ is found to be (Lasher and Stern 1964)

$$B = \frac{4\pi n e^2 E_g}{\epsilon_0 m^2 h^2 c^3} |\boldsymbol{M}_{vc}|^2 \Omega, \tag{14.19}$$

where $\boldsymbol{M}_{vc}$ is the interband momentum matrix element. Averaging $|\boldsymbol{M}_{vc}|^2$ over all directions in space and including the spin–orbit splitting of the valence band Δ, one finds that (Kane 1957)

$$\langle |\boldsymbol{M}_{vc}|^2 \rangle_{av} = \frac{m\, E_g}{12} \left[\frac{E_g + \Delta}{E_g + \frac{2}{3}\Delta} \right] \left(\frac{m}{m_c^*} - 1 \right). \tag{14.20}$$

The corresponding expression for B then becomes

$$B = \frac{\pi n e^2 E_g^2}{3\epsilon_0 m h^2 c^3} \left[\frac{E_g + \Delta}{E_g + \frac{2}{3}\Delta} \right] \left(\frac{m}{m_c^*} - 1 \right) \Omega. \tag{14.21}$$

Example 14.1: Rate constant for GaAs
Calculate $\langle|\boldsymbol{M}_{vc}|^2\rangle_{av}$ and B/Ω for GaAs.
Solution. For GaAs the material parameters are: $E_g = 1.42\,\text{eV}$, $\Delta = 0.33$ eV, $m_c^* = 0.066\,m$, $n = 3.5$.
The results are: $\langle|\boldsymbol{M}_{vc}|^2\rangle_{av} = 1.26\ mE_g$
$B/\Omega = 7.9 \times 10^8\ \text{s}^{-1}$.
The quantity $\langle|\boldsymbol{M}_{vc}|^2\rangle$ varies little from one direct-gap material to another. Consequently, the variation of B is due primarily to the variation of E_g.

The reduced density-of-states $\rho_{red}(E)$ that appears in the expression for r'_{st} is given by

$$\rho_{red}(E) = \frac{2}{2\pi^2}k^2\left[\frac{d\Delta E}{dk}\right]^{-1}\Bigg|_{\Delta E=E}, \tag{14.22}$$

where

$$\Delta E = E_g + \frac{\hbar^2 k^2}{2\bar{m}^*}$$

for simple parabolic bands with reduced mass $\bar{m}^*$. A factor of 2 has been included to account for the spin degeneracy. In III–V compounds such as GaAs, both light and heavy holes contribute.

In many semiconductor lasers there is high p-type doping in order to provide a high concentration of holes with which the injected electrons can recombine. In the presence of the acceptor impurities, the wave functions are no longer pure Bloch functions, but are linear combinations of Bloch functions with a broad range of wave vectors. Consequently, the selection rule $\boldsymbol{k}_v = \boldsymbol{k}_c$ is not applicable, so transitions can occur between any occupied conduction band state to any empty valence band state. Let us replace the reduced density-of-states by the equivalent expression $\rho_c(E'_c)\rho_v(E'_v)/[\rho_c(E'_c) + \rho_v(E'_v)]$ which involves the densities-of-states of the individual valence and conduction bands $\rho_v(E)$ and $\rho_c(E)$. The rate of spontaneous emission then takes the form

$$\begin{aligned} r_{sp}(E'_c, h\nu) = \frac{B}{\Omega}\int & \rho_c(E'_c)\rho_v(E'_v)f_c(E'_c)[1 - f_v(E'_v)] \\ & \times [\rho_c(E'_c) + \rho_v(E'_v)]^{-1}\delta(E'_c - E'_v - h\nu)dE'_v. \end{aligned} \tag{14.23}$$

We can incorporate the relaxation of the $\boldsymbol{k}$ selection rule by taking the average value of $\rho_c(E'_c) + \rho_v(E'_v)$, combining it with B, and integrating over E'_c to give

$$\begin{aligned} r_{sp}(h\nu) = \bar{B}\int\int & \rho_c(E'_c)\rho_v(E'_v)f_c(E'_c) \\ & \times [1 - f_v(E'_v)]\delta(E'_c - E'_v - h\nu)dE'_c dE'_v \end{aligned} \tag{14.24}$$

where $\bar{B} = B/[\rho_c(E'_c) + \rho_v(E'_v)]_{av}$.

14.2.2.3 Evaluation of optical gain

The optical gain $g(\nu)$ is defined by

$$dI = Ig(\nu)dx, \tag{14.25}$$

where I is the intensity of the radiation at a point x in the system. This equation is analogous to Eq. (10.41) that defines the absorption coefficient $\alpha(\nu)$, but with $g(\nu)$ replacing $-\alpha(\nu)$. Pursuing the analogy further, one can derive the following expression for the gain:

$$g(\nu) = \frac{r'_{st}}{(c/n)N(h\nu)\phi(h\nu)}, \tag{14.26}$$

where the numerator is the rate of phonon energy generation per unit volume and the denominator is the rate of flow of photon energy through unit area. The quantity $N(h\nu)$ is the number of modes per unit energy interval given by

$$N(h\nu) = [e^{\beta h\nu} - 1]\Omega P(h\nu) = \frac{8\pi n^3(h\nu)^2\Omega}{(hc)^3}. \tag{14.27}$$

Substituting this result together with those for r'_{st} and $\phi(h\nu)$ given by Eqs. (14.15) and (14.9), respectively, yields $g(\nu)$ in the form

$$g(\nu) = \frac{hc^2}{8\pi n^2\nu^2} r_{sp}\left[1 - e^{\beta(h\nu - \Delta E_F)}\right]. \tag{14.28}$$

As we have already noted, gain can effectively occur only if $E_g < h\nu < \Delta E_F$. The key quantity is ΔE_F whose excess over E_g is a measure of the degree of population inversion of the bands.

14.2.3 Transparency current density

Under steady state conditions the electrons and holes that are lost through recombination must be replenished by the current flowing through the junction. We shall focus on the situation for which $\Delta E_F \simeq h\nu$ and the spontaneous emission is much greater than the stimulated emission in accordance with Eq. (14.15). If the radiative recombination occurs in a region of width d_3 around the junction and nonradiative recombination is negligible, the current density j is given by

$$j = eR_{sp}d_3, \tag{14.29}$$

where R_{sp} is the total spontaneous emission rate obtained by integrating r_{sp} over all photon energies:

$$R_{sp} = \int r_{sp}(h\nu)d(h\nu). \tag{14.30}$$

A commonly used convention is to calculate the current density for a particular width $d_3 = 1\,\mu\text{m}$. The resulting quantity is the **nominal current**

density denoted by j_{nom}:

$$j_{nom} = e\,R_{sp}. \tag{14.31}$$

Substituting the expression for $r_{sp}(h\nu)$ from Eq. (14.24) into Eq. (14.30) and carrying out the integral over $h\nu$ yields

$$R_{sp} = \bar{B} n_c p_v, \tag{14.32}$$

where n_c and p_v are the concentrations of electrons and holes given by

$$n_c = \int \rho_c(E_c')f_c(E_c')dE_c' \tag{14.33a}$$

$$p_v = \int \rho_v(E_v')[1 - f_v(E_v')]dE_v', \tag{14.33b}$$

and $\bar{B}$ is $B/[\rho_c(E_c') + \rho_v(E_v')]_{av}$. The nominal current density then becomes

$$j_{nom} = e\bar{B} n_c p_v. \tag{14.34}$$

Under typical conditions the concentrations of the injected carriers are much greater than the dopant concentration. Consequently, charge neutrality requires that $n_c \simeq p_v = n$ and

$$j_{nom} \simeq e\bar{B}n^2. \tag{14.35}$$

The threshold value of the nominal current density is obtained by determining the carrier concentration necessary to have ΔE_F satisfy the Bernard–Duraffourg condition for the photon energy $h\nu$ of interest. Substituting this concentration into Eq. (14.34) then gives the **transparency current density** j_0. Assuming simple parabolic bands and degenerate carriers, the quasi-Fermi energies relative to the respective band edges are specified by

$$E_{Fc} = \frac{h^2}{2m_c^*}\left(\frac{3n_c}{8\pi}\right)^{2/3} \tag{14.36a}$$

$$E_{Fv} = \frac{h^2}{2m_v^*}\left(\frac{3p_v}{8\pi}\right)^{2/3}. \tag{14.36b}$$

For $n_c = p_v = n$, the difference in quasi-Fermi energies is

$$\Delta E_F = E_{Fc} + E_{Fv} + E_g = \frac{h^2}{2\mu^*}\left(\frac{3n}{8\pi}\right)^{2/3} + E_g. \tag{14.37}$$

Taking $\Delta E_F = h\nu$ and solving for n, we get

$$n = \frac{8\pi}{3}\left[\frac{2\bar{m}^*}{h^2}(h\nu - E_g)\right]^{3/2} \tag{14.38}$$

and

$$j_0 = e\bar{B}\left(\frac{8\pi}{3}\right)^2\left[\frac{2\bar{m}^*}{h^2}(h\nu - E_g)\right]^3. \qquad (14.39)$$

This expression for j_0 is restricted to zero temperature. Finite temperatures require numerical integrations that lead to results that cannot be expressed in simple analytic form. One finds from such calculations that j_0 varies with temperature approximately as $T^{3/2}$. For pure GaAs at room temperature the injected carrier concentrations are $\sim 10^{18}\,\text{cm}^{-3}$ and the quantity $\bar{B} \simeq 2 \times 10^{-10}\,\text{cm}^3\,\text{s}^{-1}$. The corresponding value of j_0 is $\sim 4000\,\text{Acm}^{-2}\,\mu\text{m}^{-1}$.

14.2.4 Current density–gain relationship

The total rate of spontaneous recombination R_{sp} that determines the current density is an integral of $r_{sp}(h\nu)$ over photon energy. As a function of $h\nu$, $r_{sp}(h\nu)$ starts at zero when $h\nu = E_g$, rises to a maximum and then declines toward zero as $h\nu \to \infty$. This behavior can be characterized by the maximum value $r_{sp}(max)$ and the full width at half-maximum ΔE. To a good approximation, $R_{sp} \simeq r_{sp}(max)\Delta E$ and

$$j_{nom} \simeq e r_{sp}(max)\Delta E. \qquad (14.40)$$

The gain $g(\nu)$ is given in convenient form by Eq. (14.28). It has a maximum value g_{max} that can be expressed approximately by

$$g_{max} = \frac{h^3c^2}{8\pi n^2 E_g^2} r_{sp}(max)\epsilon, \qquad (14.41)$$

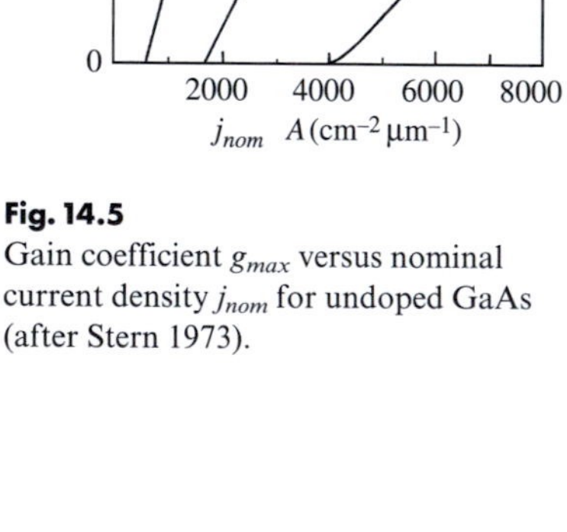

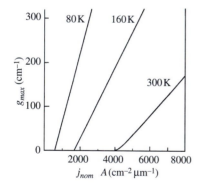

Fig. 14.5
Gain coefficient g_{max} versus nominal current density j_{nom} for undoped GaAs (after Stern 1973).

where ϵ is the value of the quantity in square brackets in Eq. (14.28) and is always less than unity. Eliminating $r_{sp}(max)$ yields the relation

$$g_{max} \simeq \frac{h^3c^2\epsilon}{8\pi e n^2 E_g^2 \Delta E} j_{nom}. \qquad (14.42)$$

At moderate temperatures the linewidth ΔE can be approximated by $\Delta E = qk_BT$ where q lies in the range $1 < q < 10$. Then

$$g_{max} \simeq \frac{h^3c^2\epsilon}{8\pi e n^2 E_g^2 q k_B T} j_{nom}. \qquad (14.43)$$

In this simple treatment the gain is proportional to the nominal current density with a slope that is inversely proportional to the temperature T. Detailed calculations (Stern 1973) support these conclusions and are presented in Fig. 14.5 which contains plots of g_{max} versus j_{nom}. Both the linearity and decrease in slope with increasing temperature are evident. The plots can be well represented by

$$g_{max} = A(j_{nom} - j_0). \qquad (14.44)$$

One finds that A varies as $1/T$ and is $\sim 0.045\,\text{cm}^{-1}/\text{A cm}^{-2}\,\mu\text{m}^{-1}$ for pure GaAs at room temperature.

14.2.5 Threshold condition in a Fabry–Perot cavity

Typical semiconductor lasers are fabricated in such a way that the active region forms a resonant cavity. A thin platelet of semiconducting material, GaAs for example, has a pair of parallel reflecting mirrors created by cleavage along parallel (110) planes. The width of a mirror is generally small, $d \sim 1\ \mu\text{m}$, compared to a height of a few millimeters. The length of the face normal to the two mirrors and connecting them is $L \gg d$. The naturally cleaved mirrors are only partially reflecting. For example, the reflectivity at the GaAs–air interface is 0.32.

We focus our attention on the relation between the threshold gain and the threshold current density. Since the longitudinal dimension of the cavity is large compared to the transverse dimensions, the modes of interest are longitudinal modes which are characterized by the number of nodes q between the two mirrors of the cavity. The number of nodes is the length of the cavity divided by one-half the wavelength λ_s of radiation in the semiconductor:

$$q = \frac{2L}{\lambda_s}. \tag{14.45}$$

Using the relation $\lambda_s = \lambda/n(\nu)$, where $\lambda = c/\nu$, we obtain

$$q = \frac{2Ln(\nu)}{\lambda}. \tag{14.46}$$

If we start with low current through the junction, there is spontaneous emission and gain. Increasing current leads to increasing gain until the threshold for lasing is reached. The gain then satisfies the condition that a light wave completely cross the cavity and be reflected without suffering attenuation,

$$R \exp[(g_{th} - \alpha)L] = 1, \tag{14.47}$$

where R is the reflectivity of the ends of the cavity, g_{th} is the threshold gain, and α is the absorption coefficient essentially due to free carriers and impurities. Solving for g_{th} we obtain

$$g_{th} = \alpha + \frac{1}{L}\log\left(\frac{1}{R}\right). \tag{14.48}$$

For $g > g_{th}$ the light output is increased and emission of coherent beams from the ends of the cavity takes place.

The threshold current density j_{th} varies linearly with g_{th} and is specified by an equation analogous to Eq. (14.44) (Sze 1981):

$$j_{th} = \frac{1}{A} g_{th} + j_0 = j_0 + \frac{1}{A}\left[\alpha + \frac{1}{L}\log\left(\frac{1}{R}\right)\right]. \tag{14.49}$$

At moderately high temperatures, $j_{th} \sim T^3$.

14.2.6 Light-emitting diodes

Even if the conditions for laser action are not attained, p–n junctions can still serve as useful light sources known as **light-emitting diodes** or LEDs. Alloys of III–V compounds can be used to produce light in the visible range. For example, $GaAs_{0.6}P_{0.4}$ produces red light. GaAs itself is a source for near infrared radiation. $GaAs_{1-x}P_x$ with $x > 0.45$ is an indirect gap material which can be used to generate orange ($x = 0.65$), yellow ($x = 0.85$), or green light ($x = 1.0$) if nitrogen impurity is added to enhanced radiative recombination.

14.3 Photodetectors

We have seen in Chapter 10 that an intrinsic semiconductor absorbs electromagnetic radiation with photon energy greater than the band gap. The electron–hole pairs thus produced increase the electrical conductivity of the sample and result in the phenomenon of **photoconductivity**. The increased conductivity can be measured by placing the sample in a suitable circuit and provides the basis for a device called a **photodetector** that can be used to detect the radiation. The band gap constitutes the wavelength threshold for the operating range of the photodetector. By exploiting the broad variety of semiconductors and their alloys that is available, one can design intrinsic photodetectors that cover a wide spectral range from the visible well into the infrared. A number of semiconductors together with their band gaps are listed in Table 3.4. A particularly interesting system is the PbTe–SnTe solid solution whose band gap as a function of composition is shown in Fig. 3.19. The band gap is zero at a composition of 34% SnTe; thus, this system can be used in detectors that operate in the far infrared.

Another candidate for a far infrared photodetector is an extrinsic semiconductor that contains shallow impurity levels, e.g., Ga-doped Ge. Carriers localized on the impurity levels can be excited by the absorption of radiation below a threshold wavelength and become free carriers. The conductivity thereby increases just as in the case of intrinsic absorption. The long-wavelength threshold for Ga-doped Ge is 115 μm.

14.3.1 Photoconductive gain

The operation of a photodetector can be characterized by various parameters. We consider first the **photoconductive gain**. Suppose that the detector material has the form of a slab as shown in Fig. 14.5. Electrodes are placed on two opposite faces a distance d apart and a voltage V is applied between them. Radiation is incident on one of the broad faces connecting the electrodes. Free carriers are generated at a rate g per unit volume by the absorption of radiation of suitable frequencies. To be specific we focus on electrons whose change in concentration as a result of the radiation is Δn. The electrons thus produced can recombine with holes or with ionized donor impurities at a rate r given by

$$r = \frac{\Delta n}{\tau_n}, \tag{14.50}$$

where τ_n is the recombination time. Under steady-state conditions,

$$g = \frac{\Delta n}{\tau_n}. \tag{14.51}$$

The change in electron concentration is associated with a change in electron flux $\Delta\phi_n$, where

$$\Delta\phi_n = \Delta n v_n = g\tau_n\mu_n\frac{V}{d}, \tag{14.52}$$

v_n is the electron drift velocity, and μ_n is the electron mobility. The number of electrons crossing the sample per second ΔN_c is given by

$$\Delta N_e = A\Delta\phi_n = Ag\tau_n\mu_n\frac{V}{d}, \tag{14.53}$$

where A is the cross sectional area of the sample. The photoconductive gain G is defined as

$$G = \frac{\Delta N_e}{\Delta N_{ph}}, \tag{14.54}$$

where ΔN_{ph} is the number of photons absorbed per second. Since the creation of one additional electron requires the absorption of one photon, ΔN_{ph} can be expressed as

$$\Delta N_{ph} = gAd. \tag{14.55}$$

Substituting Eqs. (14.50) and (14.52) into Eq. (14.51) yields

$$G = \tau_n\mu_n\frac{V}{d^2}. \tag{14.56}$$

We see that the important material parameter entering the gain is the $\tau_n\mu_n$ product. High gain is favored by long recombination time and high mobility.

An alternative expression for the gain can be obtained by introducing the **electron transit time** t_n, i.e., the time it takes an electron to move from one electrode to the other. This time is simply the distance between the electrodes divided by the drift velocity:

$$t_n = \frac{d}{v_n} = \frac{d^2}{\mu_n V}. \tag{14.57}$$

The above result for G can then be rewritten as

$$G = \frac{\tau_n}{t_n}. \tag{14.58}$$

Thus, short transit time and long recombination time favor high gain.

The mean distance that an electron travels between its generation and recombination is its **drift length** ℓ_n which is given by

$$\ell_n = v_n \tau_n = \tau_n \mu_n \frac{V}{d}. \qquad (14.59)$$

In terms of ℓ_n, the gain is

$$G = \frac{\ell_n}{d}, \qquad (14.60)$$

which shows that small electrode separation and large drift length are desirable for high gain.

14.3.2 Responsivity and detectivity

The various expressions for the photoconductive gain that have been presented are not independent, but each provides distinctive insight into the problem of enhancing the gain. In comparing the practical advantages of various photodetectors, however, it is useful to consider figures of merit such as the **responsivity** R and **detectivity** D^*. The responsivity is defined as the ratio of the change in current ΔI to the incident radiation power P_i:

$$R = \frac{\Delta I}{P_i}. \qquad (14.61)$$

For ΔI we have

$$\Delta I = e\Delta\phi_n A. \qquad (14.62)$$

Using Eq. (14.52) we get

$$\Delta I = I\Delta n/n, \qquad (14.63)$$

where I is the DC current given by $-env_nA$. For P_i we have a simple proportionality to the incident photon flux ϕ_{ph}:

$$P_i = Ah\nu\phi_{ph}. \qquad (14.64)$$

The responsivity now becomes

$$R = \frac{I\Delta n}{Ah\nu\phi_{ph}n}. \qquad (14.65)$$

In the actual operation of a photodetector, the input light signal is chopped at a frequency ν_s so that the resulting electrical signal in the circuit can be amplified. Consequently, Δn is time dependent. It can be shown that R then takes the form

$$R = \frac{I\eta\tau}{Adh\nu n} \cdot \frac{1}{[1 + (2\pi\nu_s\tau)^2]^{\frac{1}{2}}}, \qquad (14.66)$$

where τ is the decay time of the circuit and η is the fraction of the light absorbed.

Another factor that must be taken account is the competition between the output signal and the electronic noise in the circuit. The detectivity D^* is the normalized signal-to-noise ratio given by

$$D^* = \frac{R}{\mathcal{N}}(A\Delta\nu)^{\frac{1}{2}}, \tag{14.67}$$

where $\mathcal{N}$ is the root-mean-square noise current and $\Delta\nu$ is the band width. The noise is associated with random fluctuations of the current and has contributions from generation–recombination, $1/f$, Nyquist (thermal), and photon mechanisms.

Generation–recombination noise arises from the random nature of the generation and recombination processes undergone by a current carrier. The mean square fluctuation in electron current can be expressed in terms of the spectral intensity $S(\nu)$ as (van Vliet 1958)

$$\langle \Delta I^2 \rangle \simeq A^2 S(\nu)\Delta\nu$$
$$= 4\left(\frac{I}{\mathcal{N}}\right)^2 \frac{\tau_n}{1+4\pi^2\nu^2\tau_n^2} \frac{N(N_A+N)(N_D-N_A-N)}{(N_A+N)(N_D-N_A-N)+NN_D}\Delta\nu, \tag{14.68}$$

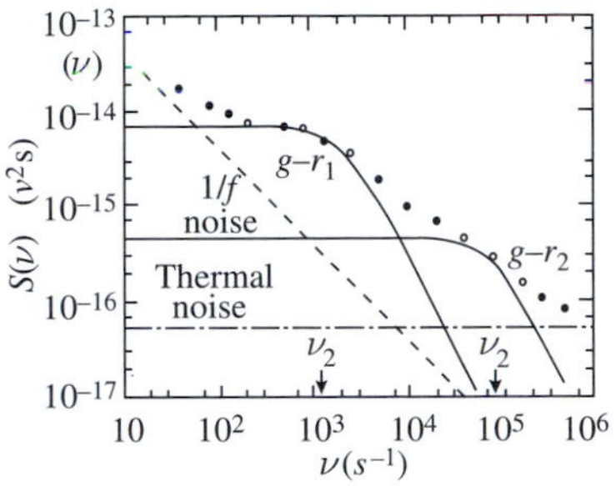

Fig. 14.6
Spectral intensity versus frequency for p-Si with two different recombination centers: theoretical, solid curves; experimental, open circles (after Bosman 1981).

where N, N_D, and N_A are the numbers of electrons, donors, and acceptors, respectively, in the detector element and τ_n is the electron lifetime in the conduction band. The frequency dependence of $S(\nu)$ is shown in Fig. 14.6 for a particular case (p-type Si) that has two types of recombination centers. Theoretical curves for the individual centers are also shown.

In $1/f$ noise the spectral intensity varies as the inverse frequency. The physical origin of $1/f$ noise is not yet well established. **Nyquist noise** or **thermal noise** arises from the random thermal motion of the carriers (Nyquist 1928). The noise power can be written as

$$\langle \Delta I^2 \rangle R = 4A^2 \frac{h\nu}{e^{\beta h\nu}-1}\Delta\nu \simeq 4A^2 k_B T \Delta\nu, \tag{14.69}$$

where R is the resistance of the semiconductor. The spectral intensity is independent of frequency at not too low temperatures. **Photon noise** is associated with the random incidence of photons on the semiconductor, thus introducing randomness into the generation of free carriers. The spectral intensity has the same form as that for generation–recombination noise.

Returning to the detectivity D^*, we focus on the generation–recombination component. In the limit of small number of electrons, we can express $\mathcal{N}$ with the aid of Eq. (14.68) as

$$\mathcal{N} = \langle \Delta I^2 \rangle^{\frac{1}{2}} \simeq 2I\left[\frac{\tau_n \Delta\nu}{ndA(1+4\pi^2\nu_s^2\tau_n^2)}\right]^{\frac{1}{2}}. \tag{14.70}$$

Substituting Eqs. (14.66) and (14.70) into Eq. (14.67) and assuming that $\tau_n = \tau$, we obtain D^* in the simple form

$$D^* \simeq \frac{\eta}{2h\nu}\left(\frac{\tau}{nd}\right)^{\frac{1}{2}}. \tag{14.71}$$

The detectivity is enhanced by a large decay time and small carrier concentration and semiconductor thickness.

14.4 Solar cells

A **solar cell** is a device that converts the electromagnetic energy of sunlight into electric energy by means of the **photovoltaic effect**. The photovoltaic effect is the production of an electric potential by the absorption of light in the region of a medium in or near a potential barrier, for example, in a p–n junction. It is necessary that the photon energy be equal to or greater than the band gap, so that electron–hole pairs are created. The built-in electric field of the junction forces the electrons and holes apart, thereby modifying the space charge in such a way that the built-in electric field is reduced from its equilibrium value $\mathcal{E}$ to the value $\mathcal{E} - \bar{\mathcal{E}}$. Correspondingly, the diffusion potential is reduced from V_d to $V_d - \bar{V}_d$. The energy barrier to the passage of carriers is reduced to $e(V_d - \bar{V}_d)$.

The above discussion describes an illuminated p–n junction under open-circuit conditions. The quantity $\bar{V}_d$ is the open-circuit voltage of the junction. The maximum value of $\bar{V}_d$ is V_d itself which leads to complete elimination of the built-in electric field.

From Eq. (12.35) we see that

$$eV_d = E_{Fn} - E_{Fp}. \tag{14.72}$$

For reasonable doping levels, $E_{Fn} \simeq E_C$ and $E_{Fp} \simeq E_V$, so

$$eV_d \simeq E_C - E_V = E_g. \tag{14.73}$$

Thus, a large band gap is favorable for producing a large open-circuit voltage.

A p–n junction can serve as an infrared detector (photovoltaic detector) or, if properly illuminated, as a source of electrical power (solar cell). In the latter application using photons from the sun, the optimum band gap is ~ 1.4 eV (Moss *et al.* 1973) which suggests InP or GaAs. However, the relatively low cost and easy availability of Si ($E_g \simeq 1.2$ eV) give it a definite advantage for practical applications. Solar cells made from single crystal silicon provide yields of up to 22%. Even if they are made from relatively cheap polycrystalline material, yields of 10% can be achieved. They have found extensive applications in space vehicles. However, to provide tens of watts of power, thousands of cells must be used.

Problems

1. State the conditions for having net stimulated emission. Assuming equal electron and hole concentrations, derive an expression for the minimum carrier concentration required to give population inversion that involves the intrinsic carrier concentration. Calculate the minimum concentration for GaAs at 300 K taking $E_g = 1.42\,\text{eV}$, $m_c^* = 0.066\,\text{m}$, and $m_v^* = 0.40\,\text{m}$.
2. Using the results of Example 14.1, calculate the net stimulated emission rate for GaAs at 300 K. Take $h\nu = 1.45\,\text{eV}$, and $\Delta E_F = 1.50\,\text{eV}$.
3. Calculate the rate of spontaneous emission for the situation in Problem 2.
4. Calculate the optical gain for the situation in Problem 2.
5. Calculate the photoconductive gain for a semiconductor infrared detector with mobility $\mu_n = 10^4\,\text{cm}^2/(\text{V s})$, recombination time $\tau_n = 10^{-4}\,\text{s}$, electrode separation $d = 1\,\text{mm}$, and voltage $V = 1\,\text{V}$.
6. Calculate the detectivity D^* for the detector of Problem 5 with carrier concentration $n = 10^{14}\,\text{cm}^{-3}$, fraction of absorbed light $\eta = 0.2$, photon wavelength $\lambda = 10\,\mu\text{m}$.

References

M. G. A. Bernard and G. Duraffourg, *Phys. Status Solidi* **1**, 699 (1961).

G. Bosman, *Thesis*, University of Utrecht, 1981.

R. Dalven, *Introduction to Applied Solid State Physics*, Second edition (Plenum Press, New York, 1990).

E. O. Kane, *J. Phys. Chem. Solids* **1**, 249 (1957).

G. Lasher and F. Stern, *Phys. Rev.* **133**, A553 (1964).

T. S. Moss, G. J. Burrell, and B. Ellis, *Semiconductor Opto-Electronics* (John Wiley, New York, 1973).

H. Nyquist, *Phys. Rev.* **32**, 110 (1928).

F. Stern, *J. Quant. Electronics* QE-**9**, 290 (1973).

S. M. Sze, *Physics of Semiconductor Devices*, Second Edition (John Wiley, New York, 1981).

G. H. B. Thompson, *Physics of Semiconductor Laser Devices* (John Wiley, New York, 1980).

K. M. van Vliet, *Phys. Rev.* **110**, 50 (1958).

Heterostructures: electronic states

15

Key ideas

A *heterostructure* consists of two different semiconductors in intimate contact. A *heterojunction* is a heterostructure with semiconductors of different type, e.g., n-type and p-type. A *modulation doped heterostructure* involves two semiconductors, one highly doped with an impurity of a certain type, say n-type, on one side and the other intrinsic or lightly doped with an impurity of opposite type, say p-type, on the other side of the interface.

The conduction band edge on the p-type side can dip below the Fermi energy and create a very narrow layer of conduction electrons at the interface that constitutes a two-dimensional electron gas (2DEG). Since it is on the side of nearly intrinsic material, very high mobility can be achieved.

The electron energies in a 2DEG form *electronic subbands* which arise from the confinement of the electron in the direction normal to the junction interface.

A square-quantum well is formed by placing a thin layer of small gap semiconductor between two thick layers of a large gap semiconductor.

The subband eigenfunctions of conduction electrons have even or odd symmetry with respect to inversion at the center of a symmetric square well.

The quantum well potential lifts the degeneracy of light and heavy hold bands at $\boldsymbol{k} = 0$. The hole effective masses characterizing motion parallel to the interface are functions of the subband quantum number n.

The density of states has a *staircase profile*. At the limit of lowest energy the quantum well density-of-states retains a finite value. All dynamical phenomena in a 2D system remain finite even at low energies.

The charge distributions of an exciton in a quantum well is *ellipsoidal* in shape and is characterized by transverse and longitudinal reduced Bohr radii. The exciton binding energy is increased compared to that in a 3D system.

Two quantum wells with a common barrier between them constitute a *double-well structure*.

Heterostructures

In a double-well structure there is a splitting of both their ground state and their excited state levels.

Coupling together a large number of quantum wells leads to the creation of a *superlattice*. The energy level diagram consists of *minibands*. The width of a miniband decreases as the barrier width increases.

15.8 Modulation doping of heterostructures

In heterostructures *charge transfer* occurs with the result that the charged impurities are not in the same region as the free carriers. The spatial separation of carriers and ionized impurities leads to greatly reduced impurity scattering and *very high mobilities*.

15.9 Self-consistent energy-level calculations

The use of an undoped spacer layer increases the separation between the ionized donor atoms and the channel electrons, thus decreasing the Coulomb interaction and therefore increasing the electron mobility. In a double heterojunction structure the active layer is a wide undoped layer of semiconductor sandwiched between two appropriately doped barriers.

15.10 N–I–P–I structures

Spatially modulated doping can produce a superlattice and causes a folding-back of the Brillouin zone. A new energy band structure along the axis of the superlattice arises. Such structures are *doping superlattices* and consist of periodic n-doped, intrinsic, p-doped, intrinsic, ... multilayers known as N–I–P–I structures. At large doping levels the effective band gap can become negative.

In the preceding three chapters we have considered junctions formed by taking a single semiconductor and doping adjacent parts of the sample with two different types of impurities, one acceptor and the other donor. We now consider junctions in which the p- and n-type parts are different semiconductors. These junctions have some properties that are similar to those of homojunctions, but other properties that are quite unique.

15.1 Heterojunctions

A **heterostructure** is obtained when two different semiconductors are placed in intimate contact. A heterostructure forms a **heterojunction** if the two semiconductors are of different type, e.g., n-type and p-type. A **modulation-doped heterojunction** (Störmer *et al.* 1979) consists of a semiconductor that is heavily doped with impurities of a certain type, e.g., n-type, and a semiconductor that is intrinsic or lightly doped with impurities of the opposite type, e.g., p-type.

Just as with a p–n junction, a heterojunction in equilibrium must have the Fermi energy constant throughout the system. Consequently, bending of the energy bands occurs in the vicinity of the interface between the two semiconductors. In a typical situation the n-type material has the larger energy band gap, and its conduction band edge is bent upward as the interface is approached, while the conduction band edge of the intrinsic or p-type material is bent downward. The situation is represented in Fig. 15.1, which shows the energy-band diagram for the modulation-doped heterojunction n-$Ga_{1-x}Al_xAs$/p-GaAs. The bending of the band edges is evident, but in addition there is a feature not found in a silicon p–n junction, i.e., a very narrow layer of conduction electrons in the GaAs at the interface which constitutes a two-dimensional electron gas (2DEG). The 2DEG arises from electron transfer across the junction.

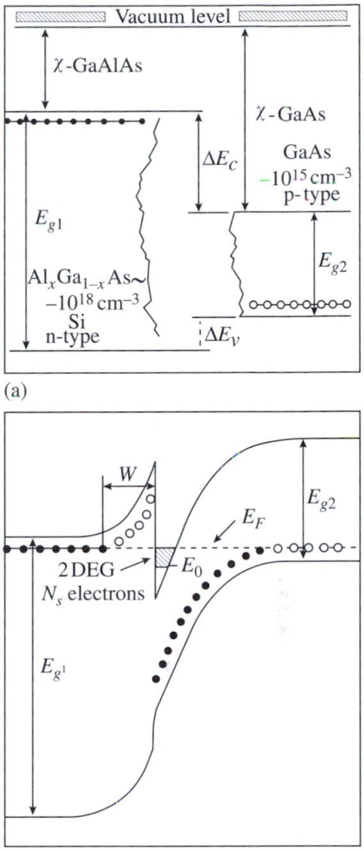

Fig. 15.1
Band edge diagram of a selectively doped GaAlAs/GaAs heterostructure (a) before equilibrium and (b) at equilibrium. χ designates electron affinity (after Weisbuch and Vinter 1991).

15.2 Free charge carrier transfer

When two different semiconductors are placed in contact, charge transfer takes place across the interface to equalize the Fermi energy (chemical potential) on both sides. Electrons from the donor impurities of the highly doped n-type $Ga_{1-x}Al_xAs$ in Fig. 15.1 are transferred to the conduction band or to acceptor impurities of the nearly intrinsic p-type GaAs until equilibrium is reached. Positively charged donor ions are therefore left near the interface on the n-type side and negatively charged conduction electrons and acceptor ions are left near the interface on the p-type side. These uncompensated charges constitute a space charge region which contains a built-in electric field similar to that found in a silicon p–n junction.

The built-in electric field causes the energy band edges to bend as seen in Fig. 15.1. Under suitable conditions the conduction band edge on the p-type side can dip below the Fermi energy and produce a region whose states are occupied by conduction electrons that form a 2DEG. Since this region is on the side with nearly intrinsic material, there are very few ionized impurities to scatter the electrons, and consequently very high mobilities can be achieved in the 2DEG. Part of the transferred electrons constitutes the 2DEG. The other part annihilates holes and produce a depletion region on the p-type side.

The transfer of charge carriers across the heterojunction is of the utmost importance in understanding the physics of two-dimensional electron systems and phenomena such as the quantum Hall effect, and for the realization of devices such as the high electron mobility transistor (HEMT) and the two-dimensional electron gas field effect transistor (TEGFET). A simple analysis of the 2DEG can be formulated by focusing on the conduction band edge as modified by the electric charge and field near the interface and the constancy of the Fermi energy across the junction. The conduction band edge as shown schematically in Fig. 15.2 has a minimum associated with a potential well that is approximately triangular in shape. The states of an electron in such a well can be determined quantum mechanically.

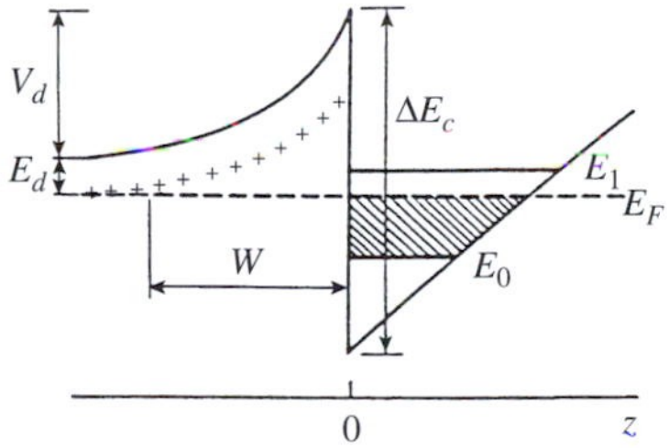

Fig. 15.2
Schematic representation of the conduction band edge (after Weisbuch and Vinter 1991).

15.3 Triangular quantum well

The potential well arises from the transfer of electrons across the heterojunction and the electric field created by the resulting space charge. If N_s is number of electrons transferred per unit area, the electric field $\mathcal{E}$ as given by Gauss's law is

$$\mathcal{E} = \frac{N_s e}{\epsilon_0 \epsilon_r}, \tag{15.1}$$

where ϵ_0 is the permittivity of vacuum and ϵ_r is the dielectric constant of the nearly intrinsic GaAs. For a triangular well the electrostatic potential $\varphi(z)$ is linear in z for $z > 0$ and is given by

$$\varphi(z) = -\mathcal{E}z. \tag{15.2}$$

An infinite potential barrier is assumed at $z = 0$.

The Hamiltonian for an electron in a triangular well in GaAs takes the form

$$H = -\frac{\hbar^2\nabla^2}{2m} + V_p(\mathbf{r}) - e\varphi(z), \tag{15.3}$$

where $V_p(\mathbf{r})$ is the periodic potential energy. Using effective mass theory as discussed in Chapter 5, we can express the eigenfunctions of H in lowest-order approximation as

$$\psi(\mathbf{r}) = F_{cn}(\mathbf{r})u_{c0}(\mathbf{r}), \tag{15.4}$$

where $u_{c0}(\mathbf{r})$ is the conduction band Bloch function for zero wave vector and $F_{cn}(\mathbf{r})$ is the envelope function satisfying the effective mass equation

$$\left[-\frac{\hbar^2\nabla^2}{2m^*} - e\varphi(z)\right]F_{cn}(\mathbf{r}) = E\,F_{cn}(\mathbf{r}). \tag{15.5}$$

The index n identifies the eigenstates, m^* is the conduction electron effective mass of GaAs and E is the energy eigenvalue. Solutions to Eq. (15.5) can be written as

$$F_{cn}(\mathbf{r}) = e^{i\mathbf{k}_\perp\cdot\mathbf{r}_\perp}\chi_n(z), \tag{15.6}$$

where $\mathbf{k}_\perp$ is the two-dimensional wave vector perpendicular to the surface normal, $\chi_n(z)$ satisfies the equation

$$\left[-\frac{\hbar^2}{2m^*}\frac{d^2}{dz^2} - e\varphi(z)\right]\chi_n(z) = E_n\chi_n(z), \tag{15.7}$$

and

$$E_n = E - \frac{\hbar^2k_\perp^2}{2m^*}. \tag{15.8}$$

The boundary conditions to be satisfied by $\chi_n(z)$ are: $\chi_n(0) = \chi_n(\infty) = 0$. A solution satisfying the boundary condition at infinity is the Airy function

$$\mathrm{Ai}\left[\left(\frac{2m^*}{\hbar^2e^2\mathcal{E}^2}\right)^{1/3}(e\mathcal{E}z - E_n)\right].$$

The boundary condition at $z = 0$ determines the allowed values of E_n:

$$E_n = -\left(\frac{e^2\mathcal{E}^2\hbar^2}{2m^*}\right)^{1/3}a_n. \tag{15.9}$$

The quantity a_n is the nth zero of the Airy function and is well approximated by

$$a_n \cong -\left[\frac{3\pi}{2}\left(n+\tfrac{3}{4}\right)\right]^{2/3}, \quad n = 0, 1, \ldots. \tag{15.10}$$

The values of E_n are then (Stern 1972)

$$E_n \cong \left(\frac{\hbar^2}{2m^*}\right)^{1/3}\left[\frac{3\pi e\mathcal{E}}{2}\left(n+\tfrac{3}{4}\right)\right]^{2/3}. \tag{15.11}$$

The energy eigenvalues are specified by Eq. (15.8),

$$E(\boldsymbol{k}_\perp, n) = E_n + \frac{\hbar^2 k_\perp^2}{2m^*}, \tag{15.12}$$

and correspond to the electronic **subbands** of the triangular well. The subbands are two-dimensional and parabolic with minima given by the E_n. The ground state energy is E_0. If the built-in field $\mathcal{E}$ is eliminated from Eq. (15.11) using Eq. (15.1), one gets

$$E_0 \simeq \left(\frac{\hbar^2}{2m^*}\right)^{1/3}\left(\frac{9\pi e^2 N_s}{8\epsilon_0\epsilon_r}\right)^{2/3}. \tag{15.13}$$

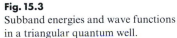

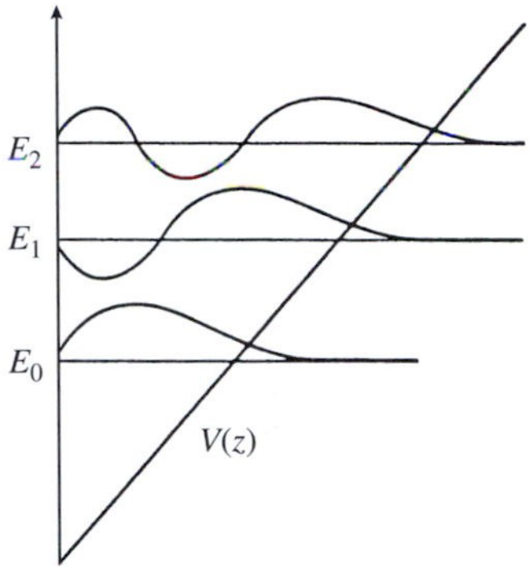

Fig. 15.3
Subband energies and wave functions in a triangular quantum well.

A schematic diagram of the three-lowest subband edges and the corresponding wave functions are shown in Fig. 15.3. It should be noted that the separation of adjacent subband edges decreases as n increases and that the ground state energy E_0 depends only on the number of electrons transferred N_s.

For small N_s and $T=0$ the transferred electrons fill states in the lowest subband up to a certain value of $k_\perp$ that is determined by the two-dimensional density-of-states $\rho_{2D}(E)$. Assuming periodic boundary conditions, the allowed values of k_x and k_y are specified by

$$k_x = \frac{2\pi p_x}{L_x}, \quad k_y = \frac{2\pi p_y}{L_y}, \tag{15.14}$$

where p_x and p_y are integers and $k_\perp = (k_x^2 + k_y^2)^{\frac{1}{2}}$. The area in k_x, k_y space associated with one allowed point is $(2\pi)^2/L_xL_y$. With the energy E given by Eq. (15.12), the area in k_x, k_y space between circles corresponding to E and $E+dE$ is $2\pi k_\perp dk_\perp$. The number of states associated with this area is

$$dN = \frac{2\pi k_\perp dk_\perp}{(2\pi)^2} L_xL_y = \frac{k_\perp dk_\perp}{2\pi} L_xL_y. \tag{15.15}$$

The energy change dE is

$$dE = \frac{\hbar^2 k_\perp dk_\perp}{m^*}. \tag{15.16}$$

The density-of-states is therefore

$$\frac{dN}{dE} = \frac{m^*}{2\pi\hbar^2} L_x L_y. \tag{15.17}$$

Introducing a factor 2 for spin and normalizing to unit area, we obtain the two-dimensional density-of-states ρ_{2D},

$$\rho_{2D} = \frac{m^*}{\pi\hbar^2}, \tag{15.18}$$

which is independent of the energy E.

The energy at the top of the filled states can be expressed as

$$E_{max} = E_0 + \frac{N_s}{\rho_{2D}} = E_0 + \frac{\pi\hbar^2 N_s}{m^*}. \tag{15.19}$$

The corresponding value of $k_\perp$ is $(2\pi N_s)^{\frac{1}{2}}$. E_{max} serves as the Fermi energy of the quantum well when the latter is at equilibrium at $T=0$. Equilibrium also requires that E_{max} be equal to the Fermi energy on the GaAlAs side of the heterojunction as shown in Fig. 15.2. The latter Fermi energy is depressed by the depletion potential V_d due to the depleted donor atoms near the interface. Assuming that the n-type part of the heterojunction is uniformly doped to concentration n_d, the depletion potential is

$$V_d = -\int_0^{-W} \mathcal{E} dz = -\int_0^{-W} \frac{en_d z}{\epsilon_0 \epsilon_r} dz = -\frac{en_d W^2}{2\epsilon_0 \epsilon_r}, \tag{15.20}$$

where W is the depletion width. The change in electrostatic potential energy of an electron at the conduction band edge at $z=0$ and $z=-W$ is

$$-eV_d = \frac{e^2 n_d W^2}{2\epsilon_0 \epsilon_r} = \frac{e^2 N_s^2}{2\epsilon_o \epsilon_r n_d}, \tag{15.21}$$

where $N_s = n_d W$ is the same as N_s introduced earlier in Eq. (15.1).

With the aid of Fig. 15.2 we can write the conduction band offset ΔE_c at $z=0$ as

$$\Delta E_c = E_0 + \frac{\pi\hbar^2 N_s}{m^*} + E_d - eV_d, \tag{15.22}$$

where E_d is the donor binding energy in GaAlAs. Eliminating V_d using Eq. (15.21) gives

$$\Delta E_c = E_0(N_s) + \frac{\pi\hbar^2 N_s}{m^*} + E_d + \frac{e^2 N_s^2}{2\epsilon_0 \epsilon_r n_d}. \tag{15.23}$$

This expression relates ΔE_c to the charge transferred N_s and has been obtained under the assumption that the Fermi energy in the GaAlAs is unaffected by the charge transfer. Since ΔE_c is also unaffected by charge transfer and can be determined from the electron affinities of the individual components (see Fig. 15.1), one can calculate N_s by solving Eq. (15.23).

15.4 Square quantum well

A square quantum well is formed by taking a sufficiently thin layer of material A and placing it between two thick layers of material B. The bandgap of B must be larger than that of A, and the band discontinuities should be such that both conduction electrons and holes are confined in material A. As an example we can take a structure consisting of GaAlAs/GaAs/GaAlAs which is shown schematically in Fig. 15.4.

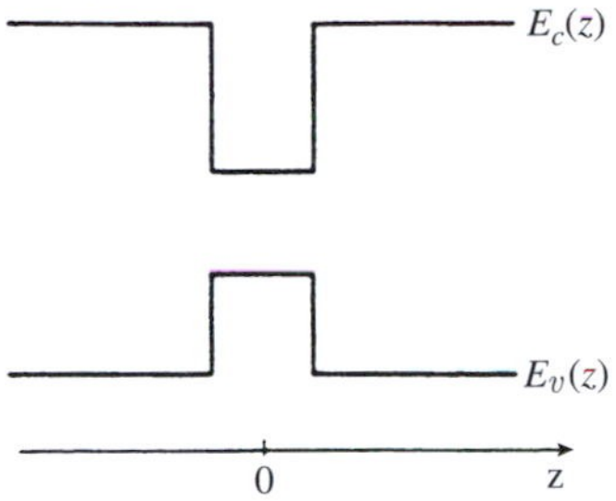

Fig. 15.4
Schematic diagram of a GaAlAs/GaAs/GaAlAs heterostructure.

15.4.1 Conduction electron energy levels

The energy levels of the conduction band in the heterostructure represented by Fig. 15.4 can be calculated using effective mass theory in the **envelope function approximation**. It is assumed that on the scale of variation of the envelope function $\chi_n(z)$ the interface potential is well localized at the geometrical interface. It is further assumed that the two constituents have the same crystal structure and the same lattice constant. The periodic parts of the Bloch function at $\boldsymbol{k} = 0$ can then be taken to be the same (Bastard 1988):

$$u_{c0}^{A}(\boldsymbol{r}) = u_{c0}^{B}(\boldsymbol{r}) = u_{c0}(\boldsymbol{r}).$$

The heterostructure wave function for a conduction electron accordingly takes the form

$$\psi_c(\boldsymbol{r}) = e^{i\boldsymbol{k}_\perp \cdot \boldsymbol{r}_\perp} u_{c0}(\boldsymbol{r}) \chi_n^{A,B}(z), \tag{15.24}$$

where $\chi_n^A(z)$ applies in material A and $\chi_n^B(z)$ in material B. The envelope function satisfies the effective mass equation

$$\left[-\frac{\hbar^2}{2m^*(z)}\frac{d^2}{dz^2} + V_c(z)\right]\chi_n^{A,B}(z) = E_n \chi_n^{A,B}(z), \tag{15.25}$$

where $V_c(z)$ designates the energy at the bottom of the conduction band throughout the structure and $m^*(z)$ is the conduction electron effective mass. The boundary conditions at the interfaces are the continuity of $\chi_n(z)$ and of $[1/m^*(z)][d\chi_n(z)/dz]$ corresponding to the continuity of charge density and current density, respectively. For the continuity of current density it is necessary to use the condition stated rather than the usual continuity of $d\chi_n(z)/dz$ alone.

15.4.1.1 Infinite quantum well

For the deepest energy levels of quantum wells that are not too narrow or shallow, it is reasonable to approximate $V_c(z)$ by an **infinite quantum well** having $V_c(z) = 0$ inside the well and $V_c(z) = \infty$ outside as shown in Fig. 15.5. Explicitly, for a well of width L,

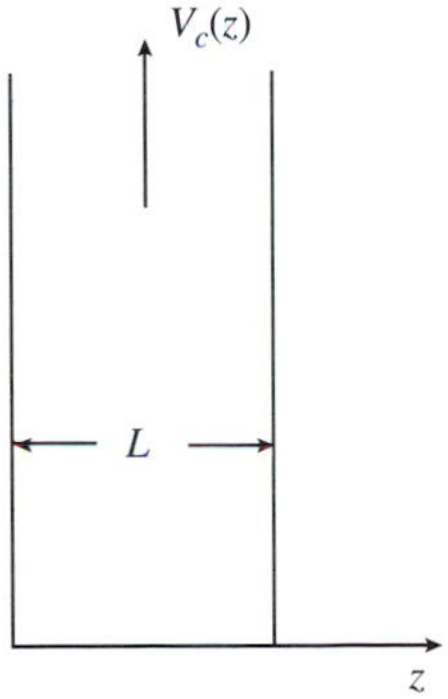

Fig. 15.5
Diagram of an infinite square well potential.

$$V_c(z) = \begin{cases} 0, & |z| < \dfrac{L}{2} \\ \infty, & |z| > \dfrac{L}{2} \end{cases}. \tag{15.26}$$

Since an electron cannot penetrate into a semi-infinite region of infinite potential energy, the envelope function must satisfy the conditions

$$\chi_n\left(\frac{L}{2}\right) = \chi_n\left(-\frac{L}{2}\right) = 0. \tag{15.27}$$

Independent solutions of the effective mass equation are

$$\chi_n(z) = (2/L)^{\frac{1}{2}} \sin k_n z \tag{15.28a}$$

$$\chi_n(z) = (2/L)^{\frac{1}{2}} \cos k_n z, \tag{15.28b}$$

where

$$k_n = \left(\frac{2m^* E_n}{\hbar^2}\right)^{\frac{1}{2}}. \tag{15.29}$$

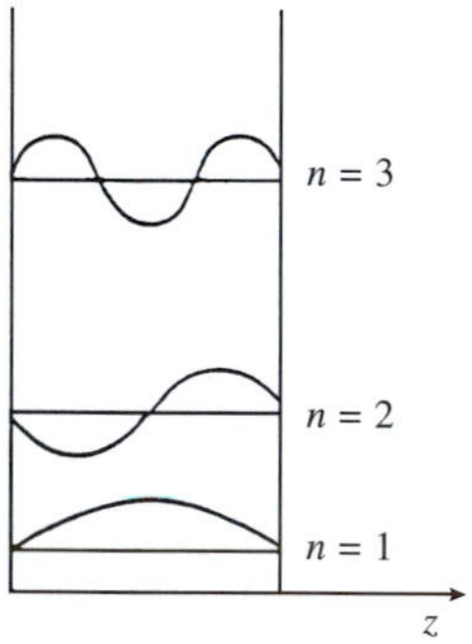

Fig. 15.6
Diagram of the lowest energy levels and envelope functions for an infinite square well potential.

The boundary conditions, Eq. (15.27), require that

$$k_n = \frac{n\pi}{L}, \tag{15.30}$$

where n is a positive integer. Even values of n apply to Eq. (15.28a) and odd values to Eq. (15.28b). The energy eigenvalues are then given by

$$E_n = \frac{\pi^2 \hbar^2 n^2}{2m^* L^2}. \tag{15.31}$$

A diagram of the first few energy levels and envelope functions is shown in Fig. 15.6. With respect to inversion at the center of the well, the even eigenfunctions involve $\cos k_n z$ and the odd eigenfunctions $\sin k_n z$.

If the translational energy parallel to the interface is included, the full energy eigenvalues are given by Eq. (15.12) with E_n specified by Eq. (15.31). Just as for a triangular well, the eigenstates can be grouped into two-dimensional subbands.

15.4.1.2 Finite quantum well

In a real heterostructure, the conduction band edge of material B is never infinitely high compared to that of material A. This fact leads us to consider the **finite quantum well** illustrated in Fig. 15.7. The conduction band edge of material B is at an energy V_0 higher than that of material A:

$$V_c(z) = \begin{cases} -V_0 & |z| < \dfrac{L}{2} \\ 0, & |z| > \dfrac{L}{2}. \end{cases} \tag{15.32}$$

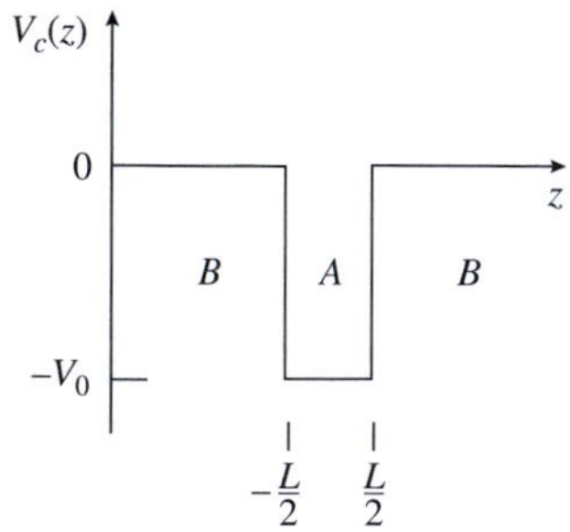

Fig. 15.7
Diagram of a finite square well potential.

The envelope functions in material B are no longer zero, but we can still classify them as even or odd with respect to inversion at the center of the well. For the even functions, we can write the solutions of the effective mass equation as

$$\chi_n(z) = C_1 \cos k_A z, \qquad |z| < \frac{L}{2} \tag{15.33a}$$

$$= C_2 \exp\left[-k_B\left(z - \frac{L}{2}\right)\right], \qquad z > \frac{L}{2} \tag{15.33b}$$

$$= C_2 \exp\left[k_B\left(z + \frac{L}{2}\right)\right], \qquad z < -\frac{L}{2}. \tag{15.33c}$$

For the odd functions,

$$\chi_n(z) = C_1 \sin k_A z, \qquad |z| < \frac{L}{2} \tag{15.34a}$$

$$= C_2 \exp\left[-k_B\left(z - \frac{L}{2}\right)\right], \qquad z > \frac{L}{2} \tag{15.34b}$$

$$= -C_2 \exp\left[k_B\left(z + \frac{L}{2}\right)\right], \qquad z < -\frac{L}{2}. \tag{15.34c}$$

The energy eigenvalues for both cases can be expressed as

$$E_n = \frac{\hbar^2 k_A^2}{2m_A^*} - V_0 \quad \text{or} \quad E_n = -\frac{\hbar^2 k_B^2}{2m_B^*}, \tag{15.35}$$

where $-V_0 < E_n < 0$. Their precise specification is provided by the boundary conditions at $z = \pm L/2$ that χ_n and $(1/m^*)d\chi_n/dz$ are continuous. For the even solutions we obtain

$$C_1 \cos(k_A L/2) = C_2 \tag{15.36a}$$

$$(k_A/m_A^*) C_1 \sin(k_A L/2) = k_B C_2 / m_B^*. \tag{15.36b}$$

A nontrivial solution of Eqs. (15.36) requires that

$$(k_A/m_A^*) \tan(k_A L/2) = (k_B/m_B^*). \tag{15.37}$$

Eliminating E_n from Eqs. (15.35) yields

$$\frac{\hbar^2 k_A^2}{2m_A^*} + \frac{\hbar^2 k_B^2}{2m_B^*} = V_0. \tag{15.38}$$

Equations (15.37) and (15.38) can be solved numerically for k_A and k_B, which upon substitution into Eq. (15.35) gives the energy eigenvalues.

For the odd solutions, Eq. (15.37) is replaced by

$$(k_A/m_A^*) \cot(k_A L/2) = -(k_B/m_B^*). \tag{15.39}$$

The calculation of k_A, k_B, and E_n now proceeds in a fashion analogous to the even parity case.

A particularly simple situation arises if $m_A^* = m_B^*$. Equations (15.37) and (15.39) then become

$$k_A \sin(k_A L/2) = k_B \cos(k_A L/2) \tag{15.40}$$

and

$$k_A \cos(k_A L/2) = -k_B \sin(k_A L/2), \tag{15.41}$$

respectively, while Eq. (15.38) becomes

$$k_A^2 + k_B^2 = \frac{2m^* V_0}{\hbar^2} \equiv k_0^2. \tag{15.42}$$

For the even states we square both sides of Eq. (15.40) and rearrange terms to give

$$k_A^2 = (k_A^2 + k_B^2)\cos^2(k_A L/2) = k_0^2 \cos^2(k_A L/2).$$

Taking the square root of both sides, we obtain

$$\cos(kL/2) = k/k_0, \tag{15.43}$$

where we have dropped the subscript A on k_A. For the odd states a similar procedure yields

$$\sin(kL/2) = k/k_0. \tag{15.44}$$

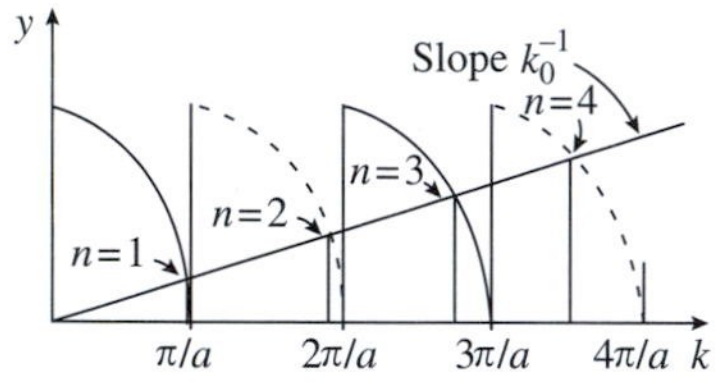

Fig. 15.8
Graphical solution of the equation that determines the energy eigenvalues for a finite square well potential. For even states $y = \cos(kL/2)$ and for odd states $y = \sin(kL/2)$. n identifies the corresponding states of the infinite square well potential.

A graphical solution of Eqs. (15.43) and (15.44) is presented in Fig. 15.8. Note that there is always at least one solution corresponding to the electron bound to the quantum well. As V_0 increases the number of bound states N_b increases according to the formula

$$N_b = 1 + \mathrm{Int}\left[\left(\frac{2m^* V_0 L^2}{\pi^2 \hbar^2}\right)^{\frac{1}{2}}\right], \tag{15.45}$$

where Int(x) denotes the integer part of x. Energies E_n greater than zero correspond to continuum states for which k_B is purely imaginary. The electron is not bound to the quantum well and behaves as a free particle. The results for the infinite well can be obtained by letting $V_0 \to \infty$.

15.4.2 Hole energy levels

In bulk semiconductors the valence bands are typically more complicated than the conduction bands and are degenerate at $\boldsymbol{k} = 0$ as discussed in Chapter 3. The hole states are described by wave functions that contain contributions from the various degenerate bands. Near $\boldsymbol{k} = 0$ the bands can be classified as heavy and light hole bands derived from states with total angular moment $J = 3/2$.

In the case of a quantum well, there are two principal effects to be analyzed: the effect of the quantum well potential V_{QW} and the effect of the $\boldsymbol{k} \cdot \boldsymbol{p}$ perturbation $H_{\boldsymbol{k}\cdot\boldsymbol{p}}$. The quantum well potential reduces the symmetry of the system and lifts the degeneracy of the bands at $\boldsymbol{k} = 0$, the $J_z = \pm\frac{3}{2}$ (heavy hole) band rising above the $J_z = \pm\frac{1}{2}$ (light hole) band as shown in Fig. 15.9a. If the axis of the quantum well is in the [001] direction, the symmetry of the system is equivalent to that arising from a uniaxial stress applied normal to the layers. From $\boldsymbol{k} \cdot \boldsymbol{p}$ perturbation theory one obtains the energies of heavy and light holes near $\boldsymbol{k} = 0$ in the forms (Chemla 1983)

$$E_{hh} = -\frac{\hbar^2}{2m}[(\gamma_1 + \gamma_2)k_\perp^2 + (\gamma_1 - 2\gamma_2)k_z^2] \tag{15.46a}$$

$$E_{\ell h} = -\frac{\hbar^2}{2m}[(\gamma_1 - \gamma_2)k_\perp^2 + (\gamma_1 + 2\gamma_2)k_z^2], \tag{15.46b}$$

where γ_1 and γ_2 are the Luttinger band parameters and $k_\perp^2 = k_x^2 + k_y^2$. This result has the remarkable feature that there are different effective masses in the z-direction and in the $(x–y)$ plane. The "heavy-hole" band has the light mass $m/(\gamma_1 + \gamma_2)$ in the $(x–y)$ plane, and the "light hole" band has the heavy mass $m/(\gamma_1 - \gamma_2)$ in that plane. This situation is represented schematically in Fig. 15.9b. It should be noted that there is a band crossing at $k_\perp \simeq \pi\sqrt{2}/L_z$ predicted by Eqs. (15.46). Actually, the bands do not cross. A more refined calculation shows that the bands repel each other and become nonparabolic as indicated in Fig. (15.9c).

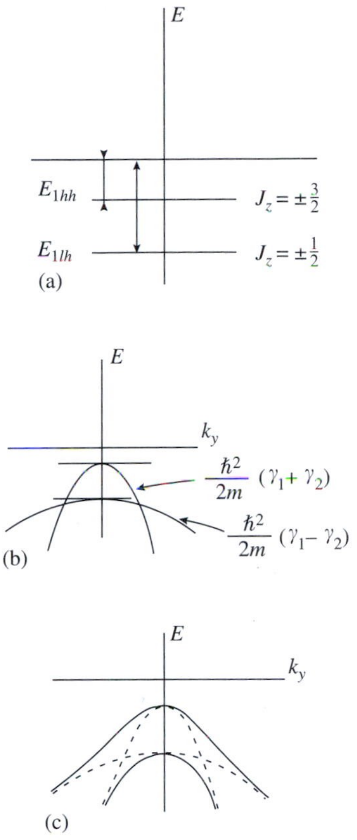

Fig. 15.9
(a) Splitting of the $J_z = \pm\frac{3}{2}$ and $J_z = \pm\frac{1}{2}$ hole levels by a quantum well potential; (b) k-dependence of the $J_z = \pm\frac{3}{2}$ and $J_z = \pm\frac{1}{2}$ bands; (c) anticrossing behavior of the $J_z = \pm\frac{3}{2}$ and $J_z = \pm\frac{1}{2}$ bands.

The energy levels of holes in an infinitely deep quantum well have been obtained analytically (Nedozerov 1970). The energies of the subband edges for heavy and light holes are given by expressions analogous to those for conduction electrons:

$$E_{hh}^{(n)} = \frac{\pi^2\hbar^2 n^2}{2mL^2}(\gamma_1 - 2\gamma_2), \tag{15.47a}$$

$$E_{\ell h}^{(n)} = \frac{\pi^2\hbar^2 n^2}{2mL^2}(\gamma_1 + 2\gamma_2). \tag{15.47b}$$

The dependence of the energy on $k_\perp$ is specified by a rather complicated relation. However, for small $k_\perp$ the subband energies can be expressed in the parabolic forms

$$E_{hh} = E_{hh}^{(n)} + \frac{\hbar^2 k_\perp^2}{2m_n^{(1)}} \tag{15.48a}$$

$$E_{\ell h} = E_{\ell h}^{(n)} + \frac{\hbar^2 k_\perp^2}{2m_n^{(2)}}, \tag{15.48b}$$

where the effective masses $m_n^{(1)}$ and $m_n^{(2)}$ are complicated functions of n. They have the characteristic that the $m_n^{(1)}$ are primarily light masses, whereas the $m_n^{(2)}$ are primarily heavy masses. They thus fit in with the effective masses that appear in the coefficients of $k_\perp^2$ in Eqs. (15.46). Some of the heavy-hole subbands have positive (electron-like) masses near $k_\perp = 0$. Very strong nonparabolic behavior of the subbands is found as $k_\perp$ increases (Weisbuch and Vinter 1991).

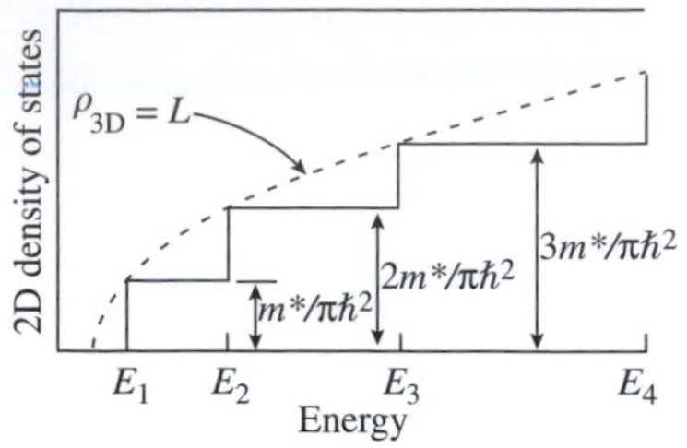

Fig. 15.10
Staircase density-of-states for a 2D quantum well. The dashed curve is the 3D case.

15.5 Density-of-states of quantum wells

We have seen in Section 15.3 that the density-of-states for a parabolic two-dimensional system is given by

$$\rho_{2D} = \frac{m^*}{\pi\hbar^2} \tag{15.49}$$

for all energies above the energy threshold corresponding to $k_\perp = 0$. In a quantum well each subband contributes this amount above its threshold. Adding up the contributions of all the subbands yields a staircase density-of-states profile as shown in Fig. 15.10 for a square quantum well. An interesting comparison with the three-dimensional case can be made by taking the 3D density-of-states

$$\rho_{3D}(E) = \frac{(2m^{*3})^{\frac{1}{2}} E^{\frac{1}{2}}}{\pi^2\hbar^3} \tag{15.50}$$

evaluated at $E = E_n$ and multiplying it by L. The result is indicated by the dashed curve in Fig. 15.10. At each subband threshold, $\rho_{3D}(E)L$ coincides with the total density-of-states of the quantum well: $\rho_{3D}(E_n)L = n\rho_{2D}$.

An important difference between the quantum well and 3D cases is evident as the lowest allowed energy is approached. In this limit ρ_{3D} goes to zero, whereas the quantum well density-of-states retains a finite value. The practical significance of this result is that all dynamical phenomena in a 2D or quantum well system remain finite even at low energies. This remark is pertinent to phenomena such as scattering processes and optical absorption at low temperature.

On the other hand, when one considers situations involving large values of n, such as in thick layers, the behaviors of 3D and quantum well systems cannot be distinguished, in line with the correspondence principle. Under these circumstances the term quantum well is not appropriate, and such systems cannot be regarded as two-dimensional.

15.6 Excitons and shallow impurities in quantum wells

We have seen in Chapter 10 that a conduction electron and a hole, which carry opposite charges, can be bound together by the Coulomb interaction to form an exciton. Excitons have a hydrogen-like energy spectrum. The binding energy of the lowest bound state is the effective Rydberg Ry* given by

$$\mathrm{Ry}^* = \frac{\mu^*}{m\epsilon^2}\mathrm{Ry}, \tag{15.51}$$

where μ^* is the reduced effective mass of the electron and hole, ϵ is the dielectric constant of the semiconductor and Ry = 13.6 eV is the ordinary Rydberg. The effective Bohr radius a_B^*, which is the measure of the separation of the electron and hole in the exciton ground state, is given by

$$a_B^* = \frac{m\epsilon}{\mu^*}a_B, \tag{15.52}$$

where $a_B = 0.529$ Å is the ordinary Bohr radius.

From our discussion of shallow donor impurities in Chapter 5 we see that the binding energy of an exciton and that of a shallow donor can be comparable. In the limit that the effective mass of the hole becomes infinite, the binding energy and effective Bohr radius of an exciton become the same as those of a shallow donor. The effective Bohr radius for a 3D exciton is typically on the order of 100 Å, since $\epsilon \simeq 10$ and $\mu^* \simeq 0.1\,\mathrm{m}$. When the quantum well thickness is on the order of or smaller than the exciton diameter, we expect the wave function and energy levels of the exciton to be significantly modified.

In the case of an infinitely deep quantum well whose width is very small, $L \ll a_B^*$, we can regard the exciton as two-dimensional with an effective Rydberg $\mathrm{Ry}_{2D}^* = 4\mathrm{Ry}^*$. The ground state wave function for the relative motion of the electron and hole can be expressed in polar coordinates ρ, φ as

$$\Psi = (8/\pi a_B^{*2})^{\frac{1}{2}} e^{-2\rho/a_B^*}, \tag{15.53}$$

where $\rho = [(x_e - x_h)^2 + (y_e - y_h)^2]^{\frac{1}{2}}$ and x_e, y_e and x_h, y_h are the coordinates of the electron and hole, respectively. The energy levels of the 2D exciton are given by

$$E_{n2D} = E_g - \frac{1}{(n - \frac{1}{2})^2}\mathrm{Ry}^*, \quad n = 1, 2, \ldots \tag{15.54}$$

where n is the principal quantum number with positive integer values and the zero for energy has been taken to be the top of the valence band. The effective Bohr radius $(\langle\rho^2\rangle)^{\frac{1}{2}} = (\langle\Psi|\rho^2|\Psi\rangle)^{\frac{1}{2}}$ is related to a_B^* by $(\langle\rho^2\rangle)^{\frac{1}{2}} = (3/8)^{\frac{1}{2}} a_B^*$.

When L is on the order of or greater than a_B^* the charge distribution of the exciton is ellipsoidal in shape and is characterized by transverse and longitudinal reduced Bohr radii. Exciton binding energies and reduced Bohr radii for this case have been calculated (Bastard *et al.* 1982) using a variation method. The results for the binding energy $Ry^*(L)$ are shown in Fig. 15.11.

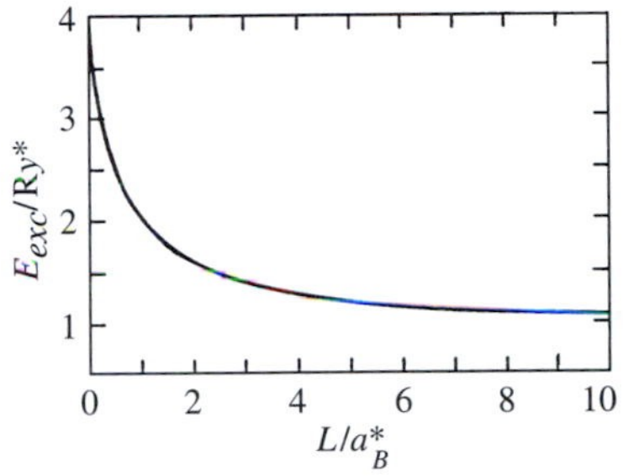

Fig. 15.11
Reduced binding energy of an exciton versus reduced well thickness for a GaAlAs/GaAs heterostructure (after Bastard *et al.* 1982).

The increase of exciton binding energy is one of the remarkable properties of a quantum well. Even at room temperature the optical properties of GaAs-based quantum wells are dominated by exciton effects. Promising applications of room temperature excitons are based on features such as optical bistability, four-wave mixing, and large electro-optic coefficients.

As the principal quantum number $n \to \infty$, the energies of the bound exciton levels approach E_g, as can be seen from Eq. (15.54). Energies greater than E_g correspond to the free-particle continuum in which the electron and hole are not bound together.

The drastic difference in binding energies of 2D and 3D excitons is strikingly revealed by the optical absorption spectra displayed in Fig. 15.12. The separation of the $n = 1$ absorption peak from the continuum due to interband transitions is four times larger for 2D excitons than for 3D excitons.

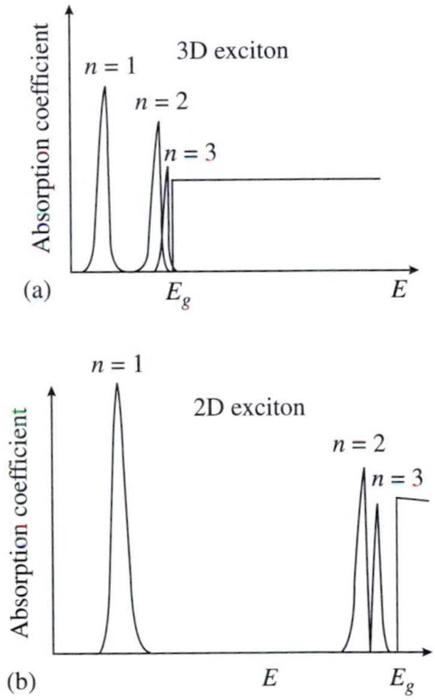

Fig. 15.12
Optical absorption coefficient versus photon energy for excitons in (a) the 3D case and (b) the 2D case (after Weisbuch and Vinter 1991).

The problem of shallow donor impurities is somewhat complicated because the binding energy of a shallow impurity is affected by the proximity of the impurity to the potential barriers at the well interfaces. For an infinite potential barrier the wave function of an impurity state must vanish at the barrier. If the impurity is located at a barrier, the wave function for the ground state should be a truncated p-like state, whereas for an impurity atom located at the center of the well the ground state should be s-like. In the large-well limit the ground state binding energies for these two cases should be $E_{imp}/4$ and E_{imp}, respectively, where E_{imp} is the 3D impurity binding energy.

Calculations of acceptor binding energies are more complicated than for donors because of the degeneracy of the valence bands.

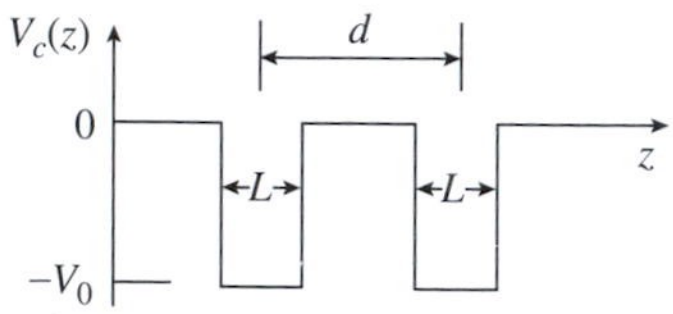

Fig. 15.13
Schematic diagram of a double-well structure.

15.7 Coupled quantum wells and superlattices

15.7.1 Double-well structure

The double-well structure results from placing two quantum wells side-by-side with a common barrier between them. A schematic diagram is given in Fig. 15.13 for the case of interest involving finite barriers. With decreasing barrier width, a wave function centered on one well will have increasing penetration through the barrier into the second well and thereby increasingly overlap a wave function centered on the second well. For not too narrow barriers we can treat the overlap as a perturbation.

The Hamiltonian for a carrier in a double well is

$$H = T + V_1(z) + V_2(z), \tag{15.55}$$

where T is the kinetic energy operator and $V_1(z)$ and $V_2(z)$ are the potential energies associated with wells 1 and 2, respectively. Let χ_1 and χ_2 be the ground-state wave functions for the isolated wells satisfying the Schrödinger equations

$$(T + V_1(z))\chi_1 = E_1\chi_1 \tag{15.56a}$$

$$(T + V_2(z))\chi_2 = E_1\chi_2, \tag{15.56b}$$

where E_1 is the ground state energy of an isolated well. By analogy with the tight-binding method of energy band theory, we write the ground-state wave function Ψ for the double well as a linear combination of χ_1 and χ_2:

$$\Psi = a_1\chi_1 + a_2\chi_2. \tag{15.57}$$

Substituting into the Schrödinger equation

$$H\Psi = E\Psi \tag{15.58}$$

and taking matrix elements with respect to χ_1 and χ_2, we obtain

$$(E_1 + \bar{V}_1 - E)a_1 + (SE_1 + \bar{V}_{12} - SE)a_2 = 0 \tag{15.59a}$$

$$(SE_1 + \bar{V}_{12} - SE)a_1 + (E_1 + \bar{V}_1 - E)a_2 = 0, \tag{15.59b}$$

where

$$S = \langle \chi_1 | \chi_2 \rangle \tag{15.60a}$$

$$\bar{V}_1 = \langle \chi_1 | V_2(z) | \chi_1 \rangle = \langle \chi_2 | V_1(z) | \chi_2 \rangle \tag{15.60b}$$

$$\bar{V}_{12} = \langle \chi_1 | V_1(z) | \chi_2 \rangle = \langle \chi_2 | V_2(z) | \chi_1 \rangle. \tag{15.60c}$$

Setting the determinant of the coefficients of a_1 and a_2 to zero, we obtain the secular equation

$$\begin{vmatrix} E_1 - E + \bar{V}_1 & S(E_1 - E) + \bar{V}_{12} \\ S(E_1 - E) + \bar{V}_{12} & E_1 - E + \bar{V}_1 \end{vmatrix} = 0, \tag{15.61}$$

whose solutions are

$$E = E_1 + \frac{\bar{V}_1 \pm \bar{V}_{12}}{1 \pm S}. \tag{15.62}$$

For $S \ll 1$, we have the approximate solutions

$$E = E_1 + \bar{V}_1 \pm \bar{V}_{12}. \tag{15.63}$$

The coupling of the two wells therefore produces a splitting of their ground-state levels by $\sim 2\bar{V}_{12}$. Similar considerations apply to excited states.

The perturbed wave functions Ψ_+ and Ψ_- are given by

$$\Psi_+ = \frac{1}{\sqrt{2}}(\chi_1 + \chi_2) \tag{15.64a}$$

$$\Psi_- = \frac{1}{\sqrt{2}}(\chi_1 - \chi_2) \tag{15.64b}$$

corresponding to the energy eigenvalues $E_1 + \bar{V}_1 + \bar{V}_{12}$ and $E_1 + \bar{V}_1 - \bar{V}_{12}$, respectively. Since we take the zero of energy at the top of the wells, $V_1(z)$ and $V_2(z)$ are negative within the wells, whereas χ_1 and χ_2 are positive in accordance with Eqs. (15.33). $\bar{V}_1$ and $\bar{V}_{12}$ are therefore negative, so Ψ_+ is the ground state wave function and Ψ_- is an excited state wave function.

15.7.2 Superlattices: periodic coupled quantum wells

15.7.2.1 Wave functions and energy levels

Coupling together larger and larger numbers of quantum wells leads to the creation of a **superlattice** with an allowed band of energy levels. For N wells, the N-degenerate zero-order states (excluding spin) of the uncoupled wells give rise to a band of N states when coupling is imposed. With spin included there are $2N$ states in the band.

In Fig. 15.14 is presented a diagram of a linear array of coupled quantum wells that are equally spaced to form a superlattice. To obtain the wave functions and energy eigenvalues, we impose periodic boundary conditions and apply the tight-binding method in a manner exactly analogous to that

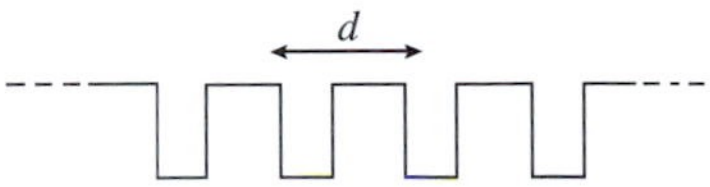

Fig. 15.14
Diagram of a superlattice composed of coupled quantum wells.

employed in Chapter 2 to derive expressions for the Bloch functions and energy bands of a periodic lattice. The tight-binding form of the Bloch function associated with a subband of quantum number n can be expressed as

$$\Psi_{k_z}(z) = \frac{1}{\sqrt{N}} \sum_{p=0}^{N-1} e^{ik_z pd} \chi_n(z - pd), \tag{15.65}$$

where k_z is the wave vector, d is the period of the superlattice, and p is an integer. The energy eigenvalues are given by

$$E_n(k_z) = E_n + V_1(n) + 2V_{12}(n) \cos k_z d \tag{15.66}$$

where

$$V_1(n) = \int_{-\infty}^{\infty} \chi_n(z - d) V(z) \chi_n(z - d) dz \tag{15.67a}$$

$$V_{12}(n) = \int_{-\infty}^{\infty} \chi_n(z - d) V(z) \chi_n(z) dz, \tag{15.67b}$$

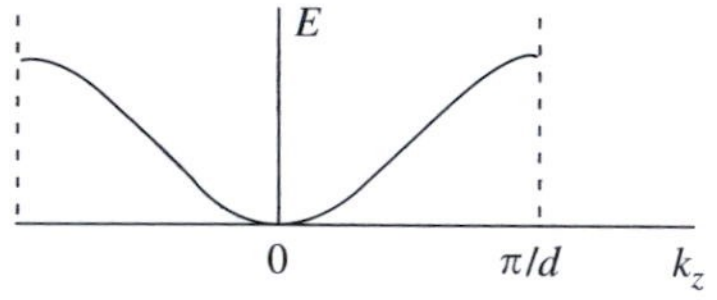

Fig. 15.15
Energy versus wave vector for a superlattice of period d.

and we have neglected the overlap integral $\int_{-\infty}^{\infty} \chi_n(z - d) \chi_n(z) dz$. $V_{12}(n)$ is called the **transfer matrix element**. The energy band described by Eq. (15.66) is a **miniband**.

A plot of allowed energies versus wave vector is given in Fig. 15.15. Its shape is typical of a carrier moving in a periodic potential. The width of the miniband is $4V_{12}(n)$ which is twice the separation of the eigenstates of a double well. The factor of two difference reflects the fact that a given well in the superlattice interacts with two neighboring wells and not just one.

The periodic boundary conditions require that the wave vector k_z take on the discrete values

$$k_z = \frac{2\pi j}{Nd}, \tag{15.68}$$

where j is an integer in the range $-\frac{N}{2} + 1 \le j \le \frac{N}{2}$ and N is assumed to be even. Values of j outside this range do not correspond to states that are physically distinct from those associated with values inside the range. Hence, there are N distinct states in a miniband, each of which can accommodate two electrons of opposite spin.

The transfer matrix element $V_{12}(n)$ decreases as the width of the barrier between wells increases. Consequently, the width of a miniband also decreases as the barrier width increases, as shown in Fig. 15.16 for the ground subband $n = 1$. On the other hand, the gaps between successive minibands increase as the barrier width increases.

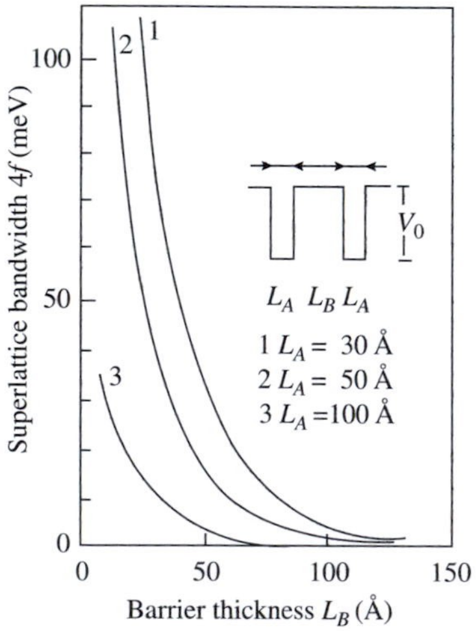

Fig. 15.16
Superlattice bandwidth $4V_{12}(1)$ versus barrier width L_B for three well widths L_A (after Bastard 1983).

15.7.2.2 Density-of-states in a superlattice

The formation of a superlattice produces a significant change in the density-of-states compared to that of a single quantum well. The dispersion in a miniband eliminates the step-function discontinuities in density-of-states characteristics of single wells.

If the kinetic energy of a carrier perpendicular to the superlattice axis is included, the total energy E is

$$E \equiv E_n(k_\perp, k_z) = \frac{\hbar k_\perp^2}{2m^*} + E_n(k_z), \tag{15.69}$$

Then the density-of-states for the superlattice can be written as (Weisbuch and Vinter 1991)

$$\rho(E) = \sum_n \rho_n(E) = \sum_n N \frac{m^*}{\pi\hbar^2} \arccos\left[\frac{E - E_n - V_1(n)}{2V_{12}(n)}\right], \tag{15.70}$$

where n is the miniband index. A plot of the density-of-states is given in Fig. 15.17 together with those for a 2D quantum well and for a 3D isotropic system. It is noteworthy that the superlattice density-of-states has singularities in its derivatives.In the limit of infinitely wide barriers between adjacent wells, the width of the miniband approaches zero, and we recover the density-of-states of isolated quantum wells.

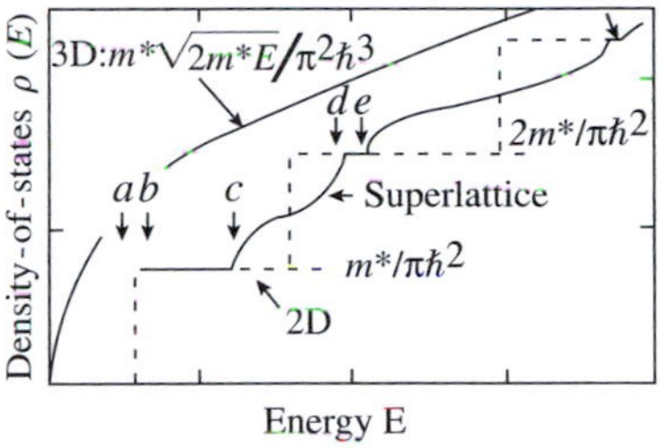

Fig. 15.17
Density-of-states versus energy for a superlattice, a 2D quantum well, and a 3D isotropic system (after Esaki 1983).

15.8 Modulation doping of heterostructures

We have seen in Section 15.2 that charge transfer can occur in heterostructures with the result that the charged impurities are not in the same region as the free carriers. Consequently, this spatial separation of the carriers and the ionized impurities can lead to greatly reduced ionized impurity scattering and very high mobilities. This is one of the principal reasons for the great interest in heterostructures.

In the structure GaAlAs/GaAs, where the wide-gap material GaAlAs is strongly n-type, electron transfer occurs to the GaAs layer until equilibrium is reached. The number of electrons transferred per unit area N_s is related to the energy difference ΔE_c of the conduction band edges of the two materials by Eq. (15.23). This result was obtained under the assumption that the wave function of an electron in the GaAs has negligible penetration into the barrier and therefore produces negligible modification of the electrostatic potential. Calculations show that the wave function penetrates no more than 20 Å into the barrier, which is much less than the width of the depletion region in the barrier. The electron wave function accordingly samples very little of the depletion region and is relatively unaffected by barrier penetration.

In deriving Eq. (15.23) it was assumed that the entire depletion lies in the n-type GaAlAs. However, there is always some residual doping in the GaAs which leads to a depletion region in the latter if it is p-type. Associated with this depletion region is an electric field that contributes to the potential difference across the junction. The depletion charge N_{dep} in the GaAs is specified by the relation (Weisbuch and Vinter 1991)

$$N_{dep} = n_a W_a \simeq n_a \left(\frac{2\epsilon_0\epsilon_r E_g}{n_a e^2}\right)^{\frac{1}{2}} = \left(\frac{2n_a\epsilon_0\epsilon_r E_g}{e^2}\right)^{\frac{1}{2}}, \tag{15.71}$$

where W_a is the width of the depletion region.

At low impurity concentration the depletion width in the GaAs is much larger than the width of the depletion region in the GaAlAs, and its associated potential can be considered triangular in the region of interest. If N_s is large compared to N_{dep}, this potential is very small and can be neglected. On the other hand, if $N_s \lesssim N_{dep}$, the residual impurity potential is very important in determining the electron excited states whose wave functions extend far from the interface.

The case of n-type residual doping in the GaAs is more difficult to analyze because the Fermi energy away from the interface depends both on the charge transferred across the interface and on the position of the donor level.

One often uses an undoped GaAlAs spacer layer to increase the separation between the ionized donor atoms and the channel electrons. The spacer layer decreases the Coulomb interaction between the ionized centers and the free electrons and therefore considerably increases the electron mobility.

15.9 Self-consistent energy-level calculations

In the quantum mechanical analysis of the energy levels of electrons in a triangular potential well presented in Section 15.3, the Coulomb interaction between electrons was neglected for simplicity. However, this interaction can significantly modify the electronic eigenstates at electron concentrations of interest. It can be conveniently taken into account using the Hartree approximation. The electron–electron interaction is described by a potential energy $V_{ee}(z)$ that satisfies the Poisson equation

$$\frac{d^2 V_{ee}(z)}{dz^2} = \frac{e^2 N_s}{\epsilon_0 \epsilon_r} |\chi_n(z)|^2. \tag{15.72}$$

Physically, an electron is assumed to move in the average electrostatic field produced by all the other electrons. The Schrödinger equation satisfied by the effective mass function $\chi_n(z)$ is

$$\left[-\frac{\hbar^2}{2m^*}\frac{d^2}{dz^2} + V_0(z) + V_{imp}(z) + V_{ee}(z) \right] \chi_n(z) = E_n \chi_n(z), \tag{15.73}$$

where $V_0(z)$ is the heterojunction potential energy and $V_{imp}(z)$ is the channel ionized impurity potential energy. Equations (15.71) and (15.72) must be solved self-consistently. This task has been accomplished using the variational method (Ando 1982) with a ground-state function of the form

$$\chi_0(z) = B b^{\frac{1}{2}} (bz + \beta) e^{-bz/2}, \quad z > 0 \tag{15.74a}$$

$$= B'(b')^{\frac{1}{2}} e^{b'z/2}, \quad z < 0 \tag{15.74b}$$

where B, B', b, b' and β are variational parameters and penetration into the barrier region $z < 0$ is accounted for. Satisfying the usual boundary and normalization conditions leaves only the two variational paramters b and b' to be determined. A good approximation is to take b' to have the standard value for penetration into a barrier of height ΔE

$$b' = (2/\hbar)(2m^*\Delta E)^{\frac{1}{2}}, \tag{15.75}$$

where the differences in effective mass inside and outside the barrier are neglected. A diagram showing the ground-state subband energy and wave function is given in Fig. 15.18. The penetration of the wave function into the barrier is evident.

In real devices at finite temperature one must use the Fermi–Dirac distribution function $f(E)$ to determine energy level occupancy. The surface concentration of channel carriers N_s is given by

$$N_s = \int_{E_0}^{\infty} \rho_{2D}(E)f(E)dE. \tag{15.76}$$

Since the density-of-states $\rho_{2D}(E)$ is a constant as stated by Eq. (15.49), use of Eq. (6.1) for $f(E)$ yields the result

$$N_s = \frac{k_B T m^*}{\pi\hbar^2}\log[1 + e^{(E_F - E_0)/k_B T}]. \tag{15.77}$$

Combining this expression with Eq. (15.23) permits one to determine the Fermi energy E_F.

The problem of creating design rules needed to obtain optimum performance of single heterostructure devices can now be tackled. If the concentration n_d of donor impurities in the barrier material is increased, the charge transfer is increased, since the depletion width W_d varies as $n_d^{-\frac{1}{2}}$ from an equation analogous to Eq. (15.71) and

$$N_s = n_d W_d \propto n_d^{\frac{1}{2}}. \tag{15.78}$$

An increased transfer of carriers leads to more effective screening of channel impurities, but produces at the same time increased scattering by the ionized impurities in the barrier. The latter scattering can be reduced by inserting an undoped spacer layer as in Fig. 15.18, but this lowers the charge transfer. Experimentally one finds that the channel carrier density is always limited; therefore, other configurations such as the double heterojunction have been studied.

In Fig. 15.19 is given a diagram of a double heterojunction in which the active layer is a wide, undoped layer of GaAs sandwiched between two appropriately doped barriers of GaAlAs. We have here a double heterojunction which leads to a doubling of the channel carrier concentration N_s. Modulation doping of multiple quantum wells offers the possibility of even higher concentrations and operation under high-current conditions.

Another difficulty that must be faced in the design of an efficient heterostructure is the great variability of donor energy levels E_d in $Ga_{1-x}Al_xAs$, which range from 6 meV at $x \leq 0.1$ to 160 meV or more for indirect gap material with $x > 0.235$. This problem can be avoided by taking the charge-transferring side of the heterojunction to be a GaAs/GaAlAs superlattice with the GaAs layers strongly n-type and the GaAlAs layers undoped and

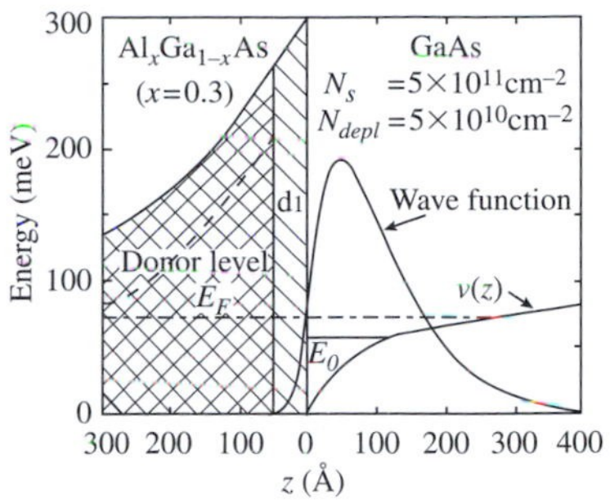

Fig. 15.18
Calculated ground state energy level and wave function for a GaAlAs/GaAs selectively doped interface with spacer thickness $d_1 = 50$ Å (after Ando 1982).

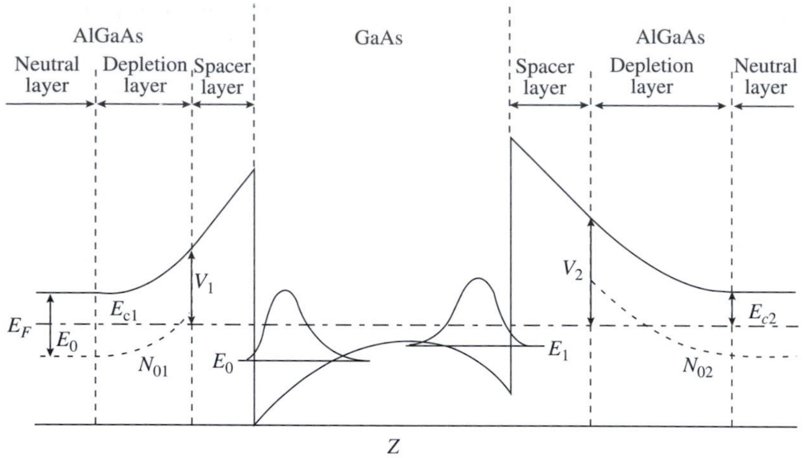

Fig. 15.19
Schematic diagram of a double heterojunction (after Miyatsuji *et al.* 1985).

thin enough to allow carrier tunneling. If the GaAs layers are very thin also, the lowest confined level is broadened into a band and raised above the bulk GaAs conduction band to the proximity of the GaAlAs barrier level. Under these circumstances electrons can flow between the GaAs confined donor levels and the GaAs channel to the right of the superlattice as shown in Fig. 15.20.

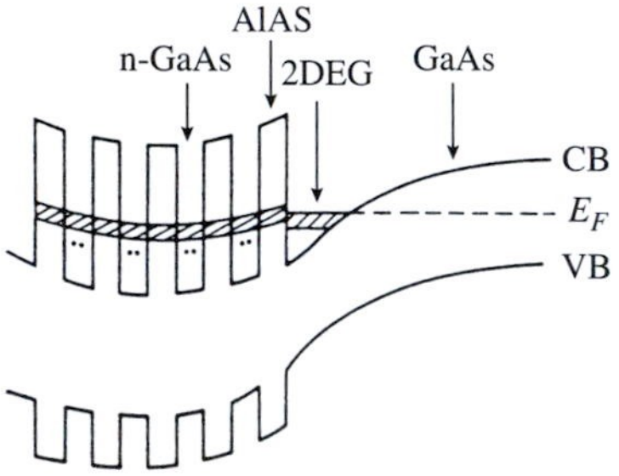

Fig. 15.20
Heterostructure of a selectively doped GaAlAs/GaAs superlattice and undoped GaAs (after Weisbuch and Vinter 1991).

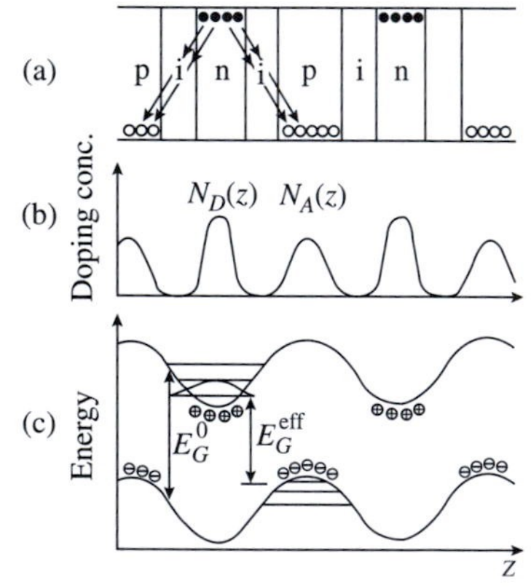

Fig. 15.21
N–I–P–I band structure formation: (a) electrons from neutral donors combine with holes on neutral acceptors to produce a net space charge shown in (b). The resulting band-gap variation and carrier confinement are shown in (c) (after Weisbuch and Vinter (1991).

15.10 N–I–P–I structures

If the doping of an otherwise homogeneous crystal is spatially modulated, a superlattice can be produced (Esaki and Tsu 1970). The spatial modulation causes a folding-back of the Brillouin zone and a new energy-band structure along the axis of the superlattice. Such systems are doping superlattices, compared to structural superlattices, and consist of periodic n-doped, intrinsic, p-doped, intrinsic, . . . , multilayers known as **N–I–P–I structures.** A diagram of such a structure is shown in Fig. 15.21 together with the doping concentration and energy profiles.

The energy levels of a N–I–P–I structure can be calculated using a self-consistent procedure similar to that employed in Section 15.9. The single-particle potential energy $V(z)$ can be written as (Weisbuch and Vinter 1991)

$$V(z) = V_{imp}(z) + V_H(z) + V_{xc}(z), \tag{15.79}$$

where $V_{imp}(z)$ is the electrostatic potential energy due to ionized impurities, $V_H(z)$ is the Hartree potential energy of electrons and holes, and $V_{xc}(z)$ is the exchange and correlation potential energy. $V_{imp}(z)$ is specified by Poisson's equation

$$\frac{d^2V_{imp}(z)}{dz^2} = \frac{e^2}{\epsilon_0\epsilon_r}[n_d(z) - n_a(z)] \tag{15.80}$$

and $V_H(z)$ by

$$\frac{d^2V_H(z)}{dz^2} = -\frac{e^2}{\epsilon_0\epsilon_r}[|\chi_e(z)|^2 - |\chi_h(z)|^2], \tag{15.81}$$

where $\chi_e(z)$ and $\chi_h(z)$ are effective mass functions for electrons and holes, respectively. $V_{xc}(z)$ can be calculated within the LDA of the density-functional formalism as (Ruden and Döhler 1983)

$$V_{xc}(z) = 0.611 \frac{e^2}{4\pi\epsilon_0\epsilon_r}\left[\frac{4\pi}{3n(z)}\right]^{1/3}, \tag{15.82}$$

where $n(z)$ is the electron concentration. Since the hole mass is relatively large, the holes tend to occupy an acceptor impurity band, which is equivalent to a neutralization of the negatively charged acceptors.

The energy levels are obtained by solving the one-dimensional effective mass equation

$$\left[-\frac{\hbar^2}{2m^*_{e,h}}\frac{d^2}{dz^2} + V(z)\right]\chi_{e,h}(z) = E\chi_{e,h}(z), \tag{15.83}$$

where m^*_e and m^*_h are the effective masses of electrons and holes, respectively. Certain conclusions can be drawn from the preceding set of equations:

1. In the case of exact compensation, no free carriers exist at low temperature without external excitation.
2. When there is unequal doping of donors and acceptors, free carriers will accumulate in the potential wells associated with the dopant of higher concentration.
3. For sufficiently large dopings and spacings, the effective bandgap can become negative, i.e., the top of the valence band in the p-type regions can be higher than the bottom of the conduction band in the n-type regions. Carrier transfer then takes place from hole wells to electron wells until a zero bandgap is attained. Associated with this transfer are band filling, reduction of the periodic superlattice potential as a result of charge neutralization by the transferred charges, and modification of the quantized energy levels.
4. When nonequilibrium conditions exist such as carrier injection or photoexcitation, the populations of electrons and holes in the wells can increase and produce charge neutralization accompanied by an effective bandgap increases. Since the electrons and holes are spatially separated, the radiative recombination rate is greatly reduced compared to the bulk rate. This situation is reminiscent of the relatively low recombination rates in indirect gap semiconductors. Interestingly, the rates of nonradiative recombination are also greatly reduced, so reasonable quantum efficiencies can be attained. One can therefore anticipate that doped superlattices can form the basis of tunable light sources and photodetectors with large gain. To overcome the relatively poor mobilities in the doped regions, undoped small-gap semiconductor layers can be inserted into the middle of the n- and p-doped layers of the superlattice, thereby creating modulation-doped heterostructures with very high mobilities.

Problems

1. Consider a triangular quantum well potential in a n-GaAlAs/i-GaAs heterostructure. Calculate the number of electrons transferred per unit area N_s using the values $\Delta E_c = 0.5$ eV, $E_d = 0.03$ eV, $n_d = 10^{18}\,\text{cm}^{-3}$, $m^* = 0.1$ m, and $\epsilon_r = 12$. You may solve the equation for N_s graphically, by iteration, or on a computer.
2. For the **finite** square well potential show that the even parity eigenfunctions are associated with odd values of n for the corresponding eigenfunctions of the **infinite** square well potential and that the odd parity eigenfunctions are associated with even values of n.
3. For a finite quantum well with $L = 100$ Å, $V_0 = 0.3$ eV, and $m_A = 0.07$ m, calculate the energies of the lowest even and odd states for $m_B = 0.21$ m.
4. A double quantum well consists of two wells whose midpoints are a distance d apart and whose parameters have the values given in problem 3. Taking $d = 150$ Å, calculate the energy of the electronic ground state and compare it with that of a single well.
5. Consider a superlattice composed of quantum wells of the type in problem 3 and having a period $d = 150$ Å. Calculate the energy E as a function of wave vector k_z for the lowest lying energy band and plot E versus k_z. Compare the lowest energy of the band with the ground state energy of the double quantum well.
6. A **quantum wire** is a semiconductor in the form of a cylinder or bar or other shape that confines the motion of a carrier to one dimension. The transverse dimension is typically on the order of 100 Å. Derive the ground-state envelope function and calculate the ground state energy for an electron in a GaAs bar of square cross section with sides of 50 Å. Assume that the bar is bounded by infinite potential barriers.
7. A **quantum dot** is a thin slice cut out of a quantum wire that confines a carrier to zero-dimensional motion. Calculate the ground state energy of an electron in a GaAs quantum dot of square cross section with sides of 50 Å and thickness 25 Å. Assume infinite potential barriers.

References

T. Ando, *J. Phys. Soc. Japan* **51**, 3900 (1982).

G. Bastard, *Acta Electronica* **25**, 147 (1983).

G. Bastard, *Wave Mechanics Applied to Semiconductor Heterostructures* (Les Editions de Physique, Les Ulis, 1988).

G. Bastard, E.E. Mendez, L. L. Chang, and L. Esaki, *Phys. Rev.* B**26**, 1974 (1982).

D. S. Chemla, *Helv. Phys. Acta.* **56**, 607 (1983).

L. Esaki, in *Recent Topics in Semiconductor Physics*, eds. H. Kamimura and Y. Toyozawa (World Scientific, Singapore, 1983)

L. Esaki and R. Tsu, *IBM J. Dev.* **14**, 61 (1970).

K. Miyatsuji, H. Hihara, and C. Hamaguchi, *Superlattices and Microstructures* **1**, 43 (1985).

S. S. Nedozerov, *Fiz. Tverd. Tela* **12**, 2269 (1970); *Sov. Phys.-Solid State* **12**, 1815 (1971).

P. Ruden and G. H. Döhler, *Phys. Rev.* B**27**, 3538 (1983).

F. Stern, *Phys. Rev.* B**5**, 4891 (1972).

H. L. Störmer, R. Dingle, A. C. Gossard, W. Wiegmann, and R. A. Logan, *Conf. Series—Inst. Phys.* **43**, 557 (1979).

C. Weisbuch and B. Vinter, *Quantum Semiconductor Structures* (Academic Press, San Diego, 1991).

Phonons in superlattices

16

Key ideas

For a superlattice of two constituents the acoustic branches overlap. The forbidden gap may overlap as well. The optical branches may or may not overlap. When there is overlap, phonons can propagate along the growth direction. These are *propagating modes*. When there is no overlap, the phonons are confined in individual layers and are called *confined modes*.

The dispersion curves are straight lines that exhibit *folding* of the acoustic branches in the first Brillouin zone. Phonons can propagate in a superlattice parallel to the interfaces. They include modes with particle displacements localized near the interfaces that are called *interface modes*.

Optical modes show folding of their dispersion curves within the first Brillouin zone and gaps at the zone center and zone boundary. *Confined optical modes* are dispersionless and behave like the bulk modes of a single layer called *guided-wave modes*.

Some normal modes of a superlattice have displacements that vary rapidly with position on an atomic scale. They include *folded acoustic*, *propagating optical*, and *confined optical modes*.

In order to properly account for Coulomb interactions in polar constituents of a superlattice, a three-dimensional model is required.

Phonons

We have seen in Chapter 7 that in a three-dimensional homogeneous crystal there are three frequency ranges of importance for the normal modes of vibration of the atoms. The low-frequency range contains the acoustic branches. In crystals containing two atoms per unit cell, the acoustic branches lie below an intermediate frequency range corresponding to a forbidden gap. The high-frequency range lies above the gap and contains the optical branches. Under certain circumstances, there may be no forbidden gap even though optical modes exist. If there are more than two atoms per unit cell, there may be several forbidden gaps and several frequency ranges containing optical modes.

16.1 Qualitative aspects of phonons in superlattices

In thin slabs of a material such as GaAs bounded by (111) free surfaces, the frequencies of the longitudinal and transverse optical modes of zero parallel wave vector very nearly coincide with those of the corresponding branches of the infinite crystal, provided the number of atomic layers is ten or more (Kanellis *et al.* 1983). One also finds several surface-mode branches whose frequencies lie within or at the edge of the gap. Reducing the number of layers below ten causes the frequencies of the bulk-like optical modes to decrease sharply while those of two of the surface modes increase until they meet the frequencies of the corresponding bulk-like modes for the two-layer case.

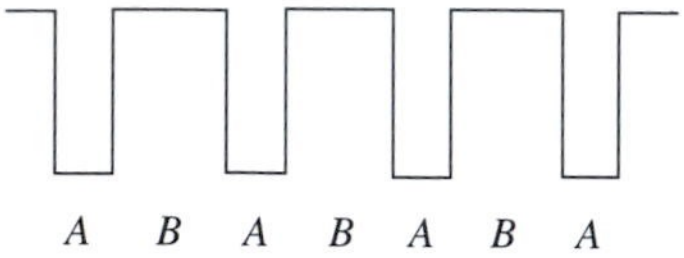

Fig. 16.1
Superlattice composed of two materials A and B.

Suppose a superlattice is created from two different materials A and B as shown in Fig. 16.1. The frequency ranges mentioned above are, in general, different for the two constituents. In the low-frequency range the acoustic branches will always overlap to some degree. A good portion of the forbidden gaps may overlap as well. The frequencies of the optical branches for different compounds are often very different, and they may overlap only partially or not at all. Two cases can be distinguished regarding the optical branches. In the first case the optical branches of the constituent compounds are in two different frequency ranges, i.e., they are well separated. In the second case the optical branches of the two compounds are in the same frequency range, and they may overlap partially or completely. When there is overlap, the lattice-vibrational wave propagates through the superlattice along the growth direction or axis. One refers to these modes as **propagating modes**. When there is no overlap of optical mode frequencies, propagation of optical modes along the superlattice axis is not possible. The modes are confined to individual layers and are referred to as confined modes. It is, of course, possible to have optical modes that propagate in directions perpendicular to the superlattice axis, i.e., parallel to the layer interfaces.

16.2 Elastic continuum theory of low-frequency modes

For the low-frequency range, where the acoustic branches overlap, elastic continuum theory can be applied to the analysis of the propagating modes. If we restrict ourselves to cubic constituents and to superlattices with axes in the [001] direction, the equation of motion for the displacement $w(z)$ of a longitudinal wave propagating along the superlattice axis is

$$\rho(z)\frac{\partial^2 w(z)}{\partial t^2} = \frac{\partial}{\partial z}\left[C_{11}(z)\frac{\partial w(z)}{\partial z}\right], \tag{16.1}$$

where $\rho(z)$ is the material density at z and $C_{11}(z)$ is the appropriate elastic modulus at z. Denoting by i the constituents A or B and assuming that the individual layers are homogeneous, we can write for each layer

$$\rho_i \frac{\partial^2 w_i(z)}{\partial t^2} = C_{11}^i \frac{\partial^2 w_i(z)}{\partial z^2}. \tag{16.2}$$

The solutions of the equations for media A and B can be written in the forms

$$w_A(z) = (W_A^+ e^{i\alpha z} + W_A^- e^{-i\alpha z})e^{-i\omega t} \tag{16.3a}$$

$$w_B(z) = (W_B^+ e^{i\beta z} + W_B^- e^{-i\beta z})e^{-i\omega t} \tag{16.3b}$$

where α and β are the effective wave vectors for propagation in media A and B, respectively, and ω is the frequency. Substitution of these expressions into Eqs. (16.2) gives the relations

$$\alpha = \omega / v_A \tag{16.4a}$$

$$\beta = \omega / v_B, \tag{16.4b}$$

where $v_A = (C_{11}^A/\rho_A)^{\frac{1}{2}}$ and $v_B = (C_{11}^B/\rho_B)^{\frac{1}{2}}$ are the velocities of longitudinal acoustic waves in the two media.

The periodicity of the superlattice can be exploited by applying Bloch's theorem to the displacements $w_i(z)$ and the normal stresses $C_{11}^i w_i'(z)$:

$$w_B(-d_B) = e^{-ik_z D} w_A(d_A) \tag{16.5a}$$

$$C_{11}^B w_B'(-d_B) = e^{-k_z D} C_{11}^A w_A'(d_A). \tag{16.5b}$$

The prime on ω indicates the first derivative of w with respect to z, d_A and d_B are the thicknesses of the constituent layers, $D = d_A + d_B$ is the superlattice period, and k_z is the Bloch wave vector. We also must satisfy the boundary conditions at the interface $z = 0$; namely, the continuity of the displacements and the normal stresses:

$$w_A(0) = w_B(0) \tag{16.6a}$$

$$C_{11}^A w_A'(0) = C_{11}^B w_B'(0). \tag{16.6b}$$

Substituting Eqs. (16.3) into Eqs. (16.5) and (16.6) leads to a set of four linear homogeneous equations in the amplitudes W_A^+, W_A^-, W_B^+, and W_B^-. To have a nontrivial solution, we set the determinant of coefficients equal to zero and obtain the dispersion relation for the propagation of longitudinal acoustic waves along the axis of a superlattice in the form (Jusserand *et al.* 1986)

$$\cos(k_z D) = \cos\left(\omega\frac{d_A}{v_A} + \omega\frac{d_B}{v_B}\right) - \frac{\epsilon^2}{2}\sin\left(\omega\frac{d_A}{v_A}\right)\sin\left(\omega\frac{d_B}{v_B}\right), \tag{16.7}$$

where

$$\epsilon = \frac{|\rho_A v_A - \rho_B v_B|}{(\rho_A v_A \rho_B v_B)^{\frac{1}{2}}}. \tag{16.8}$$

The quantity ϵ is a measure of the mismatch of the acoustic impedances of the different layers.

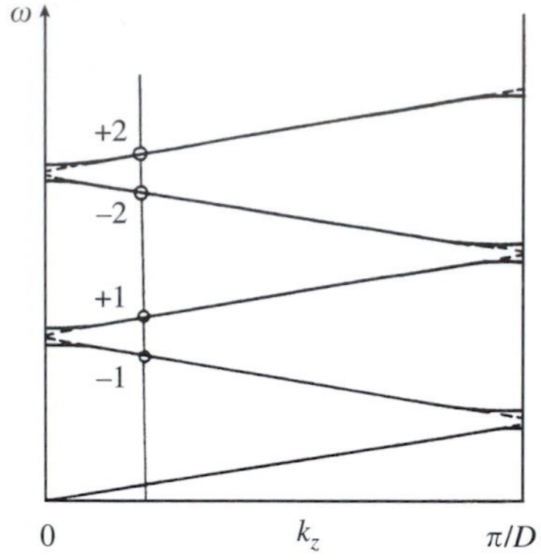

Fig. 16.2
Phonon dispersion curves of the folded acoustic branch according to elasticity theory.

In discussing the solutions of Eq. (16.7) it is convenient to use the reduced zone scheme in which k_z is restricted to the first Brillouin zone: $-\pi/D \leq k_z \leq \pi/D$. The zeroth-order solution corresponding to $\epsilon = 0$ takes the simple form

$$\omega = \pm\bar{v}k_z + \frac{2m\pi\bar{v}}{D}, \quad m = 0, 1, 2, \ldots \tag{16.9}$$

where $\bar{v}$ is the sound velocity of the average medium defined by

$$\bar{v} = D\left(\frac{d_A}{v_A} + \frac{d_B}{v_B}\right)^{-1}. \tag{16.10}$$

This approximation assumes that the two media are acoustically matched ($\rho_A v_A = \rho_B v_B$) and no reflection of a propagating elastic wave occurs at any interface.

In Fig. 16.2 the frequency specified by Eq. (16.9) is plotted versus positive wave vector. The plot consists of a series of straight lines exhibiting the exact folding of the dispersion curves in the first Brillouin zone. The lowest branch ($m = 0$) is the acoustic branch of the superlattice. The higher branches are "optical" branches that can be observed by Raman scattering. There are double degeneracies of the modes at the center of the zone and at the zone boundary with frequencies given by

$$\omega_n^{(0)} = \frac{n\pi\bar{v}}{D}. \tag{16.11}$$

The index n is $2m$ for the zone-center modes and $2m+1$ for the zone-boundary modes.

The solutions of Eq. (16.7) for the more realistic case $\epsilon \neq 0$ are only slightly modified in the interior of the first Brillouin zone. The important qualitative change is the lifting of the degeneracies at the zone center and boundary. The dispersion curves deviate from linearity for ω near each nonzero value of $\omega_n^{(0)}$, and small gaps are produced about these values as shown in Fig. 16.2. To first order in ϵ the frequencies ω_n are (Jusserand *et al.* 1987)

$$\omega_n = \omega_n^{(0)} \pm \Delta\omega_n, \tag{16.12}$$

where

$$\Delta\omega_n = \pm\frac{\bar{v}}{D}\left|\epsilon \sin\left[\frac{n\pi(1-x)v_B}{(1-x)v_B + xv_A}\right]\right| \tag{16.13}$$

and $x = d_B/D$ is the fractional concentration of constituent B. The frequency gap at either the zone center or the zone boundary is $2|\Delta\omega_n|$ and is inversely proportional to the period D of the superlattice. The zone center gaps vanish at a value of x given by $x_c = v_B/(v_A + v_B)$.

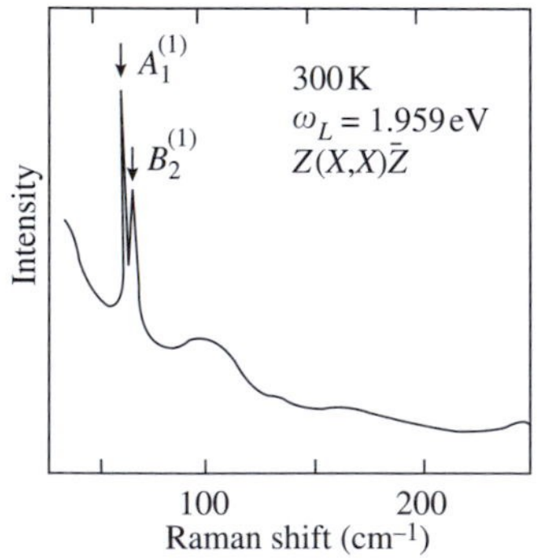

Fig. 16.3
Raman spectrum of a folded acoustic doublet in a GaAs/AlAs superlattice (after Colvard *et al.* 1980).

The folded-back modes at $k \sim 0$ can be observed experimentally by Raman scattering. In Fig. 16.3 is shown a portion of the Raman spectrum of a superlattice consisting of alternating layers of GaAs with thickness of

13.6 Å and AlAs with thickness of 11.4 Å. The two peaks around 60 cm^{-1} separated by 4 cm^{-1} correspond to the lowest folded longitudinal acoustic branch at the zone center with a gap of 4 cm^{-1}. These results are in good agreement with the predictions of the elastic continuum theory shown in Fig. 16.2.

In zeroth order approximation the frequencies of the folded acoustic modes as specified by Eq. (16.11) depend only on the average sound velocity $\bar{v}$ and the superlattice period D. The doublet splitting depends on the ratio $\bar{v}/D$ as well as on parameters associated with the inner structure of the supercell such as individual layer thicknesses and sound velocities. Thus, measurements of the average doublet frequency and wave vector enable one to determine $\bar{v}$ and D. This information coupled with measurements of the doublet splitting and knowledge of the individual sound velocities and densities allows the determination of the fractional concentration parameter x and thence the individual layer thicknesses.

Just as in a single layer, vibrational modes can propagate in a superlattice parallel to the interfaces. Since there is no compositional variation in the direction of propagation, no frequency gaps exist within the elastic continuum approximation. Some of the modes have particle displacements localized near the interfaces and are called **interface modes**, while the remaining modes are bulk-like modes with significant displacements away from the interfaces. Among the latter modes are longitudinal and transverse modes of an effective average medium.

16.3 Dielectric continuum theory of optical modes

The optical modes of vibration of a polar crystal contribute a frequency dependence to the dielectric constant as discussed in Chapter 10. This dependence leads to interesting effects on the propagation of infrared radiation in superlattices. Classical electromagnetic theory can be used to analyze the character of the normal modes, some of which are interface modes (Camley and Mills 1984).

If retardation is ignored, the electric field is specified by the electrostatic potential $\varphi(\boldsymbol{r}, t)$ which satisfies Laplace's equation in each layer $i = A$,B:

$$\nabla^2 \varphi_i(\boldsymbol{r}, t) = 0. \tag{16.14}$$

We are primarily interested in superlattice constituents that are cubic and hence optically isotropic. The superlattice possesses translational invariance parallel to its interfaces and rotational invariance about its axis. A solution to Eq. (16.14) can therefore be written without loss of generality in the form

$$\varphi_i(\boldsymbol{r}, t) = e^{i(kx-\omega t)} \Phi_i(z), \tag{16.15}$$

where $\Phi_i(z)$ satisfies

$$\frac{d^2\Phi_i(z)}{dz^2} - k^2 \Phi_i(z) = 0. \tag{16.16}$$

The general solution for $\Phi_i(z)$ is

$$\Phi_i(z) = C_+^i e^{kz} + C_-^i e^{-kz}, \tag{16.17}$$

where C_+^i and C_-^i are constants to be determined.

A partial determination of these constants can be made by invoking the periodicity of the superlattice in the z-direction and writing $\Phi_i(z)$ in the Bloch form

$$\Phi_i(z) = e^{iqz} U_q^i(z), \tag{16.18}$$

where q is the z-component of the wave vector and for any integer n,

$$U_q^i(z + nD) = U_q^i(z). \tag{16.19}$$

Within the nth layer of constituent A which extends from $z = nD$ to $z = nD + d_A$, the periodic function $U_q^A(z)$ can be written as

$$U_q^A(z) = e^{-iq(z-nD)} \left[C_+^A e^{k(z-nD)} + C_-^A e^{-k(z-nD)} \right] \tag{16.20}$$

for an interface mode. $\Phi_A(z)$ is then given by

$$\Phi_A(z) = e^{iqnD} \left[C_+^A e^{k(z-nD)} + C_-^A e^{-k(z-nD)} \right]. \tag{16.21}$$

Similarly, within the nth layer of constituent B which extends from $nD + d_A$ to $(n+1)D$, $\Phi_B(z)$ has the form

$$\Phi_B(z) = e^{iqnD} \left[C_+^B e^{k(z-nD-d_A)} + C_-^B e^{-k(z-nD-d_A)} \right]. \tag{16.22}$$

These expressions for $\Phi_A(z)$ and $\Phi_B(z)$ clearly satisfy Eq. (16.16).

The four constants C_+^A, C_-^A, C_+^B, and C_-^B are specified by the boundary conditions stating the continuity across each interface of the electrostatic potential and the normal component of the electric displacement $\mathcal{D}$. These conditions can be applied at $z = nD$ and $z = nD + d_A$; they are satisfied at all other interfaces as a result of the Bloch forms for $\Phi_A(z)$ and $\Phi_B(z)$.

Continuity of $\Phi(z)$ at $z = nD$ gives the equation

$$C_+^A + C_-^A = e^{-iqD} \left[C_+^B e^{kd_B} + C_-^B e^{-kd_B} \right]. \tag{16.23}$$

The normal component of the electric displacement is given by $-\epsilon(\omega)(d\Phi/dz)$, where $\epsilon(\omega)$ is the dielectric constant at frequency ω. Its continuity at $z = nD$ yields the equation

$$\epsilon_A(\omega) \left[C_+^A - C_-^A \right] = e^{-iqD} \epsilon_B(\omega) \left[C_+^B e^{kd_B} - C_-^B e^{-kd_B} \right]. \tag{16.24}$$

From the continuity conditions at $z = nD + d_A$ we obtain a second pair of equations:

$$C_+^A e^{kd_A} + C_-^A e^{-kd_A} = C_+^B + C_-^B \tag{16.25}$$

$$\epsilon_A(\omega) \left[C_+^A e^{kd_A} - C_-^A e^{-kd_A} \right] = \epsilon_B(\omega) (C_+^B - C_-^B) \tag{16.26}$$

The set of four equations (16.23)–(16.26) has a nontrivial solution only if the determinant of coefficients of C_+^A, C_-^A, C_+^B and C_-^B is zero. Evaluating this determinant we obtain the dispersion relation for the normal modes:

$$\cos(qD) = \cosh(kd_A)\cosh(kd_B) + \frac{1}{2}\left[\frac{\epsilon_A(\omega)}{\epsilon_B(\omega)} + \frac{\epsilon_B(\omega)}{\epsilon_A(\omega)}\right]\sinh(kd_A)\sinh(kd_B). \tag{16.27}$$

We note that the left hand side of this equation is in magnitude less than unity, whereas the first term on the right hand side is greater than unity if $k > 0$. If phonon damping is neglected, $\epsilon_A(\omega)$ and $\epsilon_B(\omega)$ are real. Consequently, their ratio $\epsilon_A(\omega)/\epsilon_B(\omega)$ must be **negative** in order for Eq. (16.27) to have a physically meaningful solution. The dielectric constant of one constituent must be positive and that of the other constituent must be negative at the frequency of interest. For polar crystals with two atoms per unit cell such as the III–V compounds, the dielectric constant is given by Eq. (10.137) and is negative only in the reststrahlen range $\omega_{TO} < \omega < \omega_{LO}$. If the reststrahlen regions of the two constituents do not overlap as in GaAs/AlAs, normal mode solutions of Eq. (16.27) occur for frequencies in both reststrahlen regions.

In Fig. 16.4 are plotted the normal mode frequencies versus wave vector k for various values of q. In each reststrahlen region the frequencies fall into two bands with a gap between them if $d_A \neq d_B$. From Eqs. (16.15), (16.21), and (16.22) we see that the modes have field amplitudes localized about each interface and are therefore **interface modes**. They are frequently referred to as **Fuchs–Kliewer modes**.

Merlin *et al.* (1980) measured Raman scattering in GaAs/AlAs superlattices and observed a peak between the TO and LO modes that may be attributed to the modes discussed above (Sood *et al.* 1985). More recent Raman scattering experiments by Sood *et al.* clearly show interface modes in both the GaAs and AlAs reststrahlen regions.

Setting $k = 0$ in Eq. (16.27) requires that $q = 2\pi p/D$, where p is an integer. Such solutions correspond to the uniform mode in which $\Phi_A(z) = \Phi_B(z) = \text{constant}$. More interesting solutions with $k = 0$ can be obtained if retardation is included and the electric vector $\mathcal{E}$ satisfies the wave equation

$$\nabla^2 \mathcal{E}_i = \frac{\epsilon_i(\omega)}{c^2}\frac{\partial^2 \mathcal{E}_i}{\partial t^2} \tag{16.28}$$

in each medium i. Taking $\mathcal{E}$ to be in the x-direction and to depend only on z and t reduces Eq. (16.28) to the form

$$\frac{\partial^2 \mathcal{E}}{\partial z^2} = \frac{\epsilon_i(\omega)}{c^2}\frac{\partial^2 \mathcal{E}_{ix}}{\partial t_2} \tag{16.29}$$

which is completely analogous to Eq. (16.2) for the elastic media case. The discussion of the latter case can be taken over to the dielectric medium case provided both $\epsilon_A(\omega)$ and $\epsilon_B(\omega)$ are positive. Folding of the dispersion curves within the first Brillouin zone and gaps at the zone center and zone

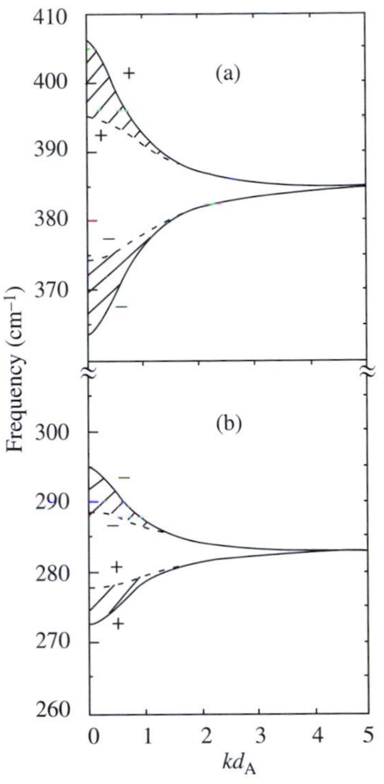

Fig. 16.4
Calculated interface modes for a GaAs/AlAs superlattice with $d_A = 20$ Å and $d_B = 60$ Å: (a) AlAs-like modes; (b) GaAs-like modes. d_A and d_B are the thicknesses of the GaAs and AlAs layers, respectively. The + and − signs indicate the parity of the electrostatic potential with respect to the center of the GaAs layers. $qD = 0$, dashed curves; $qD = \pi$, solid curves (after Sood *et al.* 1985).

boundary occur. These results provide an example of **photonic band structure**.

If one of the media has a negative dielectric constant at the frequency of interest, the electromagnetic wave cannot propagate through that medium. It is confined to the medium with the positive dielectric constant and constitutes a **confined mode**. Such modes are dispersionless and behave much like the bulk modes of a single layer, which are called **guided-wave modes**.

16.4 Microscopic theory of optical modes

The acoustic modes described by elastic continuum theory and the macroscopic optical modes described by the dielectric continuum theory do not exhaust the totality of normal modes of vibration of a superlattice. Since the individual constituents have microscopic optical modes not describable by the continuum approach, it follows that a superlattice also must have such modes. A qualitative effect of the new larger periodicity of the superlattice is to lower the symmetry of the structure and produce folding of the optical branches in the first Brillouin zone of the bulk constituents. As a result new modes appear for a given value of the wave vector in the first Brillouin zone of the superlattice that are related to bulk modes of the constituents having larger wave vector values.

Two distinct situations can arise regarding the optical branches of the two constituents. In the first case the optical branches of the two materials are in different frequency ranges and do not overlap, while in the second case there is partial or complete overlap. Examples of the two cases are GaAs/AlAs for the first and InAs/GaSb for the second.

In the GaAs/AlAs superlattice the Al atoms are much lighter than the Ga atoms, and hence the optical branches of AlAs are in a higher frequency range than those of GaAs. There is no overlap. As a result superlattice optical modes in layers having one of these frequency ranges cannot propagate into layers having the other frequency range. The modes are confined in either the GaAs or the AlAs layers. The confined modes do not show dispersion, just as with confined macroscopic optical modes. In some cases of confined modes the atomic displacements are not rigorously localized in one type of layer, but penetrate into the adjacent layers with exponentially decaying amplitude. Confinement is revealed experimentally by the dependence of the phonon frequencies only on layer thickness and not on wave vector. Decreasing the layer thickness causes the frequencies of the GaAs-like and AlAs-like LO modes to shift downward, whereas the corresponding TO frequencies shift upward. In addition to confined modes the GaAs/AlAs system has interface modes as discussed in the preceding section.

In the InAs/GaSb superlattice the frequency range of the LO phonons in GaSb lies entirely within that of InAs. Two types of interfaces occur alternately: one involving pairs of lighter atoms (Ga–As) and the other involving pairs of heavier atoms (In–Sb). Interface modes occur at both types of interfaces (Berdekas and Kanellis 1991). In addition **pseudo-confined modes** occur in which the displacements are concentrated in a layer of one material, but which, while small in the adjoining layers of the other material, do not decay exponentially going away from the interfaces with the first layer.

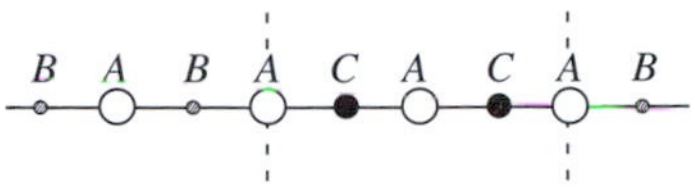

Fig. 16.5
Linear chain model of a superlattice with constituents AB and AC for $n_1 = 2$, $n_2 = 2$.

16.4.1 Linear chain model

A simple procedure for investigating the propagation of optical modes along the superlattice axis is to employ the linear chain model with nearest-neighbor interactions. We restrict our attention to the case in which both materials AB and AC have a common atom as shown in Fig. 16.5. The constituents AB and AC are labeled with an index κ taking on the values 1 and 2, respectively. The superlattice supercell contains n_1 layers of AB and n_2 layers of AC. The displacements of atom A are denoted by $u_\ell(\kappa)$ and those of atoms B or C (for $\kappa = 1$ or 2 respectively) by $v_\ell(\kappa)$, where ℓ denotes the site of an A atom. Associated with an A atom is a B or C atom which together with the A atom form a primitive unit cell in a constituent layer. For a given layer the equations of motion are

$$m_A \ddot{u}_\ell(\kappa) = \sigma_\kappa [v_\ell(\kappa) + v_{\ell-1}(\kappa) - 2u_\ell(\kappa)] \tag{16.30a}$$

$$m_{B,C} \ddot{v}_\ell(\kappa) = \sigma_\kappa [u_{\ell+1}(\kappa) + u_\ell(\kappa) - 2v_\ell(\kappa)] \tag{16.30b}$$

with σ_κ the nearest-neighbor force constant. Since the force constants differ little from one III–V compound to another, we shall take $\sigma_1 = \sigma_2 = \sigma$ in the following.

The solution of Eqs. (16.30) can be expressed as

$$u_\ell(\kappa) = U_\kappa e^{i\ell k_\kappa a_\kappa - i\omega t} \tag{16.31a}$$

$$v_\ell(\kappa) = V_\kappa e^{i\ell k_\kappa a_\kappa - i\omega t}, \tag{16.31b}$$

where k_κ is the wave vector, ω is the frequency, a_κ is the lattice constant, and U_κ, V_κ are displacement amplitudes. Substituting Eqs. (16.31) into Eqs. (16.30) yields the pair of equations

$$(\omega^2 m_A - 2\sigma) U_\kappa + \sigma(1 + e^{-ik_\kappa a_\kappa}) V_\kappa = 0 \tag{16.32a}$$

$$\sigma(e^{ik_\kappa a_\kappa} + 1) U_\kappa + (\omega^2 m_{B,C} - 2\sigma) V_\kappa = 0, \tag{16.32b}$$

whose solution can be expressed as

$$U_\kappa = W_\kappa \sigma (1 + e^{-ik_\kappa a_\kappa}) \tag{16.33a}$$

$$V_\kappa = W_\kappa (2\sigma - \omega^2 m_A), \tag{16.33b}$$

with W_κ an arbitrary amplitude.

To obtain the normal modes of the superlattice, we take linear combinations of the solutions specified by Eqs. (16.31) and (16.33) with both

$+k_\kappa$ and $-k_\kappa$,

$$u_\ell(\kappa) = W_\kappa^+ \sigma(1 + e^{-ik_\kappa a_\kappa}) e^{i\ell k_\kappa a_\kappa} + W_\kappa^- \sigma(1 + e^{ik_\kappa a_\kappa}) e^{-i\ell k_\kappa a_\kappa} \tag{16.34a}$$

$$v_\ell(\kappa) = (2\sigma - m_A\omega^2)\left[W_\kappa^+ e^{i\ell k_\kappa a_\kappa} + W_\kappa^- e^{-i\ell k_\kappa a_\kappa}\right], \tag{16.34b}$$

where the four parameters W_1^+, W_1^-, W_2^+, and W_2^- are determined by the interface boundary conditions. At the origin $\ell = 0$ where an atom of type A is shared by layers of AB and AC constituents, we must have

$$u_0(1) = u_0(2). \tag{16.35a}$$

The equation of motion of this shared atom is

$$\begin{aligned} -m_A\omega^2 u_0(1) &= \sigma[v_{-1}(1) - u_0(1) + v_0(2) - u_0(1)] \\ &= \sigma[v_{-1}(1) - u_0(1) + v_0(2) - u_0(2)]. \end{aligned} \tag{16.35b}$$

At the interface $\ell = n_2$ we impose the Bloch condition and obtain

$$u_{n_2}(2) = u_{-n_1}(2) e^{iqD},$$

where q is the Bloch wave vector. Since the atom at $\ell = -n_1$ is shared by different layers, $u_{-n_1}(2) = u_{-n_1}(1)$ and

$$u_{n_2}(2) = u_{-n_1}(1) e^{iqD}. \tag{16.35c}$$

The equation of motion of this atom can be written as

$$-m_A\omega^2 u_{n_2}(2) = \sigma\left[v_{-n_1}(1) e^{iqD} - u_{n_2}(1) + v_{n_2-1}(2) - u_{n_2}(2)\right]. \tag{16.35d}$$

Substituting Eqs. (16.34) for $\kappa = 1$ and 2 into Eqs. (16.35) yields a set of four linear homogeneous equations. Setting the determinant of the coefficients equal to zero gives the dispersion relation for the superlattice in the form (Jusserand and Paguet 1986).

$$\cos(qD) = \cos(n_1 k_1 a_1)\cos(n_2 k_2 a_2) - \eta \sin(n_1 k_1 a_1)\sin(n_2 k_2 a_2), \tag{16.36}$$

where

$$\eta = \frac{1 - \cos(k_1 a_1)\cos(k_2 a_2)}{\sin(k_1 a_1)\sin(k_2 a_2)}. \tag{16.37}$$

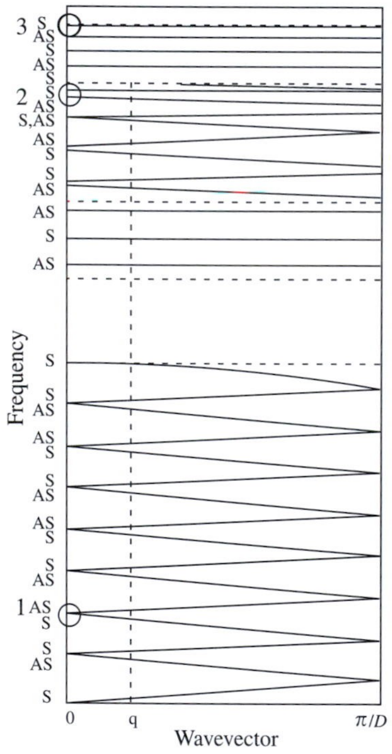

Fig. 16.6
Linear chain dispersion curves for an $(AB)_8/(AC)_8$ superlattice corresponding to LO modes. S and AS designate symmetric and antisymmetric modes of the types: (1) folded acoustic mode, (2) extended mode with frequency in the overlapping region, (3) confined mode with the highest frequency (after Jusserand 1987).

This dispersion relation describes both optical and folded acoustic vibrations. Each folded acoustic branch produces $n_1 + n_2 - 1$ additional modes at the zone center. They are "optical" modes of the superlattice and in principle are observable by spectroscopic techniques. The optical modes are confined or propagating depending on the overlap of the optical branches of the constituents. Numerical results for the normal mode frequencies as a function of wave vector are presented in Fig. 16.6 for a superlattice whose constituents have partially overlapping optical branches. All three types of modes are illustrated: folded-acoustic, propagating optical, and confined optical.

16.4.2 Three-dimensional models

The linear chain model as described in the previous section cannot fully take into account several features of the vibrational properties of real superlattices. Among the deficiencies to be overcome are the lack of interactions beyond nearest neighbors, the neglect of Coulomb interactions, and the restriction to one dimension. These deficiencies have been remedied by three-dimensional calculations for GaAs/AlAs superlattices with axis along [001] using the bond-charge model (Yip and Chang 1984) and the shell model (Richter and Strauch 1987). In the latter work force constants for bulk GaAs were used for both materials with no modifications made at the interfaces. This approximation is justified by the close similarity of the elastic constants of GaAs and AlAs.

The results of the shell-model calculations reveal folded-acoustic and confined optical modes associated with both transverse and longitudinal branches. Also present are interface modes derived from both the acoustic and optical branches. The LO confined optical modes can be approximated as standing waves with a wave vector q_j given by

$$q_j = \frac{2\pi j}{(n_p + 1)a_p}, \tag{16.38}$$

where n_p and a_p are the number of monolayers and lattice constant, respectively, of the material with positive dielectric constant. The associated displacements behave as $\cos(q_j z)$ and $\sin(q_j z)$ for $j = 1,3,5,\ldots$, and $j = 2,4,6\ldots$, respectively. Also present are guided acoustic modes propagating perpendicular to the superlattice axis and interface optical modes of the Fuchs–Kliewer type.

Problems

1. Calculate the frequency of the lowest folded longitudinal acoustic branch observed by Raman scattering in a superlattice consisting of alternating layers of GaAs with thickness of 15 Å and AlAs with thickness of 10 Å. The elastic moduli can be found in Table 7.3 and the lattice constants in Table 1.4.
2. With the aid of Table 10.1 determine what frequency ranges can support interface optical modes in a superlattice composed of ZnSe and AlAs.
3. For a ZnSe/AlAs superlattice with $d_A = d_B = d$ calculate the frequency as a function of kd for all interface optical mode branches at $q = 0$. Plot ω versus kd for each branch. Also calculate and plot the amplitudes $\Phi_A(z)$ and $\Phi_B(z)$ as functions of z.
4. Consider confined optical modes in a superlattice with amplitude $\Phi_B(z)$ given by

$$\Phi_B(z) = e^{iqnD}\{\bar{C}_s^B \sin[k(z - nD - d_A)] + \bar{C}_c^B \cos[k(z - nD - d_A)]\}, \tag{16.39}$$

but $\Phi_A(z)$ given by Eq. (16.21). Derive the dispersion relation and calculate ω as a function of kd_A for the GaAs/AlAs superlattice of Problem 1 and $q = 0$. Also calculate and plot the amplitudes $\Phi_A(z)$ and $\Phi_B(z)$ as functions of z.

References

D. Berdekas and G. Kanellis, *Phys. Rev.* B**43**, 9976 (1991).
R. E. Camley and D. L. Mills, *Phys. Rev.* B**29**, 1695 (1984).
C. Colvard, R. Merlin, M.V. Klein, and A. C. Gossard, *Phys. Rev. Lett.* **45**, 298 (1980).
B. Jusserand, Thesis, Paris (1987).
B. Jusserand and D. Paquet, *Phys. Rev. Lett.* **56**, 1752 (1986).
B. Jusserand, F. Alexandre, J. Dubard, and D. Paquet, *Phys. Rev.* B**33**, 2897 (1986).
B. Jusserand, D. Paquet, F. Mollot, F. Alexandre, and G. Le Roux, *Phys. Rev.* B**35**, 2808 (1987).
G. Kanellis, J. F. Morhange, and M. Balkanski, *Phys. Rev.* B**28**, 3390 (1983).
R. Merlin, C. Colvard, M. V. Klein, H. Morkoc, A. Y. Cho, and A. C. Gossard, *Appl. Phys. Lett.* **36**, 43 (1980).
E. Richter and D. Strauch, *Solid State Commun.* **64**, 867 (1987).
A. K. Sood, J. Menendez, M. Cardona, and K. Ploog, *Phys. Rev. Lett.* **54**, 2111, 2115 (1985).
S. Yip and Y. Chang, *Phys. Rev.* B**30**, 7037 (1984).

Optical properties of heterostructures

17

Key ideas

Electronic transitions that contribute to optical absorption include
intrasubband transitions between states of a single subband;
intersubband transitions between subbands of the same band;
interband transitions between subbands of different bands.

Optical absorption by a quasi-two-dimensional gas of free carriers in a perfect heterostructure is forbidden. Absorption can be induced by phonons, impurities, and other defects.

To have absorption the radiation must propagate parallel to the interfaces of the quantum well. If the quantum well has a plane of reflection symmetry, only transitions between subbands of opposite parity are allowed.

Optical absorption due to transitions between subbands in different bands can occur in heterostructures with *type I quantum wells* (conduction electrons and holes in the same layer) or with *type II quantum wells* (conduction electrons and holes in different layers). For type I the absorption has a staircase profile while for type II it has a smooth profile and is quite weak.

Interband optical absorption in type I and type II superlattices has properties similar to those of the individual quantum wells making up the superlattices.

The exciton effect reduces the effective energy gap in a 2D system by an amount which is four times the reduction in a 3D system due to the differences in exciton binding energy. With increasing thickness of the quantum well the effective Bohr radius increases and the exciton binding energy decreases. Strong *nonlinear absorption* observed in quantum wells is related to the large oscillator strength of the principal exciton peak.

Photoluminescence techniques include:
photoluminescence spectroscopy in which a source is at a fixed frequency and the emitted light is analyzed by a monochromator.
photoluminescence excitation spectroscopy which is the inverse of luminescence spectroscopy. The detector is set at a fixed frequency and the exciting light is scanned through a monochromator. Markedly

Optical properties

different excitation spectra indicate that the various photoluminescence lines have different physical origins.

For band-to-band recombination the polarization selection rule depends on the direction of propagation of the emitted light. The "free" exciton luminescence line has high intensity in a quantum well due to the small volume in which carriers and photons interact. Two-dimensional excitons can be localized by weak disorder.

For bulk semiconductors the determination of the optical absorption coefficient is relatively simple: one measures the attenuation of a light beam passing through a sample of suitable thickness. In the case of quasi-two-dimensional structures the measurement is complicated by the high anisotropy of these systems and their very small thicknesses.

Two-dimensional systems are prepared by the epitaxial growth of thin films of different materials on one another. A quantum well, for example, can be formed from two different semiconductors, one having a small energy gap E_A and the other having a large energy gap E_B. One starts with a relatively thick layer of single crystal semiconductor B and deposits upon it by molecular beam epitaxy (MBE) a very thin layer of semiconductor A whose thickness L is on the order of 100 Å. On top of this layer of material A is deposited by MBE a second thick layer of semiconductor B. The very thin layer of semiconductor A sandwiched between two thick layers of semiconductor B forms a quantum well as shown in Fig. 17.1. The direction of epitaxial crystal growth is taken to be the z-direction.

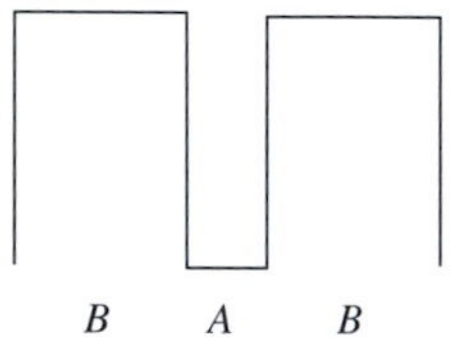

Fig. 17.1
Quantum well formed from two semiconductors A and B.

Optical experiments on quantum wells can be carried out in two different configurations. The most frequently used configuration corresponds to an electromagnetic wave propagating along the z-axis with wave vector $\boldsymbol{q}$ parallel to $\hat{z}$ and perpendicular to the plane of the heterostructure. Much more difficult to carry out are experiments in which the light beam propagates in the layer plane with $\boldsymbol{q} \perp \hat{z}$.

17.1 Optical absorption due to electronic transitions

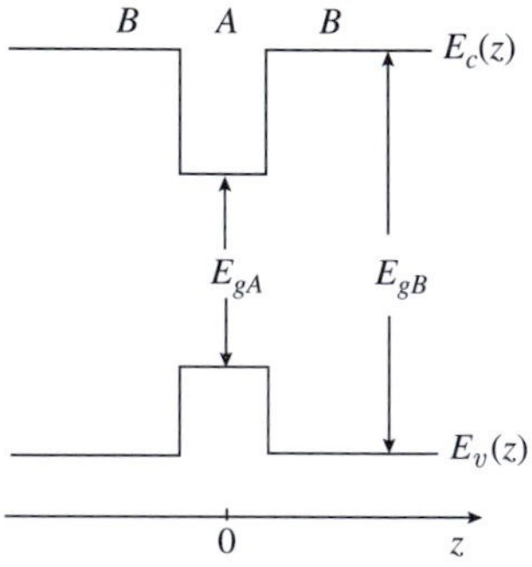

Fig. 17.2
Conduction and valence band edges in a heterostructure $B/A/B$.

In Fig. 17.2 we show the conduction and valence band edges of the heterostructure $B/A/B$ described above. Free carriers in material A are confined by the energy barriers at the interfaces between materials A and B. It was shown in Chapter 15 that the energy states of the carriers form subbands. Each band of the bulk semiconductor has a set of subbands associated with it that are distinguished by a subband index n. Three types of optical transitions can be identified immediately:

1. transitions between states of a single subband
2. transitions between subbands of the same band
3. transitions between subbands of different bands.

We shall refer to transitions of type (1) as **intrasubband transitions**, of type (2) as **intersubband transitions**, and of type (3) as **interband transitions**. These transitions are shown diagrammatically in Fig. 17.3 for the case of direct allowed interband transitions.

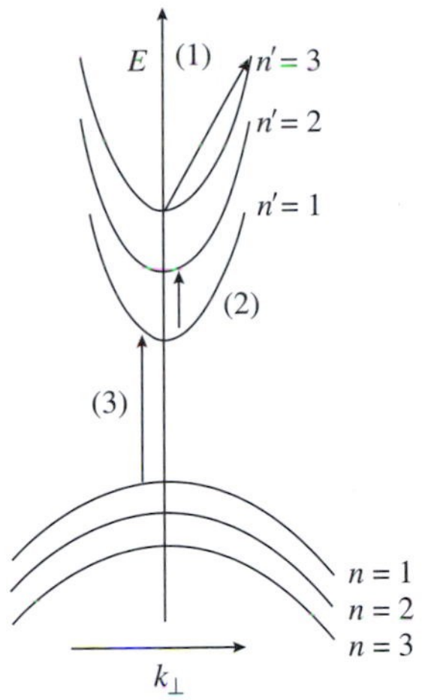

Fig. 17.3
Optical transitions in a heterostructure: (1) intrasubband transitions, (2) intersubband transitions, (3) interband transitions.

17.1.1 Intrasubband transitions

The optical absorption associated with intrasubband transitions is the two-dimensional analog of the free carrier absorption that occurs in bulk semiconductors. For perfect heterostructures the absorption by a quasi-two-dimensional gas of free carriers is forbidden for the same reason as in bulk materials, i.e., the impossibility of conserving energy and momentum simultaneously during the photon absorption process. Free carrier absorption can be induced by phonons, impurities, and other defects that are capable of providing the momentum change necessary for the carrier transition.

17.1.2 Intersubband transitions in the same band

Intersubband absorption involves the excitation of a carrier from one subband to a higher subband. We consider the specific case of conduction electrons in an infinite quantum well. The envelope function $F_{n\boldsymbol{k}_\perp}(\boldsymbol{r})$ can be written with the aid of Eqs. (15.6) and (15.28) in the form

$$F_{n\boldsymbol{k}_\perp}(\boldsymbol{r}) = e^{i\boldsymbol{k}_\perp\cdot\boldsymbol{r}_\perp}\chi_n(z), \tag{17.1}$$

where

$$\chi_n(z) = (2/L)\cos(n\pi z/L), \quad n \text{ odd} \tag{17.2a}$$

$$\chi_n(z) = (2/L)\sin(n\pi z/L), \quad n \text{ even} \tag{17.2b}$$

and n is a positive integer. In order to produce a transition between subbands the electric vector of the radiation must be in the z-direction, so the radiation must propagate parallel to the interfaces of the quantum well. Experimentally, this condition can be achieved by having the radiation incident on the sample at Brewster's angle.

An expression for the absorption coefficient $\alpha(\omega)$ can be obtained by combining Eq. (10.53) with a generalization of Eq. (10.60). It turns out that for transitions within the same band, the simple product form $F_{n\boldsymbol{k}_\perp}(\boldsymbol{r})u_{i0}(\boldsymbol{r})$ given in Eq. (11.87) as an approximation for the full wave function $\psi_i(\boldsymbol{r})$ is inadequate. Higher-order terms arising from the canonical transformations of effective mass theory must be included (Wallis 1958). The net result is that the free electron mass m is to be replaced by the effective mass m^* in Eq. (10.60). The absorption coefficient is then given by

$$\alpha(\omega) = \frac{2\pi e^2 N_W}{\epsilon_0 c n(\omega)\omega m^{*2}\Omega}\sum_{i,j}|\langle i|p_z|j\rangle|^2\delta(E_j - E_i - \hbar\omega)(f_i - f_j), \tag{17.3}$$

where N_W is the number of quantum wells in volume Ω, account has been taken of stimulated emission, f is the Fermi distribution function, the indices i, j stand for the quantum numbers $n, \boldsymbol{k}_\perp$ and $n', \boldsymbol{k}'_\perp$, respectively, and a factor of 2 has been inserted for spin. Using Eqs. (17.1) and (17.2) reduces the momentum matrix element to the form

$$\langle n\boldsymbol{k}_\perp | p_z | n'\boldsymbol{k}'_\perp \rangle = \delta_{\boldsymbol{k}_\perp,\boldsymbol{k}'_\perp} \langle n | p_z | n' \rangle. \tag{17.4}$$

Since the midplane of the quantum well is a plane of reflection symmetry, the nonvanishing matrix elements of p_z are specified by the selection rule

$$n - n' = \text{odd integer}. \tag{17.5}$$

In other words, only transitions between subbands of opposite parity are allowed. Taking n to be odd and n' to be even, we evaluate the matrix element of p_z and obtain

$$\langle n | p_z | n' \rangle = \frac{2\hbar n'}{iL}\left[\frac{\sin(n+n')\pi/2}{n+n'} + \frac{\sin(n'-n)\pi/2}{n'-n}\right]. \tag{17.6}$$

Equation (17.4) can be utilized to rewrite Eq. (17.3) for the absorption coefficient as

$$\alpha(\omega) = \frac{2\pi e^2 N_W}{\epsilon_0 c n(\omega)\omega m^{*2}\Omega} \sum_{\boldsymbol{k}_\perp,n,n'} |\langle n|p_z|n'\rangle|^2 \delta(E_{n'} - E_n - \hbar\omega) \times (f_{n\boldsymbol{k}_\perp} - f_{n'\boldsymbol{k}_\perp}). \tag{17.7}$$

If nonparabolicity of the subbands and the scattering of the carriers by defects are taken into account, the delta function in Eq. (17.7) is replaced by a peaked function with nonzero width.

Example 17.1: Intersubband absorption coefficient
Assuming all carriers are initially in the $n = 1$ subband, obtain the absorption coefficient for excitation to the $n = 2$ subband.
Solution. The matrix element of p_z is found from Eq. (17.6) to be $8\hbar/3iL$. The sum over $\boldsymbol{k}_\perp$ is simply

$$2\sum_{\boldsymbol{k}_\perp} f_{1\boldsymbol{k}_\perp} = N_1, \tag{17.8}$$

where N_1 is the number of carriers in subband 1 of a given quantum well and the factor of 2 accounts for spin. The absorption coefficient is then given by

$$\alpha(\omega) = \frac{\pi e^2 N_1 N_W}{\epsilon_0 c n(\omega)\omega m^{*2}\Omega}\left(\frac{8\hbar}{3L}\right)^2 \delta(E_2 - E_1 - \hbar\omega). \tag{17.9}$$

If the thickness of the heterostructure with N_W quantum wells is $\mathcal{L}$ then $\Omega = \mathcal{L}S$, where S is the surface area of a well. Equation (17.9) can then be rewritten as

$$\alpha(\omega) = \frac{\pi e^2 n_1 N_W}{\epsilon_0 c n(\omega)\omega m^{*2}\mathcal{L}}\left(\frac{8\hbar}{3L}\right)^2 \delta(E_2 - E_1 - \hbar\omega), \tag{17.10}$$

where n_1 is the number of carriers per unit area in subband 1. Replacing the delta function by the Lorentzian function

$$\delta(x) \to \frac{\Gamma}{\pi(x^2 + \Gamma^2)}, \tag{17.11}$$

where $x = \hbar\omega + E_1 - E_2$ and Γ is the half-width at half-maximum, we obtain

$$\alpha(\omega) = \frac{\pi e^2 n_1 N_W}{\epsilon_0 c n(\omega) \omega m^{*2} \mathcal{L}} \left(\frac{8\hbar}{3L}\right)^2 \frac{\Gamma}{\pi(x^2 + \Gamma^2)}. \tag{17.12}$$

For a typical GaAs quantum well, $n(\omega) \simeq 3.3$, $m^* \simeq 0.07\,m$, $n_1 \simeq 3 \times 10^{11}\,\text{cm}^{-2}$, $L \simeq 100\,\text{Å}$, and $\Gamma \simeq 5\,\text{meV}$. The absorption peak occurs at $\hbar\omega_{max} \simeq 160\,\text{meV}$. If $N_W = 1$ and $\mathcal{L} = 1\,\mu\text{m}$, $\alpha(\omega_{max}) \simeq 87\,\text{cm}^{-1}$. A plot of the experimental absorption versus frequency is given in Fig. 17.4.

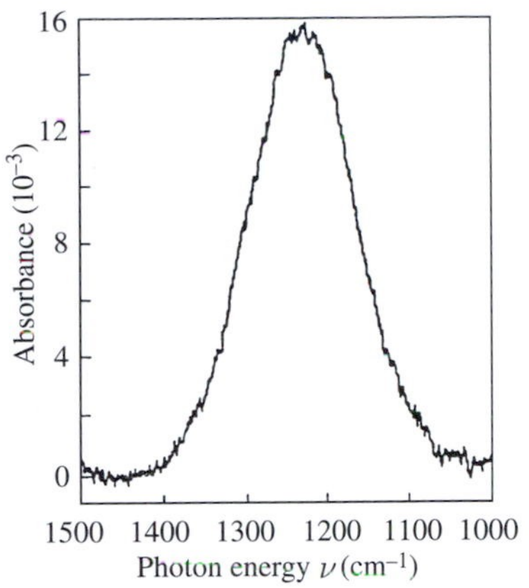

Fig. 17.4
Absorption coefficient versus frequency for the $n=1 \to n=2$ transition of a GaAs quantum well (after Levine *et al.* 1987).

17.1.3 Interband transitions

At this point we must distinguish between type I and type II quantum wells. In a type I quantum well as exemplified by the system GaAs/GaAlAs, both conduction electrons and holes are localized in the same layer, as revealed by the band-edge diagram in Fig. 17.2. However, in a type II quantum well as exemplified by InAs/GaSb and shown in Fig. 17.5, the conduction electrons and holes are spatially separated, the conduction electrons residing in the InAs layer and the holes in the adjoining GaSb.

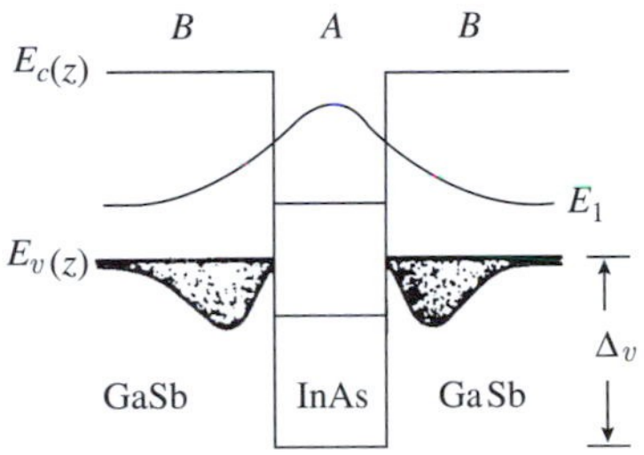

Fig. 17.5
Band-edge diagram for a type II quantum well of InAs (A) and GaSb (B). The upper and lower curves are the envelope functions for electrons and holes, respectively. Note that E_{CA} lies below E_{VB}.

17.1.3.1 Type I quantum wells

In GaAs, interband optical transitions are direct transitions. As discussed in Chapter 11, such transitions are characterized by the second term of the matrix element I_1 in Eq. (11.89),

$$I_1^{(2)} = \langle u_i | \boldsymbol{p} | u_j \rangle \langle F_i^h | F_j^e \rangle, \tag{17.13}$$

where F_i^h and F_j^e are now quantum well envelope functions of holes and conduction electrons, respectively, rather than Landau level envelope functions. The function u_i is the periodic part of the Bloch function for either the heavy hole band or the light hole band at $\boldsymbol{k} = 0$, while u_j is the periodic part of the conduction band at $\boldsymbol{k} = 0$. The matrix element $\langle u_i | \boldsymbol{p} | u_j \rangle$ is very large for direct transitions between valence and conduction bands in GaAs.

The overlap integral $\langle F_i^h | F_j^e \rangle$ can be evaluated using the expression in Eq. (17.1). The result is

$$\langle F_i^h | F_j^e \rangle \equiv \langle n\boldsymbol{k}_\perp | n'\boldsymbol{k}'_\perp \rangle = \delta_{\boldsymbol{k}_\perp, \boldsymbol{k}'_\perp} \int_{-L/2}^{L/2} \chi_n^h(z) \chi_{n'}^e(z) dz. \tag{17.14}$$

Since the functions $\chi_n^h(z)$ and $\chi_{n'}^e(z)$ must have either even or odd parity, their overlap integral can be nonzero only if they are both odd or both even

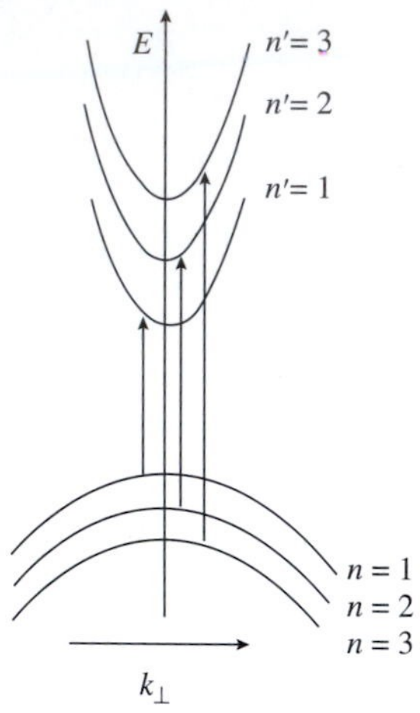

Fig. 17.6
Interband transitions with $n=n'$ for a type I quantum well.

or, alternatively,

$$n - n' = \text{even integer.} \tag{17.15}$$

Furthermore, if we restrict our attention to rectangular, infinitely deep quantum wells, the functions $\chi_n(z)$ are specified by Eqs. (17.2) and are orthonormal. Evaluation of the integral in Eq. (17.14) leads to the more restrictive selection rule

$$n = n'. \tag{17.16}$$

For wells of finite depth, the latter selection rule is relaxed, so that transitions with

$$n - n' = \text{even integer} \neq 0 \tag{17.17}$$

become allowed. Even so, transitions with $n=n'$ are typically much stronger than those with $n \neq n'$. The $n=n'$ transitions are shown schematically in Fig. 17.6.

Polarization selection rules are determined by the matrix elements $\langle u_c|\boldsymbol{p}|u_j\rangle$. In typical III–V semiconductors such as GaAs and InAs, the conduction band has s-like symmetry and the valence band p-like symmetry. Matrix elements of the form $\langle x|p_x|s\rangle = \langle y|p_y|s\rangle = \langle z|p_z|s\rangle$ are nonzero. Interband transitions can therefore be excited by radiation propagating parallel to the quantum well axis with wave vector $\boldsymbol{q}\|\hat{z}$. They can also be excited by radiation propagating perpendicular to the axis and parallel to the interfaces, in common with intersubband transitions.

Using the expressions for the valence band Bloch functions at $\boldsymbol{k}=0$ given in Chapter 3, we can evaluate the matrix elements $\langle u_i|p_x|s\rangle$, where u_i refers to either the heavy hole (hh) or light hole (ℓh) band. The results are

$$\langle hh|\, p_x|s\rangle = P/\sqrt{2} \tag{17.18a}$$

$$\langle \ell h|\, p_x|s\rangle = -P/\sqrt{6}, \tag{17.18b}$$

where $P = \langle x|p_x|s\rangle$. Since the absorption coefficient is proportional to the square of the momentum matrix element, the transitions of heavy holes are three times more intense than those of light holes.

Replacing the matrix element of p_z in Eq. (17.7) by the x-component of $I_1^{(2)}$ in Eq. (17.13), using Eq. (17.18a), and assuming that the conduction band levels are unoccupied, we can express the absorption coefficient for interband transitions involving heavy holes as

$$\alpha(\omega) = \frac{2\pi e^2 N_W}{\epsilon_0 c n(\omega)\omega m^2 \Omega} \sum_{nk_\perp} (P^2/2)|\langle \chi_n^h|\chi_n^e\rangle|^2 \delta \\ \times \left[E_{gA} + E_{en} + E_{hhn} + \frac{\hbar^2 k_\perp^2}{2}\left(\frac{1}{m_c^*} + \frac{1}{m_{hh}^*}\right) - \hbar\omega\right], \tag{17.19}$$

where E_{en} and E_{hhn} are the energies of the subband edges for conduction electrons and heavy holes, respectively, m_c^* and m_{hh}^* are the corresponding effective masses, and we have assumed $n' = n$. Transforming the sum over $\boldsymbol{k}_\perp$ to an integral and evaluating the latter yields the result

$$\alpha(\omega) = \frac{e^2 N_W P^2 \bar{m}^*}{2\epsilon_0 c n(\omega)\omega m^2 \mathcal{L}\hbar^2} \sum_n \Theta(\hbar\omega - E_{gA} - E_{en} - E_{hhn}), \qquad (17.20)$$

where $\Theta(x)$ is the Heaviside step-function, $\bar{m}^*$ is the reduced mass of the electron and heavy hole, and $\langle \chi_n^h | \chi_n^e \rangle$ has been approximated by unity. For a given value of n, the absorption is a step-function. Superposing the contributions of the various values of n produces a staircase profile to the absorption as shown in Fig. 17.7. This profile is essentially a manifestation of the joint density of states of two-dimensional subbands. The absorption threshold occurs at a photon energy $\hbar\omega_0 = E_{gA} + E_{e1} + E_{hh1}$.

A similar expression can be derived for the absorption coefficient associated with light holes. The constant coefficient is one-third that for heavy holes in Eq.(17.20).

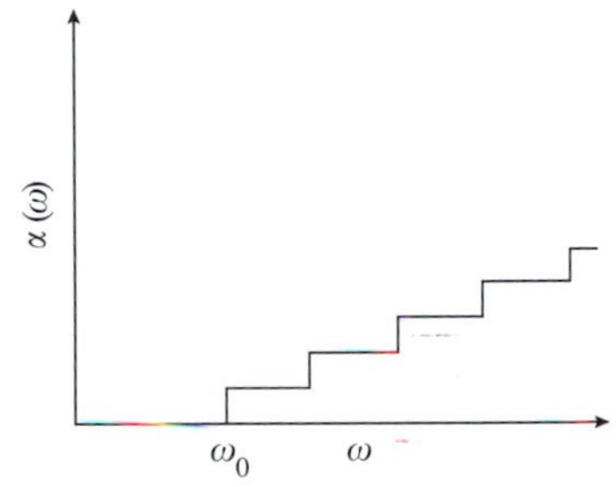

Fig. 17.7
Absorption coefficient versus frequency for interband transitions in a type I quantum well.

Example 17.2: Interband absorption in a quantum well
Discuss the interband absorption of a $GaAs/Ga_{1-x}Al_xAs$ quantum well.

Solution. As can be seen from the Heaviside step-functions, photons will start to be absorbed at an energy $\hbar\omega = E_{gA} + E_{e1} + E_{hh1}$. This corresponds to a significant blue shift of the quantum well fundamental absorption edge with respect to that of the bulk A material. This shift can be tuned by varying the quantum well thickness. Bulk GaAs starts absorbing in the infrared at a photon energy equal to the gap of 1.5192 eV at low temperature. Narrow $GaAs/Ga_{1-x}Al_xAs$ quantum wells of thickness $L = 30$ Å with $x = 0.5$ start absorbing light in the red part of the spectrum at $\hbar\omega = E_{gA} + E_{e1} + E_{hh1} \simeq 1.7275$ eV.

The magnitude of the absorption coefficient $\alpha_{hh1\to c1}$ for the case of a single quantum well embedded in a structure of total thickness $\mathcal{L} = 1\,\mu$m can be calculated using the following parameters: $\hbar\omega = 1.75$ eV, $m_c^* = 0.07\,m$, $m_{hh}^* \gg m_c^*$, $n(\omega) \simeq 3.3$, and $E_P = 2P^2/m \simeq 23$ eV. The result is $\alpha_{hh1\to e1} \simeq 64\,\text{cm}^{-1}$. Such a feeble absorption is difficult to measure, since the light beam intensity is attenuated by only $1 - \exp(-\alpha_{hh1\to e1}\mathcal{L}) \simeq 0.6\%$. Hence, multiple quantum wells must be used to enhance the absorption by the factor N_W that appears in Eq. (17.20).

17.1.3.2 Type II quantum wells

Examples of systems with type II quantum wells are InAs/GaSb and $InP/Al_{0.48}In_{0.52}As$. The spatial separation of conduction electrons and holes is illustrated schematically in Fig. 17.5. The overlap of the conduction and valence band envelope functions is only due to the exponential tails of these functions in layers B and A, respectively. In the limit of large valence and conduction band discontinuities, the overlap integrals $\langle \chi_n^{(h)} | \chi_{n'}^{(e)} \rangle$ tend to zero. On the other hand, $\langle \chi_n^{(h)} | \chi_{n'}^{(e)} \rangle$ should increase with increasing n and n', since the tails of the envelope functions into their respective barriers become more and more important.

Interband optical transitions in type II quantum wells involve both the A and B materials in a significant way, since the conduction electrons are in the A material and the holes of interest are in the B material, as shown in Fig. 17.5. The principal task in calculating the absorption coefficient is to evaluate the overlap integral between valence band envelope functions, which are extended in the z-direction in material B and labeled by a wave vector k_v, and conduction band envelope functions which are localized in the well. If one assumes that the valence band barrier Δ_v is high enough to be impenetrable by valence electrons, one can use two linearly independent valence envelope functions given by

$$\chi_{>}^{(h)}(z) = (2/\sqrt{\mathcal{L}}) \sin\left[k_v\left(z - \frac{L}{2}\right)\right]\Theta\left(z - \frac{L}{2}\right) \tag{17.21a}$$

$$\chi_{<}^{(h)}(z) = -(2/\sqrt{\mathcal{L}}) \sin\left[k_v\left(z + \frac{L}{2}\right)\right]\Theta\left(-z - \frac{L}{2}\right). \tag{17.21b}$$

Noting that $L \ll \mathcal{L}$, these functions are normalized to unity over $\mathcal{L}/2$ with $k_v = 2\pi n_v/\mathcal{L}$ and n_v an integer. The heavy hole energies are given by

$$E_{hh}(\boldsymbol{k}_\perp, k_v) = -E_{gA} + \Delta_v - \frac{\hbar^2 k_\perp^2}{2m_{hh}^*} - \frac{\hbar^2 k_v^2}{2m_{hh}^*} \tag{17.22}$$

where the zero of energy is E_C for the InAs. It is convenient for subsequent developments to introduce new functions which are even and odd with respect to reflection about $z = 0$:

$$\chi_e^{(h)}(z) = \frac{1}{\sqrt{2}}\left[\chi_{>}^{(h)}(z) + \chi_{<}^{(h)}(z)\right] \tag{17.23a}$$

$$\chi_o^{(h)}(z) = \frac{1}{\sqrt{2}}\left[\chi_{>}^{(h)}(z) - \chi_{<}^{(h)}(z)\right]. \tag{17.23b}$$

In treating the conduction band envelope functions it is necessary to include penetration of conduction electrons into the B material in order to have interband absorption under our present assumptions. For the conduction band ground state with energy E_1, we can express the envelope function as

$$\chi_1^{(e)}(z) = A_c \cos(k_c' z), |z| \leq \frac{L}{2} \tag{17.24a}$$

$$\chi_1^{(e)}(z) = B_c \exp\left[-k_c\left(z - \frac{L}{2}\right)\right] z > \frac{L}{2} \tag{17.24b}$$

$$\chi_1^{(e)}(-z) = \chi_1^{(e)}(z). \tag{17.24c}$$

The wave vectors k_c and k_c' are related by the equation

$$(k_c'/k_c)\tan(k_c' L/2) = 1 \tag{17.25}$$

which follows from the continuity of $\chi_1^{(e)}(z)$ and its derivative at $z = L/2$. For the $n = 1$ subband, the value of k_c that is appropriate is the solution of

Eq. (17.25) that gives no nodes in Eq. (17.24a). The overlap integral $\langle\chi_o^{(h)}|\chi_1^{(e)}\rangle$ vanishes, while $\langle\chi_e^{(h)}|\chi_1^{(e)}\rangle$ is given by

$$\langle\chi_e^{(h)}|\chi_1^{(e)}\rangle \simeq \frac{2\sqrt{2}B_c}{\sqrt{\mathcal{L}}}\cdot\frac{k_v}{k_c^2+k_v^2} \qquad (17.26)$$

in the limit $\mathcal{L}\to\infty$.

Exercise. Calculate the probability $P_B(E_1)$ of finding a conduction electron in the B material while in the E_1 state of the A material.
Answer. B_c^2/k_c

With the overlap integral in hand we can calculate the absorption coefficient for interband transitions from the heavy hole valence band of material B to the $n=1$ conduction subband of material A. The matrix element $\langle u_i|p_x|u_j\rangle$ involves valence and conduction band u-functions of material B. The modified form of Eq. (17.19) that is needed for type II quantum wells is

$$\alpha(\omega) = \frac{2\pi e^2 N_W}{\epsilon_0 cn(\omega)\omega m^2\Omega}\sum_{k_v\boldsymbol{k}_\perp}(P^2/2)\cdot\frac{8B_c^2}{\mathcal{L}}\frac{k_v^2}{(k_c^2+k_v^2)^2}$$
$$\times\,\delta\left[E_{gA}-\Delta_v+E_1+\frac{\hbar^2k_v^2}{2m_{hh}}+\frac{\hbar^2k_\perp^2}{2}\left(\frac{1}{m_{c\hat{A}}^*}+\frac{1}{m_{hh}^*}\right)-\hbar\omega\right]. \qquad (17.27)$$

Transforming the sums over k_v and $\boldsymbol{k}_\perp$ to integrals and evaluating the integral over $\boldsymbol{k}_\perp$ yields

$$\alpha(\omega) = \frac{2e^2\bar{m}^* N_W P^2 B_c^2}{\pi\epsilon_0 cn(\omega)\omega m^2\hbar^2\mathcal{L}}\int_0^{k_{max}}\frac{k_v^2 dk_v}{(k_c^2+k_v^2)^2}, \qquad (17.28)$$

where $\bar{m}^*$ is the reduced mass of m_{cA}^* and m_{hh}^* and

$$k_{max}^2 = \frac{2m_{hh}^*}{\hbar^2}(\hbar\omega - E_{gA}+\Delta_v-E_1). \qquad (17.29)$$

Introducing $x=\mathrm{k}_{max}/k_c$ and carrying out the integral over k_v, we obtain

$$\alpha(\omega) = \frac{e^2\bar{m}^* N_W P^2 B_c^2}{\pi\epsilon_0 cn(\omega)\omega m^2\hbar^2\mathcal{L}k_c}\left[-\frac{x}{1+x^2}+\arctan x\right]. \qquad (17.30)$$

The onset of absorption occurs at

$$\hbar\omega_0 = E_{gA}-\Delta_v+E_1. \qquad (17.31)$$

Just above the onset the absorption increases as $(\omega-\omega_0)^{\frac{3}{2}}$, as shown in Fig. 17.8. This behavior is similar to that observed in indirect optical transitions. Thus, the indirect optical transitions in real space under discussion have a similar smoothing effect on the absorption edge as the indirect transitions in reciprocal space discussed in Chapter 10 for bulk materials.

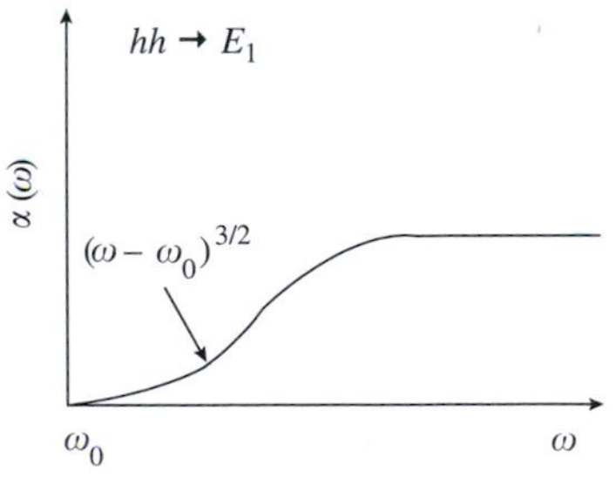

Fig. 17.8
Absorption coefficient versus frequency for a type II quantum well (after Bastard 1988).

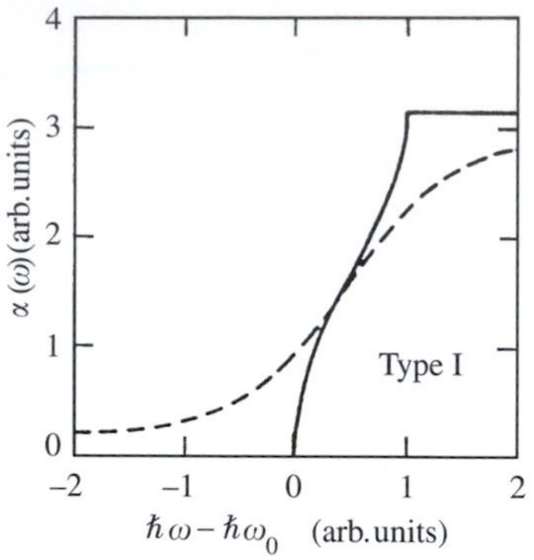

Fig. 17.9
Absorption spectrum for a type I superlattice with no broadening (solid curve) and with broadening (dashed curve) (After Voisin *et al.* 1984).

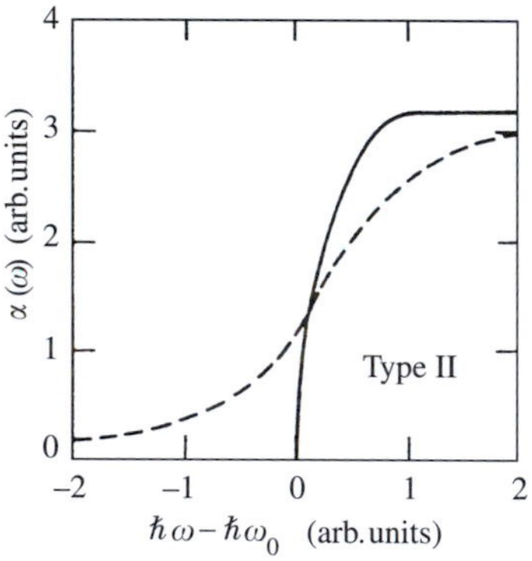

Fig. 17.10
Absorption spectrum for a type II superlattice (after Voisin *et al.* 1984).

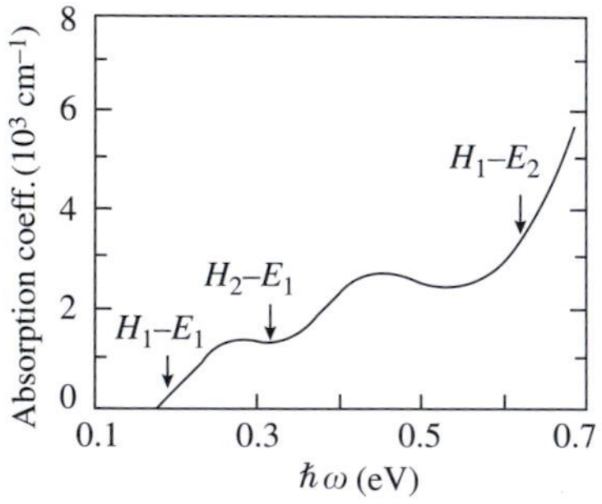

Fig. 17.11
Absorption spectrum of an InAs/GaSb superlattice at 4.2 K. *E* and *H* denote electron and heavy hole subbands. (after Chang *et al.* 1981).

Since the physical origin of this absorption is due to the exponential tail of the conduction band envelope function outside the confined layer, the absorption is expected to be very weak. Hundreds of wells may be needed in order for this absorption to be observable, since $P_B(E_1)$ is only a few percent.

17.1.4 Interband optical transitions in superlattices

When the barriers separating consecutive quantum wells become sufficiently thin, the wells become coupled together due to the tunneling of carriers through the barriers. The localized states within the wells hybridize, forming minibands along the axis of the superlattice. Due to the symmetry properties and the translational invariance of the superlattice potential, the optical absorption in typical superlattices corresponds to vertical transitions in the superlattice Brillouin zone and in the Brillouin zone of the layer plane. The relative spatial localization of electrons and holes also plays an important part in the selection rules. One finds that in type II superlattices, an allowed optical transition at $q = 0$ becomes forbidden at $q = \pi/d$, where d is the superlattice period. This rule significantly modifies the absorption lineshape. In perfect structures the absorption line shapes of type I and type II superlattices should be distinct, as shown in Figs. 17.9 and 17.10. In imperfect structures in which damping is significant, the sharp features are smeared out as indicated by the dashed lines in Figs. 17.9 and 17.10. The lineshapes then become very similar for both types of superlattices.

To distinguish between the two types of superlattices, it is preferable to employ the magnitude of the absorption coefficient rather than a lineshape analysis. As in individual quantum wells, the absorption in type II superlattices should be significantly weaker than that in type I superlattices, since the former involves indirect transitions in real space, whereas the latter involves direct transitions.

An example of an optical absorption spectrum in which important smoothing effects are observed is shown in Fig. 17.11 for an InAs/GaSb superlattice.

17.1.5 Optical absorption by excitons in heterostructures

As discussed in Chapter 10, the Coulomb interaction between an electron in the conduction band and a hole in the valence band leads to a bound state, the exciton, in which the electron and hole revolve around each other. We have seen that excitons produce observable effects on the interband absorption spectra of bulk semiconductors. Particularly important is the appearance of sharp peaks in the spectra for photon energies near E_g that are not present in the absence of the electron–hole interaction.

Exciton effects in quantum wells and superlattices are even more striking than in bulk materials. For a single quantum well the envelope function $F_{nn'\ell}(\boldsymbol{R}_\perp, \boldsymbol{r}_\perp; z_e, z_h)$ can be approximated by

$$F_{nn'\ell}(\boldsymbol{R}_\perp, \boldsymbol{r}_\perp;\, z_e, z_h) = e^{i\boldsymbol{K}_\perp \cdot \boldsymbol{R}_\perp} \Phi_p(\boldsymbol{r}_\perp) \chi_n^{(e)}(z_e) \chi_{n'}^{(h)}(z_h), \tag{17.32}$$

where $\boldsymbol{R}_\perp$ and $\boldsymbol{K}_\perp$ are the center of mass coordinate and wave vector in the plane of the quantum well, $\boldsymbol{r}_\perp$ is the relative coordinate of the electron and hole in the plane, and p is the exciton quantum number. For very narrow quantum wells we can ignore the effect of the Coulomb interaction on $\chi_n^{(e)}(z_e)$ and $\chi_{n'}^{(h)}(z_h)$. The effective Schrödinger equation satisfied by $\Phi_p(\boldsymbol{r}_\perp)$ can then be written as

$$\left\{-\frac{\hbar^2\boldsymbol{\nabla}_\perp^2}{2\bar{m}^*}-\frac{e^2}{4\pi\epsilon_0\epsilon_\infty}\int\int\frac{|\chi_n^{(e)}(z_e)|^2|\chi_{n'}^{(h)}(z_h)|^2dz_e dz_h}{[r_\perp^2+(z_e-z_h)^2]^{\frac{1}{2}}}\right\}\Phi_p^{nn'}(\boldsymbol{r}_\perp)$$
$$=E_p^{nn'}\Phi_p^{nn'}(\boldsymbol{r}_\perp), \tag{17.33}$$

$$E_p^{nn'}=E-E_{gA}-E_{en}-E_{hhn'}-\frac{\hbar^2K_\perp^2}{2(m_c^*+m_{hh}^*)}, \tag{17.34}$$

and E is the total energy eigenvalue.

Equation (17.33) can be solved using a variational function for $\Phi_p^{nn'}(\boldsymbol{r}_\perp)$. For the ground state one can take

$$\Phi_1^{11}(\boldsymbol{r}_\perp)=(2/\pi\lambda^2)^{\frac{1}{2}}\exp(-r_\perp/\lambda), \tag{17.35}$$

where λ is the effective Bohr radius. In the limit of a quantum well whose thickness approaches zero and depth approaches infinity, one has $|\chi_1^{(e)}(z_e)|^2=\delta(z_e)$ and $|\chi_1^{(h)}(z_h)|^2=\delta(z_h)$, so Eq. (17.33) becomes

$$\left[-\frac{\hbar^2\nabla_\perp^2}{2\bar{m}^*}-\frac{e^2}{4\pi\epsilon_0\epsilon_\infty r_\perp}\right]\Phi_p^{11}(\boldsymbol{r}_\perp)=E_p^{11}\Phi_p^{11}(\boldsymbol{r}_\perp), \tag{17.36}$$

which is the Schrödinger equation for a two-dimensional hydrogen atom. The energy eigenvalues for the discrete states are given by

$$E_p^{11}=-\frac{\mathrm{Ry}^*}{(p-\frac{1}{2})^2},\quad p=1,2,3,\ldots. \tag{17.37}$$

where Ry^* is the effective Rydberg for a three-dimensional hydrogen-like atom given by

$$\mathrm{Ry}^*=\frac{\bar{m}^*e^4}{2(4\pi\epsilon_0)^2\epsilon_\infty^2\hbar^2}. \tag{17.38}$$

The eigenfunctions can be characterized as $1s, 2p, \ldots$, with even or odd parity.

For the ground state $p=1$ the binding energy $E_B^{(2)}$ of a 2D exciton is

$$E_B^{(2)}=4\mathrm{Ry}^*, \tag{17.39}$$

i.e., four times that of a 3D exciton. This increase in binding energy is due to the compression of the electron–hole system in passing from the 3D to the

2D case. The total energy for the ground state with $\boldsymbol{K}_\perp = 0$ is

$$E = E_{gA} + E_{e1} + E_{hh1} - 4\mathrm{Ry}^*. \tag{17.40}$$

The exciton effect thus reduces the effective energy gap by the amount 4 Ry*, which is four times the reduction in the 3D case.

If the thickness of the quantum well is increased, the effective Bohr radius λ increases, while the binding energy of the exciton decreases. Excitonic transitions involving subbands with $n > 1$ become evident in optical absorption.

Interband optical absorption including exciton effects constitutes a rather complex many-electron problem. We shall not go into the details of the solution, but simply present the result for the absorption coefficient if the radiation has its electric vector in the x-direction. For the fundamental peak of a single well, we have (Bastard 1988)

$$\alpha(\omega) = \frac{2\pi e^2}{\epsilon_0 c n(\omega)\omega m^2 \mathcal{L}} |\langle u_{hh}|p_x|u_e\rangle|^2 |\langle \chi_1^{(h)}|\chi_1^{(e)}\rangle|^2$$
$$\times\, \delta_{\boldsymbol{K}_\perp,0} |\Phi_1^{11}(0)|^2 \delta(E_{gA} + E_{e1} + E_{hh1} - 4\mathrm{Ry}^* - \hbar\omega). \tag{17.41}$$

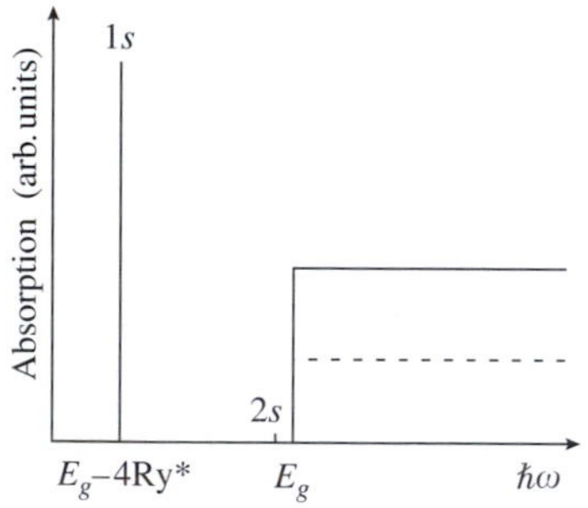

Fig. 17.12
Schematic representation of the optical absorption due to excitonic transitions (after Bastard 1988).

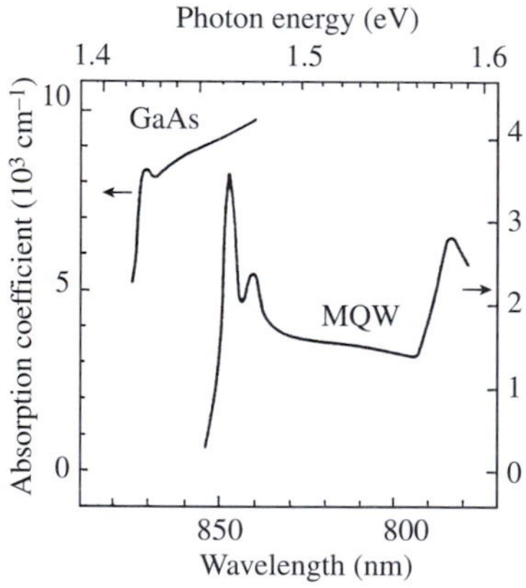

Fig. 17.13
Absorption spectra for bulk GaAs and for a GaAlAs/GaAs multiple quantum well structure (after Chemla 1985).

Similar expressions apply to transitions involving higher discrete levels specified by Eq. (17.37) and discrete levels associated with higher subbands with $n > 1$. In addition, transitions can be made to continuum levels of the exciton associated with each subband.

Several features are apparent from Eq. (17.41). The Kronecker delta, $\delta_{\boldsymbol{K}_\perp,0}$, allows only transitions to an exciton state with essentially no center-of-mass motion. The factor $|\Phi_1^{11}(0)|^2$ requires that only s-like exciton states can be created by optical absorption. States with p-like, d-like, ... character have a zero value of Φ at zero argument and hence cannot be excited. A schematic representation of the optical absorption due to excitonic transitions for the $n = 1$ subband is shown in Fig. 17.12 in the absence of broadening.

The difference between the absorption coefficient for noninteracting electron–hole systems and the experimentally observed absorption coefficient shows that even into the continuum the noninteracting picture is not correct. The long range nature of the Coulomb interaction affects the behavior of the electron and hole even when they do not form a bound state.

The fact that the binding energy for 2D excitons is four times that for 3D excitons greatly facilitates the experimental observation of exciton peaks in the optical absorption by quantum well structures. These peaks have been observed in many undoped quantum well structures of high quality, even at room temperature. In Fig. 17.13 are compared the room temperature absorption spectra near the absorption edge of bulk GaAs and of a multiple quantum well structure of GaAs/GaAlAs. The much more clearly defined exciton peaks in the latter structure are quite evident.

In quantum well structures one observes a strong nonlinear absorption that is related to the large oscillator strength of the principal exciton peak, which is well separated from the continuum absorption and is restricted to a narrow frequency range. On the other hand, at room temperature, the excitons in bulk GaAs are essentially dissociated into free electrons and holes, and strong, well-defined exciton peaks are not observed. As a result, the nonlinear optical response of GaAs quantum wells is enhanced with respect to bulk GaAs.

17.2 Photoluminescence in two-dimensional systems

Photoluminescence is of interest not only in providing fundamental physics information but also as the basis for practical light sources. Photoluminescence experiments are much easier to perform than optical absorption experiments, but they are far more difficult to interpret. To produce photoluminescence the excited states of the system must be optically populated, and they are not in equilibrium. De-excitation, involving return to the equilibrium state, may proceed through many different channels including radiative recombination of free carriers and nonradiative relaxation processes: phonon emission, Auger effect, capture by deep centers, etc. The efficiency of photoluminescence depends on the lifetimes in the excited state with respect to the radiative and nonradiative relaxation mechanisms.

17.2.1 Experimental techniques

Two similar techniques are frequently used in the experimental investigation of spontaneous radiative recombination:

1. *Photoluminescence spectroscopy*. The experimental setup consists of an intense light source at a fixed frequency, a laser for example, whose beam is focused on the sample to be studied. The emitted light from the sample is collected by a mirror and sent through a monochromator to a detector.

2. *Photoluminescence excitation spectroscopy*. Excitation spectroscopy is essentially the inverse of luminescence spectroscopy. The detection spectrometer is set at a given frequency inside the photoemission band, and the frequency of the exciting light is scanned through a monochromator. Thus, the excited discrete levels of the system under investigation are populated at rates proportional to their absorption coefficients. Once populated, the individual excited levels relax radiatively or nonradiatively.

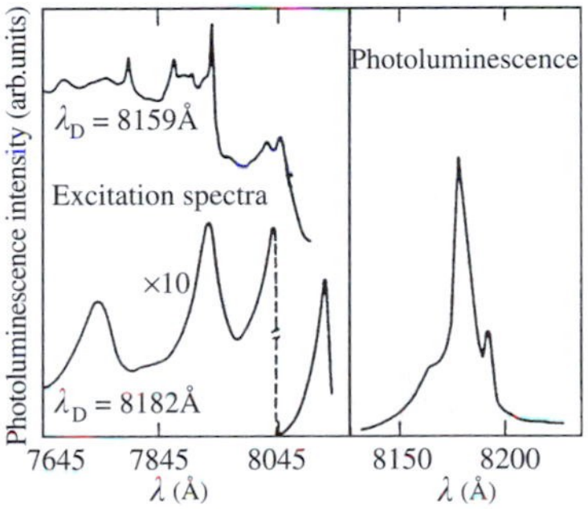

Fig. 17.14
Excitation and luminescence spectra of a GaAlAs/GaAs quantum well (after Bastard 1988).

An example of excitation and luminescence spectra is shown in Fig. 17.14. One sees that the excitation spectrum reproduces to some extent the absorption spectrum, but this is not necessarily the case. In some cases the excitation spectrum may be quite different from the absorption spectrum. The excitation spectrum may show peaks which are not found by absorption. This occurs when the excited levels have significantly different relaxation times.

Excitation spectroscopy, like photoluminescence spectroscopy, does not require thick samples, in contrast to absorption spectroscopy, and can therefore be performed on single quantum wells. Moreover, excitation spectroscopy has two frequencies which can be analyzed and exploited to probe the physical origin of the photoluminescent signal. If, for example, the photoluminescence band contains several peaks, by setting the detection frequency at each of these peaks, one obtains excitation spectra whose shape is characteristic of the various excited levels which give rise to the different photoluminescence peaks. If the excitation spectra are independent of the detection frequency, one may assume that the whole luminescence band has a single physical origin, but if the excitation spectra are markedly different, one can conclude that the various photoluminescence lines have different physical origins.

17.2.2 Quantum well luminescence

Luminescence involves optical transitions from the initial state $|i\rangle$, which is an excited state, to the final state $|f\rangle$, which is the thermal equilibrium state. This process is characterized by the same momentum matrix elements as the absorption process, i.e.,

$$\langle i|\hat{e}\cdot\boldsymbol{p}|f\rangle = \langle u_i|\hat{e}\cdot\boldsymbol{p}|u_f\rangle\langle F_i|F_f\rangle, \tag{17.42}$$

where $\hat{e}$ is a unit vector in the direction of the electric field of the radiation. The matrix element $\langle u_i|\hat{e}\cdot\boldsymbol{p}|u_f\rangle$ gives the selection rule for the polarization of the emitted light, and the overlap integral $\langle F_i|F_f\rangle$ gives the selection rule for the subband indices which governs the interband recombination. In quantum wells with band edge profiles which are symmetric in z, the only optically allowed transitions between subbands, as for absorption, preserve the parity of the z-dependent part of the envelope functions.

In many cases the photo-excited conduction or valence carriers thermalize among themselves much faster than they recombine. Thermalized electrons acquire a temperature T_c and the holes a temperature T_h which may be different from the lattice temperature T.

17.2.2.1 Band-to-band recombination

The polarization selection rule for band-to-band recombination depends on the direction of propagation of the emitted light. For propagation along the z-axis, transitions from conduction subbands to both light and heavy hole subbands are allowed provided $n-n'$ is an even integer and the quantum well potential is an even function of z. For light emitted in the layer plane with $\boldsymbol{\mathcal{E}}\|\hat{z}$, transitions from conduction subbands to light hole subbands are allowed, but to heavy hole subbands are forbidden.

If electrons and holes are assumed to be at thermal equilibrium with $T_c = T_h = T^*$ and if $k_B T^*$ is comparable to the energy separation between several nearby subbands, the photoluminescence spectrum typically displays several lines. For example, in GaAs/GaAlAs quantum wells with a GaAs layer thickness of $\sim$100 Å and at a low carrier injection rate, one observes two lines at room temperature associated with recombination of electrons with holes in both light and heavy hole subbands. At low temperature the recombination involving light hole subbands disappears.

At high injection rate and low temperature, band-to-band emission in GaAs/GaAlAs quantum wells can be observed under conditions such that excitonic effects can be disregarded.

17.2.2.2 Excitonic recombination

If the heterostructure potential is symmetric with respect to $z \to -z$, excitonic recombination fulfills the same selection rules as the excitonic absorption, i.e., $\boldsymbol{K}_\perp = 0$ and $n - n'$ an even integer. In addition, only excitons in ns states can recombine radiatively. This means that the excitonic luminescence should consist of discrete lines, most likely, a single 1 s exciton line attached to the band gap

$$\Delta E = E_{gA} + E_{e1} + E_{hh1}. \tag{17.43}$$

One of the dominant features of the optical properties of GaAs/GaAlAs quantum wells is the high intensity of "free" exciton luminescence at low temperature compared to that involving impurities. This situation is the opposite of what is usually observed in bulk GaAs. This effect may be related to the very small volume in which the carriers and the photons interact in the quantum well. Once created by recombination of an electron and a hole, the photon does not have enough space to propagate before being re-absorbed to create another exciton, as happens in the bulk. Instead, the photon almost immediately strikes a boundary and escapes from the structure.

In "good" samples the maximum of the photoluminescence spectrum coincides with the maximum of the absorption spectrum at $\hbar\omega = E_{gA} + E_{el} + E_{hh1} - 4\mathrm{Ry}^*$. The Stokes shift, which is the energy separation between the absorption peak and the luminescence peak, is zero. In such a case the observed excitons are not bound to extrinsic defects. In general, a variety of experimental results indicate some small Stokes shifts related to trapping of excitons on intrinsic interface defects.

Two-dimensional excitons can be localized by weak disorder, the localization being the result of constructive interference of the wavefunctions of excitons scattered by randomly distributed defects.

The quality of quantum well structures can be judged by the line width of the free exciton luminescence line. The Stokes shift and the line width are correlated: the broader the luminescence line, the larger the Stokes shift.

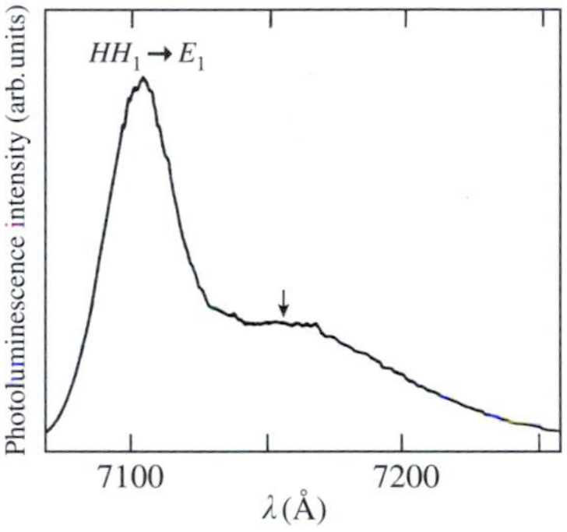

Fig 17.15
Photoluminescence spectrum of a GaAlAs/GaAs quantum well showing structure under the arrow due to recombination at acceptors (after Bastard 1988).

17.2.2.3 Extrinsic photoluminescence

When the Stokes shift becomes large, extrinsic defects may play a significant role in the radiative recombination. As can be seen in Fig. 17.15, the excitonic recombination line $e1 \to hh1$ is accompanied by a weak line which is attributed to the $e1 \to$ acceptor recombination process.

Generally speaking, the residual doping in a heterostructure is not uniform: impurities segregate near the inverted interface. Thus, there is a considerable enhancement of the luminescence from the recombination of conduction electrons with acceptors on an edge of the sample.

Problems

1. Consider a finite square quantum well of width L and depth V_0 that contains N_1 electrons in the lowest subband of energy E_1. Derive an expression for the absorption coefficient $\alpha(\omega)$ associated with transitions from the lowest subband to the continuum for photon energies satisfying $\hbar\omega \geq V_0 - E_1$.

You may take the continuum wave functions to be plane waves in a normalization length $\mathcal{L}$. Taking $L = 100$ Å, $V_0 = 0.3$ eV, $m^* = 0.07$ m, $\mathcal{L} = 10\,L$, and $E_1 = -V_0 + \pi^2\hbar^2/2m^*L^2$, calculate $\alpha(\omega)$ for a range of $\hbar\omega$ and plot $\alpha(\omega)$ versus $\hbar\omega$.

2. For the case of a type II quantum well with interband optical transitions obeying the selection rule $n = n'$, make a schematic plot of the absorption coefficient versus photon energy that includes transitions with $n = 1, 2, 3, 4$, and 5.
3. For an InAs/GaSb quantum well, set up the odd parity conduction band envelope function with one node corresponding to $n = 2$. Evaluate the appropriate overlap integral with the odd parity heavy hole envelope function and develop an expression for the absorption coefficient. At what photon energy does the onset of absorption occur?
4. Calculate the two-dimensional exciton binding energies for both light holes and heavy holes in GaAs. Take $m_c^* = 0.07\,m$, $m_{lh}^* = 0.08\,m$, $m_{hh}^* = 0.5\,m$, and $\epsilon = 13.5$. Use these results to make a rough plot of the absorption spectrum near the band edge of a GaAs/GaAlAs quantum well.

References

G. Bastard, *Wave Mechanics Applied to Semiconductor Heterostructures* (Les Editions de Physique, Les Ulis, 1988).

L. L. Chang, G. A. Sai-Halasz, L. Esaki, and R. L. Aggarwal, *J. Vac. Sci. Technol.* **19**, 589 (1981).

D. S. Chemla, *J. Lumin.* **30**, 502 (1985).

B. F. Levine, R. J. Malik, J. Walker, K. K. Choi, C. G. Bethea, D. A. Kleinman, and J. M. Vandenberg, *Appl. Phys. Lett.* **50**, 273 (1987).

P. Voison, G. Bastard, and M. Voos, *Phys. Rev.* B**29**, 935 (1984).

R. F. Wallis, *J. Phys. Chem. Solids* **4**, 101 (1958).

Transport properties of heterostructures

18

Key ideas

Heterostructures have a large *anisotropy* of their transport properties in the directions parallel and perpendicular to the growth axis.

The large spatial separation of conduction electrons and ionized donors weakens the Coulomb interaction and is largely responsible for very high electron mobilities. Intersubband scattering becomes important at high electron concentrations.

Under an applied field along the growth axis $\hat{z}$ the edges of the bands are deformed. At high electric fields *tunneling* of electrons can occur. For a potential corresponding to twice the difference between the first subband energy and the conduction band edge of the emitter, *resonant tunneling* occurs and the tunneling current is a maximum. For potentials exceeding the resonant tunneling condition the slope of the current–voltage curve is negative corresponding to a *negative differential resistance*. This produces an instability which forms the basis for high-frequency oscillators.

With a magnetic field parallel to the growth axis $\hat{z}$ the carrier motion is quantized in the z-direction by the confining potential and in the x- and y-directions by the magnetic field.

If the magnetic field is not parallel to the growth axis, *anticrossing* of the Landau levels of two different subbands occurs.

The density-of-states has point-like singularities separated by energy gaps. When the Fermi energy lies in a gap, the electron system behaves as an *insulator*, but if it lies on one of the energy eigenvalues, the system behaves a *conductor*.

The conductivity in the plane of a layer exhibits *oscillations* when the magnetic field is varied. The Hall resistivity in the layer plane at low temperatures exhibits a series of well-defined plateaus. This is the *quantized Hall effect*.

Resonance absorption can occur at cyclotron harmonics as well as at the fundamental ω_c. Oscillations appear in the cyclotron resonance lineshapes when its width is sufficiently large to permit several Landau levels to cross the Fermi energy.

Transport properties

There are two aspects of heterostructures containing two-dimensional systems that make their transport properties significantly different from those of three-dimensional structures. First, heterostructures have a large anisotropy of their transport properties in the directions parallel and perpendicular to the growth axis. Second, much higher mobilities can be achieved in such structures.

18.1 Effects of a constant electric field

We shall distinguish the effects of two different orientations of the electric field: (1) electric field applied parallel to the interfaces and (2) electric field applied perpendicular to the interfaces.

18.1.1 Electric field parallel to the interfaces: $\mathcal{E}\|\hat{x}$

The current carriers are taken to be electrons that occupy subbands with energies given by

$$E_n(\boldsymbol{k}_\|) = E_n + \frac{\hbar^2 k_\|^2}{2m_c^*}, \tag{18.1}$$

where $\boldsymbol{k}_\|$ is the component of wave vector parallel to the interfaces and E_n is the edge of the nth subband. Under the influence of the electric field $\mathcal{E}$, electrons in various subbands are accelerated, are scattered by various mechanisms, and reach a steady-state distribution. The current density $\boldsymbol{J}$ is given by Ohm's law as

$$\boldsymbol{J} = \sigma\boldsymbol{\mathcal{E}}, \tag{18.2}$$

where the conductivity σ has the form

$$\sigma = \sum_n \sigma_n \tag{18.3}$$

and

$$\sigma_n = \frac{n_n e^2 \tau_n}{m_c^*}. \tag{18.4}$$

The quantities n_n and τ_n are the concentration and scattering time of an electron in the nth subband. The corresponding mobility μ_n is given by

$$\mu_n = \frac{e\tau_n}{m_c^*}, \tag{18.5}$$

so that

$$\sigma_n = n_n e \mu_n. \tag{18.6}$$

The total conductivity can now be rewritten as

$$\begin{aligned}\sigma &= \sum_n n_n e \mu_n \\ &= ne\bar{\mu},\end{aligned} \tag{18.7}$$

where $\bar{\mu}$ is the average mobility defined by

$$\bar{\mu} = \sum_n \frac{n_n \mu_n}{n} \tag{18.8}$$

and n is the total electron concentration.

The scattering mechanisms for carriers in the conducting channel include those found in the bulk materials plus those that are associated with the presence of interfaces. Among the former are scattering by impurities, alloy disorder, and bulk phonons, while the latter arise from interface roughness, interface phonons, and intersubband transitions. In GaAs/GaAlAs quantum wells with low residual doping of the GaAs layer ($< 10^{15}$ cm^{-3}), the principal impurity scattering of electrons in the GaAs is due to their Coulomb interaction with ionized donors in the GaAlAs barriers. The large spatial separation of the electrons and ionized donors weakens the interaction and is largely responsible for the very high mobilities ($\sim 10^7$ cm^2/(V s)) that have been achieved in this type of quantum well at moderate and low temperatures. High mobilities are favored by placing undoped GaAlAs spacer layers between the GaAs layer and the doped GaAlAs barriers.

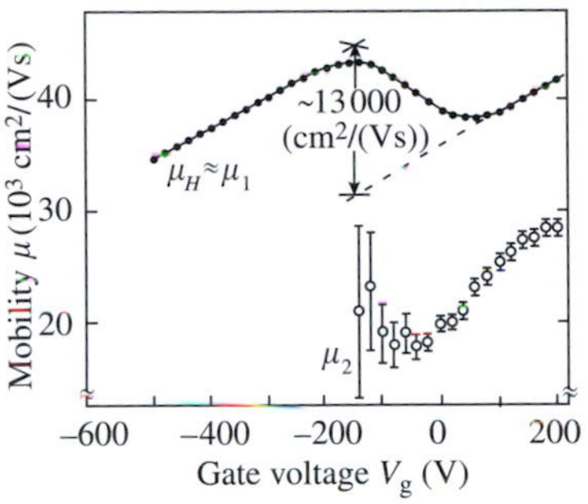

Fig. 18.1
Hall mobility drop in a GaAs/GaAlAs heterostructure at the onset of occupancy of the E_2 subband. μ_1 and μ_2 are the mobilities of electrons in the E_1 and E_2 subbands, respectively (after Störmer *et al.* 1982).

Scattering by phonons in the GaAlAs is relatively weak due to the small penetration of the electron wave functions into the spacers and barriers. The lack of penetration also leads to suppression of scattering due to alloy disorder. If the conducting channel material is itself an alloy, as in the InP/GaInAs heterostructure, alloy disorder can significantly reduce the maximum mobility. Surface roughness appears to be of less importance in III–V compound heterostructures than in metal-oxide-semiconductor structures (Weisbuch and Vinter 1991). In systems such as GaAs/GaAlAs the similar dielectric constants and densities of the two materials make interface phonon scattering relatively insignificant.

Intersubband scattering becomes important at relatively high electron concentrations involving occupancy of higher subbands ($n = 2, 3, \ldots$). A drop in mobility occurs when the $n = 2$ subband has states at the same energy as occupied states of the $n = 1$ subband, as shown in Fig. 18.1. The scattering probability increases because an electron at this energy in the $n = 1$ subband can scatter into states of the $n = 2$ subband as well as into states of the $n = 1$ subband.

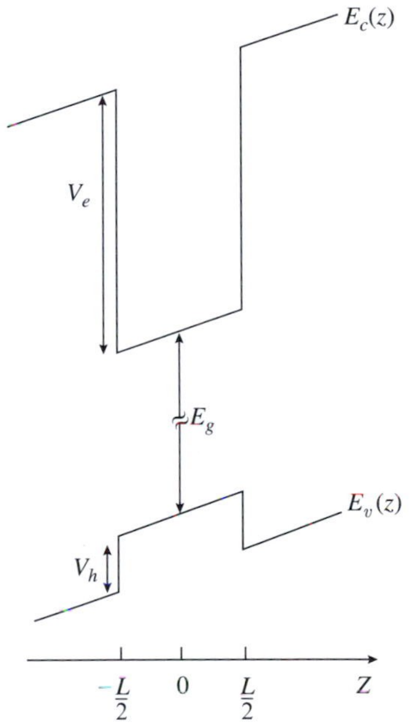

Fig. 18.2
Effect of an electric field on the conduction and valence band edges of a type I quantum well.

18.1.2 Electric field perpendicular to the interfaces: $\mathcal{E} \| \hat{z}$

When an external electric field is applied along the growth axis $\hat{z}$ perpendicular to the interfaces, the edges of the energy bands are deformed as shown in Fig. 18.2. The evolution of the ground state wave function as the electric field increases is given in Fig. 18.3.

At zero field the average electron position $\langle z_c \rangle$ in the ground state $n = 1$ is zero (Fig. 18.3a). As $\mathcal{E}$ increases, $\langle z_c \rangle$ becomes negative, and its magnitude is proportional to $\mathcal{E}$ at small fields. An electric dipole moment $\mathcal{M}$ is created,

$$\mathcal{M} = -e\langle z_c \rangle = \epsilon_0 \alpha \mathcal{E}, \tag{18.9}$$

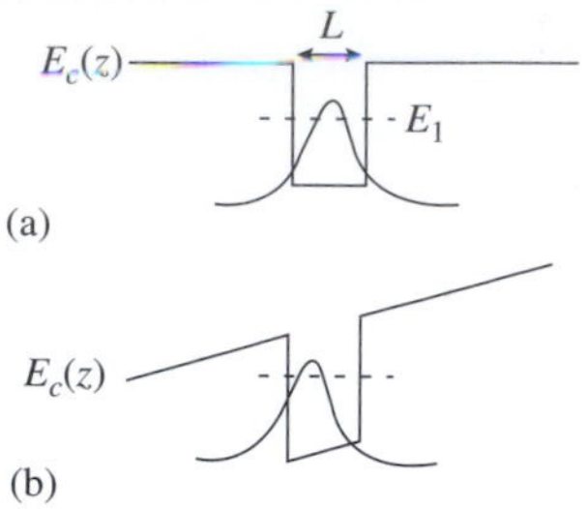

Fig. 18.3
Ground state wave function in a quantum well for (a) zero electric field and (b) large electric field.

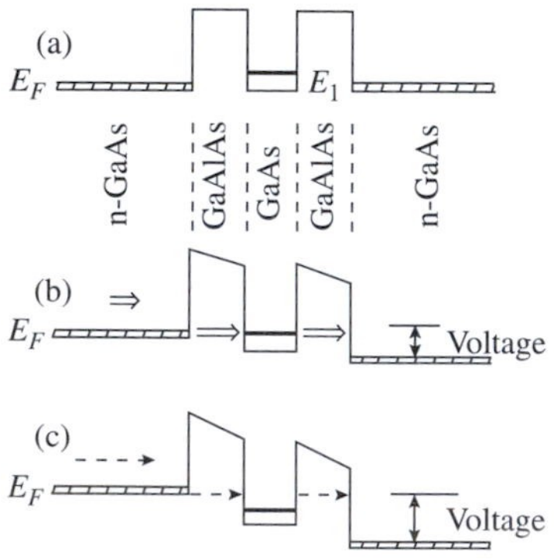

Fig. 18.4
Diagram of the resonant tunneling effect (after Sollner *et al.* 1983).

where α is the polarizability. The shift in energy ΔE_1 of the $n = 1$ subband can be written as

$$\Delta E_1 = -\int_0^{\mathcal{E}} \mathcal{M} d\mathcal{E}$$
$$= -\epsilon_0 \alpha \int_0^{\mathcal{E}} \mathcal{E} d\mathcal{E} = -\tfrac{1}{2}\epsilon_0 \alpha \mathcal{E}^2. \tag{18.10}$$

This result corresponds to the quadratic Stark effect. Further increase in the field causes the electrons to accumulate near the left-hand side of the well as shown in Fig. 18.3b. The induced dipole moment saturates along with ΔE_1. If the barrier is infinitely high, the electrons become localized in a triangular potential well. Since real barriers are neither infinitely high nor infinitely thick, the field can eventually reach a value that leads to appreciable tunneling of electrons through the barrier.

The essential ideas of electron tunneling into and out of quantum wells can be illustrated by a single well bounded by two barriers as shown in Fig. 18.4a. The regions to the left of the left barrier and to the right of the right barrier are n-GaAs and are called the emitter and collector, respectively. We assume that the quantum well is i-GaAs symmetrically bounded by GaAlAs barriers and is sufficiently narrow to place E_1 above the Fermi energy E_F of the emitter and collector. The temperature is taken to be sufficiently low that the electrons in the n-GaAs are degenerate.

In the tunneling process, the energy E and the wave vector component parallel to the interfaces $k_{\|}$ are conserved. The energy of an electron in the emitter is

$$E^e = E_C + \frac{\hbar^2 k_{\|}^2}{2m_c^*} + \frac{\hbar^2 k_z^2}{2m_c^*}. \tag{18.11}$$

For the symmetric structure in Fig. 18.4 with an applied voltage V, the energy of an electron in the well is

$$E^w = E_1 + \frac{\hbar^2 k_{\|}^2}{2m_c^*} - \frac{eV}{2}. \tag{18.12}$$

The two conservation laws then lead to the relation

$$\frac{eV}{2} = E_1 - \left(E_C + \frac{\hbar^2 k_z^2}{2m_c^*}\right). \tag{18.13}$$

The voltage has its minimum value when the quantity in parentheses has its maximum value. The maximum value of k_z is k_F corresponding to

$$E_C + \frac{\hbar^2 (k_z^{max})^2}{2m_c^*} = E_C + \frac{\hbar^2 k_F^2}{2m_c^*} = E_F. \tag{18.14}$$

Since we have taken the system to have $E_1 > E_F$, we see that Eq. (18.13) is obeyed only if the applied voltage satisfies the condition

$$eV \geq 2(E_1 - E_F), \tag{18.15}$$

as exemplified by Fig. 18.4b. When the equality sign holds, electrons with $k_z = k_F$ and $k_{\|} = 0$ can tunnel, but no others. As V increases above this value, electrons with $k_z < k_F$ and $k_{\|} > 0$ can tunnel, their number increasing as $k_{\|}^2 = k_F^2 - k_z^2$, until $k_z = 0$ and V reaches the value

$$V = \frac{2(E_1 - E_C)}{e}. \tag{18.16}$$

This is the condition for **resonant tunneling** corresponding to the tunneling current having its maximum value. For $V > 2(E_1 - E_C)/e$ as in Fig. 18.4c, Eq. (18.13) cannot be satisfied for any real value of k_z, and the tunneling current drops to zero. The qualitative behavior of the tunneling current I as a function of V is shown in Fig. 18.5, where some thermal broadening has been included.

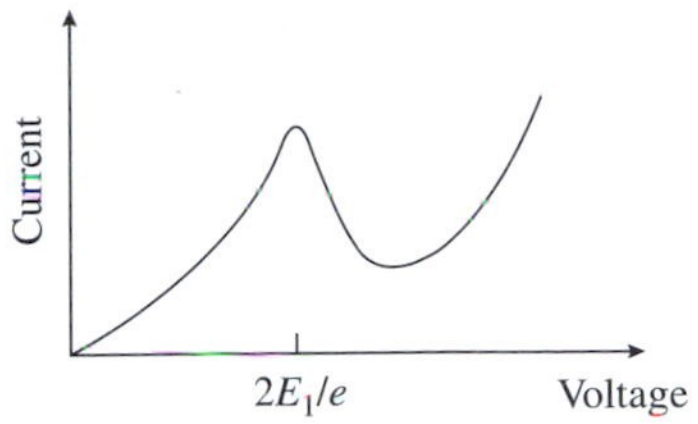

Fig. 18.5
Current–voltage characteristic for resonant tunneling. E_1 is measured relative to E_C (after Sollner *et al.* 1983).

We note that the conservation laws for tunneling from the quantum well to the collector have the form

$$E_C + \frac{\hbar^2 k_z^2}{2m_c^*} = E_1 + \frac{eV}{2}. \tag{18.17}$$

For any V satisfying Eq. (18.15) and $E_1 > E_F > E_C$ one can always find a value of k_z that satisfies Eq. (18.17). Therefore, no cutoff of the tunneling current arises from quantum well-to-collector tunneling.

Quantitative treatments of resonance tunneling have been given by a number of workers (Tsu and Esaki 1973, Bastard 1988). These treatments establish the effect of barrier thickness on the I–V curves and the detailed behavior of the transmission coefficient as a function of energy.

An important aspect of the current–voltage curve in Fig. 18.5 is the **negative differential resistance** for $V > 2(E_1 - E_C)/e$, i.e., the slope dI/dV is negative and the double barrier diode becomes unstable. This instability can be exploited to produce high-frequency oscillators (Weisbach and Vinter 1991). Bistability can arise having a high-current state in which the well level contains many electrons and a low-current state with few electrons in the well level.

18.2 Effects of a constant magnetic field

Magnetic field effects on the transport properties of bulk semi-conductors have been discussed in Section 8.6. In this section we analyze the modifications of magnetotransport properties that occur as a result of the confinement of current carriers into two dimensions. We shall focus our attention on the configuration having the external magnetic field $\mathcal{B}$ parallel to the growth axis and perpendicular to the interfaces of the heterostructure. The carrier motion is entirely quantized. In the growth (z) direction it is quantized by the confining potential $V_c(z)$, while in the individual layers it is quantized by the magnetic field.

18.2.1 Energy levels and wave functions

The eigenstates correspond to the Landau levels associated with each of the energy subbands. The Hamiltonian of an electron in a spherical parabolic energy band has the form

$$H = \frac{1}{2m^*}\left[(p_x + e\mathcal{A}_x)^2 + p_y^2\right] + g^*\mu_B\sigma_z\mathcal{B} + \frac{1}{2m^*}p_z^2 + V_c(z), \qquad (18.18)$$

where the vector potential $\mathcal{A}$ is expressed in the Landau gauge ($\mathcal{A} = (-y\mathcal{B}, 0, 0)$, g^* is the effective g-factor, μ_B is the Bohr magneton, and σ_z is the z-component of spin quantum number with values $\pm\frac{1}{2}$.

In view of the separable nature of H, we can follow the treatment of Section 11.5 and write the wave function as

$$\Psi_{\ell n}(s, y, z) = \frac{1}{\sqrt{L_x}} e^{ik_x x}\phi_\ell(y)\chi_n(z), \qquad (18.19)$$

where k_x is the x-component of wave vector, $\phi_\ell(y)$ is the magnetic wave function for Landau level ℓ specified by Eq. (11.80) and $\chi_n(z)$ is the quantum well wavefunction for subband n given by Eq. (15.33) or (15.34). The energy eigenvalues are

$$E_{\ell n\sigma_z} = (\ell + \tfrac{1}{2})\hbar\omega_c + g\mu_B\sigma_z\mathcal{B} + E_n \begin{cases} \ell = 0, 1, 2, \ldots \\ n = 1, 2, 3, \ldots \end{cases} \qquad (18.20)$$

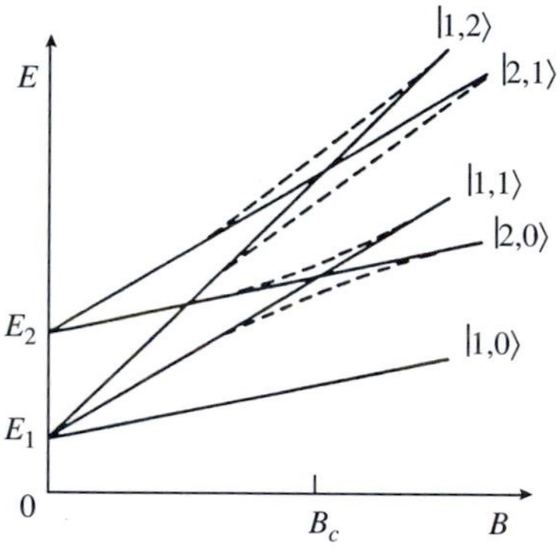

Fig. 18.6
Energy versus magnetic field for states $|n\ell\rangle$ with $n = 1,2$ and $\ell = 0,1,2$. The dashed lines indicate anticrossing. $\mathcal{B}_c = (E_2 - E_1)m^*/\hbar e$ (after Bastard 1988).

As in the case of bulk semiconductors, the energy levels are independent of k_x and are highly degenerate with degeneracy given by Eq. (11.86). The lowest lying energy levels for a particular spin orientation are plotted versus $\mathcal{B}$ as solid lines in Fig. 18.6. Pairs of levels cross at a magnetic field $\mathcal{B}_c$. If the magnetic field is not parallel to the growth axis, additional terms appear in the Hamiltonian that couple the y and z motions and lead to an anticrossing of the levels as indicated by the dashed lines in Fig. 18.6.

18.2.2 Magnetic-field-dependent density-of-states

For the case of magnetic field parallel to the growth axis with the z and x, y motions separated, the total density-of-states can be expressed as a sum of contributions from the various subbands n:

$$\rho(E) = \sum_n \rho_n(E). \qquad (18.21)$$

The subband density-of-states $\rho_n(E)$ is given by

$$\rho_n(E) = \sum_{\ell k_x \sigma_z} \delta[E - E_n - (\ell + \tfrac{1}{2})\hbar\omega_c - g^*\mu_B\sigma_z\mathcal{B}]. \qquad (18.22)$$

Since the summand is independent of k_x, the sum over k_x can be done immediately to give

$$\rho_n(E) = \frac{L_x L_y}{2\pi\lambda^2}\sum_{\ell\sigma_z} \delta[E - E_n - (\ell + \tfrac{1}{2})\hbar\omega_c - g^*\mu_B\sigma_z\mathcal{B}], \qquad (18.23)$$

where λ is the magnetic length $(\hbar/e\mathcal{B})^{\frac{1}{2}}$. In view of the complete quantization of the carrier motion, the density-of-states is zero except at the energy eigenvalues specified by Eq. (18.20), where it is infinite.

This result for the density-of-states is quite different from that for zero magnetic field. In the latter case $\rho(E)$ is nonzero for $E > E_1$, and the electron system is metallic. When $\mathcal{B} \neq 0$, the continuous behavior of $\rho(E)$ is replaced by point-like singularities separated by energy gaps. If the Fermi energy lies in a gap, the electron system behaves as an insulator, whereas if it lies on one of the energy eigenvalues, the system behaves as a conductor. In a three-dimensional system, on the other hand, there is no quantization in the direction of the magnetic field; hence, insulating regions do not exist above the zero-point energy $\hbar\omega_c/2$.

18.2.3 Magnetoconductivity in a 2D heterostructure

Consider a 2D heterostructure placed in a strong magnetic field parallel to the growth axis and subject to a weak electric field that is parallel to the layer planes ($\boldsymbol{\mathcal{E}} \perp \hat{z}$). Specializing the discussion in Chapter 8, Section 8.6 to the 2D case, the response of the electron gas consists of an electric current whose current density components satisfy the equations

$$j_x = \sigma_{xx}\mathcal{E}_x + \sigma_{xy}\mathcal{E}_y \tag{18.24a}$$

$$j_y = \sigma_{yx}\mathcal{E}_x + \sigma_{yy}\mathcal{E}_y, \tag{18.24b}$$

where the $\sigma_{\alpha\beta}$ are the elements of the conductivity tensor whose inverse is the resistivity tensor:

$$\overleftrightarrow{\rho} = \overleftrightarrow{\sigma}^{-1}. \tag{18.25}$$

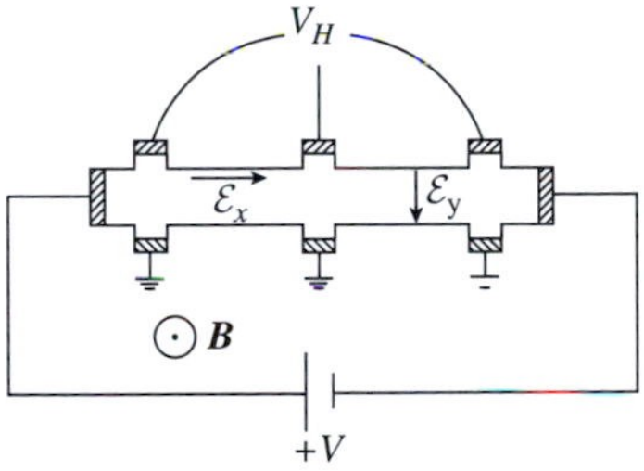

Fig. 18.7
Diagram of the sample geometry for measuring the resistivity tensor $\overleftrightarrow{\rho}$.

It must be emphasized in the present context that the current density components have dimensions charge per unit time per unit length rather than charge per unit time per unit area as in the three-dimensional case. A corresponding change occurs in the dimensions of the conductivity tensor components.

The experimental geometry is shown in Fig. 18.7. The electric current and field components can be analyzed in a fashion analogous to that in Chapter 8. Under direct current conditions in the absence of scattering of the carriers, the elements of the conductivity tensor of interest are

$$\sigma_{xx} = \sigma_{yy} = 0 \tag{18.26a}$$

$$\sigma_{xy} = -\sigma_{yx} = \frac{n_s e^2}{m_c^* \omega_c}. \tag{18.26b}$$

18.2.3.1 Magnetic field dependence of σ_{xx}: Shubnikov–de Haas effect

In order to have $\sigma_{xx} \neq 0$, it is necessary to include scattering of the carriers in the analysis. Scattering leads to broadening of the density-of-states peaks as

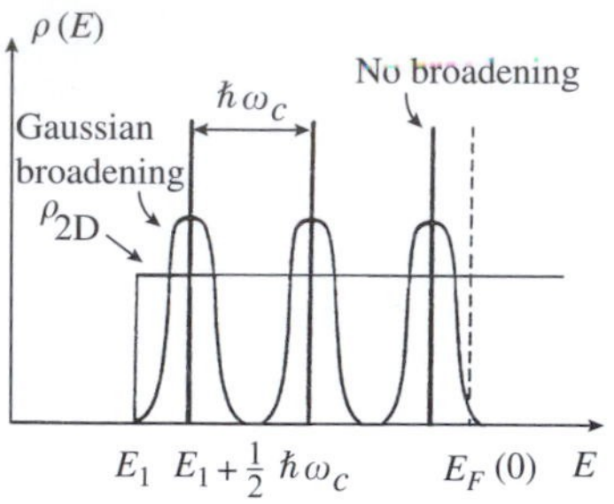

Fig. 18.8
Broadening of Landau levels in a 2D system (after Weisbuch and Vinter 1991).

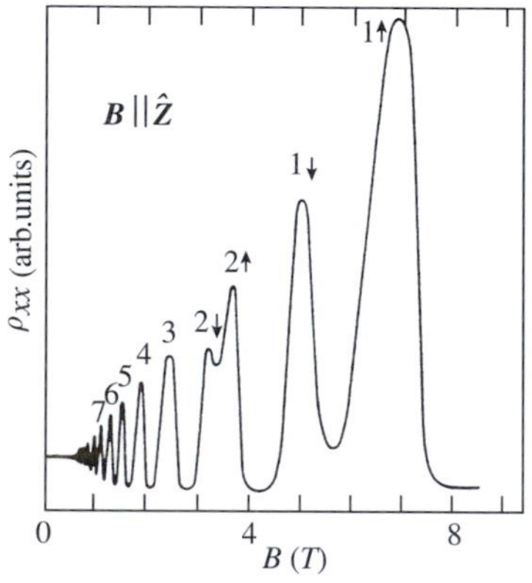

Fig. 18.9
Oscillations in a plot of ρ_{xx} versus $\mathcal{B}$ for a GaAs/GaAlAs heterostructure (after Bastard 1988).

shown in Fig. 18.8. Close to a Landau level the expression for the density-of-states given by Eq. (18.23) is replaced by (Ando and Uemura 1974)

$$\rho_n^{(E)} = \frac{L_x L_y}{\pi^2 \lambda^2 \Gamma_n} \sum_{\ell \sigma_z} \left[1 - \left(\frac{E - E_{n\ell\sigma_z}}{\Gamma_n} \right)^2 \right]^{\frac{1}{2}}, \tag{18.27}$$

where $E - E_{n\ell\sigma_z} < \Gamma_n$ and $2\Gamma_n$ is the width. A potential is associated with the scattering centers that is capable of binding carriers in localized states. In particular the states in the tails of broadened Landau levels can be expected to be localized. If the Fermi energy lies in a region of tail states, the electrons occupying these states are localized and do not contribute to the conductivity. Furthermore, at very low temperatures, electrons occupying extended states of the broadened Landau levels do not have nearby empty extended states into which to make transitions and thus contribute to the conductivity. Consequently, σ_{xx} should be very small for $T \simeq 0$ K. On the other hand, if the Fermi energy lies near the center of a broadened Landau level, empty extended states are close to occupied extended states, and σ_{xx} can be expected to be large.

If the magnetic field is varied so that successive Landau levels move through the Fermi energy, σ_{xx} is very small when E_F is in a region of localized levels and relatively large when E_F is in a region of extended levels. In other words, σ_{xx} or ρ_{xx} exhibits oscillations as $\mathcal{B}$ is varied. These are **Shubnikov–de Haas oscillations**, which are illustrated in Fig. 18.9 for a GaAs/GaAlAs heterostructure at 2 K. Note that peaks associated with spin up and spin down electrons are resolved.

A quantitative analysis based on the self-consistent Born approximation yields the following results for σ_{xx} (Ando and Uemura 1974):

$$\sigma_{xx} = 0 \qquad \text{if } |E_F - (\ell + \tfrac{1}{2})\hbar\omega_c - E_1 - g^* \mu_B \sigma_z \mathcal{B}| > \Gamma_n \tag{18.28a}$$

$$\sigma_{xx} = \frac{e^2}{\pi^2 \hbar} (\ell + \tfrac{1}{2}) \left\{ 1 - \left[\frac{E_F - (\ell + \frac{1}{2})\hbar\omega_c - E_1 - g^* \mu_B \sigma_z \mathcal{B}}{\Gamma_n} \right]^2 \right\}$$

$$\text{otherwise.} \tag{18.28b}$$

These equations clearly show the oscillatory behavior of σ_{xx} as a function of $\mathcal{B}$. It should be emphasized that the oscillations are due to the quantization of the magnetic levels and are not predicted by the classical Drude theory.

The maximum in σ_{xx} for particular values of ℓ and σ_z occurs at a magnetic field designated by $\mathcal{B}_{\ell\sigma_z}$. From Eq. (18.28b) we see that $\mathcal{B}_{\ell\sigma_z}$ is specified by the equation

$$E_F - (\ell + \tfrac{1}{2})\hbar\omega_c - E_1 - g^* \mu_B \sigma_z \mathcal{B}_{\ell\sigma_z} = 0 \tag{18.29}$$

or

$$\frac{1}{\mathcal{B}_{\ell\sigma_z}} = \frac{\hbar e}{m^*(E_F - E_1)} [\ell + \tfrac{1}{2} + g^* \mu_B \sigma_z (m^*/\hbar e)]. \tag{18.30}$$

A plot of $1/\mathcal{B}_{\ell\sigma_z}$ versus ℓ is a straight line with slope S_B given by

$$S_B = \frac{\hbar e}{m^*(E_F - E_1)}. \qquad (18.31)$$

For a degenerate two-dimensional electron gas occupying the lowest subband,

$$E_F - E_1 = \frac{\hbar^2 k_F^2}{2m^*}, \qquad (18.32)$$

where k_F is the Fermi wave vector. Using periodic boundary conditions to specify the two-dimensional wave vector $\mathbf{k}$, the area per allowed wave vector is $4\pi^2/A$, where A is the area of the 2D system. Since the area of the Fermi circle is πk_F^2, the number of occupied states N_s is given by

$$N_s = \frac{\pi k_F^2 A}{4\pi^2} = \frac{2m^*(E_F - E_1)\pi A}{4\pi^2\hbar^2} \qquad (18.33)$$

Introducing a factor of 2 for spin, we see that the areal concentration of electrons n_s is given by

$$n_s = \frac{2N_s}{A} = \frac{m^*(E_F - E_1)}{\pi\hbar^2}. \qquad (18.34)$$

Comparing this result with Eq. (18.31) yields the slope in the form

$$S_B = \frac{e}{\pi\hbar n_s}, \qquad (18.35)$$

thus establishing that the slope gives a direct measure of n_s.

18.2.3.2 Magnetic field dependence of σ_{xy}: quantum Hall effect

Measurements of the Hall resistivity ρ_{xy} versus magnetic field $\mathcal{B}$ in Si-MOSFET samples at low temperature (von Klitzing *et al.* 1980) exhibit a series of well-defined plateaus. Similar measurements have been made on a variety of heterostructures. The results for a GaAs/GaAlAs structure are shown in Fig. 18.10. The plateaus are specified by the very simple expressions

$$\rho_{xy} = \frac{1}{i}\frac{h}{e^2}$$

or

$$\sigma_{xy} = i\frac{e^2}{h}, \qquad (18.36)$$

which involve only positive integers i and fundamental constants. This is the **quantum Hall effect**. It is independent of the band structure of the particular system being investigated. The width $\Delta\mathcal{B}$ of a given plateau increases as the temperature decreases. High mobility of the carriers favors well-resolved plateaus.

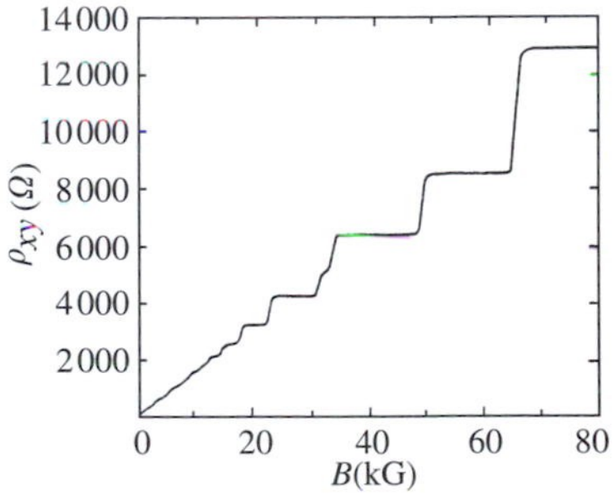

Fig. 18.10
ρ_{xy} versus $\mathcal{B}$ for a GaAs/GaAlAs heterostructure (after Paalanen *et al.* 1982).

The plateaus of the QHE arise when the Fermi energy lies in the region between broadened Landau levels where the states are localized and $\sigma_{xx} \simeq 0$. From the classical Drude theory (see Eq. 18.26b) we have, with $\sigma_{xx} = 0$ and $\omega_c = \frac{e\mathcal{B}}{m_c^* c}$,

$$\sigma_{xy} = \frac{n_s e}{\mathcal{B}}. \tag{18.37}$$

If the Fermi energy lies between Landau levels i and $i+1$ and the degeneracy of a Landau level is given by Eq. (11.82), the areal concentration n_s is given by

$$n_s = i\frac{e\mathcal{B}}{h}. \tag{18.38}$$

Combining Eqs. (18.37) and (18.38) yields the Hall conductivity in the QHE form

$$\sigma_{xy} = i\frac{e^2}{h}. \tag{18.39}$$

The above treatment, while giving the correct result, has a significant flaw: all electrons are treated as being in conducting states, whereas some of them are in nonconducting localized states. A more detailed analysis (Aoki and Ando 1981, Prange 1981) shows that the speed of localized current carriers in the presence of disorder is modified in just the right amount to exactly compensate for the lack of conduction by the localized carriers. A first-principles treatment (Laughlin 1981) has established that the QHE arises from the gauge invariance of the interaction of light with matter and the existence of a mobility gap.

A useful quantity in characterizing the QHE is the **filling factor** ν defined by $\nu = n_s/d$, where d is the degeneracy of a Landau level. So far in our discussion, ν is simply the integer i in Eq. (18.39). However, at very low temperatures less than 4.2 K, fractional values of ν expressible as p/q, where p and q are integers, have been observed for the lowest Landau level in systems such as GaAs/GaAlAs with very high mobilities (Tsui *et al.* 1983). This is the **fractional quantum Hall effect** (FQHE). The experimental data of Tsui *et al.* are shown in Fig. 18.11. The early results gave only odd values of q, but more recent results have given even values of q (Willett *et al.* 1987).

An explanation of the FQHE has been given by Laughlin (1983) based on a condensation of electrons or holes into a collective ground state due to electron–electron or hole–hole interactions. The ground state is separated from the nearest excited state by the energy $0.03\ e^2/\lambda$. The possibility of a repulsive interaction between carriers of the same charge giving rise to a condensation is related to the two-dimensional character of the system. The condensed phase consists of quasiparticles of fractional charge $e/m, m = 3,5,\ldots$, that obey statistics intermediate between Fermi–Dirac and Bose–Einstein. These quasiparticles are called **anyons**. Such a condensation does not occur in three-dimensional systems.

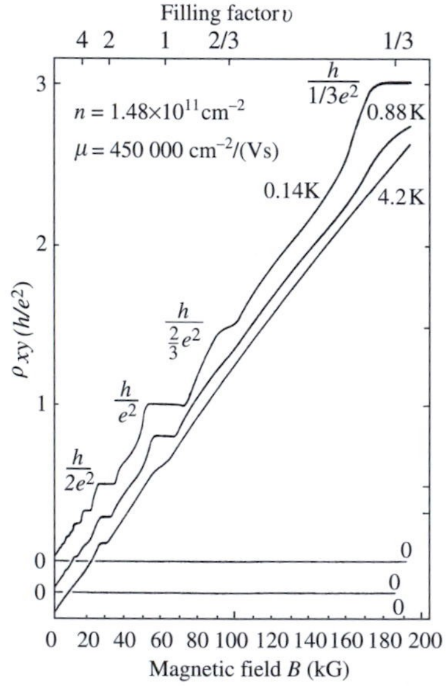

Fig. 18.11
ρ_{xy} versus $\mathcal{B}$ for a GaAs/GaAlAs heterostructure at $T = 4.2$, 0.88, and 0.14 K (after Tsui *et al.* 1983).

18.2.4 Cyclotron resonance

Cyclotron resonance has been observed for electrons and holes in a number of heterostructures including GaAs/GaAlAs, InP/GaInAs, and InAs/GaSb, when the magnetic field is parallel to the growth axis. Electron and hole masses have been obtained as well as information on polaron and screening effects in two-dimensional systems. A microscopic calculation of $\sigma_{\alpha\beta}(\omega)$ based on the self-consistent Born approximation (Ando 1975) predicts resonance absorption at cyclotron harmonics $n\omega_c$ as well as at the fundamental ω_c. Furthermore, oscillations appear in the cyclotron resonance lineshape when the width of the fundamental is sufficiently large to permit several Landau levels to cross the Fermi energy. Such oscillations are shown in Fig. 18.12.

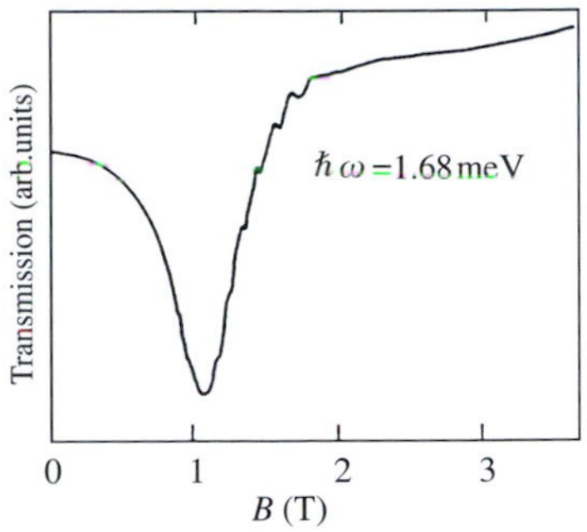

Fig. 18.12
Cyclotron resonance lineshape for a GaAs/GaAlAs heterostructure at $T = 2$ K (after Voisin *et al.* 1983).

When the effective mass of the carriers is very small and the bands are highly nonparabolic as in InAs/GaSb heterostructures, many subbands may be populated. The nonparabolicity causes the subbands to have different cyclotron resonance effective masses, with the lowest subband having the heaviest mass (Guldner *et al.* 1982). A splitting of the cyclotron resonance line occurs that reveals the absorption of the individual subbands as seen in Fig. 18.13.

If the magnetic field is perpendicular to the growth axis $\hat{z}$, it can be chosen in the y-direction with the vector potential taken as $A = (z\mathcal{B}, 0, 0)$. Neglecting spin effects the Hamiltonian becomes

$$H = \frac{1}{2m^*}[(p_x + e\mathcal{B}z)^2 + p_y^2 + p_z^2] + V_c(z). \tag{18.40}$$

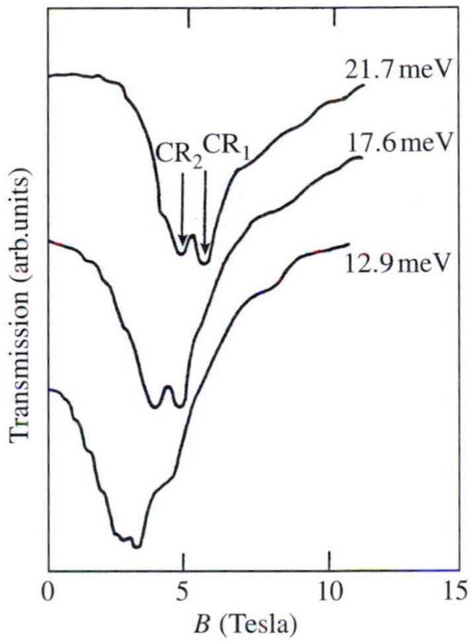

Fig. 18.13
Double cyclotron resonance absorption labeled CR_1 and CR_2 corresponding to the E_1 and E_2 subbands in an InAs/GaSb heterostructure at 2 K (after Guldner *et al.* 1982).

The confining potential $V_c(z)$ inhibits the cyclotron motion and prevents the formation of degenerate Landau levels for moderate values of the magnetic field. To lowest order in perturbation theory the energy of subband n is

$$E_n(k_x, k_y) = E_n + \frac{1}{2m^*}[(\hbar k_x + m^*\omega_c z_{nn})^2 + \hbar^2 k_y^2] + \tfrac{1}{2}m^*\omega_c^2[(z^2)_{nn} - (z_{nn})^2], \tag{18.41}$$

where

$$z_{nn} = \int_{-L_z/2}^{L_z/2} dz\, \chi_n^*(z) z \chi_n(z) \tag{18.42a}$$

$$(z^2)_{nn} = \int_{-L_z/2}^{L_z/2} dz\, \chi_n^*(z) z^2 \chi_n(z). \tag{18.42b}$$

For a square quantum well, the wave functions $\chi_n(z)$ are even or odd functions of z and therefore $z_{nn} = 0$. The shift in energy due to the magnetic field is

$$\Delta E_n = \tfrac{1}{2}m^*\omega_c^2(z^2)_{nn}, \tag{18.43}$$

which is quadratic in $\mathcal{B}$.

For a triangular quantum well, on the other hand, the wave functions do not have parity and $z_{nn} \neq 0$. The subband parabola in k_x undergoes a shift in minimum corresponding to

$$\hbar\bar{k}_x = \hbar k_x + m^* \omega_c z_{nn} = 0,$$

or

$$\hbar k_x = -m^* \omega_c z_{nn}. \tag{18.44}$$

The change in energy at the minimum is

$$(\Delta E_n)_{min} = \tfrac{1}{2} m^* \omega_c^2 [(z^2)_{nn} - (z_{nn})^2]. \tag{18.45}$$

For the $n=0$ to $n=1$ transition, the absorption line center corresponds to a vertical transition from the bottom of the shifted $n=0$ subband parabola to the $n=1$ subband parabola at the same value of k_x. This transition energy is

$$\begin{aligned}\Delta E_{01} = E_1 - E_0 + \tfrac{1}{2} m^* \omega_c^2 [&(z^2)_{11} - (z_{11})^2 - (z^2)_{00} \\ &+ (z_{00})^2 + (z_{11} - z_{00})^2].\end{aligned} \tag{18.46}$$

The shift due to the magnetic field is quadratic in $\mathcal{B}$ in agreement with experimental results on accumulation layers in Si (Beinvogl *et al.* 1976).

Problems

1. Consider variable range hopping conductivity that was discussed at the end of Chapter 8. Show that in a two-dimensional system the electrical conductivity varies with temperature as $\exp(-B'/T^{1/3})$, where B' is a constant.
2. Derive the relation $\mathcal{B}_c = (E_2 - E_1)m^*/\hbar e$ that specifies the magnetic field at which the energies of the $n=1$, $\ell=1$ and $n=2$, $\ell=0$ levels cross for $\mathcal{B} \| \hat{z}$. Generalize this result to the case of crossing of levels n, ℓ and $n+1$, $\ell-1$.
3. Derive Eq. (18.46) using Eqs. (18.41–18.45). For a magnetic field $\mathcal{B} = 2$T in the [111] direction and a triangular potential well on n-Si, evaluate ΔE_{01} using simple forms $\chi_0(z) = A_0 z \exp(-z/\alpha)$ and $\chi_1(z) = A_1(\frac{3}{2}\alpha - z)z\exp(-z/\alpha)$ for the envelope functions of the two lowest levels. A_0 and A_1 are normalization constants, α is a variational parameter for the ground state, and χ_1 is orthogonal to χ_0. The confining electric field $\mathcal{E}$ is 2×10^4 V/cm.

References

T. Ando, *J. Phys. Soc. Japan* **38**, 989 (1975).
T. Ando and Y. Uemura, *J. Phys. Soc. Japan* **36**, 959 (1974).
H. Aoki and T. Ando, *Solid State Commun.* **38**, 1079 (1981).
G. Bastard, *Wave Mechanics Applied to Semiconductor Heterostructures* (Les Editions de physique, Les Ulis, 1988).
W. Beinvogl, A. Kamgar, and J. F. Koch, *Phys. Rev.* B**14**, 4274 (1976).
Y. Guldner, J. P. Vieren, P. Voisin, M. Voos, J. C. Maan, L. L. Chang, and L. Esaki, *Solid State Commun.* **41**, 755 (1982).
R. B. Laughlin, *Phys. Rev.* B**25**, 5632 (1981).
R. B. Laughlin, *Phys. Rev. Lett.* **50**, 1395 (1983).
M. A. Paalanen, D.C. Tsui, and A.C. Gossard, *Phys. Rev.* B**25**, 5566 (1982).
R. E. Prange, *Phys. Rev.* B**23**, 4802 (1981).

T. C. L. G. Sollner, W. D. Goodhue, P. E. Tannenwald, C. D. Parker, and D. D. Peck, *Appl. Phys. Lett.* **43**, 588 (1983).

H. L. Störmer, A. C. Gossard, and W. Wiegmann, *Solid State Commun.* **41**, 707 (1982).

R. Tsu and L. Esaki, *Appl. Phys. Lett.* **22**, 562 (1973).

D. C. Tsui, H. L. Störmer, J. C. M. Huang, J. S. Brooks, and M. J. Naughton, *Phys. Rev.* B**28**, 2274 (1983).

P. Voisin, Y. Guldner, J. P. Vieren, M. Voos, J. C. Maan, P. Delescluse, and N. T. Linh, *Physica* B**117**/**118**, 634 (1983).

K. von Klitzing, G. Dorda, and M. Pepper, *Phys. Rev. Lett.* **45**, 494 (1980).

C. Weisbuch and B. Vinter, *Quantum Semiconductor Structures* (Academic Press, San Diego, 1991).

R. Willett, J. P. Eisenstein, H. L. Störmer, D. C. Tsui, A. C. Gossard, and J. H. English, *Phys. Rev. Lett.* **59**, 1776 (1987).

Metal-semiconductor devices

19

Key ideas

A *metal-oxide-semiconductor capacitor* (MOSC) is made by replacing part of the insulator layer of an ordinary capacitor by a semiconductor layer. This makes the capacitance dependent on both the applied voltage and the frequency of the impressed signal.

Under *forward bias* the energy band edges bend upward and the hole concentration increases as the oxide boundary is approached. Under *reverse bias* the band edges bend downward. At large reverse bias the intrinsic Fermi energy drops below the actual Fermi energy, so that the electron concentration exceeds the hole concentration near the boundary. An *inversion layer* of electrons is produced.

The difference between the intrinsic Fermi energy of the bulk and the intrinsic Fermi energy at an arbitrary point in the semiconductor defines a potential $\psi(x)$. Depending on the value of this potential at the surface of the semiconductor one can have an accumulation layer of holes, flat bands, depletion layer of holes, intrinsic condition, or an inversion layer of electrons.

The MOSC has a capacitance contribution from the oxide layer as well as from the space charge of the semiconductor. The relation between total capacitance and gate voltage depends on the potential at the semiconductor–oxide interface and the space charge capacitance.

The MOSC has applications as a variable capacitor, a component of the metal-oxide-semiconductor field effect transistor, and in the charge-coupled device.

The *metal-semiconductor* (MS) *diode* consists of a junction between a metal and a semiconductor.

The electric field in the space charge region and the bending of the conduction band edge produce an energy barrier called a *Schottky barrier*.

Under *forward bias* the metal is positive with respect to the semiconductor and the energy barrier for passage of an electron from the semiconductor to the metal is lowered. Under *reverse bias* the metal is negative and the energy barrier is increased.

The total current density under forward bias increases exponentially with applied voltage. Under large reverse bias the current density

Junctions involving metals

reaches a saturation value proportional to the *thermal current density*. The latter is limited by the passage of majority carriers over the potential barrier due to thermal excitation.

19.3 Metal-oxide-semiconductor field effect transistor

The metal-oxide-semiconductor transistor, or MOSFET, consists of a metal layer called the *gate* separated from a semiconductor layer called the *substrate* by a layer of insulating oxide called the *gate oxide*. The MOSFET is a *unipolar transistor*. The current flow is parallel to the principal interface of the MOSFET.

MOSFET characteristics exploit those of the MOSC. The drain current is related to both the drain voltage and the gate voltage. For sufficiently large gate voltage the drain attains the *saturation drain voltage*. For small drain voltage the drain current increases linearly with drain voltage and gate voltage. The smallest gate voltage at which a conduction channel is induced and a drain current can flow is called the *threshold gate voltage*. With increasing drain voltage the drain current levels off and reaches a maximum value at the saturation drain voltage. The *transfer characteristics* of a MOSFET are specified by the dependence of the drain current on the gate voltage at fixed drain voltage.

Among the most common heterostructures found in device applications are the metal-oxide-semiconductor (MOS) structure and the metal-semiconductor (MS) structure. The former is used in the metal-oxide-semiconductor capacitor and the metal-oxide-semiconductor field effect transistor, while the latter is the basis for the metal-semiconductor diode. These devices are referred to as **unipolar** devices, because current is transported by carriers of a single polarity, typically electrons. They are to be contrasted with bipolar devices such as the BJT which involve carriers of both polarities.

19.1 Metal-oxide-semiconductor capacitor

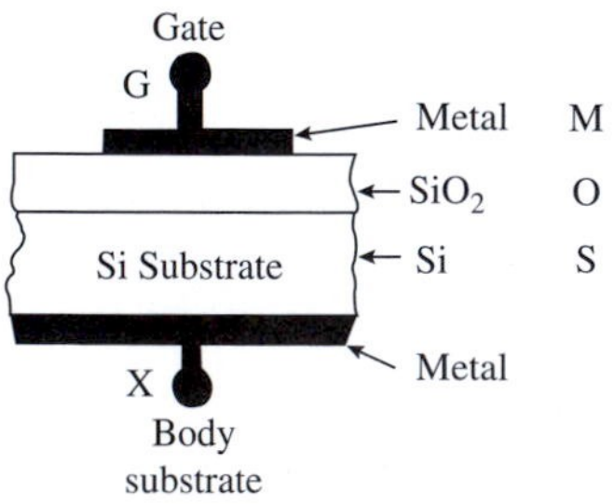

Fig. 19.1
Diagram of a MOSC on silicon (after Sah 1991).

The **metal-oxide-semiconductor capacitor** (MOSC) differs from an ordinary capacitor by having part of the insulator layer replaced by a semiconductor layer as shown in Fig. 19.1. Typically, the semiconductor is either n- or p-type Si, the oxide is SiO_2, and the metal electrodes are Al. The electrode in contact with the oxide is the **gate**. The oxide is essentially a perfect insulator and prevents any DC current from flowing between the electrodes. A qualitative effect of the semiconductor layer is to make the capacitance of a MOSC dependent upon both the applied voltage and the frequency of an impressed signal. The MOSC can therefore be used to replace mechanical variable capacitors in tunable LC circuits in AM-FM radios, television sets, and other electronic appliances.

19.1.1 Effect of applied bias on energy bands

Let us consider a MOSC fabricated with p-Si. If a negative bias voltage is applied to the gate, the MOSC is said to be under **forward bias**, whereas a positive bias voltage gives **reverse bias**. Under forward bias the energy band edges of the p-Si bend upward as the oxide boundary is approached, and holes accumulate near the boundary as shown in Fig. 19.2a. The holes are transferred from the bulk region of the p-Si to the boundary with the oxide by the action of the electric field due to the bias voltage. An **accumulation layer** of holes thus results. Its existence is consistent with the facts that the Fermi energy E_F is constant throughout the structure in the absence of a current and that the valence band edge E_V moves closer to E_F as the boundary is approached. Since the hole concentration varies as $\exp[-\beta(E_F - E_V)]$ (cf. Eq. (6.14)), the hole concentration must increase as the boundary is approached.

Under reverse bias, however, the band edges bend downward as shown in Fig. 19.2b, and the hole concentration is reduced or depleted. Minority electrons are attracted to the boundary due to the electric field that is now directed from the oxide into the semiconductor. At a sufficiently large reverse bias the intrinsic Fermi energy E_{Fi} drops below E_F. The electron concentration then exceeds the hole concentration near the boundary in accordance with Eq. (6.43). This situation constitutes **inversion** as depicted in Fig. 19.2c and is characterized by an **inversion layer** of electrons.

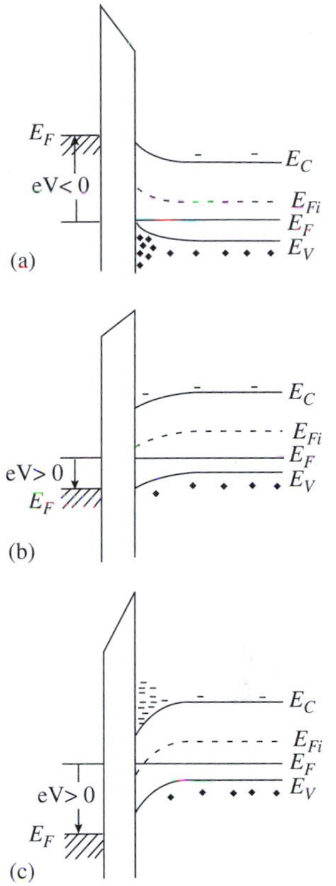

Fig. 19.2
Energy band diagrams for the ideal MOSC structure on p-Si: (a) forward bias ($V < 0$), (b) weak reverse bias ($V > 0$), (c) strong reverse bias ($V \gg 0$) (after Sze 1981).

19.1.2 Bias dependence of capacitance

The presence of an electric field in the semiconductor near its boundary with the oxide indicates that there is space charge in this region. In order to determine the bias dependence of the capacitance, it is necessary to evaluate the space charge per unit area Q_s as a function of the potential $\psi(x)$. We define $\psi(x)$ in terms of the deviation of the intrinsic Fermi energy E_{Fi} from its value E_{Fi}^{∞} far from the boundary: $e\psi(x) = E_{Fi}^{\infty} - E_{Fi}(x)$. The difference between E_{Fi}^{∞} and E_F can be used to define a potential ψ_B by

$$e\psi_B = E_{Fi}^{\infty} - E_F. \tag{19.1}$$

The various quantities just introduced are shown schematically in Fig. 19.3.

As the band edges E_C and E_V bend under the action of a bias, the intrinsic Fermi energy maintains its relative position between E_C and E_V as indicated in Fig. 19.3. The potential ψ typically increases in magnitude as the surface of the semiconductor is approached and reaches the values ψ_s at the surface. Several cases can be distinguished:

1. $\psi_s < 0$: accumulation layer of holes
2. $\psi_s = 0$: flat-band condition
3. $\psi_B > \psi_s > 0$: depletion layer of holes
4. $\psi_s = \psi_B$: intrinsic condition at the boundary
5. $\psi_s > \psi_B$: inversion layer of electrons.

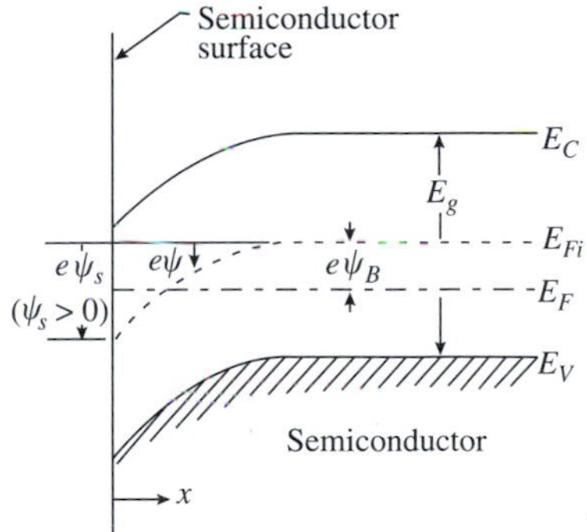

Fig. 19.3
Energy band diagram at the surface of a p-type semiconductor. The potential ψ is defined to be zero in the bulk, and is measured from the intrinsic Fermi energy E_{Fi} (after Sze 1981).

The dependence of the electron and hole concentrations on position x are specified by

$$n_p(x) = n_{p0}e^{\beta e\psi(x)} \tag{19.2a}$$

$$p_p(x) = p_{p0}e^{-\beta e\psi(x)}, \tag{19.2b}$$

where $\psi(x)$ is positive when the band edges are bent downward and the quantities n_{p0} and p_{p0} are the equilibrium concentrations of electrons and holes, respectively, far from the surface where $\psi(x)=0$. At the surface,

$$n_p = n_s = n_{p0}e^{\beta e\psi_s} \tag{19.3a}$$

$$p_p = p_s = p_{p0}e^{-\beta e\psi_s}. \tag{19.3b}$$

Returning now to the calculation of the space charge per unit area Q_s, we utilize Poisson's equation in one dimension:

$$\frac{d^2\psi(x)}{dx^2} = -\frac{\rho(x)}{\epsilon_s}, \tag{19.4}$$

where ϵ_s is the permittivity of the semiconductor and $\rho(x)$ is the total space-charge density given by

$$\rho(x) = e(n_d^+ - n_a^- + p_p - n_p). \tag{19.5}$$

The p-type semiconductor is assumed to be uniformly doped with acceptor impurities and some compensating donor impurities. At room temperature both types of impurities are fully ionized with concentrations n_d^+ and n_a^- for donors and acceptors, respectively.

In the bulk of the semiconductor there is no net space charge, so both $\rho(x)$ and ψ are zero. Hence, from Eqs. (19.3) and (19.5),

$$n_d^+ - n_a^- = n_{p0} - p_{p0}. \tag{19.6}$$

At an arbitrary position in the semiconductor we can use Eqs. (19.2) to express $p_p(x) - n_p(x)$ as

$$p_p(x) - n_p(x) = p_{p0}e^{-\beta e\psi(x)} - n_{p0}e^{\beta e\psi(x)}. \tag{19.7}$$

Poisson's equation then takes the form

$$\frac{d^2\psi}{dx^2} = -\frac{e}{\epsilon_s}\left[p_{p0}\left(e^{-\beta e\psi(x)} - 1\right) - n_{p0}\left(e^{\beta e\psi(x)} - 1\right)\right]. \tag{19.8}$$

A first integral of the last equation can be obtained by multiplying both sides by $d\psi/dx$ and integrating to yield

$$\left(\frac{d\psi}{dx}\right)^2 = -\frac{2e}{\epsilon_s}\int \left[p_{p0}\left(e^{-\beta e\psi} - 1\right) - n_{p0}\left(e^{\beta e\psi} - 1\right)\right] d\psi + c_0, \tag{19.9}$$

where c_0 is a constant of integration. Carrying out the integration over ψ and noting that in the bulk of the semiconductor, $\psi=0$ and $d\psi/dx=0$, we find that

$$c_0 = -2(p_{p0} + n_{p0})/\beta\epsilon_s. \tag{19.10}$$

Since the electric field $\mathcal{E}(x)$ is $-d\psi(x)/dx$, it follows from Eqs. (19.9) and (19.10) that

$$\mathcal{E}^2(x) = \frac{2k_BT}{\epsilon_s}\{p_{p0}[e^{-\beta e\psi(x)} + \beta e\psi(x) - 1] + n_{p0}[e^{\beta e\psi(x)} - \beta e\psi(x) - 1]\}. \tag{19.11}$$

Introducing the notation

$$G(\psi, p_{p0}, n_{p0}) = [p_{p0}(e^{-\beta e\psi} + \beta e\psi - 1) + n_{p0}(e^{\beta e\psi} - \beta e\psi - 1)]^{\frac{1}{2}} \geq 0, \tag{19.12}$$

we can express the electric field at the surface as

$$\mathcal{E}_s = \pm\left(\frac{2k_BT}{\epsilon_s}\right)^{\frac{1}{2}} G(\psi_s, p_{p0}, n_{p0}), \tag{19.13}$$

where the positive sign applies for $\psi_s > 0$ and the negative sign for $\psi_s < 0$.

It is now a simple matter to obtain the desired space charge Q_s with the aid of Gauss's law. We consider a slab with one face of area A coinciding with the surface of the semiconductor and the other lying in the bulk region where the electric field is zero. The slab thus encompasses the space charge. Gauss's law in its general form,

$$\epsilon_s \oint \boldsymbol{\mathcal{E}} \cdot d\mathbf{s} = AQ_s, \tag{19.14}$$

becomes

$$\epsilon_s[\mathcal{E}_s \cdot (-A\hat{x})] = AQ_s$$

or

$$Q_s = -\epsilon_s\mathcal{E}_s = \mp(2\epsilon_s k_BT)^{\frac{1}{2}} G(\psi_s, p_{p0}, n_{p0}). \tag{19.15}$$

The behavior of $|Q_s|$ as a function of ψ_s is shown in Fig. 19.4 for p-type silicon at room temperature with $n_a = 4 \times 10^{15}\,\mathrm{cm}^{-3}$. In the negative range of ψ_s, Q_s is positive corresponding to an accumulation layer of holes. The dominant term in G as given by Eq. (19.12) is the first term, so that $Q_s \sim \exp(\beta e|\psi_s|/2)$. At $\psi_s = 0$, the **flat-band condition**, we have $Q_s = 0$ and $\mathcal{E}_s = 0$. In the range $\psi_B > \psi_s > 0$, the hole concentration is depleted, Q_s is negative, and $\mathcal{E}_s$ is positive. The second term in G is now dominant, so $Q_s \sim -\psi_s^{1/2}$. In the range $\psi_s \gg \psi_B$, the fourth term in G dominates,

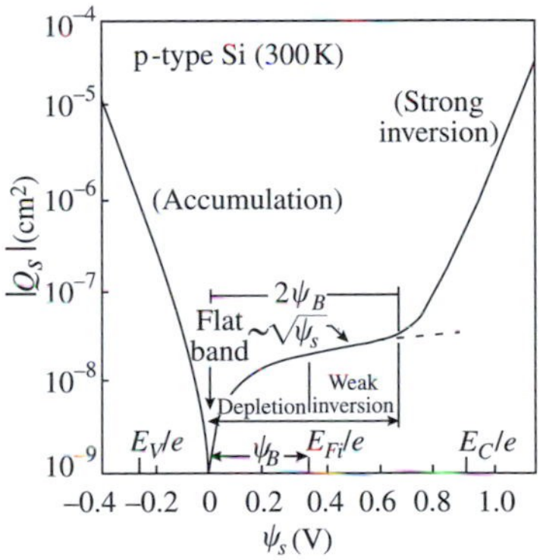

Fig. 19.4
Variation of the space charge density in the semiconductor as a function of the surface potential ψ_s for p-Si with $n_a = 4 \times 10^{15}\,\mathrm{cm}^{-3}$ at room temperature. ψ_B is defined by $e\psi_B = E_{Fi} - E_F$ in the bulk (after Sze 1981).

$Q_s \sim -\exp(\beta e\psi_s/2)$, and we have an inversion layer. The onset of strong inversion occurs when $\psi_s \simeq 2\psi_B$.

The differential capacitance per unit area of the space-charge region is defined by

$$C_s = \frac{\partial Q_s}{\partial \psi_s}. \tag{19.16}$$

Substituting Eq. (19.15) into Eq. (19.16) and using Eq. (19.12) yields the result

$$C_s = e\left(\frac{\epsilon_s}{2k_BT}\right)^{\frac{1}{2}} \frac{[p_{p0}(1 - e^{-\beta e\psi_s}) + n_{p0}(e^{\beta e\psi_s} - 1)]}{G(\psi_s, p_{p0}, n_{p0})}. \tag{19.17}$$

Under flat-band conditions ($\psi_s = 0$), C_s takes the simple form

$$C_s^{FB} = \frac{\sqrt{2}\epsilon_s}{L_D}, \tag{19.18}$$

where $L_D = (2\epsilon_s k_BT/e^2 p_{p0})^{\frac{1}{2}}$ is the extrinsic Debye length for holes and the minority carrier concentration n_{p0} has been neglected. Far from flat-band conditions, $C_s \sim \exp(\beta e|\psi_s|/2) \sim |Q_s|$, and C_s becomes large.

19.1.3 Evaluation of capacitance versus voltage curves

Up to this point we have considered only the capacitance due to the space charge in the semiconductor. In addition to this capacitance the MOSC has a contribution from the oxide layer. The oxide capacitance per unit area C_O is given by

$$C_O = \frac{\epsilon_O}{W_O}. \tag{19.19}$$

where ϵ_O is the permittivity of the oxide and W_O is its width. Since the oxide and space-charge capacitance are in series, the total capacitance C is specified by

$$\frac{1}{C} = \frac{1}{C_O} + \frac{1}{C_s}$$

or

$$C = \frac{C_O}{1 + C_O/C_s}. \tag{19.20}$$

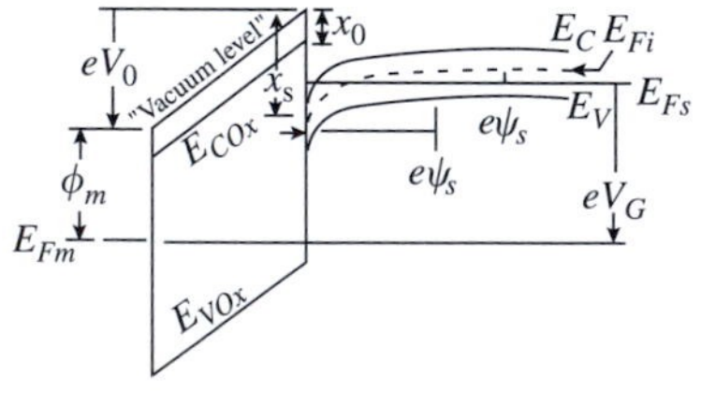

Fig. 19.5
The energy-position diagram of a MOSC on p-Si. E_{Fm} and E_{Fs} are the Fermi energies of the metal and semiconductor, respectively (after Sah 1991).

The oxide capacitance C_O is constant for a device with given oxide width W_O, whereas the space-charge capacitance C_s depends on ψ_s and hence on the applied gate voltage V_G.

The relation between V_G and ψ_s can be established with the aid of Fig. 19.5 which shows the energy versus position diagram of a MOSC as well as the important properties: metal work function ϕ_m, metal Fermi energy E_{Fm}, semiconductor electron affinity χ_s, oxide electron affinity χ_O, and voltage across the oxide V_O. The metal work function is the difference between the energy of an electron in vacuum far from the metal and the Fermi energy:

$$\phi_m = E_{vac} - E_{Fm}. \tag{19.21}$$

The semiconductor electron affinity is the energy required to raise an electron from the conduction band edge to the vacuum far from the semiconductor:

$$\chi_s = E_{vac} - E_C. \tag{19.22}$$

It can be seen from Fig. 19.5 that

$$eV_O + \phi_m = eV_G + \chi_s - (E_{Fs} - E_C(0)), \tag{19.23}$$

where E_{Fs} is the semiconductor Fermi energy and $E_C(0)$ is the conduction band edge of the semiconductor at the interface with the oxide. Since all three quantities E_C, E_{Fi}, and E_V decrease by the amount $e\psi_s$ in passing from the bulk region of the semiconductor to the oxide interface,

$$E_C(0) = E_C - e\psi_s, \tag{19.24}$$

and

$$eV_O + \phi_m = eV_G + \chi_s - e\psi_s + (E_C - E_{Fs}). \tag{19.25}$$

The work function of the semiconductor ϕ_s is given by

$$\phi_s = E_{vac} - E_{Fs}. \tag{19.26}$$

Eliminating E_{vac} from Eqs. (19.22) and (19.26), we obtain

$$\phi_s = \chi_s + (E_C - E_{Fs}). \tag{19.27}$$

Use of Eq. (19.1) yields

$$\phi_s = \chi_s + e\psi_B + (E_C - E_{Fi}^{\infty}). \tag{19.28}$$

The quantities ϕ_m, ϕ_s, and χ_s are typically on the order of a few electron volts. Eliminating χ_s from Eqs. (19.25) and (19.28), we obtain

$$V_G = V_O + \psi_s + (\phi_m - \phi_s)/e. \tag{19.29}$$

The voltage V_O across the oxide is itself a function of ψ_s. The oxide is the insulator of a capacitor with the metal gate as one electrode with charge Q_M on it. Hence, V_O is given by

$$V_O = \frac{Q_M}{C_O}. \tag{19.30}$$

Since the total charge on the capacitor is zero, we have

$$Q_M + Q_s + Q_O + Q_I = 0, \tag{19.31}$$

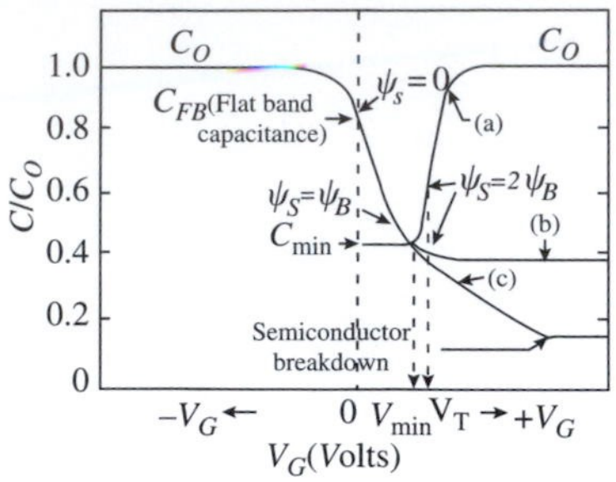

Fig. 19.6
MOSC capacitance-voltage curves: (a) low frequency, (b) high frequency, (c) deep depletion.

where Q_O is the total fixed charge in the oxide and Q_I is the charge trapped at the oxide–semiconductor interface. Eliminating Q_M from Eqs. (19.30) and (19.31) and using Eq. (19.15), we get

$$V_O = \frac{\epsilon_s \mathcal{E}_s - Q_O - Q_I}{C_O}, \tag{19.32}$$

which relates V_O to ψ_s through Eq. (19.13). The dependence of V_G on ψ_s is specified by Eqs. (19.29) and(19.32).

The space-charge capacitance C_s and the total capacitance C can now be calculated as functions of the gate voltage V_G with the aid of Eqs. (19.17) and (19.20). In doing the calculation it is simplest to treat ψ_s as the independent variable and obtain $\mathcal{E}_s$, C_s, V_O, V_G, and C in turn.

Results for a typical case are shown in Fig. 19.6. In the region of negative V_G there is an accumulation layer of holes, C_s is large, and $C \simeq C_O$ from Eq. (19.20). As V_G becomes less negative, the space-charge density decreases until a depletion layer with low carrier concentration is formed. Both C_s and C exhibit corresponding decreases to low values. When V_G becomes sufficiently positive, an inversion layer with high electron concentration is formed, and C_s becomes large again. The total capacitance rises rapidly until the limiting value C_O is reached. The curves in Fig. 19.6 were obtained by neglecting the work function difference $\phi_m - \phi_s$ and the charges Q_O and Q_I. Inclusion of these quantities produces a rigid translation of the curves parallel to the horizontal axis.

The increase in capacitance at positive V_G just described is found only if the frequency of the AC signal is sufficiently low. If the frequency is too high, the rates of recombination and generation of electrons are insufficient to enable the electron concentration to keep in phase with the applied signal and thus lead to a large differential capacitance C_s. Experimentally, one finds that frequencies in the range of 5–100 Hz (Grove *et al.* 1965) are accompanied by enhanced values of C_s. At higher frequencies C_s remains small and no rise in the total capacitance occurs at positive V_G.

An estimate of the high-frequency capacitance can be obtained as follows. We write the space charge Q_s in the form

$$Q_s = \int_{-\infty}^{\infty} \rho(x) dx = Q_n - e n_a W_d, \tag{19.33}$$

where Q_n is the total charge per unit area of the electrons in the inversion layer and W_d is the width of the depletion region. The potential distribution in the depletion region is specified by Poisson's equation

$$\frac{d^s \psi(x)}{dx^2} = \frac{e n_a}{\epsilon_s}, \tag{19.34}$$

whose solution subject to the boundary conditions

$$\psi(x) = \frac{d\psi(x)}{dx} = 0 \quad \text{at } x = W_d$$

is

$$\psi(x) = \psi_s\left(1 - \frac{x}{W_d}\right)^2, \tag{19.35}$$

where

$$\psi_s = \frac{en_a W_d^2}{2\epsilon_s}. \tag{19.36}$$

The capacitance of the semiconductor is given by

$$C_s = \frac{d\,|Q_s|}{d\psi_s} = en_a \frac{dW_d}{d\psi_s}, \tag{19.37}$$

where Eq. (19.33) has been used. Note that Q_n does not vary with ψ_s because the inversion-layer electrons do not respond to the high-frequency signal. Eliminating $dW_d/d\psi_s$ with the aid of Eq. (19.36) and substituting the result for C_s into Eq. (19.20), we obtain the total high-frequency capacitance of the MOS structure:

$$C_{hf} = \frac{C_O}{1 + (\epsilon_O W_d/\epsilon_s W_O)}. \tag{19.38}$$

As shown in Fig. 19.6, the onset of strong inversion occurs for $\psi_s \simeq 2\psi_B$, where ψ_B is given by Eq. (19.1). Continued increase of ψ_s beyond $2\psi_B$ does not lead to a significant increase in W_d, because the inversion-layer charge increases and screens the depletion region from an increase in the electric field. The maximum value of the depletion-layer width, W_{dm}, can be estimated using the equation

$$2\psi_B = \frac{en_a W_{dm}^2}{2\epsilon_s}, \tag{19.39}$$

which follows from Eq. (19.36). The Fermi energies appearing in the defining equation for ψ_B can be related to the corresponding hole concentrations by Eq. (6.14):

$$p = \bar{N}_v e^{\beta(E_V - E_F)} \tag{19.40a}$$

$$p_i = \bar{N}_v e^{\beta(E_V - E_{Fi})}. \tag{19.40b}$$

Since the acceptor impurities are essentially all ionized at room temperature, $p \simeq n_a$. Taking the logarithm of the ratio of Eqs. (19.40a) and (19.40b) and using the fact that $p_i = n_i$, we obtain

$$E_{Fi} - E_F = k_B T \log\left(\frac{n_a}{n_i}\right). \tag{19.41}$$

Substituting into Eqs. (19.1) and then (19.39) yields the result for W_{dm}:

$$W_{dm} = \left[\frac{4\epsilon_s k_B T}{e^2 n_a} \log\left(\frac{n_a}{n_i}\right)\right]^{\frac{1}{2}}. \tag{19.42}$$

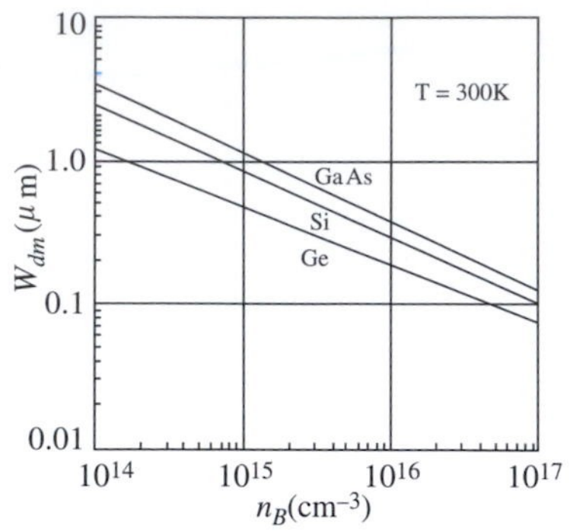

Fig. 19.7
Maximum depletion layer width versus impurity concentration for Ge, Si, and GaAs under heavy inversion. n_B is n_a for p-type and n_d for n-type semiconductors (after Sze 1981).

This equation, together with Eq. (6.17), enables one to calculate W_{dm} as a function of acceptor concentration and temperature. Results for several materials at room temperature are presented in Fig. 19.7.

19.1.4 Applications of the metal-oxide-semiconductor

At the beginning of this chapter use of the MOSC as a variable capacitor was mentioned. Another application concerns monitoring the quality of oxidized silicon surfaces during the fabrication of devices. The MOSC is itself a component of the metal-oxide-semiconductor field effect transistor, serving as the **gate** or input electrode. An application that is composed of a number of closely spaced MOSC electrodes is the **charge-coupled device** (Boyle and Smith 1970). It is used as the imaging element in video cameras and as the signal delay line in digital and analogue circuits.

19.2 Metal-semiconductor diode

As its name implies, the metal-semiconductor (MS) diode consists of a junction between a metal and a semiconductor. It is more heterogeneous than a junction involving only semiconductors, and is characterized by an interface that is imperfect on the atomic scale. Many of the semiconductor bonds are dangling rather than completed to adjacent metal atoms. Furthermore, impurity atoms tend to accumulate at the interface to form an interfacial impurity layer.

In the 1920s radio receivers employed MS diode rectifiers in the form of a "cats whisker" (pointed tungsten wire) pressed against a lead sulfide crystal surface. During this period large-area rectifiers appeared that were based on copper oxide or selenium as the semiconductor. World War II saw the use of high-purity silicon in rectifiers for radar applications.

To obtain rectification the work function of the metal ϕ_m must be larger than that of the semiconductor ϕ_s. Under these conditions, a potential barrier exists that restricts the majority carriers of the semiconductor, typically electrons in n-type material, from entering the metal and the metal electrons from entering the semiconductor. If $\phi_m < \phi_s$, one has an ohmic contact and rectification is not possible.

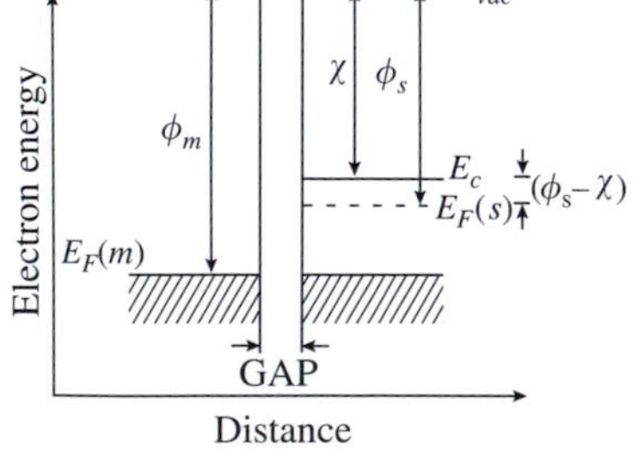

Fig. 19.8
Band diagrams of a metal and n-type semiconductor before contact is made and with a gap between them (after Dalven 1990).

19.2.1 Equilibrium characteristics of the MS diode

When a metal and an n-type semiconductor are brought close together, but before equilibrium is established, the energy level diagram has the form shown in Fig. 19.8 for $\phi_m > \phi_s$. If the metal and semiconductor are sufficiently close together that transfer of electrons can occur between the two materials, equilibrium can be achieved. From Fig. 19.8 it is clear that electrons must move from the semiconductor to the metal in order to make the chemical potentials (Fermi energies E_{F_s} and E_{F_m}) the same on both sides of the interface as shown in Fig. 19.9. A positive space charge region of width W arises in the part of the semiconductor next to the metal at the interface. An electric field $\mathcal{E}$ is therefore present in the space charge region that is directed toward the metal. In view of the definitions of the work

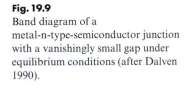

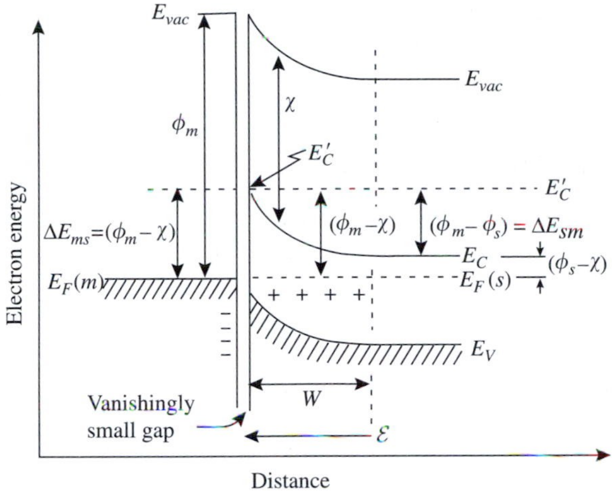

Fig. 19.9
Band diagram of a metal-n-type-semiconductor junction with a vanishingly small gap under equilibrium conditions (after Dalven 1990).

functions and the equalization of the Fermi energies, we see that the vacuum energies E_{vacm} and E_{vacs} of the metal and semiconductor, respectively, must be different after equilibration in accordance with Eqs. (19.21) and (19.26):

$$\phi_m - \phi_s = E_{vacm} - E_{vacs}. \tag{19.43}$$

As a result of the electric field in the space charge region and the concomitant bending of the conduction band edge as seen in Fig. 19.9 an energy barrier ΔE_{sm} arises that restricts the motion of electrons from the semiconductor to the metal. Since the conduction band edge is below the vacuum energy by the electron affinity χ_s, we see that at the metal-semiconductor interface $x = 0$,

$$E_C(0) = E_{vacm} - \chi_s, \tag{19.44}$$

whereas at the right-hand edge of the space charge region $x = W$,

$$E_C(W) = E_{vacs} - \chi_s. \tag{19.45}$$

Eliminating χ_s and using Eq. (19.43) give

$$\begin{aligned}\Delta E_{sm} &\equiv E_C(0) - E_C(W)\\ &= \phi_m - \phi_s.\end{aligned} \tag{19.46}$$

Thus, the energy barrier for electron motion into the metal is the difference in the work functions. On the other hand, we see from Fig. 19.9 that the energy barrier ΔE_{ms} for electron motion from the metal into the semiconductor is given by

$$\Delta E_{ms} = \phi_m - \chi_s. \tag{19.47}$$

The energy barrier is called a **Schottky barrier** if the space charge region contains donor or acceptor ions and a **Mott barrier** if the semiconductor is intrinsic.

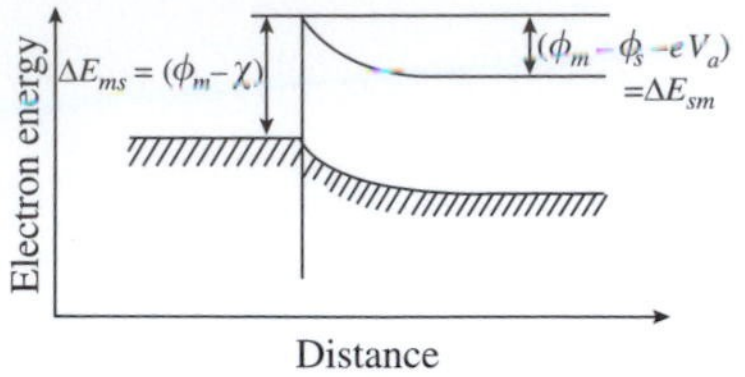

Fig. 19.10
Band diagram of the metal-n-type-semiconductor junction under an applied forward bias V_a (semiconductor negative) (after Dalven 1990).

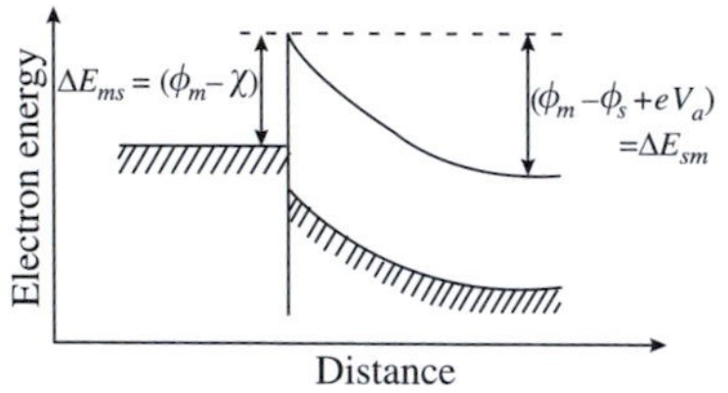

Fig. 19.11
Band diagram of the metal-n-type-semiconductor junction under an applied reverse bias V_a (semiconductor positive) (after Dalven 1990).

19.2.2 Current under applied voltage

The application of a DC voltage V_a across an MS diode produces a current arising from carriers passing over the energy barrier between the semiconductor and metal. Under an applied forward bias voltage, $V_a > 0$, the metal is made positive with respect to the semiconductor, and the energy diagram is modified as shown in Fig. 19.10. The energy barrier for passage of an electron from the semiconductor to the metal is lowered and is given by

$$\begin{aligned}\Delta E_{sm}(V_a) &= \Delta E_{sm} - eV_a \\ &= \phi_m - \phi_s - eV_a. \end{aligned} \tag{19.48}$$

Under a reverse bias voltage, $V_a < 0$, the metal is made negative, and the resulting energy diagram is shown in Fig. 19.11. The energy barrier takes the form

$$\begin{aligned}\Delta E_{sm}(V_a) &= \Delta E_{sm} + e|V_a| \\ &= \phi_m - \phi_s + e|V_a|. \end{aligned} \tag{19.49}$$

The calculation of the DC current is carried out by evaluating the net rate at which electrons flow through the metal–semiconductor interface at $x = 0$ using statistical mechanics. Since electrons flowing from the semiconductor into the metal must have a velocity component $-v_x$, the resulting current density j_{smx} can be written as (cf. Eq. (8.61))

$$j_{smx} = \frac{2(-e)}{\Omega} \sum_{\boldsymbol{k}} (-v_x) f(E_{\boldsymbol{k}}), \tag{19.50}$$

where $f(E_{\boldsymbol{k}})$ is the Fermi distribution function, a factor of 2 has been introduced for spin, and Ω is the volume. If the semiconductor is not too heavily doped and the temperature is not too high, $f(E_{\boldsymbol{k}})$ can be approximated by the Boltzmann distribution:

$$f(E_{\boldsymbol{k}}) \simeq e^{-(E_{\boldsymbol{k}} - E_F)/k_B T}. \tag{19.51}$$

Assuming a spherical energy band,

$$E_{\boldsymbol{k}} = E_C + \frac{\hbar^2 k^2}{2m^*}, \tag{19.52}$$

and taking $\boldsymbol{v}$ to be the group velocity, we have

$$v_x = \frac{1}{\hbar}\frac{\partial E_{\boldsymbol{k}}}{\partial k_x} = \frac{\hbar k_x}{m^*}. \tag{19.53}$$

Substituting these results into Eq. (19.50) and transforming the sum over $\boldsymbol{k}$ to an integral yields

$$j_{smx} = \frac{\hbar e}{4\pi^3 m^*} \int d^3k \, k_x e^{[E_F - E_C - (\hbar^2 k^2/2m^*)]/k_B T}. \tag{19.54}$$

To a good approximation the ranges of integration of k_y and k_z can be taken to be from $-\infty$ to $+\infty$. The range of k_x, however, is restricted by the requirement that the kinetic energy in the x-direction be sufficient to enable the electron to surmount the energy barrier:

$$-\infty \leq k_x \leq -k_{x0}$$

or

$$\infty \geq k_x \geq k_{x0}, \quad (19.55)$$

where

$$\begin{aligned} \frac{\hbar^2 k_{x0}^2}{2m^*} &= \Delta E_{sm}(V_a) \\ &= \phi_m - \phi_s - eV_a. \end{aligned} \quad (19.56)$$

Carrying out the integrals over k_y and k_z,

$$j_{smx} = \frac{ek_BT}{2\pi^2\hbar} e^{(E_F - E_C)/k_BT} \int_{k_{x0}}^{\infty} e^{-\hbar^2 k_x^2/2m^* k_BT} k_x dk_x. \quad (19.57)$$

The integral over k_x is elementary. The result for j_{smx} is

$$j_{smx} = \frac{em^*(k_BT)^2}{2\pi^2\hbar^3} e^{(E_F - E_C - \phi_m + \phi_s + eV_a)/k_BT}, \quad (19.58)$$

which exhibits an exponential increase in current density with increasing forward bias. Eliminating ϕ_s with the aid of Eq. (19.27) yields the simpler result

$$j_{smx} = j_{th} e^{(\chi_s - \phi_m + eV_a)/k_BT}, \quad (19.59)$$

where j_{th} given by

$$j_{th} = \frac{em^*(k_BT)^2}{2\pi^2\hbar^3} \quad (19.60)$$

is the **thermionic current coefficient**. It can be expressed in the reduced form (Sah 1991)

$$j_{th} = 10.8(m^*/m)(T/300)^2 \times 10^6 \text{ A/cm}^2 \quad (19.61)$$

and is the solid-state analogue of the coefficient characterizing the thermionic emission of electrons from a hot filament in vacuum.

To complete the calculation of the current we need to include the contribution of electrons flowing from the metal into the semiconductor, j_{msx}. In this direction the barrier height ΔE_{ms} is unaffected by the applied voltage V_a because no space charge region exists in the metal and no bending of its energy bands occurs, as is evident from Fig. 19.10. Since ΔE_{ms} is independent of V_a we can determine j_{msx} by exploiting the fact that under equilibrium conditions with $V_a = 0$, the total current j_{tot} is zero:

$$j_{tot} = j_{smx} + j_{msx} = 0 \quad \text{for } V_a = 0. \quad (19.62)$$

Replacing j_{smx} by its expression in Eq. (19.59) with $V_a = 0$ we obtain

$$j_{msx} = -j_{th} e^{(\chi_s - \phi_m)/k_B T}. \tag{19.63}$$

This equation is the **Richardson equation** that describes thermionic emission of electrons. The total current density under bias is obtained by combining Eqs. (19.59) and (19.63) to yield

$$j_{tot} = j_{sat}\left(e^{eV_a/k_B T} - 1\right), \tag{19.64}$$

where j_{sat} is the **saturation current density** defined by

$$j_{sat} = j_{th} e^{(\chi_s - \phi_m)/k_B T}. \tag{19.65}$$

Note that for large reverse bias, $V_a \ll -k_B T/e$, the exponential term in Eq. (19.64) is negligible, so $j_{tot} = -j_{sat}$. The current–voltage characteristic is qualitatively the same as that shown in Fig. 12.24.

Equation (19.64) is the **Bethe diode equation**. It has the same qualitative form as the Shockley diode equation, Eq. (12.87), for the p–n junction diode. However, the expressions for the saturation currents in the two cases, Eqs. (12.88) and (19.65), are not the same, because the physical processes involved are quite different. In the p–n junction the current is limited by minority carrier diffusion and recombination, whereas in the MS diode, the current is limited by the passage of majority carriers over the potential barrier by thermal excitation. The Bethe diode saturation current is typically five orders of magnitude larger than the Shockley diode saturation current.

The current associated with electrons flowing into the metal from the semiconductor is often called a hot electron current, because the electrons in the semiconductor must have a kinetic energy of at least $\phi_m - \chi_s$ in order to surmount the barrier. The corresponding effective temperature $T_{eff} = (\phi_m - \chi_s)/k_B$ turns out to be $\sim 12\,000\,K$ in a typical case, so the electrons initially are indeed hot. However, phonon scattering in both the metal and semiconductor is so efficient that the electrons are quickly cooled to ambient temperature.

It was noted earlier that the silicon atoms at the interface with the metal can have dangling bonds. Furthermore, oxygen is typically present in the interface region, leading to formation of SiO_2. Both the dangling bonds and the SiO_2 serve as **interface traps** that can capture free carriers. The captured carriers significantly modify the position of the Fermi energy at the interface and, thereby, the Schottky barrier height (Bardeen 1947). In fact, this pinning of the Fermi level makes the barrier height essentially independent of the specific metal used in the diode. For further details the reader is referred to the solid-state electronics literature (Sah 1991).

19.3 Metal-oxide-semiconductor field effect transistor

19.3.1 Introduction

The **metal-oxide-semiconductor field effect transistor** (MOSFET) consists of a metal layer called the **gate** separated from a semiconductor layer called the **substrate** by a layer of insulating oxide called the **gate oxide**. It is shown schematically in Fig. 19.12. The semiconductor is typically p-Si. The oxide layer is sufficiently thick that no current flows in the direction perpendicular to the interfaces when a voltage called the **gate voltage** is applied to the gate. Instead, an electric field is produced that can reach values on the order of 5×10^6 V/cm and that modulates the conductivity of the semiconductor near its boundary with the oxide. This is the **field effect** that leads to the name **field effect transistor**. Since the gate is insulated from the semiconductor, the term **insulated-gate field effect transistor** (IGFET) is sometimes used.

Up to now what we have described is a MOSC. The MOSFET exploits the modulation of the conductivity of the semiconductor that arises when the electric field is strong enough to bend the energy bands sufficiently to produce an n-type inversion layer or **inversion channel** at the interface of the p-type semiconductor with the gate oxide. Electrical contact is made to the inversion channel by means of n^+**source** and n^+**drain** electrodes that are diffused into the semiconductor as shown in Fig. 19.12. By applying voltages V_S and V_D to the source and drain, respectively, a current parallel to the interface can be made to flow in the inversion channel. This current can be modulated by varying the gate voltage and thereby varying the carrier concentration and conductivity of the inversion channel. This device is known as the **induced n-type inversion channel MOSFET**. The induced p-type inversion channel MOSFET results if the p-Si substrate is replaced by n-Si with the inversion channel becoming p-type.

If the gate voltage is zero or very low, no inversion channel exists. The current path from the source to the drain involves two n^+–p junctions back-to-back, one of which is reversed biased. Consequently, when a DC voltage is applied between source and drain, the maximum current that can flow is

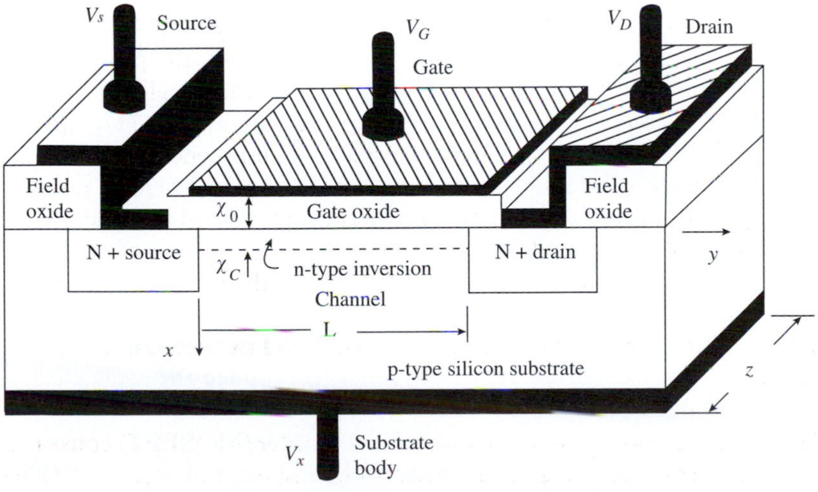

Fig. 19.12
A three-dimensional diagram of an inversion n channel silicon MOSFET (after Sah 1991).

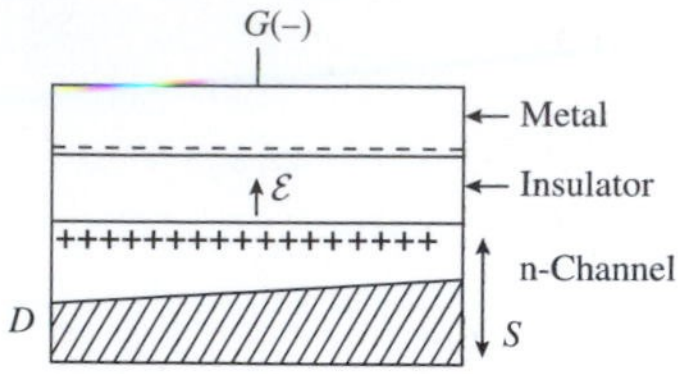

Fig. 19.13
Diagram of the metal gate, insulator, and n-type channel of a MOSFET with a negative potential applied to the gate and showing the electric field in the insulator. The undepleted part of the channel is shaded (after Dalven 1990).

the small saturation current under reverse bias. This situation corresponds to the **cutoff range** of the MOSFET current–voltage characteristics. When an inversion channel is present, however, a much larger current can flow, because the n^+–p junctions are eliminated.

An alternative way to produce an n-type inversion channel MOSFET is to dope the surface layer of the p-Si with donor impurities such as P, As, or Sb. This n-type layer constitutes the inversion channel. If a negative gate voltage is applied, electrons in the inversion channel are forced away from the insulator layer and out of the channel leaving behind positively charged donor ions as shown in Fig. 19.13. The resistance of the channel thereby increases and the source-to-drain current decreases. This is the **doped n-type inversion channel MOSFET.**

It should be emphasized that the MOSFETs just discussed are **unipolar transistors**, i.e., the current is carried by only one type of carrier, either electrons or holes, but not both. In the BJT, on the other hand, the current is carried by both majority and minority carriers; hence, inclusion of the word "bipolar" in the name is appropriate. Another difference is that the current flow is parallel to the principal interfaces in the MOSFET, but perpendicular to them in the BJT.

19.3.2 DC characteristics of the MOSFET

19.3.2.1 Surface field and surface charge

Our discussion of the MOSFET characteristics builds upon that of the MOSC, since the latter is an integral part of the former. What needs to be added are the roles of the source and drain and the current that flows between them. The procedure is to obtain a new expression for the space charge Q_s that contains the drain voltage V_D. The charge Q_n in the inversion layer is then obtained and from it an equation that relates the drain current I_D to V_D and the gate voltage V_G. The source is assumed to be grounded so that the source voltage is zero. The treatment can easily be generalized to include a nonzero source voltage (Sah 1991).

The drain voltage produces a potential $V(y)$ in the inversion channel at a distance y from the source that requires the Boltzmann distribution for electrons to take the modified form

$$n_p(x, y) = n_{p0} e^{\beta e[\psi(x) - V(y)]}. \tag{19.66}$$

At the semiconductor surface, $x = 0$,

$$n_p(0, y) = n_{p0} e^{\beta e[\psi_s - V(y)]}. \tag{19.67}$$

Under equilibrium conditions with $V(y) = 0$, we have seen that ψ_s must be $2\psi_B$ in order to produce an electron concentration corresponding to strong inversion. It is clear from Eq. (19.67) that to produce the same electron concentration with $V(y) \neq 0$, ψ_s must be given by

$$\psi_s = 2\psi_B + V(y). \tag{19.68}$$

The energy band variations in an inverted p-region with and without $V(y)$ are shown in Fig. 19.14.

We now turn to the relation between the drain current I_D and the drain voltage V_D. Consider an elemental volume of the inversion channel of length dy, thickness dx and width Z. The resistance dR_v of this volume is given by

$$dR_v = \frac{dy}{\sigma(x,y)Z\,dx}, \tag{19.69}$$

where $\sigma(x,y)$ is the conductivity. The latter depends on x and y due to the nonuniform character of the inversion channel. Since the slabs of thickness dx making up the channel constitute resistances in parallel, the total resistance dR of a section of length dy is specified by

$$\frac{1}{dR} = \frac{Z\int_0^{W_i}\sigma(x,y)dx}{dy}, \tag{19.70}$$

where W_i is the total thickness of the inversion channel. Using the relation $\sigma(x,y) = e\mu_n n_p(x,y)$, where μ_n is the effective electron mobility, we obtain

$$dR = \frac{dy}{Z\mu_n|Q_n(y)|}. \tag{19.71}$$

The quantity $Q_n(y)$ is the electron charge per unit area given by

$$Q_n(y) = -e\int_0^{W_i} n_p(x,y)dx. \tag{19.72}$$

The voltage drop dV across dR is determined by Ohm's law

$$dV = I_D\,dR = \frac{I_D\,dy}{Z\mu_n|Q_n(y)|}. \tag{19.73}$$

The electron charge $Q_n(y)$ is related to the total charge induced in the semiconductor $Q_s(y)$ by Eq. (19.29) which we rewrite in the form

$$Q_s(y) = Q_n(y) + Q_B(y), \tag{19.74}$$

where $Q_B(y)$ is the surface depletion charge per unit area given by $-en_aW_d(y)$ and $W_d(y)$ is the width of the depletion region specified by the generalization of Eq. (19.32):

$$\psi_s = \frac{en_aW_d^2(y)}{2\epsilon_s}. \tag{19.75}$$

Using the result for ψ_s appropriate for strong inversion in the presence of the drain voltage, Eq. (19.68), we obtain

$$W_d(y) = \left\{\frac{2\epsilon_s[2\psi_B + V(y)]}{en_a}\right\}^{\frac{1}{2}}. \tag{19.76}$$

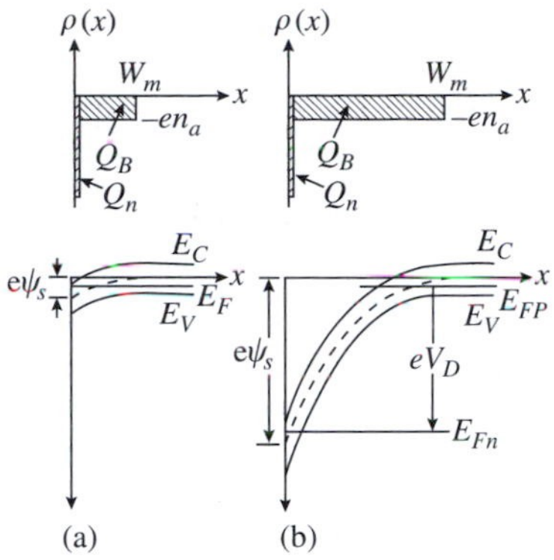

Fig. 19.14
Comparison of the charge distribution and energy band variation of an inverted p-region for: (a) equilibrium case, (b) nonequilibrium reverse bias case (after Grove and Fitzgerald 1966).

$Q_B(y)$ is then given by

$$Q_B(y) = -\{2\epsilon_s e n_a[2\psi_B + V(y)]\}^{\frac{1}{2}}. \tag{19.77}$$

Since $V(y)$ increases from source to drain due to the IR drop along the inversion channel, the energy band bending ψ_s increases from source to drain, as does the surface depletion charge Q_B.

The total charge $Q_s(y)$ can be related to the gate voltage V_G by combining Eqs. (19.15), (19.29), and (19.32). The result is

$$V_G = -\frac{Q_s(y)}{C_O} + V_{FB} + \psi_s(y), \tag{19.78}$$

where V_{FB} is the **flat-band voltage** given by

$$V_{FB} = \frac{\phi_m - \phi_s}{e} - \frac{Q_O + Q_I}{C_O}. \tag{19.79}$$

The flat-band voltage is the gate voltage required to give flat bands for which ψ_s, $\mathcal{E}_s$, and Q_s are all zero. Solving Eq. (19.78) for $Q_s(y)$ and using Eq. (19.68), we obtain

$$Q_s(y) = C_O[-V_G + V_{FB} + 2\psi_B + V(y)]. \tag{19.80}$$

Substituting this result for $Q_s(y)$ and that for $Q_B(y)$ into Eq. (19.74) yields $Q_n(y)$ in the form

$$Q_n(y) = -C_O[V_G - V_{FB} - 2\psi_B - V(y)] + \{2\epsilon_s e n_a[2\psi_B + V(y)]\}^{\frac{1}{2}}. \tag{19.81}$$

Let us suppose that V_G is sufficiently large to produce an inversion layer at the source where $V(y) = V(0) = 0$. As one proceeds toward the drain and $V(y)$ increases, we see that $|Q_n(y)|$ decreases until it reaches the value $|Q_n(L)|$ at the drain where $V(y) = V(L) = V_D$. As V_D is increased, the **saturation drain voltage** V_{Dsat} is eventually attained at which $Q_n(L) = 0$. From Eq. (19.81) we find that V_{Dsat} is given by

$$V_{Dsat} = V_G - V_{FB} - 2\psi_B + \frac{\epsilon_s e n_a}{C_O^2} \times \left\{1 - \left[1 + \frac{2C_O^2(V_G - V_{FB})}{\epsilon_s e n_a}\right]^{\frac{1}{2}}\right\}. \tag{19.82}$$

We can eliminate C_O and n_a with the aid of Eqs. (19.19) and (19.36):

$$V_{Dsat} = V_G - V_{FB} - 2\psi_B + 4\psi_B\left(\frac{\epsilon_s W_O}{\epsilon_O W_{dm}}\right)^2 \times \left\{1 - \left[1 + \left(\frac{\epsilon_O W_{dm}}{\epsilon_s W_O}\right)^2 \frac{V_G - V_{FB}}{2\psi_B}\right]^{\frac{1}{2}}\right\}. \tag{19.83}$$

In the limiting case of oxide layer width W_O small compared to the maximum depletion region width W_{dm}, the above result simplifies to

$$V_{Dsat} \simeq V_G - V_{FB} - 2\psi_B. \qquad (19.84)$$

19.3.2.2 Drain current–drain voltage relationship

With the inversion layer charge $Q_n(y)$ related to the drain voltage $V(y)$ through Eq. (19.81), we can use Ohm's law as expressed by Eq. (19.73) to obtain the drain current I_D. We integrate from the source at $y=0$, where $V=0$, to the drain at $y=L$, where $V=V_D$, and obtain

$$I_D = \frac{Z\mu_n C_O}{L}\left\{\left[V_G - V_{FB} - 2\psi_B - \frac{V_D}{2}\right]V_D - \frac{2}{3}\frac{(2\epsilon_s e n_a)^{\frac{1}{2}}}{C_O}\left[(V_D + 2\psi_B)^{3/2} - (2\psi_B)^{3/2}\right]\right\}. \qquad (19.85)$$

Eliminating C_O and n_a inside the curly brackets yields

$$I_D = \frac{Z\mu_n C_O}{L}\left\{\left[V_G - V_{FB} - 2\psi_B - \frac{V_D}{2}\right]V_D - \frac{4}{3}\frac{\epsilon_s W_O}{\epsilon_O W_{dm}}(2\psi_B)^{\frac{1}{2}}\left[(V_D + 2\psi_B)^{3/2} - (2\psi_B)^{3/2}\right]\right\}. \qquad (19.86)$$

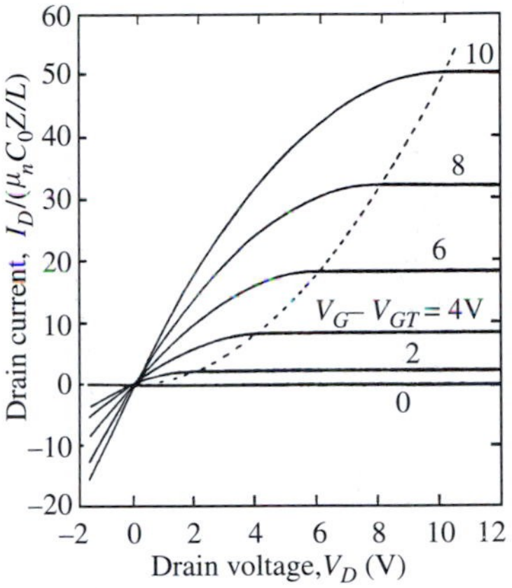

Fig. 19.15
Normalized drain current versus drain voltage for various values of the gate voltage for an n-channel MOSFET (after Sah 1991).

A family of drain current versus drain voltage curves for various gate voltages is shown in Fig. 19.15.

For small V_D, I_D increases linearly with V_D in accordance with the leading term of the small V_D expansion of Eq. (19.86):

$$I_D = \frac{Z\mu_n C_O}{L}(V_G - V_{GT})V_D. \qquad (19.87)$$

The quantity V_{GT} is the **threshold gate voltage** defined by

$$V_{GT} = V_{FB} + 2\psi_B - \frac{Q_B(0)}{C_O} = V_{FB} + 2\psi_B + \frac{(4\epsilon_s e n_a \psi_B)^{\frac{1}{2}}}{C_O}. \qquad (19.88)$$

It is the smallest gate voltage at which a conducting channel is induced and a drain current can flow. As the drain voltage increases further, the drain current levels off and reaches a maximum value at $V_D = V_{Dsat}$. From the definition of the latter, $Q_n(L)=0$ which corresponds to the **pinch-off condition**. No mobile electrons remain at the drain end of the channel. When V_D exceeds V_{Dsat}, a depletion region replaces the inversion region near the drain. The drain current then consists of electrons that flow from the source along the inversion channel and are injected into the depletion region before reaching the drain. The drain current is no longer described by Eq. (19.86), because the latter assumes the existence of an inversion layer over the entire distance from source to drain. The electrons now flow through n-depletion and depletion-n$^+$ junctions back-to-back. I_D remains constant at the value I_{Dsat} corresponding to V_{Dsat}. Replacement of V_D in Eq. (19.86) by V_{Dsat} from Eq. (19.83) yields an expression for I_{Dsat}. In the limit of small W_O, one obtains the relatively simple result

$$\begin{aligned} I_{Dsat} &\simeq \frac{Z\mu_n C_O}{2L} V_{Dsat}^2 \\ &\simeq \frac{Z\mu_n C_O}{2L}(V_G - V'_{GT})^2, \end{aligned} \tag{19.89}$$

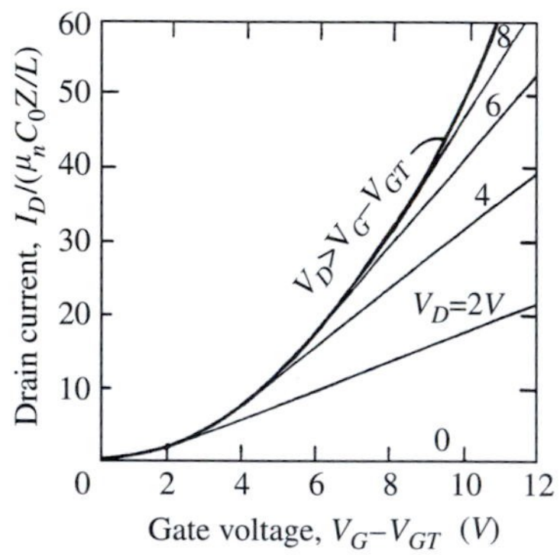

Fig. 19.16
Normalized drain current versus gate voltage for various values of the drain voltage for an n-channel MOSFET (after Sah 1991).

where

$$V'_{GT} = V_{FB} + 2\psi_B. \tag{19.90}$$

Equation (19.89) describes a parabola which is indicated by the dashed curve in Fig. 19.15. To the right of the parabola the drain current is essentially constant for given V_G.

The **transfer characteristics** of a MOSFET are provided by the dependence of I_D on V_G with V_D fixed. A typical plot showing this dependence is presented in Fig. 19.16. For small V_D and $V_G - V_{GT} \gg V_D$ the characteristics are linear in accordance with Eq. (19.87). As $V_G - V_{GT}$ is reduced at constant V_D, it eventually reaches a value at which V_D acquires its saturation value V_{Dsat}. The drain current then takes its saturation value. It is specified by Eq. (19.89) for $V_D > V_G - V_{GT} > 0$ and is represented by the left-hand envelope of the curves in Fig. 19.16.

A parameter of interest for device applications is the **transconductance** defined by

$$g_m \equiv \left.\frac{\partial I_D}{\partial V_G}\right|_{V_D}. \tag{19.91}$$

For $V_D < V_{Dsat}$ we use Eq. (19.87) and obtain

$$g_m = \frac{Z\mu_n C_O V_D}{L}. \tag{19.92}$$

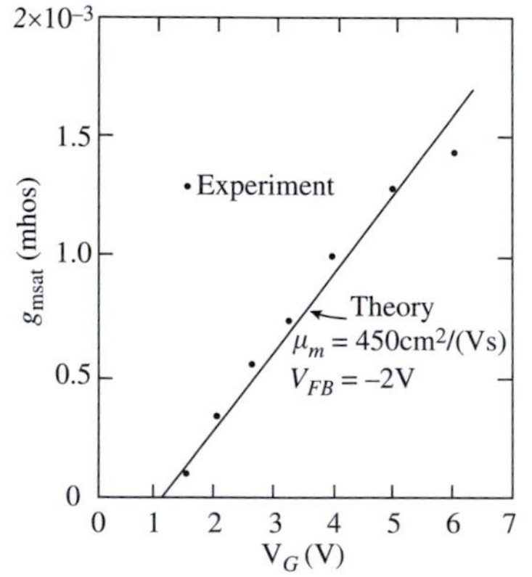

Fig. 19.17
Transconductance versus gate voltage in the saturation region ($V_D > V_{Dsat}$) for an n-channel MOSFET (after Grove 1967).

In the saturation range we replace V_D in the last equation by V_{Dsat} as specified by Eq. (19.83) or (19.84). In the latter case (very thin oxide layer),

$$g_{msat} = \frac{Z\mu_n C_O}{L}(V_G - V_{FB} - 2\psi_B). \tag{19.93}$$

We see that in the saturation range the transconductance increases linearly with gate voltage as shown in Fig. 19.17. At large gate voltages, g_{msat} levels off due to a reduction in surface mobility μ_n of the electrons.

Another parameter of interest is the **voltage amplification factor** f_a. It is defined by

$$f_a = -\left.\frac{\partial V_D}{\partial V_G}\right|_{I_D} \tag{19.94}$$

and takes the form

$$f_a = \frac{V_D}{V_G - V'_{GT} - V_D} \tag{19.95}$$

in the thin oxide layer limit. A plot of f_a versus $V_G - V'_{GT}$ at fixed V_D is shown in Fig. 19.18. Bringing V_D close to its saturation value produces large values of f_a.

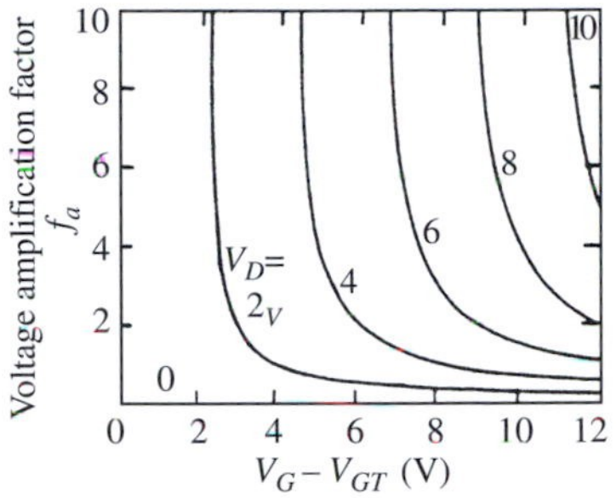

Fig. 19.18
Voltage amplification factor versus gate voltage for various values of the drain voltage for an n-channel MOSFET (after Sah 1991).

Problems

1. Calculate the space charge per unit area Q_s for a surface potential $\psi_s = -0.3$ V in p-type silicon at $T = 300$ K with $n_a = 10^{16}\,\mathrm{cm}^{-3}$. Also calculate the values of Q_s for $\psi_s = 0$ and $\psi_s = +0.3$ V. For what value of ψ_s does the onset of strong inversion occur?
2. Consider a MOSC on p-Si at 300 K with the following parameters: $n_a = 10^{16}\,\mathrm{cm}^{-3}$, $W_O = 300$ Å, $Q_O = Q_I = 0$, $\epsilon_O = 2.4\epsilon_0$, and $\phi_m = \phi_s$. For equally spaced values of ψ_s in the range -0.5 V to 1.0 V, calculate the gate voltage V_G, the oxide capacitance C_O and the total capacitance C. Make a plot of C/C_O versus V_G and compare with Fig. 19.6.
3. In connection with problem 2 explain why there is an accumulation layer of holes when V_G is negative and why there is an inversion layer of electrons when V_G is sufficiently positive.
4. Calculate the maximum value of the depletion layer width W_{dm} at 300 K for an acceptor concentration $n_a = 10^{16}\,\mathrm{cm}^{-3}$.
5. Calculate the total current density of an MS diode under bias $V_a = 0.3$ V at 300 K with $m^* = 0.1$ m, $\chi_s = 4.03$ eV, and $\phi_m = 4.08$ eV.
6. What is the physical significance of the flat-band condition $\psi_s = 0$ (see Fig. 19.5)? What are the values of $\mathcal{E}_s$ and Q_s? Obtain an expression for the flat-band voltage that involves the voltage across the oxide V_O.
7. For a MOSFET with material parameters as in problem 2, calculate the saturation drain voltage when $Q_n(L) = 0$ and $V_G = 2$ V. If the oxide layer thickness W_O is small compared to the maximum depletion region width W_{dm}, evaluate the saturation drain voltage V_{Dsat} and the threshold gate voltage V_{GT}.
8. From the results of problem 7, calculate the saturation drain current I_{Dsat} if $\mu_n = 1000\,\mathrm{cm}^2/(\mathrm{V\,s})$, $Z = 100\,\mu$m, and $L = 5\,\mu$m.

References

J. Bardeen, *Phys. Rev.* **71**, 717 (1947).
W. S. Boyle and G. E. Smith, *Bell Syst. Tech. J.* **49**, 587 (1970).
R. Dalven, *Introduction to Applied Solid State Physics*, Second edition(Plenum Press, New York, 1990).
A. S. Grove, *Physics and Technology of Semiconductor Devices* (John Wiley, New York, 1967).
A. S. Grove and D. J. Fitzgerald, *Solid-State Electron.* **9**, 783 (1966).
A. S. Grove, B. E. Deal, E. H. Snow, and C. T. Sah, *Solid-State Electron.* **8**, 145 (1965).
C. T. Sah, *Fundamentals of Solid-State Electronics* (World Scientific, Singapore, 1991).
S. M. Sze, *Physics of Semiconductor Devices*, Second edition (John Wiley, New York, 1981).

Applications of semiconductor heterostructures

20

Key ideas

The essential features of the MOSFET—source, drain, and gate which is electrically isolated from the source and drain—can be incorporated into semiconductor heterostructures to create devices such as the two-dimensional electron gas field effect transistor (TEGFET) or high electron mobility transistor (HEMT) and the modulation doped field effect transistor (MODFET) or selectively doped heterojunction transistor (SDHT).

In heterostructure devices the interfaces are very sharply defined and changes in the voltage across them can be very abrupt, thus leading to excellent control of the channel charge concentration by the externally applied gate voltage.

An important application of heterostructures is to digital devices such as switches. The high mobility in the 2D channel provides low power dissipation.

In a SISFET the metal gate of the standard MOSFET is replaced by a highly doped GaAs gate.

High-frequency operation requires that the time for carriers to cross the active region must be as short as possible. The latter should therefore be made as narrow as possible.

The operation of the double barrier diode involves the *resonant tunneling effect*. A quantum well containing electrons is separated by a barrier from a quantum well containing holes. Under an applied voltage the current is due to electrons surmounting the barrier. At resonant tunneling the current decreases, giving rise to an effective negative resistance.

The heterojunction bipolar transistor (HBT) has an emitter of n-GaAlAs, a collector of n-GaAs, and a base of p-GaAs which serves as a barrier for electrons moving from the emitter to the collector. In unipolar transistors only majority carriers with high kinetic energy (hot carriers) are involved, thus producing the *hot electron transistor* (HET) with very high-frequency operation. With a sufficiently thin GaAlAs barrier tunneling of hot electrons becomes possible giving the *tunneling hot electron transistor amplifier* (THETA).

Applications

20.3 Quantum well lasers

Lasers based on quantum wells have better performance than conventional semiconductor lasers.

The guided light wave and carrier confinement significantly lower the threshold current density of a double-heterostructure laser.

The single quantum well laser has a two-dimensional density-of-states that enhances the efficient buildup of gain.

Multiple quantum well lasers are advantageous under high loss conditions. Gain can be improved by applying strain to split the light and heavy hole bands.

20.4 One-dimensional and zero-dimensional quantum structures

One-dimensional and zero-dimensional structures lead to *quantum wires* and *quantum dots*.

In one dimension the carrier energies lie in subbands or *channels*. The channel conductance has the quantized value $2e^2/h$.

Nanoscale lithographic techniques, molecular beam epitaxy, and metalorganic chemical vapor deposition can be used to produce quantum wires and quantum dots.

If the mean free path of an electron is large compared to the dimensions of a quantum wire, the electrical transport is *ballistic*.

Devices based on 1D and 0D structures can perform digital operations involving single- or few-electron events.

The threshold for nonlinear optical effects decreases as the dimensionality decreases.

The quantum dot laser makes all electron–hole pairs efficient for gain.

20.5 Devices based on electro-optic effects in quantum well structures

An electric field applied in the distinct directions parallel and perpendicular to the layers of a quantum well structure leads to novel effects on exciton absorption.

An electric field perpendicular to the layers strongly shifts the exciton absorption peaks in the *quantum-confined Stark effect*.

The QCSE forms the basis for the *quantum well modulator* in which the optical transmission is modulated by an applied field.

A photodetector combined with a quantum well modulator creates a *self-electro-optic-effect device* (SEED).

The development of techniques such as molecular beam epitaxy has made possible the production of ultrathin semiconductor structures with precisely controlled characteristics at the atomic level. These structures have formed the basis for new electronic devices with improved performance in both standard and novel applications.

20.1 Devices with transport parallel to the interfaces: field effect transistor

20.1.1 Analysis of physical processes

The essential features of a metal-oxide-semiconductor field effect transistor (MOSFET) discussed in Chapter 19 are the electrical contact called the **source**, the contact called the **drain**, and an electrode called the **gate** which is electrically isolated from the source and drain. The same features can be incorporated into semiconductor heterostructures to create devices such as the **two-dimensional electron gas field effect transistor** (TEGFET), also known as the **high electron mobility transistor** (HEMT), the **modulation doped field effect transistor** (MODFET), or the **selectively doped hetero-junction transistor** (SDHT).

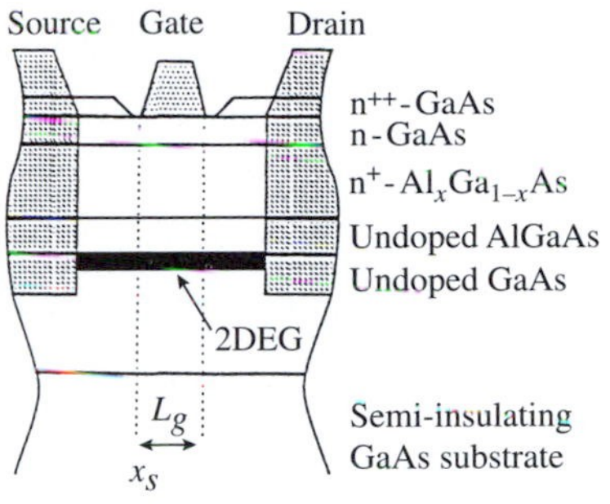

Fig. 20.1
Diagram of a TEGFET (after Weisbuch and Vinter 1991).

These devices are typically based on GaAs/GaAlAs heterostructures as shown in Fig. 20.1. A layer of undoped GaAs is grown on a semi-insulating GaAs substrate. On top of the undoped GaAs is grown a layer of $Ga_{1-x}Al_xAs$, the lower part of which is undoped and the upper part of which is heavily doped n-type ($n \sim 10^{18}\,cm^{-3}$). The gate is a metal that forms a Schottky barrier against the $Ga_{1-x}Al_xAs$. Application of a voltage to the gate leads to transfer of electrons from the n-$Ga_{1-x}Al_xAs$ to the undoped GaAs leaving the $Ga_{1-x}Al_xAs$ depleted of electrons. The transferred electrons form a 2D electron gas or channel within the GaAs at the interface with the $Ga_{1-x}Al_xAs$. The electron concentration is determined by the gate voltage. The source and drain contacts are diffused down to the 2D electron gas layer and can be used to drive a current through the latter. This source–drain current can be modulated by the gate voltage through variations induced in the 2D electron concentration.

One of the advantages of heterostructure devices is that the interfaces are very sharply defined, and changes in voltage across them can be very abrupt. This leads to excellent control of the channel charge density by the externally applied gate voltage.

The factors entering into the speed of operation of TEGFETs can be assessed using results from the analysis of MOSFETs in Chapter 19. The drain current at saturation I_{Dsat} is given by Eq. (19.89),

$$I_{Dsat} = \frac{Z\mu_n C(V_G - V'_{GT})^2}{2L}, \tag{20.1}$$

where μ_n is the mobility, C is the gate capacitance per unit area, V_G and V'_{GT} are the gate voltage and threshold voltage, respectively, and L is the gate length. We have seen in the previous chapter that the transconductance g_m defined by

$$g_m = \left.\frac{\partial I_D}{\partial V_G}\right|_{V_D} \tag{20.2}$$

is an important parameter characterizing a field effect transistor. The transconductance at saturation g_{msat} is obtained by replacing I_D in Eq. (20.2) by I_{Dsat} from Eq. (20.1):

$$g_{msat} = \frac{Z\mu_n C(V_G - V'_{GT})}{L}. \tag{20.3}$$

Introducing the electric field $\mathcal{E} = (V_G - V'_{GT})/L$ and the saturation drift velocity $v_s = \mu_n \mathcal{E}$, we obtain the simple result

$$g_{msat} = C_F \frac{v_s}{L} = \frac{C_F}{\tau}, \tag{20.4}$$

where $C_F = ZLC$ is the full capacitance and $\tau = L/v_s$ is the time required for an electron to cross the gate. For the device to have fast operation, τ should be short, which can be achieved by having L small and/or v_s large. The latter condition clearly makes high mobilities desirable.

Another important parameter is the current gain h defined by the ratio of the drain current to the gate current:

$$h = \frac{\partial I_{Dsat}}{\partial I_G}. \tag{20.5}$$

From Fig. 20.1, we see that a layer of high resistivity $Ga_{1-x}Al_xAs$ lies between the gate and the channel. This combination therefore forms a capacitor. The gate current I_G and gate voltage V_G are essentially 180° out of phase, so that, if $V_G \sim \exp(i\omega t)$,

$$I_G = i\omega C_F(V_G - V_{GT}). \tag{20.6}$$

From Eqs. (20.1)–(20.3) we obtain

$$\frac{\partial I_{Dsat}}{\partial V_G} = g_{msat}. \tag{20.7}$$

The gain under saturation conditions takes the form

$$h = \frac{g_{msat}}{i\omega C_F} = \frac{v_s}{i\omega L} = \frac{f_T}{if}, \tag{20.8}$$

where the cutoff frequency f_T is given by $v_s/2\pi L$ and $f = \omega/2\pi$. To obtain reasonable gain at high frequencies, it is clearly important to have high channel velocities. This condition is more easily satisfied in GaAs-based TEGFETs than in Si-based MOSFETs for two reasons. First, at a given value of the electric field, the average velocity of electrons in GaAs considerably exceeds that in Si. Second, the higher mobility in the undoped channel of a TEGFET compared to the mobility in a MOSFET further increases the maximum gain attainable.

The treatment we have presented is oversimplified. A more detailed treatment (Weisbuch and Vinter 1991) shows that a critical field $\mathcal{E}_c$ exists above which the electron velocity saturates, thus limiting the gain.

20.1.2 Analysis of device performance

Heterostructure transistors are particularly interesting for low-noise, very high-frequency amplification. Power gain is attainable up to a maximum frequency f_{max} defined by (Hollis and Murphy 1990)

$$f_{max} = \frac{f_T}{2[g_0(R_G + R_S) + 2\pi f_T R_G C_{DG}]^{\frac{1}{2}}}, \tag{20.9}$$

where $g_0 = \partial I_D/\partial V_{DS}$ is the output conductance, R_G is the gate resistance, R_S is the resistance between the source and the gate, and C_{DG} is a parasitic gate–drain capacitance. It is evident from this relation that a high cutoff frequency is necessary for high power gain. Also, it is critical to achieving the maximum frequency for power gain that R_G, R_S, and C_{DG} be reduced as much as possible. These features are favored by the use of heterostructures.

An important application of heterostructures is to digital devices such as switches. Factors that must be considered are the switching delay and the power consumption. To switch rapidly from the off-state to the on-state requires a small capacitance C and a high current flow. These conditions are satisfied if the ratio g_{msat}/C_F appearing in the gain is large and if the mobility of the electrons in the 2D channel is large. High mobility is also valuable in providing low power dissipation. The product of power dissipation and switching delay constitutes a useful figure of merit. Heterostructure transistors compete very favorably on the basis of this figure of merit.

20.1.3 Semiconductor–insulator–semiconductor field effect transistor (SISFET)

In the SISFET the metal gate is replaced by a highly doped GaAs gate. The gate is separated from the channel in an undoped GaAs layer by a barrier of undoped GaAlAs. The lack of a heavily n-doped part of the GaAlAs results in a lower barrier for the SISFET than the TEGFET. Consequently, the SISFET can be operated in the normal way only at relatively low temperatures in order to minimize leakage across the barrier. Indeed, the low barrier makes possible a different mode of operation in which electrons in the channel are accelerated to sufficiently high energies by the source-drain field to undergo a transfer into or through the barrier. In the **real-space transfer diode** the transfer of electrons from a hot channel to a cooler channel leads to a reduction in current with increasing electric field and hence a negative differential resistance. Taking the hot channel to be the quantum-well channel and the cool channel to be the gate electrode gives rise to the negative resistance field effect transistor (NERFET).

20.2 Devices with transport perpendicular to the interfaces

In Chapter 13 devices such as the bipolar junction transistor are discussed that involve transport of current carriers perpendicular to the interfaces of the junctions. This geometry is also useful for heterostructure devices. From Eqs. (20.4) and (20.8) we see that high-frequency operation of a TEGFET requires that the time τ for a carrier to cross the gate must be as short as possible. Consequently, the gate length L should be made as short as possible. On the other hand, molecular beam epitaxy is a particularly advantageous technique for obtaining very narrow active regions in the direction perpendicular to the interfaces. Thus, heterojunctions with perpendicular transport have desirable aspects for high-frequency, high speed devices. Another desirable feature of the perpendicular geometry is that the current is essentially three-dimensional and proportional to the area of the device. High-frequency power amplifiers with high current capacity are therefore feasible.

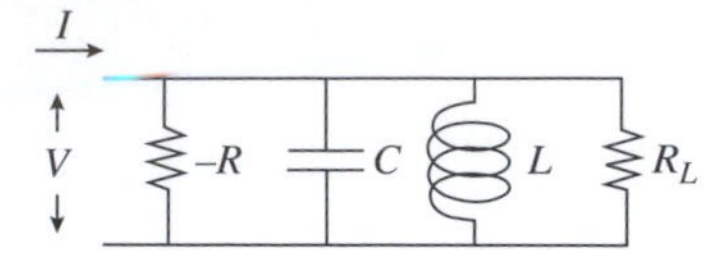

Fig. 20.2
Small-signal equivalent circuit for a double-barrier diode (after Weisbuch and Vinter 1991).

20.2.1 Heterostructure double-barrier diode

The operation of the double-barrier diode involves the resonant tunneling effect discussed in Chapter 18. The structure consists of a quantum well containing electrons separated by a barrier from a quantum well containing holes. If an external voltage is applied, a current flows associated with electrons surmounting the barrier. When the condition for resonant tunneling is attained, the current reaches a maximum and then decreases giving rise to an effective negative resistance. The diode is then in an unstable state and if incorporated into a resonating circuit, can give rise to oscillations at a frequency determined by the parameters of that circuit.

The situation can be described by the small-signal equivalent circuit shown in Fig. 20.2. The diode is represented by the negative resistance $-R$ connected to a resonating LC circuit that feeds into a load resistance R_L. The currents through the various elements are $I_{ind}(t)$, $I_{cap}(t)$, and $I_{res}(t)$ for the inductance, capacitance, and resistances, respectively. They are related to the applied voltage $V(t)$ by the equations

$$V(t) = L\frac{dI_{ind}(t)}{dt} = R_T I_{res}(t) \tag{20.10}$$

$$\frac{dV(t)}{dt} = \frac{1}{C}I_{cap}(t), \tag{20.11}$$

where

$$\frac{1}{R_T} = \frac{1}{R_L} - \frac{1}{R}. \tag{20.12}$$

Since the total current $I(t)$ is the sum of the currents through the individual elements, the following equation must be satisfied:

$$\frac{dI(t)}{dt} = \frac{V(t)}{L} + C\frac{d^2V(t)}{dt^2} + \frac{1}{R_T}\frac{dV(t)}{dt}. \tag{20.13}$$

Taking the Laplace transform defined by the equation

$$F(s) = \int_0^\infty e^{-st}F(t)dt \tag{20.14}$$

gives the relation

$$I(s) = \left(\frac{1}{sL} + sC + \frac{1}{R_T}\right)V(s). \tag{20.15}$$

Solving for $V(s)$,

$$V(s) = \frac{s}{2i\omega_0 C}\left(\frac{1}{s-\gamma-i\omega_0} - \frac{1}{s-\gamma+i\omega_0}\right)I(s), \tag{20.16}$$

where $\gamma = -1/2CR_T$ and $\omega_0^2 = (1/LC) - \gamma^2$.

Let us rewrite Eq. (20.16) as

$$V(s) = G(s)I(s). \tag{20.17}$$

It can readily be verified that $G(s)$ is the Laplace transform of the function

$$G(t) = \frac{1}{C}e^{\gamma t}\left(\frac{\gamma}{\omega_0}\sin\omega_0 t + \cos\omega_0 t\right). \tag{20.18}$$

The product form of the right-hand side of Eq. (20.17) means that $V(t)$ can be expressed as the convolution

$$V(t) = \int_0^t G(\tau)I(t-\tau)d\tau. \tag{20.19}$$

If $I(t)$ is taken to be a current pulse

$$I(t) = (I_0/\omega_0)\delta(t) \tag{20.20}$$

then

$$V(t) = \frac{I_0}{\omega_0 C}e^{\gamma t}\left(\frac{\gamma}{\omega_0}\sin\omega_0 t + \cos\omega_0 t\right). \tag{20.21}$$

If $R < R_L$ so that $R_T < 0$ and $\gamma > 0$, a fluctuation in the current will produce an oscillation in the circuit with frequency ω_0 and amplitude that initially increases exponentially with time. Eventually, nonlinearities stabilize the amplitude.

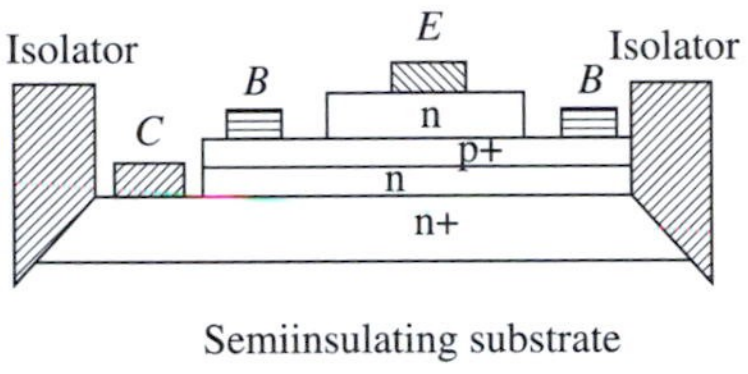

Fig. 20.3
Diagram of a heterojunction bipolar transistor. *E*, *B*, and *C* denote metallic contacts to the emitter, base, and collector, respectively (after Weisbuch and Vinter 1991).

20.2.2 Heterojunction bipolar transistor

In contrast to the double barrier diode, which is a two-terminal device, the **heterojunction bipolar transistor** (HBT) is a three-terminal device with emitter, base, and collector. A typical configuration is shown in Fig. 20.3. The emitter is n-GaAlAs, the collector is n-GaAs, and the base is p-GaAs. The latter serves as a barrier for electrons moving from the emitter to the collector. A diagram of the band edges in an HBT is given in Fig. 20.4.

An important advantage of an HBT is the smallness of the base thickness W_B that an electron must cross. The analysis of the standard bipolar transistor given in Chapter 13 can be applied to the HBT. The collector current density j_c is

$$j_c = \frac{D_e e n_{p0}}{W_B}\left(e^{eV_{BE}/k_BT} - 1\right), \tag{20.22}$$

where V_{BE} is the emitter–base voltage, D_e is the electron diffusion constant in the base, and n_{p0} is the equilibrium electron concentration in the base. The quantity $v_d = 2D_e/W_B$ is the average diffusion velocity of electrons in the base, and $n_{p0}[\exp(eV_{BE}/k_BT) - 1]$ is the average concentration of electrons injected into the base. Small base thickness and large diffusion constant clearly favor a large collector current density.

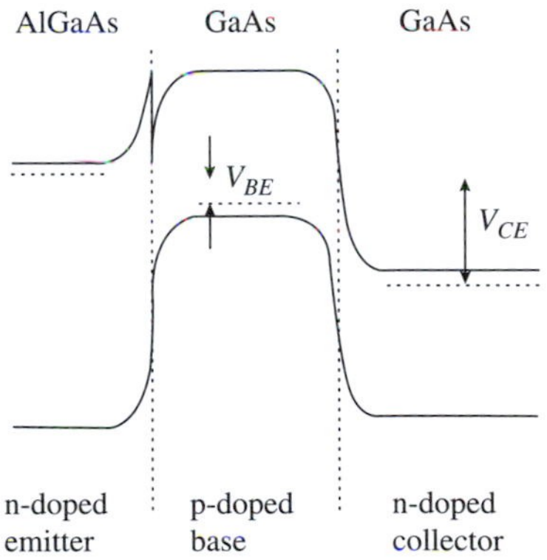

Fig. 20.4
Band diagram of an HBT for the common emitter configuration (after Weisbuch and Vinter 1991).

The transconductance g_m is given by

$$g_m = \frac{\partial I_c}{\partial V_{BE}} \equiv \frac{e}{k_B T} \cdot \frac{D_e e n_{p0}}{W_B} e^{eV_{BE}/k_B T}. \tag{20.23}$$

At large V_{BE},

$$g_m \simeq \frac{e}{k_B T} I_c. \tag{20.24}$$

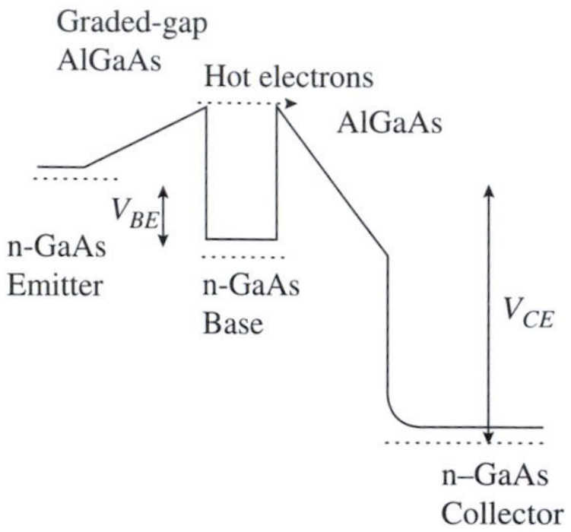

Fig. 20.5
Conduction band diagram for a hot electron transistor (after Weisbuch and Vinter 1991).

The transconductance is therefore large at large current levels. Only a small variation in V_{BE} is necessary to produce a given variation in I_c, a property that is useful in digital applications.

A parameter analogous to the transit time of a TEGFET is the time τ_B required for an electron to cross the base:

$$\tau_B = \frac{W_B}{v_d} = \frac{W_B^2}{2D_e}. \tag{20.25}$$

In order to obtain a high-speed device, τ_B and hence W_B should be made as small as possible. Ohmic losses can be reduced by having a relatively high doping level in the base. Alternatively, the diffusion velocity v_d can be augmented by doping the base with Al in such a way that its concentration decreases from the emitter junction to the collector junction. This concentration gradient gives rise to an internal field that accelerates the carriers. Ballistic electrons that cross the base without scattering also augment the average transit velocity.

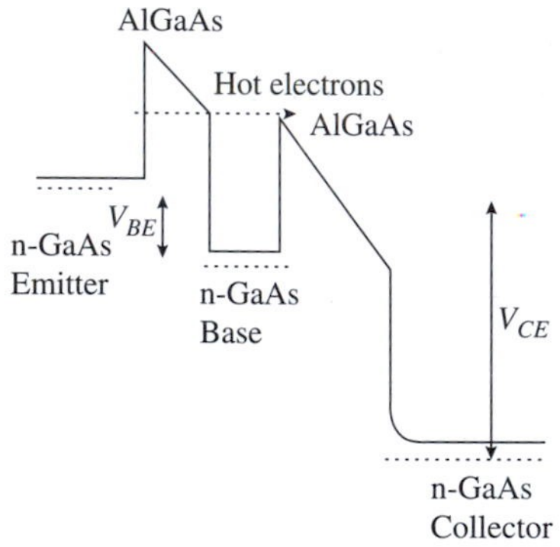

Fig. 20.6
Conduction band diagram of a tunneling hot electron transistor amplifier (after Weisbuch and Vinter 1991).

Unipolar transistors utilizing heterojunctions have been designed in which the emitter, base, and collector are all n-GaAs. Only majority carriers are involved, and when they enter the base from the emitter, they have a high kinetic energy and are therefore "hot". The band-edge diagram for the **hot electron transistor** (HET) is shown in Fig. 20.5. Very high-frequency operation should be possible with HETs, since the transit velocity and cutoff frequency are very high. If the GaAlAs barrier between emitter and base is sufficiently thin, tunneling of the hot electrons becomes possible, giving rise to the **tunneling hot electron transistor amplifier** (THETA). A band-edge diagram of a THETA is presented in Fig. 20.6.

20.3 Quantum well lasers

We have seen in Chapter 14 that a semiconductor laser contains an active region associated with a p–n junction in a material such as GaAs. The active region forms a resonant cavity bounded by a pair of parallel reflecting mirrors created by cleavage along parallel (110) planes. However, improved laser performance can be achieved if heterojunctions and quantum wells are utilized.

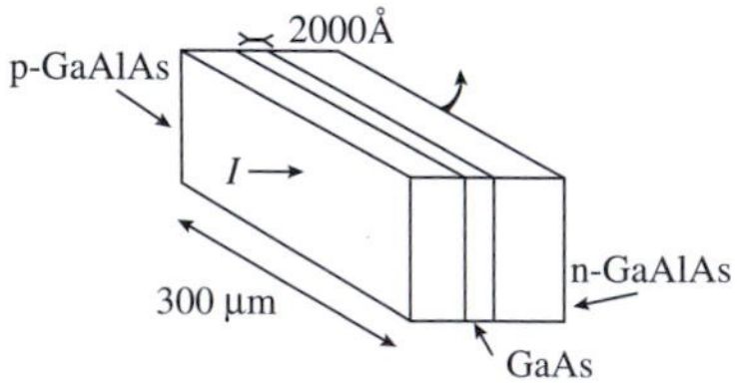

Fig. 20.7
Schematic representation of a double-heterostructure laser.

20.3.1 Double-heterostructure lasers

In a double-heterostructure laser, schematically represented in Fig. 20.7, the refractive index of the constituent materials produces an optical wave guide. The refractive index of the inner layer is high and guides the optical wave between the two outer layers of lower refractive index as shown in Fig. 20.8. The two outer layers are composed of material different from that of the inner layer. The energy gaps of the outer and inner materials are different. One of the outer layers is n-type and the other is p-type. The system exhibits energy band discontinuities which are shown schematically in Fig. 20.9. These discontinuities lead to electron and hole confinement in the active (inner) layer.

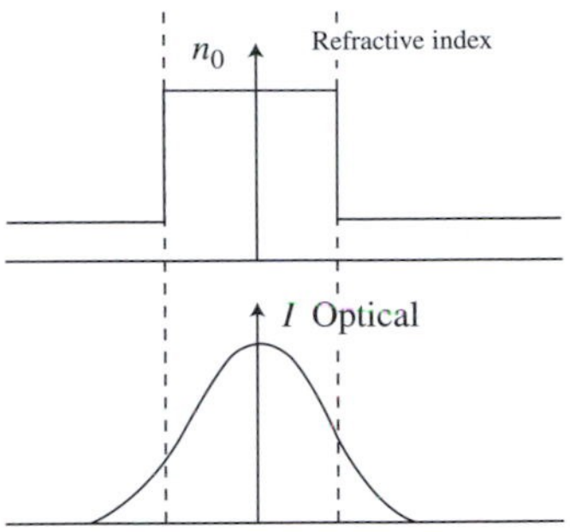

Fig. 20.8
Representation of the refractive index and guided optical wave.

The increase of light intensity due to the guided wave and the confinement of electrons and holes in the same region create a situation which facilitates the attainment of carrier population inversion. The enhanced light intensity then more readily induces the stimulated emission of radiation. One can thus achieve a reduction of a few orders of magnitude in the threshold current density. The threshold condition, Eq. (14.47), can be generalized to the case of a double-heterostructure (DH) laser to yield

$$R_1 R_2 \exp(2\Gamma g_{th} L) \exp[-2L(\Gamma\alpha_i + (1-\Gamma)\alpha_c)] = 1, \qquad (20.26)$$

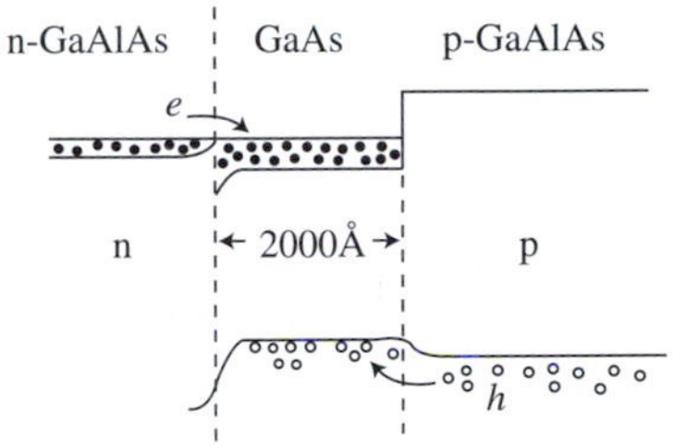

Fig. 20.9
Band structure profiles in a DH laser.

where R_1 and R_2 are the reflectivities at each end of the waveguide, Γ is the optical confinement factor, g_{th} is the threshold gain of the active layer, L is the cavity length, α_i is the internal loss coefficient of the active material, and α_c is the loss coefficient of the confining layers. Solving for Γg_{th} we have

$$\Gamma g_{th} = \Gamma\alpha_i + (1-\Gamma)\alpha_c + \frac{1}{2L}\log\left(\frac{1}{R_1 R_2}\right). \qquad (20.27)$$

The optical confinement factor is important for defining a semiconductor laser. It is given by

$$\Gamma = \int_{-d/2}^{d/2} |\mathcal{E}(z)|^2 dz \Big/ \int_{-\infty}^{\infty} |\mathcal{E}(z)|^2 dz, \qquad (20.28)$$

where d is the width of the quantum well and $\mathcal{E}$ is the electric field of the radiation. It specifies the fraction of photons in the guided wave that interacts with the active layer. Defining g to be the volume gain per unit length of the active layer, the product Γg is the amplification of the optical wave per unit length. The product $\Gamma\alpha_i$, on the other hand, is the loss per unit length of the guided wave. The difference $1-\Gamma$ is the fraction of the optical wave outside the active layer, and $(1-\Gamma)\alpha_c$ is the corresponding contribution to the loss per unit length.

20.3.2 Single quantum well (SQW) lasers

A single quantum well laser typically has a very thin active layer with quantized energy levels in the conduction and valence bands and very small values of Γ. There are several improvements of the SQW including the separate confinement heterostructure (SCH) and the graded index SCH structure (GRIN-SCH) given in Fig. 20.10 along with the associated densities of states. Both the SCH and the GRIN-SCH have optical confinement layers next to the active layer that increase the value of Γ. The energies of the

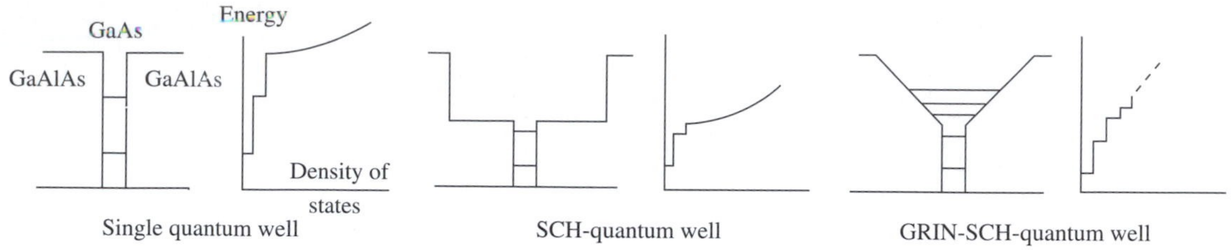

Fig. 20.10
Diagrams of the SQW, the SCH, the GRIN-SCH, and their densities of states (after Weisbuch and Vinter 1991).

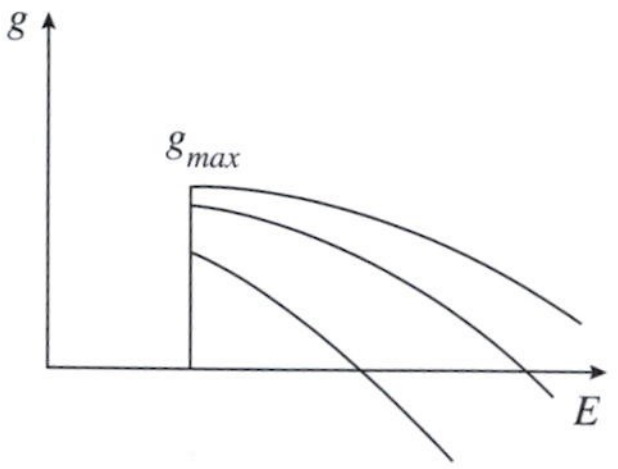

Fig. 20.11
Gain curves for a QW laser for three injected carrier concentrations.

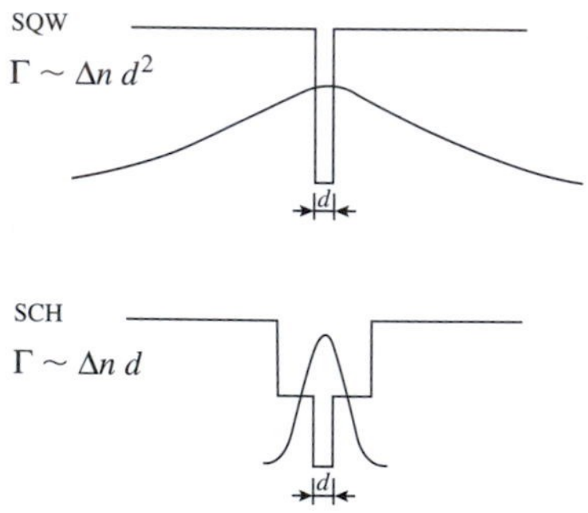

Fig. 20.12
Confinement of an optical guided wave for a narrow QW and a SCH.

photons emitted by a laser are primarily determined by the band gap and the confinement energies. As shown in Chapter 16, the density-of-states in a quantum well is two-dimensional in character. The buildup of gain is more efficient in 2D than in 3D, because additional carriers enter a 2D-band at its extremum where the gain is largest as shown in Fig. 20.11. Consequently, the slope A of the spectral gain curve specified by Eq. (14.44) is larger in 2D than in 3D, but the gain saturates at a finite value in 2D when the electron and hole states are fully inverted. The transparency current density j_0 which enters the expression for the threshold current density j_{th},

$$j_{th} = \frac{1}{\eta}\left\{ j_0 + \frac{1}{A}\left[\alpha_i + \frac{(1-\Gamma)}{\Gamma}\alpha_c + \frac{1}{\Gamma}\left(\frac{1}{2L}\right)\log\frac{1}{R_1 R_2}\right]\right\}, \qquad (20.29)$$

where η is the internal quantum efficiency, i.e., the ratio of radiative carrier recombination to total recombination, is small compared to that for a DH laser, since the density-of-states to be inverted is typically an order of magnitude smaller for a GaAs SQW laser than for a DH laser.

The confinement factor must be adjusted adequately. For very thin layers Γ is very small, varying as d^2, and the optical wave leaks into the adjacent layers. One overcomes this difficulty by introducing a pair of optical confinement layers as in the SCH structure presented in Fig. 20.12. The dependence of the confinement factor on layer thickness d is thereby removed.

The effect of layer thickness on maximum gain has been investigated by Weisbuch (1987) and Nagle (1987). Results of detailed calculations are given in Fig. 20.13. It is clear that the transparency current density, which specifies the onset of gain, is reduced by reducing the layer thickness. It is also evident, however, that there is an optimum thickness below which the transparency current density increases. Among the reasons for having an optimum in the layer thickness is that the quasi-Fermi energy is sufficiently high in very thin wells to produce a transfer of carriers into the optical confinement cavity and an increase in j_0. Also, for wide wells the population of higher-lying levels must be considered. In the case of SCH and GRIN-SCH structures optical cavity states are occupied as can be seen in the schematic representation in Fig. 20.14. Even in structures that have been optimized, however, the quasi-Fermi energy is high enough to populate the optical cavity states. At the threshold for lasing, the SCH optical cavity contains roughly the same number of electrons as the active layer. On the

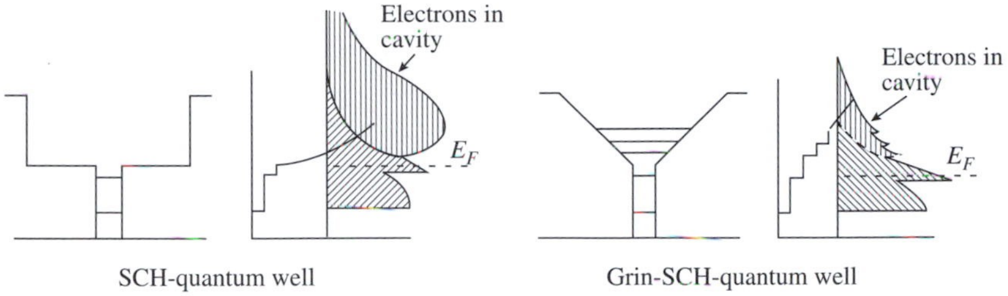

Fig. 20.14
GRIN-SCH and SCH with densities of states and occupied states (after Weisbuch and Vinter 1991).

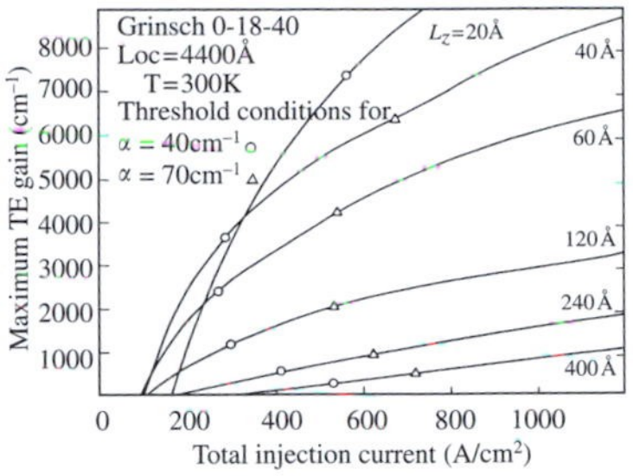

Fig. 20.13
Calculated gain–current density curves for a GRIN-SCH for various layer thicknesses (after Weisbuch 1987 and Nagle 1987).

other hand, in the GRIN-SCH optical cavity the number is only about 20% as a result of the smaller density-of-states in a triangular cavity.

20.3.3 Multiple quantum well (MQW) lasers

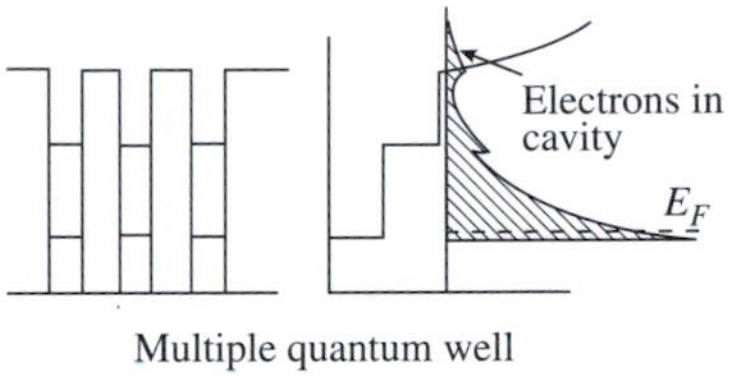

Fig. 20.15
MQW structure with density-of-states and occupied states.

Multiple quantum wells offer both advantages and disadvantages over single quantum wells for laser applications. In Fig. 20.15 diagrams of a MQW structure and the associated density-of-states and occupied states are presented. If the optical confinement factor is proportional to the total active layer width, the relation between the modal gain Γg and the current density j can be obtained by multiplying the units of both axes of the Γg versus j plot by the number of quantum wells as can be seen in Fig. 20.16. The SQW provides higher modal gain at low j, but lower gain at high j. The loss level plays an important role in determining the relative advantages and disadvantages of the SQW and MQW structures. Under low loss conditions the SQW has the advantage, since it has a lower total transparency current density j_0 and the total internal loss as measured by $\Gamma\alpha_i$ is proportional to the number of wells. At high loss, however, the MQW has the advantage, because its gain arises from the part of the gain–current density curve that has a high slope rather than the saturated part as in the SQW. Indeed, the saturated gain of the SQW may not be large enough to attain the threshold gain, and one must therefore turn to the MQW.

Lasers with ultra-low thresholds can be produced by applying reflection coatings to the ends of an SQW. Such practice leads to a lower threshold current density which can be very close to its limiting value j_0.

Another way to improve the gain is to modify the valence band structure. The presence of multiple valence bands together with a large heavy hole mass produce a high valence-band density-of-states and a quasi-Fermi energy for holes that remains above the valence band edge as shown in Fig. 20.17a. There are then too many states to be inverted to achieve gain. This difficulty can be overcome, however, by applying a strain which splits the valence band and raises the light hole band above the heavy hole band as in Fig. 20.17b. The quasi-Fermi energy of the valence band now lies well below the top of the light hole band with its small density-of-states. Inversion of these hole states is increased sufficiently at a given hole concentration to produce higher gain.

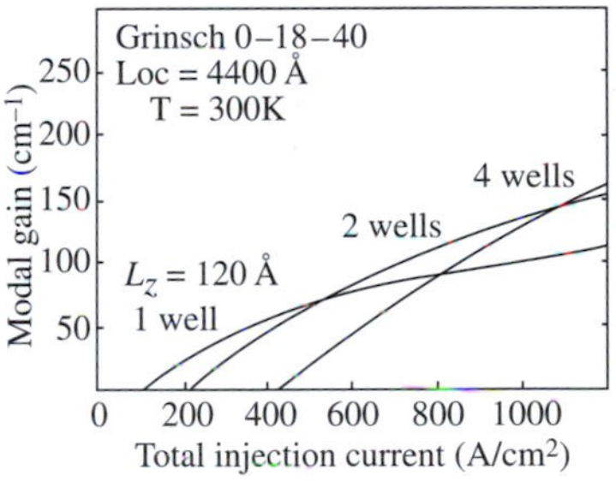

Fig. 20.16
Modal gain versus injection current density for 1-, 2-, and 4-well GRIN-SCHs (after Nagle and Weisbuch 1987).

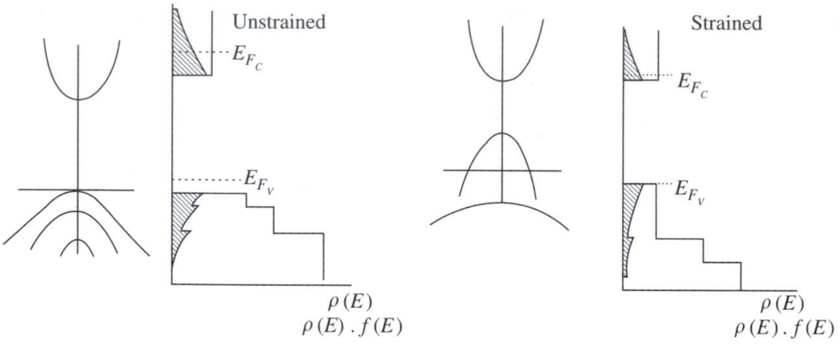

Fig. 20.17
Diagrams of conduction and valence band filling in (a) unstrained and (b) strained QWs.

This procedure has been applied with great success to strained-layer lasers such as GaInAs/GaAlAs/GaAs and GaInAsP/InP. They are characterized by a better optical confinement factor and a higher differential gain due to their higher refractive index and deeper electron/hole/quantum wells.

In addition to their low threshold current density GaAs/GaAlAs QW lasers have some other attractive properties among which are excellent manufacturability and reliability. Having the active layer thin compared to the width of the optical cavity has the advantage that only minor additional loss in the optical cavity is induced by degradation of the faces of the active layer, and collapse of laser action is unlikely.

High-speed telecommunications have greatly benefited from QW lasers. This is due to the fact that the square density-of-states profile results in a larger differential gain $A = dq/dj$ in QW lasers than in conventional semiconductor lasers. Furthermore, the best lasers for this application are MQWs having large A at large gain.

Arrays of quantum well structures provide the basis for high-power lasers that are useful in various applications. The advantage of QW lasers in this connection lies in their very high quantum efficiency due to their low transparency current density and their small internal losses associated with the confinement factor $\Gamma\alpha_i$.

There are many applications for which vertical-emitting lasers rather than edge-emitting lasers are desirable. They include 2D imaging, 2D arrays for high power with low divergence, easy access for optical accessories, and free-space direct chip-to-chip communication. An interesting realization employs integral mirrors which permit more than one million lasers to be fabricated on a single chip. A schematic representation is shown in Fig. 20.18. Such a device can have a threshold current in the milliampere range.

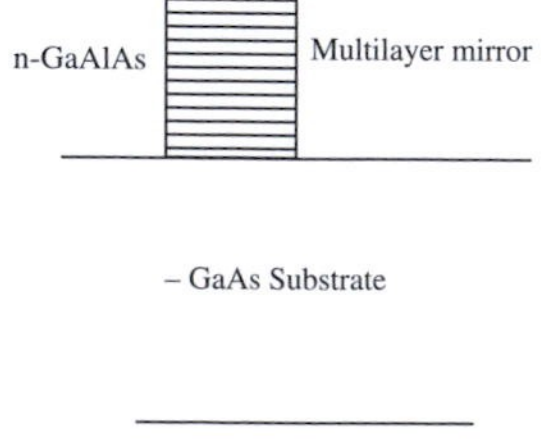

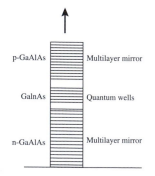

Fig. 20.18
Diagram of a mesa, vertical surface-emitting laser.

20.4 One-dimensional and zero-dimensional quantum structures

We have already noted that changing a system from 3D to 2D introduces additional quantization into the energy spectrum. Further reduction in the dimensionality leads to further quantization and produces significant modifications in the properties of the system.

20.4.1 Theoretical background

If the motion of a carrier is confined to one dimension, the carrier has only one translational degree of freedom, which we take to be in the z-direction. The carrier energies lie in subbands characterized by an index i and can be expressed as

$$E_{ik_z} = E_i + \frac{\hbar^2 k_z^2}{2m^*} \qquad (20.30)$$

for a simple energy band. Here k_z is the wave vector in the direction of motion and E_i is the energy of the ith subband at $k_z = 0$.

By imposing periodic boundary conditions over a length L_z in the z-direction, one obtains the density-of-states $D(E)$, defined as the number of states per unit energy interval, in the form (Weisbuch and Vinter)

$$D_i(E) = \frac{2L_z}{\pi\hbar}\left[\frac{m^*}{2(E - E_i)}\right]^{\frac{1}{2}}, \qquad (20.31)$$

where the factor of 2 in the numerator takes care of the spin degeneracy. The density-of-states has inverse square root singularities at the bottoms of the subbands, a behavior analogous to that associated with the Landau levels of an electron in a magnetic field as shown in Fig. 11.8. At zero temperature the number of electrons N_i per unit length in the ith subband is found to be

$$N_i = \frac{2}{\pi\hbar}[2m^*(E_F - E_i)]^{\frac{1}{2}}. \qquad (20.32)$$

In the literature on 1D structures the term channel is more often used than subband.

An electron with wave vector k_z contributes a current given by $e\hbar k_z/m^* L_z$. In thermal equilibrium this current is cancelled by that due to an electron with wave vector $-k_z$, so the net current is zero. If a weak potential difference V is applied between the points $z = 0$ and $z = L_z$, a net current flows given by

$$I_i = \frac{e^2\hbar k_F}{2m^* L_z} D_i(E_F)V. \qquad (20.33)$$

Making use of Eq. (20.31) and defining the conductance G_i of channel i by $I_i = G_{iV}$, we obtain

$$G_i = \frac{2e^2}{h}. \qquad (20.34)$$

Each populated channel contributes the same quantized amount to the conductance, the latter increasing abruptly whenever a new channel is populated. There is a similarity to the quantum Hall effect.

The motion of carriers in a real 1D structure is not restricted to a mathematical line. The carriers are confined to a very small region in x–y space by potential barriers. The barriers typically consist of a pair of barriers separated by a distance L_x and parallel to the z-axis but perpendicular to the x-axis and a second pair separated by a distance L_y and parallel to the z-axis but perpendicular to the y-axis. Such a system is called a **quantum wire**. For an infinitely deep wire the energy E_i is given by

$$E_i = \frac{\pi^2 \hbar^2}{2m^*}\left(\frac{n_x^2}{L_x^2} + \frac{n_y^2}{L_y^2}\right), \tag{20.35}$$

where n_x and n_y take on the integer values $0, 1, 2, 3, \ldots$, but not both zero. Thus, there is an infinite number of energy levels associated with subband i. Quantum wires having widths on the order of 100 Å can be fabricated.

To create a zero-dimensional (0D) system one introduces a pair of infinite potential barriers perpendicular to the z-axis, thereby confining the motion of the carriers in all three dimensions. Such a system is referred to as a **quantum dot** or **superatom**. The energy levels are entirely discrete and are given by

$$E_{n_x,n_y,n_z} = \frac{\pi^2 \hbar^2}{2m^*}\left(\frac{n_x^2}{L_x^2} + \frac{n_y^2}{L_y^2} + \frac{n_z^2}{L_z^2}\right). \tag{20.36}$$

The density-of-states consists of delta functions centered on each level with strengths containing the spin degeneracy and accidental degeneracies. For an ensemble of independent dots, the number of dots must be included as a factor in the DOS. Just as in real atoms, the probability of transitions between energy levels due to perturbations is restricted significantly by the discreteness of the levels. Consequently, very long lifetimes can be associated with occupied excited states and be determined primarily by the radiative transition probability. The combination of strongly peaked density-of-states and a very long lifetime offers the possibility of producing optical devices based on 0D structures that have superior performance.

20.4.2 Fabrication techniques for 1D and 0D structures

Nanoscale lithographic techniques can be used to pattern 2D heterostructures and obtain 1D quantum wires and 0D quantum dots. There is the problem, however, that all electronic properties at small dimensions are dominated by the damage introduced by the etching procedure employed in the pattern transfer. To alleviate this problem various alternative techniques have been proposed. They include creating patterned confining layers to provide a lateral confining potential. By varying the applied voltage of a patterned gate a variety of structures can be produced and studied. Direct fabrication methods are also useful. One of them is controlled precipitation in matrix. A supersaturated glass matrix is fabricated from supersaturated glass melts of semiconductors such as CdS, CdSe, and their solutions. Semiconductor inclusions are then precipitated through annealing. By properly regulating the temperature and duration of the annealing step, one can control the size distribution of the crystallites. Several methods have been used to fabricate free-standing nanocrystallites. Nanometer-sized clusters can be deposited by means of vapor-phase nucleation. They can also be produced by controlled reactions in solutions. Even bio-organisms have been brought to bear on the problem. Living cells are able to protect themselves from toxic heavy metals such as Cd by using their DNA-synthesized peptides to attach the heavy metal atoms. CdS microcrystallites have been extracted from cadmium salt solutions in which yeasts with their peptides have been cultured.

A method of direct semiconductor growth of quantum wire structures is based on molecular beam epitaxy (MBE) or metalorganic chemical vapor deposition (MOCVD). The semiconductor crystal is grown layer-by-layer on vicinal surfaces. The quantum wire structures arise by nucleation on well-organized steps. Also, 1D channels are provided by growth on side-walls of cleaved and etched multiquantum well structures. Yet another method is the nucleation of quantum dots during growth.

Impurity or damage induced effects can be exploited to provide lateral confinement. An example is low-energy, ion-implantation induced damage which, through the pinning of the Fermi energy by defects, results in carrier confinement. A similar process is impurity-induced interdiffusion disordering that is the basis for an index-guided laser manufacturing technique.

20.4.3 Electrical applications of 1D and 0D structures

It was shown earlier in this section that the low-field conductance of an ideal quantum wire is quantized with the value of $2e^2/h$ per occupied channel. A real quantum wire however, differs from an ideal system in a number of ways.

1. Its length is not infinite and it therefore lacks translational symmetry.
2. Potentials due to impurities in and near the channel also break the translational symmetry.
3. At finite temperatures the electron–phonon interaction leads to random scattering processes that average out interference effects. An electron can no longer remember its phase as time evolves.
4. A small number of defects, however, can produce sample-dependent interference effects that are observable and reproducible. These effects include **universal conductance fluctuations** (UCF) and **weak localization**.

All of these deviations from ideal behavior give rise to different regimes of conductance which are characterized by various scales of length. They include the length L and width W of the wire, the elastic mean free path ℓ_e associated with impurity scattering, and the inelastic mean free path ℓ_ϕ associated with phonon scattering that breaks the phase. We now discuss several regimes of interest.

If $\ell_\phi \gg \ell_e \gg L, W$, the only potential affecting the electron motion is that due to the boundaries of the wire. In this case the quantum states of interest extend over the entire length of the wire and, if occupied, carry current from one end to the other. If no voltage is applied, the currents going to the left and to the right exactly cancel. Application of a small voltage V causes the chemical potentials μ_L and μ_R at the left and right contacts, respectively, to be different with $\mu_L - \mu_R = eV$. Electrons moving from left to right occupy states up to μ_L, whereas those moving from right to left occupy states up to μ_R. This imbalance causes a net current to flow from left to right proportional to V. The quantum mechanical transition probability between rightward moving and leftward moving states determines the conductance. The wire constitutes a potential barrier between the electrons in the two contacts.

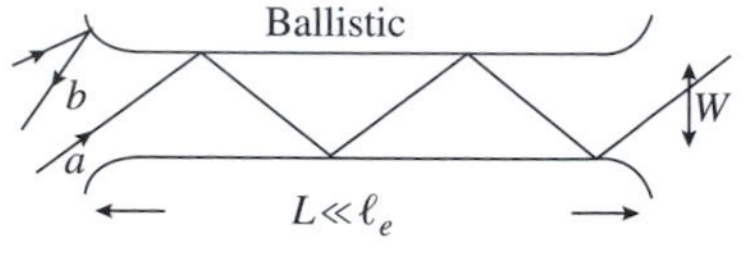

Fig. 20.19
Trajectory of an electron in ballistic transport.

One can regard an electron in a quantum wire as a particle which undergoes specular scattering off the potential walls in a ballistic fashion and whose trajectory is shown schematically in Fig. 20.19. The situation having the widths W small compared to the Fermi wavelength corresponds to **quantum ballistic transport** with only one or a few channels occupied.

If $\ell_\phi \gg L \gg \ell_e \gg W$, there are only a few impurities in the wire. The scattering introduced by the impurities increases the reflection probability of electrons entering the wire and mixes the modes, but the transport is still primarily via channels. The reflection probability is strongly affected by the precise impurity positions and the characteristic potential of the wire. Since the changes in conduction are typically on the order of e^2/h, one refers to this regime as the universal conductance fluctuation regime mentioned earlier.

If $\ell_\phi \gg L > W \gg \ell_e$, impurity scattering processes dominate, and wire modes have lost their meaning. Electrons are localized on the length scale ℓ_e in both the longitudinal and transverse directions and no states extending from one end of the wire to the other exist. This regime is referred to as weakly localized and has zero conductivity at low temperatures. Transport of electrons occurs through scattering between localized states and requires inelastic scattering. The electrons diffuse through the wire and the average concentration of impurities determines the mobility. As the temperature increases the inelastic scattering length can become smaller than the elastic length, and the mobility is determined by phonon scattering, as is often the case at room temperature.

Ballistic transport in quantum wires is at the origin of a certain number of interesting effects that are associated with the wavelike nature of electrons. The wires behave as waveguides for electron propagation and not as ordinary resistors. In the quantum ballistic regime mentioned above the conductance is quantized.

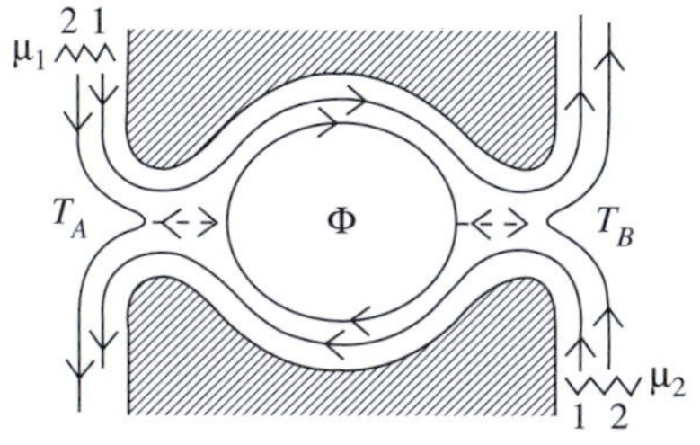

Fig. 20.20
Diagram of a ballistic quantum dot in a magnetic field.

A point contact or lateral constriction can act as a filter, since the only electrons that can be transmitted are those having the energy of a channel mode. This effect is analogous to an optical point source and is useful in creating monoenergetic electron beams. If one creates a region enclosed by two constrictions as shown in Fig. 20.20, and applies a strong magnetic field, the region behaves as a quantum dot that is coupled weakly to the outside region via the constrictions (Van Wees *et al.* 1989). The energy of a dot state is a function of the magnetic field and the constriction voltages. Consequently, the latter parameters can be manipulated to produce resonance between the dot state energy and the Fermi energies of the point contacts. This system is a lower-dimensional analogue of a double-barrier Fabry–Perot optical interferometer. Two or more dots can be coupled together to form an artificial molecule, while a periodic array of such dots forms an artificial 2D crystal.

Since the electron mean free path is on the order of tens of microns, the ballistic regime can be accessed using standard lithography. The behavior of ballistic electrons resembles optics, but with the additional possibility of modifying it with an external magnetic field. By using microfabrication techniques and three-dimensionally confined structures, it has been possible to demonstrate the tunneling properties of quantum dots in double-barrier resonant tunneling diodes (Reed *et al.* 1988). Capacitance spectroscopy has been employed to carry out studies of the electronic states in 0D systems (T.P. Smith III *et al.* 1988). With quantum ballistic electrons now available one can anticipate the development of single-electron deterministic devices. Such devices are in marked contrast with present day transistors that involve $\sim 10^4$ electrons and have properties determined by space- and time-averages of a large number of random trajectories.

20.4.4 Devices based on 1D and 0D structures

The development of techniques such as controlled growth and lithography has enabled one to produce 2D materials and devices. Further advances make possible devices that are systematic and reproducible down to the scale of nanometers. Such devices embody the **ultrasmall structure** concept for speed and integration that is the natural result of evolution from VLSI and ULSI devices. To take advantage of quantum effects one exploits the fabrication-determined energy levels of the structure to obtain devices with higher or more complex performance. Devices which perform digital operations with single- or few-electron events instead of with 10^4 electron events are the basis of **granular electronics**. The concepts behind such devices resemble the Coulomb blockade notion whereby the local potential in a structure is changed by $\Delta V = e/C$ due to the presence of a single electron. The local capacitance C is determined by the materials and the device geometry.

Under proper conditions electrons can travel without collisions in a nonquantized classical motion that yields interesting effects occurring, for example, in lateral hot-electron devices. The guiding of an electron beam through electrostatic lenses into spatially-arranged collector electrodes may provide the basis for A/D conversion in the multi-GHz range. The ballistic electron device serves as the critical element of the converter. Connections to the external system employ standard devices. The result is an integrated electron-optical component.

Some new devices dependent on electron optics employ electron-phase control of waveguided electrons. In the Mach–Zehnder interferometer the electron phase in one arm can be changed by applying an electric field.

20.4.5 1D and 0D optical phenomena

In an optical system the impact of lower dimensionality is basically due to the progressive restriction of allowed sates over the energy spectrum

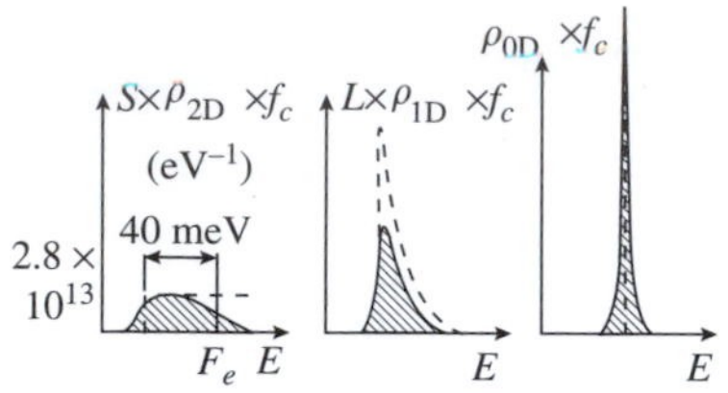

Fig. 20.21
Schematic representation of gain curves for 2D, 1D, and 0D systems (after Nagle and Weisbuch 1990).

toward the concentrated discrete atomic-like levels of quantum dots. When one deals with atomic-like levels, resonant behavior is sharpened and therefore energy selectivity. Another characteristic is the lower dispersion of optical properties over k-states due to the k-selection rule that only vertical transitions are allowed. In an injection laser above inversion, the same number of electrons in 2D, 1D, and 0D states will produce successively higher maximum gain because the electrons and holes are concentrated in successively fewer k-states as shown in Fig. 20.21. From this figure one sees that the integrated gain for fully inverted conduction and valence bands is independent of dimensionality. Consequently, the concentration of carriers in a phase space less extended in energy will produce a larger maximum in the spectral gain curve.

If one takes into account exciton effects three regimes arise for quantum dots depending on the relative magnitudes of the edge length L of the dot and the effective Bohr radius a_B^* of the exciton:

1. For $L \ll a_B^*$ the confining kinetic energy dominates the Coulomb interaction energy of the electron and hole. The latter can be regarded as a small perturbation, and the wave functions are well represented by particle-in-a-box wave functions. The oscillator strength per transition is that for ordinary interband transitions.
2. For $L \gg a_B^*$ a "giant" oscillator strength arises. The unusually large value is due to the coherent excitation of the entire volume of the quantum dot. For $L \approx a_B^*$ the exciton binding energy can be comparable to the confinement energies of electrons and holes resulting in a somewhat enhanced oscillator strength.
3. For quantum wires the excitonic effects are intermediate between those of quantum wells and quantum dots.

The threshold for nonlinear effects is lower in the lower-dimensional systems due to the smaller number of states that must be filled to reach saturation. Multiple-particle interactions are strongly enhanced in 1D and 0D. If the numbers of electron–hole pairs is increased in a confined volume, large Coulomb interactions arise leading to biexcitation and multiple-exciton effects.

20.4.6 1D and 0D optical devices

The quantum dot laser is the ultimate step in the continuous improvement of the semiconductor laser. The objective has always been to replicate the desirable optical properties of isolated atoms. The double heterostructure (DH) laser optimizes the coupling of light and matter by confining the electrons and light. In the separate confinement heterostructure (SCH) quantum well laser the number of states to be inverted is further reduced, while maintaining the light–matter interaction constant and concentrating the electrons in those radiative states. The strained quantum well laser further optimizes both the number of inverted states and the efficiency of injected carriers by splitting the light- and heavy-hole bands. The quantum dot laser makes all electron–hole pairs efficient for gain by placing the carriers in atomic-like states with very narrow band energy spread.

Quantum dot lasers should have threshold currents on the order of 10 smaller than those of quantum well lasers. In addition they should exhibit an increased modulation speed and lower spectral width arising from the much larger differential gain.

The performance of devices based on effects such as off-resonance electro-optical and nonlinear phenomena is much less enhanced by low dimensionality, because light-coupling and materials modifications produce less effect off-resonance. Quantum dots offer possibilities as infrared detectors. The noise should be diminished due to long relaxation lifetimes.

20.5 Devices based on electro-optic effects in quantum well structures

In Chapter 11 a number of electro-optic effects in bulk semiconductors were discussed. Similar effects occur in quantum well structures, but with novel features associated with the presence of two distinct directions in which one can apply the electric field, i.e., parallel to the layers or perpendicular to the layers. Of particular importance is the effect of an applied electric field on the exciton absorption (Miller 1995).

20.5.1 Quantum-confined Stark effect

If the electric field is applied in a direction parallel to the layers of the structure, the principal effect is a broadening of the exciton absorption peaks and is quite similar to that occurring in bulk semiconductors. For the case of the electric field perpendicular to the layers, however, the effect is much more dramatic due to the confinement of the electron and hole of the exciton by the walls of the wells. We have the **quantum-confined Stark effect** (QCSE) in which the electric field strongly shifts the exciton absorption peaks. The electron is forced in the direction of the positive electrode and the hole in the opposite direction, but field ionization of the exciton is prevented by the walls. The electron and hole remain relatively close to one another and still undergo strong Coulomb attraction. In the plane of the quantum wells they still orbit about one another, although the orbit is somewhat modified by the electric field. Thus, the exciton continues to exist as a bound entity, and its absorption peaks are not significantly broadened. The persistence of the exciton at high field leads to very large Stark shifts.

20.5.2 Quantum well modulators

The QCSE can be exploited as the basis for optical modulators in which the optical transmission of a system is modulated by an applied electric field. Let us consider a p–i–n diode in which the undoped intrinsic region is a quantum well region. The diode is operated under reverse bias. If the thickness of the quantum well region is on the order of 1 μm and a voltage of 1 V is applied across it, the corresponding electric field is on the order of 10^4 V/cm. Voltages of 5–10 V can produce significant QCSE shifts and changes in the absorption spectrum. The p- and n-regions of the diode are typically material that is essentially transparent at the frequency of interest.

They could be GaAlAs if the quantum wells are GaAs. The resulting structure is a **quantum well modulator** which is shown schematically in Fig. 20.22.

The quantum well modulator is characterized by two very important features:

1. the electro-absorptive mechanisms in the quantum well due to the QCSE are sufficiently strong that a modulator will work for light propagating perpendicular to the layers of the device.
2. the energy required to operate the modulator is the small amount of energy necessary to change the optical properties of the modulator. This is essentially the stored electrostatic energy associated with charging up the device capacitance.

The device that has just been described has considerable potential as a highly efficient optical output device for electronic circuits.

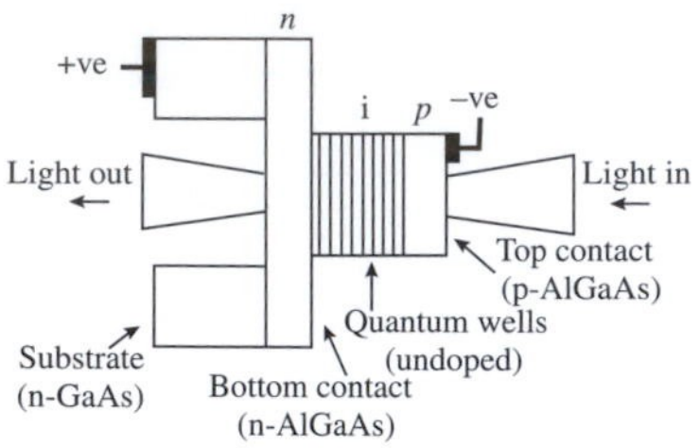

Fig. 20.22
Quantum well modulator diode structure (after Miller 1995).

20.5.3 Self-electro-optic-effect devices

Incorporating the function of a photodetector into a quantum well modulator creates an optically controlled device with an optical output that is known as the **self-electro-optic-effect device** (SEED) (Miller 1990). The advantage of such a device is the opportunity for efficient integration. If quantum well modulators are driven directly by external electrical connections, they are limited in speed and operating energy by the external electrical characteristics such as capacitance and the necessity to be driven with impedance-matched lines at high speed. If the drive is not brought in by some external connection and is electrically local, the modulators themselves can be driven with very low total energies. In addition, since they are semiconductor devices, the possibilities for integration with various opto-electronic and electronic devices are of interest.

In practice one can integrate quantum well devices quite effectively to produce two-dimensional arrays of smart opto-electronic units or "smart pixels." Devices of this sort provide new possibilities for switching architectures and information processing.

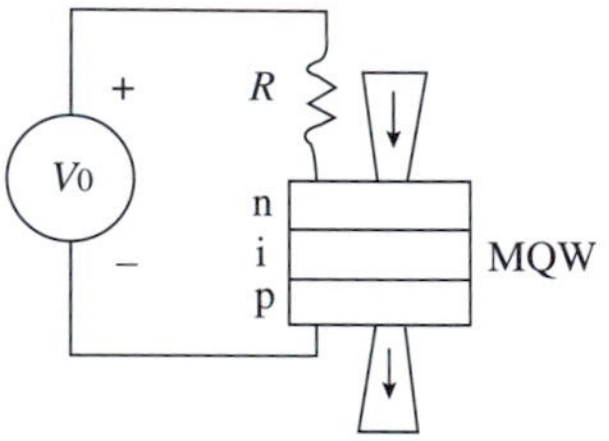

Fig. 20.23
Circuit for a simple resistive optical bistability device (R-SEED) (after Miller 1995).

SEEDs fall into two general classes: those that are based on diodes only and those that also integrate transistors. A diagram of a very simple SEED is presented in Fig. 20.23. It is a **resistor-biased SEED** or **R-SEED** and is optically bistable. It is operated at the wavelength λ_0 where the absorption increases as the reverse bias voltage decreases.

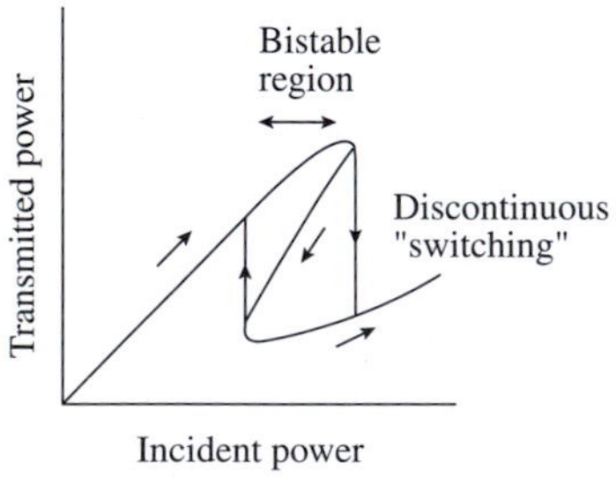

Fig. 20.24
Characteristic bistable curve for an R-SEED (after Miller 1995).

The bistability of a simple SEED is produced by a positive feedback mechanism and is illustrated in Fig. 20.24 in terms of the incident power-transmitted power curve. If the light incident on the diode is initially of low intensity, there is little photocurrent and therefore little voltage drop across the resistor. Essentially all of the voltage drop occurs across the diode, and the absorption by the diode is quite low. As the light intensity on the diode is increased, the photocurrent increases also. The voltage drop across the resistor therefore becomes larger together with a smaller voltage drop across the diode, more absorption by the diode, more photocurrent, and so forth. As a result one can switch into a state of high absorption. The diode is now so highly absorbing that the device requires less input power to maintain it in this state. Eventually, when some lower power is attained, a similar but opposite mechanism causes the diode to switch back to its high voltage state as shown in Fig. 20.24.

Instead of using a resistor as the lead in the bistable circuit, one could replace it with another reverse-biased photodiode. One photodiode then performs the function of the local resistor with an effective "resistance" whose value is controlled by the light illuminating the second "load" photodiode. This circuit is a **diode-based SEED** or **D-SEED**. This concept can be extended to include a second quantum well diode in the load giving the **symmetric SEED** or **S-SEED** as shown in Fig. 20.25. Each load diode has an incident light beam, and the S-SEED is bistable with respect to the ratio of the powers of the two incident beams rather than with respect to some absolute power. Switching occurs when the photocurrent in one diode is higher than that in the other. The S-SEED can therefore be operated over a very wide range of power.

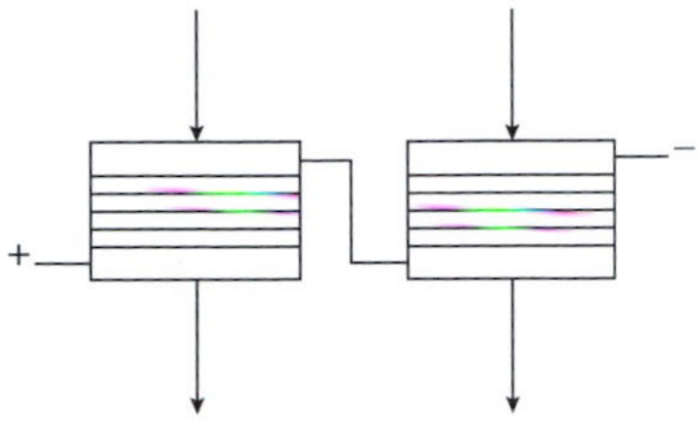

Fig. 20.25
Diagram of a symmetric SEED (S-SEED) (after Miller 1995).

As a switch the S-SEED is useful in logic systems. Elementary logic functions can be performed with the S-SEED itself. More complex situations can be handled by connecting more than two diodes in series to form the **multistate SEED** or **M-SEED** (Lentine 1989). Under voltage biasing of N diodes in series there can be N states, each having only one diode on. One can proceed even further by constructing families of circuits that embody the **logic SEED** or **L-SEED** (Lentine 1990).

Quantum well diodes can be integrated with transistors. Since electronics performs complex logic functions very well, there are significant advantages to combining electronics with the capabilities of optical devices for interconnections. Thus, a good integration technology is highly desirable.

Comment. With 100 fJ of input energy and a factor of 10 loss overall in an optical system, a 1 W laser has sufficient power to drive 1 Tb/s of information through the system. This is a very large rate and one that is difficult to achieve with purely electronic systems.

Problems

1. With the aid of the discussion of the Gunn effect in Chapter 8 develop a physical explanation for the critical electric field $\mathcal{E}_c$ that exists in a GaAs/GaAlAs TEGFET. If the saturation velocity is 10^7 cm/s, and the mobility of electrons in the GaAs is $4 \times 10^4 \text{ cm}^2/\text{V s}$, what is the value of $\mathcal{E}_c$?
2. If the gate length is 0.5 μm and the saturation drift velocity is 1.5×10^7 cm/s, what is the transit time of an electron across the gate? What is the cutoff frequency?
3. Compare the physical processes occurring in the Gunn oscillator and in the heterostructure double-barrier diode. Focus on which quantities decrease when negative resistance appears.
4. Compare the electron and hole wave functions for the first few exciton states in an "infinite" quantum well with and without an applied electric field.
5. Calculate the shift of the excitonic absorption peaks due to the quantum confined Stark effect
6. In a quantum well modulator the quantum well region containing 100 quantum wells is operated under 1 V reverse bias. What is the applied field and the QCSE shift?

References

M. A. Hollis and R. A. Murphy, in *High-Speed Semiconductor Devices*, ed. S. M. Sze (John Wiley, New York, 1990).

A. L. Lentine, D. A. B. Miller, J. E. Henry, J. E. Cunningham, and L. M. F. Cirovsky, *IEEE J. Quantum Electron.* **25**, 1921 (1989).

A. L. Lentine, D. A. B. Miller, J. E. Henry, J. E. Cunningham, L. M. F. Cirovsky, and L. A. D'Asaro, *Appl. Opt.* **29**, 2153 (1990).

D. A. B. Miller, *Optical and Quantum Electron.* **22**, S61 (1990).

D. A. B. Miller, in *Confined Electrons and Photons*, eds. E. Burstein and C. Weisbach (Plenum Press, New York, 1995).

J. Nagle, *Thesis*, Université Paris VI, 1987.

J. Nagle and C. Weisbuch, *Technical Digest*, ECOC '87, Helsinki, part II, p. 25 (1987).

J. Nagle and C. Weisbuch, in *Science and Engineering of 1 and 0 Dimensional Semiconductor Systems*, eds. C. M. Sotomayer-Torres and S. P. Beaumont (Plenum, New York, 1990).

M. A. Reed, J. N. Randall, R. J. Aggarwal, R. J. Matyi, T. M. Moore, and A. E. Wetzel, *Phys. Rev. Lett.* **60**, 1535 (1988).

T. P. Smith III, K. Y. Lee, C. M. Knoedler, J. M. Hong, and D. P. Kern, *Phys. Rev.* B**38**, 2172 (1988).

B. J. Van Wees, L. P. Kouvenhoven, C. J. P. M. Harmans, J. G. Williamson, C. E. Timmering, M. E. I. Broekaart, C. T. Foxon, and J. J. Harris, *Phys. Rev. Lett.* **62**, 2523 (1989).

C. Weisbuch, in *Technologies for Optoelectronics*, eds. F. Potter and J. M. Bulabois, *Proc. SPIE.* **869**, 155 (1987).

C. Weisbuch and B. Vinter, *Quantum Semiconductor Structures* (Academic Press, San Diego, 1991).

Index